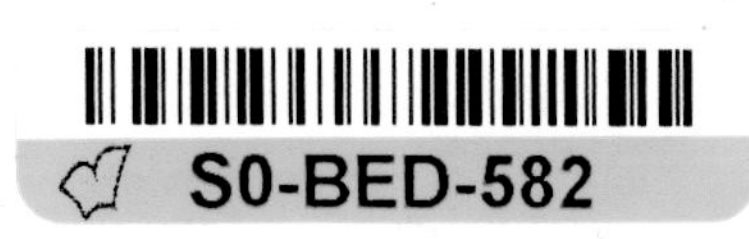

Triangles

θ (degrees)	$\sin\theta$	$\cos\theta$	$\tan\theta$	$\cot\theta$	$\sec\theta$	$\csc\theta$
30	$\frac{1}{2}$	$\frac{\sqrt{3}}{2}$	$\frac{\sqrt{3}}{3}$	$\sqrt{3}$	$\frac{2\sqrt{3}}{3}$	2
45	$\frac{\sqrt{2}}{2}$	$\frac{\sqrt{2}}{2}$	1	1	$\sqrt{2}$	$\sqrt{2}$
60	$\frac{\sqrt{3}}{2}$	$\frac{1}{2}$	$\sqrt{3}$	$\frac{\sqrt{3}}{3}$	2	$\frac{2\sqrt{3}}{3}$

Right Triangles

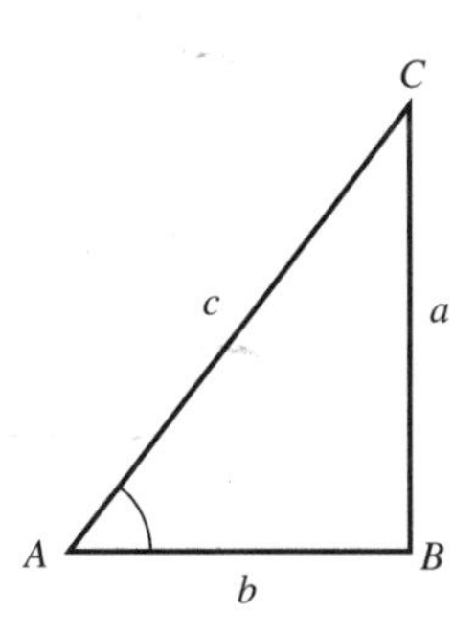

$$\sin A = \frac{a}{c} \qquad \cot A = \frac{b}{a}$$

$$\cos A = \frac{b}{c} \qquad \sec A = \frac{c}{b}$$

$$\tan A = \frac{a}{b} \qquad \csc A = \frac{c}{a}$$

Oblique Triangles

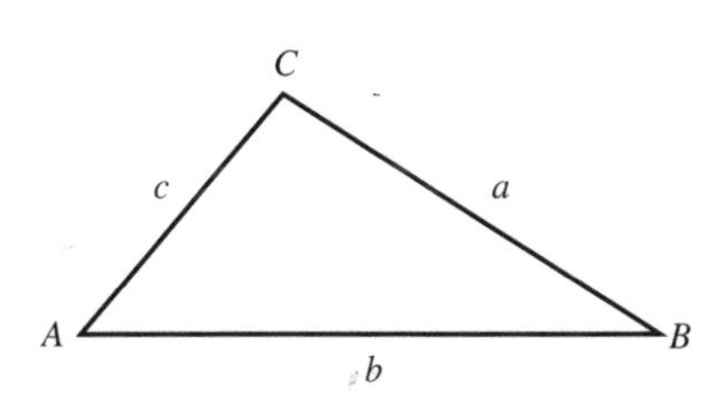

Law of Sines

$$\frac{\sin A}{a} = \frac{\sin B}{b} = \frac{\sin C}{c}$$

Law of Cosines

$$a^2 = b^2 + c^2 - 2bc\cos A$$
$$b^2 = a^2 + c^2 - 2ac\cos B$$
$$c^2 = a^2 + b^2 - 2ab\cos C$$

viviana_e@
yahoo.com

Plane Trigonometry

Seventh Edition

Bernard J. Rice
Jerry D. Strange
University of Dayton

PWS Publishing Company

I(T)P **An International Thomson Publishing Company**

Boston ▲ Albany ▲ Bonn ▲ Cincinnati ▲ Detroit ▲ London
Madrid ▲ Melbourne ▲ Mexico City ▲ New York ▲ Paris
San Francisco ▲ Singapore ▲ Tokyo ▲ Toronto ▲ Washington

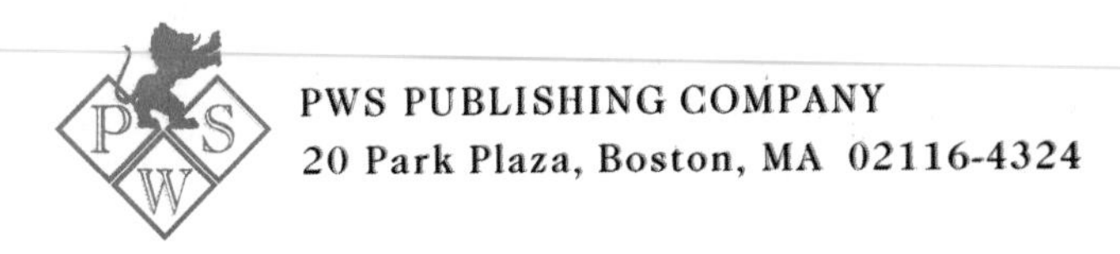

For more information contact:

PWS Publishing Company
20 Park Plaza
Boston, MA 02116

International Thomson Publishing
Europe
Berkshire House I68-I73
High Holborn
London WC1V 7AA
England

Thomas Nelson Australia
102 Dodds Street
South Melbourne, 3205
Victoria, Australia

Nelson Canada
1120 Birchmount Road
Scarborough, Ontario
Canada M1K 5G4

International Thomson Editores
Campos Eliseos 385, Piso 7
Col. Polanco
11560 Mexico D.F., Mexico

International Thomson Publishing
GmbH
Königswinterer Strasse 418
53227 Bonn, Germany

International Thomson Publishing
Asia
221 Henderson Road
#05-10 Henderson Building
Singapore 0315

International Thomson Publishing
Japan
Hirakawacho Kyowa Building, 31
2-2-1 Hirakawacho
Chiyoda-ku, Tokyo 102
Japan

Sponsoring Editor: David Dietz
Editorial Assistant: Julia Chen
Production Editor: Patricia Adams
Marketing Manager: Marianne C.P. Rutter
Manufacturing Coordinator: Marcia A. Locke
Interior Illustrator: Scientific Illustrators
Interior Designer: Kathleen Wilson
Cover Designer: Jay Shippole
Cover Photo: © Pete McArthur/Tony Stone Images
Compositor: Better Graphics Inc.
Cover Printer: John P. Pow Company, Inc.
Text Printer and Binder: Quebecor Printing/Martinsburg

Library of Congress Cataloging-in-Publication Data

Rice, Bernard J.
Plane trigonometry / Bernard J. Rice, Jerry D. Strange—7th ed.
p. cm.
Includes index.
ISBN 0-534-94824-3 (hardcover)
1. Trigonometry, Plane. I. Strange, Jerry D. II. Title.
QA533.R5 1995 95–38468
516.24′2—dc20 CIP

This book is printed on recycled, acid-free paper.

Printed and bound in the United States of America
95 96 97 98 99—10 9 8 7 6 5 4 3 2 1

Contents

10 Exponential and Logarithmic Functions 333

11 Analytic Geometry 356

APPENDIX A Calculators 399

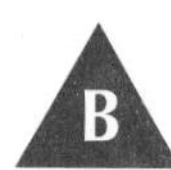

APPENDIX B Review of Elementary Geometry 409

Answers to Odd-Numbered Exercises 419

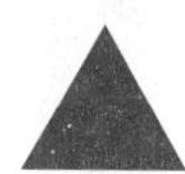

Index 473

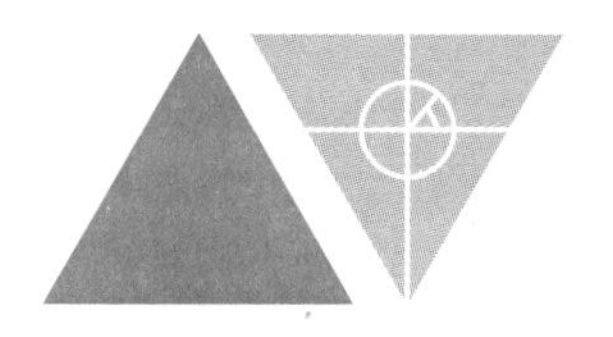

Preface

Since its conception over 20 years ago, our *Plane Trigonometry* text has undergone a number of major revisions, but we have retained our basic belief that trigonometry is best learned and understood by introducing triangle trigonometry before proceeding to analytic trigonometry. The early use of the trigonometric functions as ratios, combined with a multitude of modern applications, shows the student that trigonometry is an important tool in solving problems. The step to analytic trigonometry is then a natural extension of an already familiar subject.

The calculator is the primary tool for the evaluation of trigonometric functions and special attention is given in Chapters 1 and 2 to the problem of rounding the calculator answer to one that is consistent with accepted procedures for accuracy and precision. Although the calculator is used to evaluate the trigonometric functions, the student is expected to be able to find the values of the trigonometric functions of certain special angles (0°, 30°, 45°, 60°, 90°, etc.) without the use of a calculator.

In the previous edition, we incorporated the use of the graphing calculator in a few well-chosen places, usually in the form of calculator comments. Here, we have expanded the graphing calculator examples and exercises and have included a brief introduction to the graphing calculator in Chapter 1. The graphing calculator greatly enhances the teaching of analytic trigonometry and we wholeheartedly endorse its use as a teaching aid. However, the text is written to permit the omission of the graphing calculator if students are not required to have them.

In addition to the more extensive use of the graphing calculator, there are several other major changes to this edition of *Plane Trigonometry*. Some of the specific changes are:

- An expanded early discussion of the function concept in Chapter 1.
- A major revision and expansion of Chapter 10 to include a discussion of the exponential function and its relation to the logarithm.
- A new Chapter 11 on analytic geometry.

There are also some relatively minor changes that were recommended by users to enhance the usability of the text. These include:

- Additional historical comments.
- A short overview to begin each chapter.
- Suggested examination-type problems at the end of each chapter.

As with most books there are more topics than even the most

ambitious instructor can comfortably cover in a one-semester or one-quarter course. Not all applications need to be or should be covered. For example, the discussion of vectors in Chapter 3 can easily be omitted and some of the sinusoidal applications in Chapter 5 can be lengthy and are not required for the continuity of the text. Certain formulas in Chapter 3 such as Heron's formula are used sparingly in the sequel. The material in Chapters 9, 10, and 11 may be covered independently or omitted completely.

Numerical answers to the odd-numbered exercises were done with a calculator, but since decimal approximations often depend upon how numbers are carried through the process, your answers may differ slightly from those given in the book. A separate booklet containing answers to the even-numbered exercises is available, as is a solutions manual for the odd-numbered exercises.

Ancillaries

- An ***Instructor's Answer Manual*** provides answers to all even-numbered exercises.
- A printed and bound ***Test Bank*** is available free to adopters.
- ***EXPTest*** and ***ExamBuilder*** are improved computer-generated test systems for Windows, DOS, and Macintosh, and are available to instructors without cost.
- A ***Student's Solutions Manual*** contains detailed solutions to odd-numbered exercises.

Acknowledgments

Over the years the book has been critically reviewed by mathematicians and users from a wide variety of institutions throughout the country. In particular, the seventh edition has benefited from reviews and comments from the following:

John P. Bibbo
Southwestern College

Nancy Brannen
Lake City Community College

John Bunn
St. Louis Community College—Forest Park

Cynthia L. Coulter
Catawba Valley Community College

Gene Garza
University of Montevallo

Guy W. Hinman
Brevard Community College

Norma James
New Mexico State University

Jianzhong Wang
San Houston State University

Lamar Norwood
Campbell University

Sharon F. Welker
Catawba Valley Community College

We again wish to thank Paul Foerster for allowing us to use his excellent modeling problems as a source to draw on in Chapter 5.

Special thanks again to Carroll Schleppi of the University of Dayton for her work in preparing a solutions manual for this edition. Finally, it is a pleasure to acknowledge the fine cooperation and considerable effort of those responsible for the production of this edition: our production editor, Patty Adams, our editorial assistant, Julia Chen, our artist, George Morris, and last, but not least, our editor David Dietz.

Bernard J. Rice
Jerry D. Strange

Some Fundamental Concepts

*I*n this chapter we give a brief history of trigonometry and discuss some background topics that will be helpful to you as you begin the study of trigonometry. Because calculators are an essential part of trigonometry, special attention is given to interpreting calculator answers so that the results obtained are reasonable. In Section 1.2 we discuss the distinction between accuracy and precision as well as the effect of rounding off decimal numbers. The concept of function, which is required in later chapters, is covered in Sections 1.3 through 1.5. Included in these sections is a discussion of how to use a graphing calculator to draw graphs of functions. Angles and triangles are discussed in Sections 1.6 and 1.7 in anticipation of the trigonometry of angles in Chapter 2 and their many applications to triangles. ▼

Historical Background

Trigonometry is one of the oldest branches of mathematics. An ancient scroll called the Ahmes Papyrus, written about 1550 B.C., contains problems that are solved using similar triangles, the heart of the trigonometric idea. Historical evidence shows that by about 1100 B.C., the Chinese were making measurements of distance and height using what is essentially right-triangle trigonometry. The subject eventually became intertwined with the study of astronomy. In fact the Greek astronomer Hipparchus (180–125 B.C.) is credited with compiling the first trigonometric tables and thus has earned the right to be known as "the father of trigonometry." The trigonometry of Hipparchus and the other astronomers was strictly a tool of measurement. Thus it is difficult to classify the early uses of the subject as either mathematics or astronomy.

In the fifteenth century trigonometry was developed as a discipline within mathematics by Johann Muller (1436–1476). This development created an interest in trigonometry throughout Europe and thus placed Europe in a position of prominence with respect to astronomy and trigonometry.

In the eighteenth century trigonometry was systematically developed in a completely different direction, highlighted by the publication in 1748 of the now-famous "Introduction to Infinite Analysis" by Leonhard Euler (1707–1783). From this new viewpoint trigonometry did not necessarily have to be considered in relation to a right triangle. Rather, its analytic or functional properties became paramount. As this wider outlook on the subject evolved, many new applications arose, especially for describing physical phenomena that are "periodic."

To benefit from reading this book, you should have some ability with elementary algebra, particularly manipulative skills. Some of the specific background knowledge you will need is presented in this chapter.

1.1 Calculators and Computations

Calculators are important tools in trigonometry, and throughout this book you will be expected to use a calculator to perform arithmetic calculations. Calculators eliminate much of the computational drudgery in problem solving; however, if you wish to avoid incorrect answers, you must take time to learn how your calculator works. This section includes a general discussion of the use of calculators. Your owner's manual will give you the specific instructions for your calculator.

Order of Operations

For arithmetic operations such as $5 \cdot 4 + 6$, there might seem to be two possible answers. If we first multiply 5 and 4 and then add 6, the answer is 26. On the other hand, if we first add 4 and 6 and then multiply by 5, the answer is 50. Clearly two different answers to this problem is unacceptable. There are two ways to remove this kind of ambiguity when performing a sequence of arithmetic operations.

1. **The ambiguity can be removed by the use of parentheses.** Parentheses are used to indicate the order in which a series of operations is to be performed. For instance, the problem given above would be written as $(5 \cdot 4) + 6$ to indicate that 5 was to be multiplied by 4 and then 6 was to be added to this product to give 26; $5 \cdot (4 + 6)$ would indicate that 4 was to be added to 6 and then this sum was to be multiplied by 5 to give 50. The rule when parentheses are used is: *Compute what is inside the parentheses first and then continue with the remaining operations.*

In some cases parentheses inside of parentheses (called **nested** parentheses) are used to show the sequence in which the operations are to be performed. (In complicated expressions, brackets and braces may be used in addition to parentheses.) The rule when nested parentheses, brackets, or braces are used is: *Begin with the innermost grouping symbols first.* For example, in the expression $2 + (3 - (7 - 2))$, we do $7 - 2$ first. Thus we have

$$2 + (3 - (7 - 2)) = 2 + (3 - 5) = 2 - 2 = 0$$

Fortunately, calculators with [( ] and [ )] keys are programmed to follow the indicated rules for parentheses; all you need to do is enter the appropriate parentheses.

2. **The ambiguity can be removed by agreement.** The sequence of operations can be agreed upon so that everyone will interpret a sequence of operations in the same way. The agreement defines a **priority of operations** to be used when parentheses are not included. The following priority of operations assumes that the operations are performed starting on the *left* and moving to the *right*; it is used in most calculators and computers.

PRIORITY OF OPERATIONS In any series of arithmetic operations:

1. Special operations such as squaring [x^2] reciprocating [$1/x$] taking a root [$\sqrt{x}$], and so on are performed immediately as they occur.
2. Multiplication [×] and division [÷] are completed as they occur.
3. Addition [+] and subtraction [−] are completed in the order in which they occur after the operations listed above are completed.

COMMENT Note that the priority of operations is an agreement to mentally insert parentheses around the multiplications and divisions in order to separate these operations from the additions and subtractions. For example, the indicated priority of operations requires that $5 \cdot 4 + 6$ be interpreted as $(5 \cdot 4) + 6 = 26$. If we use the priority of operations, there is no possibility of coming up with two answers to this problem.

Most modern calculators have the indicated priority of operations built into their logic circuits. The user simply enters the sequence of numbers and operations as they occur from left to right, and the logic circuit of the calculator follows the priority of operations. To override the priority of operations, the user must insert the appropriate parentheses by using the [( ] and [ )] keys.

Example 1 Insert the implied parentheses of the priority of operations in the following arithmetic computations: **(a)** $5 + 6 \cdot 7$, **(b)** $2 - 9 + 10$, **(c)** $9 - 2 \cdot 3 - 4 \div 2$, **(d)** $3^2 + 8 - 21 \div 7$.

SOLUTION

(a) $5 + 6 \cdot 7 \rightarrow 5 + (6 \cdot 7) = 47$
There are implied parentheses around $6 \cdot 7$, since multiplication precedes addition (Rule 2).

(b) $2 - 9 + 10 \rightarrow (2 - 9) + 10 = 3$
There are implied parentheses around $2 - 9$, since additions and subtractions are done in sequence from left to right (Rule 3).

(c) $9 - 2 \cdot 3 - 4 \div 2 \rightarrow (9 - (2 \cdot 3)) - (4 \div 2) = (9 - 6) - 2 = 1$
There are implied parentheses around $2 \cdot 3$ and $4 \div 2$, since multiplication and division precede addition and subtraction. The parentheses around $9 - (2 \cdot 3)$ indicate that after the multiplication and division are completed, the additions and subtractions are done in sequence from left to right.

(d) $3^2 + 8 - 21 \div 7 \rightarrow (9 + 8) - (21 \div 7) = 17 - 3 = 14$
First we square 3 to get 9. There are implied parentheses around $9 + 8$ and $21 \div 7$. Since there are no multiplications or divisions in the first two terms, we add 9 to 8 to get 17. However, before performing the subtraction we must divide 21 by 7 to get 3. Finally, subtracting 3 from 17 yields 14.

Check the arithmetic computations given in (a) through (d) with your calculator. Enter the numbers and operations from left to right as they appear in the given expression. Do not use the [( ] and [ )] keys. If your answers do not agree with those in the book, either you have made an error or the priority of operations in your calculator is not the same as described above. ▲

Approximate Numbers

The number in the display register of a calculator is often only an approximation of the actual result of a calculation. This is because calculators can display only a finite number of digits, typically eight to twelve. Thus to display π, the calculator will show [3.141592654] to nine decimal places, which is only an approximation of the actual value of π. The following examples illustrate the way calculator results are displayed as approximate numbers. We assume that the calculator has a ten-digit display.

- When you compute 130/3, the display register shows

  [43.33333333]

- When you compute, $\sqrt{2}$, the display register shows

  [1.414213562]

In both of these cases, as with the other calculator displays in this book, the number of digits displayed and the value of the last digit may differ from those on your calculator. Check your calculator to see how it displays these results.

What decimal representation of 130/3 or $\sqrt{2}$ is considered to be acceptable? That is, how many of the digits shown in the register of a calculator should be included in an answer? As noted earlier, when $\sqrt{2}$ is evaluated using a calculator, the display shows `1.414213562`. We express this result as $\sqrt{2} \approx 1.414213562$, and we say that the approximation of $\sqrt{2}$ is accurate to nine decimal places. The digits in a numerical calculation that are known to be accurate are called **significant digits**. Thus $\sqrt{2} \approx 1.414213562$ is accurate to nine decimal places and has ten significant digits.

Example *2*

(a) $130/3 \approx 43.3333$ is accurate to four decimal places and has six significant digits.

(b) $5/7 \approx 0.714$ is accurate to three decimal places and has three significant digits. ▲

Note that when specifying the number of significant digits for numbers that are greater than zero and less than 1, we do not consider zeros that are required to locate the decimal point to be significant digits.

Example *3*

(a) $1/59 \approx 0.016949$ is accurate to six decimal places and has five significant digits.

(b) $5/10{,}937 \approx 0.000457$ is accurate to six decimal places and has three significant digits. ▲

The process of reducing the digits in a number to a specified number of significant digits or decimal places is called **rounding off**. The following method is used in this book.

ROUNDING OFF NUMBERS

1. If the digit to be dropped is less than 5, drop the digit and all digits to the right of it and use the remaining digits.
   Example: If we wish to round off 8.13457 to three significant digits, we drop the last three digits and write 8.13.
2. If the digit to be dropped is 5 or greater, drop the digit and all digits to the right of it and increase the last remaining digit by 1. This process is referred to as **rounding up**.
   Example: If we wish to round off 0.0235197 to two significant digits, we drop the last four digits, increase the last remaining digit by 1, and write 0.024.

In this book we will always round up; that is, if the digit to be dropped is 5 or greater, we will increase the last significant digit by 1. Thus

84.582983 can be approximated by rounding as follows, depending on the number of significant digits to be used.

Number of Significant Digits	Representation of 84.582983
1	80 ⟵ This zero is not a significant digit.
2	85
3	84.6
4	84.58
5	84.583
6	84.5830 ⟵ This zero is one of the significant digits.
7	84.58298

Sometimes instead of specifying accuracy by requiring a certain number of significant digits, we ask that results be accurate to a certain number of decimal places. The same rules of rounding off are used in both cases. In this text, instructions for rounding off numbers will specify decimal places in some problems and significant digits in others. However, *significant-digit accuracy and decimal-place accuracy are not the same concept; they can lead to different results when you round off a calculator readout.*

Example 4

(a) 84.5678 rounded off to three decimal places in 84.568.
84.5678 rounded off to three significant digits is 84.6.
In this case accuracy to three significant digits is equivalent to accuracy to one decimal place.

(b) 104.1538 rounded off to the nearest hundredth is 104.15.
104.1538 rounded off to five significant digits is also 104.15.

(c) 0.0002371 rounded off to five decimal places is 0.00024. This is the same as rounding off to an accuracy of two significant digits.

(d) 26,479 rounded off to three significant digits is 26,500. Notice that the two zeros are not considered to be significant digits because they are used to locate the decimal point. ▲

COMMENT The numbers 0.285 and 0.00285 both have three significant digits. Sometimes 0.00285 is said to be a more *precise* number than 0.285 since it is accurate to five decimal places and 0.285 is accurate to only three decimal places. In the case of numbers with no digits to the right of the decimal point, the number with the fewest zeros between the last significant digit and the decimal point is the more precise number. Thus 1350 and 135,000 both have three significant digits, but 1350 is considered to be the more precise of the two numbers. Precision of numbers will not play a role in the rules for rounding below. You need only be concerned with significant-digit accuracy and decimal-place accuracy.

Calculations with Approximate Numbers

There is a great temptation, especially when using a calculator, to simply use the answer for an arithmetic calculation that appears on the display register. If you do so you may make the answer seem more accurate than it really is. For example, if you multiply 8.4 (two significant digits) and 12.137 (five significant digits), the result on the display register is

$$8.4 \times 12.137 = \boxed{101.9508}$$

If we were to use 101.9508 as the answer, the implication would be that the product is accurate to seven significant digits, although the numbers being multiplied are only accurate to two and five digits, respectively. **A guiding principle in making calculations with approximate numbers is that the result cannot be more accurate than the least accurate number used in the calculation.** Using this principle, we establish the following conventions for arithmetic operations on approximate numbers.

Rules for Rounding

- In calculations involving addition or subtraction of approximate numbers, round the calculator readout to the *decimal-place accuracy* of the least accurate number.

  Example: Consider the computation

  $$0.74 + 0.0515 - 0.3329 = 0.4586.$$

  0.74 is accurate to two decimal places.
  0.0515 is accurate to four decimal places.
  0.3329 is accurate to four decimal places.
  Therefore 0.4586 should be rounded to 0.46, with two-decimal-place accuracy.

- In calculations involving multiplication, division, powers, and roots, round the calculator result to the *significant-digit accuracy* of the least accurate number.

  Example: Consider the computation 1.93(13.77) = 26.5761.
  1.93 is accurate to three significant digits.
  13.77 is accurate to four significant digits.
  Therefore 26.5761 should be rounded to 26.6, with three-significant-digit accuracy.

  Example: $\sqrt{29.14} \approx 5.398147831$, which should be rounded to 5.398, with four significant digits, because 29.14 has four significant digits.

- For computations involving a mixture of arithmetic operations, round the answer to the least accurate number, following the above rules. All such computations should be done in the calculator before rounding. *Only the final result should be rounded off.*

Example 5

(a) Consider the computation $R = 1.90(63.21) + 4.9072$. We note that 1.90 has three significant digits, 63.21 has four, and 4.9072 has five. The calculator result is **125.0062**, which we round to three significant digits. Thus $R = 125$ is the correct way to express the answer.

(b) $\sqrt{2}$, π, and 3 are examples of *exact numbers*. When used in computations, an exact number is considered to have as many significant digits as any other number in the computation. Consider $Q = 3.005\sqrt{2}$. The number 3.005 has four significant digits and $\sqrt{2}$ is exact; hence the calculator display of **4.249711755** should be rounded off to four significant digits—that is, $Q = 4.250$. ▲

Example 6 The length of a rectangle is measured with a meter stick to be 95.7 cm, and the width is measured with a vernier caliper to be 8.426 cm. What is the area of the rectangle?

SOLUTION The area is given by $A = 8.426 \times 95.7$, which gives a calculator readout of **806.3682**. Since the length has only three significant digits, this result must be rounded to three significant digits. Therefore the proper answer to the problem is $A = 806\ \text{cm}^2$. Notice that if we had rounded 8.426 to 8.43 *before* performing the computation, the calculator readout would have been **806.751** which rounds to $807\ \text{cm}^2$. ▲

COMMENT Chapter 3 discusses calculations that involve the relationship between the angles and sides of triangles. In that chapter we will add a rule for rounding off triangle calculations.

▲▼ Exercises for Section 1.1

In Exercises 1–10 insert the parentheses implied by the *priority of operations*.

1. $7 + 2 - 3$

2. $7 - 2 + 3$

3. $6 \times 5 \div 3$

4. $9 \div 3 \div 3$

5. $6 \times 7 - 8$

6. $6 - 7 \times 8$

7. $5 + 4 - 3 + 2$

8. $5 + 4 \times 3 - 2$

9. $5 \div 4 + 3 \times 2$

10. $6 \times 7 \times 3 + 2$

In Exercises 11–16 use a calculator with parentheses keys to evaluate each expression.

11. $-2[7.1 - (1.2 - 2.8)]$

12. $3[\{(5.6 - 9.2) + 5.3\} - 5.8]$

13. $[-(-2) - (4 - 6)][3 + (7 - 3)]$

14. $[3 - 7(5 - 6)][-3(-2 - 5(-4))][-6 + (5 - 7)]$

15. $7.5 + 3.2[1.5 - 6.6(8.1 - 4.5)]$

16. $\{[2.3 - 7(1.3 + 2.7)] + 5(2.0 - 5.8) + 8\}$

In Exercises 17–26 indicate the number of significant digits and the number of decimal places of accuracy.

17. 3.37 **18.** 2.002 **19.** 56.31 **20.** 204,000
21. 5700 **22.** 0.3342 **23.** 0.1030 **24.** 0.25
25. 0.000011 **26.** 0.0003751

In Exercises 27–36 round off the given number to three significant digits.

27. 9818 **28.** 72,267 **29.** 54.745 **30.** 0.06583
31. 24.95 **32.** 39.75 **33.** 0.4896 **34.** 0.9997
35. 900,498 **36.** 1.002

In Exercises 37–51 use a calculator as required. Round off and express the result with the appropriate number of digits.

37. $1.07 + 2.1 - 3.145$
38. $2.6 + 0.0064 + 55.7$
39. $62.4 - 75.3 + 5.008$
40. $6285.3 \div 25.1 + 820$
41. $765 - 42.3 + 525.1 \div 5.2$
42. $-1395 - [748.6 - (4096 - 2716)]10.903$
43. $4.07 - \{5.87 - [0.95 + 2.53(1.65 - 3.02)]\}$
44. $[0.395 - (1.761 - 2.059)][5.379 + 2.13(0.943 - 1.687)]$
45. $2.176\sqrt{3}$
46. $\sqrt{2.4^2 + 1.93^2}$
47. $\sqrt{2.14^2 + 3.9^2}$
48. $\dfrac{5.0887(2.20)}{8813}$
49. $\dfrac{0.9917(771.33)}{\sqrt{30.04}}$
50. $(6.982)^3$
51. $950 \div 25 \div 763 \div 17.2$

1.2 The Rectangular Coordinate System

We often wish to make an association between points on a line (or in a plane) and numbers, a process called coordinatization. The number (or numbers) assigned to a point is called the **coordinate** (or coordinates) of the point.

To associate points with real numbers, choose any straight line, and then choose any point on the line to be the starting point, or origin. Take any unit distance and measure that distance to the right of the origin. Then the number 0 is associated with the origin, the number 1 with the point a unit distance to the right of the origin, the number 2 with the point two units to the right of the origin, and so on. Similarly, the point one unit to the left of 0 is -1, the point two units to the left of 0 is -2, and so on. In this way the so-called **integral points** are determined. Coordinates of points in between integral points represent noninteger real numbers. The line, illustrated in Figure 1.1, is called a **real number line.**

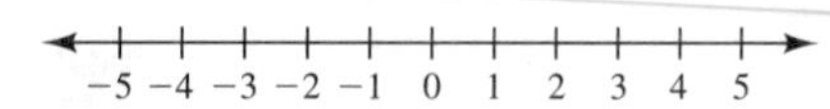

Figure 1.1
Real number line

For purposes of elementary trigonometry, the most important type of coordinatization is the association of each point in a plane with a pair of numbers. In this case we choose two mutually perpendicular intersecting lines, as shown in Figure 1.2. Normally the horizontal line is called the ***x*-axis**, the vertical line is called the ***y*-axis**, and their intersection is called the **origin**. When considered together, the two axes form a rectangular coordinate system.* As you can see, the coordinate axes divide the plane into four zones, or **quadrants**. The upper right quadrant is called *the first quadrant*, and the others are numbered consecutively in a counterclockwise direction from the first quadrant, as in Figure 1.2. The coordinate axes are not considered to be in any quadrant.

To locate points in the plane, use the origin as a reference point and mark off a suitable scale on each of the coordinate axes. The displacement of a point in the plane to the right or left of the y-axis is called the ***x*-coordinate**, or **abscissa**, of the point and is denoted by x. Values of x measured to the right of the y-axis are *positive* and values to the left of it are *negative*. The displacement of a point in the plane above or below the x-axis is called the ***y*-coordinate**, or **ordinate**, of the point and is denoted by y. Values of y above the x-axis are *positive* and values below it are *negative*. Together the abscissa and the ordinate of a point are called the **coordinates** of the point. The coordinates of a point are conventionally written in parentheses, with the abscissa written first and separated from the ordinate by a comma—that is, (x, y).

We see that a point (x, y) lies

- in quadrant I if both coordinates are positive;
- in quadrant II if the x-coordinate is negative and the y-coordinate is positive;
- in quadrant III if both coordinates are negative;
- in quadrant IV if the x-coordinate is positive and the y-coordinate is negative.

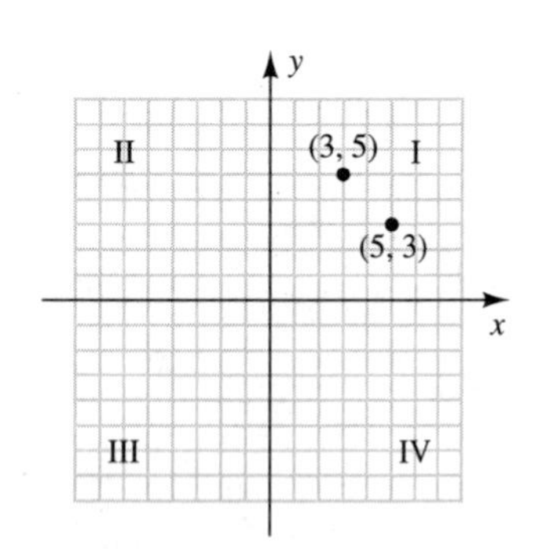

Figure 1.2
Cartesian coordinate system

Since the first number represents the horizontal displacement and the second the vertical displacement, order is significant. For example, the ordered pair (3, 5) represents a point that is displaced three units to the right of the origin and five units up, whereas the ordered pair (5, 3) represents a point that is five units to the right and three units up. The association of points in the plane with ordered pairs of real numbers is an obvious extension of the concept of the real number line.

* This system is also called the Cartesian coordinate system in honor of René Descartes, who invented it.

To be precise, we should always distinguish between the point and the ordered pair; however, it is common practice to blur the distinction and say "the point (x, y)" instead of "the point whose coordinates are (x, y)."

Each point in the plane can be described by a unique ordered pair of numbers (x, y), and each ordered pair of numbers (x, y) can be represented by a unique point in the plane called the **graph** of the ordered pair.

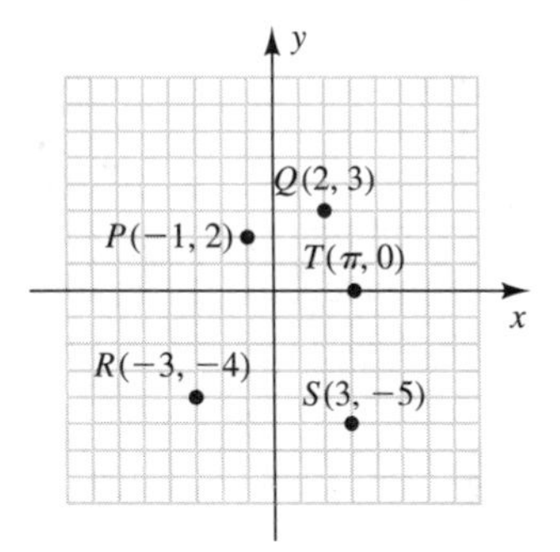

Figure 1.3
Locating points in the plane

Example 1 Locate the points **(a)** $P(-1, 2)$, **(b)** $Q(2, 3)$, **(c)** $R(-3, -4)$, **(d)** $S(3, -5)$, and **(e)** $T(\pi, 0)$ in the plane.

SOLUTION

(a) $P(-1, 2)$ is in quadrant II because the x-coordinate is negative and the y-coordinate is positive.

(b) $Q(2, 3)$ is in quadrant I because both coordinates are positive.

(c) $R(-3, -4)$ is in quadrant III because both coordinates are negative.

(d) $S(3, -5)$ is in quadrant IV because the x-coordinate is positive and the y-coordinate is negative.

(e) $T(\pi, 0)$ is not in any quadrant but lies on the positive x-axis.

The points are plotted in Figure 1.3. ▲

When an entire set of ordered pairs is plotted, the corresponding set of points in the plane is called the **graph** of the set.

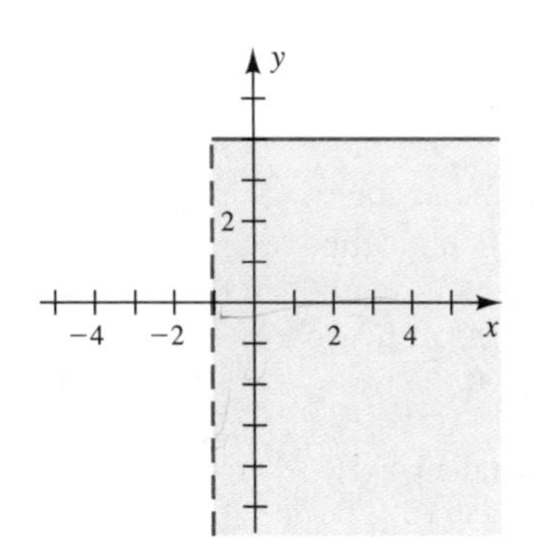

Figure 1.4

Example 2 Graph the set of points whose abscissas are greater than -1 and whose ordinates are less than or equal to 4.

SOLUTION This set is described by the two inequalities

$$x > -1$$
$$y \leq 4$$

The shaded region in Figure 1.4 is the graph of the set. The solid line is part of the region, whereas the broken line is not. ▲

Graphing calculators can display a rectangular coordinate system, and the view of the system can be altered by adjusting the *scale* on the axes. This is done by defining a minimum and maximum value for each axis, which establishes a *range*. The point $(2, -3)$ is shown below in two different situations:

- **The figure on the left shows the point graphed with a scale of $-10 \leq x \leq 10$, $-10 \leq y \leq 10$, a range of values used as a *default* setting on many calculators.**
- **The figure on the right shows the same point graphed with a scale of $-1 \leq x \leq 4$, $-4 \leq y \leq 1$.**

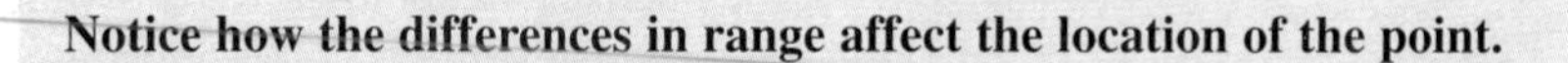

Notice how the differences in range affect the location of the point.

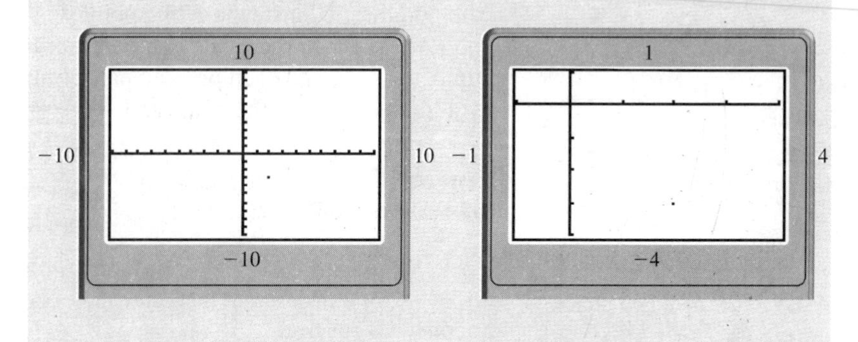

See Section 1.5 for a discussion of the graphing calculator.

Sometimes we want to find the distance between two points in the plane. Consider two points P_1 and P_2 on the x-axis, as shown in Figure 1.5. The distance between these two points can be found by counting the units between them or, algebraically, by subtracting their coordinates. To ensure that the distance will be positive, we define it in terms of the absolute value of the difference between the coordinates of P_1 and P_2. Thus

$$d(P_1, P_2) = |x_2 - x_1| \tag{1.1}$$

Figure 1.5
Distance between two points on the x-axis

Computing the distance in Figure 1.5, we have

$$d(-2, 3) = |3 - (-2)| = |5| = 5$$

A similar scheme is followed if the points lie on the y-axis.

Now consider two points $P_1(x_1, y_1)$ and $P_2(x_2, y_2)$ that determine a slanted line segment, as shown in Figure 1.6. Draw a line through P_1 parallel to the x-axis and a line through P_2 parallel to the y-axis. These two lines intersect at the point $M(x_2, y_1)$, thus forming a right triangle. Hence, by the Pythagorean theorem,*

* The Pythagorean theorem states that in a right triangle, the square of the length of the hypotenuse is equal to the sum of the squares of the lengths of the other two sides: $c^2 = a^2 + b^2$. Conversely, a triangle is a right triangle if the sum of the squares of the lengths of two of its sides is equal to the square of the length of the third side.

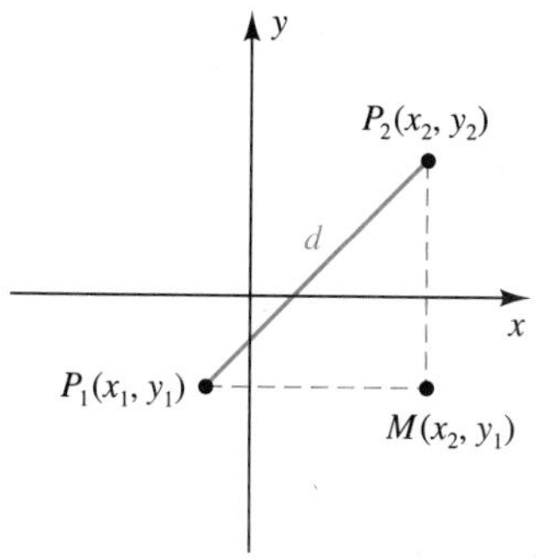

Figure 1.6
Distance between two points

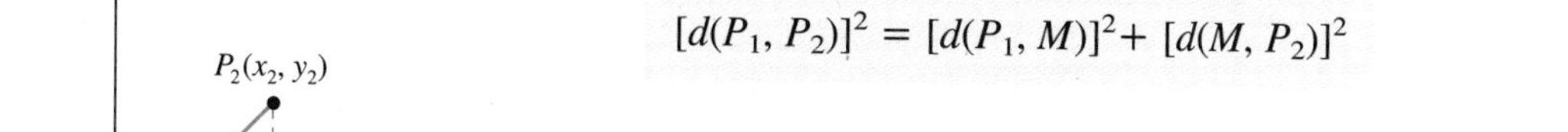

$$[d(P_1, P_2)]^2 = [d(P_1, M)]^2 + [d(M, P_2)]^2 \tag{1.2}$$

We see from Figure 1.6 that $d(P_1, M)$ is the horizontal distance between P_1 and P_2. Therefore, the distance $d(P_1, M)$ is given by

$$d(P_1, M) = |x_2 - x_1|$$

Likewise, the vertical distance $d(M, P_2)$ is given by

$$d(M, P_2) = |y_2 - y_1|$$

Recalling that $|A|^2 = A^2$ and denoting $d(P_1, P_2)$ by d, we make these substitutions into equation (1.2):

$$d^2 = (x_2 - x_1)^2 + (y_2 - y_1)^2$$

Taking the square root of both sides yields a formula for the distance d.

THE DISTANCE FORMULA The length of the line segment connecting $P_1(x_1, y_1)$ and $P_2(x_2, y_2)$ is

$$d = \sqrt{(x_2 - x_1)^2 + (y_2 - y_1)^2} \tag{1.3}$$

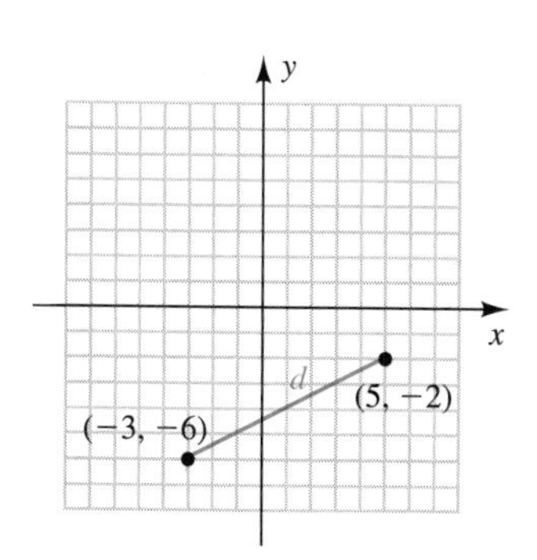

Figure 1.7

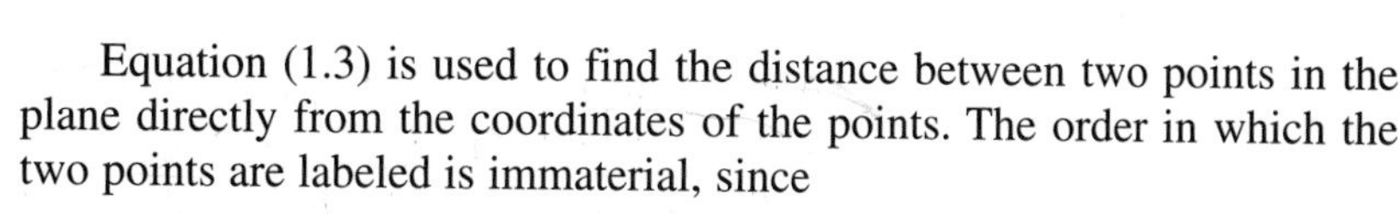

Equation (1.3) is used to find the distance between two points in the plane directly from the coordinates of the points. The order in which the two points are labeled is immaterial, since

$$(x_2 - x_1)^2 = (x_1 - x_2)^2 \qquad \text{and} \qquad (y_2 - y_1)^2 = (y_1 - y_2)^2$$

Example 3 Find the distance between $(-3, -6)$ and $(5, -2)$. (See Figure 1.7.)

SOLUTION Let $(x_1, y_1) = (-3, -6)$ and $(x_2, y_2) = (5, -2)$. Substituting these values into the distance formula, we have

$$\begin{aligned} d &= \sqrt{(x_2 - x_1)^2 + (y_2 - y_1)^2} \\ &= \sqrt{[5 - (-3)]^2 + [-2 - (-6)]^2} \\ &= \sqrt{64 + 16} = \sqrt{80} = 4\sqrt{5} \approx 8.9 \end{aligned}$$

Notice that the numerical sign of each number is included when the values are substituted into the distance formula. ▲

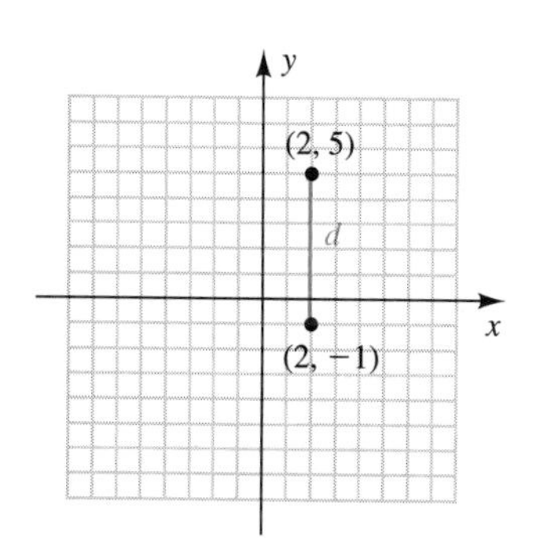

Figure 1.8

Example 4 Find the distance between $(2, 5)$ and $(2, -1)$.

SOLUTION In this case the two given points lie on a vertical line since they have the same abscissa. (See Figure 1.8.) The distance between the two points, therefore, can be found directly.

$$d = |5 - (-1)| = |5 + 1| = 6 \text{ units}$$

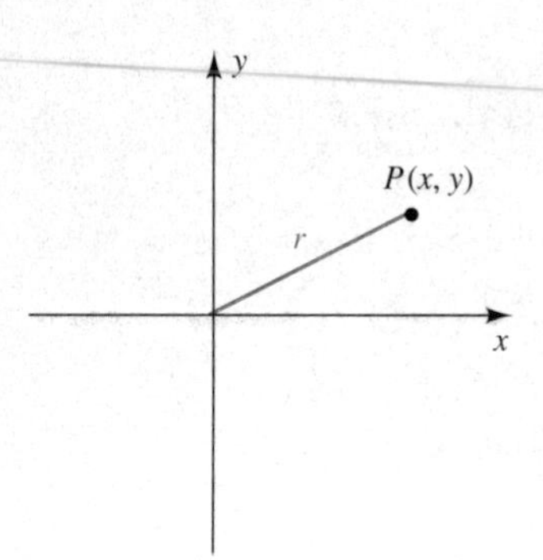

Figure 1.9

The distance can also be found from the distance formula [Equation (1.3)]. Letting $(x_1, y_1) = (2, 5)$ and $(x_2, y_2) = (2, -1)$, we have

$$d = \sqrt{(2-2)^2 + (-1-5)^2} = \sqrt{36} = 6 \text{ units}$$

▲

Example 5 Find the distance from the origin to any point (x, y).

SOLUTION From the distance formula, the length of r is given by

$$r = \sqrt{(x-0)^2 + (y-0)^2} = \sqrt{x^2 + y^2}$$

where r is always nonnegative. (See Figure 1.9.) ▲

Line segments can be displayed on the screen of a calculator in a number of ways. The figure shows the line segment from $(-5, 3)$ to $(2, -1)$ on the screen. If you have a graphing calculator, try graphing this line segment. Then check the endpoints of the line segment with the cursor.

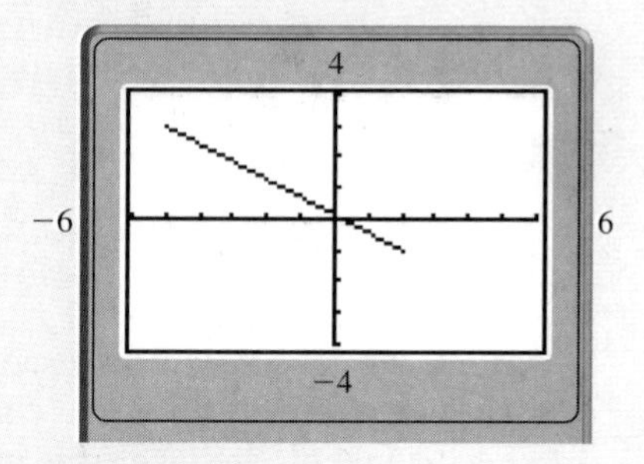

▲▼ Exercises for Section 1.2

In Exercises 1–6 plot the ordered pairs on graph paper.

1. $(3, 2)$ **2.** $(4, 6)$ **3.** $(-2, \frac{1}{2})$
4. $(-6, -5)$ **5.** $(\frac{1}{4}, -\frac{1}{2})$ **6.** $(-2.5, 1.7)$

7. In what two quadrants do points have positive abscissas?
8. In what two quadrants do points have negative ordinates?
9. In what quadrant are the abscissa and the ordinate both negative?
10. In what quadrant is the ratio y/x negative?
11. What is the ordinate of a point on the x-axis?
12. What is the abscissa of a point on the y-axis?

In Exercises 13–20 plot the pairs of points and find the distance between the points.

13. $(0, 4), (-1, 3)$ **14.** $(1, 2), (5, 4)$
15. $(\frac{1}{2}, \frac{1}{2}), (\frac{1}{2}, -\frac{3}{4})$ **16.** $(-1, 5), (-1, -6)$

17. (0.5, 1.6), (6.2, 7.5)
18. (−5, 3), (2, −1)
19. (2, −6), $(-\sqrt{3}, -3)$
20. (−3, 4), (0, 4)

In Exercises 21–28 graph the set of points whose coordinates satisfy the given condition(s).

21. $x = 0$
22. $y > 0$
23. $x > 0$
24. $y = 2$
25. $x = y$
26. $x > -1$ and $y > 0$
27. $x > -1$ and $y < -1$
28. $x > 0$ and $y > 0$

29. The point $(x, 3)$ is 4 units from (5, 1). Find x.
30. Find the distance between the points $(\sqrt{x}, \sqrt{y})$ and $(-\sqrt{x}, -\sqrt{y})$.
31. Find the distance between the points (x, y) and $(-x, y)$.
32. Find the point on the x-axis that is equidistant from (0, −1) and (3, 2).

In Exercises 33–36 find the distance from the origin to the given point.

33. (1, 2)
34. (−1, −5)
35. (2, −3)
36. (−4, 1)

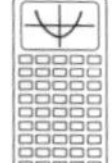

Graphing Calculator Exercises

If you have a graphing calculator, use it to display the following line segments. Choose a scale for the axes that will show the entire line segment and will give a clear indication for the endpoints.

1. (1, 2), (5, 4)
2. (−3, 4), (4, −2)
3. (0, −0.5), (0.1, 0.3)
4. (−1, 1), (0.3, 0.2)
5. (−2, 15), (1, 3)
6. (0, 20), (−4, 10)
7. (13, 25), (18, 30)
8. (−8, 15), (−15, −7)

1.3 Functions

A very important idea in mathematics is that of pairing a number x with a number y by some specified rule called a **rule of correspondence**. The rule of correspondence indicates how a value of x is used to obtain a value of y. We use this idea in the study of **direct variation** when we pair each number x with a number y by the rule of correspondence $y = kx$. Direct variation is an example of a more general idea called **functional pairing**. The pairings in the following table were generated using $y = 3x^2$ as the rule of correspondence.

x	−2	−1	0	1	2	3
y	12	3	0	3	12	27

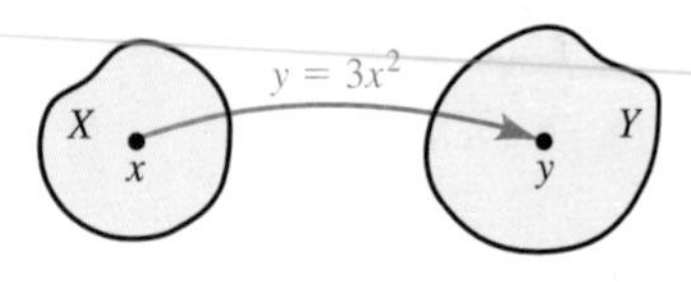

Figure 1.10

If we let X represent the set of values that we can assign to x, and Y the corresponding set of values obtained by the rule $y = 3x^2$, then Figure 1.10 represents the general pairing. The figure emphasizes the idea that the pairings are **ordered;** that is, the value of y is obtained *from* the value of x. The figure is also intended to show that the functional pairing is **unique**; that is, only one value of y is paired with each value of x.

Rules of correspondence that assign a unique value of y in Y to each x in X are of particular interest in mathematics, and we make the following definition.

DEFINITION 1.1 A correspondence f that assigns to each x in X exactly one element y in Y is called a **function**. The set X is called the **domain** of the function and the set of ys is called the **range** of the function. We say that "y is the image of x" or that "y is a function of x."

Functions are frequently expressed by a formula such as $y = 3x^2$, but other means may be used. For instance, a set of ordered pairs $\{(x, y)\}$ is a function if no two distinct pairs have the same first element. This follows from the fact that a function assigns to each x in X in a unique y in Y. The following remarks should clarify the important points of the definition of a function.

- ▸ A formula such as $y = \pm\sqrt{x}$ does *not* define y as a function of x since it assigns two values to each positive value of x. For example, ± 2 are both images of $x = 4$. However, we note that each of the formulas $y = \sqrt{x}$ and $y = -\sqrt{x}$ taken separately does define a function.
- ▸ The expression $y = 8$ defines a function since y has the value 8 for every value of x. The definition does not require that y have a different value for each x, only that it be unique.
- ▸ The expression $x = 5$ does not represent a function because many values of y correspond to $x = 5$. For example, $(5, -1)$, $(5, 0)$, $(5, 3)$ are some of the ordered pairs that satisfy the expression $x = 5$.
- ▸ The set $\{(2, 3), (-1, 4), (0, -5), (3, 4)\}$ is a function since each first element is paired with a unique second element.
- ▸ The set $\{(-3, 4), (2, 5)\ (2, -6), (9, 7)\}$ is not a function because the distinct pairs $(2, 5)$ and $(2, -6)$ have the same first element. Therefore, two different values of y are assigned to the same x.

COMMENT Rules of correspondence that permit more than one value of y to be paired with a value of x are called **relations**. Hence $y = \pm\sqrt{x}$, $x = 5$, and $\{(-3, 4), (2, 5), (2, -6), (9, 7)\}$ are relations. Note that every function is a relation but a relation is not necessarily a function.

The domain of a function can be quite arbitrary. For example, we could limit the domain of $y = 3x^2$ to $x = 0, 1, 2$. In this case the range elements are 0, 3, 12 and the functional pairings are $\{(0, 0), (1, 3), (2, 12)\}$. **If the domain is not specified, we assume that it consists of all real numbers for which the rule of correspondence will yield a real number.** Examples 2 and 3 show functions with restricted domains.

Example 1 The equation $y = x^2 + 5$ defines a function since each value of x determines only one value of y. The domain consists of all real numbers and the range consists only of those real numbers greater than or equal to 5 (since the smallest possible value of x^2 is 0). ▲

Example 2 Find the domain and range of the function $y = \sqrt{x}$.

SOLUTION If we substitute a negative real number for x in $y = \sqrt{x}$, we do not get a real number for y. However, each nonnegative real number substituted for x yields a nonnegative real number for y. Therefore, both the domain and range of this function consist of all nonnegative real numbers. ▲

Example 3 Find the domain and range of the function $y = \dfrac{4}{x - 3}$.

SOLUTION Since division by zero is not allowed, we must exclude 3 as a domain element; we conclude that the domain consists of all real numbers except 3.

To find the range of this function we solve for x and note any restrictions on y. Thus

$$y = \frac{4}{x - 3}$$

$$xy - 3y = 4 \qquad \text{Multiplying both sides by } x - 3$$

$$xy = 3y + 4 \qquad \text{Adding } 3y \text{ to both sides}$$

$$x = \frac{3y + 4}{y} \qquad \text{Dividing both sides by } y$$

The only limitation on value of y is that it cannot equal 0. Therefore the range consists of all real numbers except 0. ▲

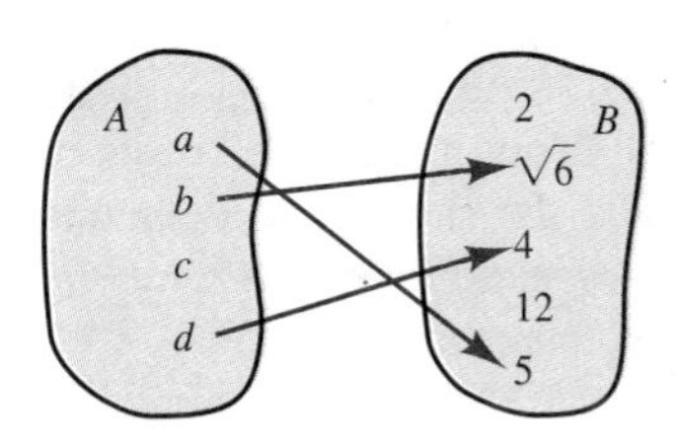

Figure 1.11

Functional rules are not restricted to formulas or mathematical expressions. The next two examples show functions that are described by diagrams and tables.

Example 4 Let A denote the set $\{a, b, c, d\}$ and B denote the set $\{5, 2, 4, \sqrt{6}, 12\}$. Figure 1.11 diagrams a function in which 5 is the image of a, $\sqrt{6}$ is the image of b, and 4 is the image of d. Notice that no number is paired with c. We say that the function is **undefined** for the element c. The domain of this function is the set $\{a, b, d\}$ and the range is the set $\{5, 4, \sqrt{6}\}$. We note that the function defined in Figure 1.11 may also be written as the set of ordered pairs $\{(a, 5), (b, \sqrt{6}), (d, 4)\}$. ▲

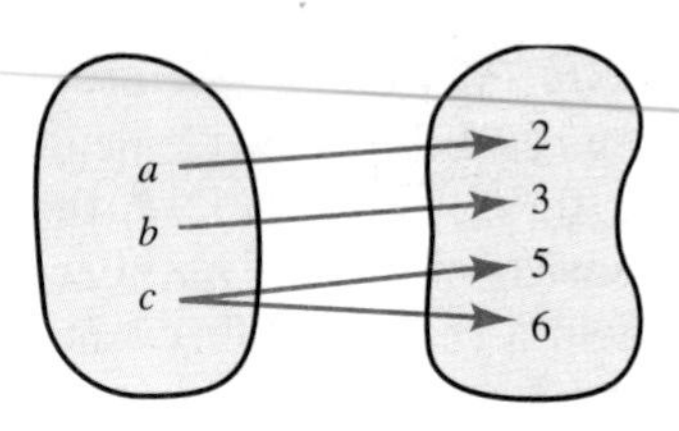

Figure 1.12

Example 5 A table is a listing of ordered pairs. The following table defines a function.

x	-3	-2	-1	0	1	2	3
y	4	0	1	2	0	5	-4

In this case the domain consists of the numbers $-3, -2, -1, 0, 1, 2, 3$. The range consists of the numbers $-4, 0, 1, 2, 4, 5$. The table is the rule of correspondence for this function. ▲

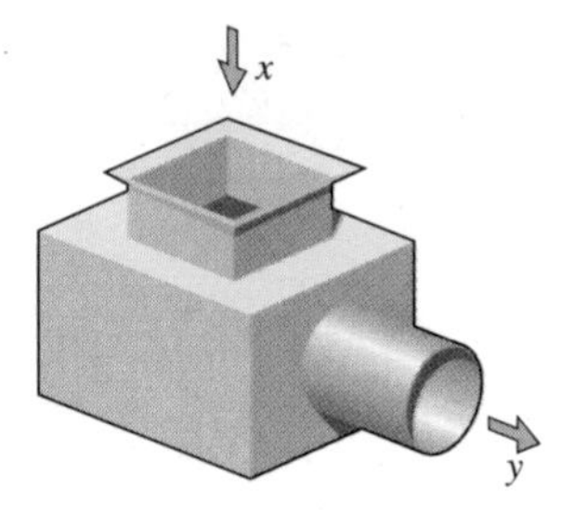

Figure 1.13

Although a functional relation can be given in a variety of forms, it must never assign more than one range element to the same domain element. Figure 1.12 does not describe a function, since c is assigned to the two different elements 5 and 6.

In a useful analogy, we can compare a function to a machine that has an input and an output. (See Figure 1.13). When an element x enters the machine, it is transformed by the machine (function) into a new element, y. The set of elements that can be put into the machine represents the domain of the function and the output represents the elements of the range.

Example 6 An ice-vending machine dispenses a 5-lb bag of ice for \$1, a 10-lb bag for \$2, and a 15-lb bag for \$3. This machine represents a function in which the money is input and the corresponding output is 5 lb of ice, 10 lb of ice, or 15 lb of ice. (See Figure 1.14.) If we ignore the units of measurement (dollars and pounds), the function can be represented by the ordered pairs $\{(1, 5), (2, 10), (3, 15)\}$. ▲

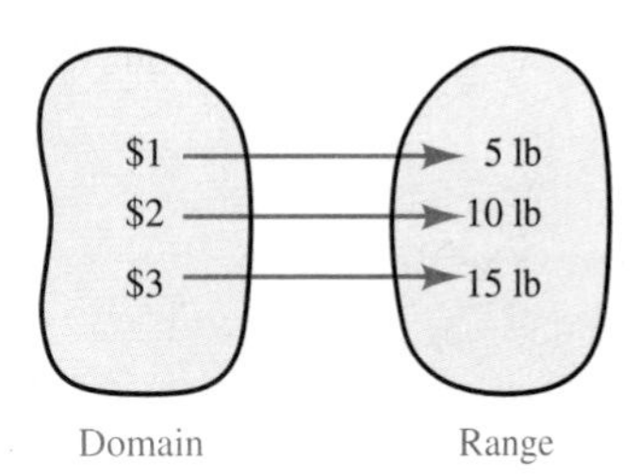

Figure 1.14

Functional Notation

Single letters or symbols customarily denote functions; for instance f, F, G, g, h, or H. If x is an element of the domain of a function f, then the corresponding value in the range is denoted by $f(x)$, which is read "f of x," or the "function evaluated at x." Referring back to Figure 1.11, we note that $f(a) = 5$, $f(b) = \sqrt{6}$, $f(c)$ is undefined, and $f(d) = 4$. The domain element x is sometimes called the **argument** of f and is usually called the **independent variable**. The corresponding value of y is called the **dependent variable**.

WARNING The symbol $f(x)$ does NOT mean the product of f times x.

The function f may or may not have a definite mathematical expression describing the rule of correspondence. However, if such a specific expression is available, functional notation offers a convenient way of indicating numbers in the range. To find a number in the range that is associated with any number in the domain, merely replace the letter x wherever it appears in the expression for $f(x)$ by the number from the domain.

It is sometimes helpful to think of x as representing a blank so that the functional notation tells us what to put in the blank. For example, if $f(x) = x^2 + 3x$, we may think:

$$f(\ ) = (\ )^2 + 3(\ )$$

Then anything may be placed in the blank. For example,

$$f(a^2) = (a^2)^2 + 3(a^2) = a^4 + 3a^2$$

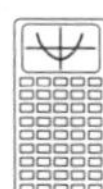

Calculators take the drudgery out of evaluating complex functions for specified values of the domain. Thus with a calculator it is just as easy to evaluate $f(x) = \pi x^3 + \sqrt{x + 2}$ for $x = 3.105$ as for $x = 2$.

Example 7 Let $f(x) = x^2 + 3x$.
(a) Find $f(2)$. **(b)** Find $f(a - 4)$.

SOLUTION **(a)** Substitute 2 for x, so that

$$f(2) = 2^2 + 3 \cdot 2 = 10$$

(b) Substitute $a - 4$ for x, so that

$$f(a - 4) = (a - 4)^2 + 3(a - 4) = a^2 - 8a + 16 + 3a - 12 = a^2 - 5a + 4$$

▲

WARNING Do not confuse functional notation with the distributive law. Thus, $f(a - 4) \neq f(a) - f(4)$.

Although functional notation does not follow the distributive law, we can combine the values of two functions by the arithmetic operations of addition, subtraction, multiplication, and division. For example, if $f(x) = 2x$ and $g(x) = x + 3$, the arithmetic functions are

$$f(x) + g(x) = 2x + x + 3 = 3x + 3$$
$$f(x) - g(x) = 2x - (x + 3) = x - 3$$
$$f(x) \cdot g(x) = 2x(x + 3) = 2x^2 + 6x$$
$$\frac{f(x)}{g(x)} = \frac{2x}{x + 3}$$

In combining two functions to form a new function, the domain of the new function need not be the same as either of the original functions. Thus the function formed above by $f(x)/g(x)$ is not defined at $x = -3$, but the original functions $f(x) = 2x$ and $g(x) = x + 3$ are both defined at $x = -3$.

Example 8 If $f(x) = 2x + 3$, find and expand the product $f(x + 1) \cdot f(5x)$.

SOLUTION Note that $f(x + 1) = 2(x + 1) + 3 = 2x + 5$ and $f(5x) = 2(5x) + 3 = 10x + 3$. Therefore

$$\begin{aligned} f(x + 1) \cdot f(5x) &= (2x + 5)(10x + 3) \\ &= 20x^2 + 56x + 15 \end{aligned}$$

▲

The function $f(x) = x^2$ has the property that $f(2) = 2^2 = 4$ and $f(-2) = (-2)^2 = 4$, so $f(-2) = f(2)$. We say that $f(x) = x^2$ is an even function. The function $g(x) = x^3$ has the property that $g(2) = 2^3 = 8$ and $g(-2) = (-2)^3 = -8$, so $g(-2) = -g(2)$. We say that this function is an odd function. In general we define even and odd functions as follows.

DEFINITION 1.2

- A function f is **even** if

$$f(-x) = f(x)$$

for every x in the domain of f.

- A function f is **odd** if

$$f(-x) = -f(x)$$

for every x in the domain of f.

Example 9

(a) Show that $f(x) = 5 - x^4$ is an even function.

(b) Show that $g(x) = 2x^3 + x$ is an odd function.

(c) Show that $h(x) = x^2 - 3x$ is neither even nor odd.

SOLUTION

(a) To show that f is an even function, we must show that $f(-x) = f(x)$ for any real number x.

$$f(-x) = 5 - (-x)^4 = 5 - x^4 = f(x)$$

This shows that f is an even function.

(b) For any real number x, we have

$$g(-x) = 2(-x)^3 + (-x) = -2x^3 - x = -(2x^3 + x) = -g(x)$$

This shows that g is an odd function.

(c) For any real number x, we have

$$h(-x) = (-x)^2 - 3(-x) = x^2 + 3x$$

Since $x^2 + 3x$ is not $h(x)$ or $-h(x)$, we conclude that h is neither even nor odd.

▲

Exercises for Section *1.3*

Which of the expressions in Exercises 1–8 define functions? (Assume that y is dependent on x.)

1. $y = 2x + 5$ **2.** $y = x^2$ **3.** $y = 10$ **4.** $y < 3x$

5. $y = \sqrt[3]{x}$ **6.** $y^2 = x^3$ **7.** $y^2 = 5x$ **8.** $y = 1/x$

Which of the sets or tables in Exercises 9–12 define y as a function of x?

9. $\{(2, 3), (-1, 4), (3, 0), (0, 4)\}$ **10.** $\{(-1, 0), (2, 3), (2, -2)\}$

11.

x	1	1	2	3
y	2	3	4	7

12.

x	2	7	8
y	3	9	3

13. Which of the following diagrams define functions?

(a)

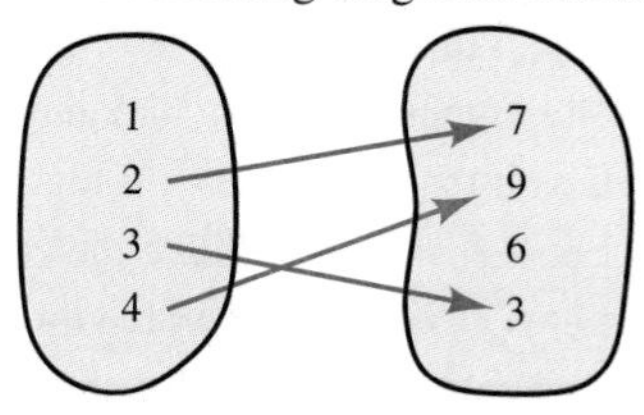

(b)

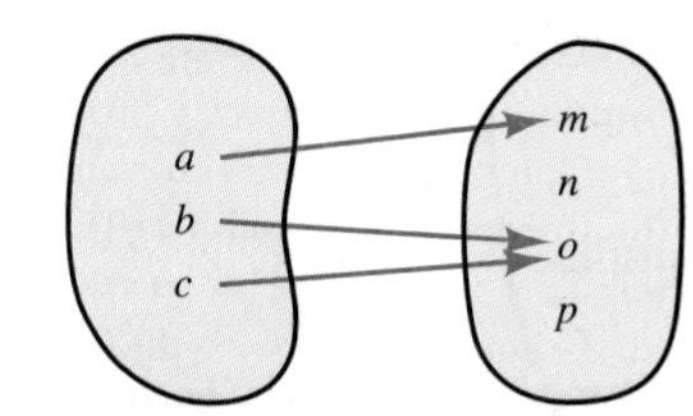

(c)

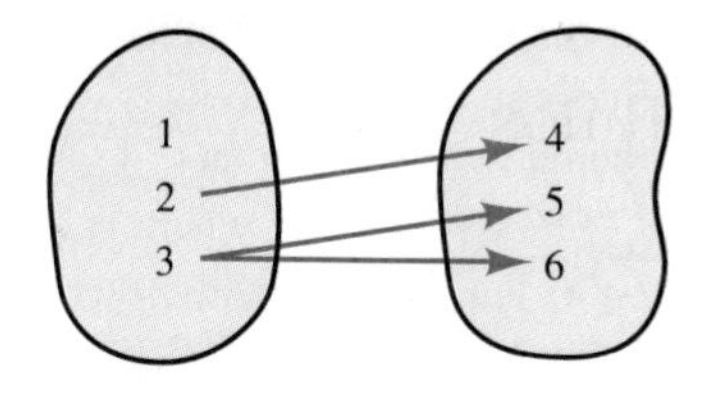

Find the domain and the range of each of the real valued functions in Exercises 14–24.

14. $y = 2x$ **15.** $y = 4 - 3x$

16. $y = 3x^2 + 5$ **17.** $y = \sqrt{x}$

18. $y = 1/x^2$ **19.** $y = x^{1/3}$

20. $\{(1, 2), (3, 5), (7, 1), (12, -2)\}$

21. $f(x) = \sqrt{-x}$ **22.** $f(x) = \sqrt{-x^2}$

23. $f(x) = \sqrt{x - 1}$ **24.** $f(x) = \sqrt{1 - x}$

25. Suppose that the function f has the rule $f(x) = 3x + 1$. Compute the following:
(a) $f(3)$ (b) $f(\pi)$ (c) $f(z)$ (d) $f(x - h)$
(e) The element in the domain that maps onto 10
(f) The range of the function

26. Suppose that the function G has the rule $G(t) = t^2 - 2t + 1$. Compute the following:
(a) $G(2)$ (b) $G(-1)$ (c) $G(x^2)$ (d) $G(x - h)$ (e) $G(\sqrt{t})$

In Exercises 27–34 identify the given function as even, odd, or neither even nor odd.

27. $f(x) = -3x^5$
28. $f(x) = 2x + 5$
29. $f(x) = 2x - x^3$
30. $f(x) = 5x^4$
31. $f(x) = x^4 - 3x$
32. $f(x) = x^2 + 10$
33. $y = \sqrt{x^2 + 4}$
34. $y = (x^2 - 1)/x^2$

35. Given that $g(x) = 2x - 7$, find and expand $g(x - 4) \cdot g(x - 1)$.
36. Given that $h(t) = t^2$, find and expand $h(t + 2) \cdot h(\sqrt{t})$.
37. Given that $F(x) = 3x^2$, find and expand $F(x) \cdot F(x - 3)$.
38. Let $f(x) = x - 3$. Solve the equation $f(x + 2) \cdot f(x - 1) = 0$.
39. Let $f(x) = x^2 + 2x$. Solve the equation $f(x) - 3 = 0$.
40. Let $h(t) = t^2 - 9$. Solve the equation $h(t - 1) + 9 = 0$.
41. Let $f(x) = 2x - x^2$. Solve the equation $f(x) - f(2x) = 0$.
42. Suppose that y is directly proportional to x; that is, $y = f(x) = kx$.
(a) Compute $f(x_1)/f(x_2)$ for any two numbers x_1 and x_2. ($f(x_2) \neq 0$).
(b) Compare $f(1/x)$ with $1/f(x)$.
(c) Compare $f(x^2)$ with $[f(x)]^2$.
(d) Compare $f(x) + 1$ with $f(x + 1)$.
(e) Compare $f(x_1 + x_2)$ with $f(x_1) + f(x_2)$.
(f) Compare $af(x)$ with $f(ax)$, where a is a constant.
43. Suppose that y is inversely proportional to x; that is, $y = f(x) = k/x$. Answer the questions in Exercise 42 for this function.
44. Let $f(x) = mx + b$, where m and b are constants.
(a) Compare $f(ax)$ with $af(x)$.
(b) Compare $f(x) + 1$ with $f(x + 1)$.

Use a calculator to find the indicated range value in Exercises 45–48.

45. Given $f(s) = 3.101s^2 + 29.46s$, find $f(-0.415)$.
46. Given $g(t) = 4356t(3.7 + t)$, find $g(287.7)$.
47. The current i in a resistor is found to vary with time according to

$$i(t) = 1.196t^2 + 0.076t$$

where i is in amperes and t is in seconds. Find $i(1.375)$.
48. The equation of motion of a rocket is given by

$$h(t) = 150.6t - 16.1t^2$$

where h is altitude in feet when t is measured in seconds. Find $h(0.45)$.

49. Jim must pay taxes of 12% to the IRS, 10% to Social Security, 6% to the state, and 1.5% to the city. In addition, \$150 is withheld from each paycheck for retirement. Write a function that gives Jim's take-home pay, P, as a function of his gross wages, w.

50. The insurance premium for a company health plan is presently \$25,000. The company controller estimates that this cost will increase each year by an amount equal to the square root of gross revenues. Write an expression for insurance premium, P, as a function of gross revenue, U.

1.4 The Graph of a Function

The **graph of a function** f is the set of points (x, y) in the plane whose coordinates satisfy the equation $y = f(x)$. The first number in the ordered pair is the domain element and the second number is the range element. Domain values are plotted along the horizontal axis, and range values along the vertical axis.

Example 1 The graph of the function defined by $f = \{(1, 3), (-2, 2), (5, -1)\}$ is shown in Figure 1.15.

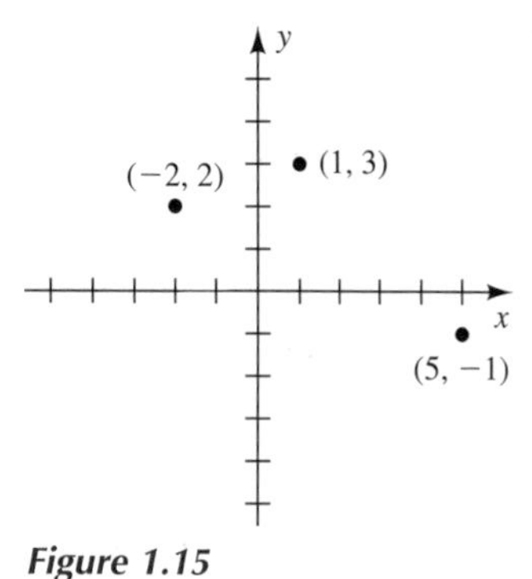

Figure 1.15 ▲

Example 2 Draw the graph of the function g if the domain of g is $\{0, 1, 4, 9\}$ and the rule of correspondence is $g(x) = -\sqrt{x}$.

SOLUTION From the given conditions, we have $g = \{(0, 0), (1, -1), (4, -2), (9, -3)\}$. The graph of g is shown in Figure 1.16.

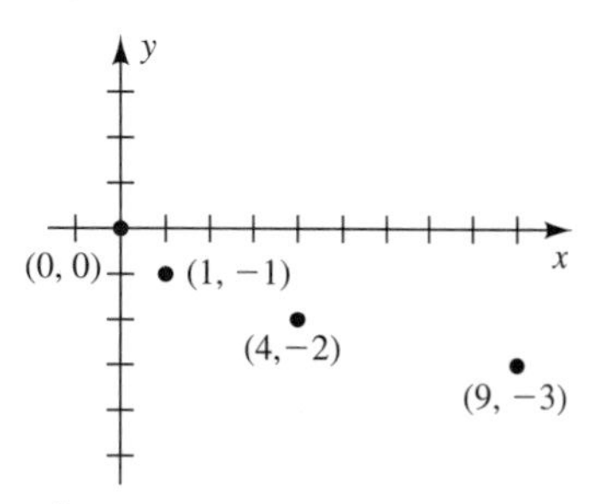

Figure 1.16 ▲

The domain of a function is usually an interval of real numbers. The graphs of such functions can be constructed by plotting all points in the plane whose coordinates satisfy the function. However, this approach is impractical since there is no limit to the number of such points that can be plotted. In practice the graph of such a function is constructed by plotting a few selected points and connecting these points with a smooth curve. As an illustration, consider the function defined by $y = x^2$, whose domain is the entire set of real numbers. By assigning values to x, we obtain the following set of ordered pairs.

x	−3	−2	−1	0	1	2	3
y	9	4	1	0	1	4	9

Now plot the set of ordered pairs in the Cartesian plane, as shown in Figure 1.17(a). The representation of the graph of $y = x^2$ can then be obtained by connecting the points with a smooth curve, as in Figure 1.17(b). This graph is, of course, only an approximation of the actual graph of the function. Its accuracy depends on the number of points plotted and the care taken in drawing the smooth curve connecting the points.

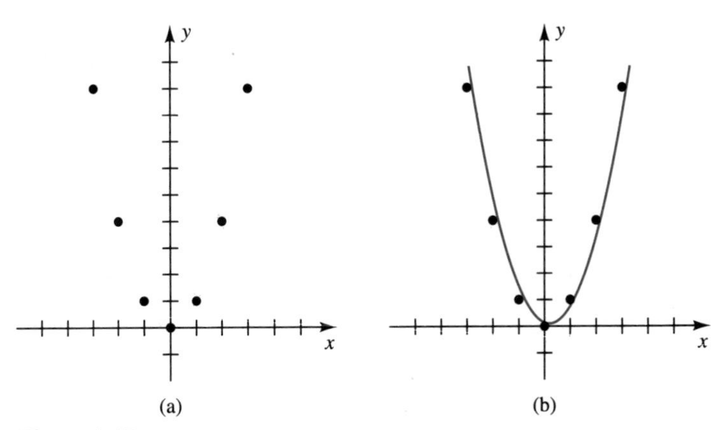

Figure 1.17

Usually only a segment of the graph in the vicinity of the origin is plotted, even though the graph may extend much farther. For instance, in Figure 1.17 only the portion of the curve from $x = -3$ to $x = 3$ is plotted, although the domain of the function is the set of all real numbers.

When graphing a function it is always helpful to determine the domain beforehand, since the domain determines the extent of the graph with respect to the x-axis.

Example 3 Graph the function $y = \sqrt{4 - x}$.

SOLUTION The domain consists of all real numbers x for which $4 - x \geq 0$. Otherwise the range values would not be real numbers. Hence the domain is the set of all real numbers x such that $x \leq 4$. This means that the graph must be to the left of the vertical line $x = 4$, except for the point $(4, 0)$, which *is* part of the graph, as you can see in Figure 1.18.

$y = \sqrt{4 - x}$

x	y
4	0
2	$\sqrt{2}$
0	2
−2	$\sqrt{6}$

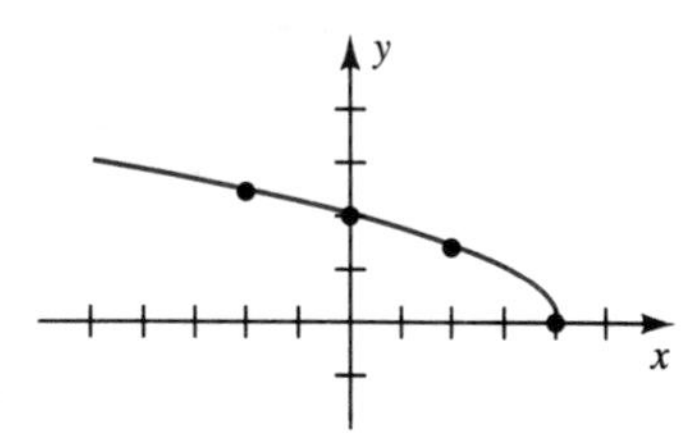

Figure 1.18

▲

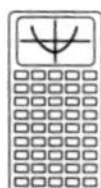

A graphing calculator may be used to display the graphs of functions. Simply insert the expression for the function and use the graphing capability of the calculator. The figure shows the display of the graph of $y = \sqrt{4 - x}$ on the interval $-4 \leq x \leq 7$.

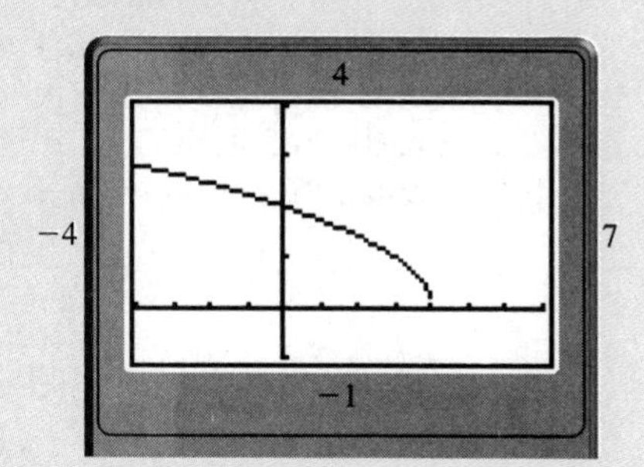

See Section 1.5 for a discussion of the graphing calculator.

In many physical situations, the nature of the problem restricts the domain of the function. For instance, the height above ground of a ball thrown upward from ground level with an initial velocity of 32 ft/sec is described by the function $h = 32t - 16t^2$, where t is the elapsed time in seconds and h is the vertical height in feet. The ball will strike the ground when $h = 0$—that is, when $32t - 16t^2 = 0$, or $t = 2$ sec after the ball is thrown upward. The domain of this function is $0 \leq t \leq 2$, since the equation has no meaning before the ball is thrown or after it hits the ground.

Example 4 Draw the graph of $h = 32t - 16t^2$ on $0 \le t \le 2$.

SOLUTION The table and graph in Figure 1.19 result from computing some convenient values of h in the interval $0 \le t \le 2$. The graph shows that the maximum height the ball reaches is 16 ft.

$h = 32t - 16t^2$

t	h
0	0
.5	12
1.0	16
1.5	12
2.0	0

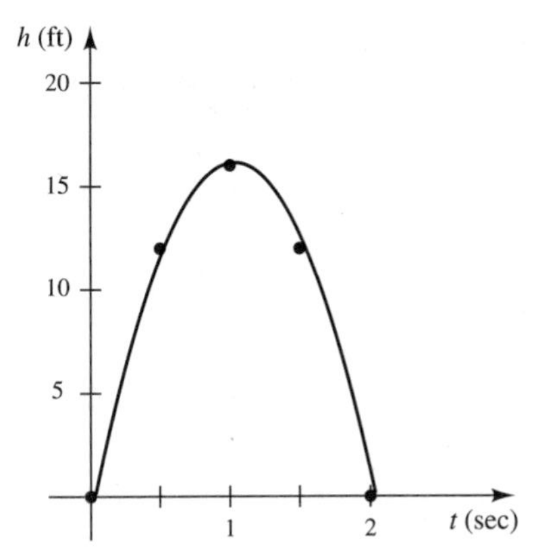

Figure 1.19

▲

In business, the relationship between the price of an item and the number of items purchased by the consumer (in a fixed period of time) is called the **demand function**. The relation between the price and the number of items supplied by the producer is called the **supply function**. If n is the number of items and p is the price, we can express either a supply or demand function in the form

$$p = f(n)$$

Typical demand and supply curves are shown in the next example.

Example 5

(a) Draw the demand curve corresponding to $p = 8000 - \frac{1}{2}n^2$ for $0 \le n \le 100$.

(b) Draw the supply curve corresponding to $p = 100 + 5\sqrt{n}$ for $0 \le n \le 900$.

SOLUTION Some convenient values of n are used to compute the values in the tables. The curves are drawn from these tabled values. Although the graphs of demand and supply functions consist of sets of discrete points, it is common practice to show a smooth curve through these points, as in Figure 1.20. ▲

The two curves shown in Figure 1.20 characterize the conditions in a free market: The demand curve shows that as price decreases, the demand for the item will increase; and the supply curve shows that as the price increases, the supply will also increase.

$p = 8000 - \frac{1}{2}n^2$

n	p
0	8000
20	7800
40	7200
60	6200
80	4800
100	3000

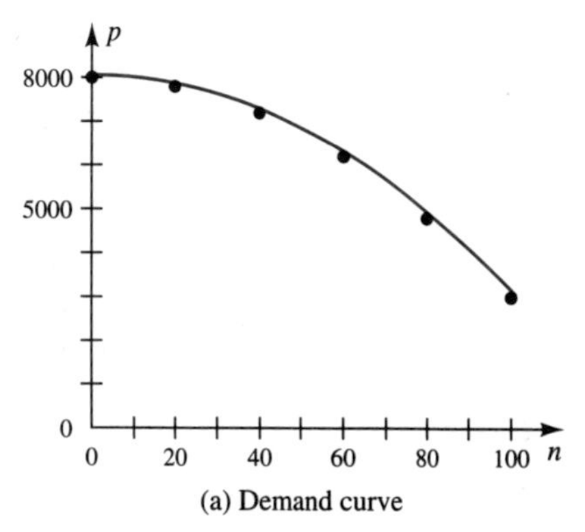

(a) Demand curve

$p = 100 + 5\sqrt{n}$

n	p
0	100
100	150
400	200
900	250

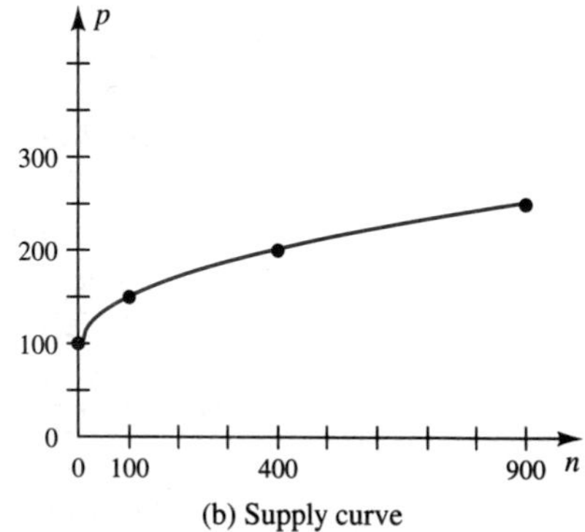

(b) Supply curve

Figure 1.20

Even and Odd Functions

The concept of even and odd functions was presented in Definition 1.2 of the previous section. The graphs of even and odd functions have an interesting kind of symmetry with respect to the coordinate axes.

- Given that f is an even function, then $y = f(x) = f(-x)$ for every x in the domain of f. This means that if (x, y) is on the graph of the function f, then so is $(-x, y)$ See Figure 1.21. A graph that has this property is said to be **symmetric with respect to the y-axis.**

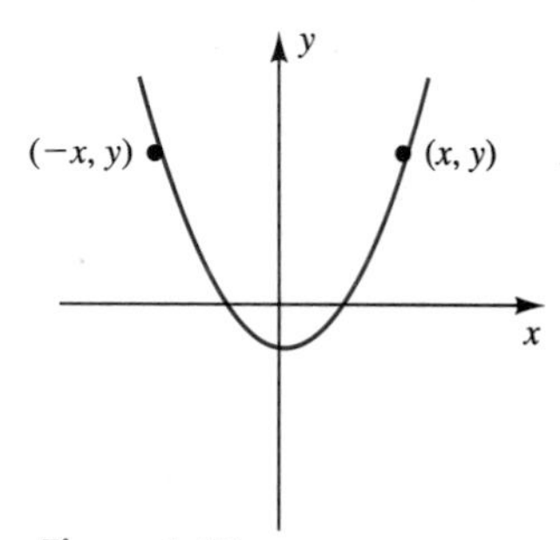

Figure 1.21
Even function

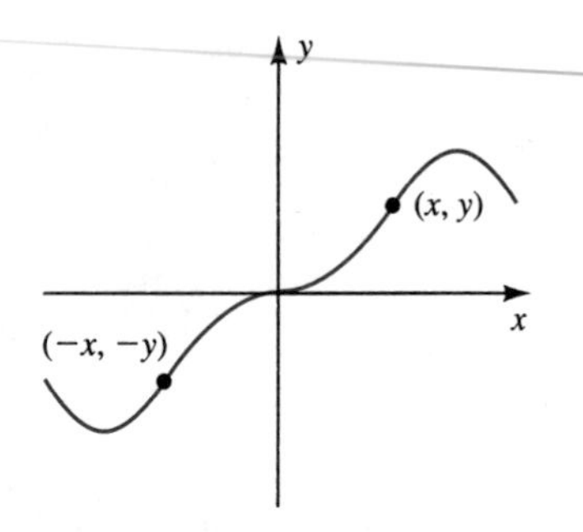

Figure 1.22
Odd function

▸ Given that f is an odd function, then if (x, y) is on the graph of the function f, so is $(-x, -y)$. See Figure 1.22. A graph that has this property is said to be **symmetric with respect to the origin.**

We can use our knowledge of the symmetry of a graph to assist us in the sketching process. For example, the function given in Figure 1.17 is even; so we could have drawn the graph by plotting points from $x = 0$ to $x = 3$ and then used its symmetry with respect to the y-axis to obtain the segment from $x = -3$ to $x = 0$.

Using a Graph to Define a Function

A set of ordered pairs of real numbers is represented graphically by points in the Cartesian plane. Conversely, a graph determines a set of ordered pairs of real numbers corresponding to the coordinates of the points on the graph. The set of ordered pairs determined by a graph may or may not define a function.

There is a simple test, called the **vertical line test**, to tell whether a graph defines a function of x. Draw vertical lines on the graph. If no vertical lines can be drawn to intersect the graph in more than one point, the graph defines a function because for each x there is exactly one y. Figures 1.23(a) and (b) define y as a function of x. On the other hand, Figures 1.23(c) and (d) do not define functions since some vertical lines intersect these graphs in two or more points, indicating that there is more than one value of y for some values of x.

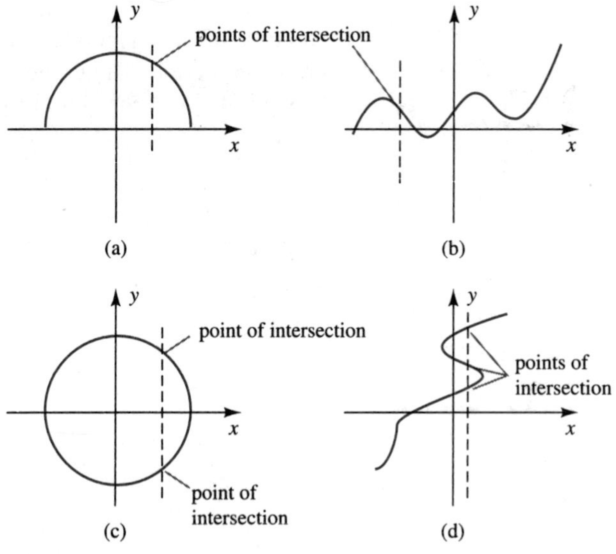

Figure 1.23

Increasing and Decreasing Functions

A function f whose graph rises from left to right is said to be an increasing function. See Figure 1.24(a). This means that as x increases, $f(x)$ also increases. A function f whose graph falls from left to right is said to be a decreasing function. See Figure 1.24(b). In this case $f(x)$ decreases as x increases. In general we speak of a function increasing or decreasing on some interval as in the following definition.

DEFINITION 1.3

1. A function f is **increasing** on the interval $[a, b]$ if $f(x_1) < f(x_2)$ whenever $x_1 < x_2$ in $[a, b]$.
2. A function f is **decreasing** on the interval $[a, b]$ if $f(x_1) > f(x_2)$ whenever $x_1 < x_2$ in $[a, b]$.

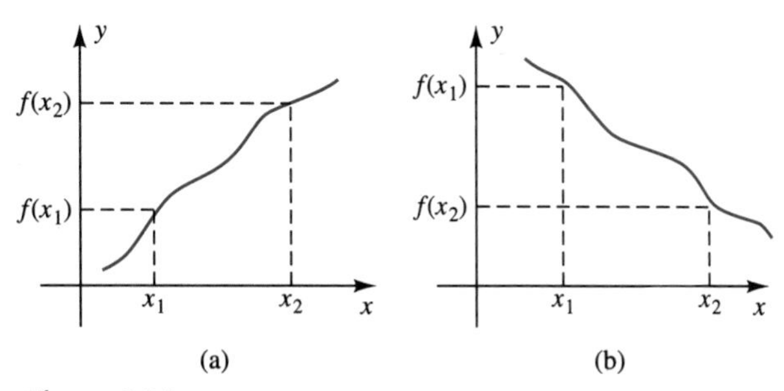

Figure 1.24

Example 6 Increasing and decreasing functions are illustrated in Figure 1.25.

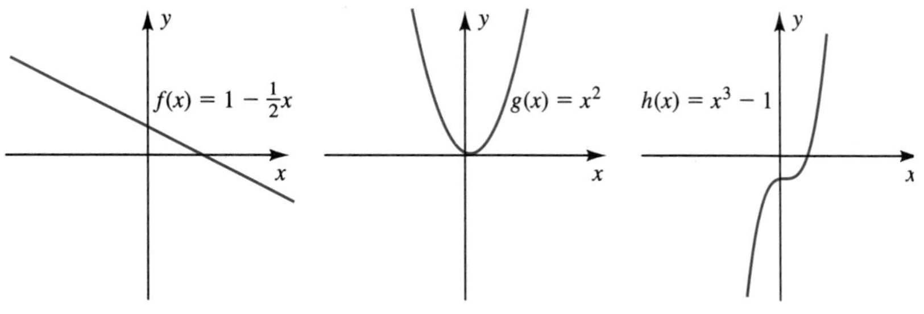

(a) f is decreasing for all x in the domain

(b) g is decreasing on the interval $(-\infty, 0]$ and increasing on the interval $[0, \infty)$

(c) h is increasing for all x in the domain

Figure 1.25 ▲

The Zeros of a Function

A **zero** of a function is any domain value whose corresponding range value is 0. Thus if $f(C) = 0$, where C is a number in the domain of the function f, then C is called a zero of f. The concept of a zero of a function has a simple graphical analog: *The zeros of a function correspond to the abscissas of the points at which the graph of the function intersects the x-axis.* The abscissas of these points are called ***x*-intercepts.** In Figure 1.26(a) the values x_1, x_2, and x_3 are zeros of the function. Figure 1.26(b) shows a function that has no zeros. The zeros should be indicated on the graph, at least approximately.

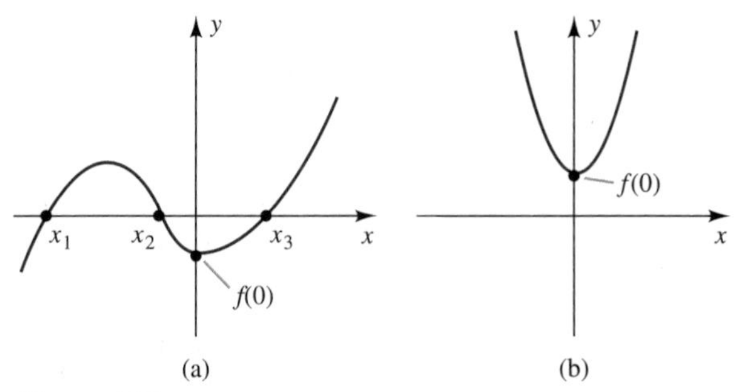

Figure 1.26

COMMENT Be sure to distinguish carefully between the zeros of a function and the value of the function when $x = 0$. The x-intercepts are the zeros, whereas the value of the y-intercept is $f(0)$.

Example 7 Draw the graph of $f(x) = x^2 - 3x + 1$ and indicate the zeros.

SOLUTION Computing $f(x)$ for some convenient values of x, we get the graph shown in Figure 1.27.

$f(x) = x^2 - 3x + 1$

x	$f(x)$
−1	5
0	1
1	−1
2	−1
3	1
4	5

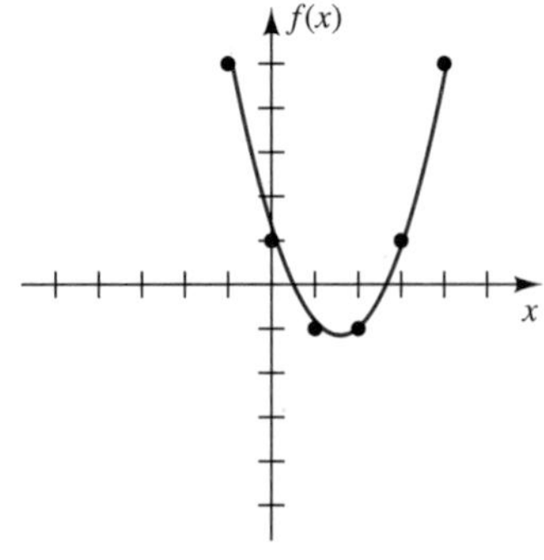

Figure 1.27

Here the graph crosses the x-axis at approximately $x = 0.4$ and $x = 2.6$. Therefore these points are the zeros of the function. By using a graphical method, we can estimate the zeros. Analytically the zeros are obtained by setting $f(x) = 0$. Therefore

$$x^2 - 3x + 1 = 0$$

whose roots, using the quadratic formula, are found to be

$$x = \frac{3 \pm \sqrt{5}}{2}$$

These two roots of the quadratic equation give the exact zeros of the function. ▲

▲▼ Exercises for Section *1.4*

In Exercises 1–4 indicate those graphs that define functions of x.

1. (a)

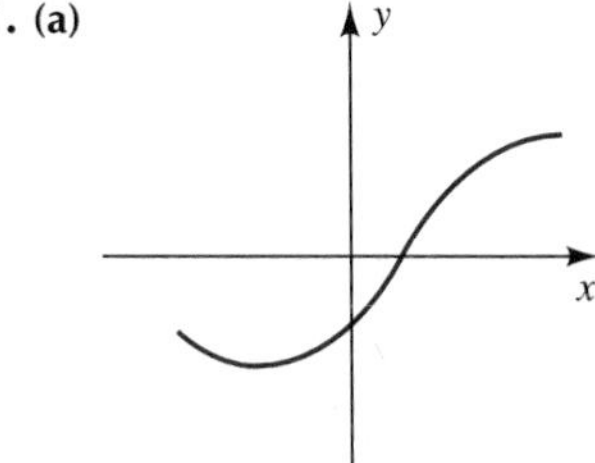

(b)

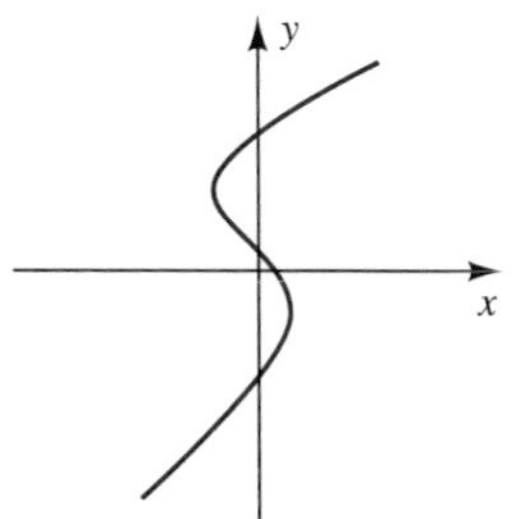

2. (a)

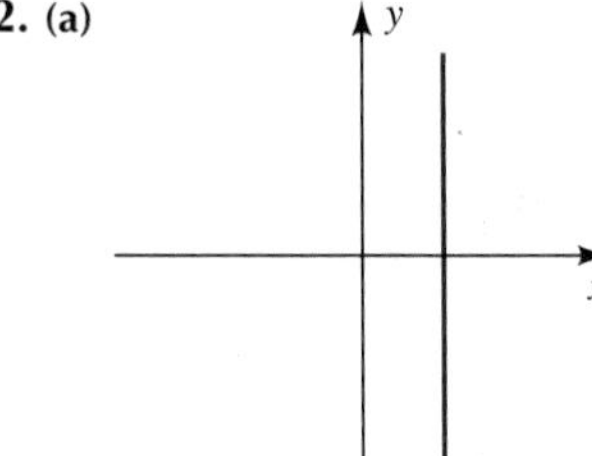

(b)

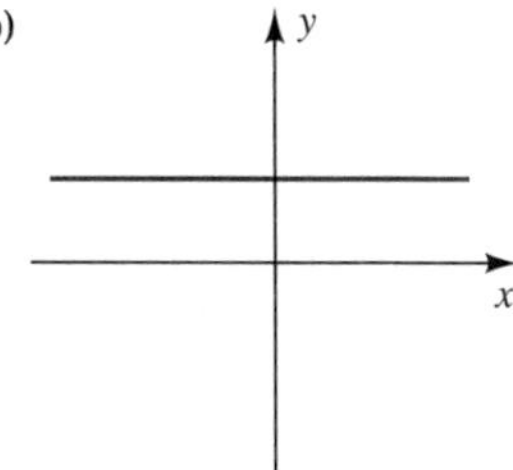

3. (a)

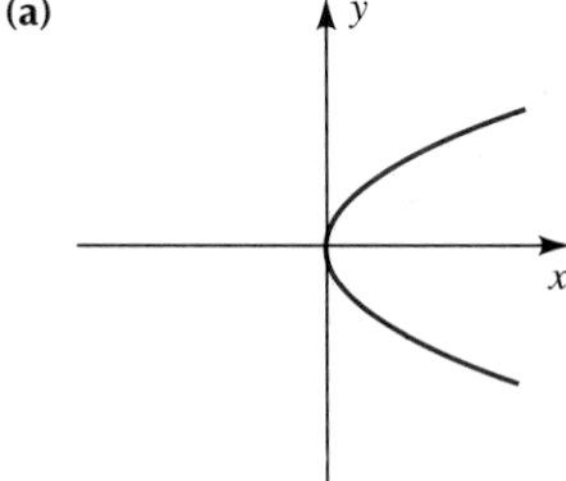

(b)

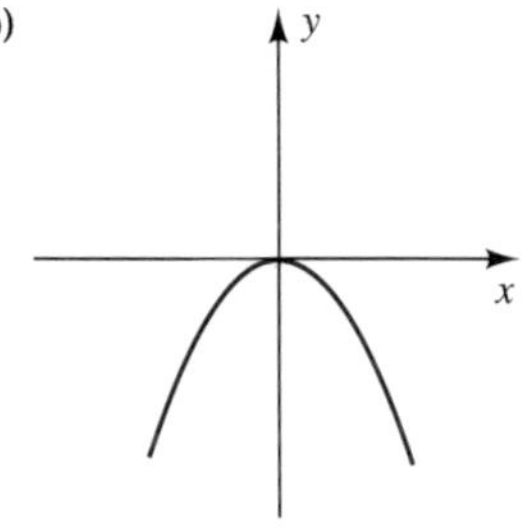

4. **(a)**

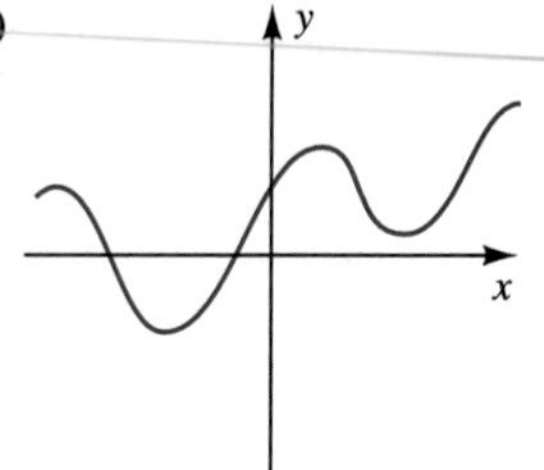

(b)

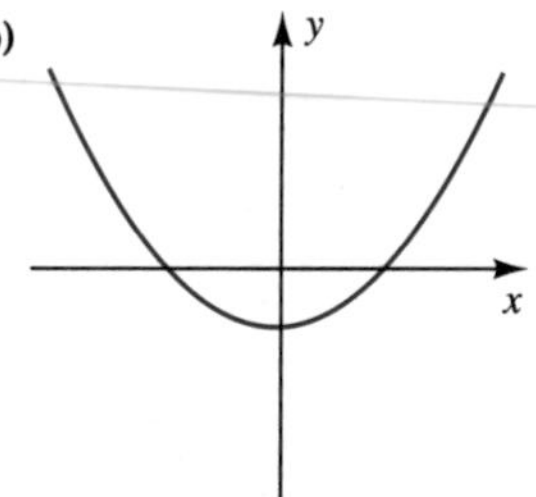

5. Indicate the intervals on which f is increasing and the intervals on which it is decreasing.

(a)

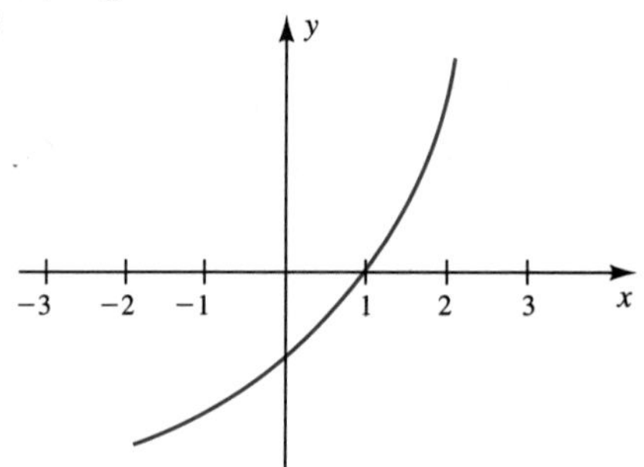

(b)

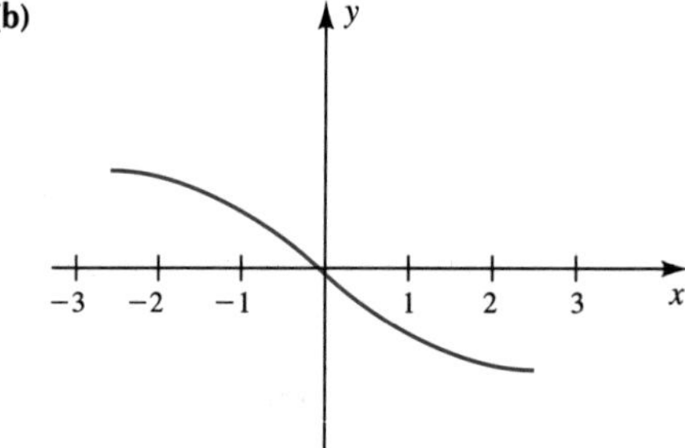

(c)

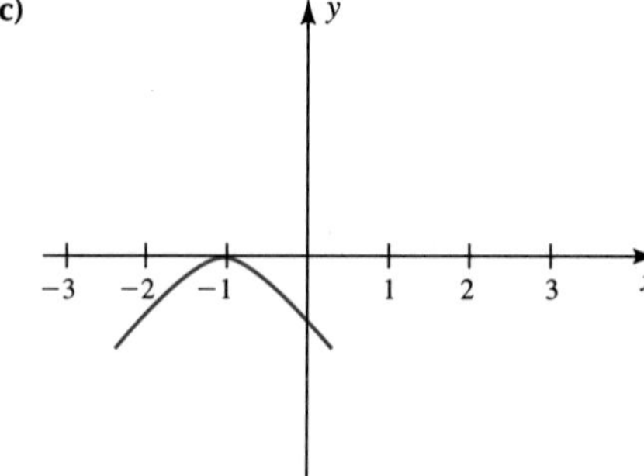

6. Indicate the intervals on which f is increasing and the intervals on which it is decreasing.

(a)

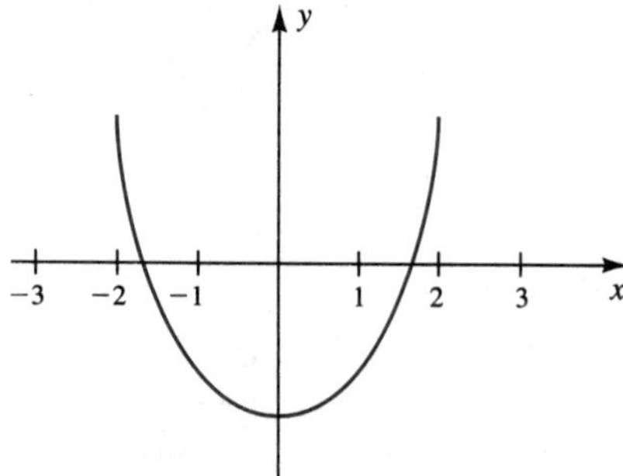

(b)

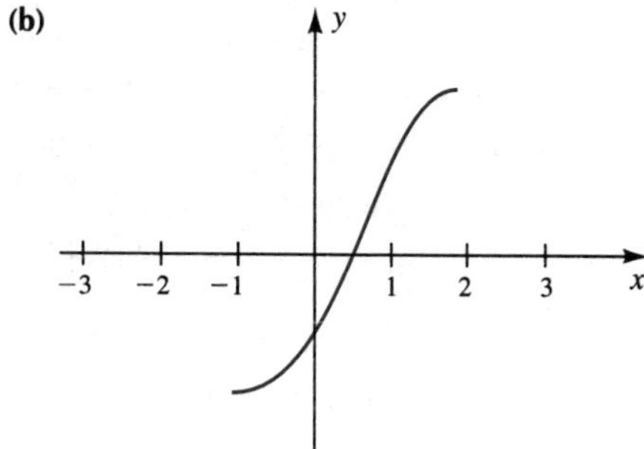

(c)

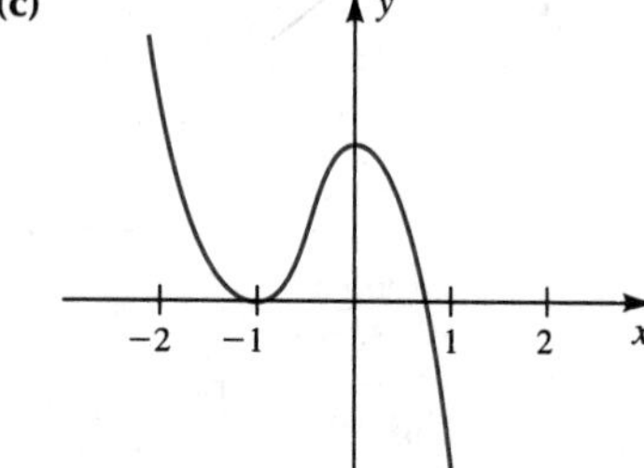

Graph the functions in Exercises 7–28. Indicate the domain and range. Where appropriate, find the value of the zeros.

7. $\{(-2, 3), (0, 4), (2, 5)\}$
8. $\{(-1, 0), (0, 0), (1, 2), (3, -4)\}$
9. $f(x) = 3x + 5$
10. $y = 2 - \frac{1}{2}x$
11. $y = x^3$
12. $f(x) = -x^2$
13. $z = t^2 + 4$
14. $i = r - r^2$

15. $y(x) = \sqrt{x}$

16. $\phi = \dfrac{w^2}{2}$

17. $p = z^2 - z - 6$

18. $v = 10 + 2t$

19. $y = \sqrt{16 - 4x^2}$

20. $y = \sqrt{25 - x^2}$

21. $\alpha = \theta^{1/3}$

22. $z = \sqrt[3]{t^2}$

23. $y(x) = \sqrt{x - 2}$

24. $y = \sqrt[3]{2 - x}$

25. $v = s^3 - 4s^2$

26. $a = \frac{1}{4}b^4$

27. $y = -x^3$

28. $\beta = -\alpha^{1/2}$

29. The work w done in moving an object varies with the distance s according to $w = \sqrt[3]{2s}$. Show this relationship graphically.

30. The path of a certain projectile is described by the function $h = 100x - 2x^2$, where h is the vertical height in feet and x is the horizontal displacement in feet. Draw the path of the projectile.

31. An office machine is supposed to be serviced once a month. If it is not serviced, the cost of repairs is \$20 plus five times the square of the number of months the machine goes unserviced. Express the cost of repairs as a function of the number of months the machine goes unserviced and draw the graph.

32. A typical demand curve in economics is given by

$$D(x) = \frac{50}{x + 10}$$

Sketch this function for $0 \le x \le 10$.

33. A typical supply function in economics is given by

$$S(x) = \frac{1}{2}x^2 + 20$$

Sketch this function for $0 \le x \le 6$.

34. The following table describes the functional relationship between the safe speed at which a car can round a curve and the degree of the curve. Draw the graph of safe speed as a function of the degree of the curve.

Degrees of curve	5	10	15	20	25	30	35
Safe speed (mph)	70	68	66	63	58	50	34

35. A thermocouple generates a voltage in millivolts when its two ends are kept at different temperatures. In laboratory experiments the cold end of the thermocouple is usually kept at 0°C while the temperature of the other end varies. The next table represents the results of an experiment in which the output voltage of the thermocouple was recorded for various thermocouple temperatures. Draw the graph of voltage versus temperature.

$T(°C)$	0	50	100	150	200	250
V(mv)	0	2.1	4.1	5.9	7.0	7.8

36. The following table shows men's height and corresponding normal weight in pounds. Draw the graph of weight versus height.

Ht.	5′2″	5′4″	5′6″	5′8″	5′10″	6′0″	6′2″	6′4″
Wt.	130	136	144	152	161	170	184	196

1.5 The Graphing Calculator

In the previous section we discussed graphs of functions, which we obtained by connecting a few plotted points with a smooth curve. Several calculators are presently available that will generate and display graphs of functions: for example, Texas Instruments models 81, 82, and 85; Hewlett Packard model 48G, Casio model FX7700GE, and Sharp model EL9300. Each of these graphing calculators works similarly, but they are different enough so that we can give only general instructions on their use. In this section we will describe the general features of graphing calculators. You will have to refer to the manual for your specific brand to use it to graph functions.

Features of the Graphing Calculator

All graphing calculators have a **viewing screen,** which is normally several times larger than the screens of calculators that do not have a graphing capability. The screen is a rectangular grid of small regions, called **pixels,** that can be energized to generate figures such as graphs of functions. A typical graphing calculator is shown in Figure 1.28. Most graphing calculators have a grouping of four keys that control the movement of the cursor around the viewing screen. These keys are usually grouped as shown in Figure 1.28 and are used to move the cursor left and right as well as up and down. A row of keys below the viewing screen is provided to perform specific operations such as [Y =], [GRAPH], [RANGE], [TRACE], and [ZOOM]. As we indicated previously, you will need to refer to the calculator manual for your brand to see how it works.

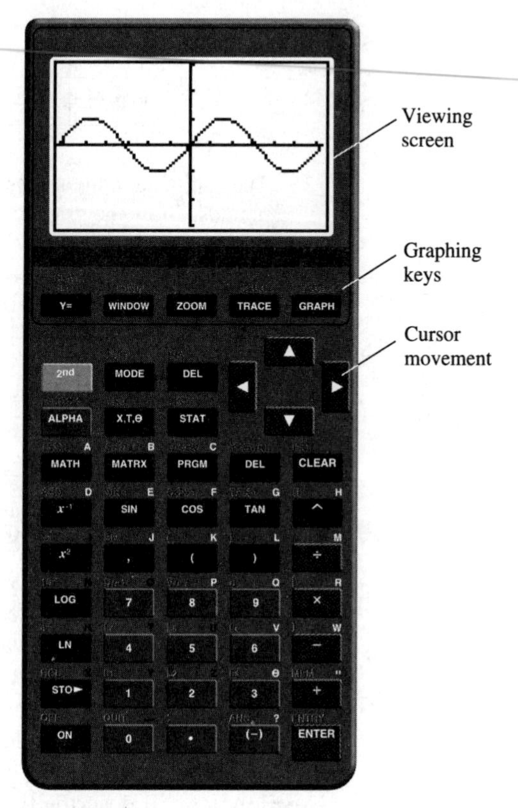

Figure 1.28

With our generic calculator, the function to be graphed is entered by pressing the Y = key and then keying in the appropriate function. Usually you will have the capability to key in several functions and have their graphs displayed either one at a time on the viewing screen or all at the same time (simultaneously).

Ranges of Values for the Viewing Screen

To display the graph of a function that has been keyed into the calculator, press the GRAPH key. The calculator has a default setting for the range of values used for the x- and y-axes; typically, the default range is from -10 to 10 on both axes. The user has the option of selecting the range of values to be displayed for both axes by pressing the RANGE key.

Example 1 Figure 1.29(a) shows the graph of $y = x^3 - x$ with viewing screen ranges of $-10 \leq x \leq 10$ and $-10 \leq y \leq 10$; Figure 1.29(b) shows the same graph with a viewing screen of $-3 \leq x \leq 3$ and $-2 \leq y \leq 2$. The details of the graph of $y = x^3 - x$ are clearly more visible in Figure 1.29(b). Notice that the interval chosen for the x-axis is independent of that chosen for the y-axis and the choice of values for the viewing screen is up to you.

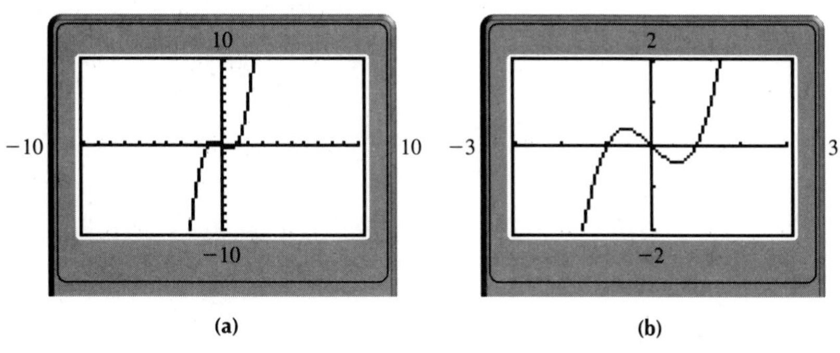

Figure 1.29 ▲

Example 2 As noted earlier, selecting an appropriate range of values for the viewing screen is an important part of graphing functions. For instance, when graphing $y = x^2 + 12x + 3$ using the default range settings $-10 \leq x \leq 10$ and $-10 \leq y \leq 10$, you get the graph shown in Figure 1.30(a). The same graph is shown in Figure 1.30(b) using range settings of $-15 \leq x \leq 5$ and $-40 \leq y \leq 20$. Only in the latter graph do we see the distinctive U shape, characteristic of a *parabola*. This shows why the selection of range values for the viewing screen is such an important part of using a graphing calculator.

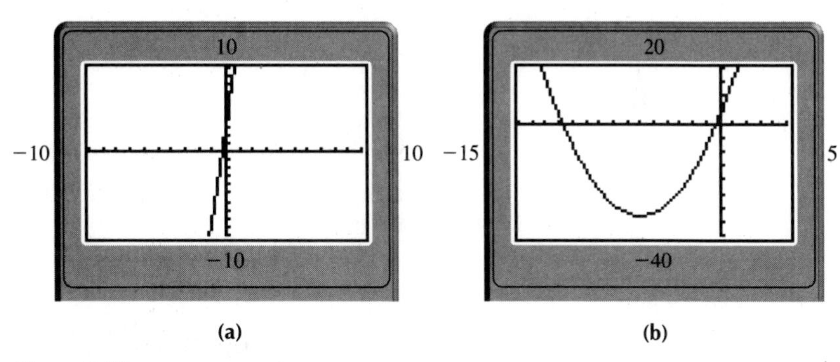

Figure 1.30 ▲

Tracing a Graph

Suppose that in the previous example we wished to find the minimum value of $y = x^2 + 12x + 3$ as shown in Figure 1.30(b). To do this graphically, we use the **TRACE** key. When a graph is displayed and the **TRACE** key is pressed, a blinking cursor will appear somewhere near the middle of the graph, and the x- and y-coordinates of the cursor will be displayed at the bottom of the viewing screen. By using the [←] and [→] keys, the cursor can be moved along the curve. Keeping track of the y-coordinate allows you to locate and estimate the minimum value of the function. Use your calculator and the [TRACE] key to verify that $y = x^2 + 12x + 3$ has a minimum value of $y = -32$ and that this value occurs at $x = -6$. How close is your estimate?

Example 3 Draw the graph of $y = |x - 3| + 2$ and estimate the minimum value of y.

SOLUTION The graph of the given function is shown in Figure 1.31. Notice how your calculator handles the absolute value operation. The [TRACE] key yields a minimum of $y = 2$, which occurs at $x = 3$.

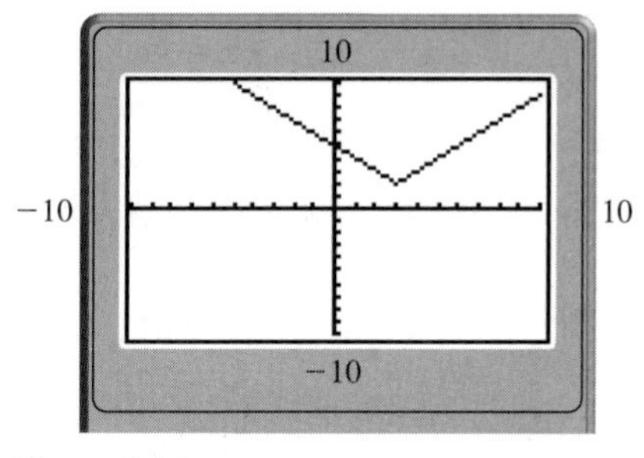

Figure 1.31 ▲

Distortion of a Graph

The viewing screen of a graphing calculator is not square; consequently if standard range values for x and y are used, the resulting graph will be somewhat distorted. For example, a graph that should be a semicircle may look more like half of an ellipse. The reason for this is that because the viewing screen is not square, one unit on the x-axis spans more pixels than one unit on the y-axis does. Your graphing calculator probably has a command that will eliminate this distortion.

Example 4 The relation $x^2 + y^2 = 9$ can be shown to describe a circle with its center at the origin and a radius of 3. Since the graphing calculator will only graph functions, the graph of $x^2 + y^2 = 9$ is obtained by graphing both $y = \sqrt{9 - x^2}$ and $y = -\sqrt{9 - x^2}$ on the same coordinate axes with a viewing screen range $-5 \leq x \leq 5$ and $-5 \leq y \leq 5$. The resulting graph, shown in Figure 1.32(a), should be the circle shown in Figure 1.32(b), but it looks more like an ellipse. Check to see whether your calculator has a command to remove this kind of distortion.

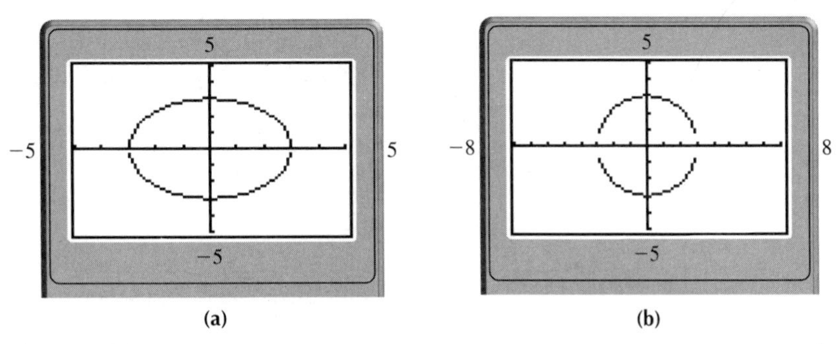

Figure 1.32 ▲

The ZOOM Key

Another way to establish range values for the viewing screen is to use the ZOOM key on your calculator. This key allows you to redefine the range of the viewing screen by marking the corners of a rectangle of interest on the present viewing screen.

Example 5 Suppose that you have graphed $y = x^3 - 3x^2 - 4x - 1$, as shown in Figure 1.33(a), and you are interested in the number of times the graph crosses the x-axis. It is clear that the graph crosses the x-axis at about $x = 4$, but it is not clear from Figure 1.33(a) how many zeros are in the vicinity of the origin. By using the ZOOM key on the calculator, you can define a new viewing screen as shown by the rectangle in Figure 1.33(b). When you press the ENTER key, the

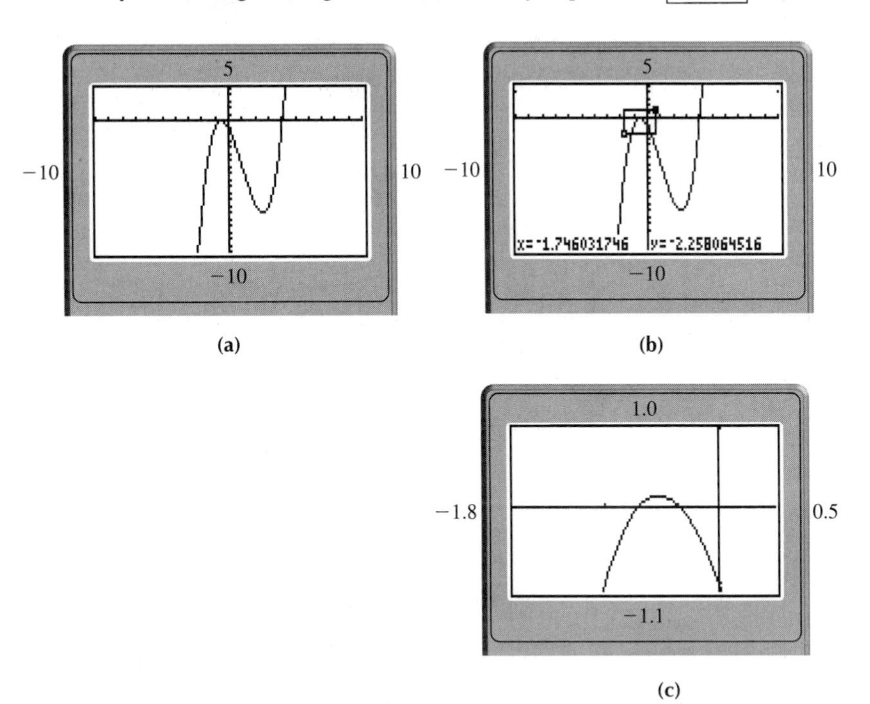

Figure 1.33 ▲

indicated rectangle fills the viewing screen, as shown in Figure 1.33(c). The graph can now be seen to have two additional crossings of the x-axis. By using the TRACE key, you can verify that these crossings occur at about $x = -0.4$ and $x = -0.7$.

Exercises for Section 1.5

In Exercises 1–20 use a graphing calculator with standard range settings to sketch the graph of the given equation. Use the TRACE key to estimate the x-intercepts of the graph.

1. $y = 3x + 4$
2. $y = 2 - 3x$
3. $2x + 4y = 5$
4. $3y + 5 = 4x$
5. $y = x^2 - x - 6$
6. $y = -x^2 + 2x - 4$
7. $a = -2t^2 - 3t + 5$
8. $m = z^3 + 1$
9. $y = |x + 3| - 1$
10. $y = 2 + |x - 2|$
11. $y = \sqrt{x + 7}$
12. $y = \sqrt{x - 2}$
13. $y = -\sqrt{x + 5}$
14. $y = -\sqrt{2x + 3}$
15. $s = t^2 + 8t$
16. $q = -|3 - t|$
17. $y = \dfrac{1}{x}$
18. $y = \dfrac{2}{x - 1}$
19. $y = x^3 + 2x^2$
20. $y = x^4 - 4x^2$

In Exercises 21–26 set the range of values for the viewing screen so that your graph agrees with the graph shown.

21. $y = \sqrt{x - 12}$

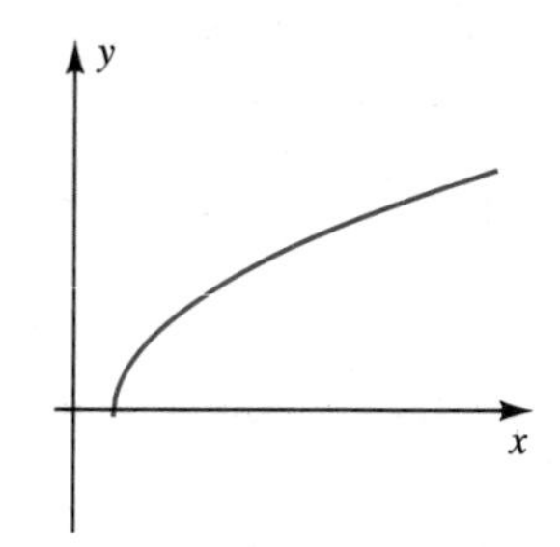

22. $y = x^2 + x + 20$

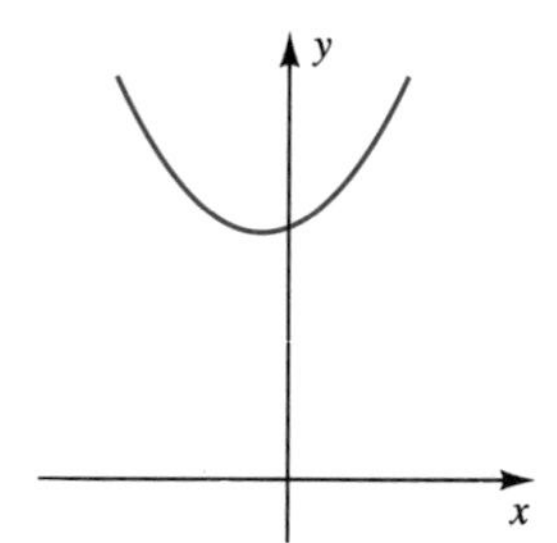

23. $y = x^3 - 11x^2 - 12x$

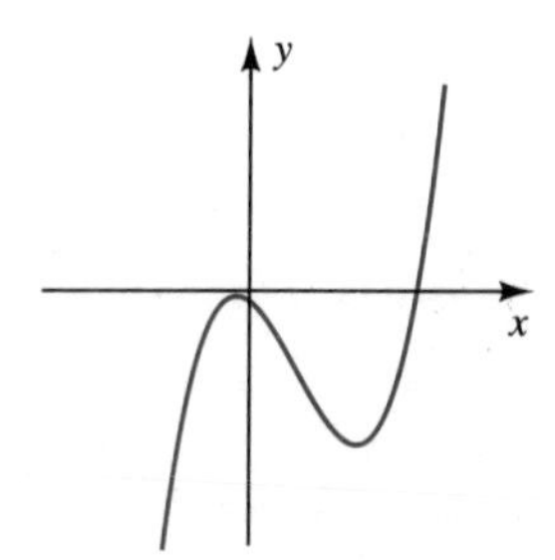

24. $y = -x^3 + 9x^2 - 60$

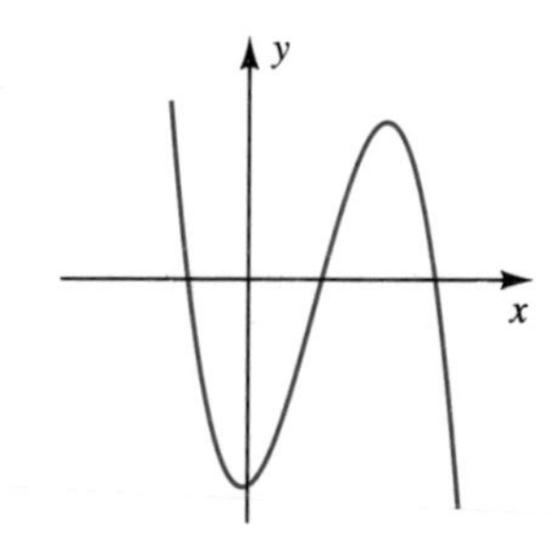

25. $y = x^2 + \dfrac{54}{x}$

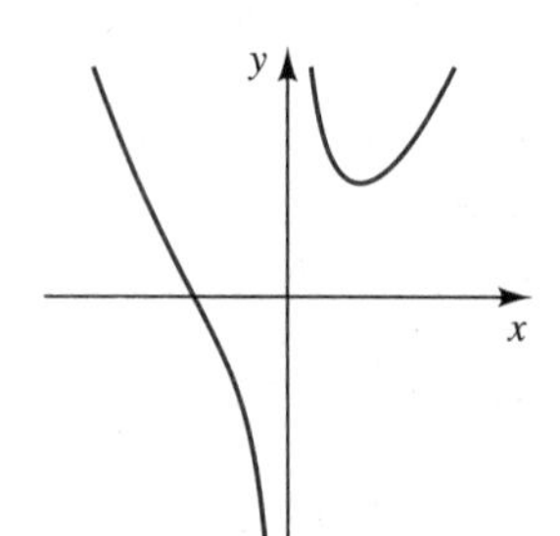

26. $y = 2x^4 - 5x^2 - 7$

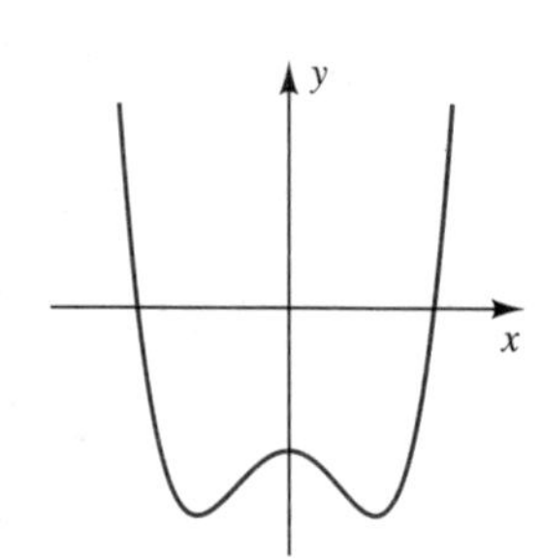

In Exercises 27–30 use a graphing calculator to sketch the graph and use the TRACE key to find the coordinates of the minimum value of the curve.

27. $y = x^2 + x - 6$

28. $y = 7x + 2x^2$

29. $y = x^2 + \dfrac{3}{x}$ for $x > 0$

30. $y = 2x^4 - 5x^2$ for $x > 0$

In Exercises 31–36 use the ZOOM feature of your calculator to find the *x*-intercept of the graph to two decimal places.

31. $y = 5 - x^2$ for $x > 0$

32. $y = 2 - x^3$

33. $s = t^2 + \dfrac{4}{t}$

34. $r = \dfrac{2x + 3}{x - 1}$

35. $y = |x + 2| - \sqrt{2}$

36. $y = |x - \sqrt{3}|$

In Exercises 37–40 sketch the graph of each relation using the standard range of values for the viewing screen. (*Hint:* In each case solve the relation for *y* and then plot both resulting functions on the same viewing screen.) If your calculator has a command to produce a square viewing screen, use it to eliminate distortion of the graph.

37. $x^2 + y^2 = 16$

38. $x^2 + y^2 = 25$

39. $y^2 - x - 3 = 0$

40. $2x - y^2 = 3$

1.6 Angles

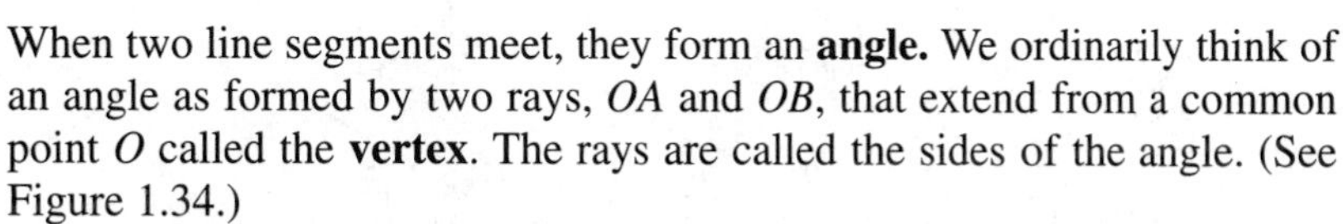

When two line segments meet, they form an **angle.** We ordinarily think of an angle as formed by two rays, OA and OB, that extend from a common point O called the **vertex**. The rays are called the sides of the angle. (See Figure 1.34.)

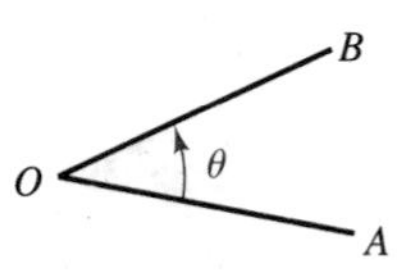

Figure 1.34
An angle

We refer to an angle by mentioning a point on each of its sides and the vertex. Thus the angle shown in Figure 1.34 is called "the angle AOB," which is written $\angle AOB$. If there is only one angle under discussion whose vertex is at O, we sometimes simply say "the angle at O" or, more simply, "angle O." It is also customary to use Greek letters to designate angles. For example, $\angle AOB$ might also be called the angle θ (read "theta").

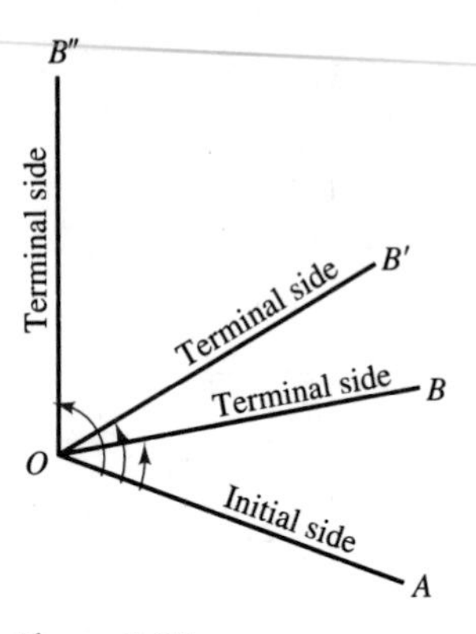

Figure 1.35
Generated angles

Sometimes it is useful to think of an angle as being "formed" when one of the sides is rotated about its vertex while the other side remains fixed, as shown in Figure 1.35. If we think of OA as being fixed and of OB as rotating about the vertex, OA is called the **initial side** and OB the **terminal side** of the generated angle. Other terminal sides such as OB' and OB'' result in different angles. The *size* of the angle depends on the amount of rotation of the terminal side. Thus $\angle AOB$ is smaller than $\angle AOB'$, which in turn is smaller than $\angle AOB''$. Two angles are equal in size if they are formed by the same amount of rotation of the terminal side. The notation $m(A)$, read "measure of A," is sometimes used to represent the measure of an angle A. The distinction between the measure of the angle and the angle itself is often blurred, and we write $A = 60°$ instead of $m(A) = 60°$.

The Degree

The most common unit of angular measure is the **degree.** We will define a degree to be $\frac{1}{360}$ of the measure of an angle formed by one complete revolution of the terminal side about its vertex. The measure of an angle formed by one complete revolution is then 360 degrees, written 360°. One-half of this angle, 180°, is called a **straight angle,** and one-fourth of it, 90°, is called a **right angle.** (See Figure 1.36.)

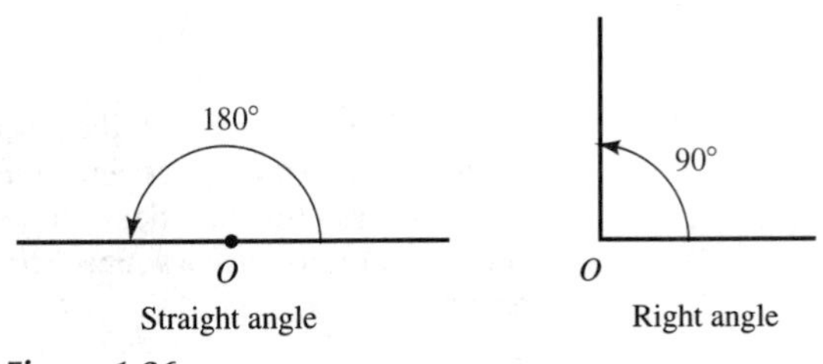

Figure 1.36

A positive angle is **acute** if it is smaller in size than a right angle. It is **obtuse** if it is larger than a right angle but smaller than a straight angle. (See Figure 1.37.)

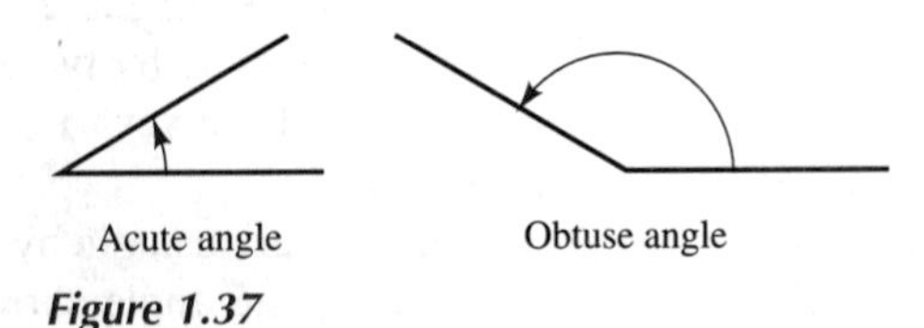

Figure 1.37

Figure 1.38 shows two angles that are larger than a straight angle.

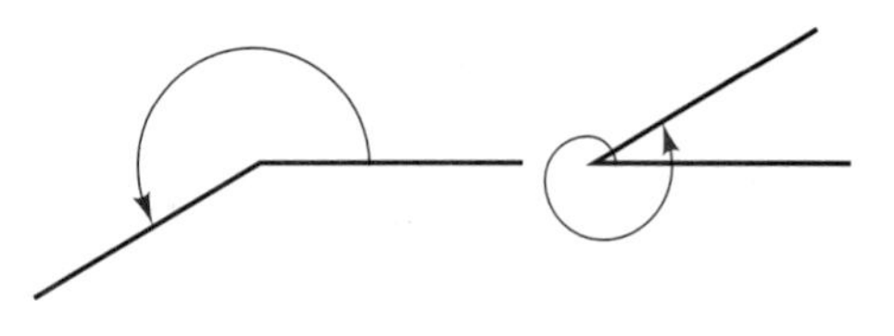

Figure 1.38

Angles with the same initial and terminal sides are said to be **coterminal.** The two angles A and A' shown in Figure 1.39 are coterminal. In this case $m(A') = m(A) + 360°$. In general coterminal angles differ in measure by a multiple of 360°. There are many important considerations, both practical and theoretical, that make it necessary to distinguish between coterminal angles formed in different ways.

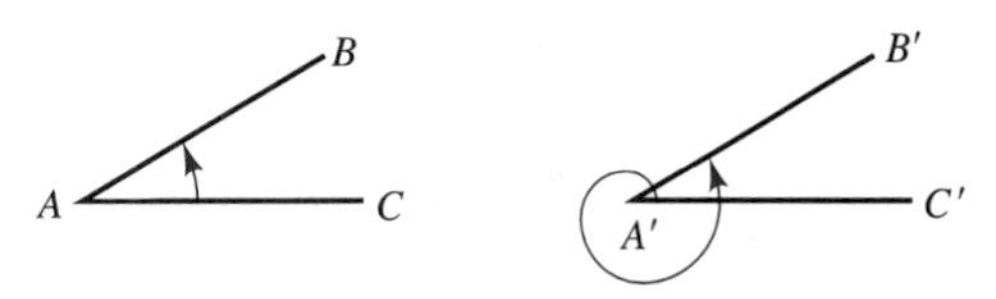

Figure 1.39

The *sign* of a generated angle indicates the direction in which the terminal side is rotated to form the angle: A counterclockwise rotation of the terminal side is **positive** and a clockwise rotation is **negative,** as shown in Figure 1.40. The measure of an angle has no numerical limit, since a terminal side may be rotated as much as desired.

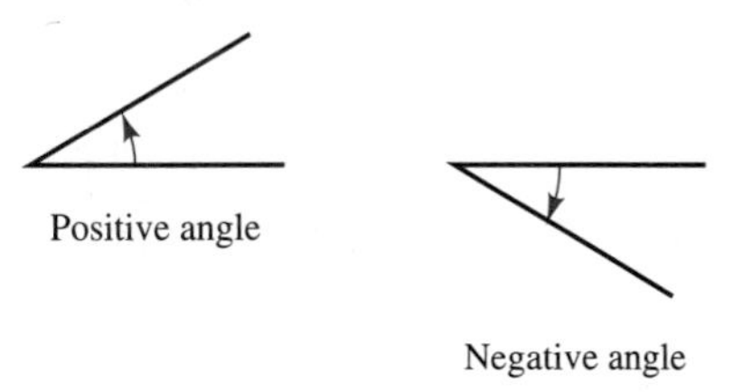

Figure 1.40

Example 1 Draw the following angles: **(a)** θ (theta) of measurement 42°, **(b)** ϕ (phi) of −450°, **(c)** β (beta) of 1470°, and **(d)** α (alpha) of − 675°. Indicate a coterminal angle between −90° and 90° for each generated angle.

SOLUTION See Figure 1.41.

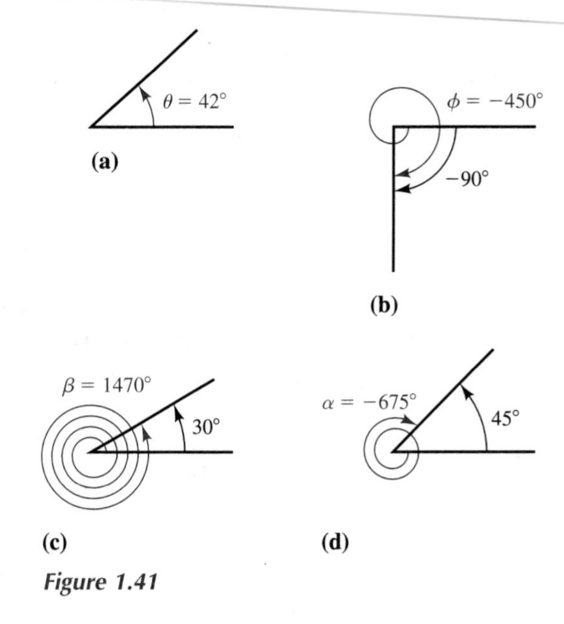

Figure 1.41 ▲

Example 2 Determine an angle between 0° and 360° that is coterminal with an angle of 1650°.

SOLUTION To find a coterminal angle between 0° and 360° of an angle greater than 360°, subtract the number of times 360° is contained in the given angle. In this example we note that $4(360) = 1440$ and $5(360) = 1800$, which means we should subtract 1440° from 1650°. Thus

$$1650° - 1440° = 210°$$

is the desired coterminal angle. ▲

Example 3 A fan blade rotates 150 times per minute. Through how many degrees does a point on the tip of one of the blades move in 7 sec?

SOLUTION Since the blade rotates 150 times per minute, it will rotate

$$150 \frac{\text{rev}}{\text{min}} \times \frac{1 \text{ min}}{60 \text{ sec}} = 2.5 \frac{\text{rev}}{\text{sec}}$$

Therefore in 7 sec the blade will rotate

$$2.5 \frac{\text{rev}}{\text{sec}} \times 7 \text{ sec} = 17.5 \text{ rev}$$

Since each rotation is 360°, the angle generated by the fan blade is

$$17.5 \times 360° = 6300°$$

▲

If the sum of the measures of two positive angles is 90°, the two angles are **complementary**. If the sum of the measures of two positive angles is 180°, they are **supplementary**. (See Figure 1.42.)

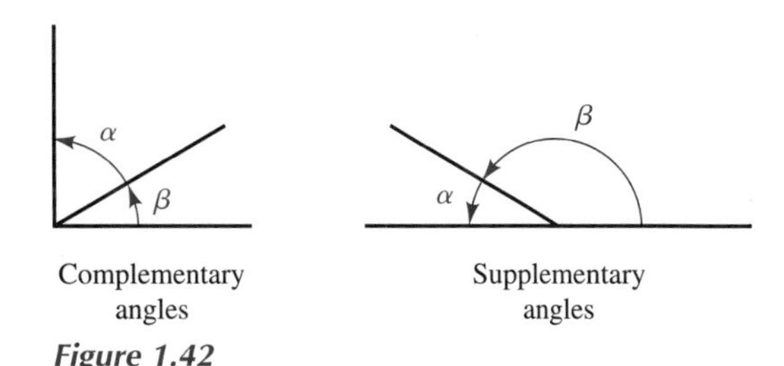

Complementary angles

Supplementary angles

Figure 1.42

The basic angular unit, the degree, is subdivided into 60 equal parts, each of which is called a **minute** and denoted by the symbol ′. The minute is further subdivided into 60 equal parts, each of which is called a **second** and denoted by the symbol ″. As the next example shows, arithmetic calculations are sometimes a bit more cumbersome with these subdivisions than with the decimal system.

Example 4 Find the sum and difference of the two angles whose measurements are 45°41′09″ and 32°52′12″.

SOLUTION The sum of the two angles is found by adding the corresponding units—that is, degrees to degrees, minutes to minutes, and seconds to seconds. Thus

$$45°41'09'' + 32°52'12'' = (45 + 32)°(41 + 52)'(09 + 12)'' = 77°93'21''$$

Since 93′ = 1°33′, we write the answer as 78°33′21″.

To find the difference of the two angles, we write 45°41′09″ in the following form:

$$45°41'09'' = 45°40'69'' = 44°100'69''$$

Thus

$$45°41'09'' - 32°52'12'' = (44 - 32)°(100 - 52)'(69 - 12)''$$
$$= 12°48'57''$$

▲

The use of minutes and seconds for angular subdivisions (and, indeed, the degree measurement itself) is based on an ancient Babylonian numeral system. Although the decimalization of angular measurement has been accelerated with the widespread use of the calculator, both systems will continue to be used for the foreseeable future. Therefore you should know how to make conversions between the two systems.

To convert an angle measured in degrees, minutes, and seconds to a decimal representation in degrees, simply divide the minutes by 60 (since

$60' = 1°$) and the seconds by 3600 (since $3600'' = 1°$) and then add the results. To ensure that the decimal form does not imply more accuracy than the given degree form, we adopt the following convention:

- An angle measured to the nearest minute should contain two decimal places in the converted form.
- An angle measured to the nearest second should contain four decimal places in the converted decimal form.

Example 5 Convert 15°35′ to decimal degrees.

SOLUTION

$$\begin{aligned} 15°35' &= 15° + \left(\frac{35}{60}\right)^\circ \\ &\approx 15° + (0.583\ldots)^\circ \\ &\approx 15.58° \end{aligned}$$

▲

Example 6 Convert 37°47′23″ to decimal degrees.

SOLUTION

$$\begin{aligned} 37°47'23'' &= 37° + \left(\frac{47}{60}\right)^\circ + \left(\frac{23}{3600}\right)^\circ \\ &\approx 37° + (0.78333\ldots)^\circ + (0.006388\ldots)^\circ \\ &\approx 37.7897° \end{aligned}$$

▲

To convert decimal notation to degrees, minutes, and seconds, multiply the fractional part of a degree by 60 to obtain minutes and multiply the fractional part of this result by 60 to obtain seconds.

Example 7 Convert 67.8235° to degrees, minutes, and seconds.

SOLUTION

$$\begin{aligned} 67.8235° &= 67° + (0.8235 \times 60)' \\ &= 67° + (49.41)' \\ &= 67°49' + (0.41 \times 60)'' \\ &\approx 67°49'25'' \quad \text{To the nearest second} \end{aligned}$$

▲

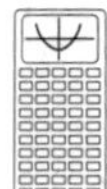

Some calculators have a built-in feature that permits easy conversion from angles given in degrees, minutes, and seconds to decimal degrees. Check your calculator for specific instructions. Or, if your calculator is programmable, write a program to carry out this conversion.

In trigonometry we often locate angles in the Cartesian plane. An angle is said to be in **standard position** in the plane if its vertex is at the origin and its initial side is along the positive half of the x-axis, as shown

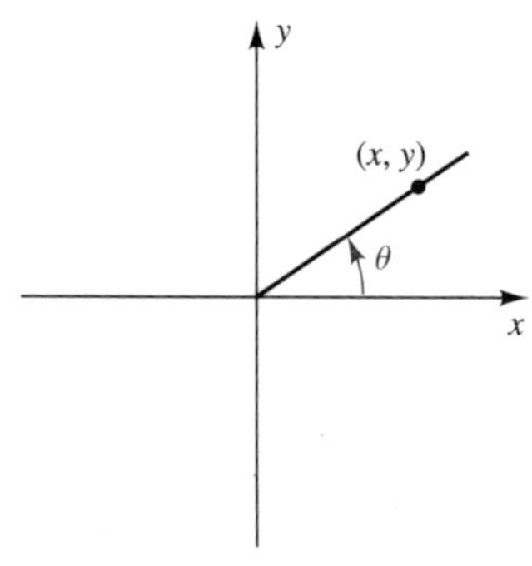

Figure 1.43
An angle in standard position

in Figure 1.43. The magnitude of an angle in standard position is measured from the positive x-axis to the terminal side.

Example 8 Draw an angle in standard position whose terminal side passes through $(-2, -2)$. What is the measure of this angle in degrees?

SOLUTION See Figure 1.44. The terminal side of θ obviously bisects the third quadrant, and therefore $\theta = 180° + 45° = 225°$. You should also observe that the measure of this angle is not unique, since there are many angles with the indicated side as the terminal side. Each of these angles differs by a multiple of 360°. Thus $\theta = 225° + m \cdot 360°$, where m is any integer. ▲

Any angle in standard position is coterminal with one whose measure is between 0° and 360°. For example, −45° is coterminal with an angle of 315°.

An angle is called a **first-quadrant angle** if the terminal side is in the first quadrant, a **second-quadrant angle** if the terminal side is in the second quadrant, and so on for the other quadrants. Angle θ in Figure 1.44 is a third-quadrant angle. If the terminal side of an angle is on one of the axes, the angle is called a **quadrantal angle;** that is, the angle does not lie in any quadrant.

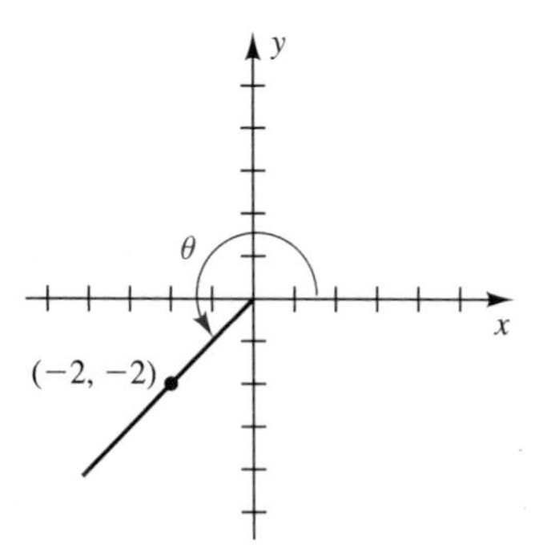

Figure 1.44

Measuring Instruments

Figure 1.45 shows a simple form of a **protractor**, the simplest instrument used to measure angles. It is marked in degrees around its rim.

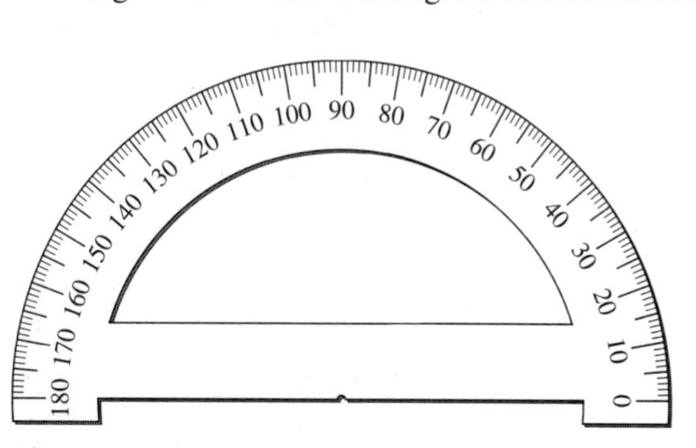

Figure 1.45 A protractor

Engineers and surveyors use more accurate devices for angle measurements. Figure 1.46(a) on page 48 shows a transit, which measures angles by locating two different line-of-sight objects. Figure 1.46(b) shows a modern laser measuring device that replaces the transit.

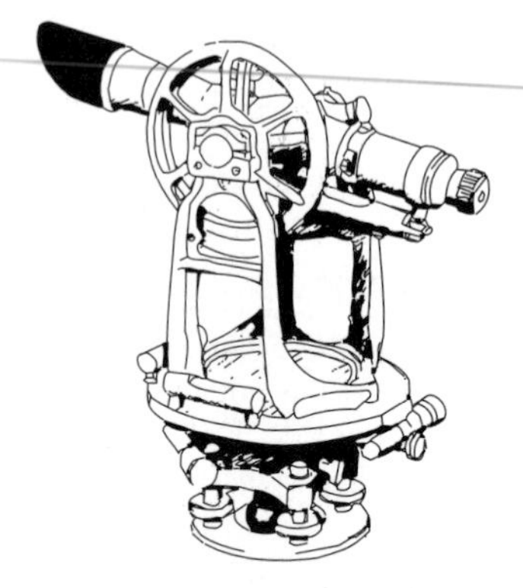

(a) A surveying transit

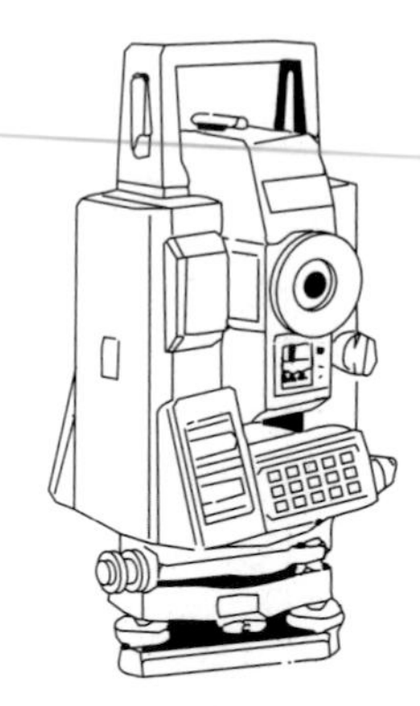

(b) An electronic survey station

Figure 1.46

Exercises for Section *1.6*

1. What is the sum of two complementary angles?
2. What is the difference in degree measure of two coterminal angles?

In Exercises 3–12 find the sum $A + B$ and the difference $A - B$ of the two given angles.

3. $A = 45°10'$, $B = 30°5'$
4. $A = 72°12'$, $B = 30°38'$
5. $A = 58°35'40''$, $B = 50°34'20''$
6. $A = 42°40'10''$, $B = 65°50'50''$
7. $A = 60°10'15''$, $B = 70°45'$
8. $A = 138°40'20''$, $B = 23°52'30''$
9. $A = 240°45'40''$, $B = 333°25'14''$
10. $A = 320°50'20''$, $B = -30°55'10''$
11. $A = -40°42'57''$, $B = -80°18'13''$
12. $A = -90°0'49''$, $B = 269°57'1''$

In Exercises 13–20 convert the given angle measure to decimal degree representation. Express angles to four decimal places.

13. $18°25'36''$
14. $54°50'16''$
15. $94°17'08''$
16. $-90°5'48''$
17. $283°36'30''$
18. $480°45'45''$
19. $183°14'40''$
20. $71°12'20''$

In Exercises 21–28 convert the given angle measure to degree/minute/second representation.

21. $48.2572°$
22. $-34.5618°$
23. $-235.4500°$
24. $30.5052°$
25. $45.7575°$
26. $234.5831°$
27. $15.2575°$
28. $68.3040°$

In Exercises 29–44 draw the angle and name the initial and terminal sides. If the given angle is greater than 360° or is negative, indicate an angle between 0° and 360° that is coterminal with the given one.

29. $420°$
30. $-300°$
31. $-317.5°$
32. $500°$
33. $-225°$
34. $-270°$
35. $590°$
36. $489.1°$

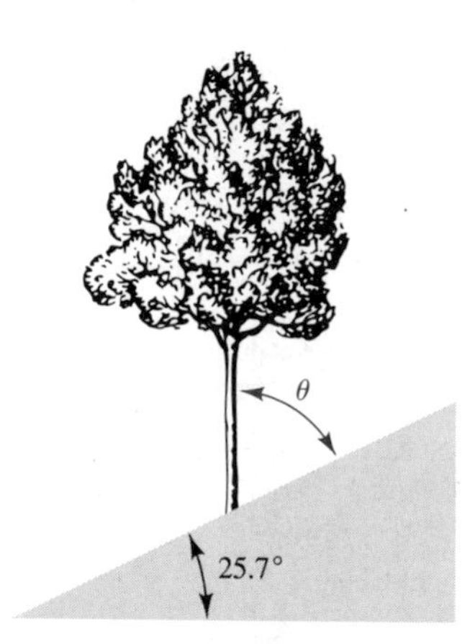

Figure 1.47

Figure 1.48

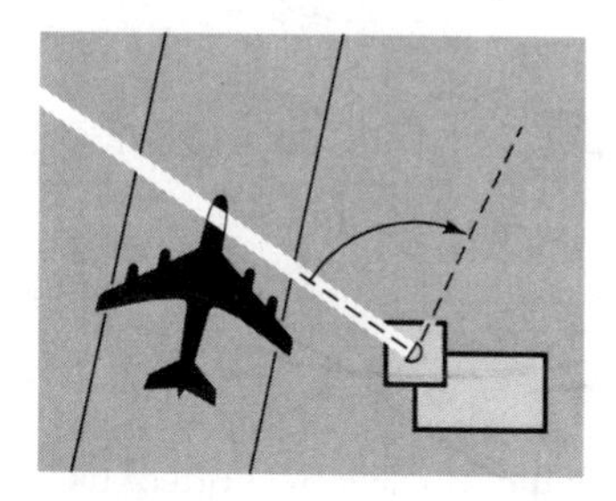

Figure 1.49

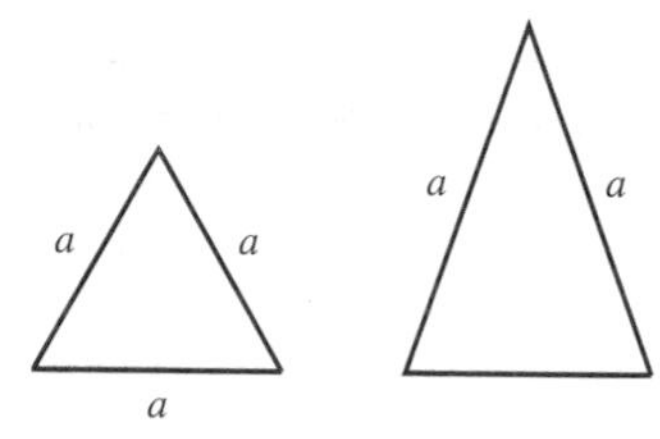

Equilateral triangle

Isosceles triangle

Figure 1.50

37. 720°	**38.** 780°	**39.** 840°	**40.** 765°
41. 1485°	**42.** 2000°	**43.** −1290°	**44.** −1205°

In Exercises 45–50 draw an angle in standard position whose terminal side passes through the given point. Give a degree measure of the angle.

45. $(-1, 1)$	**46.** $(5, 0)$	**47.** $(0, -3)$
48. $(4, -4)$	**49.** $(-4, 0)$	**50.** $(1000, -1000)$

51. A contractor surveying a building site records the angle 79.473° in the log. Convert this angle to degrees, minutes, and seconds.

52. A road that makes an angle of 30°40′ north of east intersects a road that makes an angle of 76°45′ south of east. What is the angle between the two roads?

53. During a lab experiment, a student measures an angle as 16°50′. Another member of the group measures the same angle as 16.75°. What is the difference, to the nearest one-hundredth of a degree, between the two measurements?

54. A pine tree grows vertically on a hillside that makes an angle of 25.7° with the horizontal. (See Figure 1.47.) What angle θ does the tree make with the hillside above it?

55. A bicycle wheel rotates 50 times in 1 min. (See Figure 1.48.) Through how many degrees does a point on the rim of the wheel move in 15 sec?

56. A searchlight at an airport makes 5 revolutions in 1 min. (See Figure 1.49.) Through how many degrees does the searchlight rotate in 15 sec?

1.7 Some Facts About Triangles

Much of Chapters 2 and 3 is devoted to a discussion of triangles and how trigonometry is used to compute unknown parts of a triangle. Therefore this section summarizes some geometrical facts about triangles.

A triangle is said to be **equiangular** if the measures of each of its three angles are exactly the same; it is said to be **equilateral** if all three sides have the same length. A theorem of geometry tells us that a triangle is equiangular if and only if it is equilateral. A triangle is said to be **isosceles** if two of its sides are equal; in such a triangle, the angles opposite the two equal sides are also equal. (See Figure 1.50.)*

A **right** triangle is one in which one of the angles is a right angle. See Figure 1.51(b) on page 50. An **oblique** triangle is one without a right angle. In any triangle in the plane, the sum of the measures of the angles is 180°.

* Technically we should distinguish between the lengths of the sides and the sides themselves. Thus the lengths of sides are equal and the sides themselves are congruent. Common usage often blurs this distinction, and in this text the context of the problem will determine whether "equal" is used in the sense of congruence or of equality of measurement.

Thus in an equilateral triangle, each of the angles measures 60°. In a right triangle each of the two non-right angles is acute, and the sum of their measures is 90°.

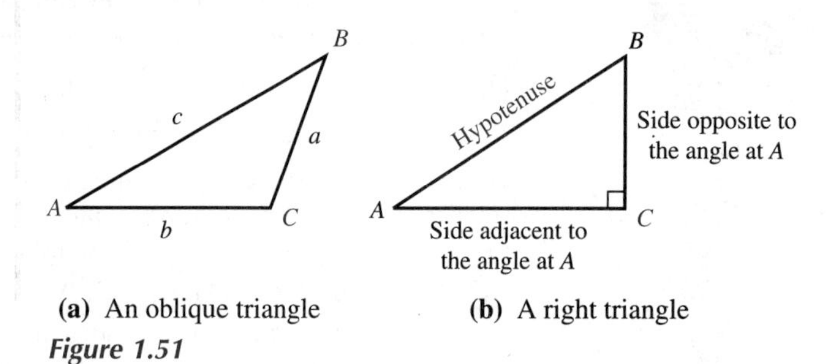

(a) An oblique triangle **(b)** A right triangle

Figure 1.51

There is a standard notation for labeling the angles and sides of a triangle. The vertices are denoted by capital letters and the sides by the corresponding lowercase letters. As shown in Figure 1.51(a), side a is opposite vertex A, side b is opposite vertex B, and side c is opposite vertex C. Sides b and c are said to be **adjacent** to the angle at A. Similarly, sides a and c are adjacent to angle B, and sides a and b are adjacent to angle C. In the right triangle shown in Figure 1.51(b), the side opposite the right angle C is called the **hypotenuse.**

Using the Pythagorean theorem (Section 1.2), the third side of a right triangle may be found if any two of the other sides are known. For instance, if a and b are given, the hypotenuse c can be computed from the relationship

$$c = \sqrt{a^2 + b^2}$$

Similarly, if lengths of sides a and c are given, then the length of side b is computed from

$$b = \sqrt{c^2 - a^2}$$

Example 1 The line-of-sight distance to the top of an antenna attached to the chimney of a house is known to be 15 m. If the sighting is taken 6.0 m from the house, how high is the top of the antenna above the ground?

SOLUTION A diagram of the situation is shown in Figure 1.52. As you can see, the unknown measurement is the third side of a right triangle in which two of the sides are known. Hence from the Pythagorean theorem,

$$h^2 + 6^2 = 15^2$$

so

$$h^2 = 225 - 36$$

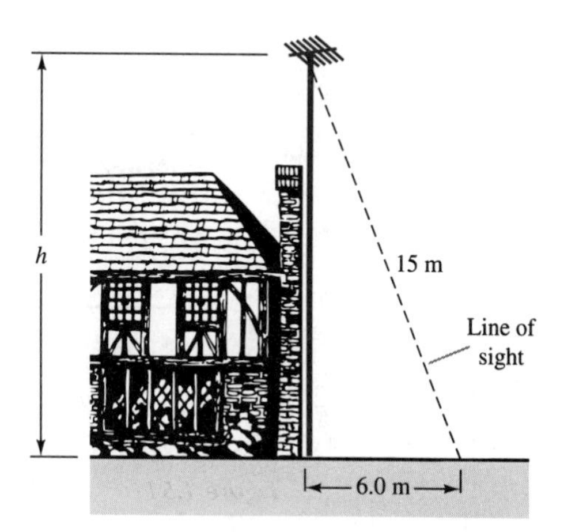

Figure 1.52

and

$$h = \sqrt{189} \approx \boxed{13.74772708} \approx 14 \text{ m} \qquad \text{To two significant digits}$$

You may wonder why the value of h was rounded off to two significant digits when it is the square root of a three-digit number. The reason is that the measurements given in the problem are accurate to two significant digits, so the answer cannot be more accurate than that. ▲

From the converse of the Pythagorean theorem, we can determine whether a triangle with given sides is a right triangle. For example, a triangle with sides 2.40, 3.10, and 3.92 is a right triangle because $2.40^2 + 3.10^2 = 3.92^2$. On the other hand, a triangle with sides 7, 8, and 9 is not a right triangle because $7^2 + 8^2 = 113$, which is not equal to 81, the square of the third side.

Special Right Triangles

Right triangles are at the heart of trigonometry. We make special note of the properties of 30°-60° and 45°-45° right triangles.

1. If one of the acute angles of a right triangle is 30°, then the other must be 60°, since the sum of the acute angles of a right triangle is equal to 90°. In a 30°-60° right triangle, the length of the side opposite the 30° angle is equal to half the length of the hypotenuse. To verify this fact, we divide an equilateral triangle into two 30°-60° right triangles, as shown in Figure 1.53(a). The bisector of the 60° angle also bisects the side opposite that angle. This construction

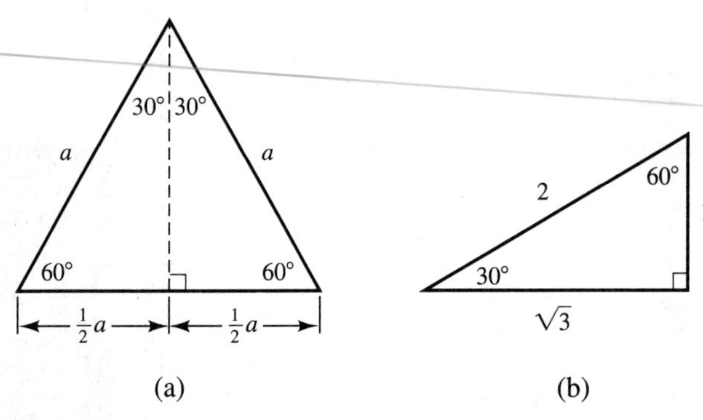

Figure 1.53

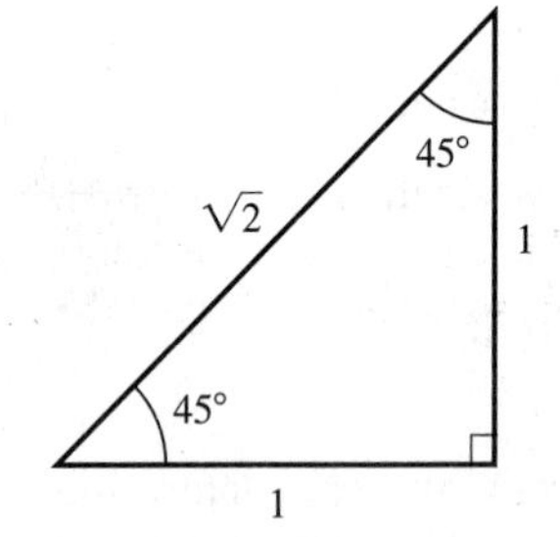

Figure 1.54

shows that the side opposite the 30° angle in the 30°-60° right triangle is $\frac{1}{2}a$, where a is the length of the hypotenuse. If the length of the hypotenuse is 2, the length of the side opposite the 30° angle is 1, and the length of the side adjacent the 30° angle is $\sqrt{2^2 - 1^2} = \sqrt{3}$. [See Figure 1.53(b).]

2. The acute angles of an isosceles right triangle are equal and each measures 45°. The sides opposite the 45° angles are equal, so if each has a length of 1 unit, the length of the hypotenuse is $\sqrt{1^2 + 1^2} = \sqrt{2}$. (See Figure 1.54.)

Another group of right triangles that is of interest to us are those right triangles whose side lengths are integers; for instance, a triangle with sides of 3, 4, and 5 is a right triangle.

COMMENT Carpenters often use the fact that a 3-4-5 triangle is a right triangle to square the walls of a building. They mark a point 3 ft from the corner along the base of one wall and 4 ft from the corner along the base of the other wall. Then by making the distance between these two points equal 5 ft, the walls will be square.

HISTORICAL COMMENT Three integers that have the property that the sum of the squares of two of the integers is equal to the square of the third are called **Pythagorean triples.** For example 3, 4, and 5 form a Pythagorean triple, since $3^2 + 4^2 = 5^2$. Do you know any other Pythagorean triples? Could you generate ten different sets of these triples? The ancient Babylonians were aware of these triples and understood them well enough to be able to generate them by a formula. It should be obvious that any integer multiple of a set of

Pythagorean triples is also a Pythagorean triple. What is not so obvious are the following formulas the Babylonians developed to generate these triples.

THEOREM 1.1 Let m and n be positive integers with $m > n$. Then $a = m^2 - n^2$, $b = 2mn$, and $c = m^2 + n^2$ form a Pythagorean triple.

Example Use Theorem 1.1 to generate the Pythagorean triple for $m = 3$ and $n = 2$.

SOLUTION The three integers are

$$a = 3^2 - 2^2 = 5,\ b = 2(3)(2) = 12, \text{ and } c = 3^2 + 2^2 = 13$$

Now notice that

$$5^2 + 12^2 = 13^2$$

▲

Similar Triangles

Another basic concept of trigonometry is that of similar triangles. Generally two triangles are **similar** if they have the same shape (not necessarily the same size). Thus similar triangles have equal angles but not necessarily equal sides. See Figure 1.55. The relationship between the sides of similar triangles has been known for centuries; it is given in the following theorem of Euclid.

THEOREM 1.2 If two triangles are similar, their sides are proportional.

COMMENT The fact that triangles ABC and $A'B'C'$ are similar is written

$$\Delta ABC \sim \Delta A'B'C'$$

Note that it is incorrect to denote the similarity by

$$\Delta ABC \sim \Delta B'A'C'$$

for this would incorrectly state that angles A and B' and angles B and A' were congruent.

In terms of the triangles shown in Figure 1.55, Theorem 1.3 can be written as

$$\frac{a}{a'} = \frac{b}{b'} = \frac{c}{c'}$$

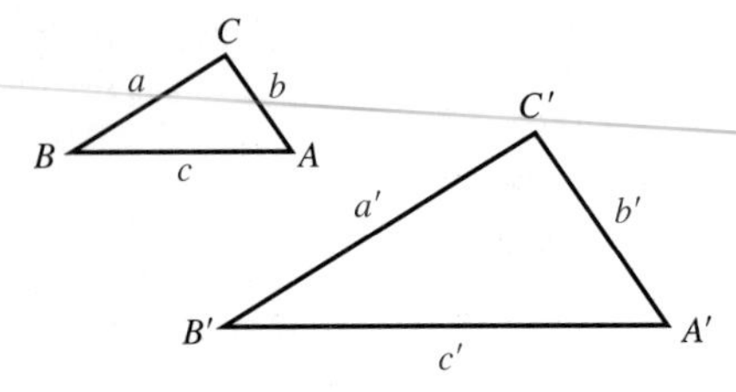

Figure 1.55
Two similar triangles

Combinations of any two of the three ratios will yield an equation with four parts. If we know three of these parts, we can find the fourth. Similar triangles are commonly used to compute distances that are difficult to measure by direct means. The next two examples show how this is done.

Example 2 A spectator is watching divers dive from an unusually high platform. The spectator, who knows he is 600 ft from the diving site, notes that his pencil of length 6 in. is just large enough to cover the diving height when he holds the pencil about 3 ft from his eye. How high is the diving platform? (See Figure 1.56. Note that it is not drawn to scale.)

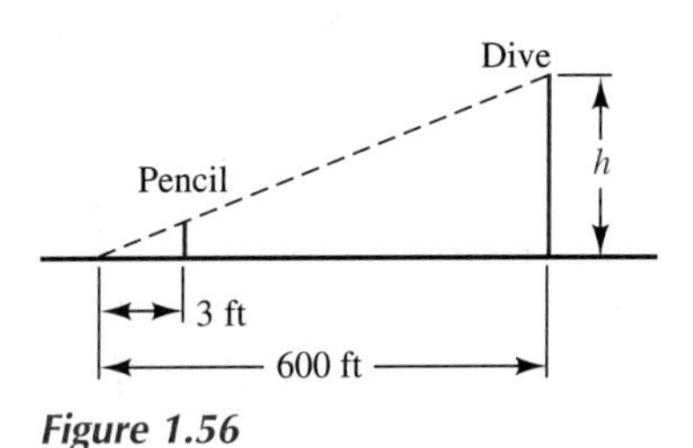

Figure 1.56

SOLUTION Because of the relatively large distances involved, we may, for the sake of approximation, ignore the fact that the sighting is taken 5 ft or so above ground level. Then since the small triangle and the large triangle are obviously similar, we have

$$\frac{3 \text{ ft}}{600 \text{ ft}} = \frac{\frac{1}{2} \text{ ft}}{h \text{ ft}}$$

and thus

$$h = \frac{1}{6}(600) = 100 \text{ ft}$$

▲

Example 3 A group of physics students was provided with a pencil and a 12-in. ruler and told to determine the height of the steeple on the campus chapel. Describe how they did it.

SOLUTION First the students used the ruler to measure the length of the shadow of the steeple on the ground and found it to be 28 ft. Then holding the ruler vertically with one end on the sidewalk, they marked the end of its shadow. (See Figure 1.57.) Finding the length of the shadow of the ruler to be 5.3 in., they used the pencil to write

$$\frac{h \text{ ft}}{12 \text{ in.}} = \frac{28 \text{ ft}}{5.3 \text{ in.}}$$

$$h = \frac{12}{5.3}(28) \approx \boxed{63.39622642} \approx 63 \text{ ft} \qquad \text{To two significant digits}$$

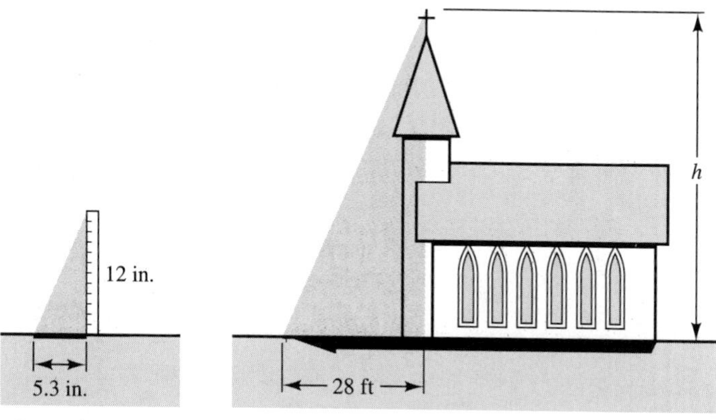

Figure 1.57 ▲

HISTORICAL NOTE The early Greek mathematician Eratosthenes (240 B.C.) closely approximated the circumference of the earth by simple arc length and shadow measurements. He noticed that at Syene (now the site of the Aswan dam in Egypt), the rays of the noon sun on the summer solstice shone straight down a deep well. At Alexandria, approximately 500 miles due north, the sun's rays were simultaneously measured to be 7.5° from the vertical. Using this information, Eratosthenes determined that the circumference of the earth must be 360/7.5 times the distance between Syene and Alexandria, or approximately 24,000 miles. (The actual value is 24,875 miles.)

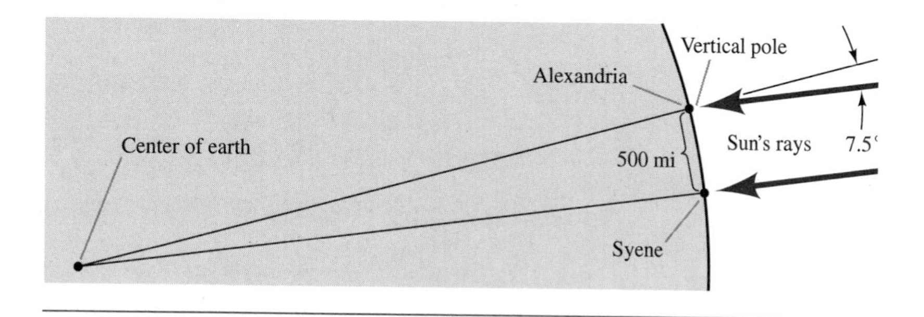

Exercises for Section 1.7

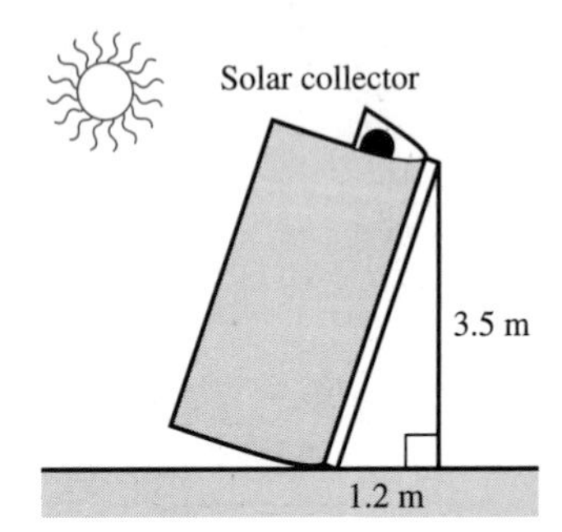

Figure 1.58

1. If the angle between the equal sides of an isosceles triangle is 32°, how large is each of the other angles?
2. How many degrees are there in each angle of an equilateral triangle?
3. A baseball diamond is a square 90.0 ft on a side. What is the distance across the diamond from first to third base?
4. An airplane flying at an altitude of 5000 ft above ground level is directly over Tiger Stadium in Detroit. Calculate the line-of-sight distance between the airplane and an observer who is 3 miles from the stadium.
5. The solar collector of a water heating system is held in a right-angle bracket, as shown in Figure 1.58. Calculate the length of the solar collector.
6. One side of a rectangle is half as long as the diagonal. The diagonal is 4 m long. How long is the other side of the rectangle?
7. What is the length of the diagonal of a cube that is 5.0 cm on an edge?
8. If a room is 21 ft long, 15 ft wide, and 10 ft high, what is the length of the diagonal of **(a)** the floor, **(b)** an end wall, and **(c)** a side wall of the room?

In Exercises 9–14 show that triangles with sides having the given measures are right triangles.

9. 3, 4, 5
10. 5, 12, 13
11. 7, 24, 25
12. 9, 40, 41
13. 11, 60, 61
14. 10, 24, 26

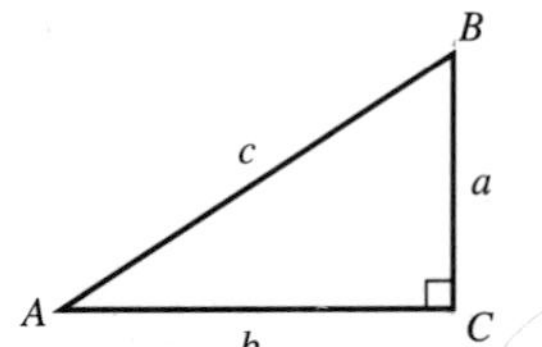

Figure 1.59

In Exercises 15–19 identify the 30°-60° right triangles and indicate the 30° angle. A typical right triangle is shown in Figure 1.59.

15. $a = 10, b = 10\sqrt{3}, c = 20$
16. $a = \sqrt{3}, b = 1, c = 2$
17. $a = 3, b = 3\sqrt{3}, c = 5$
18. $a = \sqrt{2}, b = \sqrt{3}, c = \sqrt{5}$
19. $a = 5\sqrt{3}, b = 5, c = 10$

In Exercises 20–25 determine whether the triangles whose vertices lie at the given points are right triangles.

20. $(-1, 2), (2, 2)\ (2, 6)$
21. $(1, 0), (0, 0), (0, 1)$
22. $(-1, 2), (1, 5), (3, 1)$
23. $(\sqrt{2}, 1 + \sqrt{2}), (-1 + \sqrt{2}, \sqrt{2}), (0, 2)$
24. $(-7, 3), (4, -6), (2, -8)$
25. $(-1, 1), (-1, -2), (5, 0)$
26. Show that an integer multiple of any Pythagorean triple is also a Pythagorean triple.
27. Show that if a, b, and c are Pythagorean triples and if $a = m^2 - n^2$ and $b = 2mn$, then it must follow that $c = m^2 + n^2$.

28. Use Theorem 1.1 to generate the Pythagorean triple for $m = 4$ and $n = 3$.

29. Use Theorem 1.1 to generate the Pythagorean triple for $m = 6$ and $n = 5$.

In Exercises 30–34 use the similar triangles shown in Figure 1.60 to find the unknown sides for the conditions given.

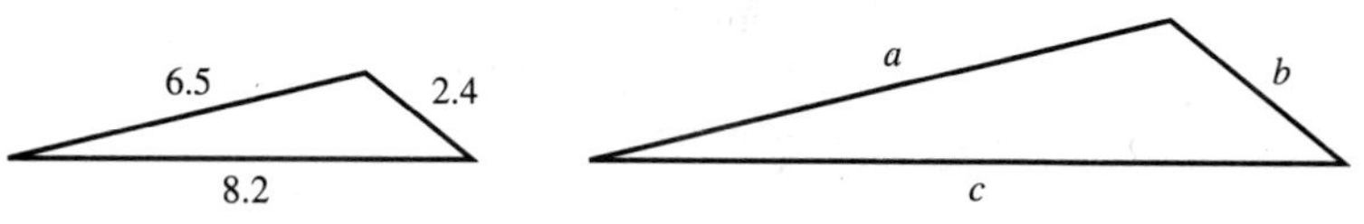

Figure 1.60

30. Given $a = 9.2$, find b and c.

31. Given $b = 3.0$, find a and c.

32. Given $b = 2.5$, find a and c.

33. Given $c = 12$, find a and b.

34. Given $c = 8.7$, find a and b.

35. If the sides of a triangle are 2, 4, and 5 cm, what is the perimeter of a similar triangle in which the longest side is 15 cm?

36. A snapshot is 7.60 cm wide and 10.1 cm long. It is enlarged so that it is 25.3 cm wide. How long is the enlarged picture? What is its area? its perimeter?

37. Is every equilateral triangle similar to every other equilateral triangle? Is every isosceles triangle similar to every other isosceles triangle? Give reasons.

38. At the same time that a yardstick held vertically casts a 5.0-ft shadow, a vertical flagpole casts a 30-ft shadow. How high is the flagpole?

39. At a certain time of day, a television relay tower casts a shadow 100 m long and a nearby pole 12 m tall casts a shadow 15 m long. How tall is the tower?

40. Assume that the three triangles in Figure 1.61 are similar. Find the measure of the unknown sides.

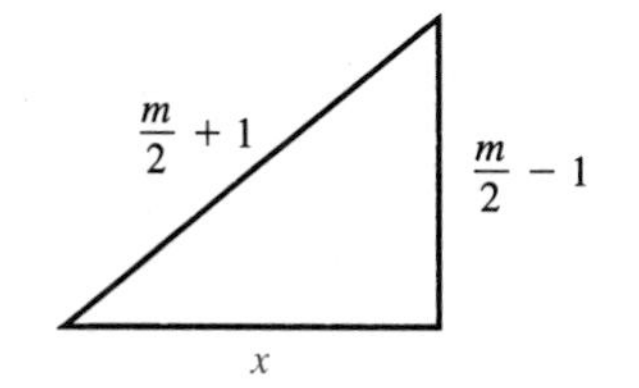

Figure 1.62

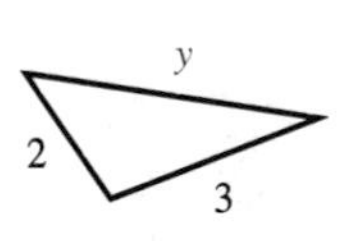

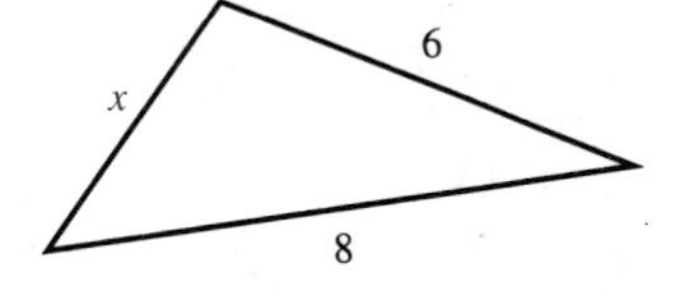

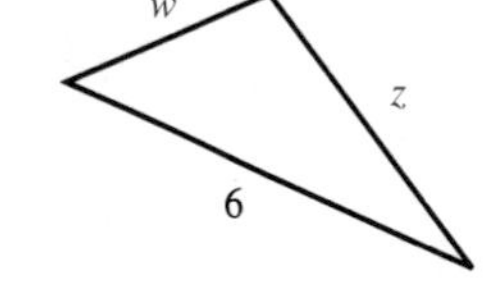

Figure 1.61

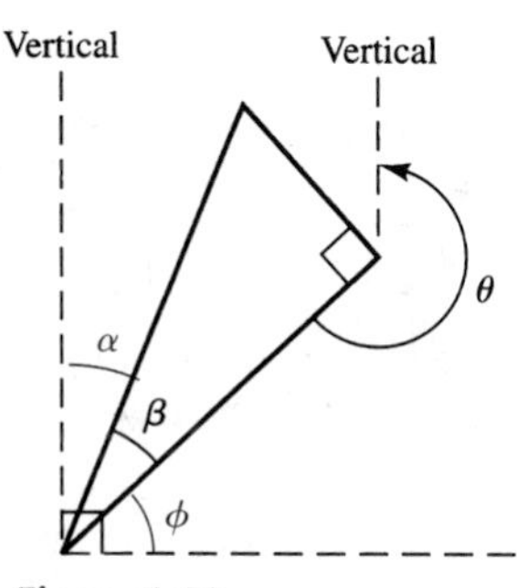

Figure 1.63

41. If the measure of the hypotenuse of the right triangle in Figure 1.62 is $(m/2) + 1$ and one leg has measure $(m/2) - 1$, find the measure of the other leg.

42. Find angles α and ϕ in Figure 1.63 if $\theta = 215°$ and $\beta = 26.6°$.

43. Find angles α and ϕ in Figure 1.63 if $\theta = 211°14'$ and $\beta = 24°12'$.

Key Topics for Chapter 1

Define and/or discuss each of the following.

Priority of Operations
Significant Digit
Rounding Off
Cartesian Coordinate System
Pythagorean Theorem
Function
Angles
Coterminal Angles
Degree Measure
Complementary Angles
Supplementary Angles
Standard Position of an Angle
Similar Triangles

Review Exercises for Chapter 1

1. If $f(x) = 2x^3 - x^2 + 1$, find $f(1)$, $f(0)$, and $f(-1)$.
2. If $f(x) = 3x - 2$, find $f(r + s)$.
3. Graph the set $\{(2, -1), (4, 1), (-3, -3), (-2, 5)\}$.
4. Graph the set $\{(0, 2), (\frac{1}{2}, 1), (1, 0)\}$.
5. Graph $y = f(x)$ if $f(x) = 2x - 3$.
6. Graph $y = f(x)$ if $f(x) = x^2 - 5$.

In Exercises 7–15 tell whether the given equation, table, or diagram defines a function.

7. $y = x^4$
8. $y = x \pm 5$
9. $y < 2$
10. $y = 1/x^2$
11. $y = x - x^{-1}$

12.

x	2	−1	4	7
y	3	0	7	−3

13.

x	0	5	7	8	10	5
y	0	3	4	2	−1	−3

14.

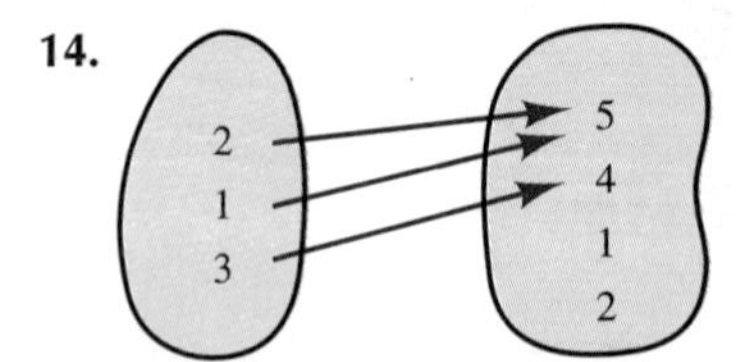

15.

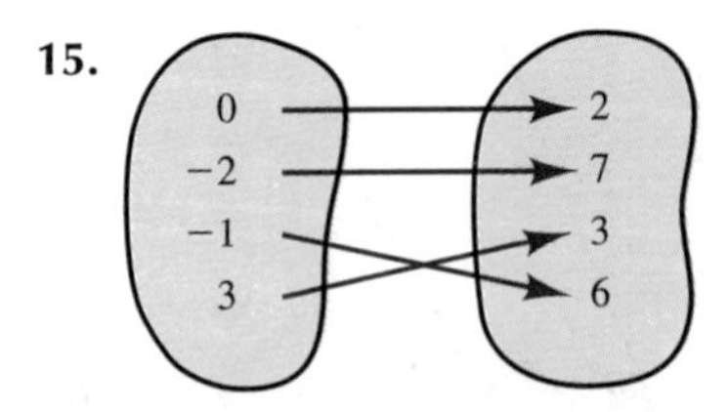

16. Find the distance from (0, 5) to (3, 2).
17. Determine whether the points (0, 1), (4, 2), and (3, 5) are vertices of a right triangle.

18. Let $y = \frac{3 - x}{x + 1}$. Find the distance from $(1, f(1))$ to $(0, f(0))$.

19. Determine the angles between 0° and 360° that is coterminal with an angle whose measure is 3400°.

20. Determine the angle between 0° and 360° that is coterminal with an angle whose measure is 800°.

21. Change 38°43′23″ to decimal degree form.

22. Change 67.5428° to degree/minute/second form.

23. Give the degree measure of an angle in standard position whose terminal side passes through $(-1, 1)$.

24. Give the degree measure of an angle in standard position whose terminal side passes through $(-\sqrt{2}, 0)$.

25. Subtract: 45°23′14″ − 35°35′54″.

26. The Gateway Arch in St. Louis is approximately 670 ft high. If you are 500 ft away from a point on the ground directly below the highest point of the arch, what is your line-of-sight distance to the top?

27. Determine the angle in standard position between 0° and 360° that is coterminal with an angle whose measure is 10,000°.

28. The hypotenuse and one leg of a right triangle are 54.6 ft and 34.9 ft, respectively. Find the other leg.

29. A 4.6-ft arc of the circumference of a circle subtends* an angle of 32°. Find the circumference of the circle. (*Hint:* See the Historical Note in Section 1.7, page 55.)

30. Find your line-of-sight distance to the top of a 100-ft tower if you are standing 50 ft from its base.

31. A woman walks 1.8 km due east and then turns and walks 2.4 km due north. How far is she from her initial starting point?

32. Find x and y if the two triangles in Figure 1.64 are similar triangles.

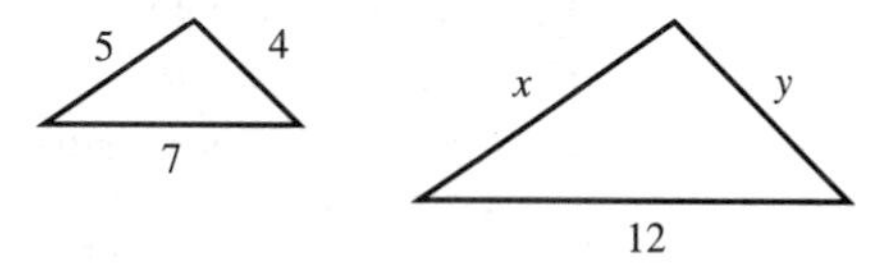

Figure 1.64

33. Under certain conditions the path of a thrown ball can be represented mathematically by the equation $h = a + bx + cx^2$, where x is the horizontal displacement of the ball, h is the corresponding height, and a, b, and c are constants. Draw the path of a ball for the interval $x = 0$ to $x = 7$, if $h = 7x - x^2$.

* An angle **subtends** a line segment or arc if the line segment or arc extends between the sides of the angle.

Practice Test Questions for Chapter 1

In Exercises 1–10 answer *true* or *false*.

1. Coterminal angles have the same degree measure.
2. Any angle larger than a right angle is said to be an obtuse angle.
3. Angles generated by a clockwise rotation of the terminal side are considered to be positive.
4. One degree is equal to $\frac{1}{300}$ of one revolution of the terminal side of an angle in standard position.
5. A function is even if $f(-x) = -f(x)$.
6. An equilateral triangle has three equal angles.
7. The number 0.027 has two significant digits.
8. The zeros of a function correspond to the x-intercepts of its graph.
9. The answer to 0.231(74.58) should have four significant digits.
10. A point in the plane is in quadrant IV if both coordinates are positive.

In Exercises 11–20 fill in the blank to make the statement true.

11. The digits in a calculation that are known to be accurate are called ____________.
12. The computation 9.5(3.471) is accurate to ____________ significant digits.
13. Rules of correspondence that allow more than one value of y to be paired with each x are called ____________.
14. $f(x)$ is the symbol for the ____________ value of the function f.
15. The ____________ of a function f is the set of points in the plane whose coordinates satisfy the equation $y = f(x)$.
16. Two positive angles are ____________ if their sum is 90°.
17. A function is ____________ on the interval $[a, b]$ if $f(x_1) < f(x_2)$ whenever $x_1 < x_2$.
18. A positive angle is ____________ if it is smaller than a right angle.
19. Two triangles are ____________ if their corresponding angles are equal.
20. An angle in standard position has its ____________ at the origin of the Cartesian coordinate system.

Solve the following exercises. Show all your work.

21. Find $\alpha + \beta$ if $\alpha = 145°14'56''$ and $19°47'23''$.
22. Find the angle between 0° and 360° that is coterminal with 825°.
23. Find the angle between 0° and 360° that is coterminal with 700°.
24. Express 125.3172° as an angle in degrees, minutes, and seconds.
25. Express $5°14'36''$ as an angle in decimal parts of a degree.
26. Given $f(x) = x^3 - 2x + 2$, evaluate $f(-1)$, $f(0)$, and $f(t + 2)$.
27. Plot $(-2, 5)$ and $(3, 1)$ and determine the distance between the two points.
28. Two sides of a right triangle are 4.51 cm and 3.90 cm. Calculate the length of the hypotenuse.

29. A person on an observation tower that is 350 ft above ground level spots a friend on the ground who is standing 240 ft from the base of the tower. Calculate the line of sight distance between the two people.

30. Centerville is 12 miles due north of Midland, and Lewis is 7.5 miles due east of Midland. Calculate the line of sight distance from Centerville to Lewis.

31. Draw the graph of $y = \sqrt{x + 2}$ on the interval $-3 \le x \le 3$.

32. Draw the graph of $y = 6 + 3x - x^2$ on the interval $-2 \le x \le 5$. Estimate the zeros of the function.

33. A person who is 6.0 ft tall casts a shadow that is 7.5 ft long. At the same time a flagpole casts a shadow that is 78.2 ft long. Calculate the height of the flagpole.

34. A scale drawing of a triangular field has sides of 3 in., 5 in., and 7 in. If the longest side of the actual field is 102 ft, calculate the perimeter of the field.

The Trigonometric Functions

*T*rigonometry is introduced in this chapter through angles in standard position in the Cartesian plane. The six basic trigonometric functions of an angle θ are defined in Section 2.1, and some of the interrelationships that exist among these six functions are discussed in Section 2.2. Although the trigonometric values can all be obtained from a calculator, it is instructive and helpful to know how to obtain the values of the functions for some special angles, such as 0°, 30°, 45°, 60°, and 90°, without the aid of a calculator. These so-called special angles are discussed in Section 2.3. In Section 2.4 we introduce the concept of a reference angle in order to obtain all angles that correspond to a given trigonometric value. We conclude the chapter with some hints and warnings on using a calculator to obtain trigonometric values for known angles and to obtain angles corresponding to known trigonometric values. ▼

2.1 Definitions of the Trigonometric Functions

Trigonometry was invented as a means of using triangles to calculate distances that could not be measured. Although triangle applications are still an important part of the study of trigonometry, many modern applications have nothing to do with triangles. For this reason we state the basic definitions of trigonometry in terms of a generated angle in standard position in the plane. [See Figure 2.1(a).]

Referring to Figure 2.1(a), we note that there are three important numbers relative to any point on the terminal side of angle θ; namely, the x- and y-coordinates of the point and the distance r from the origin to the point. For a given angle, θ, we are interested in the ratios of the numbers x, y, and r. By inspection we can see that the following six ratios can be formed

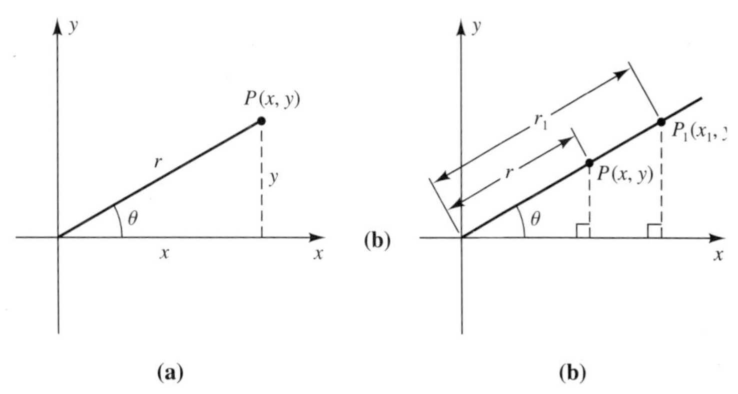

Figure 2.1
An angle in standard position

from the three numbers:

$$\frac{y}{r}, \frac{x}{r}, \frac{y}{x}, \frac{x}{y}, \frac{r}{x}, \frac{r}{y}$$

For a given angle θ, these six ratios are independent of the point chosen on the terminal side. The following argument proves that this is true. Choose two distinct points on the terminal side of angle θ, as shown in Figure 2.1(b), and then draw a line through each point perpendicular to the x-axis. Two similar triangles with a common vertex at the origin are formed. From our knowledge of similar triangles, we know that corresponding sides of the two triangles are proportional. Therefore

$$\frac{y}{y_1} = \frac{r}{r_1} \quad \text{and} \quad \frac{x}{x_1} = \frac{r}{r_1} \quad \text{and} \quad \frac{y}{y_1} = \frac{x}{x_1}$$

or, rearranging terms,

$$\frac{y}{r} = \frac{y_1}{r_1} \quad \text{and} \quad \frac{x}{r} = \frac{x_1}{r_1} \quad \text{and} \quad \frac{y}{x} = \frac{y_1}{x_1}$$

Each proportion says that the ratio of the two numbers associated with the smaller triangle is equal to the ratio of the corresponding numbers in the larger triangle. Recognizing that these are the first three of the six ratios mentioned earlier and that the other three could be handled in the same manner, we conclude that the six ratios are independent of the point chosen on the terminal side of θ.

Although the six ratios are independent of the point selected on the terminal side, they are dependent on the generated angle θ. For instance, if in Figure 2.1(b) angle θ increased, the x-coordinate of P would decrease and the y-coordinate would increase, while r remained constant. Consequently the ratio x/r would decrease as θ increased, and the ratio y/r would increase as θ increased. The six ratios are functions of the angle θ and have come to be called the **trigonometric functions**. Definition 2.1 gives the names of the six ratios.

DEFINITION 2.1 The six trigonometric functions of angle θ in Figure 2.1 are as follows:

$$\text{sine } \theta = \frac{y}{r} \qquad \text{Abbreviated } \sin \theta$$

$$\text{cosine } \theta = \frac{x}{r} \qquad \text{Abbreviated } \cos \theta$$

$$\text{tangent } \theta = \frac{y}{x} \qquad \text{Abbreviated } \tan \theta$$

$$\text{cotangent } \theta = \frac{x}{y} \qquad \text{Abbreviated } \cot \theta$$

$$\text{secant } \theta = \frac{r}{x} \qquad \text{Abbreviated } \sec \theta$$

$$\text{cosecant } \theta = \frac{r}{y} \qquad \text{Abbreviated } \csc \theta$$

COMMENT You should take time to learn the definition of each of the trigonometric functions since they are the building blocks of trigonometry. You should know them so well that when someone mentions $\sin \theta$ you automatically think "y divided by r."

COMMENT The angle θ may be in any of the four quadrants; the definition of the trigonometric function is exactly the same. Furthermore, the values of the trigonometric functions are independent of how the angle is generated, whether by more than 2000 rotations around the origin or by just part of a single rotation. Only the final position of the terminal side matters.

HISTORICAL NOTE The word *sine* has an interesting origin. The first trigonometric tables were of chords of a circle corresponding to an angle θ, as shown in the figure below. As you can see, if the radius of the circle is 1, then $\sin \theta$ is just one-half the chord length. The Hindus gave the name *jiva* to the half chord, and the Arabs used the word *jiba*. In the Arabic language there is also a word *jaib*, meaning "bay," whose Latin translation is "sinus." A medieval translator inadvertently confused the words *jiba* and *jaib* and thus the word *sine* is used instead of *half chord*.

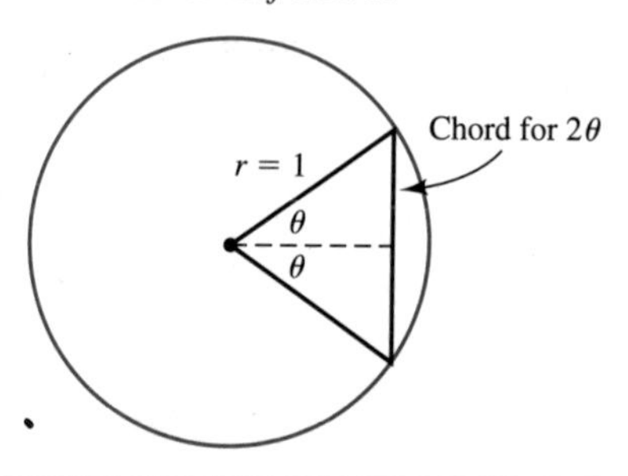

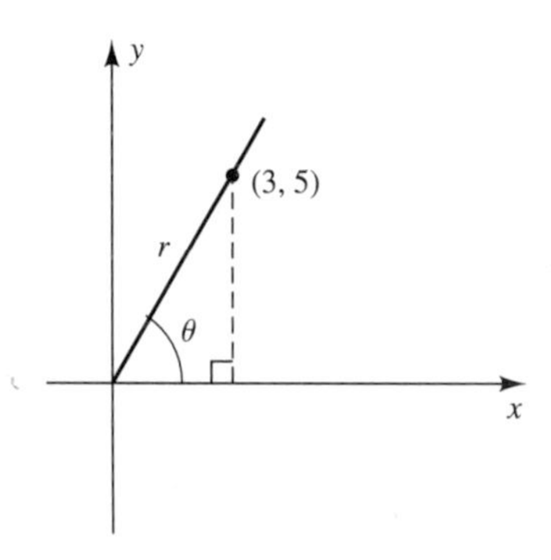

Figure 2.2

Example 1 Determine the trigonometric functions of the angle θ in standard position whose terminal side passes through the point (3, 5), as shown in Figure 2.2.

SOLUTION Note from Figure 2.2 that r is given by $r = \sqrt{x^2 + y^2}$. Therefore $r = \sqrt{3^2 + 5^2} = \sqrt{34}$. Using $x = 3$, $y = 5$, and $r = \sqrt{34}$ in Definition 2.1, we find that

$$\sin\theta = \frac{5}{\sqrt{34}} \qquad \csc\theta = \frac{\sqrt{34}}{5}$$
$$\cos\theta = \frac{3}{\sqrt{34}} \qquad \sec\theta = \frac{\sqrt{34}}{3}$$
$$\tan\theta = \frac{5}{3} \qquad \cot\theta = \frac{3}{5}$$

Of course, $\sin\theta$ and $\cos\theta$ can be written in rationalized form as $\sin\theta = 5\sqrt{34}/34$ and $\cos\theta = 3\sqrt{34}/34$, respectively. ▲

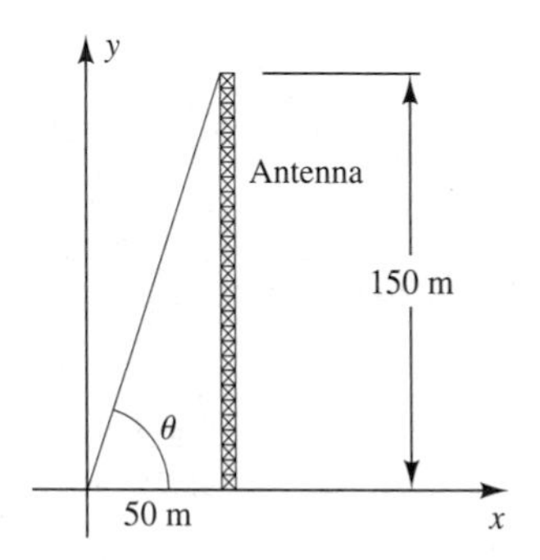

Figure 2.3

Example 2 A support line from the top of a 150-m antenna is anchored at a spot 50 m from the base of the antenna. Find the tangent of the angle of elevation of the cable. (The angle of elevation is the angle between the horizontal and one's line of sight when looking up at the object.)

SOLUTION In Figure 2.3 the support line and the antenna are drawn in the Cartesian plane, with the anchor point at the origin. From the figure we see that $x = 50$ and $y = 150$, so the tangent of the angle of elevation of θ is

$$\tan\theta = \frac{y}{x} = \frac{150}{50} = 3.0$$ ▲

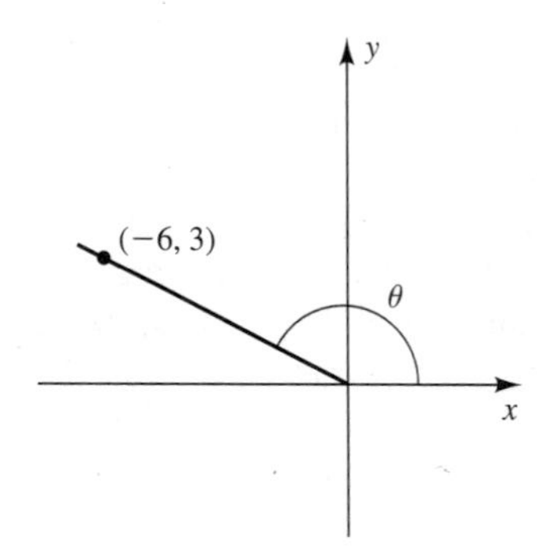

Figure 2.4

Example 3 An angle θ in standard position has the point $(-6, 3)$ on its terminal side. Find the values of the six trigonometric functions of θ. (See Figure 2.4.)

SOLUTION Using Definition 2.1 with $x = -6$, $y = 3$, and $r = \sqrt{(-6)^2 + 3^2} = \sqrt{45} = 3\sqrt{5}$, we get

$$\sin\theta = \frac{y}{r} = \frac{3}{3\sqrt{5}} = \frac{1}{\sqrt{5}} \qquad \csc\theta = \frac{r}{y} = \frac{3\sqrt{5}}{3} = \sqrt{5}$$
$$\cos\theta = \frac{x}{r} = \frac{-6}{3\sqrt{5}} = -\frac{2}{\sqrt{5}} \qquad \sec\theta = \frac{r}{x} = \frac{3\sqrt{5}}{-6} = -\frac{\sqrt{5}}{2}$$
$$\tan\theta = \frac{y}{x} = \frac{3}{-6} = -\frac{1}{2} \qquad \cot\theta = \frac{x}{y} = \frac{-6}{3} = -2$$ ▲

Example 4 The terminal side of an angle θ in standard position passes through the point $(3, -4)$. Find the six trigonometric functions of the angle. (See Figure 2.5 on page 66.)

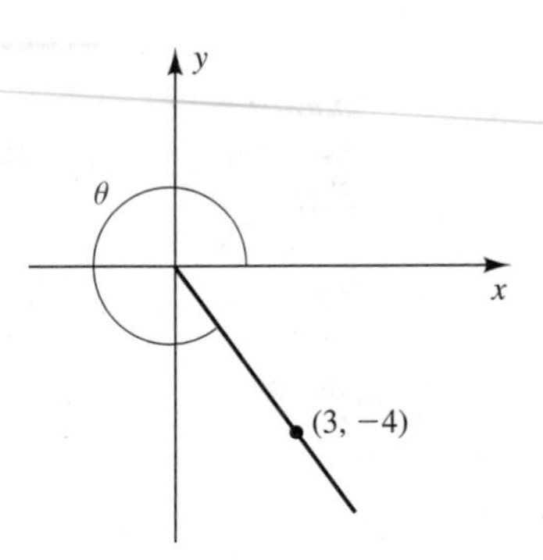

Figure 2.5

SOLUTION Using Definition 2.1 with $x = 3$, $y = -4$, and $r = \sqrt{3^2 + (-4)^2} = 5$, we get

$$\sin\theta = \frac{y}{r} = \frac{-4}{5} = -\frac{4}{5} \qquad \csc\theta = \frac{r}{y} = \frac{5}{-4} = -\frac{5}{4}$$

$$\cos\theta = \frac{x}{r} = \frac{3}{5} \qquad \sec\theta = \frac{r}{x} = \frac{5}{3}$$

$$\tan\theta = \frac{y}{x} = \frac{-4}{3} = -\frac{4}{3} \qquad \cot\theta \frac{x}{y} = \frac{3}{-4} = -\frac{3}{4}$$

▲

In each of these examples, four of the six trigonometric values are negative. This follows from the fact that since r is positive, the signs of the functional values depend on the signs of x and y.

The sine function is the ratio of y to r, which means that it is positive for angles in the first and second quadrants and negative for angles in the third and fourth quadrants. This is because y is positive above the x-axis and negative below.

The cosine function, which is the ratio of x to r, is positive for angles in the first and fourth quadrants and negative for angles in the second and third quadrants. This is because x is positive to the right of the y-axis and negative to the left.

The tangent function, which is the ratio of y to x, is positive in the first and third quadrants because y and x have the same signs in these quadrants. The tangent is negative in the second and fourth quadrants. The signs of the remaining three functions can be analyzed in the same way. Table 2.1 summarizes the results for all six functions.

Table 2.1

Quadrant	sin θ	cos θ	tan θ	cot θ	sec θ	csc θ
I	+	+	+	+	+	+
II	+	−	−	−	−	+
III	−	−	+	+	−	−
IV	−	+	−	−	+	−

COMMENT Table 2.1 is best learned by knowing the definitions of the trigonometric functions.

Example 5

(a) $\sin\theta > 0$ in quadrants I and II.

(b) $\tan\theta < 0$ in quadrants II and IV.

(c) $\sec\theta < 0$ in quadrants II and III. ▲

Example 6 Show that the terminal side of θ (in standard position) is in quadrant II if $\sin\theta > 0$ and $\cos\theta < 0$.

SOLUTION From Table 2.1, $\sin\theta > 0$ in quadrants I and II, and $\cos\theta < 0$ in quadrants II and III. Since both of the given conditions are satisfied in quadrant II, we conclude that the terminal side of θ must be in this quadrant. ▲

Example 7

(a) If $\sec\phi > 0$ and $\sin\phi < 0$, the terminal side of angle ϕ is in quadrant IV.

(b) If $\csc\phi > 0$ and $\cot\phi < 0$, the terminal side of angle ϕ is in quadrant II. ▲

COMMENT If an angle is in standard position, you can determine the values of all six trigonometric functions if you know the value of one of them and the quadrant of terminal side of the angle. If the quadrant is not given, two sets of values are possible.

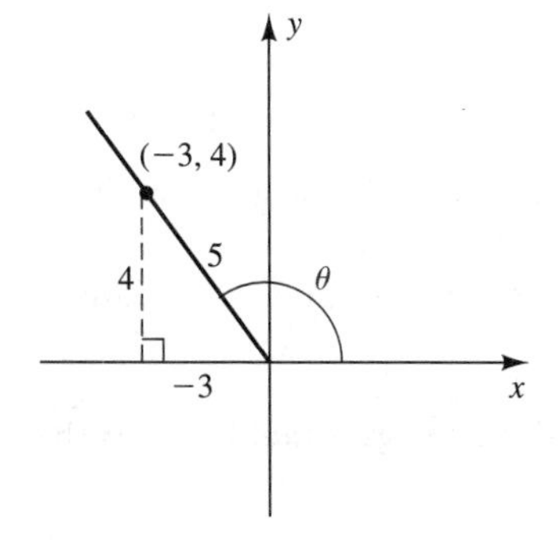

Figure 2.6

Example 8 Given that $\tan\theta = -\frac{4}{3}$ and that θ in standard position has its terminal side in quadrant II, find the values of the other five trigonometric functions.

SOLUTION We choose a convenient point on the terminal side—in this case (−3, 4), as shown in Figure 2.6. [If we had been told to locate the point in quadrant IV, we would have chosen the point (3, −4).] The desired trigonometric functional values for the given angle are

$$\sin\theta = \frac{4}{5} \qquad \cos\theta = \frac{-3}{5} \qquad \cot\theta = \frac{-3}{4} \qquad \sec\theta = \frac{5}{-3} \qquad \csc\theta = \frac{5}{4}$$ ▲

Example 9 Given that $\cos\theta = -\frac{5}{13}$, find the values of the other trigonometric functions. (See Figure 2.7 on page 68.)

SOLUTION Since the quadrant is not specified, two angles between 0° and 360° will satisfy the given condition. One is in the second quadrant and the other is in the third quadrant. For the second-quadrant angle,

$$\sin\theta = \frac{12}{13} \qquad \tan\theta = -\frac{12}{5} \qquad \cot\theta = -\frac{5}{12} \qquad \sec\theta = -\frac{13}{5} \qquad \csc\theta = \frac{13}{12}$$

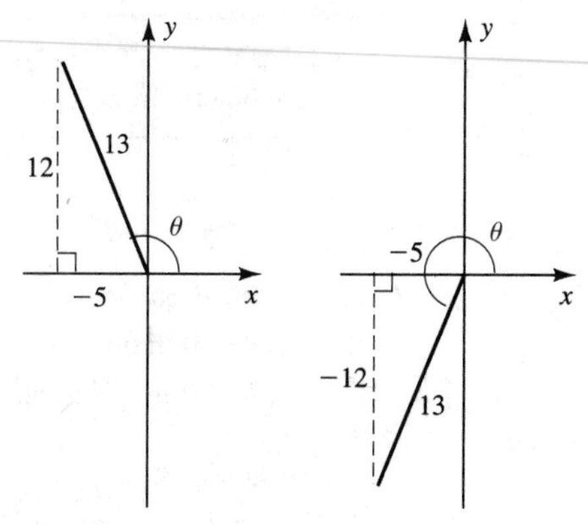

Figure 2.7

For the third-quadrant angle,

$$\sin\theta = -\frac{12}{13} \qquad \tan\theta = \frac{-12}{-5} = \frac{12}{5} \qquad \cot\theta = \frac{-5}{-12} = \frac{5}{12}$$

$$\sec\theta = -\frac{13}{5} \qquad \csc\theta = -\frac{13}{12}$$

▲

▲▼ Exercises for Section 2.1

In Exercises 1–9 find the exact values of the trigonometric functions of an angle in standard position whose terminal side passes through the points given.

1. $(2, 4)$
2. $(-1, 5)$
3. $(-9, 16)$
4. $(3, 1)$
5. $(2, -7)$
6. $(-1, -1)$
7. $(3, -1)$
8. $(1, -1)$
9. $(-\sqrt{3}, -1)$
10. Show that the values of the trigonometric functions of an angle θ are independent of the choice of the point P *on its terminal side.*
11. In which quadrants must the terminal side of θ lie for **(a)** $\sin\theta$ to be positive? **(b)** $\cos\theta$ to be positive? **(c)** $\tan\theta$ to be positive?

In Exercises 12–18, for the given conditions, indicate the quadrant in which the terminal side of θ lies.

12. $\sin\theta > 0$ and $\tan\theta < 0$
13. $\sec\theta < 0$ and $\cot\theta < 0$
14. $\cos\theta > 0$ and $\sin\theta < 0$
15. $\tan\theta > 0$ and $\csc\theta < 0$
16. $\sin\theta < 0$ and $\cos\theta < 0$
17. $\sin\theta < 0$ and $\sec\theta > 0$
18. $\csc\theta < 0$ and $\sec\theta < 0$

In Exercises 19–38 find the exact values of all the trigonometric functions of an angle θ that satisfies the conditions.

19. $\tan\theta = \frac{3}{4}$ in quadrant I
20. $\sec\theta = -3$

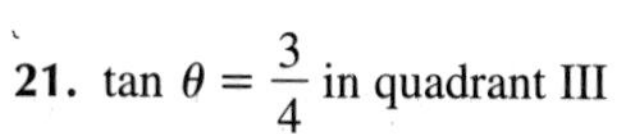

21. $\tan\theta = \dfrac{3}{4}$ in quadrant III

22. $\tan\theta = \dfrac{1}{2}$ in quadrant III

23. $\cos\theta = \dfrac{\sqrt{3}}{2}$

24. $\cot\theta = -3$ in quadrant IV

25. $\sin\theta = \dfrac{2}{3}$ in quadrant II

26. $\sin\theta = \dfrac{\sqrt{3}}{2}$

27. $\sin\theta = -\dfrac{1}{2}$

28. $\sin\theta = \dfrac{1}{5}$

29. $\tan\theta = 10$ in quadrant I

30. $\csc\theta = 2$

31. $\cos\theta = \dfrac{12}{13}$

32. $\cos\theta = -\dfrac{\sqrt{3}}{2}$ in quadrant II

33. $\tan\theta = -\sqrt{3}$

34. $\cot\theta = -\sqrt{2}$

35. $\sin\theta = \dfrac{u}{v}$

36. $\tan\theta = \dfrac{u}{v}$

37. $\cos\theta = u$

38. $\sin\theta = \dfrac{1}{v}$

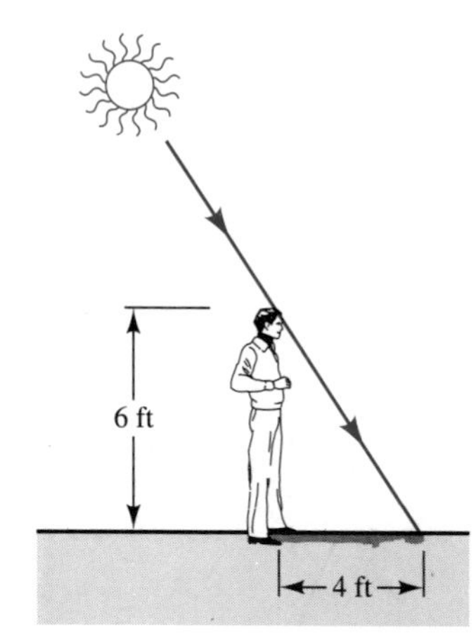

Figure 2.8

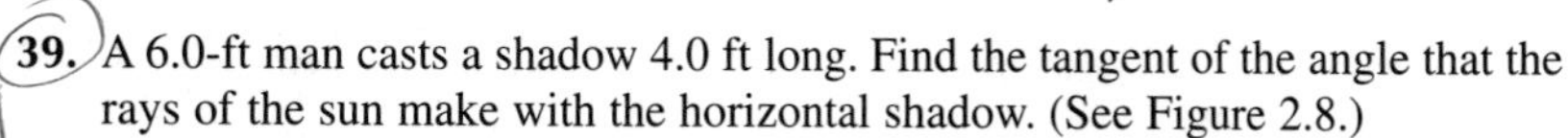

39. A 6.0-ft man casts a shadow 4.0 ft long. Find the tangent of the angle that the rays of the sun make with the horizontal shadow. (See Figure 2.8.)

40. A wire 30 ft long is used to brace a flagpole. If the wire is attached to the pole 25 ft above the level ground, what is the cosine of the angle made by the wire with the ground?

41. The line-of-sight distance to the top of a 128-ft-high building is 456 ft. (See Figure 2.9.) What is the tangent of the angle of elevation? (*Note:* Recall the angle of elevation is defined as the angle between the horizontal and the line of sight up to the object in question.)

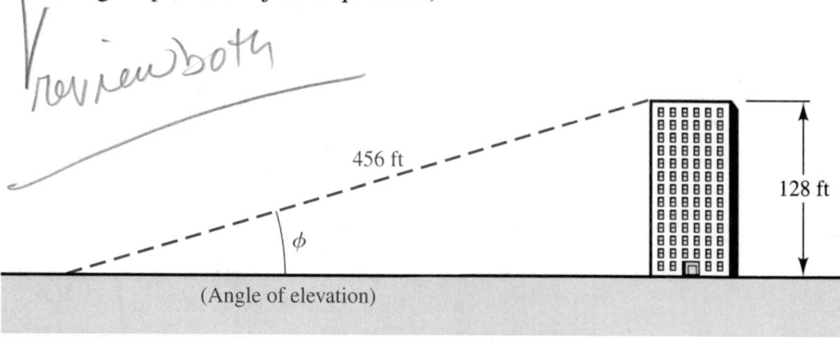

Figure 2.9

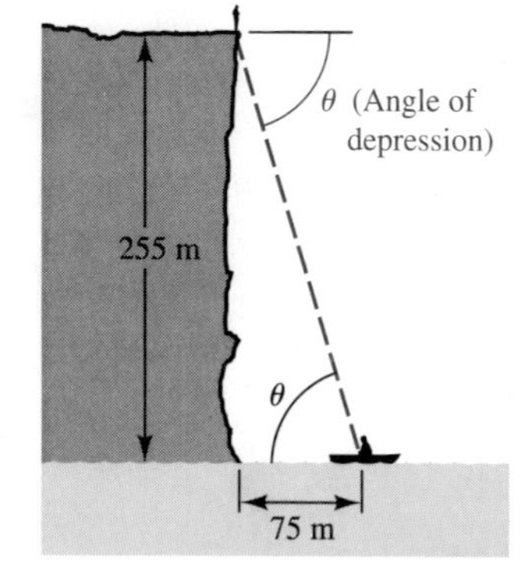

Figure 2.10

42. Suppose that a girl is flying a kite at the end of a 100-m string that makes an angle of 45° with the ground. Find the cosine of the angle that the string makes with the ground.

43. A man on a 255-m cliff looks down on a rowboat known to be 75.0 m from the base of the cliff. (See Figure 2.10.) What is the sine of the angle of depression? (The angle of depression is defined as the angle between the horizontal and one's line of sight when looking down on an object.)

2.2 Fundamental Relations

The trigonometric functions are interrelated by some interesting and useful formulas. From Definiton 2.1, $\sin\theta = y/r$ and $\csc\theta = r/y$; consequently

$$\sin\theta \csc\theta = \frac{y}{r}\cdot\frac{r}{y} = 1$$

Similarly,

$$\cos\theta \sec\theta = \frac{x}{r}\cdot\frac{r}{x} = 1$$

and

$$\tan\theta \cot\theta = \frac{y}{x}\cdot\frac{x}{y} = 1$$

Rearranging each of these formulas, we get

$$\sin\theta = \frac{1}{\csc\theta}, \quad \cos\theta = \frac{1}{\sec\theta}, \quad \tan\theta = \frac{1}{\cot\theta} \qquad \textbf{(2.1)}$$

The three relations in (2.1) are called the **reciprocal relations** for the trigonometric functions. Of course, they can also be written in the forms $\csc\theta = 1/\sin\theta$, $\sec\theta = 1/\cos\theta$, and $\cot\theta = 1/\tan\theta$.

CAUTION As with all mathematics formulas, there are values of θ for which these equations are invalid. For example, the relation $\sin\theta = 1/\csc\theta$ has no meaning for any angle θ for which $\theta = 0$. It would become very cumbersome to keep listing the values of θ for which relations are invalid, so we won't always do it. Just remember that if some value of θ yields a 0 denominator, then the relation is invalid for that value of θ.

Example 1 Given $\sin\theta = \frac{2}{3}$, use Relations (2.1) to find $\csc\theta$.

SOLUTION

$$\csc\theta = \frac{1}{\sin\theta} = \frac{1}{2/3} = \frac{3}{2}$$ ▲

Example 2 Given $\sec\alpha = 2$, use Relations (2.1) to find $\cos\alpha$.

SOLUTION

$$\cos\alpha = \frac{1}{\sec\alpha} = \frac{1}{2}$$ ▲

Two other relations that are of considerable importance are

$$\tan\theta = \frac{\sin\theta}{\cos\theta} \quad \text{and} \quad \cot\theta = \frac{\cos\theta}{\sin\theta} \tag{2.2}$$

The first of these is verified by the following sequence of operations:

$$\begin{aligned} \tan\theta &= \frac{y}{x} && \text{Definition of } \tan\theta \\ &= \frac{y/r}{x/r} && \text{Divide numerator and denominator by } r \\ &= \frac{\sin\theta}{\cos\theta} && \text{Definition of } \sin\theta \text{ and } \cos\theta \end{aligned}$$

The fact that $\cot\theta = \cos\theta/\sin\theta$ follows immediately from the relation $\cot\theta = 1/\tan\theta$.

Example 3

(a) Given $\sin\phi = 1/\sqrt{5}$ and $\cos\phi = 2/\sqrt{5}$, use Relation (2.2) to find $\tan\phi$.

(b) Indicate the quadrant in which the terminal side of ϕ lies.

SOLUTION

(a) $\tan\phi = \dfrac{\sin\phi}{\cos\phi} = \dfrac{1/\sqrt{5}}{2/\sqrt{5}} = \dfrac{1}{2}$

(b) The terminal side of ϕ lies in quadrant I since $\sin\phi > 0$ and $\cos\phi > 0$. ▲

Finally, with the aid of the Pythagorean theorem, we derive an important relation between the sine and cosine functions. For any angle θ whose terminal side passes through the point $P(x, y)$, the Pythagorean theorem requires that

$$y^2 + x^2 = r^2$$

Divide both sides of this equation by r^2,

$$\left(\frac{y}{r}\right)^2 + \left(\frac{x}{r}\right)^2 = 1$$

or, in terms of the trigonometric functions,

$$(\sin\theta)^2 + (\cos\theta)^2 = 1$$

It is customary to write $(\sin\theta)^2$ as $\sin^2\theta$ and $(\cos\theta)^2$ as $\cos^2\theta$. (A similar convention holds for expressing powers of the other trigonometric functions.) Thus the equation reads

$$\sin^2\theta + \cos^2\theta = 1 \tag{2.3}$$

which is often called the **Pythagorean relation** of trigonometry.

COMMENT Be sure to notice the distinction between $\sin^2\theta$ and $\sin \theta^2$, since $\sin^2\theta \neq \sin \theta^2$. For now, the quantity $\sin \theta^2$ will not be used, but it will be used in later chapters.

Two alternative forms of Relation (2.3) often prove useful. By dividing both sides by $\cos^2\theta$, we obtain

$$\frac{\sin^2\theta}{\cos^2\theta} + \frac{\cos^2\theta}{\cos^2\theta} = \frac{1}{\cos^2\theta}$$

Then since $\sin \theta/\cos \theta = \tan \theta$ and $1/\cos \theta = \sec \theta$, we have

$$\tan^2\theta + 1 = \sec^2\theta \tag{2.3a}$$

In a similar manner we can show that

$$\cot^2\theta + 1 = \csc^2\theta \tag{2.3b}$$

Example 4 Given that $\sin \theta = \frac{1}{4}$ and that θ is a second-quadrant angle, use the Pythagorean relation to find the exact value of $\cos \theta$. Then find the exact value of $\tan \theta$.

SOLUTION Solving the Pythagorean relation for $\cos \theta$, we have $\cos \theta = -\sqrt{1 - \sin^2\theta}$. The negative square root is chosen because $\cos \theta < 0$ when θ is in quadrant II. Therefore,

$$\cos \theta = -\sqrt{1 - \left(\frac{1}{4}\right)^2} = -\sqrt{1 - \left(\frac{1}{16}\right)} = -\frac{\sqrt{15}}{4}$$

Finally, using $\sin \theta = \frac{1}{4}$ and $\cos \theta = -\sqrt{15}/4$ in Relation (2.2), we get

$$\tan \theta = \frac{\sin \theta}{\cos \theta} = -\frac{1/4}{\sqrt{15}/4} = -\frac{1}{\sqrt{15}}$$ ▲

Any combination of trigonometric functions such as $2 \sin x + \tan x$ or $\cos 2x - \sin 3x$ is called a **trigonometric expression.** The fundamental relations can often be used to simplify or alter the form of trigonometric expressions. The next two examples illustrate the process.

Example 5 Use a fundamental relation to show that $\sin x \cot x$ can be reduced to $\cos x$, when $\sin x \neq 0$.

SOLUTION By Relation (2.2) we have, if $\sin x \neq 0$, $\cot x = \dfrac{\cos x}{\sin x}$. Thus

$$\sin x \cot x = \sin x \frac{\cos x}{\sin x} = \cos x$$

This shows that $\sin x \cot x$ is equivalent to $\cos x$, when $\sin x \neq 0$. ▲

Example 6 Use a combination of fundamental relations to show that $(1 + \tan^2\theta)\cos^2\theta$ is equivalent to 1.

SOLUTION By Relation (2.3a) we have $1 + \tan^2\theta = \sec^2\theta$. Thus

$$(1 + \tan^2\theta)\cos^2\theta = \sec^2\theta \cos^2\theta$$

By Relations (2.1) we know that $\sec\theta = 1/\cos\theta$. From this it follows that $\sec^2\theta = 1/\cos^2\theta$, and we have

$$\sec^2\theta \cos^2\theta = \frac{1}{\cos^2\theta}\cos^2\theta = 1$$

This shows that $(1 + \tan^2\theta)\cos^2\theta$ is equivalent to 1. ▲

▲▼ Exercises for Section 2.2

In Exercises 1–25 use the fundamental relations (2.1)–(2.3) to find the exact value of the indicated trigonometric function.

1. $\sin\theta = \dfrac{1}{2}$, find $\csc\theta$
2. $\cos\phi = \dfrac{2}{3}$, find $\sec\phi$
3. $\sec\beta = 3$, find $\cos\beta$
4. $\tan\theta = \dfrac{10}{7}$, find $\cot\theta$
5. $\sin A = \dfrac{\sqrt{3}}{2}$, $\tan A < 0$, find $\cos A$
6. $\sin\alpha = \dfrac{\sqrt{2}}{2}$, $\sec\alpha > 0$, find $\cos\alpha$
7. $\cot\theta = \sqrt{2}$, find $\tan\theta$
8. $\csc\theta = \dfrac{2}{\sqrt{3}}$, find $\sin\theta$
9. $\csc\alpha = 2$, $\tan\alpha > 0$, find $\cos\alpha$
10. $\cos x = -\dfrac{1}{2}$, $\tan x > 0$, find $\sin x$
11. $\sin\phi = -\dfrac{5}{13}$, $\cos\phi = \dfrac{12}{13}$, find $\tan\phi$
12. $\sin\beta = -\dfrac{2}{\sqrt{7}}$, $\cos\beta = \sqrt{\dfrac{3}{7}}$, find $\tan\beta$

13. $\tan\theta = \frac{1}{2}$, $\cos\theta = -\frac{2}{\sqrt{5}}$, find $\sin\theta$

14. $\tan\theta = \frac{2}{3}$, $\sin\theta = \frac{2}{\sqrt{13}}$, find $\cos\theta$

15. $\sin x = -\frac{1}{\sqrt{10}}$, $\cos x = -\frac{3}{\sqrt{10}}$, find $\tan x$

16. $\sec\theta = \frac{13}{12}$, $\tan\theta = \frac{5}{12}$, find $\sin\theta$

17. $\csc B = \frac{\sqrt{5}}{2}$, $\sec B > 0$, find $\tan B$

18. $\sin\gamma = \frac{2}{3}$, $\sec\gamma < 0$, find $\cot\gamma$

19. $\cos\phi = -\frac{\sqrt{2}}{3}$, $\csc\phi > 0$, find $\cot\phi$

20. $\sec\alpha = -2$, $\sin\alpha < 0$, find the other five functions

21. $\csc\theta = 3$, $\cos\theta > 0$, find the other five functions

22. $\sin\theta = \frac{2}{3}$, $\sec\theta > 0$, find the other five functions

23. $\cos\theta = -\frac{2}{\sqrt{5}}$, θ in quadrant II, find the other five functions

24. $\tan\alpha = 1$, α in quadrant III, find the other five functions

25. $\tan\beta = \sqrt{2}$, β in quadrant III, find the other five functions

In Exercises 26–30 use Relations (2.1) and a calculator to find the value of the indicated trigonometric functions.

26. $\sin\theta = 0.4313$, find $\csc\theta$

27. $\cos x = 0.1155$, find $\sec x$

28. $\tan\phi = 2.397$, find $\cot\phi$

29. $\csc A = 1.902$, find $\sin A$

30. $\sec t = 2.030$, find $\cos t$

31. The slope of a line is equal to the tangent of its angle of elevation. Use the fundamental relations to find the slope of a line for which the cosecant of the angle of elevation is $\sqrt{2.6}$ in quadrant I.

32. In a problem involving the rotation of coordinate axes, a student calculates the tangent of the rotation angle to be 2. Use the fundamental relations to find the sine and cosine of the rotation angle. Assume the rotation angle is acute.

33. Rework Exercise 25 in Section 2.1 (page 69), but this time use the fundamental relations.

34. Prove Relation (2.3b).

In Exercises 35–44 use the fundamental relations to show that the trigonometric expression on the left is equivalent to that on the right.

35. $\cos x \tan x = \sin x$

36. $\cos x \csc x = \cot x$

37. $\sec\phi \sin\phi = \tan\phi$

38. $\sin\theta \cos\theta \sec^2\theta = \tan\theta$

39. $\csc^2 x \cos x \sin x = \cot x$

40. $(1 + \cot^2\delta)\sin\delta = \csc\delta$

41. $\dfrac{\cos^2 x + \sin^2 x}{\sec x} = \cos x$

42. $(1 - \sin^2 t)\sec^2 t = 1$

43. $\dfrac{\sec^2 y - \tan^2 y}{\sin y} = \csc y$

44. $\dfrac{\sin z \sec z}{\csc^2 z - \cot^2 z} = \tan z$

2.3 The Values of the Trigonometric Functions for Special Angles

In the previous sections we discussed and computed values of the trigonometric functions from known points on the terminal side of an angle or by using the fundamental relations. No attempt was made to relate the measure of the angle to the values of its trigonometric functions. In practice it is important to know how to obtain the trigonometric functions for a specified angle. The values of the trigonometric functions of certain angles can be found geometrically, as illustrated in the next two examples.

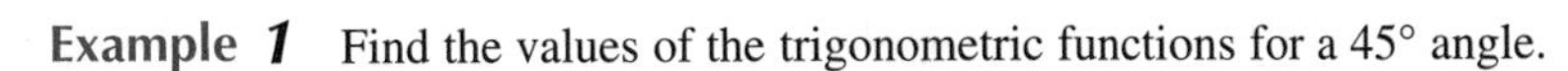

Example 1 Find the values of the trigonometric functions for a 45° angle.

SOLUTION Drawing a 45° angle in standard position, we observe that the terminal side bisects the first quadrant. Consequently the x-coordinate of any point on the terminal side of a 45° angle equals the y-coordinate. For convenience we choose the point (1, 1), as shown in Figure 2.11. Then $r = \sqrt{1^2 + 1^2} = \sqrt{2}$. Using $x = 1$, $y = 1$, and $r = 2$ in the definitions, we get

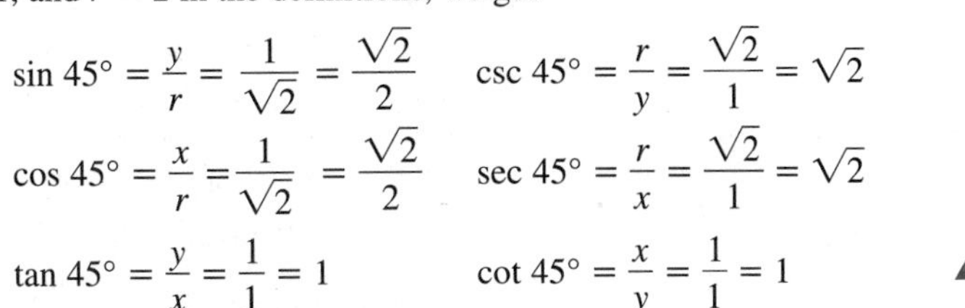

$$\sin 45^\circ = \frac{y}{r} = \frac{1}{\sqrt{2}} = \frac{\sqrt{2}}{2} \qquad \csc 45^\circ = \frac{r}{y} = \frac{\sqrt{2}}{1} = \sqrt{2}$$

$$\cos 45^\circ = \frac{x}{r} = \frac{1}{\sqrt{2}} = \frac{\sqrt{2}}{2} \qquad \sec 45^\circ = \frac{r}{x} = \frac{\sqrt{2}}{1} = \sqrt{2}$$

$$\tan 45^\circ = \frac{y}{x} = \frac{1}{1} = 1 \qquad \cot 45^\circ = \frac{x}{y} = \frac{1}{1} = 1$$

▲

Figure 2.11

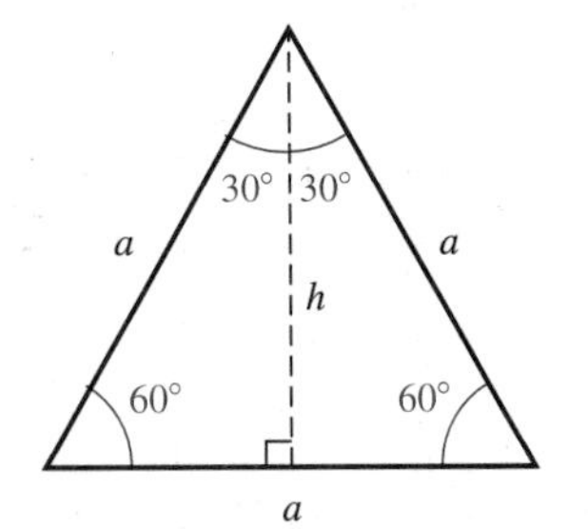

Figure 2.12

Example 2 Find the values of the trigonometric functions of a 60° angle.

SOLUTION Consider the equilateral triangle shown in Figure 2.12. The bisector of any one of the angles divides the equilateral triangle into two congruent right triangles. Since a line that bisects an angle of an equilateral triangle also bisects the side opposite that angle, the length of the side opposite the 30° angle is one-half the length of the hypotenuse. By the Pythagorean theorem, the length of h is $h = \sqrt{a^2 - (\frac{1}{2}a)^2} = a\sqrt{3}/2$. If $a = 2$, then $h = \sqrt{3}$ and $\frac{1}{2}a = 1$. From this we can conclude that the terminal side of a 60° angle in standard position passes through the point $(1, \sqrt{3})$ if r is 2. (See Figure 2.13.) Hence, by definition,

$$\sin 60^\circ = \frac{y}{r} = \frac{\sqrt{3}}{2} \qquad \csc 60^\circ = \frac{r}{y} = \frac{2}{\sqrt{3}} = \frac{2\sqrt{3}}{3}$$

$$\cos 60^\circ = \frac{x}{r} = \frac{1}{2} \qquad \sec 60^\circ = \frac{r}{x} = \frac{2}{1} = 2$$

$$\tan 60^\circ = \frac{y}{x} = \frac{\sqrt{3}}{1} = \sqrt{3} \qquad \cot 60^\circ = \frac{x}{y} = \frac{1}{\sqrt{3}} = \frac{\sqrt{3}}{3}$$

▲

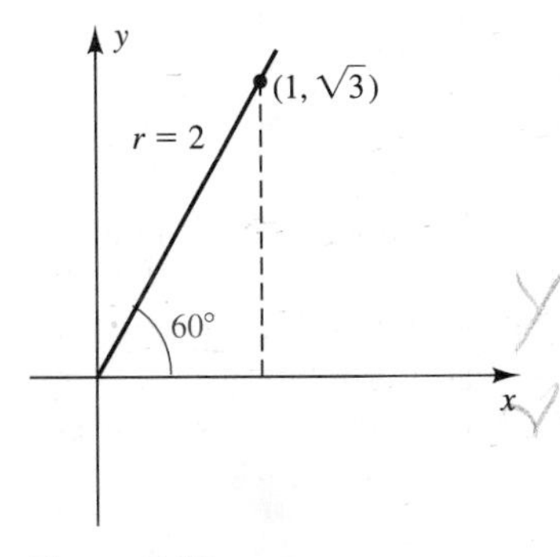

Figure 2.13

The values of the trigonometric functions for a 30° angle are found by the same right-triangle relationship used for a 60° angle. Table 2.2 summarizes the trigonometric functions for 30°, 45°, and 60° angles. Study it carefully. You should know how to *derive* each of the values in the table. (Keep in mind Figures 2.11 and 2.13.)

Table 2.2 Values for some special angles

θ (degrees)	$\sin\theta$	$\cos\theta$	$\tan\theta$	$\cot\theta$	$\sec\theta$	$\csc\theta$
30	$\frac{1}{2}$	$\frac{\sqrt{3}}{2}$	$\frac{\sqrt{3}}{3}$	$\sqrt{3}$	$\frac{2\sqrt{3}}{3}$	2
45	$\frac{\sqrt{2}}{2}$	$\frac{\sqrt{2}}{2}$	1	1	$\sqrt{2}$	$\sqrt{2}$
60	$\frac{\sqrt{3}}{2}$	$\frac{1}{2}$	$\sqrt{3}$	$\frac{\sqrt{3}}{3}$	2	$\frac{2\sqrt{3}}{3}$

COMMENT A calculator can be used to obtain approximate decimal equivalents of the values in Table 2.2. For example,

$$\tan 30^\circ = \frac{\sqrt{3}}{3} \approx \boxed{0.5773502692} \qquad \text{(to ten decimal places)}$$

As you will soon learn, the calculator is a very important tool in trigonometry, but knowing the exact values for certain special angles such as those listed in Table 2.2 is also important. It is best to remember the values for these special angles as exact values written in terms of radicals and simple fractions.

An interesting and useful observation that can be made about the values in Table 2.2 is that, for the complementary angles 30° and 60°,

$$\sin 30^\circ = \frac{1}{2} = \cos 60^\circ$$

$$\tan 30^\circ = \frac{\sqrt{3}}{3} = \cot 60^\circ$$

$$\sec 30^\circ = \frac{2\sqrt{3}}{3} = \csc 60^\circ$$

Two trigonometric functions that have equal values for complementary angles are called **cofunctions**. The various cofunction relationships are apparent from the names of the trigonometric functions; for example, the names "sine" and "cosine" reflect the cofunction relationship. Similarly, the tangent and the cotangent are cofunctions, as are the secant and the cosecant. Making use of the fact that θ and $90^\circ - \theta$ are complementary angles, we have

$$\begin{Bmatrix}\text{trigonometric function of} \\ \text{an acute angle } \theta\end{Bmatrix} = \{\text{cofunction of } (90° - \theta)\} \quad \textbf{(2.4)}$$

Example 3

(a) $\sin 40° = \cos 50°$

(b) $\tan 5.6° = \cot 84.4°$

(c) $\cos 13°15' = \sin 76°45'$ ▲

Quadrantal Angles

Recall that an angle in standard position whose terminal side lies on a coordinate axis is called a **quadrantal angle.** Angles of 0°, ±90°, and ±180° are examples. For these angles, one of the coordinates of a point on the terminal side must be zero. Since division by zero is undefined, two of the six trigonometric functions are undefined at each quadrantal angle.

y
180°
(−1, 0)
x

Example 4 Find the trigonometric functions of an angle $\theta = 180°$ whose terminal side passes through the point $(-1, 0)$, as shown in Figure 2.14.

SOLUTION In this case $x = -1$, $y = 0$, and $r = 1$. Since $\theta = 180°$ we write

$$\sin 180° = \frac{y}{r} = \frac{0}{1} = 0 \qquad \csc 180° = \frac{r}{y} = \frac{1}{0} \quad \text{(undefined)}$$

$$\cos 180° = \frac{x}{r} = \frac{-1}{1} = -1 \qquad \sec 180° = \frac{r}{x} = \frac{1}{-1} = -1$$

$$\tan 180° = \frac{y}{x} = \frac{0}{-1} = 0 \qquad \cot 180° = \frac{x}{y} = \frac{-1}{0} \quad \text{(undefined)}$$ ▲

Example 4 gives the values for a quadrantal angle of 180° or one coterminal with it. The values of the other quadrantal angles can be found by a similar procedure; they are listed for your reference in Table 2.3.

Table 2.3 Values for the quadrantal angles

θ (degrees)	$\sin\theta$	$\cos\theta$	$\tan\theta$	$\cot\theta$	$\sec\theta$	$\csc\theta$
0	0	1	0	undefined	1	undefined
90	1	0	undefined	0	undefined	1
180	0	−1	0	undefined	−1	undefined
270	−1	0	undefined	0	undefined	−1

COMMENT You should be able to verify all the values in Tables 2.2 and 2.3.

Example 5 Use Tables 2.2 and 2.3 to find the exact value of the expression

$$\sin 60° + 3 \cos 180° + \cos^2 45°$$

SOLUTION From Table 2.2 we have

$$\sin 60° = \frac{\sqrt{3}}{2} \quad \text{and} \quad \cos 45° = \frac{\sqrt{2}}{2}$$

and from Table 2.3 we have

$$\cos 180° = -1$$

Therefore

$$\begin{aligned} \sin 60° + 3 \cos 180° + \cos^2 45° &= \frac{\sqrt{3}}{2} + 3(-1) + \left(\frac{\sqrt{2}}{2}\right)^2 \\ &= \frac{\sqrt{3}}{2} - 3 + \frac{1}{2} \\ &= \frac{\sqrt{3} - 5}{2} \end{aligned}$$

▲

▲▼ Exercises for Section 2.3

1. Verify the values of the six trigonometric functions of 30° given in Table 2.2.
2. Verify the values of the six trigonometric functions of 90° given in Table 2.3. Note that (0, 1) lies on the terminal side of the angle of 90° when it is placed in standard position.
3. Verify the values of the six trigonometric functions of 0° given in Table 2.3. Use (1, 0) as a point on the terminal side.
4. Verify the values of the six trigonometric functions of 270° given in Table 2.3. Use (0, −1) as a point on the terminal side.

In Exercises 5–14 express the value of the given trigonometric function in terms of a cofunction of a complementary angle.

5. sin 20°
6. cos 50°
7. tan 20°15′
8. cot 72°
9. sec 81°
10. cos 30°40′
11. cos 17.5°
12. sin 37.5°
13. tan 60.4°
14. csc 12.6°

Mathematical expressions frequently contain combinations of trigonometric functions like those in Exercises 15–24. Use Tables 2.2 and 2.3 if necessary to find the exact value of each of these expressions.

15. $2 \sin 30° + 5 \cos 45° - 2$
16. $3 \sin 90° + 4 \cos 90° - 10$
17. $\tan 45° \sin 270°$
18. $\cos 0° \cot 45°$
19. $\csc^2 60° + 3 \cot^2 270°$
20. $\tan^2 30° + 1$
21. $\dfrac{\sec^2 180°}{\sin 180°}$
22. $\dfrac{\tan 45° \sec 45°}{\cos 270°}$
23. $(2 \sin 60° + \sqrt{3} \cos 180°)^2$
24. $(\sec^2 45° - \tan^2 45°)^2 \cos 60°$

In Exercises 25–34 use Tables 2.2 and 2.3 to determine the value of the angle θ, $0° \le \theta \le 90°$, for which the given statement is true.

25. $\sin \theta = \frac{1}{2}$ **26.** $\cos \theta = \frac{1}{2}$ **27.** $\tan \theta = \sqrt{3}$ **28.** $\sec \theta = 2$

29. $\cot \theta = \sqrt{3}$ **30.** $\sin \theta = 0$ **31.** $\cos \theta = 0$ **32.** $\tan \theta = 0$

33. $\cot \theta = 1$ **34.** $\sec \theta = 1$

35. For which acute angle θ is $\sin \theta = \cos \theta$?

36. For which acute angle θ is $\tan \theta = \cot \theta$?

37. The distance x, in feet, between two buildings is given by $x = 15 \tan 30°$. Calculate the distance x.

38. The horizontal and vertical components of the velocity of a rocket are given by $v_H = 1000 \cos 30°$ and $v_V = 1000 \sin 30°$, both in feet per second. Approximate the value of the two components to the nearest 10 ft/sec.

2.4 Reference Angles

The values of the trigonometric functions for 30°, 45°, and 60° can be used to find the values of the trigonometric functions for related angles in other quadrants. To accomplish this we use the concept of a reference angle. We noted in Section 1.6 that any angle in standard position is coterminal with an angle whose measure is between 0° and 360°. Consequently we only need to discuss reference angles for angles between 0° and 360°.

> DEFINITION 2.2 The **reference angle** of a given angle $\theta(0° \le \theta < 360°)$ is the positive acute angle α between the terminal side of θ in standard position and the x-axis. (See Figure 2.15.)

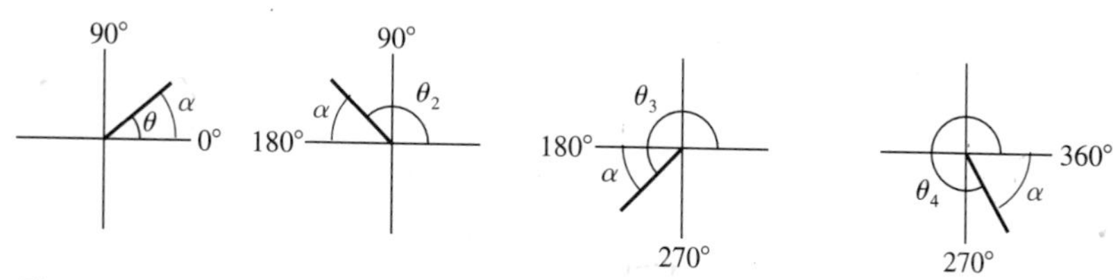

Figure 2.15
Reference angles

The reference angle α of a given angle θ in standard position can be found by using the following rules—θ_1 represents an angle in quadrant I, θ_2 an angle in quadrant II, and so forth.

REFERENCE ANGLE RULES The reference angle α of an angle θ in standard position is found as follows:

1. If θ_1 is a first-quadrant angle, then $\alpha = \theta_1$.
2. If θ_2 is a second-quadrant angle, then $\alpha = 180° - \theta_2$.
3. If θ_3 is a third-quadrant angle, then $\alpha = \theta_3 - 180°$.
4. If θ_4 is a fourth-quadrant angle, then $\alpha = 360° - \theta_4$.

Example 1 Find the reference angle for **(a)** 196° and **(b)** 98°.

SOLUTION

(a) $\alpha = 196° - 180° = 16°$ [See Figure 2.16(a).]

(b) $\alpha = 180° - 98° = 82°$ [See Figure 2.16(b).]

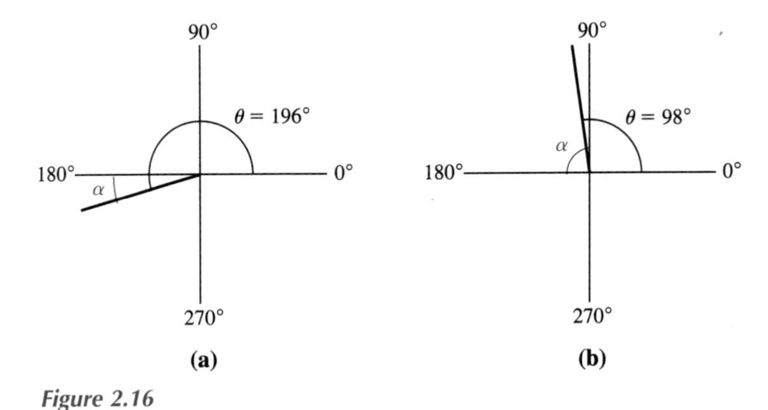

Figure 2.16

▲

COMMENT The reference angle is always computed with respect to the x-axis.

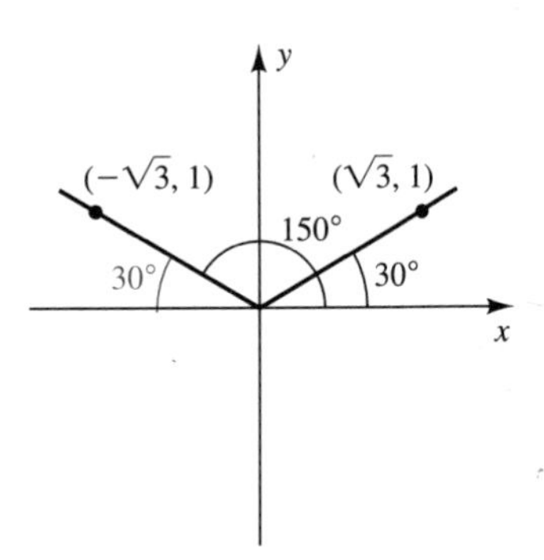

Figure 2.17

Trigonometric Functions of Angles Greater Than 90°

The next example shows how the to evaluate tan 150°. The example illustrates the general procedure used to evaluate trigonometric functions of angles greater than 90° in terms of trigonometric functions of acute angles.

Example 2 Find the exact value of tan 150°. Show that tan 150° = −tan 30°.

SOLUTION A generated angle of 150° is shown in Figure 2.17. The reference angle is 30° and is also shown as a generated angle in the first quadrant. From our knowledge of the 30°-60° right triangle, we know that the terminal side of a 30° angle must pass through the point $(\sqrt{3}, 1)$. Thus $(-\sqrt{3}, 1)$ is a point on the terminal side of 150°. Therefore by definition

$$\tan 150° = \frac{1}{-\sqrt{3}} = -\frac{1}{\sqrt{3}} = -\tan 30° \qquad ▲$$

Example 2 shows that tan 150° is equal to −tan 30°, where 30° is the reference angle of 150°. We generalize this result in the following rule.

TRIGONOMETRIC FUNCTIONS OF ANY ANGLE θ To express the value of a trigonometric function of an angle θ in standard position in terms of its reference angle, proceed as follows.

1. Determine the quadrant in which the terminal side of θ lies.
2. Use the Reference Angle Rules to find the reference angle for θ.
3. Add the appropriate sign (+ or −) to the trigonometric function of the reference angle. Use Table 2.1 to determine the sign.

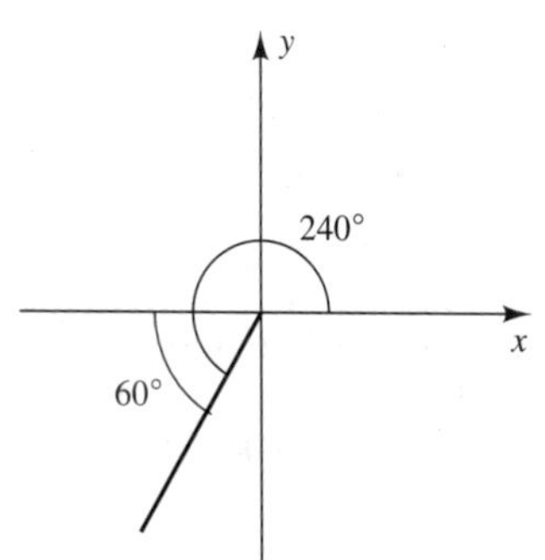

Figure 2.18

The following examples show how to use this procedure for angles that are greater than 90° and whose reference angles are the special angles in Table 2.2. General angles are discussed in Section 2.5.

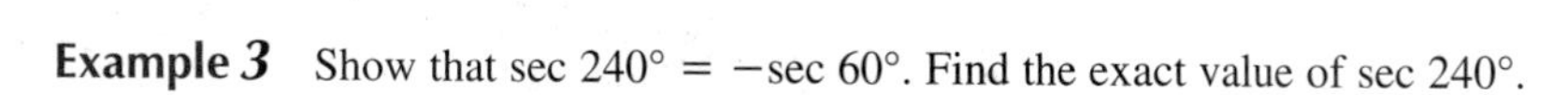

Example 3 Show that sec 240° = −sec 60°. Find the exact value of sec 240°.

SOLUTION We see from Figure 2.18 that 240° is a third-quadrant angle, so the reference angle is

$$\alpha = 240° - 180° = 60°$$

The reference angle is also shown in Figure 2.18. Since the secant function is negative in the third quadrant and sec 60° = 2, we have

$$\sec 240° = -\sec 60° = -2 \qquad ▲$$

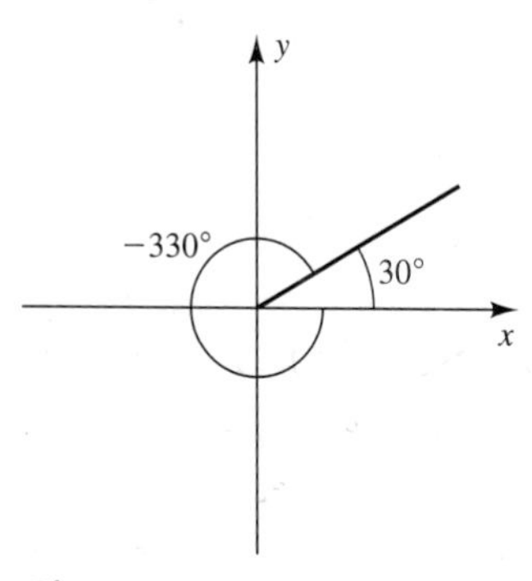

Figure 2.19

Example 4 Show that sin(−330°) = sin 30° and find the exact value of sin(−330°).

SOLUTION We see from Figure 2.19 that the required reference angle is

$$\alpha = 360° - 330° = 30°$$

The reference angle is also shown in Figure 2.19. Since the sine function is positive in the first quadrant and $\sin 30° = \frac{1}{2}$, we have

$$\sin(-330°) = \sin 30° = \frac{1}{2} \qquad ▲$$

The definitions of the trigonometric functions of any angle show that the functional values are completely determined by the location of the terminal side when the angle is in standard position. Thus **coterminal angles have equal functional values.** For instance, since 30° and 390° are coterminal, $\sin 30° = \sin 390°$, $\cos 30° = \cos 390°$, $\tan 30° = \tan 390°$, and so on. Thus in finding values of trigonometric functions, we need only consider angles between 0° and 360°.

Example 5 Show that $\sin 945° = -\sin 45°$. Find the exact value of $\sin 945°$.

SOLUTION Note that $945° = 2(360°) + 225°$, so 945° is coterminal with 225°. The reference angle for 225° is 45°. Since 225° is a third-quadrant angle and the sine function is negative in the third quadrant, we have

$$\sin 945° = \sin 225° = -\sin 45° = -\frac{\sqrt{2}}{2}$$ ▲

By considering cases in which the angle θ has its terminal side in each of the four quadrants in turn, you can verify the following general relations.

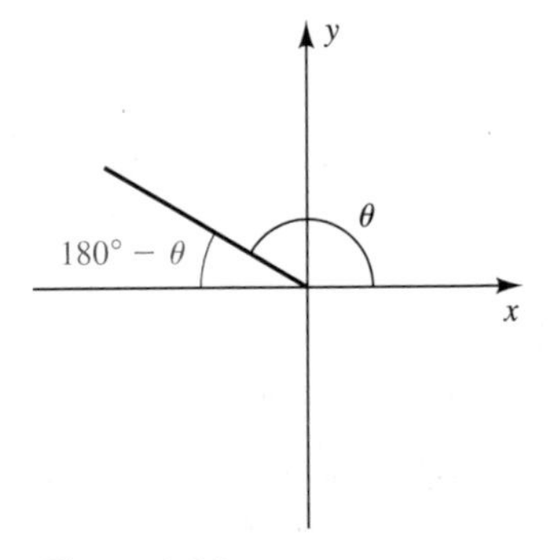

Figure 2.20

$$\begin{aligned} \sin(-\theta) &= -\sin\theta & \cos(-\theta) &= \cos\theta \\ \sin(180° - \theta) &= \sin\theta & \cos(180° - \theta) &= -\cos\theta \\ \sin(180° + \theta) &= -\sin\theta & \cos(180° + \theta) &= -\cos\theta \\ \sin(360° - \theta) &= -\sin\theta & \cos(360° - \theta) &= -\cos\theta \end{aligned} \qquad (2.5)$$

You need not memorize these relations, but be ready to work out any of the results above using the methods of this section. Figure 2.20 gives the idea for a second-quadrant angle. From Figure 2.20 we see that $180° - \theta$ is the reference angle for θ. Since θ is in the second quadrant,

$$\sin\theta = \sin(180° - \theta) \quad \text{and} \quad \cos\theta = -\cos(180° - \theta)$$

▲▼ Exercises for Section 2.4

In Exercises 1–14 give the reference angle for each of the given angles.

1. 100°	**2.** 200°	**3.** 300°	**4.** 400°
5. 210°	**6.** 315°	**7.** 480°	**8.** 570°
9. −135°	**10.** −225°	**11.** 225°	**12.** 120°
13. 930°	**14.** 1020°		

In Exercises 15–34 use the values of the trigonometric functions of the special angles (Table 2.2) to find the exact value of the indicated expression.

15. cos(−45°)	**16.** tan 240°
17. sin 150°	**18.** sec 135°

19. $\tan 300°$

20. $\cos(-240°)$

21. $\csc 210°$

22. $\sin 330°$

23. $\sec 315°$

24. $\cot 150°$

25. $\sin 210°$

26. $\tan 225°$

27. $\cot(-510°)$

28. $\cos 840°$

29. $\cos 1050°$

30. $\csc 600°$

31. $\tan 930° - 2 \sin 120°$

32. $\sin 1020° + \cot 135°$

33. $\sin 945° + \tan 240°$

34. $\cot 1140° - \sqrt{2} \sec 315°$

35. Indicate all values of θ for which $\sin \theta = 1$.

36. Indicate all values of θ for which $\cos \theta = 1$.

37. Indicate all values of θ for which $\cos \theta = 0$.

38. Indicate all values of θ for which $\tan \theta = 0$.

39. Indicate all values of θ for which $\sin \theta = \frac{1}{2}$.

40. Construct a table showing the exact values of $\sin \theta$, $\cos \theta$, and $\tan \theta$ for angles of 30°, 45°, 60°, 120°, 135°, 150°, 210°, 225°, 240°, 300°, 315°, and 330°.

41. The current in an a.c. circuit is given by $i = 3 \sin \theta$. Find the value of the current when $\theta = 0°, 30°, 60°, 120°, 150°$, and 180°.

42. The azimuth error in a radar measurement involves the term $\sec \theta$. Evaluate $\sec \theta$ when $\theta = 225°$.

2.5 Finding Values of Trigonometric Functions with a Calculator

Tables 2.2 and 2.3 in Section 2.3 are obviously incomplete listings of values of the trigonometric functions; Table 2.2 gives the functional values for only 30°, 45°, and 60°, and Table 2.3 is restricted to the quadrantal angles 0°, 90°, 180°, 270°, and 360°. The trigonometric functions for the other angles can be found with a calculator.

COMMENT Most calculators operate in either a "degree" mode or a "radian" mode (another unit of angular measure). We will discuss the radian in Chapter 4, so for now just keep your calculator in the degree mode.

Calculators have keys for $\sin \theta$, $\cos \theta$, and $\tan \theta$. The values of these three trigonometric functions for any angle θ are found by pushing the desired key and then entering the angle. (*Note:* In some calculators the order of entry is reversed.)

Example 1 Use a calculator to find **(a)** $\sin 37°$ and **(b)** $\tan 122°$ to four decimal places.

SOLUTION

(a) To find sin 37°, push the [sin] key, and then enter 37 and push the [=] key to obtain [0.601815023]. The result, when rounded to four decimal places, is

$$\sin 37° \approx 0.6018$$

(b) To find tan 122°, push the [tan] key, and then enter 122 and push the [=] key to obtain [−1.600334529]. Thus

$$\tan 122° \approx -1.6003$$ ▲

COMMENT Recall that the number of digits displayed in the register will depend on the brand of calculator. Keep in mind that calculator computations are often approximate.

Calculators are not designed with keys for csc θ, sec θ, and cot θ because these functions can be obtained from sin θ, cos θ, and tan θ by using the relations

$$\csc \theta = \frac{1}{\sin \theta} \qquad \sec \theta = \frac{1}{\cos \theta} \qquad \cot \theta = \frac{1}{\tan \theta}$$

which were introduced in Section 2.2.

Example 2 Use a calculator to evaluate **(a)** cot 17°, **(b)** csc 310°, and **(c)** sec 500° to four decimal places.

SOLUTION

(a) To find cot 17°, push the [tan] key, and then enter 17 and push the [=] key to obtain [0.305730681]. Now push [1/x] [=] to obtain [3.270852619]. Thus

$$\cot 17° \approx 3.2709$$

(b) To find csc 310°, push the [sin] key, and then enter 310 and push the [=] key to obtain [−0.766044443]. Now push [1/x] [=] to obtain [−1.305407289]. Thus

$$\csc 310° \approx -1.3054$$

(c) To find sec 500°, we use the following sequence of keystrokes.

sec 500° = [cos] 500 [=] [1/x] [=] ≈ [−1.305407289] ≈ −1.3054 ▲

Angles are commonly expressed in units of degrees, minutes, and seconds ($1° = 60' = 3600''$). Some calculators allow angles to be entered directly in this format, but others accept only decimal equivalents of these units in degrees. If this is the case with your calculator, you must change degrees, minutes, and seconds into the equivalent decimal form before entering the angle in the register.

Example 3 Use a calculator to evaluate cos 204°15′23″ to four decimal places.

SOLUTION First express 204°15′23″ in degrees by dividing 15′ by 60 and 23″ by 3600. Thus

$$204°15'23'' = 204° + \left(\frac{15}{60}\right)^{\circ} + \left(\frac{23}{3600}\right)^{\circ}$$

The required sequence of keystrokes for the right-hand side is

204 [+] 15 [÷] 60 [+] 23 [÷] 3600 [=] [204.2563889]

[Store] 204.2563889 and then push the [cos] key followed by [RCL] [=] to obtain [−0.911716240]. Thus

$$\cos 204°15'23'' \approx -0.9117$$

▲

Calculators are also used to find angles corresponding to a given trigonometric function. The procedure for finding an angle corresponding to a given functional value varies depending on the brand of calculator, but most calculators require you to enter the given number and then push an [inv] or [2nd] button prior to pushing the trigonometric function button. Other models have single buttons for this purpose labeled [$\sin^{-1}$], [$\cos^{-1}$], or [$\tan^{-1}$].

COMMENT There are two angles θ between 0° and 360° for which cos θ = 0.5. However, if you enter 0.5 in a calculator and push [inv] [cos], only 60° is displayed in the register. To obtain the other angle, which is 300°, you must use the concept of a reference angle.

CAUTION The notation $\sin^{-1}x$ means the angle whose sine is x; consequently, it *cannot* mean the reciprocal of sin x. Because $\sin^{-1}x$ is reserved for the inverse sine function, you must write the reciprocal of sin x as

$$\frac{1}{\sin x} = (\sin x)^{-1}$$

A similar caution applies to the notation for the other inverse trigonometric functions.

Example 4 Find, to the nearest hundredth of a degree, the angles θ between $0°$ and $360°$ for which $\tan \theta = 1.5$.

SOLUTION To find the first angle, push [inv] [tan], and then enter 1.5 and push [=] to obtain [56.30993247], which rounds to 56.31°. The other angle for which $\tan \theta = 1.5$ is in the third quadrant. Specifically it is the third-quadrant angle whose reference angle is 56.31°. Thus

$$\theta_3 = 180° + 56.31° = 236.31°$$

The desired angles are 56.31° and 236.31°. ▲

WARNING The angle displayed in the register for a given trigonometric function is not always a positive acute angle. For example, if $\sin \theta = -0.5$, your calculator will give only one angle, $\theta = -30°$. However, if $\cos \theta = -0.5$, your calculator will give the angle $\theta = 120°$. The reason we get a negative acute angle in the first case and a positive obtuse angle in the second is explained in Chapter 8. To avoid confusion at this time, we suggest that you use your calculator to find the reference angle for a given trigonometric function. *You can obtain the reference angle from your calculator if you enter the absolute value of the given function.* The values given by the inverse functions are a result of some carefully agreed upon definitions of the inverse trigonometric functions. More about this in Chapter 8.

Example 5 Find, to the nearest tenth of a degree, angles θ $(0° \leq \theta < 360°)$ if $\sin \theta = -\,0.5664$.

SOLUTION Enter the absolute value of the given function and push [inv] [sin]. The resulting reference angle is

$$\text{[inv] [sin] } 0.5664 = \text{[34.49956639]} \approx 34.5°$$

Now, since the sine is negative in both the third and fourth quadrants, we get

$$\theta_3 \approx 180° + 34.5° = 214.5°$$
$$\theta_4 \approx 360° - 34.5° = 325.5°$$

▲

Example 6 Find, to the nearest hundredth of a degree, angle $\theta (0° \leq \theta < 360°)$ if $\cot \theta = -2.573$ and $\cos \theta < 0$.

SOLUTION To obtain the reference angle for θ, enter 2.573, the absolute value of the given function. Next, push the [1/x] key to obtain [0.38865138], which is $\tan \theta$. (Recall that $\tan \theta = 1/\cot \theta$.) Now, store 0.38865138 and push [inv] [tan] followed by [rcl] [=] to obtain the reference angle [21.23868397] $\approx 21.24°$. Another more direct sequence of steps for finding this reference angle is

$$\text{[inv] [tan] } \underbrace{2.573 \text{ [1/x]}}_{\text{converts } \cot x \text{ to } \tan x} \text{ [=] [21.23868397]} \approx 21.24°$$

Both cot θ and cos θ must be negative. Since cot θ is negative in the second and fourth quadrants and cos θ is negative is the second and third quadrants, the desired angle must be in the second quadrant. Thus

$$\theta \approx 180° - 21.24° = 158.76°$$

▲

Example 7 Find, to the nearest second, angle ϕ ($0° \le \phi < 360°$) if $\sec \phi = 1.33327$ and $\csc \phi < 0$.

SOLUTION To obtain the reference angle for ϕ, use the following steps:

[inv] [cos] 1.33327 [1/x] [=] [41.40653592]°

(1.33327 [1/x]: converts sec x to cos x)

We require $\sec \phi > 0$ and $\csc \phi < 0$, which means that angle ϕ is in the fourth quadrant. Thus

$$\phi \approx 360° - 41.40653592° = 318.5934641°$$

To convert the decimal part of a degree to minutes, we multiply 0.5934641 by 60. Thus the equivalent minutes are

$$60(0.5934641)' = 35.607846'$$

Finally, we convert the decimal part of a minute to seconds by multiplying 0.607846 by 60. Thus the equivalent seconds are

$$60(0.604846) = 36.47076'' \approx 36''$$

The required angle is

$$\phi \approx 318°35'36''$$

Notice that we do not round off, but retain all digits in the calculator until we round off to seconds.

▲

Applications of the trigonometric functions are discussed in depth in Chapter 3. The next example shows a typical expression containing a trigonometric function.

Example 8 Find the acute angle A to the nearest tenth of a degree if $\cos A$ is given by

$$\cos A = \frac{15^2 + 18^2 - 17^2}{2(15)(18)}$$

SOLUTION Using a calculator we perform the indicated arithmetic operations to obtain

$\cos A \approx$ [0.481481481]

Now use [inv] [cos] to find acute angle A; that is,

$A \approx$ [inv] [cos] 0.481481481 [=] [61.21779532]

Rounding off to one decimal place, we conclude that $A \approx 61.2°$.

▲

Exercises for Section 2.5

In Exercises 1–24 use a calculator to find the values of the trigonometric functions to four decimal places.

1. $\sin 13°$
2. $\cos 78°$
3. $\tan 17.3°$
4. $\cos 5.45°$
5. $\cos 100.2°$
6. $\cos 211.1°$
7. $\cos 399°$
8. $\sin 105.7°$
9. $\tan 1540°$
10. $\sec 50.8°$
11. $\sec 142°$
12. $\cot 305°$
13. $\csc 111.9°$
14. $\csc(-32.6°)$
15. $\cot(-43.3°)$
16. $\sin 54°32'$
17. $\cos 213°31'$
18. $\cos 335°56'$
19. $\tan 950°52'$
20. $\csc 3°46'$
21. $\sec 59°38'$
22. $\sec 720°58'$
23. $\cot 100°3'$
24. $\cot 470°24'$

In Exercises 25–40 use a calculator to find θ, where $0° \le \theta < 360°$. Express θ to the nearest tenth of a degree.

25. $\sin \theta = 0.5567$ and $\cos \theta > 0$
26. $\tan \theta = 1.802$ and $\sec \theta > 0$
27. $\tan \theta = 0.4414$ and $\cos \theta < 0$
28. $\cos \theta = 0.9002$ and $\sin \theta < 0$
29. $\sin \theta = -0.4253$ and $\tan \theta < 0$
30. $\sin \theta = 0.4331$ and $\cos \theta > 0$
31. $\cot \theta = 3.0326$ and $\csc \theta > 0$
32. $\cos \theta = -0.8635$ and $\cot \theta < 0$
33. $\sec \theta = 1.345$ and $\tan \theta < 0$
34. $\csc \theta = 2.026$ and $\cos \theta < 0$
35. $\tan \theta = -0.3378$
36. $\sin \theta = 0.8279$
37. $\csc \theta = 1.505$
38. $\cos \theta = -0.4642$
39. $\cot \theta = -0.5137$
40. $\sec \theta = 1.104$

Exercises 41–49 show some typical expressions that involve the trigonometric functions. In Chapter 3 we will develop similar expressions; for now simply evaluate each expression.

41. A civil engineer finds that the distance between two points on a new bridge is given by

$$x = 352.7 \cos 17.9°$$

where x is measured in feet. Calculate this distance to the nearest tenth of a foot.

42. The velocity of a rocket is given by the expression

$$v = 1200 \sin 78°15'$$

where v is measured in feet per second. Calculate v to the nearest ft/sec.

43. Suppose that in calculating the height in meters of a radio antenna we encounter the expression

$$x = \frac{155 \sin 27.4°}{\sin 68.6°}$$

Calculate x to the nearest tenth of a meter.

44. The length of the side of a certain triangle is given by

$$b = \frac{5.6 \sin 13.1°}{\sin 47.9°}$$

Calculate b to the nearest tenth.

45. To determine the course for a ship, a navigator must determine the value of an acute angle A, where $\sin A$ is given by

$$\sin A = \frac{247.3 \sin 30.6°}{208.7}$$

Calculate angle A to the nearest tenth of a degree.

46. The angle θ between two sides of a triangle can be found if $\sin \theta$ is known. Find θ to the nearest tenth of a degree if $\sin \theta$ is given by

$$\sin \theta = \frac{2.3 \sin 10.3°}{3.4}$$

47. The resultant velocity of a projectile fired from a moving airplane is given by

$$v = \sqrt{250^2 + 720^2 - 2(250)(720) \cos 100°}$$

Assume v is measured in feet per second. Calculate v to the nearest ft/sec.

An airplane flying faster than the speed of sound rides ahead of its own sound wave. The sound wave emitted by the plane can be represented as a cone with its vertex at the plane's tail. (See Figure 2.21.). Aeronautical engineers have shown that the angle θ in degrees between the sides of the sound cone is given by

$$\sin \tfrac{1}{2}\theta = \frac{v_S}{v_P}$$

where v_S is the speed of sound and v_P is the speed of the airplane. Notice that if $v_S > v_P$, then angle θ is not defined because $\sin \frac{1}{2}\theta$ must be less than or equal to 1.

48. Find the angle of the sound cone if $v_S = 1120$ ft/sec and $v_P = 1500$ ft/sec.

49. Find the angle of the sound cone if $v_S = 1120$ ft/sec and $v_P = 2000$ ft/sec.

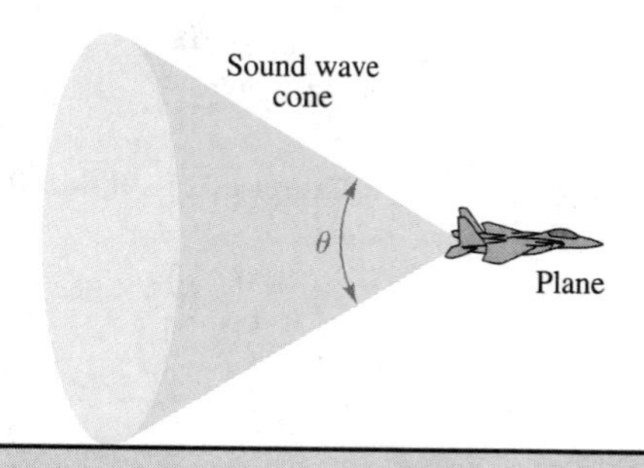

Figure 2.21

Key Topics for Chapter 2

Define and/or discuss each of the following.

$\sin\theta$, $\cos\theta$, $\tan\theta$, $\cot\theta$ $\sec\theta$, $\csc\theta$
The Reciprocal Relations
The Pythagorean Relation
The Special Angles
Quadrantal Angles
Reference Angle
Finding the exact values of the trigonometric functions at special angles
Use of a calculator to approximate values of the trigonometric functions

Review Exercises for Chapter 2

1. Find the exact values of the six trigonometric functions of the angle in standard position whose terminal side passes through $(-2, 5)$.
2. Determine the quadrant in which the terminal side of θ lies if $\cot\theta < 0$ and $\sec\theta < 0$.
3. Find the exact values of the other five trigonometric functions of θ if $\cos\theta = -\frac{4}{5}$ and $\tan\theta > 0$.
4. Find the exact values of the other five trigonometric functions of α if $\csc\alpha = \frac{13}{5}$ and $\cos\alpha < 0$.

In Exercises 5–10 indicate the reference angle for the given angle.

5. $315°18'$
6. $109°41'$
7. $241.9°$
8. $185.3°$
9. $272.2°$
10. $156.4°$

In Exercises 11–16 find the exact value of each trigonometric function. Do not use a calculator.

11. $\tan 150°$
12. $\sin -870°$
13. $\cos -135°$
14. $\sec 405°$
15. $\csc 225°$
16. $\cot 660°$

In Exercises 17–24 evaluate the function to four decimal places.

17. $\tan 203°$
18. $\sin 310°20'$
19. $\cos(-112.6°)$
20. $\csc(-9°15')$
21. $\sec 98.3°$
22. $\cot 189.6°$
23. $\sin 975°$
24. $\tan(-1053°)$

In Exercises 25–30 find the exact values of θ $(0 \leq \theta < 360°)$. Do not use a calculator.

25. $\sin\theta = \dfrac{\sqrt{3}}{2}$
26. $\cos\theta = \frac{1}{2}$
27. $\tan\theta = -\sqrt{3}$
28. $\sec\theta = -2$
29. $\csc\theta = -1$
30. $\cos\theta = -\dfrac{\sqrt{2}}{2}$
31. Given $\sin x = -0.8102$, find x $(0° \leq x < 360°)$ to the nearest minute.
32. Given $\tan\theta = 1.202$, find θ $(0° \leq \theta < 360°)$ to the nearest minute.
33. Given $\sec\phi = 2.603$ and $\sin\phi < 0$, find ϕ $(0° \leq \phi < 360°)$ to the nearest tenth of a degree.
34. Given $\csc\theta = -1.118$ and $\cot\theta > 0$, find θ $(0° \leq \theta < 360°)$ to the nearest tenth of a degree.

35. Given $\tan \beta = -0.7761$ and $\cos \beta > 0$, find β $(0^\circ \le \beta < 360^\circ)$ to the nearest tenth of a degree.

36. Surveyors laying out a new road must calculate the distance between two points by evaluating

$$x = \frac{1500 \sin 25.7^\circ}{\sin 105.3^\circ} \text{ ft}$$

Find x to the nearest tenth of a foot.

37. In filing a flight plan, a pilot must calculate the ground speed of the plane by taking into account the speed and direction of the wind. The ground speed in miles per hour can be expressed as

$$v = \frac{210 \sin 135^\circ}{\sin 40^\circ}$$

What is the numerical value to the nearest mph?

38. In evaluating the resultant of two forces, a student encounters the expression

$$F = \sqrt{5^2 + 9^2 - 2(5)(9) \cos 133^\circ 18'}$$

Evaluate F to two decimal places.

Practice Test Questions for Chapter 2

In Exercises 1–10 answer *true* or *false*.

1. For angle θ in standard position, $\cos \theta = x/y$.
2. For any acute angle β, $\cos \beta = 1/\sec \beta$.
3. $\operatorname{Sec} \phi > 0$ in quadrant IV.
4. If θ is an acute angle and $\tan \theta = \cot \theta$, then $\theta = 45^\circ$.
5. If α and β are complementary angles, then $\sin^2\alpha + \cos^2\beta = 1$.
6. For acute angles, if $\theta_1 < \theta_2$, then $\cos \theta_1 < \cos \theta_2$.
7. $\operatorname{Sin} \theta < 0$ in quadrants III and IV.
8. If $\sin \alpha < 0$ and $\cos \alpha < 0$, the terminal side of angle α is in quadrant III.
9. $\cot^2\theta + 1 = \sec^2\theta$.
10. The reference angle for 135° is 35°.

In Exercises 11–20 fill in the blank to make the statement true.

11. $\sin \alpha \csc \alpha =$ ______.
12. If θ is an angle in standard position, the ratio x/y defines the ______.
13. $\operatorname{Cos} \theta$ is positive in quadrants ______ and ______.
14. The terminal side of angle ϕ is in quadrant ______ if $\csc \phi > 0$ and $\tan \phi < 0$.
15. $\tan^2\theta +$ ______ $= \sec^2\theta$.
16. Express $\cot \theta$ as the ratio of two other trigonometric functions: $\cot \theta =$ ______.
17. The exact value of $\sin 60^\circ$ is ______.
18. An angle in standard position whose terminal side lies on a coordinate axis is called a ______ angle.

19. The ____________ angle for a given angle θ in standard position is the positive acute angle between the terminal side of θ and the x-axis.
20. Express sec 238° in terms of the cosine of an acute angle. sec 238° = ____________.

Solve the stated problem in the following exercises. Show all your work.

21. Find the exact values of the other five trigonometric functions of θ in quadrant IV if $\sec\theta = 5/2$.
22. Find the exact values of the six trigonometric functions of an angle in standard position whose terminal side passes through $(-2, 3)$.
23. Use Relations (2.1)–(2.3) to find the exact value of $\cos\theta$ if $\sin\theta = 2/3$, $\tan\theta > 0$.
24. Use Relations (2.1)–(2.3) to find the exact value of $\tan\alpha$ if $\cos\alpha = -1/3$, $\csc\alpha < 0$.
25. Use Relations (2.1)–(2.3) to show that $\sin x \cot x = \cos x$.
26. Use Relations (2.1)–(2.3) to show that $\sec^2\beta \sin\beta \cos\beta = \tan\beta$.
27. Find the exact value of $\csc 60° + 3 \tan 135° - \cos 210°$.
28. Find the exact value of $\sec^2 30° - \tan^2 330°$.
29. Give reference angle for each of the following angles.
    (a) 100° **(b)** 200° **(c)** 300°
30. Given that $\sin\theta = -0.1537$ and $\cos\theta < 0$, use a calculator to find θ, where $0° \le \theta \le 360°$.

31. Given that $\sec\phi = 1.2252$, use a calculator to find the acute angle ϕ.
32. Given that $\cot\beta = -0.5635$, use a calculator to find all angles β, where $0° < \beta < 360°$.

The Solution of Triangles

*T*rigonometry is an old subject with many applications, both old and new. The first applications of trigonometry were to right triangles. Even at this limited stage the variety of applications is astonishing, as you will see in Section 3.1. The solution of more general triangles and problems connected with this requires two very important laws, the law of cosines and the law of sines. Armed with these, the solution to many interesting problems involving triangles is within our grasp. Sections 3.6 and 3.7 show some important triangle applications encountered when using vectors.

A triangle is composed of six parts, the three sides and the three angles. A principal use of the trigonometric functions is for solving triangles. To *solve a triangle* means to find the value of each of the six parts. Since the six parts of a triangle are not independent, unknown values can be calculated if certain values are known. For example, if two of the angles are known, the other one is obtained by subtracting the sum of the two known angles from 180°. If the triangle is a right triangle and two of the sides are known, the third side may be obtained using the Pythagorean theorem.

Initially we will assume that one of the angles of the triangle is a right angle. We will look at a few of the numerous applications associated with the solution of a right triangle. Later in the chapter we will generalize the discussion to oblique triangles.

3.1 Solving Right Triangles

In discussing right triangles it is customary to designate the vertices and corresponding angles by the capital letters A, B, and C, as shown in Figure 3.1(a) on page 94. The right angle is usually denoted by the letter C, and the lowercase letters a, b, and c designate the sides opposite angles A, B, and C, respectively. Side c is then the hypotenuse.

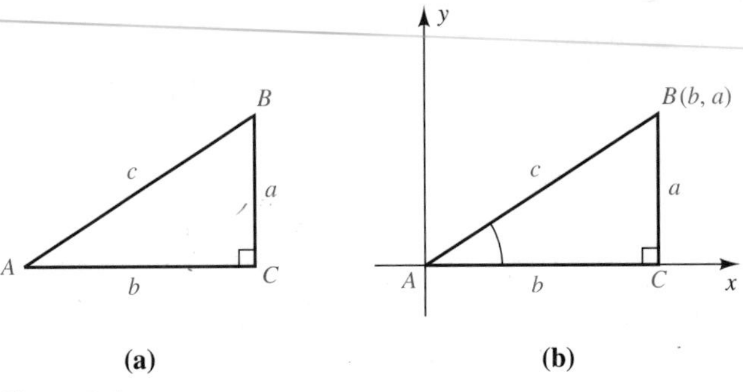

Figure 3.1

To see how the trigonometric functions are related to the parts of a right triangle, consider the right triangle shown in Figure 3.1(b). With angle A in standard position, the coordinates of the vertex at B are (b, a), and the distance from A to B is c. Thus the six trigonometric functions of angle A in terms of the sides of the standard right triangle are

$$\sin A = \frac{a}{c} \qquad \csc A = \frac{c}{a}$$
$$\cos A = \frac{b}{c} \qquad \sec A = \frac{c}{b} \tag{3.1}$$
$$\tan A = \frac{a}{b} \qquad \cot A = \frac{b}{a}$$

The sides of a right triangle are often referred to in terms of one of the two acute angles. For example, the side of length a is called the **side opposite** angle A, the side of length b is called the **side adjacent to** angle A, and the side of length c is called the **hypotenuse**. When we use this terminology, the six trigonometric functions defined in Equation 3.1 become

$$\sin A = \frac{a}{c} = \frac{\text{opposite side}}{\text{hypotenuse}} \qquad \csc A = \frac{c}{a} = \frac{\text{hypotenuse}}{\text{opposite side}}$$
$$\cos A = \frac{b}{c} = \frac{\text{adjacent side}}{\text{hypotenuse}} \qquad \sec A = \frac{c}{b} = \frac{\text{hypotenuse}}{\text{adjacent side}} \tag{3.2}$$
$$\tan A = \frac{a}{b} = \frac{\text{opposite side}}{\text{adjacent side}} \qquad \cot A = \frac{b}{a} = \frac{\text{adjacent side}}{\text{opposite side}}$$

The definitions in (3.2) are convenient for solving right triangles.

When you are given a right triangle, you can determine all six parts if you know two parts other than the right angle, at least one of which is a side.

- If an angle and one of the sides is given, the third angle is simply the complement of the one given. The other two sides can be obtained from the values of the known trigonometric functions.
- If two sides are given, the value of the third side is obtained from the Pythagorean theorem. The angles may then be determined by taking ratios of the sides. Each ratio will uniquely determine the value of some trigonometric function.

Thus in solving right triangles we make use of the trigonometric functions, the Pythagorean theorem, and the fact that the two acute angles are complementary. You will usually find it to your advantage to make a rough sketch of the triangle. This will help you to determine what is given and which trigonometric functions must be used to find the unknown parts.

The relationship between the accuracy of the sides and that of the angles is given in Table 3.1. We will use this convention in the examples and the answers to the exercises. A general discussion of rounding off approximate numbers was given in Sections 1.1 and 1.6. You may wish to review this material before you begin the examples and exercises of this section.

Table 3.1

Accuracy of Sides	Accuracy of Angles
2 significant digits	nearest degree
3 significant digits	nearest 0.1° or 10′
4 significant digits	nearest 0.01° or 1′

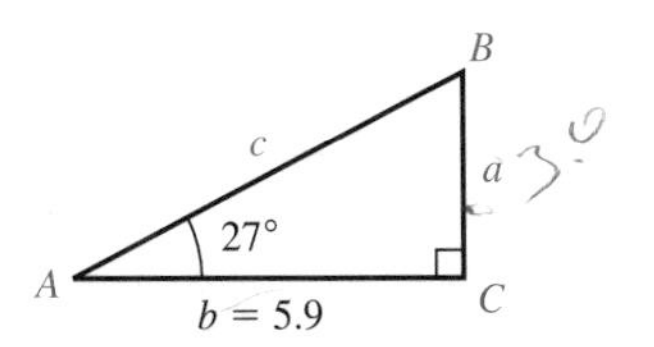

Figure 3.2

Example 1 Solve the triangle in Figure 3.2.

SOLUTION Since A and B are complementary angles, $B = 90° - 27° = 63°$. Also, $\tan A =$ opposite side/adjacent side, so

$$\tan 27° = \frac{a}{5.9}$$

Solving for a, we get

$$a = 5.9 \tan 27° \approx 3.0$$

The calculator keystrokes for the computation of a are

$$a = 5.9 \boxed{\textbf{tan}}\ 27 \boxed{=} \boxed{\textbf{3.006200152}}$$

which rounds off to 3.0. Similarly,

$$\cos 27° = \frac{5.9}{c}$$

so

$$c = \frac{5.9}{\cos 27°} \approx 6.6$$

The calculator keystrokes for c are

$c = 5.9$ [÷] [cos] 27 [=] [6.621724802]

which rounds off to 6.6 ▲

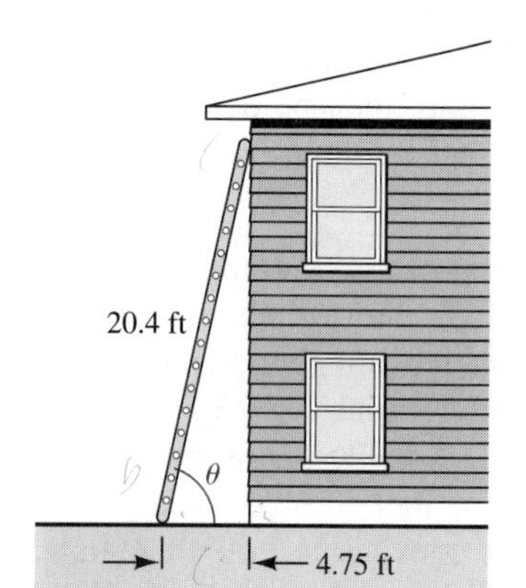

Figure 3.3

Example 2 A 20.4-ft-long ladder is placed against a building so that its lower end is 4.75 feet from the base of the building. What angle does the ladder make with the ground?

SOLUTION The desired angle is designated θ in Figure 3.3. From the figure we see that

$$\cos\theta = \frac{\text{adjacent side}}{\text{hypotenuse}} = \frac{4.75}{20.4}$$

Using a calculator, we proceed as follows:

$\theta =$ [inv] [cos] [(] 4.75 [÷] 20.4 [)] [=] [76.53548287]

Then we round off to obtain $\theta \approx 76.5°$. ▲

Example 3 A radar station tracking a missile determines the missile's angle of elevation to be 20.7° and the line-of-sight distance (called the **slant range**) to the missile to be 38.2 km. Determine the altitude and horizontal range of the missile.

SOLUTION From Figure 3.4 we see that

$$\sin 20.7° = \frac{h}{38.2}$$

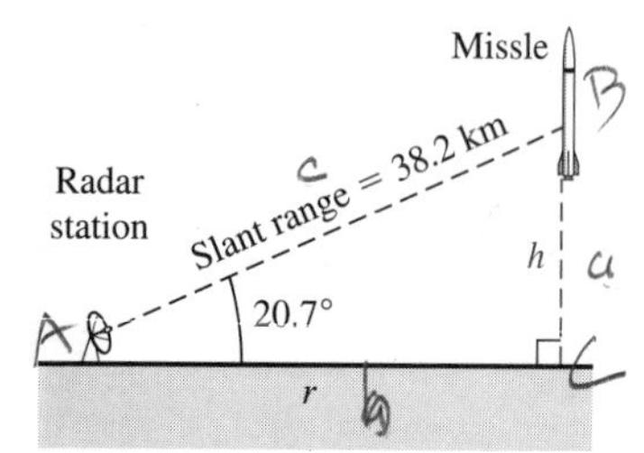

Figure 3.4

Solving for the altitude, h, we have

$h = 38.2 \sin 20.7° \approx$ [13.50273903] ≈ 13.5 km

Similarly, we have

$$\cos 20.7° = \frac{r}{38.2}$$

So the horizontal range is

$r = 38.2 \cos 20.7° \approx$ [35.73396198] ≈ 35.7 km ▲

Example 4 An airplane flying at a speed of 185 ft/sec starts to descend to the runway on a straight-line glide path that is 7.5° below the horizontal. If the plane is at a 2980-ft altitude at the start of the glide path, how long will it take for the plane to touch down?

SOLUTION Figure 3.5 shows that the length of the glide path is the hypotenuse of a right triangle. Therefore we can write

$$\sin 7.5° = \frac{2980}{d}$$

$d = \dfrac{2980}{\sin 7.5°} \approx$ [22830.66678] $\approx 22{,}800$ ft   Three significant places

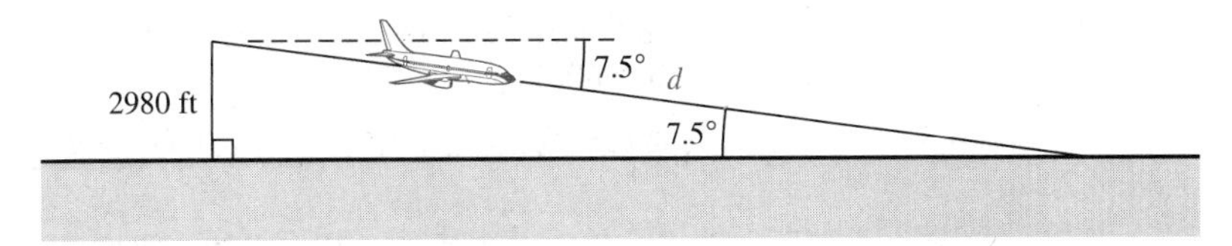

Figure 3.5

Recall that

$$\text{velocity} = \frac{\text{distance}}{\text{time}}$$

Thus, solving for time, we have

$$\text{time} = \frac{\text{distance}}{\text{velocity}} \approx \frac{22{,}800 \text{ ft}}{185 \text{ ft/sec}} \approx \boxed{123.2432432} \approx 123 \text{ sec}$$

The descent takes 123 sec = 2.05 min. ▲

The next example shows how right-triangle trigonometry can be used to find vertical dimensions of objects that appear in reconnaissance and satellite photographs. The analyst needs to know only the angle of elevation of the sun and the scale of the photograph.

Example 5 A representation of an aerial photograph of a building complex is shown in Figure 3.6(a). If the sun was at an angle of 26.5° when the photograph was taken, how high is the rectangular building?

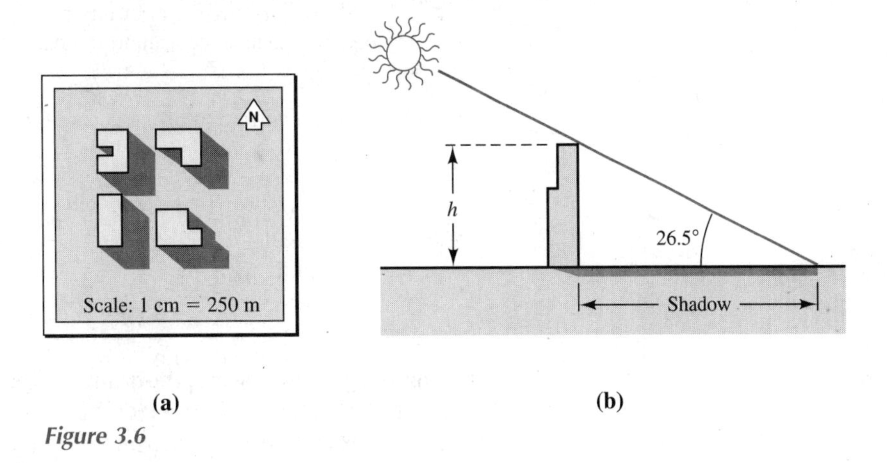

Figure 3.6

SOLUTION Assume the length of the shadow in the photograph measures 0.48 cm. To get the real length of the shadow, multiply 0.48 cm by the scale factor given in the photograph. Thus

$$\text{shadow length} = 0.48(250) = 120 \text{ m}$$

From Figure 3.6(b) we see that $\tan 26.5° = h/120$. Solving for h gives us the height of the building.

$$h = 120 \tan 26.5° \approx \boxed{59.82979297} \approx 59.8 \text{ m}$$ ▲

A **parallel of latitude** on the earth's surface is a circle around the earth in a plane parallel to the equatorial circle. The **latitude angle** is made by two radii of the earth, one from the center of the earth to the equator and one from the center to the parallel of latitude. (See Figure 3.7.)

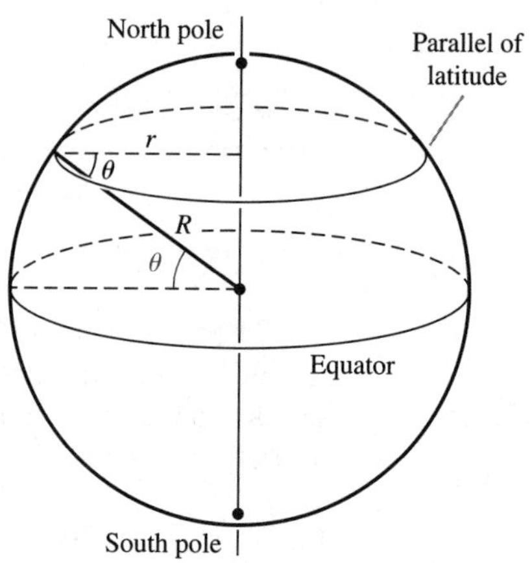

Figure 3.7

Example 6 Show that the length of any parallel of latitude around the earth is equal to the equatorial distance around the earth times the cosine of the latitude angle.

SOLUTION By the definition of the cosine function,

$$\cos \theta = \frac{r}{R} \quad \text{or} \quad r = R \cos \theta$$

The length of the parallel of latitude is $C_p = 2\pi r$. If $C_e = 2\pi R$ denotes the circumference of the earth, then

$$C_p = 2\pi r = 2\pi R \cos \theta = C_e \cos \theta$$ ▲

Example 7 shows how right-triangle trigonometry can be used to find the height of an object when a side of the right triangle cannot be determined. The procedure is to measure the angle of elevation to the top of the object at two different locations and the distance between the two locations.

Example 7 Two observers who are 4250 ft apart measure the angle of elevation to the top of a mountain to be 18.7° and 25.3°, respectively. (See Figure 3.8.) Find the height of the mountain.

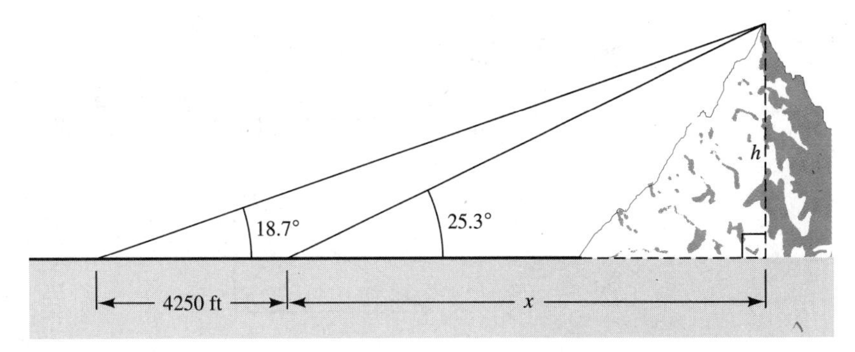

Figure 3.8

SOLUTION From Figure 3.8 we can write the two equations

$$\tan 18.7^\circ = \frac{h}{x + 4250} \qquad \textbf{(1)}$$

$$\tan 25.3^\circ = \frac{h}{x} \qquad \textbf{(2)}$$

The only two unknowns in these two equations are x and h. Solving Equation (2) for x, we get

$$x = \frac{h}{\tan 25.3^\circ} = h \cot 25.3^\circ$$

Substituting the expression into Equation (1), we have

$$\tan 18.7^\circ = \frac{h}{h \cot 25.3^\circ + 4250}$$

To solve this equation for h, we proceed as follows:

$$(h \cot 25.3^\circ + 4250)\tan 18.7^\circ = h$$

$$h \cot 25.3^\circ \tan 18.7^\circ + 4250 \tan 18.7^\circ = h$$

$$(1 - \cot 25.3^\circ \tan 18.7^\circ)h = 4250 \tan 18.7^\circ$$

$$h = \frac{4250 \tan 18.7^\circ}{1 - \cot 25.3^\circ \tan 18.7^\circ} \approx \boxed{\mathbf{5066.421507}}$$

This value rounds off to 5070 to three significant digits. Thus the top of the mountain is 5070 ft above the observers.

Notice that in keying this expression into your calculator, cot 25.3° tan 18.7° can be represented as

[**tan**] 18.7 [÷] [**tan**] 25.3

or as

[( ] [**tan**] 25.3 [ )] [x^{-1}] [**tan**] 18.7 ▲

Exercises for Section 3.1

In Exercises 1–5 solve each right triangle. Use decimal degrees for angles unless the given angle is stated in degrees and minutes.

1.

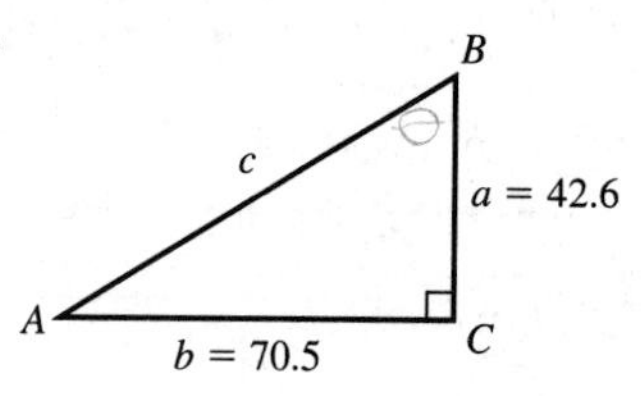

2.

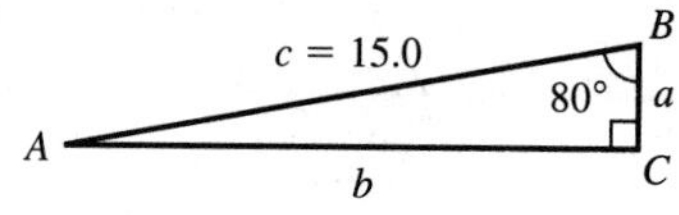

3.

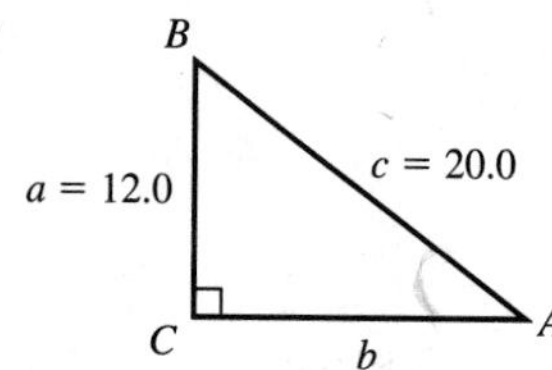

4.

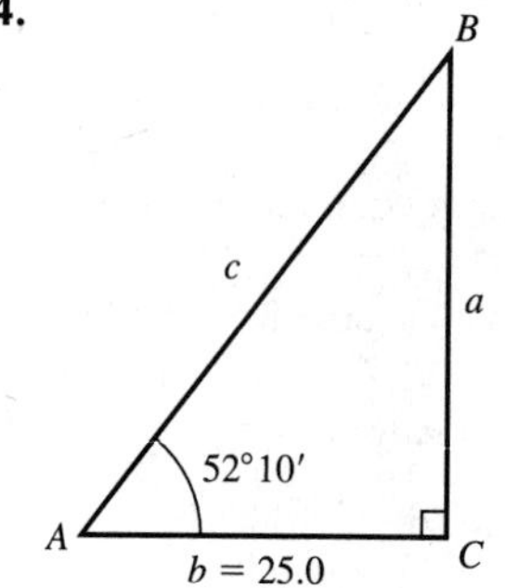

5.

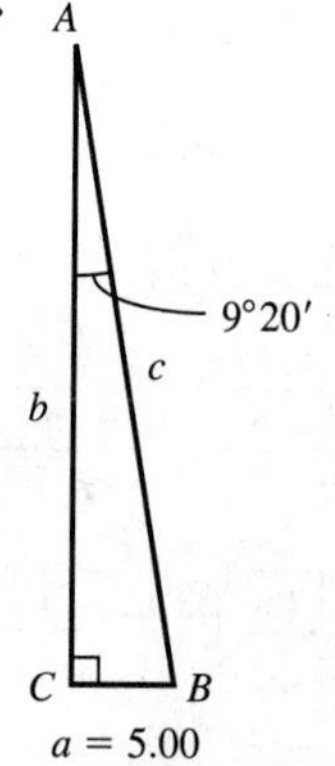

6. One side of a rectangle is half as long as the diagonal. The diagonal is 5.0 m long. How long are the sides of the rectangle? Solve without using the Pythagorean theorem.
7. A person measures the shadow of a flagpole to be 15 ft when the angle of elevation of the sun is 70°. Calculate the height of the flagpole.
8. A child flies a kite at the end of a 200-ft string. Calculate the height of the kite if the string makes an angle of 42° with the ground.
9. A 50-ft support wire on a TV antenna makes an angle of 65° with the ground. Calculate the height above the ground of the point where the wire is attached to the antenna.
10. A 24-ft ladder leans against a house. Calculate the angle that the ladder makes with the ground if the foot of the ladder is 6.5 ft from the side of the house.
11. An engineer wishing to know the width of a river walks 100 yd downstream from a point that is directly across from a tree on the opposite bank. If the angle between the riverbank and the line of sight to the tree at this second point is 55.1°, what is the distance d across the river? (See Figure 3.9.)

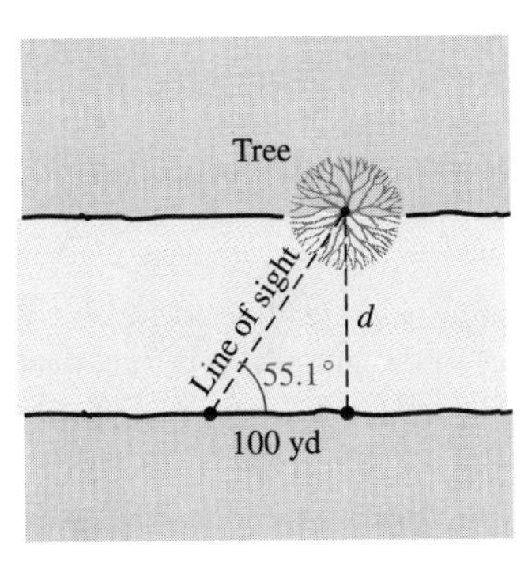

Figure 3.9

12. A solar collector is placed on the roof of a house, as shown in Figure 3.10. What angle does the collector make with the vertical?
13. A solar panel is to be tilted as shown in Figure 3.11 so that angle $\phi = 100°$ when the angle of elevation of the sun is 27°. Find h if the length of the panel is 6.4 m. = b

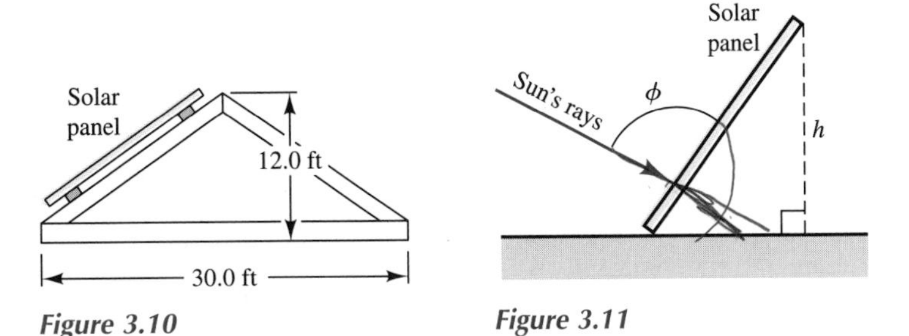

Figure 3.10 *Figure 3.11*

14. A television antenna stands on top of a house that is 20 ft tall. The angle subtended by the antenna from a point 30 ft from the base of the building is 15°. Find the height of the antenna.
15. Civil engineers designing a steel truss for the bridge shown in Figure 3.12 want $\overline{BC}$ to be 11.0 m and $\overline{AC}$ to be 7.00 m. What angle will $\overline{AB}$ make with $\overline{AC}$? with $\overline{BC}$?

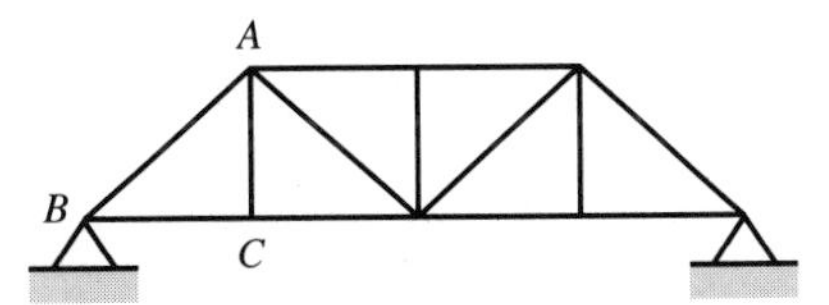

Figure 3.12

16. At noon in the tropics, when the sun is directly overhead, a fisherman holds his 5.0-m pole inclined 30° to the horizontal. How long is the shadow of the pole? How high is the tip of the pole above the level of the other end?

17. A hexagonal bolt head measures 12.0 mm from one edge to the opposite edge. Find the distance c from one corner to the opposite corner. (See Figure 3.13.)

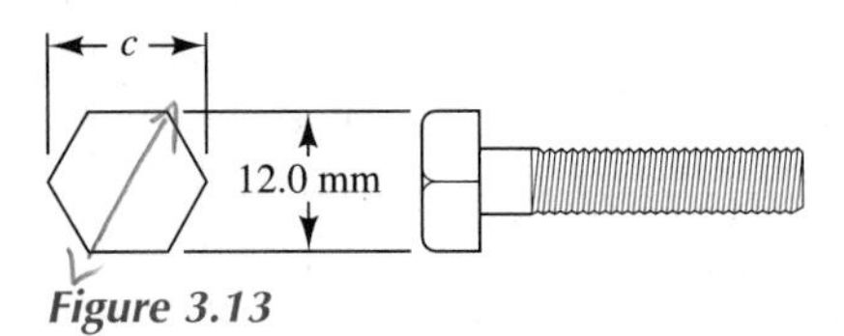

Figure 3.13

18. A 24-in.-wide sheet of aluminum is bent along its center line to form a V-shaped gutter. Find the angle between the sides of the gutter if it is 7 in. deep.

19. A cylindrical steel bar rests in a V-shaped groove, as shown in Figure 3.14. Find the radius of the bar if $a = 1.5$ in. and $\theta = 45°$.

20. The length of each blade of a pair of shears from the pivot to the point is 6.0 in. (See Figure 3.15.) When the points of the open shears are 4.0 in. apart, what angle do the blades make with each other?

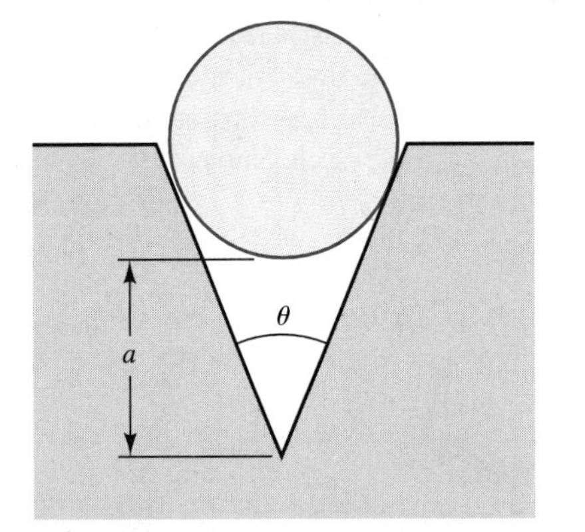

Figure 3.14

4.0 in.
6.0 in.

Figure 3.15

21. From the top of a building that is 220.0 ft high, an observer looks down on a parking lot. If the observer's lines of sight to two different cars in the lot make angles of 28°15′ and 36°20′ below the horizontal, what is the distance between the two cars?

22. A certain automatic landing system, used at airports when visibility is poor, locks on to a plane when the airplane is 4.0 mi (slant range) from the runway and at an altitude of 3800 ft. If the glide path is a straight line to the runway, what angle does it make with the horizontal?

23. An airplane takes off with an airspeed of 265 ft/sec and climbs at an angle of 8.7° with the horizontal until it reaches an altitude of 5800 ft. (See Figure 3.16.) How long does it take the plane to reach this altitude?

Figure 3.16

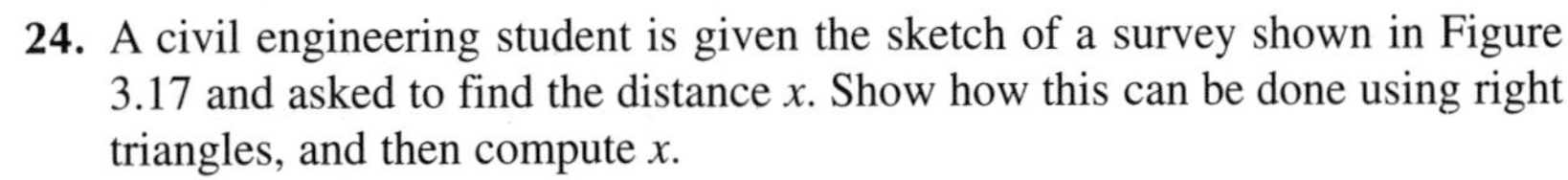

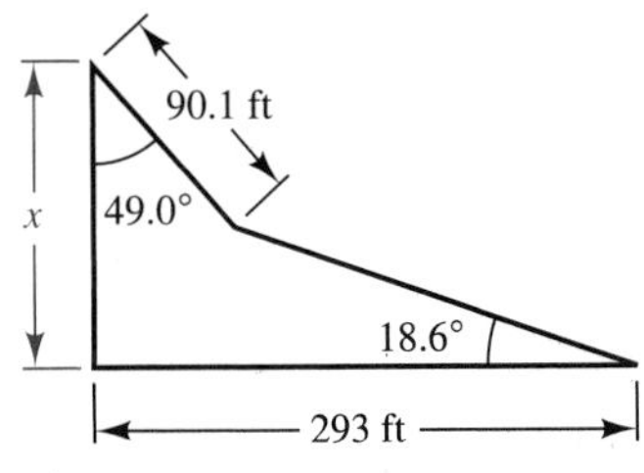

Figure 3.17

24. A civil engineering student is given the sketch of a survey shown in Figure 3.17 and asked to find the distance x. Show how this can be done using right triangles, and then compute x.

25. Ten holes are to be drilled in a circular cover plate of a rocket motor. The holes are equally spaced on a circle of radius 12.9 cm, as shown in Figure 3.18. What is the straight-line center-to-center distance between the holes?

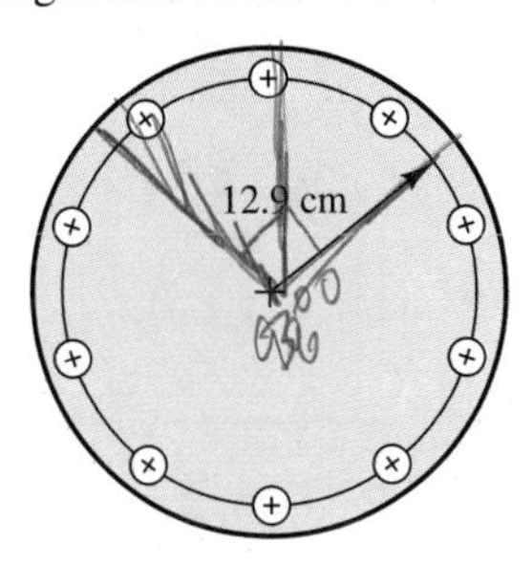

Figure 3.18

26. A 78.0-ft rocket with its base on the ground is elevated at an angle of 69°40′. What is the height of the nose of the rocket above the ground?

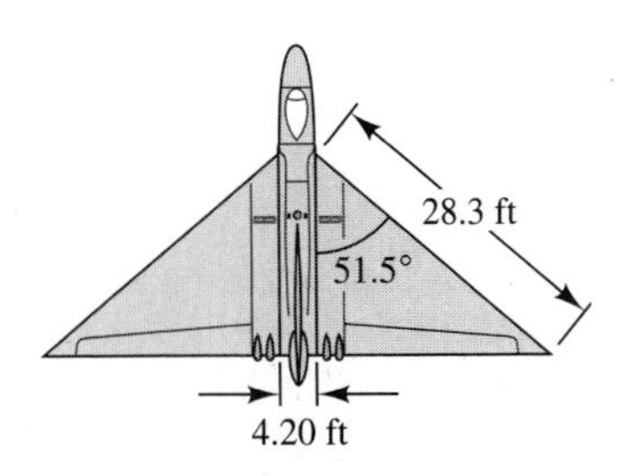

Figure 3.19

27. The triangular wing of a delta-wing airplane is swept back at an angle of 51.5° to the center line of the fuselage. If the leading edge of the wing is 28.3 ft long and the fuselage is 4.20 ft wide, what is the wingspan of the airplane? (See Figure 3.19.)

28. The shroud of a nose cone is shown in cross section in Figure 3.20. What is the diameter of the rocket using this nose cone?

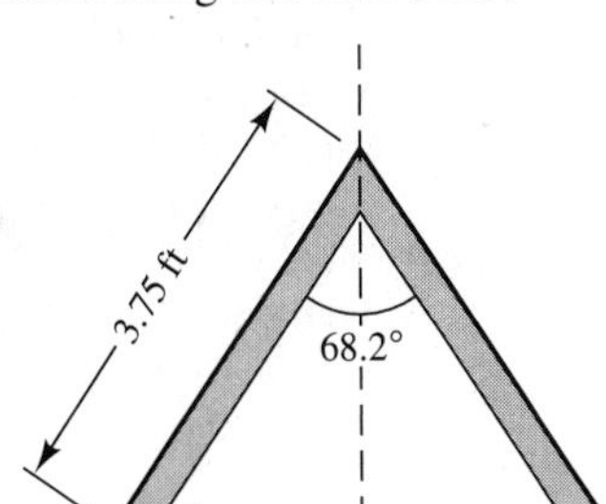

Figure 3.20

29. An astronomer measures the shadow of a crater on the moon in a photograph and finds its length to be 0.32 cm. If the sun was at an angle of 49.1° to the horizontal when the photograph was taken, how deep is the crater? (See Figure 3.21.) Assume the map has a scale of 1 cm = 2500 m.

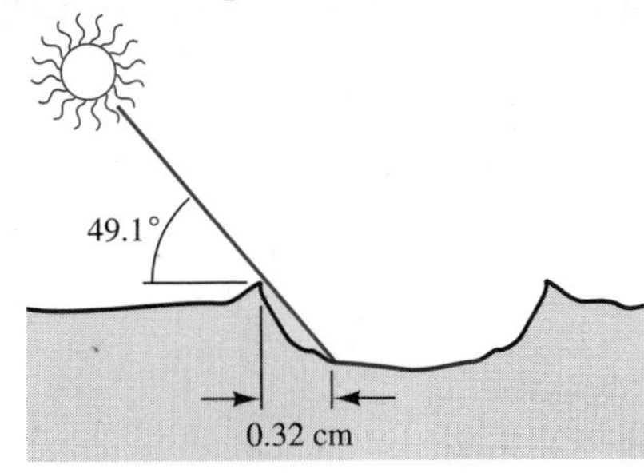

Figure 3.21

30. Figure 3.22 represents an aerial photograph of a cliff in a remote region of Antarctica. Compute the height of the cliff if the elevation angle of the sun was 19.0° when the photograph was taken.

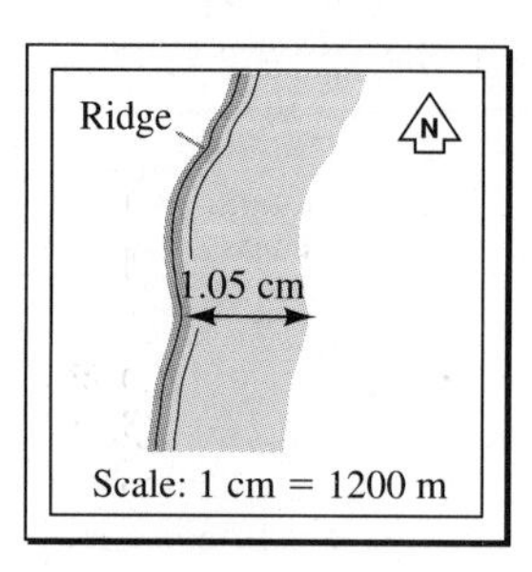

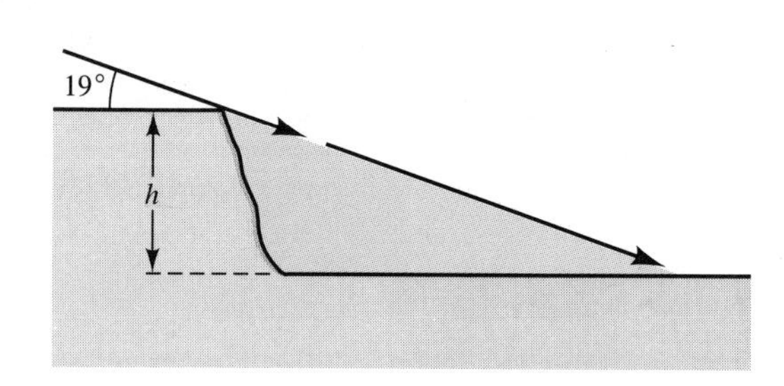

Figure 3.22

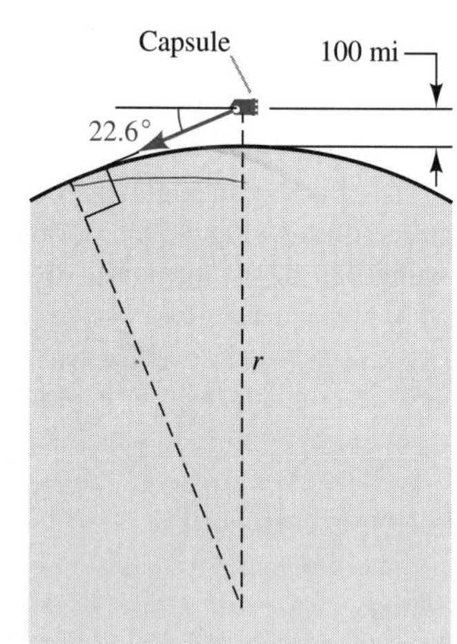

Figure 3.23

31. A space capsule orbits the moon at an altitude of 100 mi. As shown in Figure 3.23, a sighting from the capsule to the moon's horizon has an angle of depression of 22.6°. Find the radius of the moon.

32. Determine the length of the Arctic Circle (66°33′N). Assume the circumference of the earth is 25,000 mi.

33. The 40° parallel of latitude passes through the United States. If a citizen of the United States were to travel due east along the 40° parallel, how far would he or she travel before returning home? Assume the circumference of the earth is 25,000 mi.

An interesting method of measuring the height of clouds is shown in Figure 3.24. A sweeping light beam is placed at point A, and a light source detector is placed at point B. The axis of the detector is kept vertical, and the light beam is made to sweep from the horizontal ($\alpha = 0°$) to the vertical ($\alpha = 90°$). When the beam illuminates the base of the clouds directly above the detector, as shown in the figure, the detector is activated and it measures the angle α. Since d is known, the height h of the clouds can be computed. Use the diagram of the cloud altitude detector to solve Exercises 34–36.

34. Compute the height of the cloud if the light source is 100 ft from the detector and the angle is 82°.

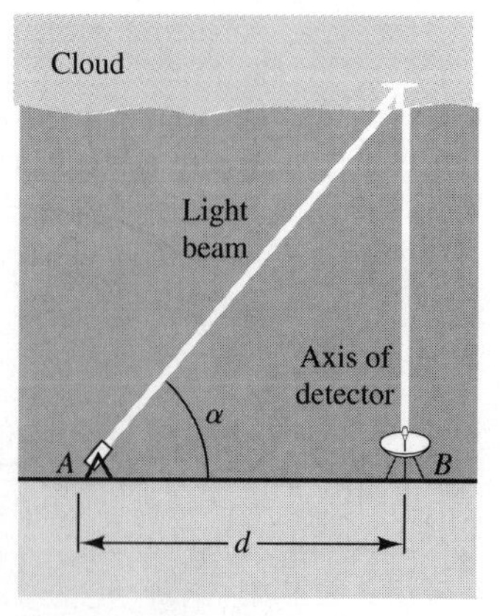

Figure 3.24

35. Find the angle α when the clouds are 1000 ft high and the light source is located 100 ft from the detector.

36. Suppose that a detector that is 1 km from the light beam detects a cloud layer when $\alpha = 60°$ and another layer when $\alpha = 75°$. What is the vertical separation of the cloud layers?

37. An observer at the base of a hill knows that the television antenna on top of the hill is 550 ft high. If the angle of elevation from the observer to the base of the antenna is 16.4° and to the top of the antenna is 29.1°, how high is the hill?

38. As shown in Figure 3.25, a fishing boat sailing due north records the location of a lighthouse as 16.2° east of north. If 12.7 miles later the boat records the location of the lighthouse as 43.7° east of north, how close will the boat come to the lighthouse?

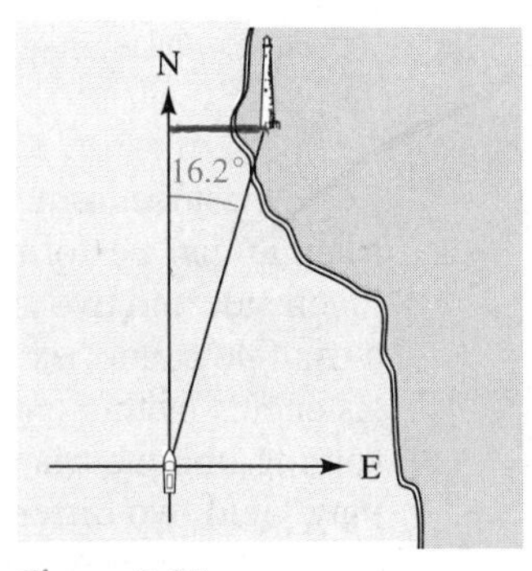

Figure 3.25

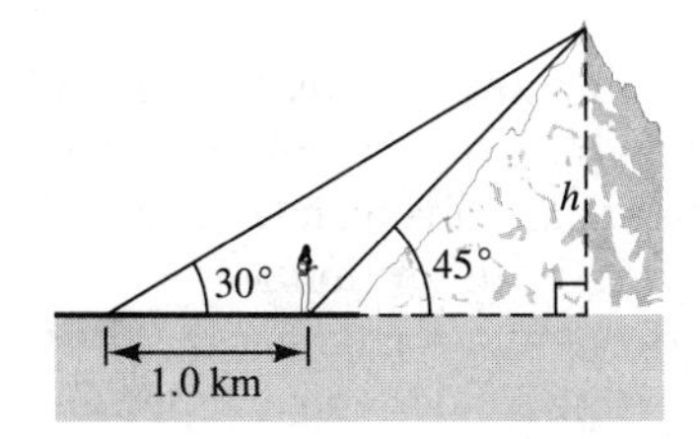

Figure 3.26

39. A woman walking along the prairie stops to measure the angle of elevation to a high mountain. It measures 30.0°. The woman then walks 1.0 km toward the mountain and measures again. This time the angle of elevation is 45.0°. (See Figure 3.26.) How high is the mountain?

40. A meteorological rocket is launched 4000 ft down range from a radar station. Ten seconds after the rocket is launched, the radar reports an elevation angle of 35.4°. Find the speed of the rocket in ft/sec if it is rising vertically at a constant rate.

41. An airplane flying horizontally at a constant rate and at an altitude of 5400 ft flies directly over a radar station at 2:14 P.M. Fifteen seconds later the radar reports an elevation angle to the plane of 53.5°. Find the speed of the plane in ft/sec.

3.2 Oblique Triangles: The Law of Cosines

Any triangle that is not a right triangle is called **oblique**. Hence in an oblique triangle no angle is equal to 90°. In the rest of this chapter we will consider conditions under which oblique triangles can be solved. As noted in the discussion of right triangles, knowledge of a least three of six parts is necessary to solve a triangle, but they cannot be just any three parts. For example, three angles will not define a unique solution, because many similar triangles can be constructed with these angles but different side lengths. There are four different combinations of side lengths and angles that will describe a triangle.

CASE 1 Two sides and the included angle are given.

CASE 2 Three sides are given.

CASE 3 Two angles and one side are given.

Of course, as noted earlier, the given information must be such that a triangle can be formed. For instance, no triangle can be formed from a given side and two angles of 88° and 95°, because the sum of the angles of a triangle cannot exceed 180°. It is important that you be aware of inconsistencies of this type when solving triangles. A fourth case that arises in solving oblique triangles is important, even though the information given may yield two different triangles, one triangle, or no triangle.

CASE 4 Two sides and an angle opposite one of the sides are given.

This last case is sometimes referred to as the **ambiguous** case, since two triangles, one triangle, or no triangles may be formed from this information. For instance, Figure 3.27 shows two triangles that can be obtained from the information $a = 15$, $b = 20$, and $A = 20°$. In Section 3.4 we will discuss this case in detail. For the present you should remember that it is possible for two noncongruent triangles to have two sides and an opposite angle the same.

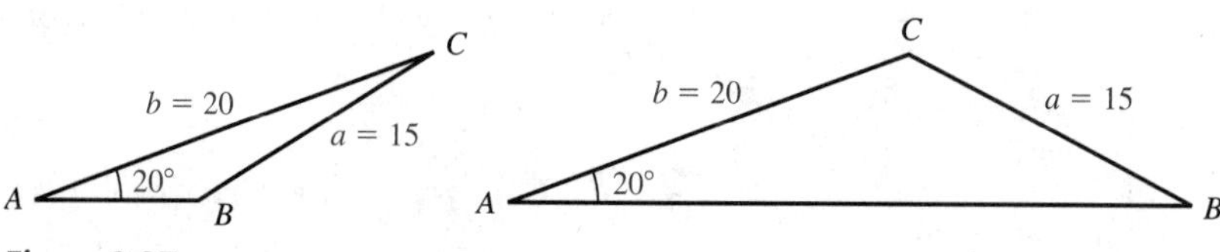

Figure 3.27

The **law of cosines** enables us to solve an oblique triangle when two sides and the included angle are given, as in Case 1, or when three sides are given, as in Case 2. To derive the law of cosines, we subdivide a general oblique triangle into two right triangles.

Consider any oblique triangle ABC; either of the triangles shown in Figure 3.28 will do. Drop a perpendicular from the vertex B to side AC or its extension. Call the length of this perpendicular h. In either case we obtain $h = c \sin A$, and hence

$$h^2 = c^2 \sin^2 A$$

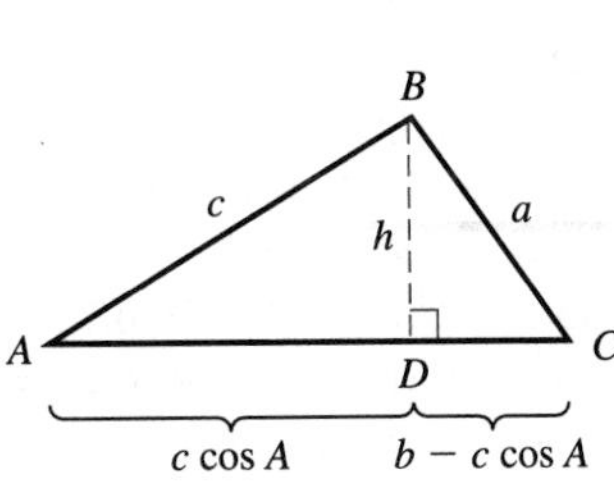

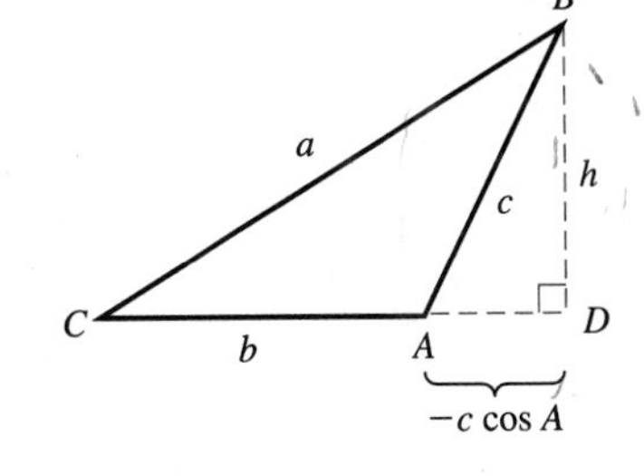

Figure 3.28
Law of cosines

Using

$$\sin^2 A + \cos^2 A = 1$$

we have

$$h^2 = c^2(1 - \cos^2 A)$$

Referring to Figure 3.28, we also know from right triangle BCD that

$$h^2 = a^2 - (b - c \cos A)^2$$

Equating the right-hand sides of the two preceding equations, we get

$$c^2(1 - \cos^2 A) = a^2 - (b - c \cos A)^2$$
$$c^2 - c^2 \cos^2 A = a^2 - b^2 + 2bc \cos A - c^2 \cos^2 A$$

Simplifying this expression, we arrive at

$$a^2 = b^2 + c^2 - 2bc \cos A \qquad \textbf{(3.3)}$$

In a similar manner it can be shown that

$$b^2 = a^2 + c^2 - 2ac \cos B$$

and that

$$c^2 = a^2 + b^2 - 2ab \cos C$$

Each of these formulas is a statement of the law of cosines. This law says that the square of any side of a triangle is equal to the sum of the squares of the other two sides minus twice the product of these sides times the cosine of the angle between them. If the angle is 90°, the law of cosines reduces to the Pythagorean theorem, so it is quite properly considered an extension of that famous theorem.

The law of cosines gives the relationship between three sides and one of the angles of any triangle. Thus, if any three of these parts are given, you can compute the remaining parts or show that such a triangle cannot exist. For example, if three sides of a triangle are given, any angle may be found from Equation (3.3) by solving for $\cos A$, $\cos B$, or $\cos C$. Alternative forms of the law of cosines are given in Equation (3.4).

$$\begin{aligned} \cos A &= \frac{b^2 + c^2 - a^2}{2bc} \\ \cos B &= \frac{a^2 + c^2 - b^2}{2ac} \\ \cos C &= \frac{a^2 + b^2 - c^2}{2ab} \end{aligned} \qquad \textbf{(3.4)}$$

Note that in this form the law says that the cosine of an angle may be found by computing a fraction whose numerator is the sum of squares of the adjacent sides minus the square of the opposite side and whose denominator is twice the product of the adjacent sides.

B

$a = 5.18$ c

60.0°

C $b = 6.00$ A

Figure 3.29

Example 1 Solve the triangle with side $a = 5.18$, side $b = 6.00$, and angle $C = 60.0°$, as shown in Figure 3.29.

SOLUTION
$$c^2 = a^2 + b^2 - 2ab \cos C$$
$$c = \sqrt{5.18^2 + 6.00^2 - 2(5.18)(6.00) \cos 60.0°}$$

The calculator sequence for c is

$c \approx$ [√] [(] 5.18 [x^2] [+] 6.00 [x^2] [−] 2 [×] 5.18 [×] 6.00 [×] [cos] 60 [)] [=] [5.634926796]

Therefore

$$c \approx 5.63$$

To solve for angle A, we use Equation (3.4) of the law of cosines.

$$\cos A = \frac{b^2 + c^2 - a^2}{2bc}$$
$$= \frac{6.00^2 + 5.63^2 - 5.18^2}{2(6.00)(5.63)}$$

The sequence of calculator keystrokes for angle A is

$A =$ [inv] [cos] [(] [(] 6.00 [x^2] [+] 5.63 [x^2] [−] 5.18 [x^2] [)] [÷] 2 [÷] 6.00 [÷] 5.63 [)] [=] [52.78106305]

Therefore angle A to the nearest tenth of a degree is

$$A \approx 52.8^\circ$$

The remaining angle could also be found from the law of cosines, but since the sum of the angles is 180°,

$$B \approx 180^\circ - 60^\circ - 52.8^\circ = 67.2^\circ$$ ▲

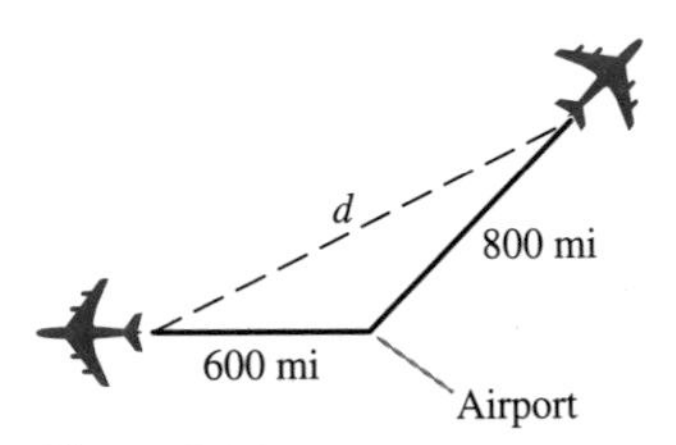

Figure 3.30

Example 2 Two airplanes leave an airport at the same time, one going northeast at 400 mph and the other going directly west at 300 mph. (See Figure 3.30.) How far apart are they 2 hr after leaving?

SOLUTION From Figure 3.30 and the law of cosines,

$$d = \sqrt{600^2 + 800^2 - 2(600)(800)\cos 135.0^\circ} \approx \boxed{1295.693833}$$

Thus $d = 1300$ mi when rounded off to three significant digits. ▲

COMMENT The [inv] [cos] keys on your calculator are designed to give angles between 0° and 90° when the cosine of the angle is positive and to give angles between 90° and 180° when the cosine of the angle is negative. Since the interior angles of a triangle are greater than 0° and less than 180°, you may calculate angles by the law of cosines without resorting to the use of reference angles. For example, if θ is an interior angle of a triangle and $\cos\theta = 0.22435$, then θ is found as follows:

$\theta \approx$ [inv] [cos] 0.22435 [=] [77.03534122] °

On the other hand, if θ is an interior angle of a triangle and $\cos\theta = -0.22435$, then

$\theta \approx$ [inv] [cos] −0.22435 [=] [102.9646588] °

We make special note that this property applies only to the [inv] [cos] keys; it does not apply to the [inv] [sin] and [inv] [tan] keys.

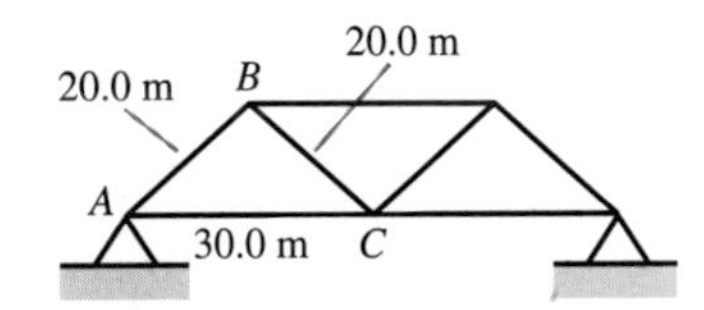

Figure 3.31

Example 3 In a steel bridge, one part of a truss is in the form of an isosceles triangle, as shown in Figure 3.31. At what angles do the sides of the truss meet?

SOLUTION $\cos A = \dfrac{20^2 + 30^2 - 20^2}{(2)(20)(30)} = \dfrac{900}{1200} = 0.75$

Hence $A \approx$ [inv] [cos] 0.75 [=] [41.40962211]° or, rounding off to the nearest tenth of a degree,

$$A \approx 41.4°$$

This is also the value of angle C, since the triangle is isosceles. Then

$$B \approx 180° - 2(41.4°) = 97.2°$$

Suppose we had decided to use the law of cosines to find angle B. Then

$$\cos B = \frac{20^2 + 20^2 - 30^2}{(2)(20)(20)} = \frac{800 - 900}{800} = -\frac{1}{8} = -0.125$$

And angle B is

$B =$ [inv] [cos] −0.125 [=] [97.18075578] $\approx 97.2°$

as we found above. ▲

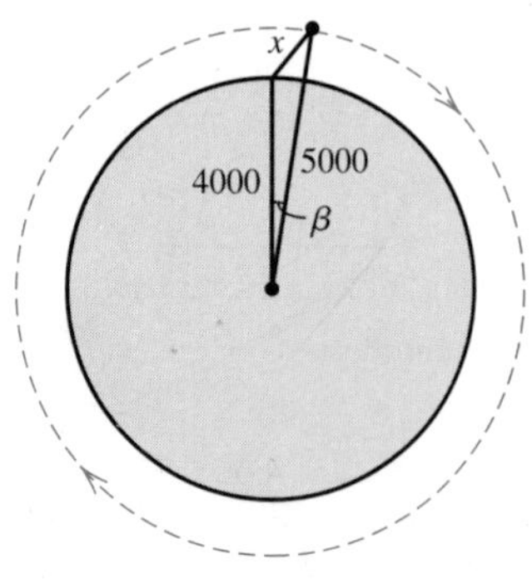

Figure 3.32

Example 4 A satellite traveling in a circular orbit 1000 mi above the Earth passes directly over a tracking station at noon. Assume that the satellite takes 2.0 hr to make an orbit and that the radius of the Earth is 4000 mi. Find the distance between the satellite and tracking station at 12:03 P.M.

SOLUTION From Figure 3.32 we see that we must compute the angle β. Since the satellite takes 2 hr (or 120 min) for an orbit of 360°, it moves 360°/120 = 3° during each minute. Thus the satellite travels a total of 9.0° in 3 minutes. Hence $\beta = 9.0°$. By the law of cosines,

$$\begin{aligned} x &= \sqrt{(4000)^2 + (5000)^2 - 2(4000)(5000)\cos 9.0°} \\ &\approx \sqrt{1{,}492{,}466.376} \\ &\approx \boxed{1221.665411} \end{aligned}$$

The distance between the satellite and the tracking station at 12:03 P.M. is about 1220 mi. ▲

COMMENT The form of the law of cosines in Equation (3.4) confirms that the sides of triangles cannot have just any lengths. For example, for $a = 1$, $b = 3$, and $c = 1$, Equation (3.4) gives

$$\cos A = \frac{(3)^2 + (1)^2 - (1)^2}{2(3)(1)} = \frac{9}{6} = \frac{3}{2}$$

which is impossible, since $|\cos A| \le 1$. (Why?) Hence no such triangle exists. The same conclusion can be reached by noting that $a + c < b$.

Note that if you use the calculator keystrokes described after Equation (3.4) when no triangle exists, you will get an error message.

Exercises for Section 3.2

In Exercises 1–6 use the law of cosines to find the unknown side.

1. $a = 45.0$, $b = 67.0$, $C = 35.0°$
2. $a = 20.0$, $b = 40.0$, $C = 28.0°$
3. $a = 10.5$, $b = 40.8$, $C = 120.0°$
4. $b = 12.9$, $c = 15.3$, $A = 104.2°$
5. $b = 38.0$, $c = 42.0$, $A = 135.3°$
6. $a = 3.49$, $b = 3.54$, $C = 5°24'$

In Exercises 7–12 use the law of cosines to find the largest angle to the nearest tenth of a degree.

7. $a = 7.23$, $b = 6.00$, $c = 8.61$
8. $a = 16.0$, $b = 17.0$, $c = 18.0$
9. $a = 18.0$, $b = 14.0$, $c = 10.0$
10. $a = 300$, $b = 500$, $c = 600$
11. $a = 170$, $b = 250$, $c = 120$
12. $a = 56.0$, $b = 67.0$, $c = 82.0$

In Exercises 13–18 use the law of cosines to solve the triangles.

13. $a = 4.21$, $b = 1.84$, $C = 30.7°$
14. $a = 5.92$, $b = 7.11$, $C = 60.6°$
15. $a = 120.0$, $b = 145.0$, $C = 94°25'$
16. $a = 900$, $b = 700$, $c = 500$
17. $a = 2.00$, $b = 3.00$, $c = 4.00$
18. $a = 5.01$, $c = 5.88$, $B = 28°40'$

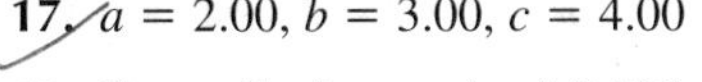

19. Jim walks from point A 2.50 km due west to point B. He then takes a path that is 25.4° south of west and walks 1.40 km to point C. What is the distance between A and C?
20. If the sides of the triangular sections of a geodesic dome as shown in Figure 3.33 are 1.65, 1.65, and 1.92 m, what are the interior angles of the triangular sections?

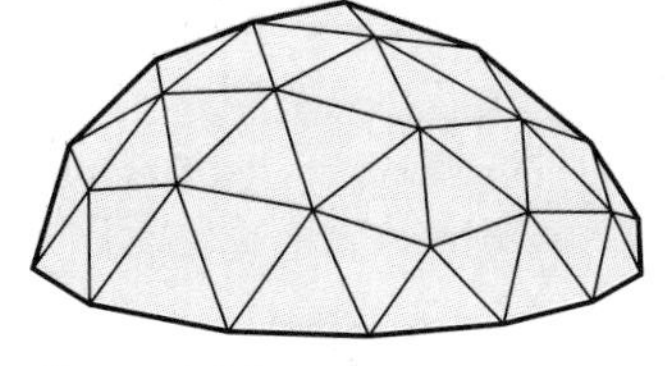

Figure 3.33

HISTORICAL NOTE The architectural advantages of the geodesic dome were publicized in the 1950s by architect R. Buckminster Fuller.

---

21. A baseball diamond is a square with 90.0-ft sides. The pitcher's mound is 60.5 ft from home plate on a line from home plate to second base.
    (a) How far is the pitcher's mound from second base?
    (b) How far is the pitcher's mound from first base?
22. As we mentioned in Chapter 1, carpenters often use the fact that a 3-4-5 triangle is a right triangle to square the walls of a building. They mark a point 3 ft from the corner along the base of one wall and 4 ft from the same corner along the base of the other wall. If the distance between the two points is 5 ft, the walls will be square. If a carpenter measures the distance between the two points to be 5 ft. 3 in., at what angle do the walls meet?
23. The stake that marked corner C of a triangular lot ABC has been lost. Consulting her deed to the property, the owner finds that $\overline{AB} = 80.0$ ft, $\overline{BC} = 50.0$ ft, and $\overline{CA} = 40.0$ ft. At what angle with $\overline{AB}$ should she run a line so that by laying off 40.0 ft along this line she can locate corner C?
24. A solar collector is placed on a roof that makes an angle of 24.0° with the horizontal. If the upper end of the collector is supported as shown in Figure 3.34, how long is the collector?

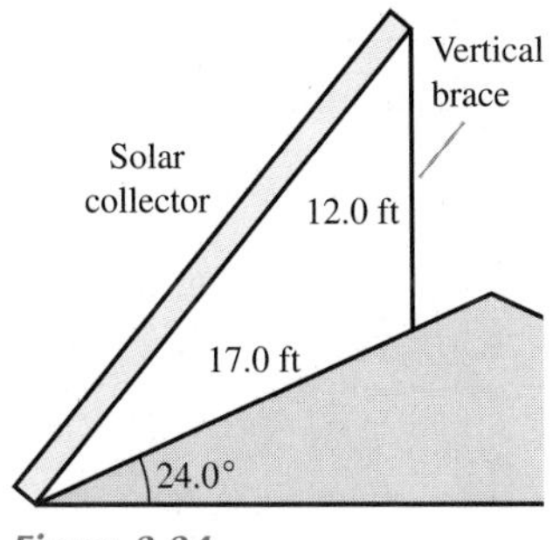

Figure 3.34

25. A reflector used in a solar furnace is composed of triangular sections having sides of lengths 5.50, 5.50, and 1.30 ft. Find the interior angle between the sides of equal length.

26. An airplane flying directly north toward a city C alters its course toward the northeast at a point 100 km from C and heads for city B, approximately 50.0 km away. If B and C are 60.0 km apart, what course should the airplane fly to get to B? (*Note:* Northeast means a course halfway between north and east.)

27. In planning a tunnel under a hill, a civil engineer lays out the triangle ABC shown in Figure 3.35, in order to determine the course of the tunnel. If $\overline{AB} = 3500$ m, $\overline{BC} = 4000$ m, and angle $B = 60°00'$, determine the sizes of the angles A and C and the length of $\overline{AC}$. Assume that the lengths of $\overline{AB}$ and $\overline{BC}$ are accurate to the nearest meter.

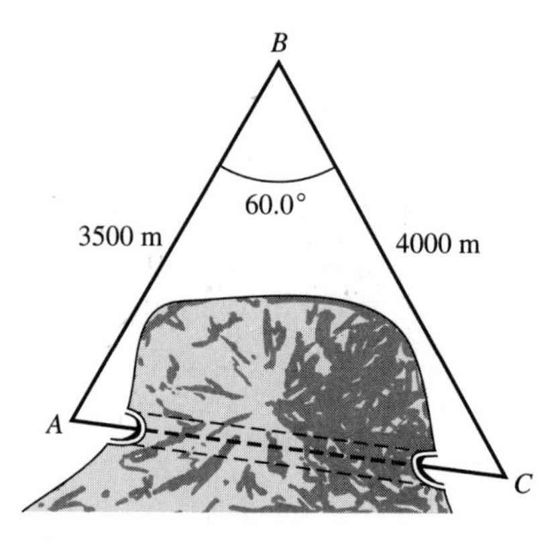

Figure 3.35

28. A construction worker places a 20-ft ladder against a 15-ft scaffold, as shown in Figure 3.36. A 12-ft ladder is then placed on the scaffold at B to reach the building at C. Point B is 4 ft from the wall. Calculate the distance between points A and C.

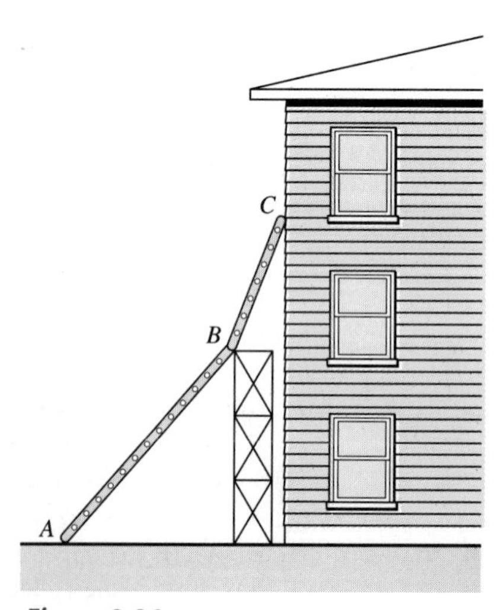

Figure 3.36

29. To measure the distance between two points A and B on opposite sides of a building, a third point C is chosen from which the following measurements are made (the lengths are accurate to the nearest meter): $\overline{CA} = 200$ m, $\overline{CB} = 400$ m, and angle $ACB = 60.0°$. What is the distance between A and B?

30. Show that for any triangle ABC,

$$\frac{a^2 + b^2 + c^2}{2abc} = \frac{\cos A}{a} + \frac{\cos B}{b} + \frac{\cos C}{c}$$

31. Two hikers leave camp at noon. One walks due west at 3.2 mph, and the other walks a line 75° north of east at 2.4 mph. Approximately how far apart will the two hikers be at 1:30 P.M.? Assume that they walk in straight lines and at constant rates.

3.3 Oblique Triangles: The Law of Sines

To solve triangles for which two angles and one side are given, we use the **law of sines**. The law of sines, in conjunction with the law of cosines, enables us to solve any triangle for which we are given three parts, one of which is a side, or at least to declare that no solution is possible.

Consider either triangle shown in Figure 3.37. We see from Figure 3.37 that if we draw a perpendicular h from the vertex B to side b or its extension,

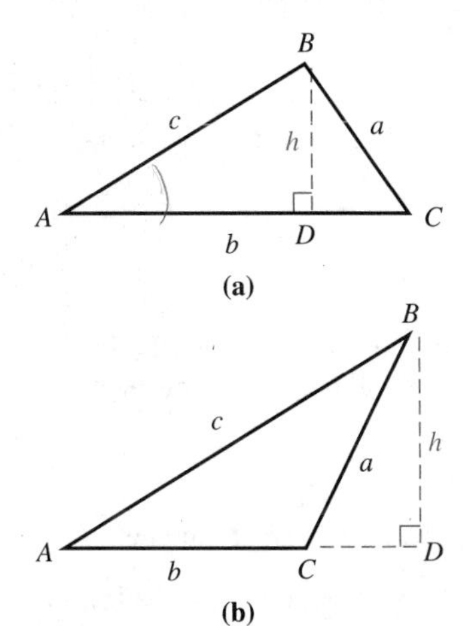

Figure 3.37

$$h = c \sin A \qquad \text{and} \qquad h = a \sin C$$

Equating these two expressions, we get

$$c \sin A = a \sin C$$

Rearranging factors gives

$$\frac{a}{\sin A} = \frac{c}{\sin C}$$

In a similar manner we can show that

$$\frac{a}{\sin A} = \frac{b}{\sin B}$$

Hence

$$\frac{a}{\sin A} = \frac{b}{\sin B} = \frac{c}{\sin C} \tag{3.5}$$

Together the set of equations in (3.5) is called the **law of sines**. The law of sines states that in any triangle the ratios formed by dividing each side by the sine of the angle opposite it are equal.

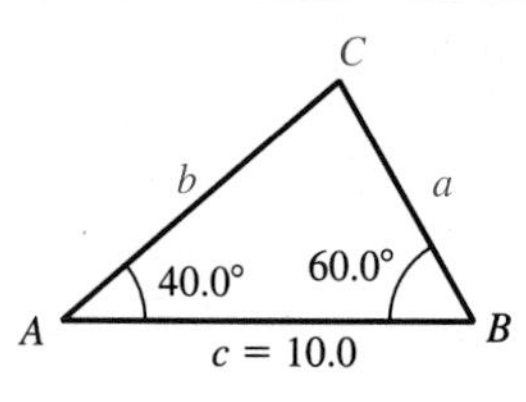

Figure 3.38

Combinations of any two of the three ratios given in Equation (3.5) yield an equation with four parts. Obviously, if we know three of these parts, we can find the fourth.

Example 1 In Figure 3.38, $c = 10.0$, $A = 40.0°$, and $B = 60.0°$. Find a, b, and C.

SOLUTION We begin by observing that

$$C = 180° - (A + B) = 180° - (40.0° + 60.0°) = 80.0°$$

Using the law of sines, we have

$$\frac{a}{\sin 40°} = \frac{10.0}{\sin 80°} \quad \text{or} \quad a = \frac{10.0 \sin 40°}{\sin 80°} \approx 6.53$$

and

$$\frac{b}{\sin 60°} = \frac{10.0}{\sin 80°} \quad \text{or} \quad b = \frac{10.0 \sin 60°}{\sin 80°} \approx 8.79$$

Note that the calculator keystrokes for a and b are the same. The keystrokes for a are as follows:

$a \approx$ 10.0 [sin] 40 [÷] [sin] 80 [=] [6.527036447] ▲

WARNING You should be aware that since $\sin A = \sin(180° - A)$, the formula

$$\sin A = \frac{a \sin B}{b}$$

gives two possible values for angle A. Therefore you must be careful when using this form of the law of sines. The next example illustrates the problem.

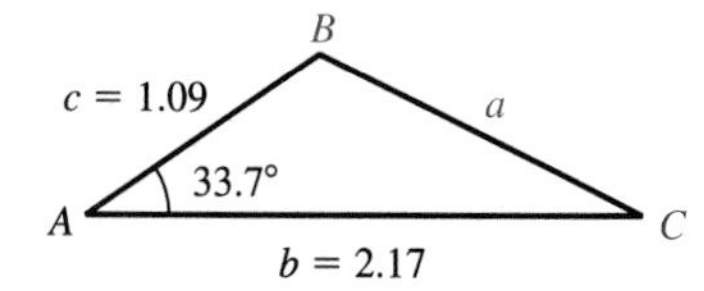

Figure 3.39

Example 2 In Figure 3.39, $A = 33.7°$, $b = 2.17$, and $c = 1.09$. Find angle B.

SOLUTION Before we can solve for angle B, we must find side a. By the law of cosines, we have

$$a = \sqrt{2.17^2 + 1.09^2 - 2(2.17)(1.09)\cos 33.7°} \approx 1.40$$

Now if we use the law of sines to find angle B, we get

$$\sin B = \frac{2.17 \sin 33.7°}{1.40}$$

$$B \approx 59.3°$$

The keystrokes for angle B are

$B \approx$ [inv] [sin] [(] 2.17 [sin] 33.7 [÷] 1.40 [)] [=] [59.31757799]

We might stop here and conclude that $B = 59.3°$. However, there is another possibility. Recall that the sine function is positive in both the first and the second quadrants, so another answer is

$$B \approx 180° - 59.3° = 120.7°$$

We can verify that this is in fact the correct value for angle B by using the law of cosines to compute B. Thus

$$\cos B = \frac{1.40^2 + 1.09^2 - 2.17^2}{2(1.40)(1.09)} \approx \boxed{-0.511402359}$$

$$B \approx \boxed{\text{inv}}\ \boxed{\cos}\ -0.511402359\ \boxed{=}\ \boxed{120.7572854}$$

We note that the correct value 120.7° is the supplement of 59.3°. However, the law of sines alone gives us no clue that the supplement of 59.3° is the required answer. A rough drawing of the triangle is usually sufficient to tell you which of the two angles to use. ▲

COMMENT One way to avoid the difficulty described in Example 2 is always to use the law of cosines when two sides and an included angle are known. Another way to avoid this difficulty is to use the law of sines to solve for the angle opposite the shorter of the two sides and then obtain the third angle by substitution. (If there is an obtuse angle, it is opposite the longest side.)

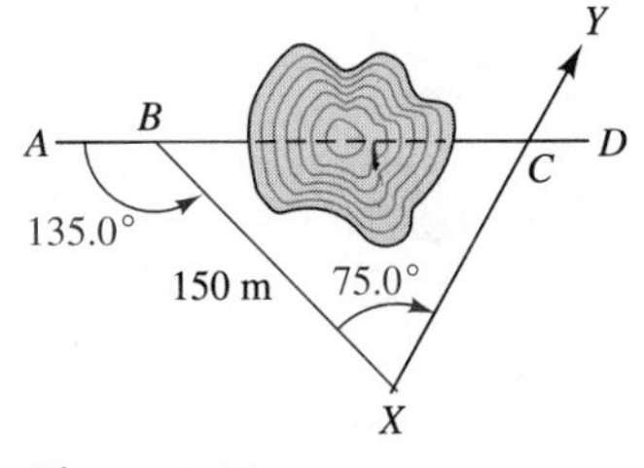

Figure 3.40

Example 3 A surveyor wants to run a straight line from A in the direction AB, as shown in Figure 3.40, but finds that an obstruction interferes with the line of sight. Therefore, the crew lays off the line segment $\overline{BX}$ for a distance of 150.0 m in such a way that angle $ABX = 135.0°$, and runs $\overline{XY}$ at an angle of 75.0° with $\overline{BX}$. At what distance from X on this line should a stake be placed so that A, B, and C are on the same straight line?

SOLUTION Since angle $CBX = 45.0°$ and angle $BCX = 60.0°$, the law of sines tells us that

$$\frac{\overline{CX}}{\sin 45.0°} = \frac{150}{\sin 60.0°}$$

Hence

$$\overline{CX} = \frac{150 \sin 45.0°}{\sin 60.0°} \approx \boxed{122.4744871} \approx 122 \text{ m}$$ ▲

The next example involves a triangle in which two sides of a triangle and the angle opposite one of them are given (Case 4, see p. 106). You will recall that such data may define one, two, or no triangles. For simplicity we restrict the following discussion to examples in which a unique triangle is defined by the given data. The discussion of the ambiguous case is presented in the next section.

Example 4 A satellite traveling in a circular orbit 1000 mi above the earth is due to pass directly over a tracking station at noon. Assume that the satellite takes 2 hr to make an orbit and that the radius of the earth is 4000 mi. If the tracking antenna is aimed 30.0° above the horizon, at what time will the satellite pass through the beam of the antenna? (See Figure 3.41.) Assume that the beam is directed to intercept the satellite after it is overhead.

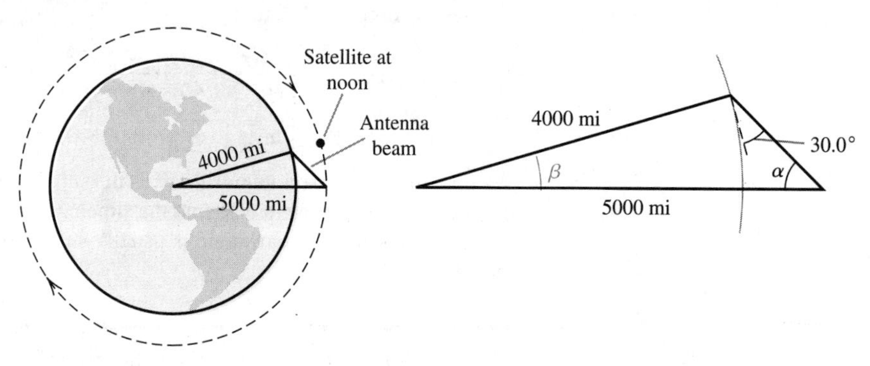

Figure 3.41

SOLUTION From the law of sines,

$$\frac{\sin \alpha}{4000} = \frac{\sin 120°}{5000}$$

$$\sin \alpha = \frac{4000 \sin 120°}{5000} \approx \boxed{0.692820323}$$

$$\alpha \approx \boxed{\text{inv}}\ \boxed{\text{sin}}\ 0.692820323\ \boxed{=}\ \boxed{43.85377861}$$

Hence

$$\alpha \approx 43.85°$$

Angle β is given by

$$\beta \approx 180° - (120° + 43.85°) = 16.15°$$

The time involved in the change from $\beta = 0.0°$ to $\beta = 16.15°$ is given by $(16.15°/360°)(120 \text{ min}) \approx 5.38 \text{ min} \approx 5 \text{ min } 23 \text{ sec}$. Thus the satellite will pass through the beam of the antenna at 12:5:23 P.M. ▲

Exercises for Section 3.3

In Exercises 1–15 solve the given oblique triangles.

1. $A = 32.0°, B = 48.0°, a = 10.0$

2. $A = 60.0°, B = 45.0°, b = 3.00$

3. $A = 45.2°, a = 8.82, b = 5.15$

4. $A = 75.0°, a = 27.7, b = 11.8$

5. $A = 35.6°, b = 12.2, a = 17.5$

6. $A = 82.1°, b = 7.21, a = 29.0$

7. $A = 120°50', a = 6.61, b = 5.09$

8. $A = 51°10', a = 59.2, b = 53.5$

9. $C = 53.0°, b = 18.3, a = 30.2$

10. $C = 58.0°, c = 83.0, b = 51.0$

11. $B = 122.0°, b = 30.0, a = 25.0$

12. $B = 63.0°$, $b = 5.00$, $c = 4.00$
13. $C = 110.0°$, $B = 50.0°$, $b = 40.0$
14. $C = 73.2°$, $A = 13.7°$, $c = 20.5$
15. $B = 48.0°$, $A = 43.4°$, $c = 61.3$

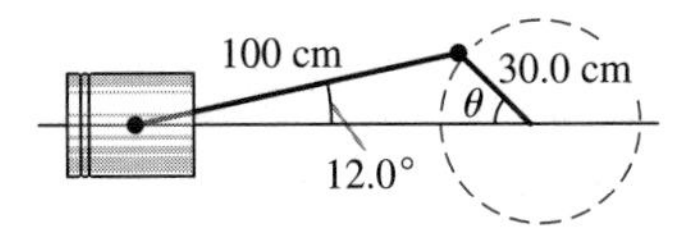

Figure 3.42

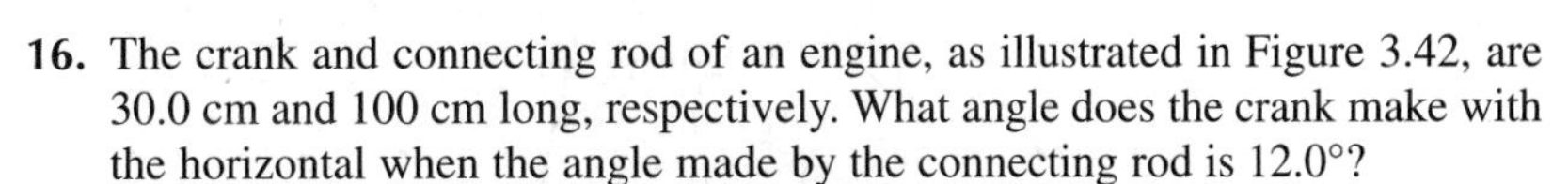

16. The crank and connecting rod of an engine, as illustrated in Figure 3.42, are 30.0 cm and 100 cm long, respectively. What angle does the crank make with the horizontal when the angle made by the connecting rod is 12.0°?

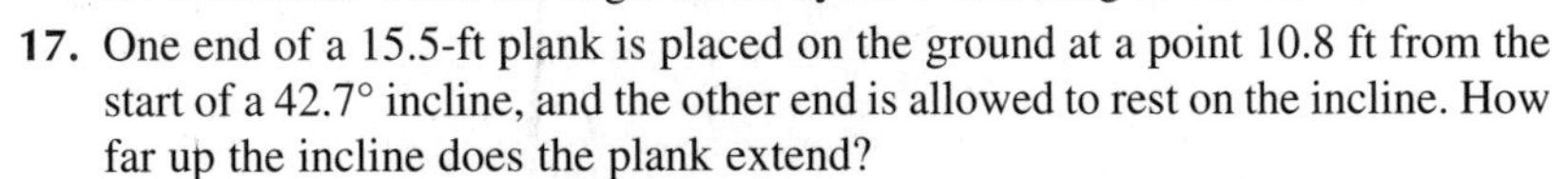

17. One end of a 15.5-ft plank is placed on the ground at a point 10.8 ft from the start of a 42.7° incline, and the other end is allowed to rest on the incline. How far up the incline does the plank extend?

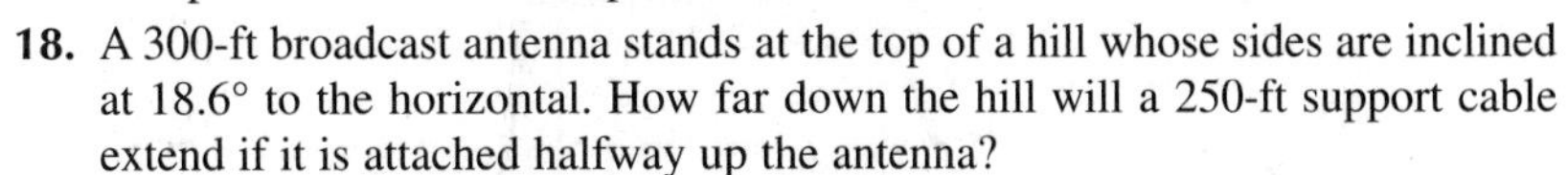

18. A 300-ft broadcast antenna stands at the top of a hill whose sides are inclined at 18.6° to the horizontal. How far down the hill will a 250-ft support cable extend if it is attached halfway up the antenna?

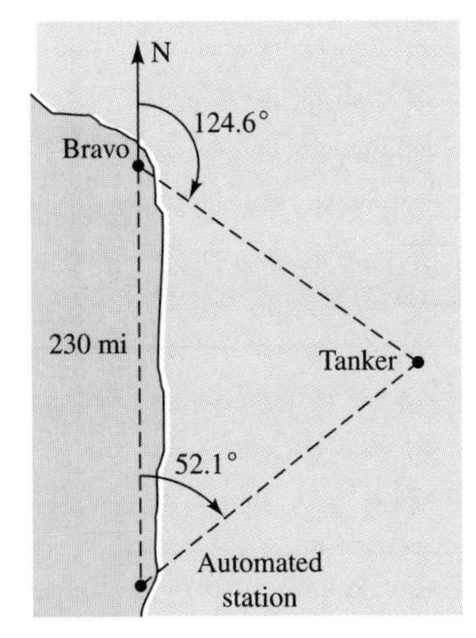

Figure 3.43
Bearing angles

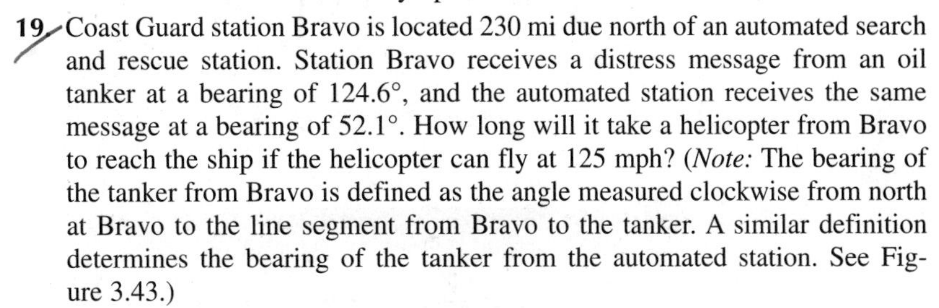

19. Coast Guard station Bravo is located 230 mi due north of an automated search and rescue station. Station Bravo receives a distress message from an oil tanker at a bearing of 124.6°, and the automated station receives the same message at a bearing of 52.1°. How long will it take a helicopter from Bravo to reach the ship if the helicopter can fly at 125 mph? (*Note:* The bearing of the tanker from Bravo is defined as the angle measured clockwise from north at Bravo to the line segment from Bravo to the tanker. A similar definition determines the bearing of the tanker from the automated station. See Figure 3.43.)

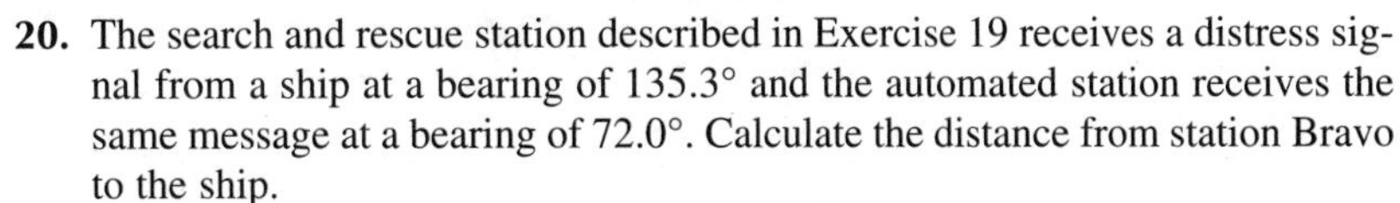

20. The search and rescue station described in Exercise 19 receives a distress signal from a ship at a bearing of 135.3° and the automated station receives the same message at a bearing of 72.0°. Calculate the distance from station Bravo to the ship.

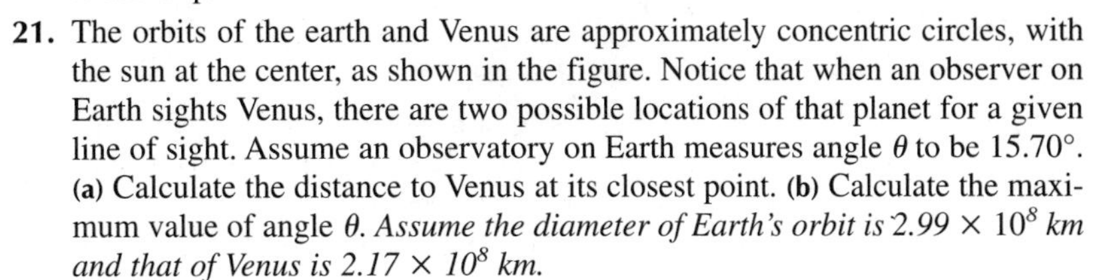

21. The orbits of the earth and Venus are approximately concentric circles, with the sun at the center, as shown in the figure. Notice that when an observer on Earth sights Venus, there are two possible locations of that planet for a given line of sight. Assume an observatory on Earth measures angle θ to be 15.70°. **(a)** Calculate the distance to Venus at its closest point. **(b)** Calculate the maximum value of angle θ. *Assume the diameter of Earth's orbit is* 2.99×10^8 *km and that of Venus is* 2.17×10^8 *km.*

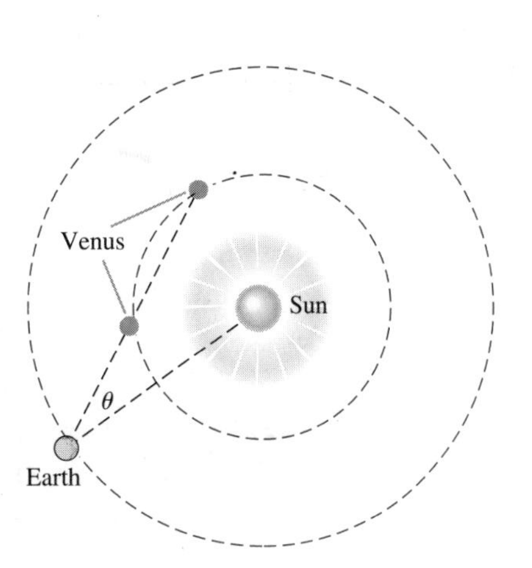

22. A balloon is tethered above a bridge by a cord. To find the height of the balloon above the surface of the bridge, a girl measures the length of the bridge and the angles of elevation to the balloon at each end of the bridge. If she finds the length of the bridge to be 263 ft and the angles of elevation to be 64.3° and 74.1°, what is the height of the balloon? (See Figure 3.44.)

23. A weight is attached to a rope strung between two vertical poles, as shown in Figure 3.45. How far is the weight from the left pole if $\theta = 41.0°$ and $\phi = 75.0°$?

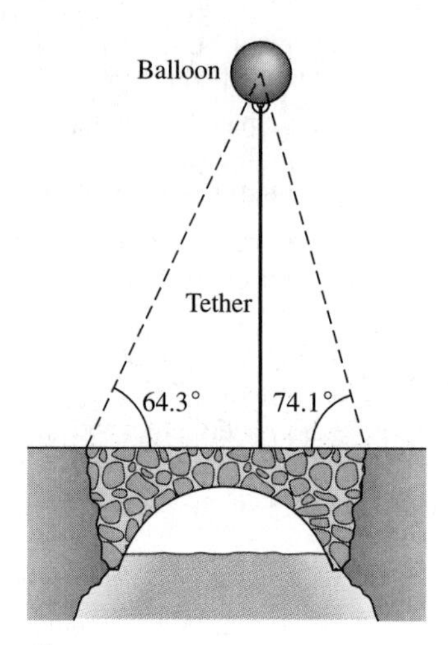

Figure 3.44

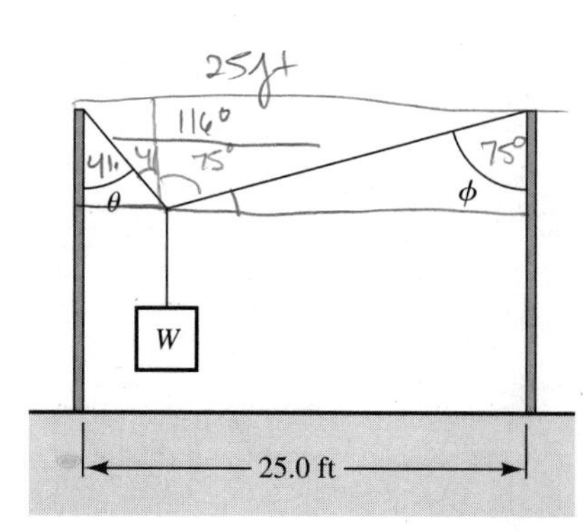

Figure 3.45

24. How far from the left pole will the weight in Exercise 23 be if $\theta = 38.7°$ and $\phi = 69.1°$?

25. From a position at the base of a hill, an observer notes that the angle of elevation to the top of an antenna is 43.5°. (See Figure 3.46.) After walking 1500 ft toward the base of the antenna up a slope of 30.0°, the observer finds the angle of elevation to be 75.4°. Find the height of the antenna and the height of the hill.

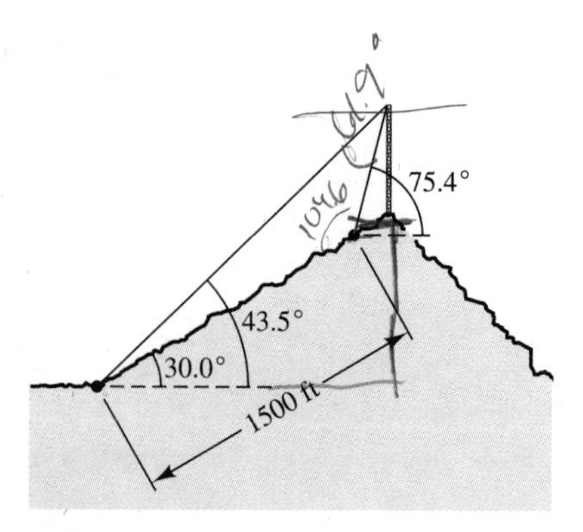

Figure 3.46

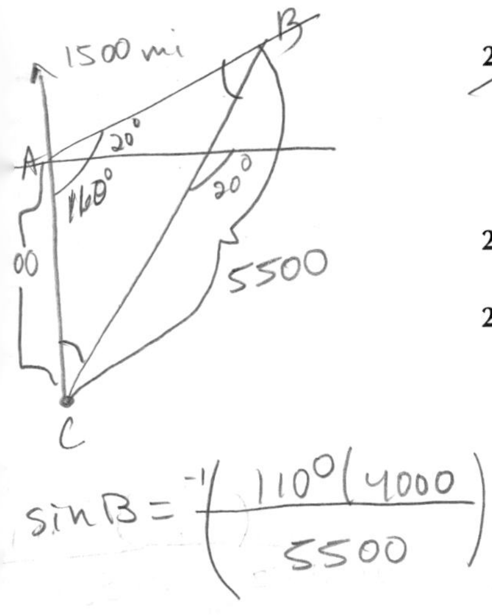

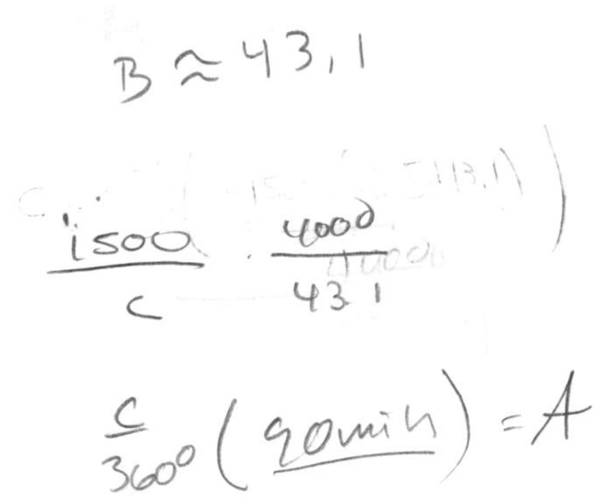

26. A satellite traveling in a circular orbit 1500 mi above the earth is due to pass directly over a tracking station at noon. Assume that the satellite takes 90 min to make an orbit and that the radius of the earth is 4000 mi. If the tracking antenna is aimed 20.0° above the horizon, at what time will the satellite pass through the beam of the antenna? See Example 4.

27. Determine at what angle above the horizon the antenna in Exercise 26 must be pointed in order for its beam to intercept the satellite at 12:05 P.M.

28. In a laser-beam experiment, mirrors are used to change the direction of a beam of light, as shown in Figure 3.47. Suppose the distance between mirror 1 and mirror 2 is 10.75 cm. Furthermore, the angle between the beam and its reflection at mirror 1 is 68.45°, and the angle between the beam reflected from mirror 1 and its reflection at mirror 2 is 32.25°. Calculate the distance from mirror 2 to the intersection of the original beam and its reflection at point A.

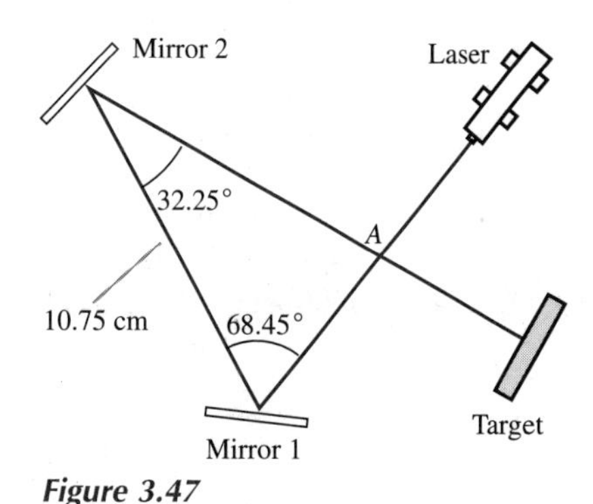

Figure 3.47

▲▼ 3.4 The Ambiguous Case

We will now analyze how to solve triangles for which we are given the measures of two sides and the angle opposite one of them. Suppose that two sides a and b and an angle A are given. As you will see, there may be one, two, or no triangles with these measurements.

$\mathbf{A < 90°}$**:** Perhaps the best way to clarify the situation in which angle A is acute is to draw a figure. To locate the vertex C, we construct a line segment having a length of b units along one side of angle A. Then it is obvious from Figure 3.48 on page 120 that the length of side a will determine whether there are two, one, or no triangles. In Figure 3.48(a), one and only one triangle is possible, since $a \geq b$.

There are three possible outcomes when $a < b$. To understand how this happens, consider the line segment h that forms a right triangle with side b and angle A; this segment is shown as a dashed line in Figure 3.48. If $a < b$ and $a < h$, as in Figure 3.48(b), no triangle is possible, because side a is too short to intersect side c. If $a < b$ and $a = h$, as in Figure

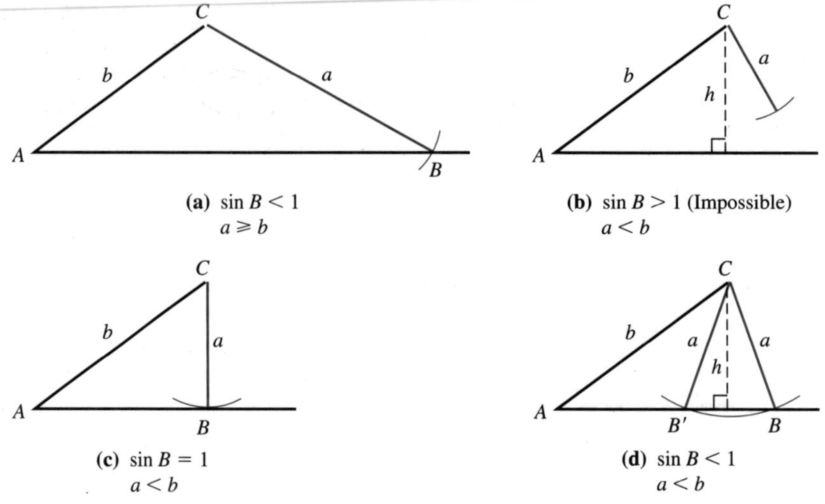

Figure 3.48
Ambiguous case, $A < 90°$

3.48(c), one right triangle is formed. Finally, if $a < b$ and $a > h$, as in Figure 3.48(d), two different triangles can be formed, because side a can intersect side c at either B or B'.

The three possible outcomes shown in Figure 3.48 for $a < b$ can be identified analytically by solving

$$\frac{a}{\sin A} = \frac{b}{\sin B}$$

for $\sin B$, to obtain

$$\sin B = \frac{b \sin A}{a}$$

When this expression is used to compute $\sin B$, the three possible outcomes are as follows:

1. If $a < b$ and $\sin B > 1$, then angle B cannot exist, so there is no triangle. [See Figure 3.48(b).]
2. If $a < b$ and $\sin B = 1$, then $B = 90°$, so there is one right triangle. [See Figure 3.48(c).]
3. If $a < b$ and $\sin B < 1$, then two angles B are possible: the acute angle B and the obtuse angle $B' = 180° - B$. [See Figure 3.48(d).]

COMMENT If there is no triangle for the given measures, your calculator will give an error message or an unrealistic answer when you try to solve for angle B. Further, your calculator will not tell you when there are two solutions. You must determine this by identifying the conditions given in outcome 3 above.

$A \geq 90°$: The case in which angle A is greater than or equal to 90° is much easier to analyze than the case in which $A < 90°$. As you can see in Figure 3.49(a), if $a \leq b$, there is no triangle. If $a > b$, there is one triangle, as shown in Figure 3.49(b).

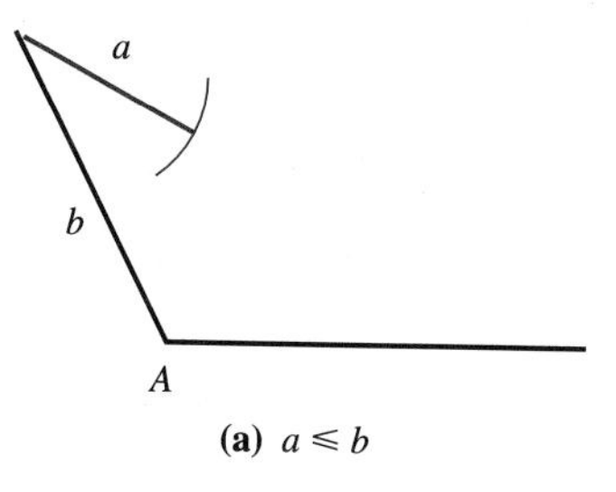

(b) $a > b$

Figure 3.49
Ambiguous case, $A > 90°$

Example 1 How many triangles can be formed if $a = 4.0$, $b = 10$, and $A = 30°$?

SOLUTION Using the law of sines, we have

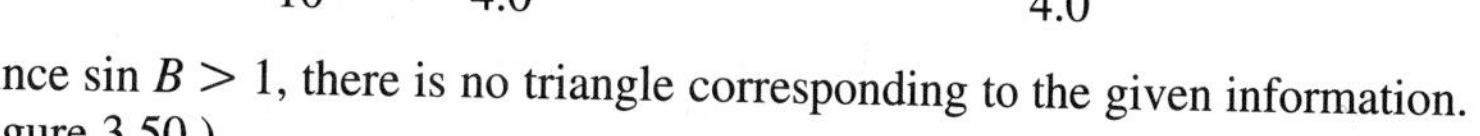

$$\frac{\sin B}{10} = \frac{\sin 30°}{4.0} \qquad \text{or} \qquad \sin B = \frac{10(0.5)}{4.0} = 1.25$$

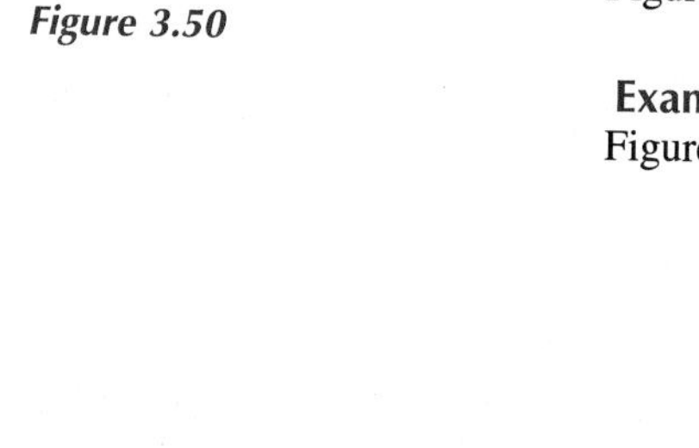

Figure 3.50

Since $\sin B > 1$, there is no triangle corresponding to the given information. (See Figure 3.50.) ▲

Example 2 Solve the triangle with $B = 134.7°$, $b = 526$, and $c = 481$. (See Figure 3.51.)

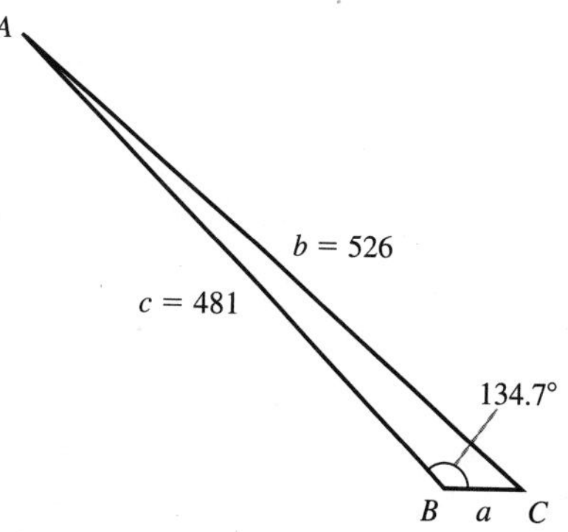

Figure 3.51

SOLUTION In this triangle the given angle is obtuse and the side opposite the given angle is greater than the side adjacent to it. This means that there is only one triangle that can be formed with the given information. Using the law of sines we have

$$\frac{\sin C}{481} = \frac{\sin 134.7°}{526} \qquad \text{or} \qquad \sin C = \frac{481 \sin 134.7°}{526} \approx \boxed{0.649989633}$$

Since B is obtuse, both A and C are acute. Consequently angle $C \approx 40.5°$, and $C' \approx 180° - 40.5° = 139.5°$. The fact that $B' + C' > 180°$ verifies that there is only one solution. Angle A is then given by

$$A \approx 180° - (134.7° + 40.5°) = 4.8°$$

Finally, side a is given by

$$\frac{a}{\sin 4.8°} = \frac{526}{\sin 134.7°} \quad \text{or} \quad a = \frac{526 \sin 4.8°}{\sin 134.7°} \approx \boxed{61.92259169} \approx 61.9$$ ▲

Example 3 Verify that there are two triangles with measurements $a = 9.00$, $b = 10.0$, and $A = 60.0°$, and then solve each triangle.

SOLUTION Substituting the given values into the law of sines, we get

$$\frac{\sin B}{10.0} = \frac{\sin 60°}{9.00} \quad \text{or} \quad \sin B = \frac{10.0 \sin 60°}{9.00} \approx \boxed{0.962250449}$$

Consequently $B \approx 74.2°$ and $B' \approx 180° - 74.2° = 105.8°$. Since $A + B' < 180°$, there are two possible solutions, shown in Figure 3.52. One triangle is ABC and the other is $AB'C'$.

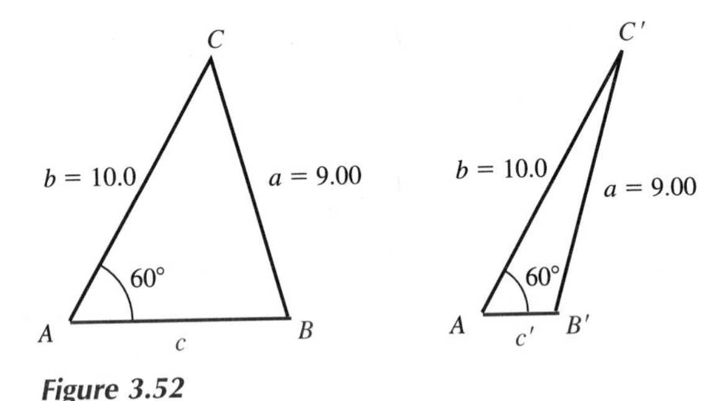

Figure 3.52

To solve triangle ABC, we note that angle C is given by

$$C = 180° - (A + B) \approx 180° - (60° + 74.2°) = 45.8°$$

Hence

$$\frac{c}{\sin 45.8°} = \frac{9.00}{\sin 60°} \quad \text{or} \quad c = \frac{9.00 \sin 45.8°}{\sin 60°} \approx \boxed{7.450353582} \approx 7.45$$

Similarly, in triangle $AB'C'$ we have

$$C' = 180° - (A + B') \approx 180° - (60° + 105.8°) = 14.2°$$

So

$$\frac{c'}{\sin 14.2°} = \frac{9.00}{\sin 60°} \quad \text{or} \quad c' = \frac{9.00 \sin 14.2°}{\sin 60°} \approx \boxed{2.549309135} \approx 2.55$$

▲

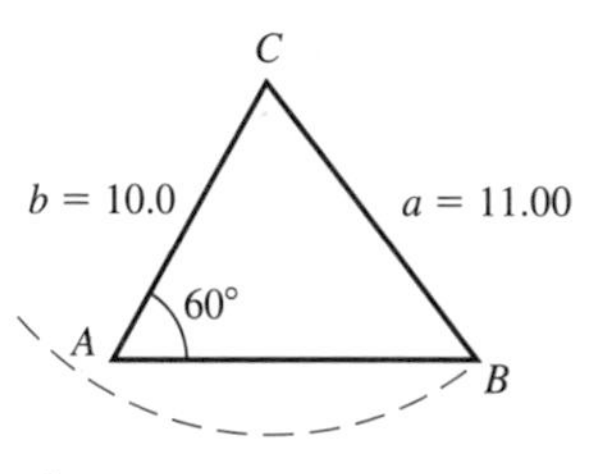

Figure 3.53

Example 4 Verify that only one triangle can be drawn for the case in which $b = 10.0$, $A = 60.0°$, and $a = 11.0$.

SOLUTION Figure 3.53 depicts the information given. Applying the law of sines, we get

$$\frac{11.0}{\sin 60°} = \frac{10.0}{\sin B}$$

from which we determine that $\sin B \approx 0.787295822$. Since the side opposite the given angle is greater than the side adjacent to it, there is only one solution. Notice that had we missed the fact that $a > b$, we would have gotten

$$B \approx 51.9° \qquad \text{and} \qquad B' \approx 128.1°$$

However, the "solution" $B \approx 128.1°$ is impossible because a triangle cannot have angles of 60.0° and 128.1°. Thus we reach the same conclusion as before. ▲

COMMENT This last example shows how important it is to continually check that both given and calculated data are consistent and do not produce impossible situations.

The various subcases within the ambiguous case need not be memorized, since if a given problem is ambiguous or has no solution, that will become evident as you try to solve it. For example, if in the course of solving a triangle you find that $\sin B > 1$, this means that no triangle corresponding to the given data exists. If $\sin B < 1$, there are two angles that may be solutions if they do not cause the sum of the angles in the triangle to exceed 180°.

▲▼ Exercises for Section 3.4

In Exercises 1–5 state whether the triangle has one solution, two solutions, or no solution, given that $A = 30°$ and $b = 4$.

1. $a = 1$ **2.** $a = 2$ **3.** $a = 3$ **4.** $a = 4$ **5.** $a = 5$

For the triangles in Exercises 6–19 find all the unknown measurements or show that no such triangle exists.

6. $a = 20.0$, $b = 10.0$, $A = 35°40'$
7. $a = 2.00$, $b = 6.00$, $A = 26°20'$
8. $a = 4.00$, $b = 8.00$, $A = 30.0°$
9. $a = 15.0$, $c = 8.00$, $A = 150.0°$
10. $a = 50.0$, $b = 19.0$, $B = 22°30'$
11. $b = 60.0$, $c = 74.0$, $B = 140.0°$
12. $a = 50.0$, $c = 10.0$, $A = 48.0°$
13. $C = 28.0°$, $a = 20.0$, $c = 15.0$
14. $B = 40.0°$, $a = 12.0$, $b = 10.0$
15. $A = 30.04°$, $b = 400.5$, $a = 325.1$
16. $B = 100.0°$, $a = 10.0$, $b = 12.0$
17. $C = 70.0°$, $b = 100$, $c = 100$
18. $B = 41.2°$, $a = 4.20$, $b = 3.20$
19. $a = 0.900$, $b = 0.700$, $A = 72°10'$

20. If $b = 12$ and $A = 30.0°$, for what values of a will there be two triangles?

21. If $c = 15$ and $B = 25.0°$, for what values of b will there be two triangles?

3.5 Analysis of the General Triangle

In Section 3.2 we mentioned that most of the time you could expect three parts of a triangle to be sufficient to determine it uniquely. With the aid of the two fundamental laws derived in Sections 3.2 and 3.3, we are now in a position to summarize the various approaches to solving a triangle.

CASE 1

TWO SIDES AND AN INCLUDED ANGLE ARE GIVEN Use the law of cosines to obtain the third side. A second angle may be obtained using either the law of cosines or the law of sines. The third angle is computed by subtracting the sum of the other two from 180°.

CASE 2

THREE SIDES ARE GIVEN Use the law of cosines to obtain one of the angles, preferably the largest one. A second angle may be obtained using either the law of cosines or the law of sines, and the third one may be obtained from the fact that the sum of the angles must be 180°. The sum of any two sides must exceed the length of the third side for a triangle to exist.

CASE 3

TWO ANGLES AND A SIDE ARE GIVEN The two given angles must have a sum of less than 180°; otherwise no triangle is possible. Use the law of sines to determine the two unknown sides.

CASE 4

TWO SIDES AND A NONINCLUDED ANGLE ARE GIVEN If two sides a and b and an angle A are given, there may be two, one, or no triangles with these measurements. A carefully drawn figure will usually make the situation clear. The following rules express this case analytically.

1. If a is acute, there is no solution, one solution, or two solutions, depending on whether $a < b \sin A$, $a = b \sin A$, or $a > b \sin A$, unless $a \geq b$, in which case there is only one solution.
2. If A is obtuse, there is either no solution or one solution, depending on whether $a \leq b$ or $a > b$.

Exercises for Section 3.5

In Exercises 1–17 solve each triangle or show that no such triangle exists.

1. $A = 60.0°$, $B = 75.0°$, $a = 600$

2. $A = 75.0°$, $a = 120$, $b = 75.0$

3. $B = 15.0°$, $C = 105.0°$, $a = 4.00$

4. $A = 30.0°$, $b = 60.0$, $c = 50.0$

5. $a = 8, b = 2, c = 6$
6. $C = 15.0°, b = 15.0, c = 10.0$
7. $C = 30.0°, a = 300, b = 500$
8. $a = 2000, b = 1000, c = 2500$
9. $B = 120.0°, a = 60.0, b = 25.0$
10. $B = 30.0°, a = 500, b = 400$
11. $B = 58.3°, a = 11.2, b = 9.75$
12. $A = 20.0°, a = 2.00, b = 3.00$
13. $A = 60.0°, B = 100.0°, b = 2.00$
14. $B = 125.2°, a = 2.20, b = 1.30$
15. $A = 100.0°, a = 2.00, b = 1.00$
16. $A = 42.3°, a = 20.0, c = 30.0$
17. $a = 6.50, b = 4.00, c = 8.00$

18. The angle of depression from a window 35.0 ft above a street to the curb on the far side of the street is 15.0°, and the angle of depression to the curb on the near side is 45.0°. How wide is the street?

19. At successive milestones on a straight road leading to a mountain, the angle of elevation to the top of the mountain is measured as 30.0° and 45.0°. What is the line-of-sight distance to the top of the mountain from the milestone nearest to it?

20. A driveway slopes from a garage to the street at an angle of 15.5° with the horizontal. A 15-ft ladder with one end on the driveway and the other end against the garage makes an angle of 60.0° to the driveway. Calculate how far the foot of the ladder is from the garage.

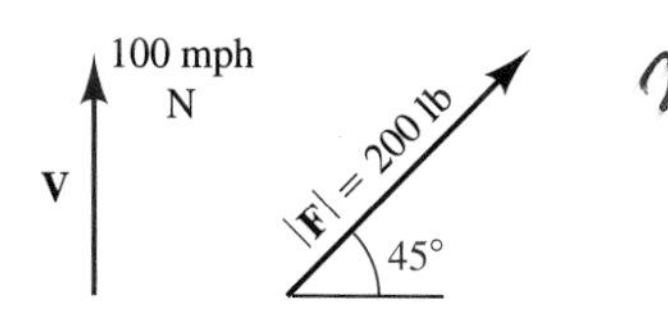

Figure 3.54

3.6 Vectors

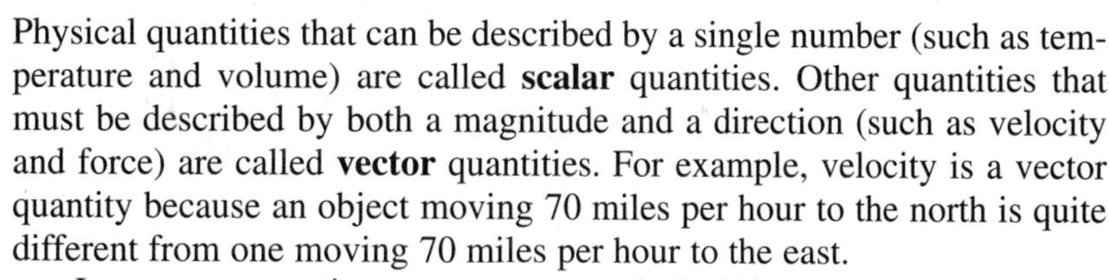

Physical quantities that can be described by a single number (such as temperature and volume) are called **scalar** quantities. Other quantities that must be described by both a magnitude and a direction (such as velocity and force) are called **vector** quantities. For example, velocity is a vector quantity because an object moving 70 miles per hour to the north is quite different from one moving 70 miles per hour to the east.

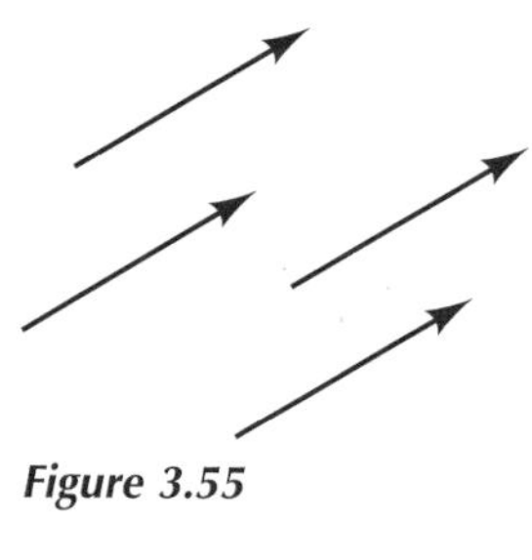

Figure 3.55

Letters representing vectors are set in boldface type to distinguish them from scalars. The magnitude of a vector **F** is denoted by $|\mathbf{F}|$ and is always a positive number. The direction of a vector is given in a variety of ways, depending on the application. A vector is represented graphically by an arrow, with the tip of the arrowhead at the **terminal point** of the vector. The length of the arrow corresponds to the magnitude of the vector, and the direction of the arrowhead gives its direction. Thus in Figure 3.54 one arrow represents a velocity of 100 mph due north and the other a force of 200 lb acting at 45° above the horizontal.

For mathematical purposes two vectors are considered equal if they have the same direction and length, regardless of the location of the initial point of the vector. Thus all the vectors in Figure 3.55 are mathematically equivalent.

Note that this type of vector equality may not always be the kind you need. For example, if **F** in Figure 3.56 is a 10-lb force pointing down, it does make a difference in which of the three places it is applied. Vectors for which you may ignore the actual point of application are said to be *free*. Physical situations such as those shown in Figure 3.56 cannot be described using free vectors and hence will not be discussed here.

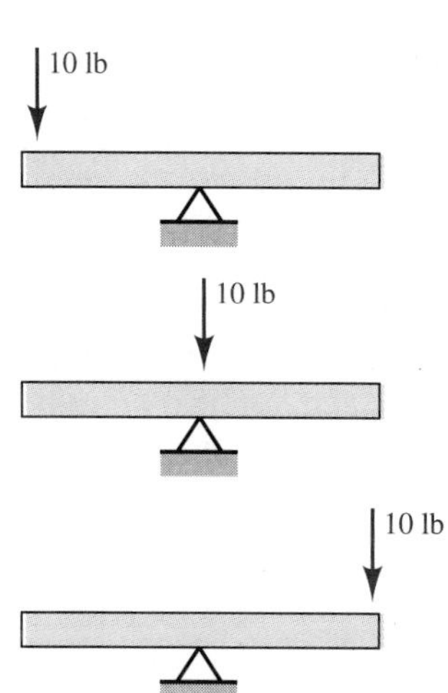

Figure 3.56

If we first impose a coordinate system on a plane, we can move all the vectors in this plane so that their initial points are at the origin. (See Figure 3.57.) Such vectors are said to be in **standard position**. In effect, a vector at the origin represents all other vectors with the same direction and length.

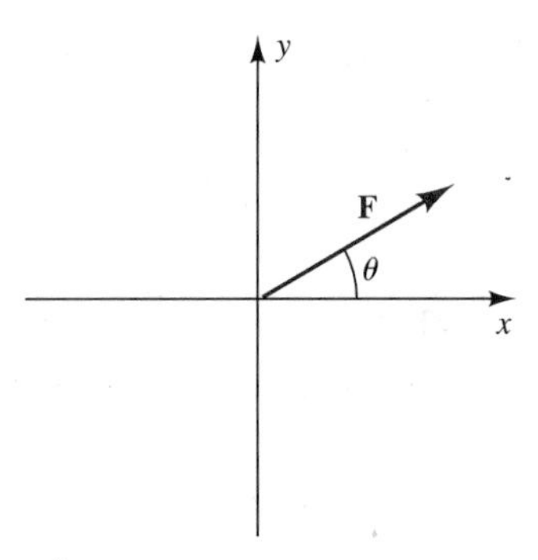

Figure 3.57

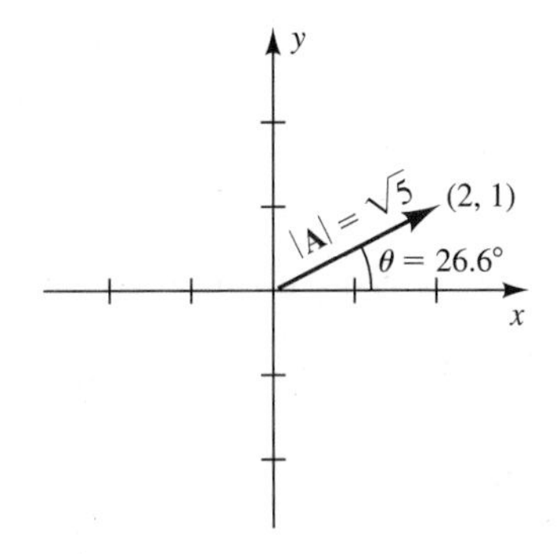

Figure 3.58

If a vector is placed in standard position, its length can be found from the Pythagorean theorem. The direction of the vector is the angle it makes with the positive x-axis.

Example 1 Find the magnitude and direction of the vector **A** in standard position whose terminal point is at (2, 1).

SOLUTION Figure 3.58 shows that the length of the vector is

$$|\mathbf{A}| = \sqrt{2^2 + 1^2} = \sqrt{5}$$

and the angle θ between **A** and the positive x-axis is given by $\tan\theta = \frac{1}{2}$. Thus

$$\theta = \boxed{\textbf{inv}}\ \boxed{\textbf{tan}}\ 0.5\ \boxed{=}\ \boxed{\textbf{26.56505118}} \approx 26.6^\circ$$ ▲

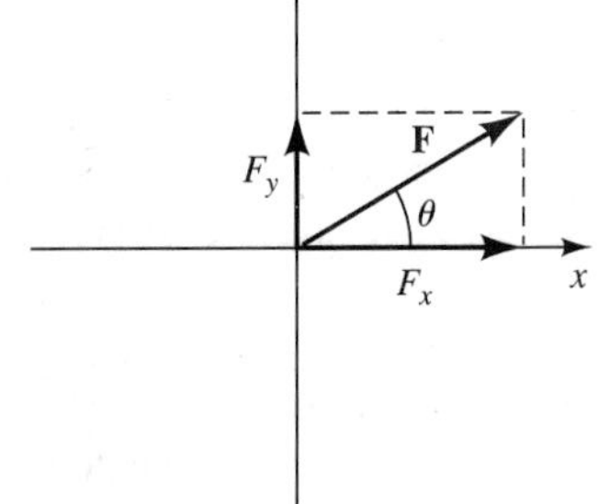

Figure 3.59

The perpendicular projections of the vector onto the x- and y-axes are called the **components** of the vector. We say that a vector **F** is resolved into its x and y components, called the **horizontal** and **vertical components of F**, respectively. Resolving a vector into its x and y components is a simple problem of trigonometry. Figure 3.59 shows that

$$F_x = |\mathbf{F}|\cos\theta \qquad \text{Horizontal component of F}$$
$$F_y = |\mathbf{F}|\sin\theta \qquad \text{Vertical component of F}$$
$$F_x^2 + F_y^2 = |\mathbf{F}|^2$$

Example 2 Find the horizontal and vertical components of a force vector of magnitude 15 lb acting at an angle of 38° to the horizontal.

SOLUTION $F_x = |\mathbf{F}|\cos\theta = 15\cos 38° \approx$ [11.8201613] ≈ 12 lb

$F_y = |\mathbf{F}|\sin\theta = 15\sin 38° \approx$ [9.23492213] ≈ 9.2 lb ▲

Example 3 Find the magnitude and direction of the vector whose components are shown in Figure 3.60.

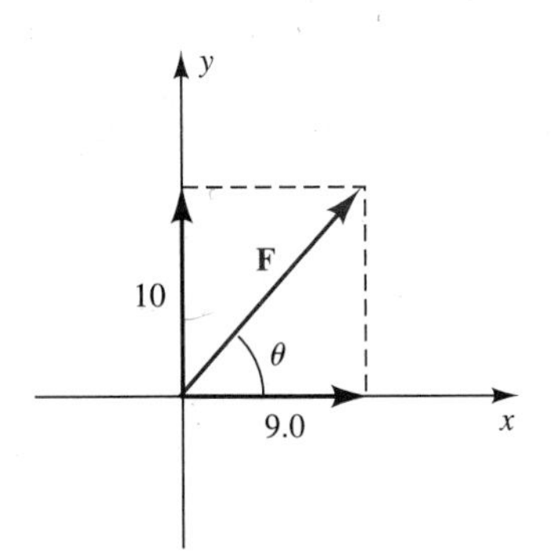

Figure 3.60

SOLUTION The magnitude is

$$|\mathbf{F}| = \sqrt{10^2 + 9^2} = \sqrt{181} \approx \boxed{13.45362405} \approx 13.5$$

The angle that the vector makes with the horizontal is determined as follows:

$$\tan\theta = \frac{10}{9.0}$$

$\theta \approx$ [inv] [tan] [(] 10 [÷] 9.0 [)] [=] [48.0127875] $\approx 48°$

Angle θ is rounded off to the nearest degree. ▲

Scalar Multiplication

Given any vector **A**, we can obtain other vectors in the same direction as **A** (or the direction opposite to that of **A**) by multiplying **A** by a real number c. The resulting vector, denoted by $c\mathbf{A}$, is a vector that points in the same direction as **A** if $c > 0$ or in the direction opposite that of **A** if $c < 0$. The magnitude of $c\mathbf{A}$ is $|c|\,|\mathbf{A}|$; that is, it is larger than $|\mathbf{A}|$ if $|c| > 1$ and smaller than $|\mathbf{A}|$ if $|c| < 1$. The vector $c\mathbf{A}$ is called a **scalar multiple** of the vector **A**. Figure 3.61 shows some scalar multiples of a given vector **A**.

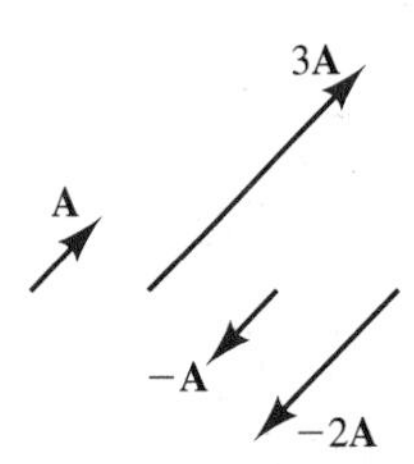

Figure 3.61

The particular scalar multiple of a vector obtained by multiplying a vector **A** by the reciprocal of its magnitude is a vector in the direction of **A** whose length is 1. This is the **unit vector** in the direction of **A**, denoted by $\mathbf{u_A}$. Thus,

$$\mathbf{u_A} = \frac{\mathbf{A}}{|\mathbf{A}|} \tag{3.6}$$

Vector Addition

A rule called the **parallelogram rule** gives the procedure for adding two vectors.

THE PARALLELOGRAM RULE To add two vectors **A** and **B**, place them with their initial points together. Then form a parallelogram with these vectors as sides. The vector that starts from the initial point of **A** and **B** and forms the diagonal of the parallelogram is called the **sum**, or **resultant**, of **A** and **B**. Figure 3.62 depicts the resultant of two vectors.

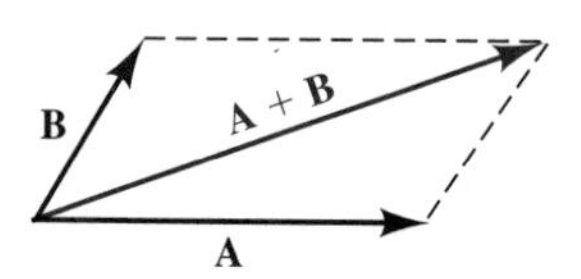

Figure 3.62

Addition of vectors is frequently accomplished by using the component method of representation for each vector in the sum. For instance, the vector sum **A** + **B** is shown in Figure 3.63, along with the components A_x, A_y, B_x, and B_y. The figure shows that the horizontal component of **A** + **B** is $A_x + B_x$ and the vertical component is $A_y + B_y$. In words, the horizontal component of the sum of two vectors is the sum of the individual horizontal components, and the vertical component is the sum of the individual vertical components. Thus finding the sum of two vectors can be reduced to finding a vector from its x and y components.

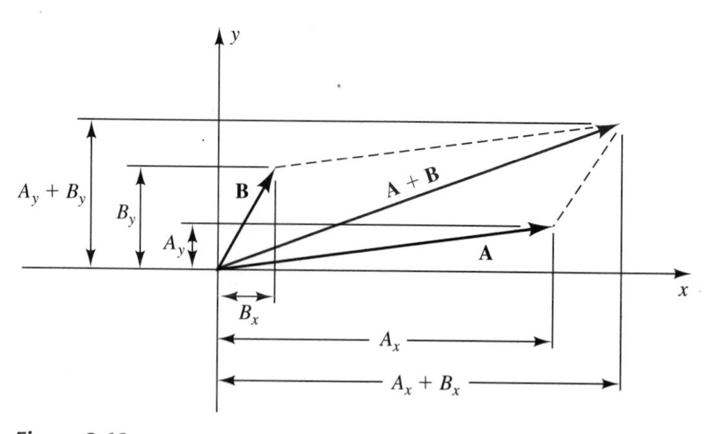

Figure 3.63

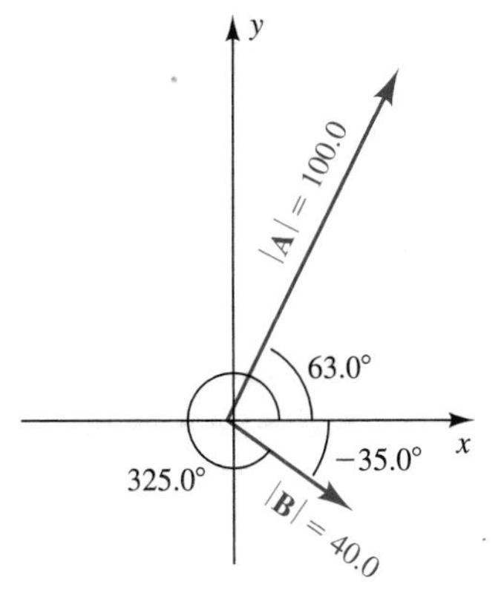

Figure 3.64

Example 4 Find the sum of the two vectors given in Figure 3.64; vector **A** of magnitude 100 and direction 63.0° and vector **B** of magnitude 40.0 and direction 325.0°.

SOLUTION First we find A_x, A_y, B_x, and B_y. (See Figure 3.65.)

$$A_x = 100 \cos 63.0° \approx \boxed{45.39904997} \approx 45.4$$
$$A_y = 100 \sin 63.0° \approx \boxed{89.10065242} \approx 89.1$$
$$B_x = 40 \cos 325.0° \approx \boxed{32.76608177} \approx 32.8$$
$$B_y = 40 \sin 325.0° \approx \boxed{-22.94305745} \approx -22.9$$

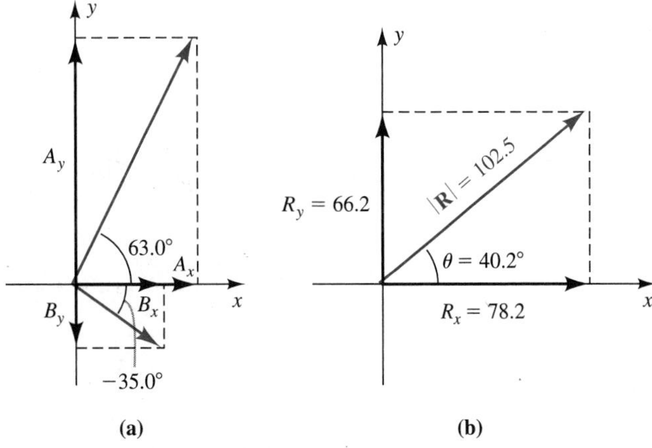

Figure 3.65

Let **A** + **B** = **R**. Then

$$R_x = A_x + B_x \approx 45.4 + 32.8 = 78.2$$
$$R_y = A_y + B_y \approx 89.1 + (-22.9) = 66.2$$

Finally, the magnitude of **A** + **B** is

$$|\mathbf{R}| = \sqrt{R_x^2 + R_y^2} \approx \sqrt{78.2^2 + 66.2^2} \approx 102.5$$

and the direction is given by

$$\tan \theta = \frac{R_y}{R_x} \approx \frac{66.2}{78.2}$$

$$\theta \approx \boxed{\text{inv}}\ \boxed{\text{tan}}\ \boxed{(}\ 66.2\ \boxed{\div}\ 78.2\ \boxed{)}\ \boxed{=}\ \boxed{40.24949352} \approx 40.2°$$

▲

An Alternative Form of the Parallelogram Rule

Vectors **A** and **B** can also be added by placing the initial point of **B** at the terminal point of **A**. Then **A** + **B** is the arrow whose initial point is at the initial point of **A** and whose terminal point is at the terminal point of **B**.

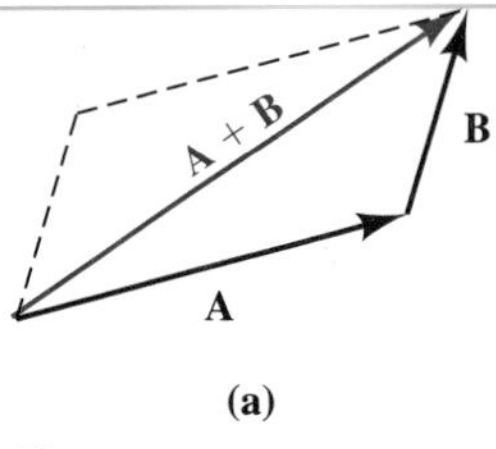

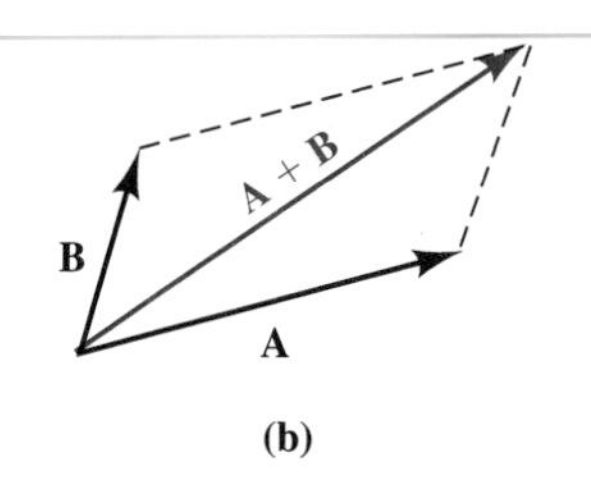

Figure 3.66

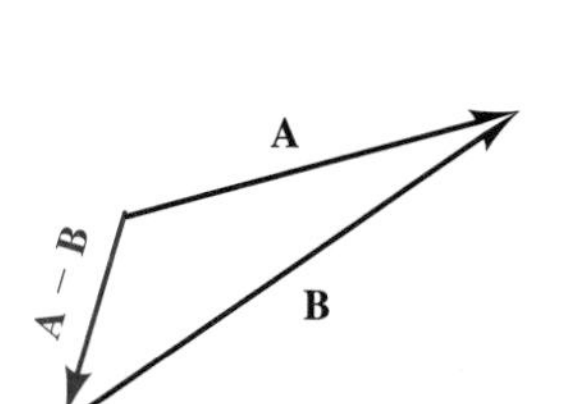

Figure 3.67

[See Figure 3.66(a).] Figure 3.66(b) shows that this method is just another way of applying the parallelogram rule.

For vector subtraction, we reverse the procedure for addition by placing the terminal point of **B** at the terminal point of **A**. Then **A** − **B** is the arrow drawn from the initial point of **A** to the initial point of **B**. (See Figure 3.67.)

One advantage of using this form of the parallelogram rule is that we can use the law of sines and the law of cosines to obtain the resultant without using the components of each vector. The next example illustrates the process.

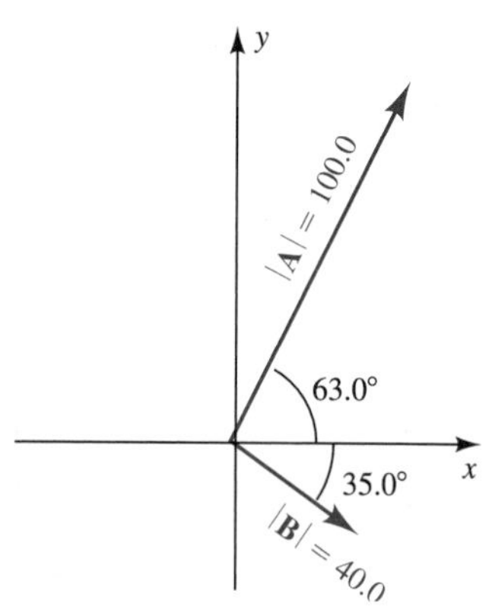

Figure 3.68

Example 5 Find the resultant for the two vectors in Example 4 using the parallelogram rule. See Figure 3.68.

SOLUTION We see in Figure 3.69 that the angle opposite the resultant vector **R** is 82.0°. This measure is obtained from the given angles based on the fact that the sum of the angles that make up a straight angle must equal 180°. Since the given vectors are the adjacent sides of the 82.0° angle, we can compute the value of **R** by using the law of cosines. Thus

$$\mathbf{R} = \sqrt{100^2 + 40^2 - 2(100)(40)\cos 82.0°} \approx \boxed{102.4041757} \approx 102.4$$

Next we use the law of sines to compute the angle α between **B** and **R**.

$$\frac{\sin \alpha}{100} = \frac{\sin 82.0°}{102.4}$$

$$\sin \alpha = \frac{100 \sin 82.0°}{102.4} \approx \boxed{0.967058661}$$

$$\alpha \approx \boxed{\text{inv}}\ \boxed{\sin}\ 0.967058661\ \boxed{=}\ \boxed{75.25286324} \approx 75.3°$$

Finally, the angle θ that **R** makes with the x-axis is

$$\theta \approx 75.3° - 35.0° = 40.3°$$

The slight differences in **R** and θ, as seen in Example 4, can be attributed to the rounding-off process. ▲

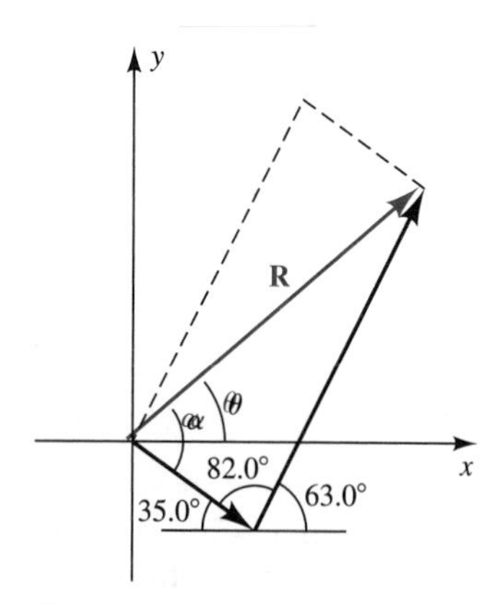

Figure 3.69

Another advantage of using the alternative form of the parallelogram rule for vector addition is that we can add several vectors sequentially without computing intermediate resultants. For instance, to add the three

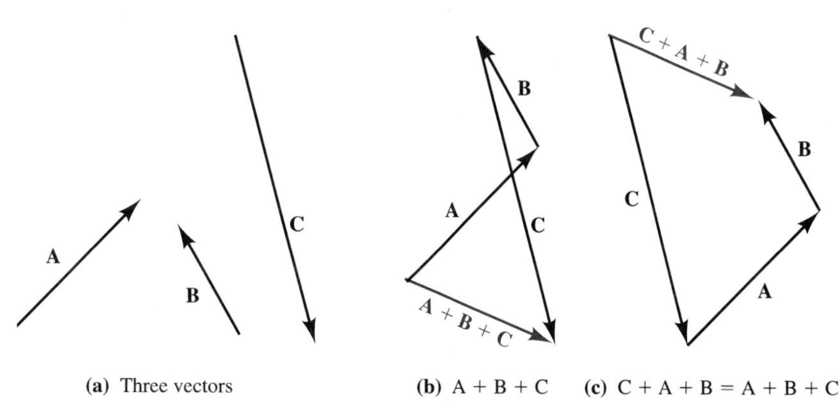

(a) Three vectors (b) A + B + C (c) C + A + B = A + B + C

Figure 3.70

vectors shown in Figure 3.70, place the initial point of **B** at the terminal point of **A** and then place the initial point of **C** at the terminal point of **B**. The resultant vector **A** + **B** + **C** is the vector drawn from the initial point of **A** to the terminal point of **C**. This method is particularly useful when vector addition is being done graphically. Notice that since addition is a commutative operation for vectors (**A** + **B** = **B** + **A**), the order of the addition is unimportant. In Figure 3.70(b) and (c) the order of addition is different, but the resultant vector is the same.

Exercises for Section *3.6*

In Exercises 1–10 draw the vector whose initial point is at the origin and whose terminal point is at the indicated point. Calculate the magnitude and direction of each vector. Express magnitudes to three significant digits and directions as positive angles to the nearest tenth of a degree.

1. $(1, 2)$ **2.** $(-3, 2)$ **3.** $(\sqrt{2}, \sqrt{7})$ **4.** $(1, -1)$
5. $(-4, -4)$ **6.** $(5, 8)$ **7.** $(4, -3)$ **8.** $(-3, -7)$
9. $(-\sqrt{3}, 6)$ **10.** $(\sqrt{5}, 3)$

In Exercises 11–20 find the horizontal and vertical components of each vector.

11. $|\mathbf{F}| = 10.0, \theta = 50.0°$ **12.** $|\mathbf{F}| = 25.0, \theta = 75.0°$
13. $|\mathbf{F}| = 13.7, \theta = 34°10'$ **14.** $|\mathbf{F}| = 0.751, \theta = 56°30'$
15. $|\mathbf{F}| = 158, \theta = 125.0°$ **16.** $|\mathbf{F}| = 875, \theta = 145.0°$
17. $|\mathbf{F}| = 43.5, \theta = 220.0°$ **18.** $|\mathbf{F}| = 9.41, \theta = 195.0°$
19. $|\mathbf{F}| = 10.4, \theta = 335.0°$ **20.** $|\mathbf{F}| = 0.05, \theta = 280°$

In Exercises 21–30 find the magnitude and direction of the vector whose components are given.

21. $F_x = 20.0, F_y = 15.0$ **22.** $F_x = 56.0, F_y = 13.0$
23. $F_x = 17.5, F_y = 69.3$ **24.** $F_x = 0.012, F_y = 0.200$

25. $F_x = 0.130, F_y = 0.280$
26. $F_x = 1930, F_y = 565$
27. $F_x = 8.06, F_y = -7.05$
28. $F_x = -3.17, F_y = 5.22$
29. $F_x = -2.11, F_y = -7.75$
30. $F_x = 159, F_y = 306$

The vectors in Exercises 31–40 are defined in terms of a magnitude and a direction. Use the component method to find the sum of A and B.

31. $|\mathbf{A}| = 20.4, \theta_{\mathbf{A}} = 15.6°$
$|\mathbf{B}| = 25.9, \theta_{\mathbf{B}} = 49.3°$
32. $|\mathbf{A}| = 165, \theta_{\mathbf{A}} = 25.7°$
$|\mathbf{B}| = 223, \theta_{\mathbf{B}} = 70.1°$
33. $|\mathbf{A}| = 153, \theta_{\mathbf{A}} = 0.0°$
$|\mathbf{B}| = 255, \theta_{\mathbf{B}} = 58.4°$
34. $|\mathbf{A}| = 9.51, \theta_{\mathbf{A}} = 90.0°$
$|\mathbf{B}| = 5.03, \theta_{\mathbf{B}} = 40.2°$
35. $|\mathbf{A}| = 2.56, \theta_{\mathbf{A}} = 35.7°$
$|\mathbf{B}| = 3.00, \theta_{\mathbf{B}} = 96.0°$
36. $|\mathbf{A}| = 29.2, \theta_{\mathbf{A}} = 15.6°$
$|\mathbf{B}| = 82.6, \theta_{\mathbf{B}} = 150.0°$
37. $|\mathbf{A}| = 125, \theta_{\mathbf{A}} = 145.2°$
$|\mathbf{B}| = 92.5, \theta_{\mathbf{B}} = 215.0°$
38. $|\mathbf{A}| = 550, \theta_{\mathbf{A}} = 140.4°$
$|\mathbf{B}| = 925, \theta_{\mathbf{B}} = 310.1°$
39. $|\mathbf{A}| = 0.332, \theta_{\mathbf{A}} = 32°12'$
$|\mathbf{B}| = 1.06, \theta_{\mathbf{B}} = 100°42'$
40. $|\mathbf{A}| = 0.0221, \theta_{\mathbf{A}} = 17°24'$
$|\mathbf{B}| = 0.0988, \theta_{\mathbf{B}} = 208°12'$

Exercises 41–50: Use the law of sines and the law for cosines to calculate the magnitude and direction of the resultant vectors found in Exercises 31–40.

In Exercises 51–57 perform each indicated operation graphically, using the vectors given below.

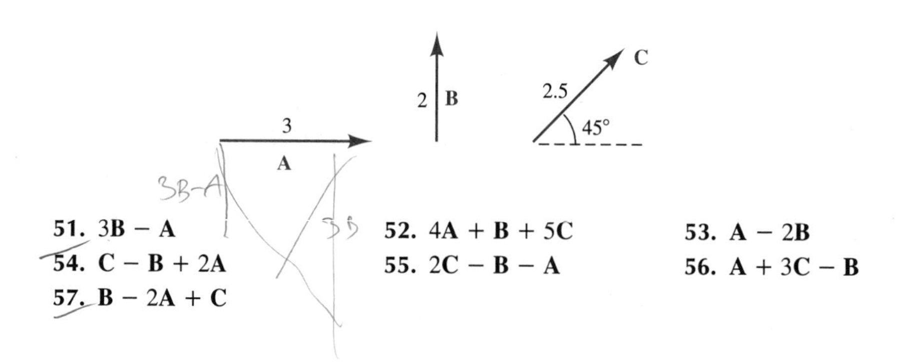

51. **3B − A**
52. **4A + B + 5C**
53. **A − 2B**
54. **C − B + 2A**
55. **2C − B − A**
56. **A + 3C − B**
57. **B − 2A + C**

3.7 Applications of Vectors

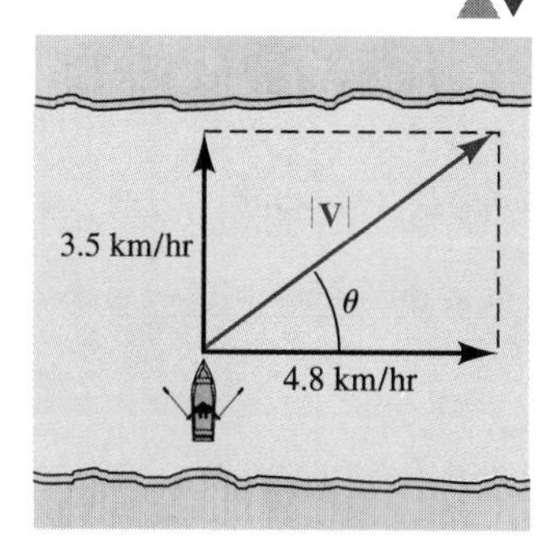

Figure 3.71

Vectors are important tools for solving a wide range of physical problems. We have already mentioned that velocity and force are vector quantities. The examples and exercises in this section illustrate how vectors are used to solve problems involving these quantities.

Consider a boat moving across a river, as shown in Figure 3.71. If the pilot heads for a point directly on the other shoreline, the boat will end up downstream from that point, because as the boat moves across the river it will be carried downstream by the current. The true motion of the boat is given by the vector sum of the velocity vector of the boat and the velocity vector of the current.

Example 1 A boat that can travel at a rate of 3.5 km/hr in still water is pointed directly across a river with a current of 4.8 km/hr. Calculate the actual velocity of the boat relative to the shoreline.

SOLUTION Referring to the diagram in Figure 3.71, we see that the velocity vector is the vector sum of the velocity of the boat and that of the current. Since these vectors are at right angles, we use the Pythagorean theorem to get

$$|\mathbf{V}| = \sqrt{3.5^2 + 4.8^2} \approx 5.9 \text{ km/hr}$$

The angle θ that the velocity vector makes with the shoreline is then given by

$$\tan\theta = \frac{3.5}{4.8} \approx \boxed{0.729166667}$$

The keystrokes for angle θ are

$$\theta \approx \boxed{\text{inv}}\ \boxed{\tan}\ 0.729166667\ \boxed{=}\ \boxed{36.09828397} \approx 36^\circ$$ ▲

Navigation

In navigation it is customary to measure the direction in which a ship or an airplane is moving by an angle called the **course** of the craft. Specifically the course of a ship or an airplane is the angle, measured clockwise, between north and the direction in which the ship or plane is moving. (See Figure 3.72.)

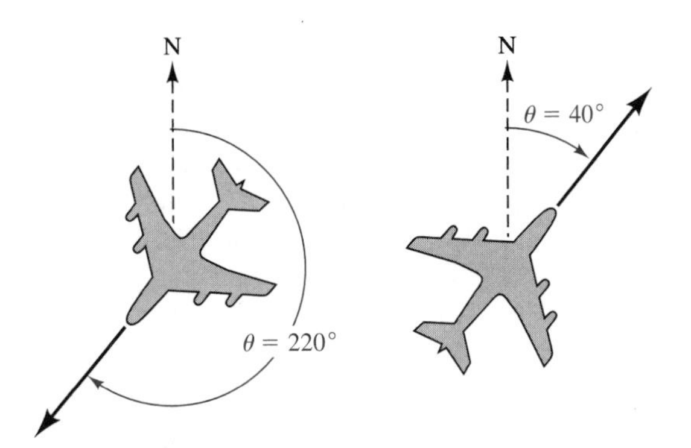

Figure 3.72 Course angle

To maintain a certain course in a crosswind, a pilot must point a plane slightly into the wind. The wind has the same effect on the plane as the current of the river had on the boat in Example 1. The direction in which a craft is pointed in order to maintain a certain course is called its **heading.** The relationship between heading and course is shown in Figure 3.73. The triangle formed by the three vectors shows that the course vector is the vector sum of the heading vector and the wind vector.

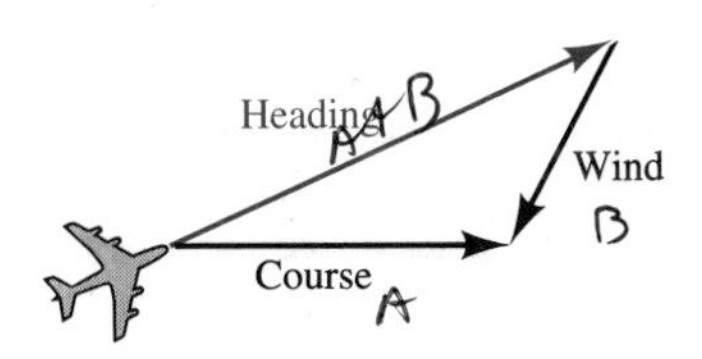

Figure 3.73
Heading vector + wind vector = course vector

Example 2 The pilot of an airplane with a cruising speed of 220 mph wishes to fly a course of 270° in a wind of 15 mph from the south. Calculate the heading the pilot should set for this course.

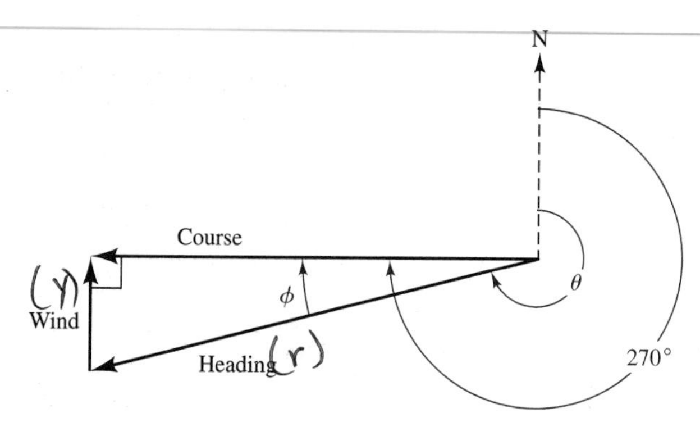

Figure 3.74

SOLUTION Since the wind is from the south and the desired course is due west, the vector sum is the right triangle shown in Figure 3.74. The angle ϕ between the course vector and the heading vector is given by

$$\sin \phi = \frac{15}{220}$$

$$\phi = \boxed{\text{inv}}\ \boxed{\text{sin}}\ \boxed{(}\ 15\ \boxed{\div}\ 220\ \boxed{)}\ \boxed{=}\ \boxed{3.909563525} \approx 3.9^\circ$$

Finally, from the figure we see that the desired heading is

$$\theta = 270^\circ - \phi \approx 270^\circ - 3.9^\circ = 266.1^\circ$$

▲

An airplane's ground speed depends on the wind field in which the airplane is flying. For instance, if a plane with a cruising speed of 200 mph is flying into a 10-mph head wind, its ground speed is 190 mph; if it is flying with a 10-mph tail wind, its ground speed is 210 mph. In general the **ground speed** is the vector sum of the airspeed and the wind speed, as shown in Figure 3.75.

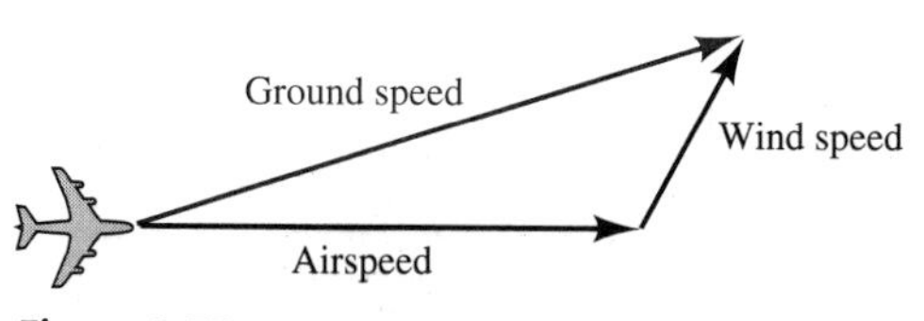

Figure 3.75

Example 3 Consider a round-trip flight from Chicago to Boston, with a one-way airline distance of 870 mi. A light plane having an airspeed of 180 mph is to make the round trip. How many flying hours will it take for the round trip if there is a constant southwest wind* of 23.0 mph? What headings will the pilot use for the two parts of the trip?

* Wind direction is conventionally specified by the direction from which the wind is blowing. Thus a "southwest wind" is a wind blowing from the southwest.

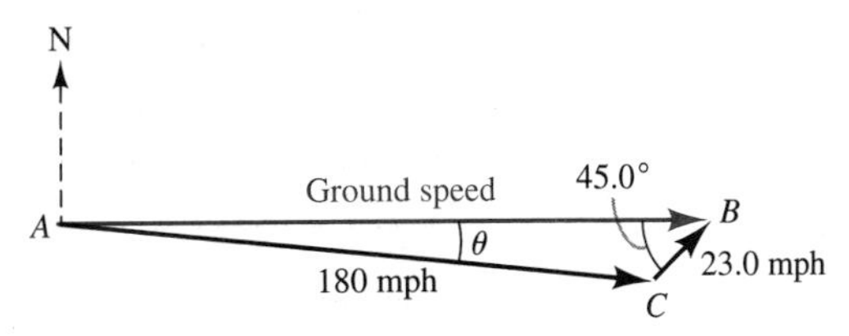

Figure 3.76
Eastbound trip

SOLUTION For a eastbound trip, shown in Figure 3.76, the law of sines is applied to determine θ.

$$\sin\theta = \frac{23.0}{180}\sin 45.0° \approx \boxed{0.090352533}$$

$$\theta \approx \boxed{\text{inv}}\ \boxed{\text{sin}}\ 0.090352533\ \boxed{=}\ \boxed{5.183888382} \approx 5.2°$$

(Note that $\theta = 174.8°$, the complement of $5.2°$, is impossible.) Applying the law of sines again, we determine the ground speed, represented by $\overline{AB}$:

$$\frac{\overline{AB}}{\sin C} = \frac{180}{\sin B}$$

Solving for $\overline{AB}$ with $C \approx 180° - (5.2° + 45°) = 129.8°$, we get

$$\overline{AB} \approx \frac{\sin 129.8°}{\sin 45.0°}\cdot 180 \approx \boxed{195.5730562} \approx 196 \text{ mph}$$

Thus the time required for the eastbound trip is

$$\text{time} = \frac{\text{distance}}{\text{velocity}} \approx \frac{870 \text{ mi}}{196 \text{ mph}} \approx \boxed{4.43877551} \approx 4.44 \text{ hr}$$

For the westbound trip, shown in Figure 3.77, θ is again $5.2°$, and the ground speed is found using the law of sines. Here $C \approx 180° - (5.2° + 135.0°) = 39.8°$, so

$$\overline{AB} \approx \frac{\sin 39.8°}{\sin 135.0°}\cdot 180 \approx \boxed{162.9453273} \approx 163 \text{ mph}$$

The time required is

$$\text{time} \approx \frac{870 \text{ mi}}{163 \text{ mph}} \approx \boxed{5.337423313} \approx 5.34 \text{ hr}$$

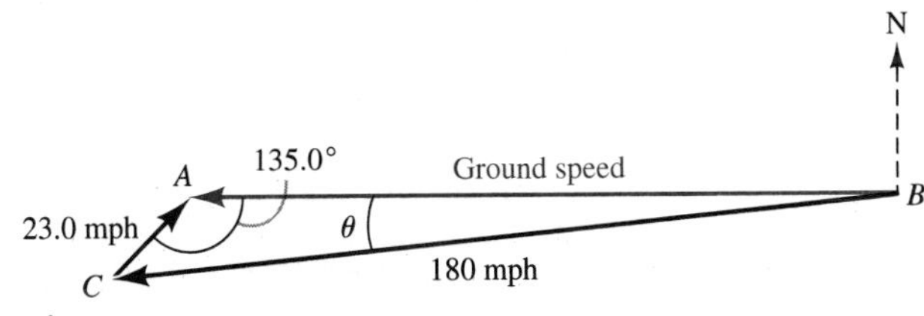

Figure 3.77
Westbound trip

Thus the total time for the round trip is 9.78 hr, or 9 hr 47 min. The heading for the eastbound trip is $90° + 5.2°$, or $95.2°$, and the heading for the westbound trip is $270° - 5.2°$, or $264.8°$. ▲

If the vector quantities to be added are not at right angles, the sum may also be obtained by the method of addition of components discussed in Section 3.6.

Example 4 A ship with a heading of 72° and a speed of 12 knots moves through an ocean current of 2 knots from the north. Calculate the course of the ship.

SOLUTION Figure 3.78 shows the conditions.

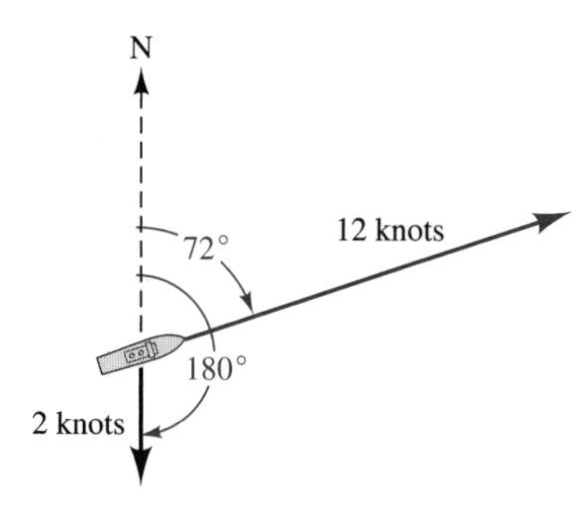

Figure 3.78

Since the heading vector and the current vector are not at right angles, we use the addition-of-components methods to get the course vector. In Figure 3.79 the north and east components of the ship's speed are

$$V_n = 12 \cos 72° \approx \boxed{3.708203933} \approx 3.7 \text{ knots}$$

$$V_e = 12 \sin 72° \approx \boxed{11.4126782} \approx 11.4 \text{ knots}$$

The north and east components of the course vector are then

$$C_n \approx -2 + 3.7 = 1.7 \text{ knots}$$

$$C_e = V_e \approx 11.4 \text{ knots}$$

The course angle is given by

$$\tan \theta = \frac{C_e}{C_n} \approx \frac{11.4}{1.7}$$

$$\theta \approx \boxed{\text{inv}}\ \boxed{\tan}\ \boxed{(}\ 11.4\ \boxed{\div}\ 1.7\ \boxed{)}\ \boxed{=}\ \boxed{81.51839421} \approx 81.5°$$

V_n $|\mathbf{V}| = 12$ knots V_e -2 knots θ C_n C_e

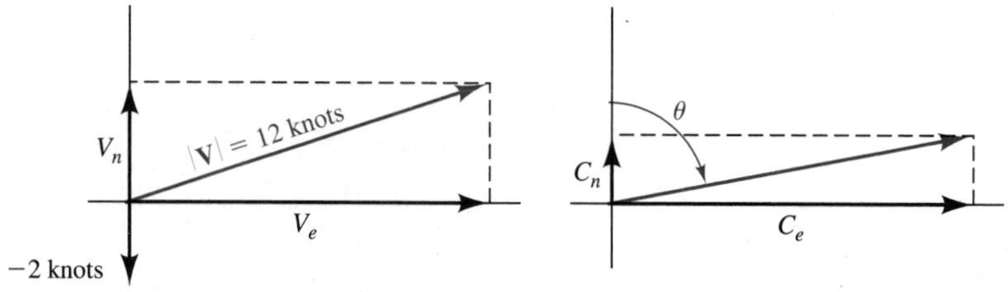

Figure 3.79 ▲

Force Analysis

Perhaps the most common use of vectors is in representing forces. Physicists and mechanical engineers spend considerable time analyzing the effects of forces. An important step in this process is representing force vectors in component form.

Example 5 A physics lab experiment on components of forces is shown schematically in Figure 3.80. Calculate the force parallel to the inclined plane.

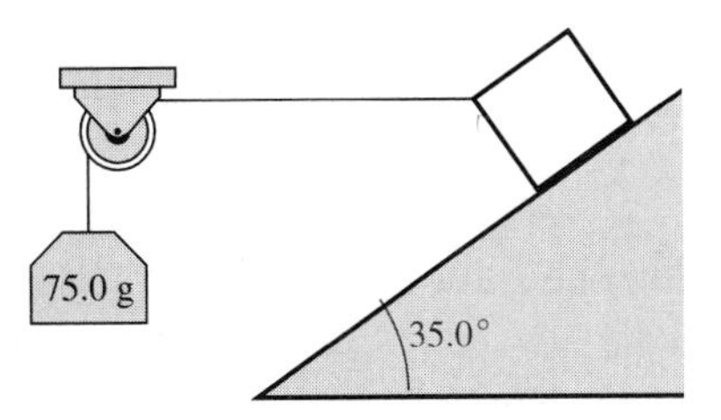

Figure 3.80

Figure 3.81

SOLUTION The vector representation of the 75.0-g force is shown in Figure 3.81. Since we want the component of force parallel to the plane, we must use

$$\cos 35.0° = \frac{F_p}{75.0}$$

Solving for F_p, we get

$$F_p = 75.0 \cos 35.0° \approx \boxed{61.43640332} \approx 61.4 \text{ g}$$

as the parallel component of the applied force. ▲

The weight of an astronaut on the moon is one-sixth his or her weight on earth. This fact has a marked effect on such simple acts as walking, running, jumping, and the like. To study these effects and to train astronauts to work under lunar gravity conditions, scientists at NASA's Langley Research Center designed an inclined plane apparatus to simulate reduced gravity.

The apparatus consists of an inclined plane and a sling that holds the astronaut in a position perpendicular to the inclined plane, as shown in Figure 3.82. The sling is attached to one end of a long cable that runs parallel to the inclined plane. The other end of the cable is attached to a trolley that runs along a track high overhead. This device allows the astronaut to move freely in a plane perpendicular to the inclined plane.

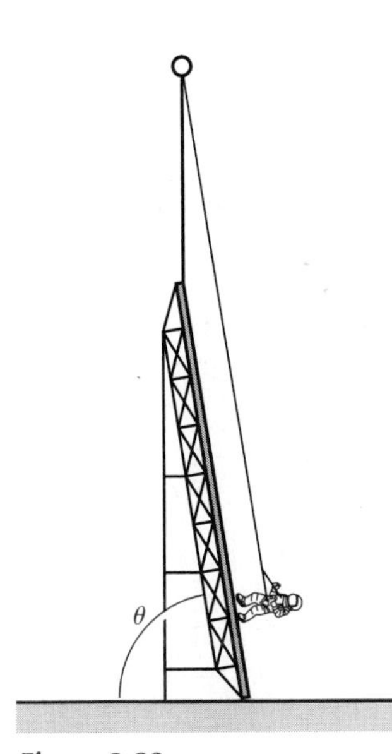

Figure 3.82

Example 6 Make a vector diagram to show the components of the astronaut's weight that are parallel to the plane and those that are perpendicular to the plane. The perpendicular component represents the force exerted by the feet of the astronaut against the plane.

SOLUTION To draw an accurate vector diagram, it is necessary to know that weight is a force that always acts vertically downward—that is, toward the center

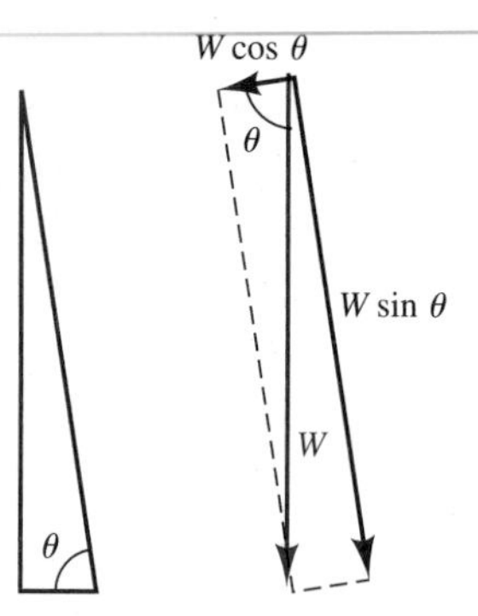

Figure 3.83

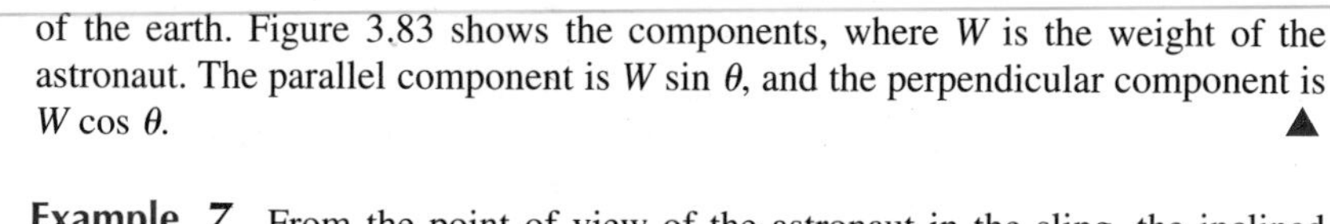

of the earth. Figure 3.83 shows the components, where W is the weight of the astronaut. The parallel component is $W \sin \theta$, and the perpendicular component is $W \cos \theta$. ▲

Example 7 From the point of view of the astronaut in the sling, the inclined plane is the ground, and the downward force against the inclined plane is $W \cos \theta$. What is the value of θ required to simulate lunar gravity?

SOLUTION To simulate lunar gravity we must have $W \cos \theta = W/6$. Thus,

$$\cos \theta = \frac{1}{6}$$

$$\theta \approx \boxed{\text{inv}}\ \boxed{\cos}\ 6\ \boxed{1/x}\ \boxed{=}\ \boxed{80.40593177} \approx 80.4°$$ ▲

The next example deals with the design of an automated spacecraft such as the one shown in Figure 3.84. Surprisingly it is an application of trigonometry that gives the design requirements for construction of the legs of the lander. Since three points determine a plane, our lunar lander is designed with three legs.

Figure 3.84

The spacecraft problem in Example 8 makes use of the fact that an object is said to be in **equilibrium** if the forces acting on the object have a vector sum equal to 0. Thus to be in equilibrium a force in one direction must be balanced by an equal force in the opposite direction. For instance, in Example 6 the component of weight parallel to the plane ($W \sin \theta$) must be balanced by the tension in the cable, and the perpendicular component ($W \cos \theta$) must be balanced by the force exerted by the plane.

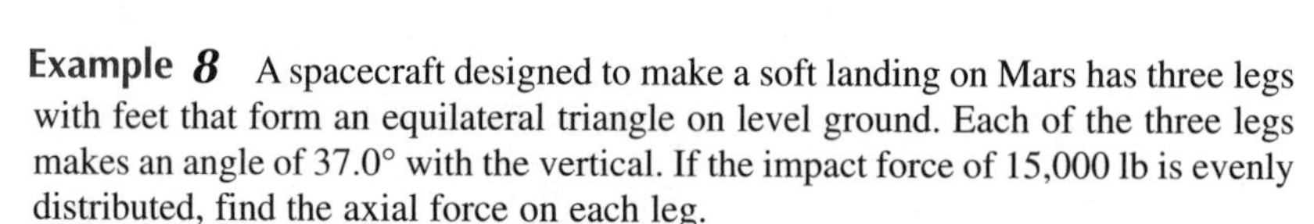

Example 8 A spacecraft designed to make a soft landing on Mars has three legs with feet that form an equilateral triangle on level ground. Each of the three legs makes an angle of 37.0° with the vertical. If the impact force of 15,000 lb is evenly distributed, find the axial force on each leg.

SOLUTION We assume that each of the three legs shares the impact force equally. Thus each leg must withstand a vertical force of 5000 lb. Since the legs are in equilibrium, each leg must have an internal force whose vertical resultant balances the impact force. The force diagram in Figure 3.85 shows that

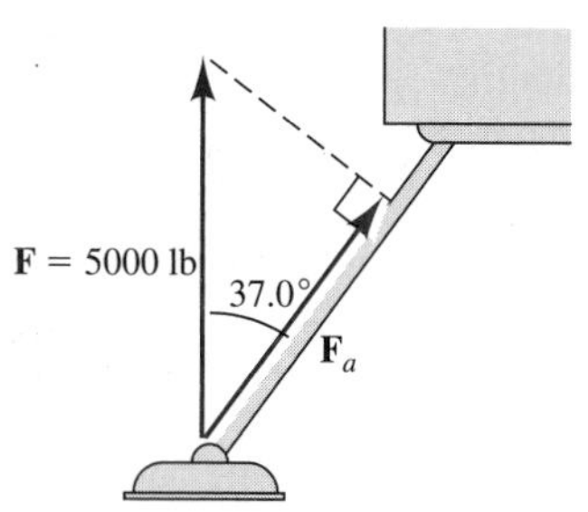

Figure 3.85

$$\cos 37.0° = \frac{\mathbf{F}_a}{5000}$$

$$\mathbf{F}_a = 5000 \cos 37.0° \approx \boxed{3993.17755} \approx 3990 \text{ lb}$$ ▲

Exercises for Section 3.7

1. What are the horizontal and vertical components of the velocity of a ball thrown 100 ft/sec at an angle of 40.0° with the horizontal?
2. A boat that travels at the rate of 5.0 mph in still water is pointed directly across a stream with a current of 3.0 mph. What will be the actual speed of the boat, and in which direction will the boat go?

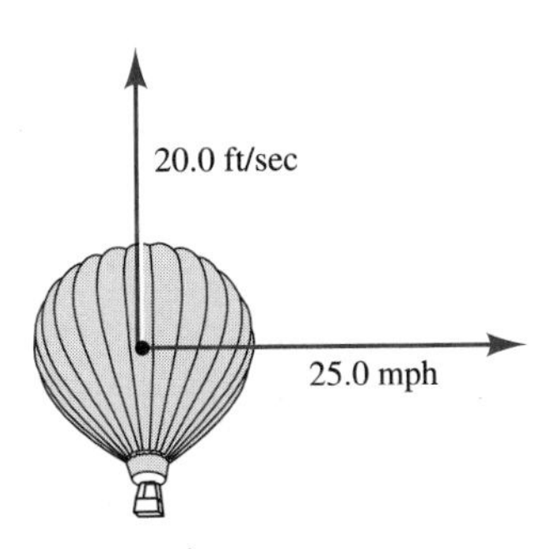

Figure 3.86

3. In which direction must the boat in Exercise 2 be pointed if it is to go straight across the stream?
4. An airplane flying at a speed of 190 knots starts to climb at an angle of 13.0° above the horizontal. How fast is the airplane rising vertically?
5. A balloon rising at the rate of 20.0 ft/sec is being carried horizontally by a wind with a velocity of 25.0 mph. (See Figure 3.86). Find the actual velocity of the balloon and the angle that its path makes with the vertical (60 mph = 88 ft/sec).
6. An object is thrown vertically downward with a speed of 50.0 ft/sec from a plane moving horizontally with a speed of 250 ft/sec. What is the velocity of the object as it leaves the plane?
7. A bullet is fired from a plane at an angle of 20.0° below the horizontal in the direction in which the plane is moving. If the bullet leaves the muzzle of the gun with a speed of 1200 ft/sec and the plane is flying 500 ft/sec, what is the resultant velocity of the bullet?
8. A rocket is launched from a ship at an elevation angle of 42.0° in the direction in which the ship is moving. The speed of the rocket is 540 ft/sec, and that of the boat is 45.0 ft/sec. Find the resultant velocity of the rocket.
9. A plane is headed due north at 300 mph. If the wind is from the east at 50.0 mph, what is the ground speed of the plane? What is its course?
10. A plane is headed due west with an airspeed of 150 mph. If the wind is from the southwest at 35 mph, what is the ground speed of the plane? What is the course of the plane?
11. An airplane with a heading of 140.0° and a cruising speed of 185 mph is flying in an 18.0-mph wind field from due south. Calculate the ground speed of the airplane.
12. Calculate the ground speed of the airplane in Exercise 11 if the wind field is from due west.
13. A pilot is preparing a flight plan from the airport in Juliette to Denver, which is 280 miles due north. If the airplane cruises at 160 mph and there is a constant 18.0-mph wind from the west, what heading should the pilot fly? How long will it take to make the trip?
14. If the airplane in Exercise 13 makes the return trip in a west wind of 32.0 mph, determine the heading the pilot should fly.
15. Find the horizontal and vertical components of the force in Figure 3.87.
16. What are the horizontal and vertical components of the force in Figure 3.87 if $|\mathbf{F}| = 250$ lb and $\theta = 21.7°$?
17. The weight of an object is represented by a vector acting vertically downward. (See Figure 3.88.) If a 15.3-lb block rests on a plane inclined at 28.4° above the horizontal, what is the component of force acting perpendicular to the plane?
18. Find the component of force acting parallel to the plane in Exercise 17.
19. A 75.0-lb block rests on an inclined plane. If the block exerts a force of 67.0 lb perpendicular to the plane, what is the angle of inclination of the plane?
20. A horizontal force of 750 lb is applied to a block resting on a plane inclined at 17.9° above the horizontal. Find the component of the force parallel to the plane.

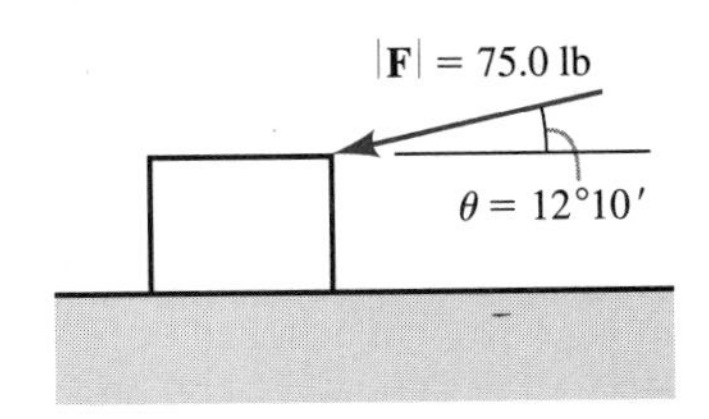

Figure 3.87

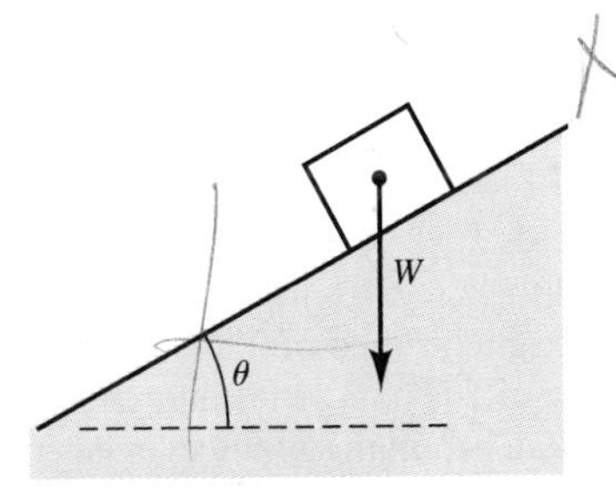

Figure 3.88

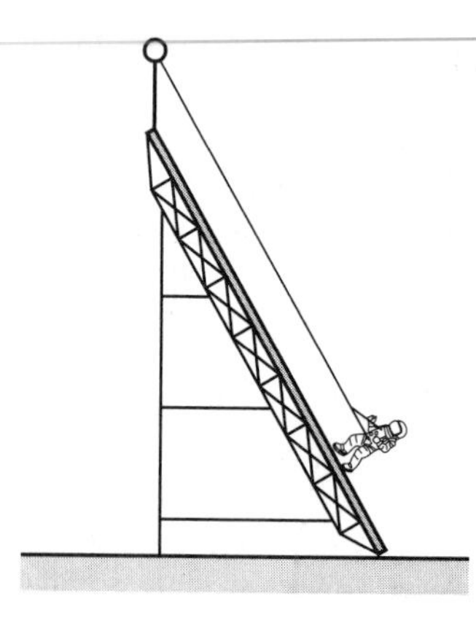

Figure 3.89

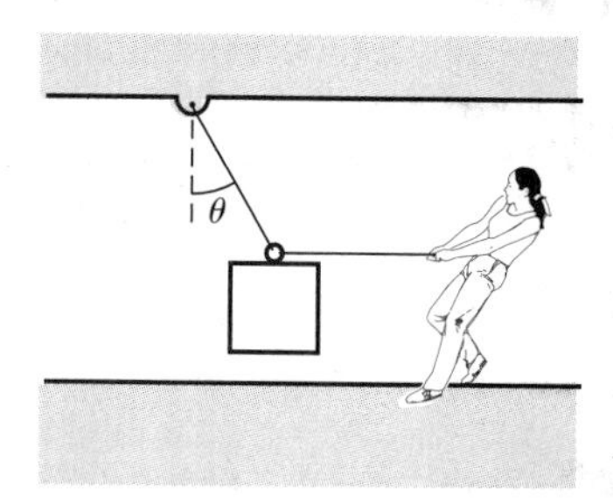

Figure 3.90

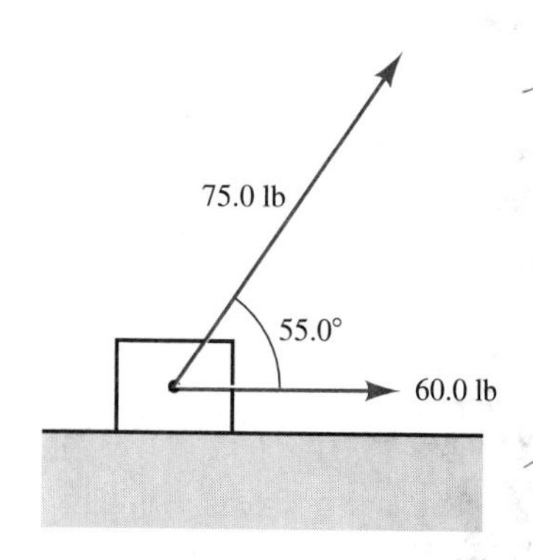

Figure 3.91

21. What is the angle of the inclined plane needed to simulate the gravity of an asteroid whose gravity is $\frac{1}{8}$ that of earth in order to train a 200-lb astronaut?

22. If the inclined plane used to simulate gravity were inclined at 60.0° (see Figure 3.89), what percentage of the astronaut's weight would bear against the plane?

23. Suppose that each of the three legs of a lunar lander will withstand an axial load of 1200 lb. What is the maximum angle that the legs can make with the vertical and still support a total impact force on the lander of 4500 lb? (See Example 8.)

24. Suppose that a lunar lander is to be designed for a total impact force of 3500 lb and that each of the three legs makes an angle of 28.0° with the vertical. Find the axial force each leg is designed to withstand.

25. In Figure 3.90, an object weighing 120 lb hangs at the end of a rope. The object is pulled sideways by a horizontal force of 30.0 lb. What angles does the rope make with the vertical? (*Hint:* As noted in Example 6, weight is a force that is always considered to be acting vertically downward, and the system is in equilibrium.)

26. A force of 60.0 lb acts horizontally on an object. Another force of 75.0 lb acts on the object at an angle of 55.0° with the horizontal. What is the resultant of these forces? (See Figure 3.91.)

27. Find the resultant force in Figure 3.91 if the 75.0-lb force is replaced by a 115-lb force.

28. If two forces, one of 500 lb and the other of 400 lb, act from a point at an angle of 58.3° with each other, what is the size of the resultant?

29. Two forces, one of 75.0 lb and the other of 100 lb, act at a point. If the angle between the forces is 60.0°, find the magnitude and direction of the resultant force. Give the direction as an angle between the resultant and the 100-lb force.

30. The airspeed of a plane is 400 mph, and there is a 75.0-mph wind from the northeast at a time when the heading of the plane is due east. Find the ground speed and the direction of the path of the plane.

31. Consider a flight from Miami to New York to be along a north-south axis with an airplane distance of approximately 1000 mi. A jet with an airspeed of 500 mph makes the round trip. If there is a constant northwest wind of 100 mph, how long will the round trip take? What headings will the pilot use for the two parts of the trip?

32. A plane is flying on a heading of 153° at an airspeed of 150 knots. There is a 12-knot wind blowing from a direction of 28.5°. Calculate the ground speed and the resulting course of the plane.

33. Calculate the heading and airspeed required for a pilot to fly to an airport 52 miles due east of Des Moines in 15 minutes if there is a 18-mph wind blowing from a direction of 200°.

34. An oil tanker on a heading of 250° is advised to change course to a heading of 235° to avoid a large iceberg. After following a heading of 235° for eight hours, the captain changes course to a heading of 274° until the ship reaches its original course, a point that is 150 miles from beginning of the avoidance manuever. Calculate the additional distance the ship traveled to avoid the iceberg.

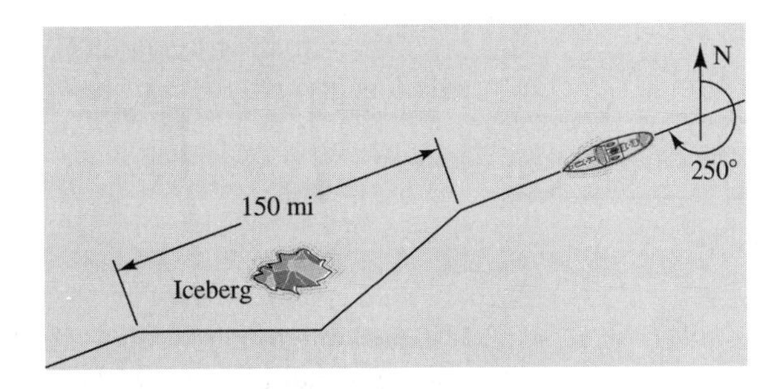

35. A line of flight distance from Chicago to St. Louis is 290 miles on a heading of 210°. Because of a thunderstorm the pilot decides to take a heading of 250° from Chicago and then change to a heading of 180° into St. Louis. Calculate the additional mileage flown by the plane to avoid the thunderstorm.

Figure 3.92(a) shows a schematic diagram of an electric circuit containing an alternating current generator, a resistance R, and an inductive reactance X_L, connected in series. A quantity Z, called the impedance of the circuit and used in a.c. circuit theory, is related to the resistance and the inductive reactance by the right-triangle relationship shown in Figure 3.92(b). The angle θ is called the phase angle. The following exercises illustrate the use of the impedance triangle in a.c. circuit theory. In Exercises 36–41 solve for the missing components in the impedance triangle.

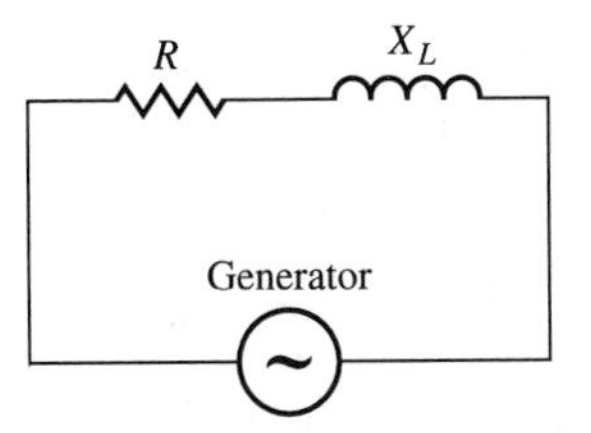

Figure 3.92

36. $R = 30.0$, $\theta = 60.0°$

37. $R = 30.0$, $Z = 60.0$

38. $Z = 100$, $\theta = 20.0°$

39. $R = 200$, $X_L = 100$

40. $Z = 1000$, $X_L = 800$

41. $X_L = 25.0$, $\theta = 30.0°$

3.8 Area Formulas

Recall that the area of a triangle is equal to one-half the product of any base and the corresponding altitude. Using the law of cosines and the law of sines, we can derive some equivalent expressions for area for which the height need not be specifically computed. In this section we will examine three such formulas.

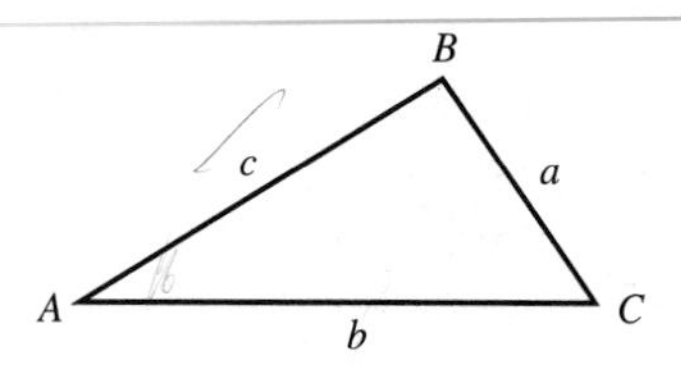

Consider the triangle in Figure 3.93. The altitude from the vertex B to side b is given by $h = c \sin A$. Hence the area S is given by

$$S = \tfrac{1}{2}bc \sin A \qquad (3.7)$$

It is possible to derive equivalent formulas for each vertex. In general *the area of a triangle is equal to one-half the product of the lengths of any two sides and the sine of the included angle.*

Example 1 Find the area of a triangle with $a = 3.8$ in., $b = 5.1$ in., and $C = 48°$.

SOLUTION Since C is the angle between sides a and b, the area is given by

$$S = \tfrac{1}{2}(3.8)(5.1)\sin 48° \approx \boxed{7.201073359} \approx 7.2 \text{ in.}^2$$ ▲

The expression for area may be altered further by using the law of sines. Since $c = b \sin C/\sin B$, we can substitute this value into Equation (3.7) to get

$$S = \frac{b^2}{2}\frac{\sin A \sin C}{\sin B} \qquad (3.8)$$

Thus the area may be calculated if one of the sides and two of the angles are given (since the third angle may then be found easily).

Example 2 Find the area of a triangle with two angles and an included side of 30°, 45°, and 2.7 cm, respectively.

SOLUTION The remaining angle is 105°. Hence

$$S = \frac{(2.7)^2}{2}\frac{\sin 30° \sin 45°}{\sin 105°} \approx \boxed{1.334162597} \approx 1.3 \text{ cm}^2$$ ▲

We have saved the most important area formula for last. For the most part, area formulas such as (3.7) and (3.8) or the more well-known formula of one-half the base times the altitude to that base is unusable because of the impracticality of measuring the angles of the triangle or the altitude. We now derive a formula for the area in terms of the side lengths a, b, and c. This formula was derived around 75 A.D. by Heron of Alexandria. The formula is of course called **Heron's formula**.

Recall from Section 3.2 that three sides uniquely determine a triangle, although a word of caution is necessary: Not *any* three numbers may be used. For example, there does not exist a triangle with sides 1, 1, and 3. This will be made clearer when we derive Heron's formula. The derivation we give is not due to Heron; his ingenious proof can be found in Professor William Dunham's book *Journey Through Genius*.

Our derivation comes from Equation (3.7) by first squaring both sides to obtain:

$$\begin{aligned} S^2 &= \tfrac{1}{4}b^2c^2\sin^2 A \\ &= \tfrac{1}{4}b^2c^2(1 - \cos^2 A) \\ &= (\tfrac{1}{2}bc)(\tfrac{1}{2}bc)(1 - \cos A)(1 + \cos A) \end{aligned}$$

Using the law of cosines, we have

$$\begin{aligned} \tfrac{1}{2}bc(1 + \cos A) &= \tfrac{1}{2}bc\left(1 + \frac{b^2 + c^2 - a^2}{2bc}\right) \\ &= \frac{2bc + b^2 + c^2 - a^2}{4} \\ &= \frac{(b + c)^2 - a^2}{4} \\ &= \frac{(b + c - a)(b + c + a)}{4} \end{aligned}$$

Similarly, we could obtain

$$\tfrac{1}{2}bc(1 - \cos A) = \frac{(a - b + c)(a + b - c)}{4}$$

Therefore the expression S^2 becomes

$$S^2 = \frac{(b + c - a)(a + b + c)(a - b + c)(a + b - c)}{16}$$

$$S = \tfrac{1}{4}\sqrt{(b + c - a)(a + b + c)(a - b + c)(a + b - c)}$$

Heron expressed this formula in terms of the semiperimeter, s: $s = \tfrac{1}{2}(a + b + c)$, but we will instead use the more familiar concept of the perimeter, $P = a + b + c$ to obtain:

Heron's Formula $\quad S = \tfrac{1}{4}\sqrt{P(P - 2a)(P - 2b)(P - 2c)}$ **(3.9)**

Example 3 Find the area of the triangle whose sides are 2, 2, and 3.

SOLUTION Since the perimeter is 7, we have

$$S = \tfrac{1}{4}\sqrt{7(7-4)(7-4)(7-6)} = \tfrac{1}{4}\sqrt{63} = \tfrac{3}{4}\sqrt{7}$$

$$S \approx \boxed{1.984313483} \approx 1.98 \text{ square units}$$

▲

Note that Equation (3.9) clarifies the inherent limitations on the lengths of the sides of a triangle. For example, there is no triangle whose sides are 1, 2, and 3.

Exercises for Section 3.8

In Exercises 1–10 find the area of the triangles with the given measurements.

1. $a = 15.0$, $b = 5.00$, $C = 30.0°$
2. $a = 12.0$, $b = 10.0$, $c = 5.00$
3. $a = 10.0$, $A = 60.0°$, $B = 45.0°$
4. $a = 10.0$, $A = 120°$, $B = 30.0°$
5. $a = 3.0$, $b = 4.0$, $c = 5.0$
6. $a = 1.0$, $b = 4.0$, $c = 5.0$
7. $a = 1.22$, $b = 1.39$, $c = 2.51$
8. $b = 4.30$, $B = 37.2°$, $C = 68.3°$
9. $c = 2.42$, $A = 108.3°$, $B = 31.4°$
10. $b = 25.6$, $A = 100.3°$, $B = 30.6°$
11. Why is it impossible to express the area of a triangle in terms of only its angles?
12. How are the areas of similar triangles related?
13. Workers need a triangular steel plate with sides of 12.0, 16.0, and 24.0 in. What is the area of such a plate?
14. If the plate mentioned in Exercise 13 is made of sheet steel weighing 0.90 oz/in.^{2} of surface, how much does the plate weigh?
15. A farmer has a field shaped as shown in Figure 3.94. Find the area of the field.
16. A home is built on a lot in the shape of a quadrilateral, with measurements 35, 100, 28, and 83 m. If the diagonal measures 110 m, what is the area of the lot in hectares? (1 hectare = 10,000 m^2.) (See Figure 3.95.)
17. Find the perimeter of a triangle if two of the sides are 100 m and 150 m and the area is 600 m^2.
18. Find the area of a quadrilateral whose sides are 3, 5, 6, and 4 m if the angle between the sides of length 3 and 5 is 100°.
19. Find all parts of a triangle whose area is 25.0 m^2 and two of whose sides are 15.3 and 11.3 m.
20. Find all possible triangles with an area of 96.5° m^2 and angles of 68.2° and 58.3°.
21. A solar reflector is composed of 36 triangular sections, the lengths of whose sides are 6.2, 6.2, and 1.1 m. Find the total area of the reflector.
22. A triangular mirror used in a solar furnace has sides of lengths 25, 25, and 3.5 cm. What is the area of the mirror?

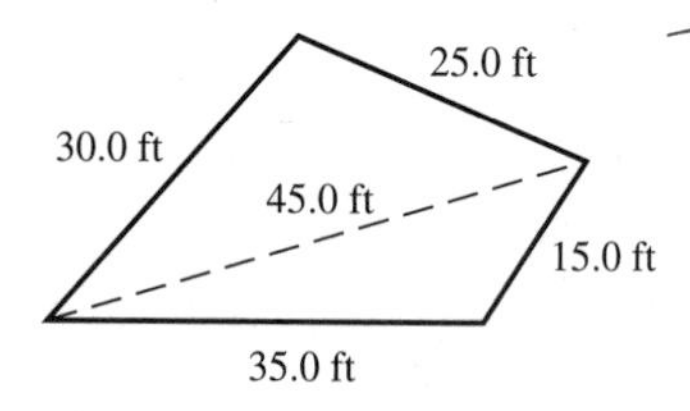

Figure 3.94

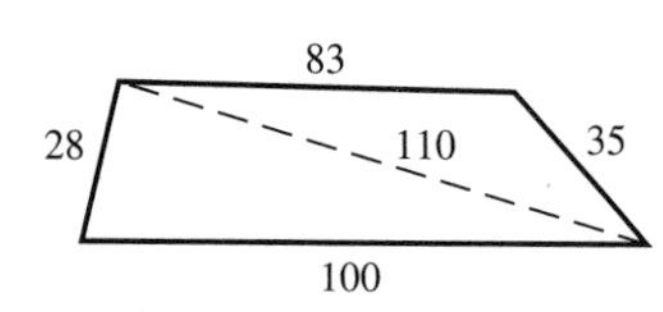

Figure 3.95

Key Topics for Chapter 3

Define and/or discuss each of the following.

Right Triangles
Law of Cosines
Law of Sines
Ambiguous and Impossible Cases
Solution to the General Triangle
Vectors
Components of a Vector
Resultant
Course
Heading
Area of a Triangle
Heron's Formula

Review Exercises for Chapter 3

1. Solve the right triangle ABC if $A = 32°$ and $a = 3.0$.
2. Solve the right triangle ABC if $A = 17°23'$ and $b = 5.8$.
3. Solve the right triangle ABC if $a = 29.0$ and $c = 41.0$ (c is the hypotenuse).
4. Solve the right triangle ABC if $b = 7.50$ and $c = 11.3$.
5. Find the horizontal and vertical components of $\mathbf{F}$ if $|\mathbf{F}| = 50.0$ lb and $\theta = 36.7°$.
6. Find the horizontal and vertical components of $\mathbf{F}$ if $|\mathbf{F}| = 8.00$ lb and $\theta = 17.9°$.
7. Find the resultant force for components $F_x = 43.0$ lb and $F_y = 72.0$ lb.
8. Find the resultant force for components $F_x = 9.4$ lb and $F_y = 3.7$ lb.
9. An engineer's drawing of a component of a trimming die is shown in Figure 3.96. Find the measure of angle θ.

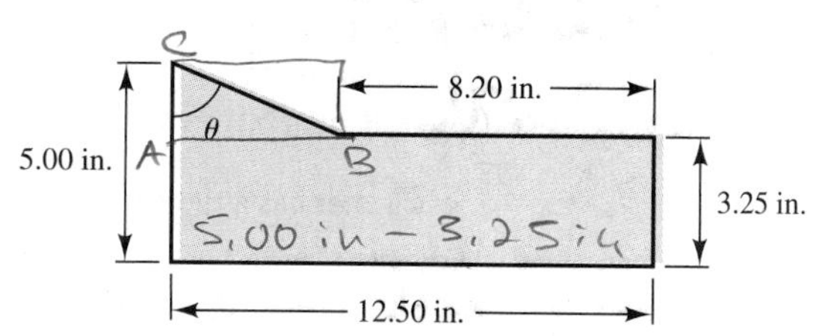

Figure 3.96

10. An airplane heading due east at 350 mph experiences a 40-mph crosswind out of the north. In what direction is the plane moving relative to an observer on the ground?
11. A tunnel through a mountain ascends at an angle of $5°25'$ with the horizontal. If the tunnel is 5000 ft long, what is its vertical rise?
12. Eight holes are equally spaced on the circumference of a circle. If the center-to-center distance between the holes is 3.8 cm, what is the radius of the circle? (See Figure 3.18 on page 103.)
13. An ant starts at the origin of the coordinate plane and travels on a straight line that makes an angle of 23.0° with the positive x-axis. What are the ant's coordinates after it has traveled 15.0 in.?

14. The angle of elevation to a weather balloon from one tracking station is 30°, and from another it is 45°. If the balloon is at an altitude of 50,000 ft, how far apart are the two tracking stations? Assume that the balloon and the tracking station are in the same vertical plane.

In Exercises 15–19 use the law of cosines to solve each triangle or show that no triangle exists.

15. $A = 29.0°, b = 17.0, c = 28.0$

16. $a = 7.80, c = 9.10, B = 38°18'$

17. $a = 11.2, b = 7.90, c = 15.4$

18. $a = 210, b = 175, c = 78.0$

19. $a = 23.0, b = 5.88, c = 17.8$

In Exercises 20–25 use the law of sines to solve each triangle or show that no triangle exists.

20. $A = 39°12', B = 17°42', c = 20.8$

21. $C = 27.6°, A = 112.2°, a = 3120$

22. $b = 75.0, B = 11.5°, C = 40.0°$

23. $a = 15.0, b = 12.0, A = 25.0°$

24. $A = 42.0°, c = 25.0, a = 17.0$

25. $B = 63°50', a = 23.0, b = 12.0$

In Exercises 26–28 find the area of the indicated triangle.

26. $a = 15.6, b = 19.2, c = 27.8$

27. $b = 7.53, c = 3.98, A = 42.5°$

28. $A = 72°, B = 37°, b = 29$

29. Solve the triangle $a = 9.06$, $c = 6.68$, $B = 138.0°$ and find its area.

30. Two light planes leave Kennedy airport at the same time, one flying a course of 265.0° at 175 mph and the other flying a course of 300.0° at 190 mph. How far apart will the two planes be at the end of $2\frac{1}{2}$ hr?

31. From a certain point the angle of elevation to the top of a tower that stands on level ground is 30.0°. At a point 100 m nearer the tower, the angle of elevation is 58.0°. How high is the tower?

32. From a helicopter the angles of depression to two successive milestones on a straight level road below are 15.0° and 30.0°. Find the altitude of the helicopter.

33. In measuring the height of a bell tower with a transit set 5.00 ft above the ground, a student finds that from one point the angle of elevation to the top of the tower is 45.0°. After moving the transit 50.0 ft in a straight line toward the tower, the student finds the angle to be 60.0°. Find the height of the bell tower.

34. A plane flies due east from Atlanta at 250 mph for 1 hr and then turns and flies 35.0° north of east at 300 mph for 1.5 hr. How far is the plane from Atlanta at the end of 2.5 hr?

35. An airplane heading due east at 375 mph experiences a 31.5-mph crosswind out of the northeast. In what direction is the plane moving relative to an observer on the ground?

36. How much would it cost to lay a 4-in.-thick slab of concrete in the shape of a triangle whose sides are 15, 20, and 22 ft if the cost of the concrete is $\$30/yd^3$?

37. A 25-ft ladder leans against a building built on a slope. If the foot of the ladder is 11 ft from the base of the building and the angle between the side of the building and the ground is 128°, how high up the side of the building does the ladder reach? (See Figure 3.97.)

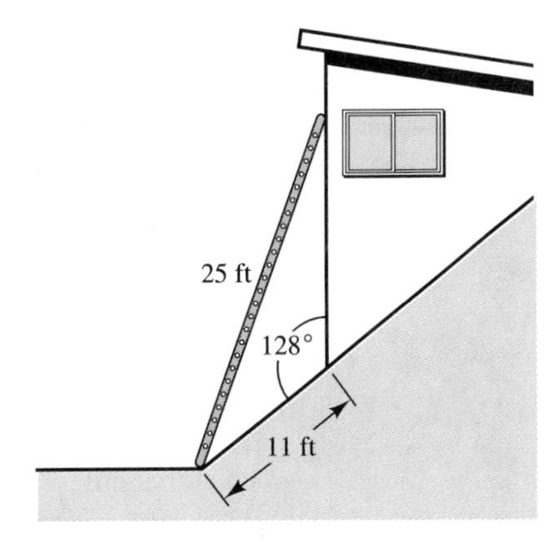

Figure 3.97

Practice Test Questions for Chapter 3

In Exercises 1–6 answer *true* or *false*.

1. A triangle with sides of 2, 3, and 4 is a right triangle.
2. The sum of the acute angles of a right triangle is 100°.
3. Only one angle of a triangle can be obtuse.
4. The law of sines can be used to find the angles of a triangle when three sides are given.
5. More than one triangle is possible when two sides and an angle opposite one of the sides are given.
6. A scalar quantity can be completely described by a single number.

In Exercises 7–10 fill in the blank to make the statement true.

7. The side opposite the 90° angle in a right triangle is called the ____________.
8. A(n) ____________ triangle is a triangle that does not have a right angle.
9. It requires both a magnitude and a direction to describe a(n) ____________ quantity.
10. A triangle whose two sides and the included angle are known can be solved using the law of ____________.

Solve the stated problem in the following exercises. Show all your work.

11. Solve the right triangle ABC if $A = 34.2°$ and $b = 175$.
12. Solve the right triangle ABC if $a = 12.1$ and $c = 33.7$.
13. Solve the triangle ABC if $a = 205$, $b = 189$, and $c = 122$.
14. Solve the triangle ABC if $b = 4.00$, $c = 5.00$, and $A = 60.0°$.
15. Solve the triangle ABC if $B = 73°20'$, $C = 15°10'$, and $c = 25.0$.
16. A straight tunnel through a mountain rises 252 ft. If the tunnel is 3000 ft long, calculate the angle of elevation of the roadway.

17. A block weighing 125 lb rests on a plane inclined at an angle of 10.3°. Calculate the components of weight parallel and perpendicular to the inclined plane.

18. From the top of a mountain, the angles of depression to two successive milestones on the plane below are 15.4° and 28.9°. Calculate the height of the mountain in feet. (Assume the milestones and the top of the mountain are in the same vertical plane.)

19. A triangular lot has dimensions of 125 ft, 185 ft, and 270 ft. **(a)** Find the largest angle. **(b)** Find the area of the lot.

20. A force of 25.0 lb at an angle of 5.30° above the horizontal is applied at a point. A second force of 30.0 lb at an angle of 20.7° above the horizontal is applied at the same point. Calculate the direction and magnitude of the resultant force.

Radian Measure

Many applications of the trigonometric functions require the use of an angle measurement called the radian. The radian measure occurs throughout the use of trigonometry and is the unit of choice in more advanced mathematics such as calculus. Section 4.1 shows how the radian is related to the degree and how by a simple flip of a switch or press of a button on a calculator you can be in "radian mode." Sections 4.2 and 4.3 give some applications in which the radian measure of angles is used to compute angular displacement and angular velocity.

4.1 The Radian

In Chapter 1 we defined the degree measure for comparing angles of different sizes. In this section we will define another unit of angular measure called the **radian**. Although less familiar than the degree, the radian is used extensively in advanced mathematics and is the standard unit of angular measurement in the International System.

Figure 4.1
One radian

> **DEFINITION 4.1** One **radian** is the measure of an angle whose vertex is at the center of a circle and whose sides subtend an arc on the circle equal in length to the radius of the circle. (See Figure 4.1.)

Thus the radian measure of a central angle can be found by determining how many times the length of the radius r is contained in the corresponding circular arc of length s. From this definiton we can write:

$$(\text{measure of } \theta \text{ in radians}) = \frac{\text{length of arc on the circle, } s}{\text{radius of circle, } r}$$

or, in abbreviated form,

$$\theta = \frac{s}{r} \tag{4.1}$$

From Definition 4.1 we can make two important observations:

- The radian measure is independent of the size of the circle.
- Since the units of the arclength and the radius are the same, they cancel, and hence *the radian measure of an angle is a pure number.* (We sometimes say that the radian is "unitless.")

To establish the relationship between the degree measure and the radian measure, we observe that the circumference, C, of a circle of radius r is given by $C = 2\pi r$. The number of radians in one circumference is thus $C/r = 2\pi$. Since there are 360° in one circumference, it follows that there are 2π radians in 360°; that is,

$$2\pi \text{ radians} = 360°$$

from which we get the following important conversion formulas.

$$\begin{aligned} 1 \text{ degree} &= \frac{\pi}{180} \text{ radians} \approx 0.0175 \text{ radian} \\ 1 \text{ radian} &= \frac{180}{\pi} \text{ degrees} \approx 57.3 \text{ degrees} \end{aligned} \tag{4.2}$$

Sometimes Formulas (4.2) are combined as

$$\frac{\text{degree of measure of an angle}}{180°} = \frac{\text{radian measure}}{\pi}$$

RULE 1 To convert from radian measure to degree measure, multiply the radian measure by $180/\pi$.

RULE 2 To convert from degree measure to radian measure, multiply the degree measure by $\pi/180$.

The next two examples show how to use these rules to convert from degrees to radians and vice versa.

Example 1 Express **(a)** 60°, **(b)** 225°, and **(c)** 24.8° in radian measure.

SOLUTION

(a) $60° = 60\left(\dfrac{\pi}{180}\right)$ radians $= \dfrac{\pi}{3}$ radians

(b) $225° = 225\left(\dfrac{\pi}{180}\right)$ radians $= \dfrac{5\pi}{4}$ radians

(c) $24.8° = 24.8\left(\dfrac{\pi}{180}\right)$ radians $\approx$ **0.432841654** ≈ 0.433 radian

When the radian measure is a convenient multiple of π, you will usually find it beneficial *not* to convert it to a decimal fraction. ▲

Example 2 Express **(a)** $\dfrac{\pi}{6}$ radians, **(b)** $\dfrac{3\pi}{4}$ radians, and **(c)** 1.130 radians in degrees.

SOLUTION

(a) $\dfrac{\pi}{6}$ radians $= \left(\dfrac{\pi}{6}\right)\left(\dfrac{180}{\pi}\right)$ degrees $= 30°$

(b) $\dfrac{3\pi}{4}$ radians $= \left(\dfrac{3\pi}{4}\right)\left(\dfrac{180}{\pi}\right)$ degrees $= 135°$

(c) 1.130 radians $= 1.130\left(\dfrac{180}{\pi}\right) \approx$ **64.74423085** $\approx 64.7°$ ▲

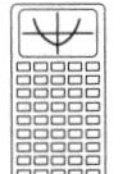

Some calculators have a d↔r key that allows for direct conversion from degrees to radians and from radians to degrees. Thus the entry 180 followed by d↔r will yield 3.141592654 radians. If you do not have a d↔r key, you will have to multiply by $\pi/180$ to convert from degrees to radians and by $180/\pi$ to convert from radians to degrees.

Most calculators have a special button for the number π. If yours does not have a π button, you can use 3.1416 as an approximation, but your answers may be slightly different from those given in the answer section.

Table 4.1 is a conversion table showing frequently occurring angles with both their degree measure and their exact radian measure. *You should memorize the entries in this table so that you don't have to use a calculator to make the approximate conversion.*

Table 4.1 Table of degree and radian measures for commonly occurring angles

Angle in Degrees	Angle in Radians
0	0
30	$\pi/6$
45	$\pi/4$
60	$\pi/3$
90	$\pi/2$
120	$2\pi/3$
135	$3\pi/4$
150	$5\pi/6$
180	π
270	$3\pi/2$
360	2π

The word *radian* is understood; it need not be written. This is not the case with degree measurement; its units must always be included. The next example emphasizes the difference between these two angular measures.

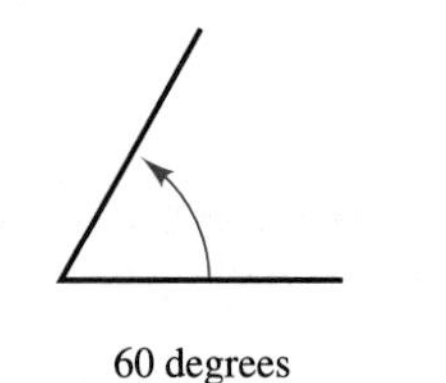

60 degrees

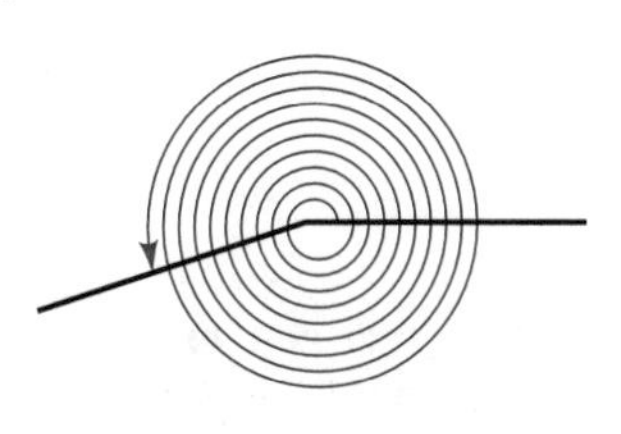

60 radians

Figure 4.2

Example 3 Compare an angle of 60° with one of 60 radians. (See Figure 4.2.)

SOLUTION The angle of 60 radians is obtained by 9 revolutions of the terminal side (each revolution being 2π radians) plus an additional 3.45 radians. To obtain the value 3.45, divide 60 by 2π; the remainder is 0.549, which when multiplied by 2π is 3.45. Thus the angle of 60 radians is coterminal with an angle of 3.45 radians. ▲

Trigonometric Functions for Angles in Radians

The six trigonometric functions may be found for angles measured in radians by using a calculator. When using a calculator, remember to use the radian mode. The functional values in the following example were evaluated using a calculator.

Example 4 Evaluate **(a)** sin 1.38, **(b)** sec 3, and **(c)** $\tan \frac{1}{2}$.

SOLUTION

(a) $\sin 1.38 \approx \boxed{0.98185353}$

(b) $\sec 3 \approx \boxed{-1.01010867}$ (*Note:* $\sec 3 = 1/\cos 3$.)

(c) $\tan \frac{1}{2} \approx \boxed{0.54630249}$ ▲

Example 5 Find angle θ, where $0 \le \theta < 2\pi$ radians, if **(a)** $\sin \theta = -0.5113$ and $\cos \theta > 0$ and **(b)** $\sec \theta = -2.2332$ and $\tan \theta > 0$. Round off to three decimal places.

SOLUTION

(a) We note that θ is a fourth-quadrant angle, since $\sin\theta < 0$ and $\cos\theta > 0$. The reference angle for θ is the acute angle α whose sine is 0.5113. Using a calculator, we get

$$\alpha = \boxed{\textbf{inv}}\ \boxed{\textbf{sin}}\ 0.5113\ \boxed{=}\ \boxed{\textbf{0.53669679}}\ \text{radian}$$

The desired fourth-quadrant angle (see Figure 4.3) is

$$\theta_4 \approx 2\pi - 0.53669679 \approx 5.746 \text{ radians}$$

(b) In this case angle θ is a third-quadrant angle, since $\sec\theta < 0$ and $\tan\theta > 0$. The reference angle for θ is the acute angle α whose secant is 2.2332. Using a calculator, we get

$$\alpha = \boxed{\textbf{inv}}\ \boxed{\textbf{cos}}\ 2.2332\ \boxed{\mathbf{1/x}}\ \boxed{=}\ \boxed{\textbf{1.106506492}}\ \text{radians}$$

The value of θ (see Figure 4.4) is then

$$\theta_3 \approx \pi + 1.106506492 \approx 4.248 \text{ radians}$$

▲

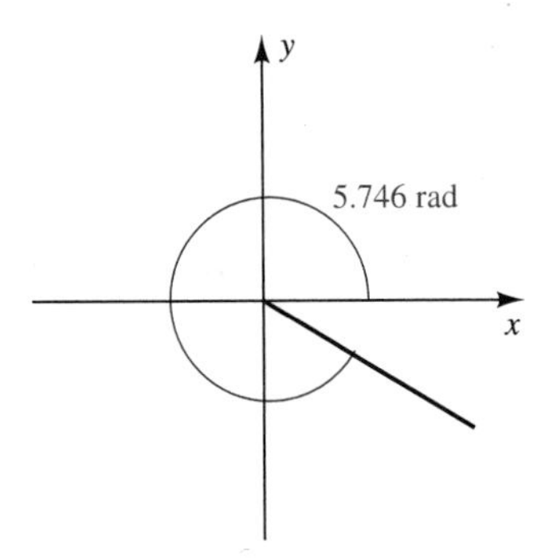

Figure 4.3

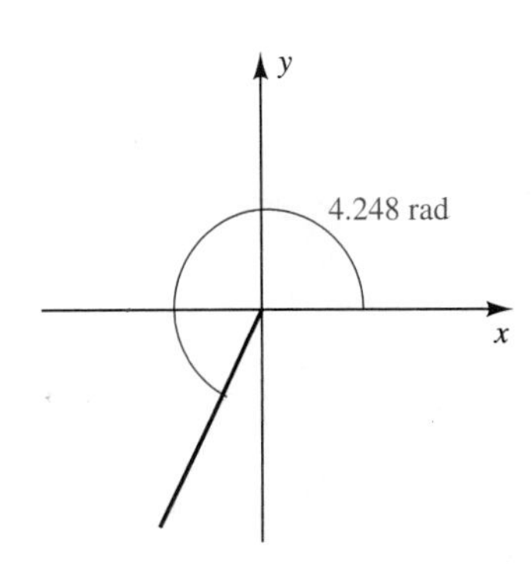

Figure 4.4

Example 6 The interior angles of a triangle can be stated in radians as well as in degrees. In Figure 4.5, a = 17.5 cm, B = 0.95 radian, and C = 1.22 radians. Find angle A and sides b and c.

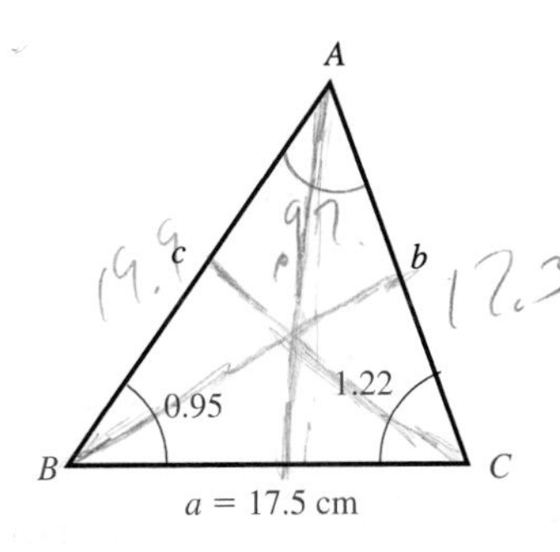

Figure 4.5

SOLUTION The sum of the interior angles of a triangle is 180°, or π radians. Using 3.14 to approximate π, we have

$$A \approx 3.14 - (0.95 + 1.22) = 0.97 \text{ radian}$$

Now, using the law of sines and a calculator, we have

$$\frac{b}{\sin 0.95} = \frac{17.5}{\sin 0.97} \quad \text{or} \quad b = \frac{17.5 \sin 0.95}{\sin 0.97}\ \boxed{=}\ \boxed{\textbf{17.25665884}} \approx 17.3 \text{ cm}$$

$$\frac{c}{\sin 1.22} = \frac{17.5}{\sin 0.97} \quad \text{or} \quad c = \frac{17.5 \sin 1.22}{\sin 0.97}\ \boxed{=}\ \boxed{\textbf{19.92304930}} \approx 19.9 \text{ cm}$$

▲

Exercises for Section 4.1

1. What is the difference in radian measure of two coterminal angles?
2. What is the radian measure of a right angle? of a straight angle?

In Exercises 3–20 express each angle in radian measure. Express the measure in multiples of π when convenient.

3. $75°$	**4.** $-30°$	**5.** $480°$	**6.** $210°$
7. $-240°$	**8.** $42°$	**9.** $95°$	**10.** $1485°$
11. $750°$	**12.** $-300°$	**13.** $92.1°$	**14.** $105.7°$
15. $0.092°$	**16.** $-34°39'$	**17.** $253°36'$	**18.** $311°48'$
19. $0°27'$	**20.** $400°40'$		

In Exercises 21–28 give the radian measure of an angle in standard position between $-\pi$ and π whose terminal side passes through the given point.

21. $(2, 2)$	**22.** $(0, 3)$	**23.** $(-\pi, 0)$	**24.** $(5, -5)$
25. $(-1, -1)$	**26.** $(-25, 0)$	**27.** $(0, -1)$	**28.** $(-3, -3)$

In Exercises 29–38 draw the angle with the given radian measure and name the initial and terminal sides. If the given angle is negative or is greater than 2π, indicate an angle between 0 and 2π that is coterminal with the one given. Express each of the given angles in degrees.

29. 4	**30.** 2π	**31.** π	**32.** $-\frac{7\pi}{6}$
33. -3π	**34.** -100	**35.** 10	**36.** 30
37. 100π	**38.** -100π		

In Exercises 39–46 use a calculator to find the value to four decimal places of each trigonometric function.

39. $\tan 1.23$	**40.** $\sin 0.54$	**41.** $\cot(-3)$	**42.** $\sec 0.02$
43. $\cot 3.78$	**44.** $\csc 0.5$	**45.** $\cos 2$	**46.** $\tan \pi$

In Exercises 47–54 use a calculator to find the values of θ to two decimal places, where $0 \le \theta < 2\pi$.

47. $\sin \theta = 0.4331$ and $\cos \theta < 0$	**48.** $\sin \theta = -0.4253$ and $\tan \theta < 0$
49. $\cos \theta = -0.8675$ and $\cot \theta > 0$	**50.** $\cos \theta = 0.3326$ and $\csc \theta < 0$
51. $\cos \theta = 0.9012$ and $\tan \theta > 0$	**52.** $\tan \theta = -6.8269$ and $\sin \theta > 0$
53. $\cot \theta = -1.0502$ and $\sin \theta < 0$	**54.** $\sec \theta = 1.113$ and $\cot \theta < 0$

In Exercises 55–66 find the exact value of each trigonometric function without using a calculator.

55. $\sin \frac{\pi}{4}$	**56.** $\cos \frac{\pi}{3}$	**57.** $\tan \frac{5\pi}{3}$	**58.** $\cos 7\pi$
59. $\tan \frac{7\pi}{4}$	**60.** $\sin \frac{-5\pi}{6}$	**61.** $\cos \frac{13\pi}{6}$	**62.** $\sin \frac{7\pi}{3}$

63. $\csc \frac{3\pi}{4}$

64. $\cot \frac{5\pi}{3}$

65. $\sec 5\pi$

66. $\cot \frac{19\pi}{6}$

In Exercises 67–70 use the law of sines to solve each triangle.

67. $A = 0.88$ radian, $B = 1.29$ radians, $c = 20.7$

68. $A = 1.31$ radians, $C = 1.00$ radian, $b = 0.652$

69. $B = 0.59$ radian, $C = 1.27$ radians, $a = 274$

70. $B = 0.72$ radian, $C = 0.93$ radian, $a = 13{,}500$

71. Construct a table showing the exact values of the trigonometric functions for angles with radian measures of 0, $\pi/6$, $\pi/4$, and $\pi/3$ and for angles around the unit circle having these angles as reference angles. (*Note:* A "unit circle" is a circle with radius 1. It is sometimes assumed that the unit circle is centered at the origin.)

72. Through how many radians does a person standing on the equator rotate in 8 hr?

73. Determine an angle with a measure between 0 and 2π that is coterminal with an angle whose measure is 80 radians.

74. A roulette wheel consists of 38 equal compartments. Those numbered from 1 to 36 are alternately black and red. There are also two green pockets numbered 0 and 00 on opposite sides of the wheel. (See Figure 4.6.)
   (a) How many radians are represented by red or black?
   (b) How many radians are represented by green?
   (c) How many radians are represented by any one number?

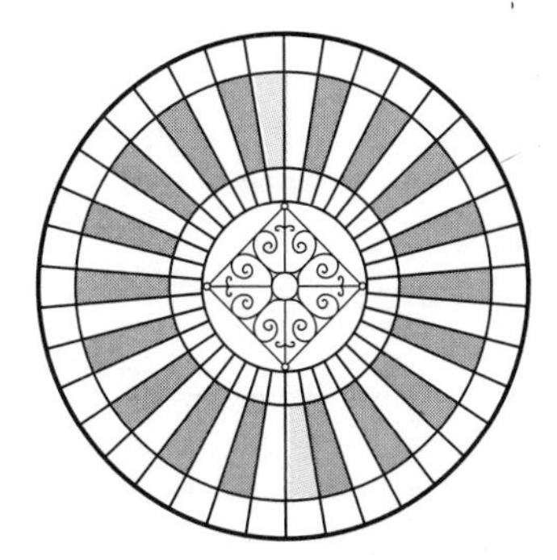

Figure 4.6

4.2 Arc Length and the Area of a Sector

Whether to measure an angle in degrees or radians is sometimes a matter of personal preference, but more often than not the unit is dictated by the circumstances of the problem. In this section we present several problems that are best solved using radian measure.

Arc Length

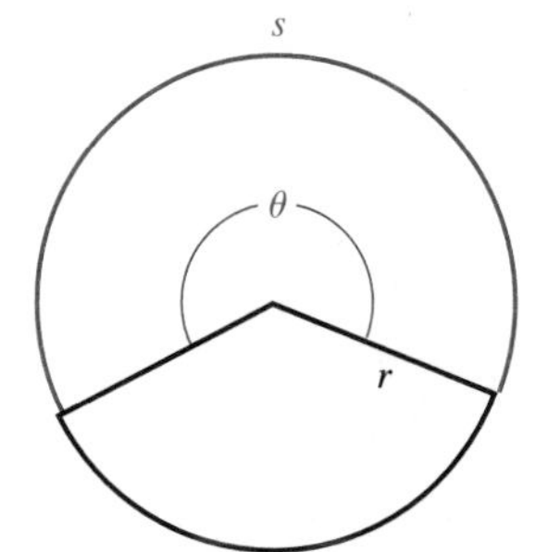

Figure 4.7

Figure 4.7 shows a central angle subtending an arc of length s. Recall that the radian measure of angle θ is found by determining how many times the radius of the circle is contained in the length of the corresponding arc—that is, $\theta = s/r$. If we solve this formula for s, we get a formula for the length of an arc.

ARC LENGTH FORMULA The length of an arc, s, subtended by a central angle θ of a circle of radius r is given by

$$s = r\theta \qquad (4.3)$$

where θ is *measured in radians.*

Example 1 Find the length of an arc on a circle of radius 5.0 cm that subtends a central angle of 38°.

SOLUTION To use Formula (4.3) you must first convert the degree measure to radian measure. Thus,

$$38°\left(\frac{\pi}{180}\text{ radians}\right) = \frac{19\pi}{90}\text{ radians}$$

Therefore $s = 5.0 \times 19\pi/90$ cm $\approx$ $\boxed{3.316125579}$ ≈ 3.3 cm. ▲

The formula $s = r\theta$ is used in physics to relate linear displacement, s, to angular displacement, θ. The next example illustrates this application of the arc length formula.

Example 2 As the drum in Figure 4.8 rotates counterclockwise, a cord is wound around the drum. How far will the weight on the end of the cord be moved when the drum is rotated through an angle of 53.8° if $r = 4.5$ ft?

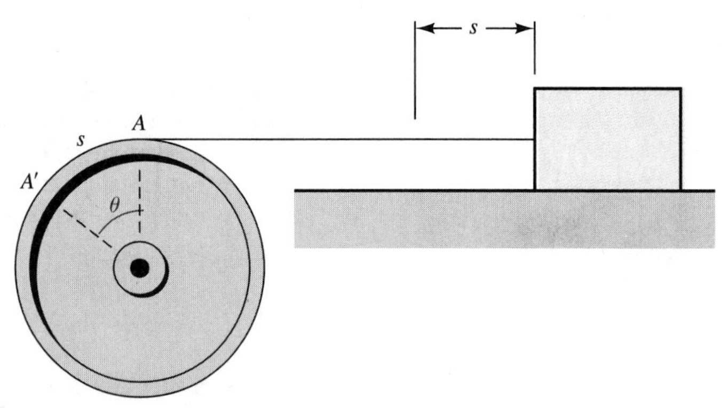

Figure 4.8

SOLUTION As the drum rotates the cord is wound around the drum, so the distance that the weight moves is equal to the arc length along the edge of the drum formed by a rotation of 53.8°. To use Formula (4.3) we must specify the angle of rotation in radians. Thus

$$\theta = 53.8\left(\frac{\pi}{180}\right)$$

The distance s is then given by

$$s = r\theta = 4.5\left[53.8\left(\frac{\pi}{180}\right)\right] \approx \boxed{4.225442119} \approx 4.2 \text{ ft}$$ ▲

Area of a Sector of a Circle

Another application of radian measure is in finding areas of sectors of circles (see Figure 4.9). To obtain a formula for computing the area of a

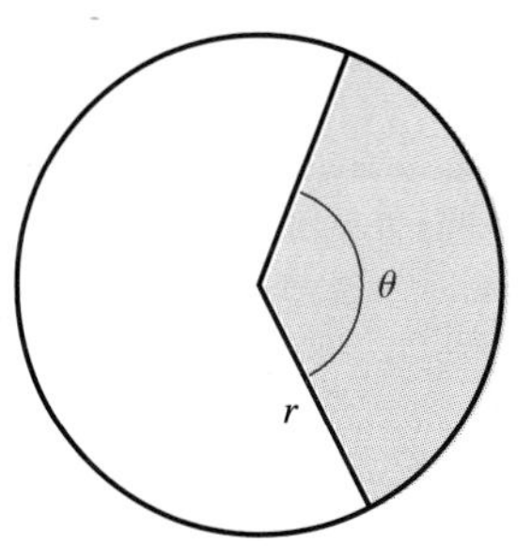

Figure 4.9

circular sector with central angle θ, we note that the area of the entire circle is given by $A = \pi r^2$. This formula can be written as $A = \frac{1}{2}(2\pi)r^2$, where 2π is the central angle of the entire circle. We interpret this to mean that the area of a circle can be expressed in terms of its central angle and the square of its radius. The area of a sector of a circle is the same proportion of the total area of the circle as its central angle is of 2π. For instance, a circular sector with a central angle of $\frac{1}{4}\pi$ has an area that is $\frac{1}{4}\pi/2\pi = \frac{1}{8}$ of the total area. This relationship is the basis for the formula for the area of a circular sector.

AREA OF A CIRCULAR SECTOR The area of a circular sector with a central angle θ and radius r is given by

$$A = \tfrac{1}{2}\theta r^2 \qquad \textbf{(4.4)}$$

where θ is measured in radians.

Example 3 Determine the area of the sector of a circle with a radius of 15.0 ft if the central angle of the sector is 52.7°.

SOLUTION To use Formula (4.4) we must convert 52.7° to radians. Thus

$$\theta = 52.7°\left(\frac{\pi}{180}\right)$$

The area of the sector is then

$$A = \frac{1}{2}\left(52.7 \times \frac{\pi}{180}\right)(15.0)^2 \approx \boxed{103.4762081} \approx 103 \text{ ft}^2$$ ▲

Exercises for Section 4.2

1. A circle has a radius of 25.3 cm. Calculate the length of the arc on the circumference subtended by a central angle of 78.6°.
2. Calculate the length of an arc on the circumference of a circle of radius 9.75 in. if the arc is subtended by a central angle of 187°.
3. Calculate the central angle of a circle with a 4.5-ft radius if the angle subtends an arc of 2.0 ft.
4. Calculate the central angle of a circle with a 21.4-cm radius if the angle subtends an arc of 1.9 cm.
5. A 92°15′ central angle of a circle subtends an arc of length 33.0 ft on the circumference of the circle. Find the radius of the circle.
6. Find the radius of a circle if a 310°30′ central angle subtends an arc of 0.720 m on the circumference.
7. A 10.0-ft-long pendulum swings through an arc of 30.0°. How long is the arc described by its midpoint?

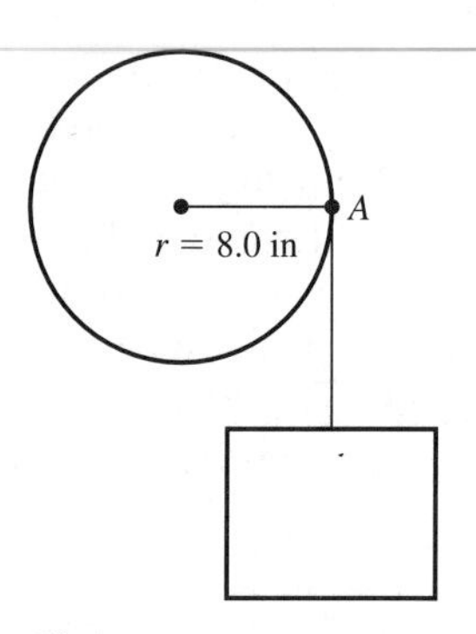

Figure 4.10

Figure 4.11

8. A racing car travels a circular course about the judge's stand. If the angle subtended by the judge's line of sight while the car travels 1.0 mi is 120°, how large is the entire track?
9. How high will the weight in Figure 4.10 be lifted if the drum is rotated through an angle of 81.5°?
10. Through what angle, in degrees, must the drum in Figure 4.10 be rotated to raise the weight 10 in.?
11. Compute the radius of a drum like that in Figure 4.10 if the weight is raised 5.2 in. by a rotation of 70.7°.
12. The scale on an ammeter is 8.0 cm long. If the scale is an arc of a circle having a radius of 3.2 cm, what angle in degrees will the needle make between the zero reading and a full scale reading? (See Figure 4.11.)
13. A voltage of 6.2 V causes the needle on a voltmeter to deflect through an angle of 48.0°. If the needle is 1.75 in. long, how far does the tip of the needle move in indicating the applied voltage?
14. The diameter of the earth is approximately 8000 mi. Find the distance between two points on the equator whose longitude differs by 3°.
15. The rotation of either of the wheels shown in Figure 4.12 causes the other wheel to rotate also. How much rotation in degrees occurs in the larger wheel when the smaller one rotates through an angle of 100.0°?

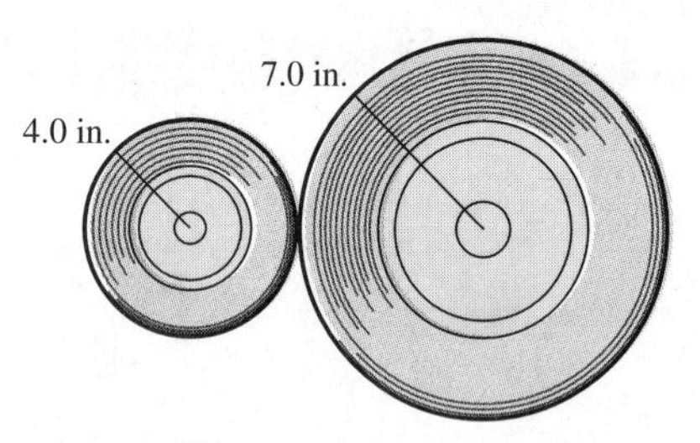

Figure 4.12

16. How much rotation must occur in the smaller wheel in Figure 4.12 to cause the larger one to rotate a quarter of a revolution?
17. Figure 4.13 shows a schematic drawing of a typical bicycle drive chain. How far will the bicycle move forward for each complete revolution of the drive sprocket?

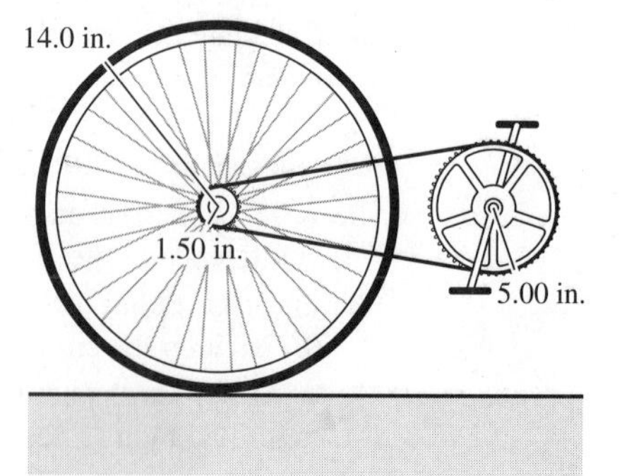

Figure 4.13

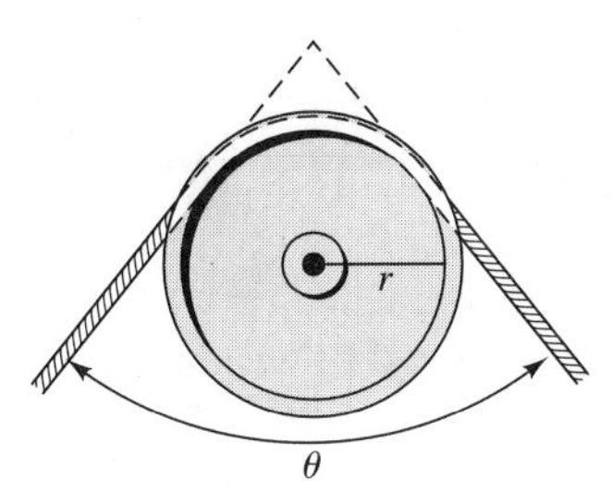

Figure 4.14

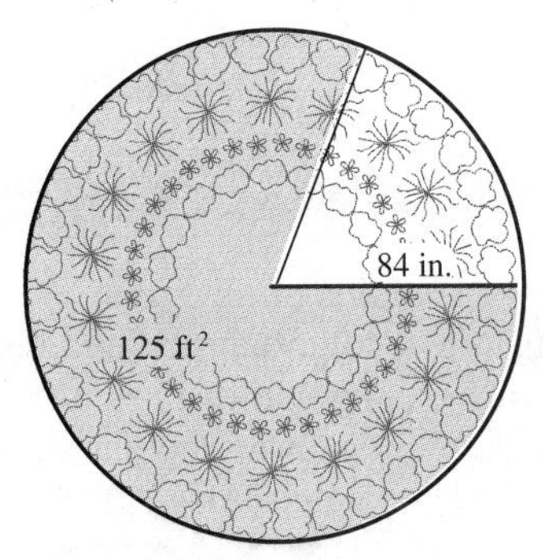

Figure 4.15

18. How much rotation of the drive sprocket must occur to move the bicycle in Figure 4.13 a total of 5.00 ft forward?
19. A cable is looped over a pulley, as shown in Figure 4.14. Given that the radius of the pulley is 7.0 in. and angle θ is 78.6°, determine how much of the cable is in contact with the pulley.
20. Find angle θ in Figure 4.14 if 5.8 in. of cable contacts the pulley.
21. Find the area of a sector of a circle, given that the central angle is 85.9° and the diameter is 24.0 ft.
22. Find the area of a sector of a circle, given that the central angle is 46° and the radius is 30.25 m.
23. The area of a circular sector is 4.05 in.^{2} and its central angle is 15.2°. Find the radius of the circle.
24. The area of a circular sector is 44.0 cm^2 and its radius is 12.0 cm. Find the central angle of the sector. Express the answer to the nearest tenth of a degree.
25. The area of a sector of a circle is 100 ft^2 and its bounding arc has a length of 5.75 ft. Find the central angle of the sector in degrees and the radius of the circle in feet.
26. The area of a circular sector is 75.4 mi^2 and its bounding arc has a length of 1.25 mi. Find the radius of the circle.
27. The area of a sector of a circular flower bed is found to be about 125 ft^2. If the radius of the bed is 84 in, what is the central angle of the sector in degrees? (See Figure 4.15.)
28. A circle of radius 3 ft is divided into a number of red and blue sectors. What is the total central angle for the blue sectors if their total area is 100 in.^{2}?
29. A cylindrical tank with a horizontal axis is filled with oil to a depth of 10 in. Given that the tank is 4 ft long and has a diameter of 3.0 ft, find the volume of oil in the tank. (See Figure 4.16.)

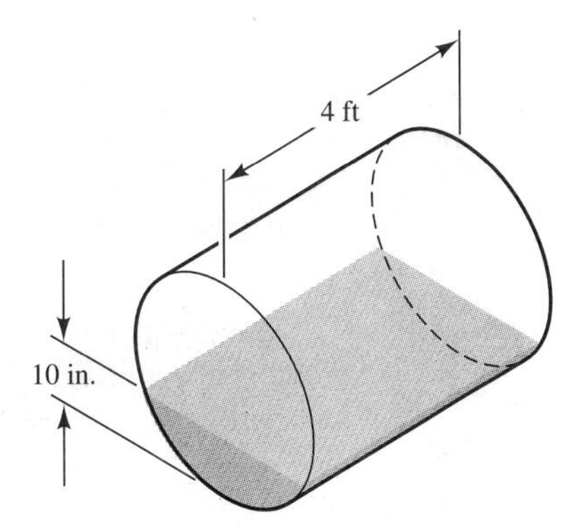

Figure 4.16

30. Derive a formula for the radius r of a circle that has an arc of s units bounding a sector of area A.

4.3 Angular and Linear Velocity

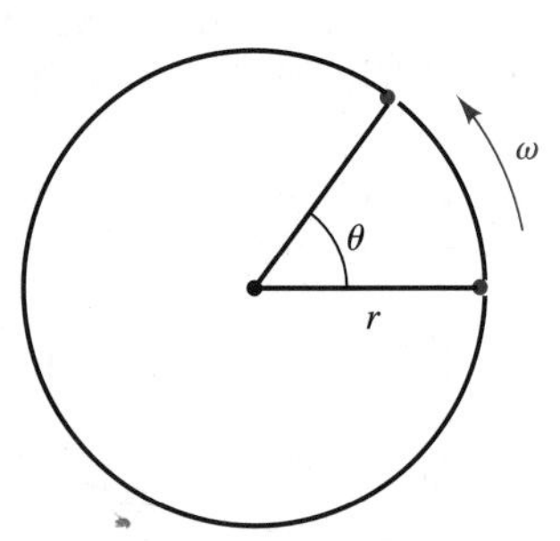

Figure 4.17

Consider a point on the circumference of a circle of radius r, as shown in Figure 4.17. If the circle rotates at a constant rate about its center, the point is said to have an angular velocity.* Angular velocity, which is a measure of the rate of change of angular displacement, is denoted by ω and defined as follows.

> DEFINITION 4.2 The **angular velocity** of a rotating object is the change in its angular displacement divided by the time required for the change to occur. Symbolically,
>
> $$\omega = \frac{\theta}{t} \qquad \textbf{(4.5)}$$

Notice that angular velocity is independent of the radius of the circle.

Two common units of angular velocity are revolutions per minute (rpm) and radians per second, but any ratio of angular displacement units to time units can be used. Conversion from one unit of angular velocity to another is sometimes required. The next example shows how this is done.

Example 1

(a) Change 80 rpm to radians per second.

(b) Change 100 rad/sec to revolutions per minute.

SOLUTION

(a) $80\,\frac{\text{rev}}{\text{min}} \times 2\pi\,\frac{\text{rad}}{\text{rev}} \times \frac{1}{60}\,\frac{\text{min}}{\text{sec}} = \frac{8\pi}{3}\text{ rad/sec} \approx \boxed{8.37758041} \approx 8.4\text{ rad/sec}$

(b) $100\,\frac{\text{rad}}{\text{sec}} \times \frac{1}{2\pi}\,\frac{\text{rev}}{\text{rad}} \times 60\,\frac{\text{sec}}{\text{min}} = \frac{3000}{\pi}\text{ rpm} \approx \boxed{954.9296586} \approx 955\text{ rpm}$ ▲

The angular velocity of a particle on the circumference of a rotating circle is not the same as its linear velocity. Linear velocity is denoted by v and defined as follows.

* Velocity is a vector quantity, meaning it has magnitude and direction. However, the word *velocity* is used here without regard to direction since there are only two ways to go around a circle, clockwise or counterclockwise, and only two ways to go on a straight line. Thus the direction part of an angular or linear "velocity vector" is not usually specified but is implied.

DEFINITION 4.3 The **linear velocity** of an object is the change in displacement divided by the time required for the change to occur. Symbolically,

$$v = \frac{s}{t} \tag{4.6}$$

where s is the distance traveled by the particle and t is the elapsed time. Some typical units for v are feet per second, meters per second, and miles per hour.

If s is the distance measured along the circumference of a circle of radius r, we can relate ω to v as follows: From the previous section we know that $s = r\theta$ if θ is measured in radians. Thus we can substitute $r\theta$ for s in Formula (4.6) to obtain

$$v = \frac{s}{t} = \frac{r\theta}{t}$$

Since $\omega = \theta/t$, the relationship between linear velocity and angular velocity is

$$v = r\omega \tag{4.7}$$

WARNING Formula (4.7) is valid only if ω is measured in radians/unit time. If ω is given in other units, such as rpm, the units must be converted to rad/unit time before Formula (4.7) is used.

COMMENT Formula (4.7) shows that linear velocity is dependent on the radius of the circle of rotation. For a given angular velocity, the larger the radius, the larger the linear velocity of a point on the circle. Thus if P_1 and P_2 are points on a rotating wheel, as shown below, P_2 travels faster in the linear sense than P_1, even though both have the same angular velocity.

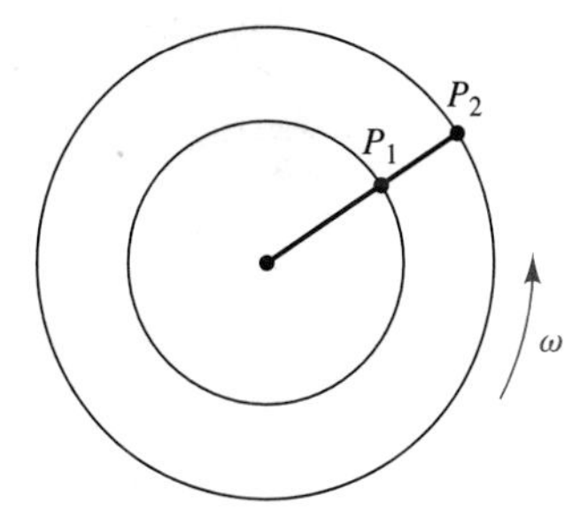

Example 2 Consider a point P on the circumference of a wheel whose radius is 18 cm. Suppose the wheel rotates at $\pi/9$ rad/sec. Determine **(a)** the linear velocity of point P and **(b)** the linear distance traveled by P in 10 sec.

SOLUTION

(a) Substituting $\omega = \pi/9$ rad/sec and $r = 18$ cm into Formula (4.7), we get

$$v = r\omega = 18\left(\frac{\pi}{9}\right) = 2\pi \text{ cm/sec} \approx 6.28 \text{ cm/sec}$$

(b) The distance traveled by the point in 10 sec is found by using the formula $s = vt$.

$$s = 2\pi(10) = 20\pi \text{ cm} \approx 62.8 \text{ cm}$$

▲

Example 3 A belt runs a pulley of radius 5 cm at 100 rpm. Find the angular velocity of the pulley in radians per second and the linear velocity of a point on the belt. (See Figure 4.18.)

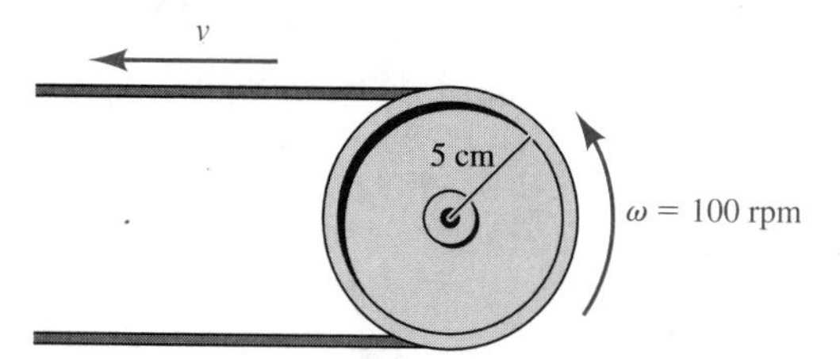

Figure 4.18

SOLUTION To write the angular velocity of the pulley in radians per second, change revolutions to radians and minutes to seconds. Thus,

$$\begin{aligned}\omega &= 100\,\frac{\text{rev}}{\text{min}} \times 2\pi\,\frac{\text{rad}}{\text{rev}} \times \frac{1}{60}\,\frac{\text{min}}{\text{sec}}\\ &= \frac{10\pi}{3}\text{ rad/sec} \approx \boxed{10.47197551} \approx 10.5 \text{ rad/sec}\end{aligned}$$

The linear velocity of a point on the belt will be the same as that of a point on the circumference of the pulley, which is given by

$$v = r\omega = 5\left(\frac{10\pi}{3}\right) = \frac{50\pi}{3}\text{ cm/sec} \approx \boxed{52.35987756} \approx 52.4 \text{ cm/sec}$$

▲

Example 4 A bicycle travels 15 mi at a constant speed for 1 hr. What is the angular velocity of a point on the wheel in revolutions per minute and in radians per second? Assume that the bicycle has 26-in. diameter wheels.

SOLUTION In 1 hr a point on the tip of a wheel will travel 15 mi. Thus

$$\begin{aligned}s &= 15 \text{ mi} \times 5280\,\frac{\text{ft}}{\text{mi}} \times 12\,\frac{\text{in.}}{\text{ft}}\\ &= 950{,}400 \text{ in.}\end{aligned}$$

(Note that we had to change either the 15 mi to inches or the 26-in. diameter to miles.)

The circumference of the wheel is $\pi d = 26\pi$ in. Thus the number of revolutions made by the wheel of the bicycle in 1 hr is

$$\frac{950{,}400}{26\pi} \approx \boxed{11635.45061} \text{ rev/hr}$$

To find ω in revolutions per minute, we have

$$\omega \approx 11{,}635.45061 \frac{\text{rev}}{\text{hr}} \times \frac{1 \text{ hour}}{60 \text{ min}} \approx \boxed{193.9241768} \approx 194 \text{ rpm}$$

Finally, to find ω in radians per second, we have

$$\omega \approx 194 \frac{\text{rev}}{\text{min}} \times 2\pi \frac{\text{rad}}{\text{rev}} \times \frac{1 \text{ min}}{60 \text{ sec}} \approx \boxed{20.31563249} \approx 20.3 \text{ rad/sec}$$ ▲

Exercises for Section 4.3

In Exercises 1–5 find the linear velocity in feet per second of a point on the rim of a wheel of the given radius with the given angular velocity.

1. $r = 1$ ft, $\omega = 12.5$ rad/sec
2. $r = 3$ ft, $\omega = 25$ rpm
3. $r = 2.45$ ft, $\omega = 100$ rpm
4. $r = 4.3$ m, $\omega = 5$ rpm
5. $r = 7$ in., $\omega = 250$ rpm

In Exercises 6–10 find the angular velocity in radians per second that corresponds to the given linear velocity.

6. $r = 10$ in., $v = 150$ in./sec
7. $r = 3.0$ ft, $v = 2.0$ ft/min
8. $r = 10$ ft, $v = 30$ mph
9. $r = 1.30$ m, $v = 140$ km/hr
10. $r = 10$ cm, $v = 100$ km/hr

In Exercises 11 and 12 find the distance traveled by a point on the rim of a wheel in the given period of time.

11. $r = 10.0$ cm, $\omega = \pi$ rad/sec, $t = 20.0$ sec
12. $r = 2$ ft, $\omega = 5$ rpm, $t = 3$ min
13. A 10-in.-diameter pulley rotates with an angular velocity of 200 rpm. What is the linear velocity in feet per second of a belt attached to the pulley? Approximately how many miles does a point on the belt travel in 8 hr?
14. **(a)** What is the angular velocity in radians per second of the minute hand of a clock?

    (b) How far does the tip of the minute hand move in 1 hr if it is 8 in. long?
15. How far does the tip of a 4.0-in. minute hand move in 90 sec?
16. What is the linear velocity in feet per second of the tip of a 6-in. second hand?
17. A bicycle is ridden at a constant speed of 12 mph for 2 hr. What is the wheel's angular velocity in radians per second if its diameter is 26 in.?
18. A bicycle travels 10 mi in 1 hr. Find the angular velocity of the drive sprocket in revolutions per minute if the drive sprocket has a 5.00-in. radius and is con-

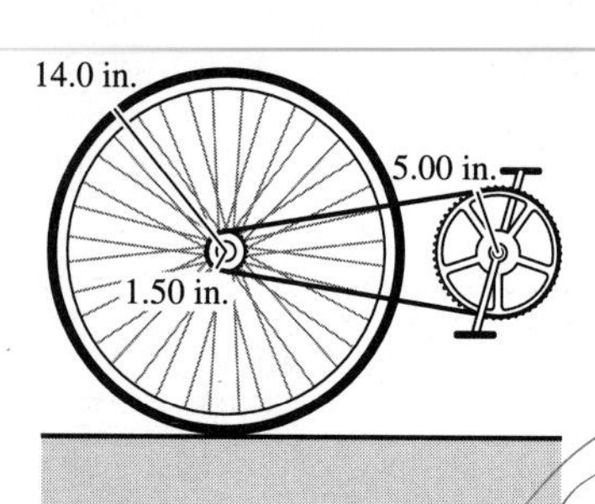

Figure 4.19

nected to a rear wheel with a 14.0-in. radius and a sprocket with a radius of 1.50 in. (See Figure 4.19.)

19. Two pulleys, one with a radius of 1 ft and the other with a radius of 2 ft, are connected by a belt. If the larger pulley has an angular velocity of 100 rpm, find the speed of the belt in feet per second and the angular velocity of the other pulley in revolutions per minute. (See Figure 4.20.)

For Exercises 20 and 21 refer to Figure 4.21.

20. Find the linear velocity in miles per hour of a person standing on the equator.

21. Find the linear velocity in miles per hour of a person standing at a latitude of 35°.

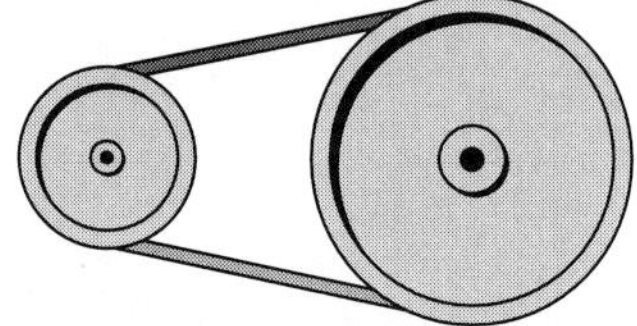

Figure 4.20

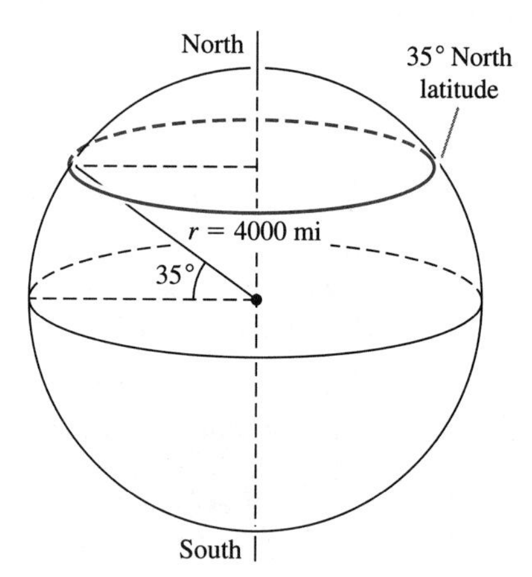

Figure 4.21

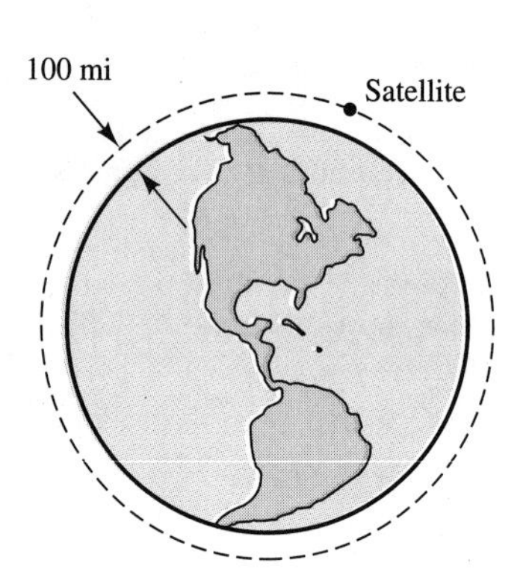

Figure 4.22

22. A satellite circles the earth at a height of 100 mi. If it circles the earth every 90 min, what is the satellite's linear velocity in miles per hour? (See Figure 4.22.)

▲▼ Key Topics for Chapter 4

Define and/or discuss each of the following.

Radian Measure
Trigonometric Functions Using Radians
Arc Length
Area of a Sector
Angular Velocity
Linear Velocity

▲▼ Review Exercises for Chapter 4

In Exercises 1–5 express each angle in radian measure. Use multiples of π.

1. 36° **2.** 25° **3.** 110° **4.** 3000° **5.** −85°

In Exercises 6–10 express each angle in degrees.

6. $\frac{5\pi}{12}$ **7.** $\frac{\pi}{18}$ **8.** $\frac{13\pi}{6}$ **9.** 435 **10.** 180π

In Exercises 11–20 evaluate the given function. If the radian measure is given as a multiple or fractional part of π, give the exact value. Otherwise use a calculator.

11. $\sin \frac{7\pi}{6}$ **12.** $\cos \frac{16\pi}{3}$ **13.** $\tan \frac{25\pi}{6}$ **14.** $\cot \frac{-5\pi}{4}$

15. $\sec \pi$ **16.** $\sin 4.16$ **17.** $\cos 2.73$ **18.** $\tan 5.27$

19. $\sin -3.07$ **20.** $\sec 0.3$

In Exercises 21–30 find the value of x for $0 \leq x < 2\pi$. When appropriate, give the exact value.

21. $\sin x = -\frac{\sqrt{3}}{2}, \cos x > 0$ **22.** $\cos x = \frac{1}{2}, \sin x > 0$

23. $\tan x = 3, \cos x > 0$ **24.** $\cos x = \frac{\sqrt{2}}{2}, \tan x < 0$

25. $\sec x = -2, \sin x > 0$ **26.** $\sin x = -0.8102, \cos x < 0$

27. $\tan x = 1.202, \sin x < 0$ **28.** $\sin x = 0.6032, \tan x > 0$

29. $\cot x = 2, \cos x < 0$ **30.** $\csc x = -1.118, \tan x > 0$

31. Determine θ if $\tan \theta = 2$, $2\pi < \theta < 4\pi$.

32. Determine θ if $\sin \theta = 0.6773$, $-\pi < \theta < \pi$.

33. A central angle of 47° subtends an arc of 5.5 cm on the circumference of a circle. Determine the radius of the circle.

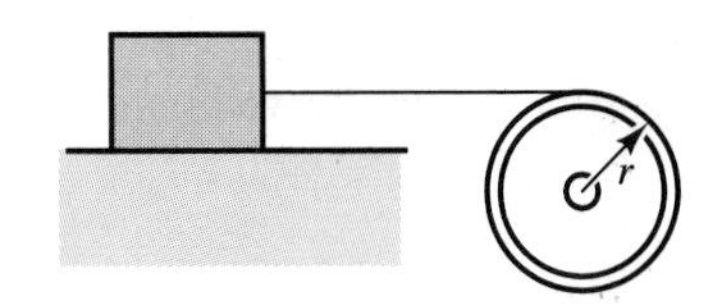

Figure 4.23

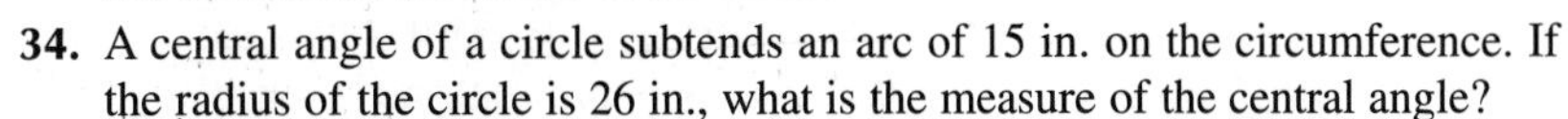

34. A central angle of a circle subtends an arc of 15 in. on the circumference. If the radius of the circle is 26 in., what is the measure of the central angle?

35. The drum shown in Figure 4.23 rotates clockwise through an angle of 200°. Determine how far the block moves if the radius of the drum is 0.75 ft.

36. Compute the radius of the drum in Figure 4.23 if the weight is moved 3.5 m by a rotation of 35.2°.

37. Determine the area of a sector of a circle if the angle of the circular sector is 40.0° and the diameter of the circle is 28.0 cm.

38. The area of a circular sector is 100 ft^2 and the subtended arc is 9.5 ft. Find the radius of the circle.

39. A bicycle wheel moves along the pavement at 14.0 mph. Determine the angular velocity of the wheel in radians per second and in revolutions per minute if it has a diameter of 27.0 in.

40. Compute the velocity at which the weight in Figure 4.23 moves if the drum is rotating with an angular velocity of 2.5 rpm and the radius of the drum is 8.0 in. Express the velocity in feet per second.

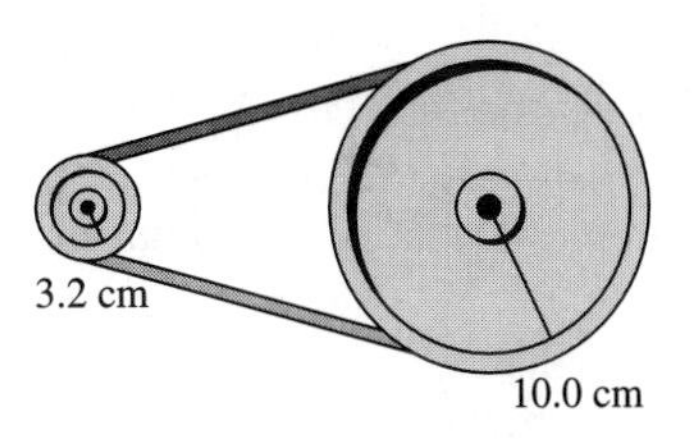

Figure 4.24

41. A conveyor belt rolls on cylindrical bearings that are 2.5 in. in diameter. If the belt has a velocity of 4.6 ft/sec, what is the angular velocity of the roller bearings in radians per second?

42. The two wheels shown in Figure 4.24 are connected by a belt. Determine the angular velocity of the smaller wheel if the angular velocity of the larger one is 25 rev/sec. What is the velocity of the belt?

43. Assume that the radius of the larger wheel in Figure 4.24 is R and the radius of the smaller one is r. If the angular velocity of the larger wheel is ω, show that the angular velocity of the smaller one is $\omega R/r$.

Practice Test Questions for Chapter 4

In Exercises 1–6 answer *true* or *false*.

1. One radian is equal to 180°.

2. To convert degrees to radians, multiply the degree measure by $\pi/180$.

3. The sum of the measures of the interior angles of a triangle is π radians.

4. The formula to convert angular velocity to linear velocity requires that the angular velocity be given in rpm.

5. The radian measure of the central angle of a circle depends on the radius of the circle.

6. A straight angle measures π radians.

In Exercises 7–14 fill in the blank to make the statement true.

7. An angle whose measure is one radian subtends an arc on a circle equal in length to the ____________ of the circle.

8. To convert radian measure to degree measure, multipy the radian measure by ____________.

9. The area of a circular sector with central angle θ in radians and radius r is given by $A =$ ____________.

10. A right angle is equal to ____________ radians.

11. Linear velocity divided by ____________ equals the radius of revolution.

12. An angle of 3 radians in standard position has its terminal side in quadrant ____________.

13. Given that $\sin\theta = -0.2271$ and $\cos\theta < 0$, then $\theta =$ ____________ radians, $0 \le \theta \le 2\pi$.

14. Given that $\sec\alpha = 1.911$ and $\tan\alpha < 0$, then $\alpha =$ ____________ radians, $0 \le \alpha \le 2\pi$.

Solve the stated problem in the following exercises. Show all your work.

15. Express 80° in radian measure using multiples of π for the answer.

16. Find the degree measure of $7\pi/30$ radians.

17. Given triangle ABC with $A = 0.36$ radian, $B = 1.25$ radians, and $a = 32.7$ cm:
(a) Find angle C.
(b Calculate the length of side b.

18. In moving along the roadway, a bicycle wheel with a radius of 14 in. rotates through an angle of 852°. Calculate the linear distance that the wheel moved.

19. The area of a circular sector is 304 ft^2 with a radius of 15.7 ft. Calculate the central angle of the sector. Express your answer to the nearest tenth of a degree.

20. A pulley with a radius of 3 in. revolves at 275 rpm.
(a) Calculate the angular velocity of the pulley in rad/sec.
(b) Calculate the linear velocity of a point on the circumference of the pulley.
(c) Calculate the angular distance traveled by the point on the circumference of the pulley in 2 min.

Analytic Trigonometry

The use of trigonometry in some physical applications and in advanced mathematics courses such as calculus require that you think of the argument of the trigonometric functions as being a real number. In Section 5.1 we show how angles measured in radians are related to real numbers. By so doing we open the rich field of analytic trigonometry, which highlights nontriangular concepts such as amplitude, period, and phase shift. The interpretation of the trigonometric functions in terms of real numbers permits us to give these functions a graphical representation, and in Sections 5.3 and 5.4 we introduce the graphs of $y = \sin x$ and $y = \cos x$. Section 5.5 gives some applications that involve the sine and cosine of real numbers. In Sections 5.6–5.8 we construct graphs of the remaining four trigonometric functions. Throughout the chapter the graphing calculator plays an important role in helping us generate and understand graphs of the trigonometric functions. ▼

▲▼ 5.1 Trigonometric Functions of Real Numbers

Modern trigonometry consists of two more or less distinct branches. The study of the six ratios and their applications to problems involving triangles is one branch, called **triangle trigonometry**. The other branch, which is concerned with the general functional behavior of the six ratios, is called **analytic trigonometry**.

In analytic trigonometry we consider the six trigonometric functions to be functions both of real numbers and of angles. The discussion in this section shows that the reinterpretation of the domain of the trigonometric functions to include real numbers is a matter of matching real numbers with the radian measures of angles.

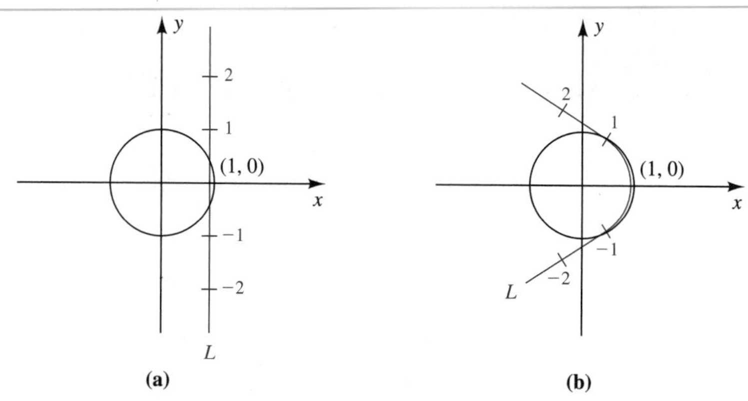

Figure 5.1

We begin by locating the unit circle in the rectangular plane so that its center is at the origin and it passes through (1, 0), as shown in Fig. 5.1(a). The circumference of the unit circle is 2π, or approximately 6.28.

Let L denote a real number line that is parallel to the y-axis through the point (1, 0). Now if L is wrapped around the unit circle, each point on the line is mapped onto a point on the circle, as in Figure 5.1(b). The positive half-line is wound in a counterclockwise direction, whereas the negative half-line is wound in a clockwise direction.

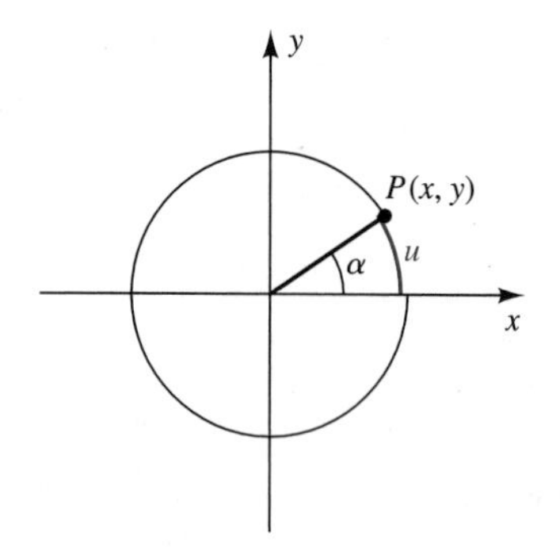

Figure 5.2

Each real number u mapped onto the unit circle determines both a point P in the plane and an angle α in standard position, as shown in Figure 5.2. The point P and the angle α are said to be *associated with* the real number u.

Real Numbers and Angles Measured in Radians

A relationship can be established between the number u and the angle α on the basis of the fact that the radian measure of an angle is the ratio of the length of the arc of a circle subtended by the angle to the radius of the circle. That is,

$$\alpha \text{ (radians)} = \frac{u \text{ (arc length)}}{r \text{ (radius)}}$$

Since $r = 1$ for the unit circle, the measure of angle α is numerically equal to the arc length u. Symbolically

$$\alpha \text{ (radians)} = u$$

Example 1 illustrates this fact.

Example 1 Sketch the points and angles associated with the real numbers 2, 10, −3.6, and 6.

SOLUTION See Figure 5.3.

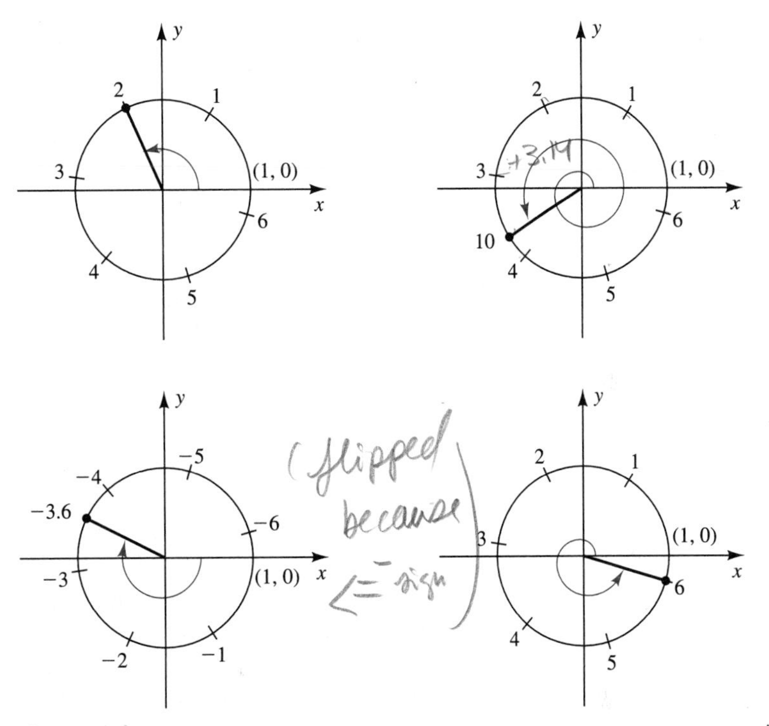

Figure 5.3 ▲

Using the natural association of real numbers with angles in standard position, we can define the trigonometric functions for any real number u. If α, in radians, is the angle associated with the real number u, then

$$\sin u = \sin \alpha \qquad \cos u = \cos \alpha \qquad \tan u = \tan \alpha$$

$$\sec u = \sec \alpha \qquad \csc u = \csc \alpha \qquad \cot u = \cot \alpha$$

To clearly understand what has been accomplished by these definitions, you must keep in mind that u is a real number and not the measure of an angle. These definitions permit discussion of trigonometric functions without reference to the concept of angle.

The trigonometric functions of angles were defined in Chapter 2. The extension to real numbers shows how the same functions are related to a circle. Because of their relationship to a circle, the trigonometric functions are often called the *circular functions*.

WARNING Since it is customary to delete the dimension of radians, it will be impossible to distinguish between an argument that is a real number and an argument that is an angle in radians. Take care to indicate the correct units of measurements if the argument is an angle measured in degrees, minutes, and seconds. Otherwise, by convention, the argument is an angle measured in radians or a real number. For example sin 30° means the sine of an angle of 30°, but sin 30 means either of the following, both of which have the same numerical value:

1. the sine of an angle of 30 radians.
2. the sine of the real number 30.

When the unit circle is located in the rectangular plane, the point P associated with the number u has coordinates (x, y). This association is shown in Figure 5.2. There is an interesting and important relationship between the coordinates of the point associated with u and $\cos u$ and $\sin u$. Since we are using the unit circle, we have

$$\cos u = \frac{x}{r} = \frac{x}{1} = x \qquad \sin u = \frac{y}{r} = \frac{y}{1} = y$$

That is, **the x- and y-coordinates of the point associated with the real number u are equal to $\cos u$ and $\sin u$, respectively.**

Figure 5.4 shows a unit circle with real numbers from 0 to 2π marked off on the circumference. Using this circle we can approximate values of $\cos u$ and $\sin u$ for any real number u in the interval $[0, 2\pi]$. As u increases,

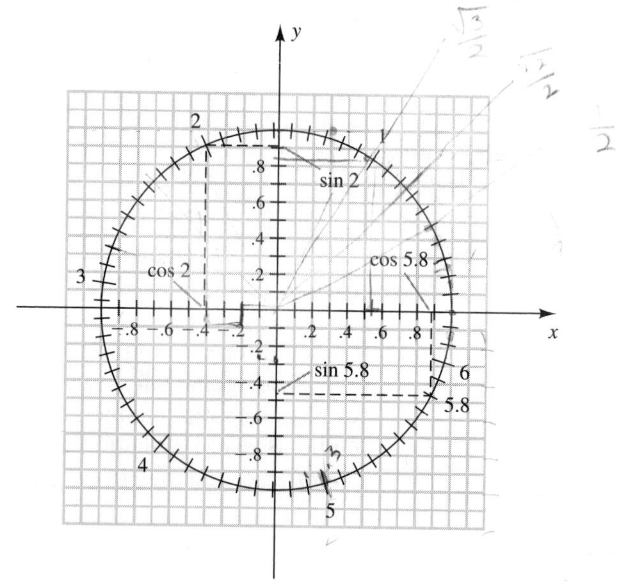

Figure 5.4

the values of $\sin u$ and $\cos u$ repeat themselves every circumference, or 2π units. Thus

$$\cos u = \cos(u + 2n\pi) \qquad \sin u = \sin(u + 2n\pi)$$

where n is an integer.

Example 2 Using Figure 5.4, estimate the cosine and sine of **(a)** 2, **(b)** 5.8, **(c)** 8.28, and **(d)** −4.28.

SOLUTION Using Figure 5.4 we estimate as follows.

(a) $\cos 2 \approx -0.42$, $\sin 2 \approx 0.91$.

(b) $\cos 5.8 \approx 0.89$, $\sin 5.8 \approx -0.46$.

(c) To find $\cos 8.28$ and $\sin 8.28$, note that $8.28 = 6.28 + 2$. Thus $\cos 8.28 = \cos 2 \approx -0.42$, and $\sin 8.28 = \sin 2 \approx 0.91$.

(d) To find $\cos(-4.28)$ and $\sin(-4.28)$, note that $-4.28 = 2 - 6.28$. Thus $\cos(-4.28) = \cos 2 \approx -0.42$, and $\sin(-4.28) = \sin 2 \approx 0.91$. ▲

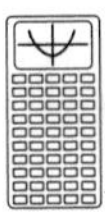

The values of the trigonometric functions for real numbers may be estimated using a calculator. When numbers are entered into a calculator in *radian mode*, the numbers may be considered either real numbers or radian measures of angles.

Example 3 Use a calculator to calculate cos 0.5, sin 1.4, and tan 0.714 to four decimal places. Also, check the results in Example 2 using your calculator.

SOLUTION A calculator in radian mode gives the following:

$$\cos 0.5 \approx \boxed{0.877582562} \approx 0.8776$$

$$\sin 1.4 \approx \boxed{0.98544973} \approx 0.9854$$

$$\tan 0.714 \approx \boxed{0.866507856} \approx 0.8665$$

▲

For angles whose reference angle is a special angle, the exact value of the trigonometric function may be found by using Table 2.2 (page 76).

Example 4 Let $f(x) = \sin x$ and $g(x) = \cos x$. Find the *exact* values of **(a)** $f(\pi/3)$, **(b)** $g(5\pi/6)$, and **(c)** $f(11\pi/4)/g(11\pi/4)$.

SOLUTION

(a) $f(\pi/3) = \sin(\pi/3) = \sqrt{3}/2$.

(b) Since $5\pi/6$ is in the second quadrant, the cosine function is negative. The reference angle for $5\pi/6$ is $\pi/6$. Thus

$$g(5\pi/6) = \cos(5\pi/6) = -\cos(\pi/6) = -\sqrt{3}/2$$

(c) The reference angle for $11\pi/4$ is $\pi/4$. Note that

$$f(11\pi/4)/g(11\pi/4) = \sin(11\pi/4)/\cos(11\pi/4) = \tan(11\pi/4).$$

Since the tangent is negative in the second quadrant,

$$\tan(11\pi/4) = -\tan(\pi/4) = -1.$$ ▲

In addition to being able to find the values of the trigonometric functions for a given domain value, we also need to be able to find the domain value(s) if the range value is given. For example, let $f(x) = \sin x$ and consider the equation $f(x) = \frac{1}{2}$. We solved problems of this type in Chapters 2 and 4, but then the domain values were angles; here the domain is the set of real numbers, so the solutions to the equation will be real numbers. Recall from Chapter 4 that $\sin(\frac{1}{6}\pi) = \frac{1}{2}$, where $\frac{1}{6}\pi$ was considered to be the radian measure of an angle. But there is no difference between $\sin(\frac{1}{6}\pi$ radians) and $\sin(\frac{1}{6}\pi)$ when $\frac{1}{6}\pi$ is considered to be a real number. Similarly, $\sin(\frac{5}{6}\pi)$ is also equal to $\frac{1}{2}$. Furthermore, for angular measure, $\sin(x + 2n\pi) = \sin x$; this is also true for real numbers. Consequently the real-number solutions of $f(x) = \frac{1}{2}$ are $x = \frac{1}{6}\pi + 2n\pi$ and $x = \frac{5}{6}\pi + 2n\pi$, where n is any integer value.

Example 5 Given that $f(x) = \sin x$, without using a calculator find the exact solutions of $f(x) = \sqrt{2}/2$.

SOLUTION From our knowledge of the trigonometric functional values at the elementary angles, we know that one solution of $\sin x = \sqrt{2}/2$ is $x = \pi/4$. Since the values of the sine function are positive in the first and second quadrants, we obtain a second value of $3\pi/4$. But these values are repeated every 2π units. For instance, $\pi/4 + 2\pi$, $\pi/4 + 4\pi$, $\pi/4 + 6\pi$, and so on are also solutions. In general the total solution set is written as $\pi/4 + 2n\pi$ and $3\pi/4 + 2n\pi$, where n is any integer. ▲

Example 6 Use a calculator to find all solutions of the equation $\cos x = 0.61$. Estimate x to two decimal places.

SOLUTION With the calculator in the radian mode, we find the angle or number whose cosine is 0.61; that is, we find $x =$ [inv] [cos] 0.61. The readout is [0.9147357359], so $x \approx 0.91$. But there are many other solutions to this equation. Since the cosine repeats itself every 2π units, $0.91 + 2n\pi$ are also solutions. Further, the positive values of the cosine are repeated in the fourth quadrant. Hence $2\pi - 0.91 \approx 5.37$ is also a solution, as are those numbers formed by adding $2n\pi$ to this number. In summary, the solutions of $\cos x = 0.61$ are:

$$0.91 + 2n\pi \quad \text{and} \quad 5.37 + 2n\pi$$ ▲

The names of the trigonometric functions are the same whether they are used in the sense of ratios or in a functional sense. However, by writing $y = \sin x$ or $f(x) = \sin x$, we emphasize the functional concept of the sine function. Any convenient letter may be used for the argument of the trigonometric functions. Thus $\sin x$, $\sin u$, $\sin \theta$, and $\sin y$ all mean the same thing. Only the application reveals whether the argument is to be interpreted as an angle or as a real number.

The following list gives some practical nonangular applications of trigonometric functions.

- A weight hanging on a vibrating spring has a velocity described by $\sin 3t$. In this case the argument is $3t$, where t is the time in seconds.
- The instantaneous voltage for certain electrical systems is given by $156 \sin 377t$, where t is the value of the time in seconds.
- The equation of motion of a shaft with flexible bearings is given by

$$x_0 \sin \frac{\pi}{2L} x$$

where x and L are given in centimeters.

Exercises for Section 5.1

Sketch the point on the unit circle associated with each real number. Find the cosine and sine of each number to two decimal places using Figure 5.4, and check the answer using a calculator.

1. (a) 1 (b) -2 (c) 3 (d) 10
(e) 3π (f) -4 (g) -4π (h) $\frac{1}{3}\pi$
(i) $\frac{1}{3}$ (j) $\frac{1}{2}$ (k) $\sqrt{7}$ (l) 5.15

In calculus the ratio $\frac{\sin x}{x}$ is important. Use a calculator in the radian mode to find the value of the ratio for each of the following values of x.

2. (a) 0.3 (b) 0.2 (c) 0.1 (d) 0.05 (e) 0.01

What value do you think $\frac{\sin x}{x}$ approaches as x approaches 0? What is $\frac{\sin x}{x}$ when $x = 0$?

3. If $f(x) = \sin x$ and $g(x) = \cos x$, show that $[f(x)]^2 + [g(x)]^2 = 1$.

In Exercises 4–10 let $f(x) = \sin x$ and find the *exact* functional values. Do not use a calculator.

4. $f(\frac{1}{2}\pi)$ 5. $f(\pi)$ 6. $f(3\pi)$ 7. $f(\frac{1}{3}\pi)$
8. $f(\frac{4}{3}\pi)$ 9. $f(\frac{1}{6}\pi)$ 10. $f(2\pi)$

In Exercises 11–16 let $f(x) = \sin x$ and approximate the indicated values to four decimal places using a calculator.

11. $f(1)$ 12. $f(-2)$ 13. $f(-5.5)$
14. $f(50)$ 15. $f(-10)$ 16. $f(0.01)$

In Exercises 17–22 let $f(x) = \sin x$ and find the *exact* values of all solutions of the equation.

17. $f(x) = 1$ 18. $f(x) = 0$ 19. $f(x) = 0.5$
20. $f(x) = -1$ 21. $f(x) = 2$ 22. $f(x) = \sqrt{3}/2$

In Exercises 23–26 let $f(x) = \sin x$ and use a calculator to solve the equation. Round off answers to two decimal places.

23. $f(x) = 0.1$ **24.** $f(x) = 1.1$

25. $f(x) = -0.3$ **26.** $f(x) = 0.88$

In Exercises 27–32 let $g(x) = \cos x$ and find the *exact* values without using a calculator.

27. $g(0)$ **28.** $g(\frac{1}{2}\pi)$ **29.** $g(\frac{1}{3}\pi)$

30. $g(3\pi)$ **31.** $g(\frac{5}{6}\pi)$ **32.** $g(-\frac{2}{3}\pi)$

In Exercises 33–38 let $g(x) = \cos x$ and approximate the indicated values to four decimal places using a calculator.

33. $g(1)$ **34.** $g(2.5)$ **35.** $g(\frac{1}{2})$

36. $g(5)$ **37.** $g(-\frac{1}{3})$ **38.** $g(-3)$

In Exercises 39–44 let $g(x) = \cos x$ and find the *exact* value of all solutions of the equation.

39. $g(x) = 1$ **40.** $g(x) = 0.5$ **41.** $g(x) = -\dfrac{\sqrt{3}}{2}$

42. $g(x) = 0$ **43.** $g(x) = 2$ **44.** $g(x) = -\frac{1}{2}$

In Exercises 45–50 let $g(x) = \cos x$ and use a calculator to solve the equation. Round off answers to two decimal places.

45. $g(x) = 0.2$ **46.** $g(x) = -0.85$ **47.** $g(x) = 0.86$

48. $g(x) = -\frac{1}{3}$ **49.** $g(x) = \frac{1}{3}$ **50.** $g(x) = -0.2$

51. Recall that a zero of a function $f(x)$ is a value $x = x_0$ such that $f(x_0) = 0$.
(a) What are the zeros of $\sin x$? **(b)** What are the zeros of $\cos x$?

52. Which pairs of trigonometric functions have the same zeros?

53. The velocity of a certain weight attached to a spring is given by $v = \sin 2t$, where v is in feet per second when t is measured in seconds. Find the velocity for $t = 0.3$ sec, to the nearest hundredth of a ft/sec.

54. The instantaneous voltage applied to a circuit is given by $V = 220 \sin 377t$, where t is measured in seconds. Find the voltage being applied when $t = 0.01$ sec, to the nearest tenth of a volt.

55. In calculus we show that if an object has motion described by $y = \tan t$, its velocity is given by $v = \sec^2 t$. Find the time between 0 and 1 sec for which the velocity is 2 m/sec, to the nearest hundredth of a second.

5.2 Analytic Properties of the Sine and Cosine Functions

We begin this section with a discussion of periodicity and boundedness. We will define each property and discuss each in a general setting, and then relate the properties to the trigonometric functions.

Periodicity

Something is *periodic* when it repeats itself at regular intervals. Examples of periodic phenomena occur throughout the physical sciences—the motion of a pendulum, alternating current, the vibration of a tuning fork. The trigonometric functions play an important role in describing periodic phenomena. The following is a precise but general definition of a periodic function.

DEFINITION 5.1 A function f is said to be *periodic* if there exists a number p, where $p > 0$, such that

$$f(x) = f(x + p)$$

for all x in the domain of the function. If there is a smallest number p for which this expression is true, then p is called the **period** (or the *fundamental period*) of the function.

Figure 5.5 shows graphs of several periodic functions. A periodic function is completely described by its period and the functional values over one period interval, as Example 1 illustrates.

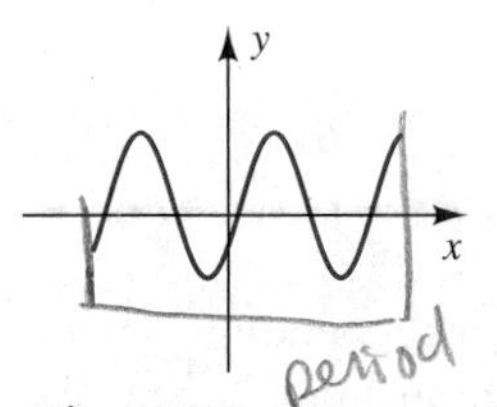

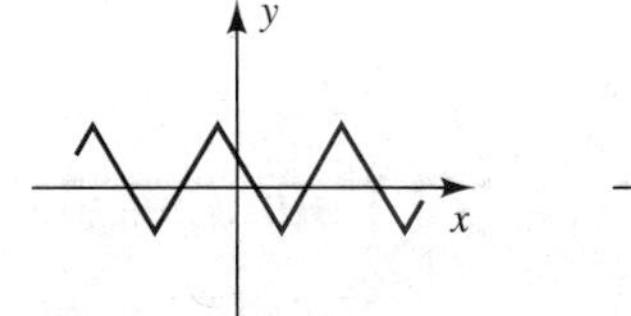

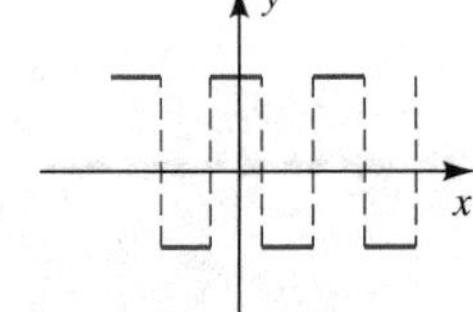

Figure 5.5
Periodic functions

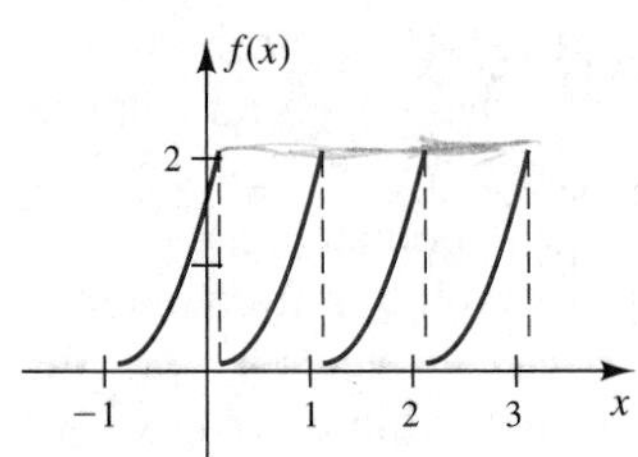

Figure 5.6

Example 1 Sketch the graph of the function that is defined over $0 \leq x < 1$ by $f(x) = 2x^2$ and is periodic with period one.

SOLUTION We sketch the graph of the function on $0 \leq x < 1$ and then repeat this pattern for all other intervals. (See Figure 5.6.) ▲

If a function is periodic with period p, then it is also periodic with periods $2p, 3p, \ldots, np, \ldots$; that is,

$$f(x) = f(x + p) = f(x + 2p) = f(x + 3p) = \cdots = f(x + np) = \cdots$$

However, only the value p is considered to be *the* period of the function.

The periodic nature of the trigonometric functions is of particular interest to us. Recall from Section 5.1 that for any real number u, the values of $\cos u$ and $\sin u$ are equal to the x- and y-coordinates, respectively, of the point on a unit circle corresponding to the number u. (See Example 2 in Section 5.1.) Furthermore, the values of sin u and cos u repeat themselves every 2π units. Thus sin u and cos u are periodic functions with period 2π. Since csc $u = 1/\sin u$ and sec $u = 1/\cos u$, it follows that csc u and sec u are also periodic with period 2π. We indicate this periodicity by writing

$$\begin{aligned} \sin u &= \sin(u + 2\pi) \\ \cos u &= \cos(u + 2\pi) \\ \sec u &= \sec(u + 2\pi) \\ \csc u &= \csc(u + 2\pi) \end{aligned}$$

The values of tan u in the first quadrant are repeated in the third quadrant. Similarly, the values of tan u in the second quadrant are repeated in the fourth quadrant. Thus tan u is a periodic function with period π. The same is true for cot u, and thus we write

$$\begin{aligned} \tan u &= \tan(u + \pi) \\ \cot u &= \cot(u + \pi) \end{aligned}$$

Boundedness

Many real-world physical phenomena tend to be constrained within certain limits or bounds. Such phenomena are described mathematically by functions called bounded functions. A function f is **bounded** if there is a number M for which $|f(x)| \leq M$ for all x in the domain of the function. For example, the functions graphed in Figure 5.7(a) are bounded, whereas those in Figure 5.7(b) are unbounded.

If a function is bounded, it has many bounds. But in describing the property of boundedness, we usually give only the lower and upper extremes of the functional values. For example, function f in Figure 5.7(a) varies between 0 and 4, so we write $0 \leq f(x) \leq 4$; the boundedness of g in Figure 5.7(a) is described by $-1 \leq g(x) \leq 2$. The graph of a function that is bounded between c and d is constrained between the lines $y = c$ and $y = d$, whereas the graph of an unbounded function is unlimited in its vertical extent.

Since the sine and cosine functions may be conceived of as coordinates of points on a unit circle, and since these coordinates are constrained to lie between -1 and 1, the sine and cosine functions are bounded by -1 and 1. Thus the graphs of the sine and cosine lie between the lines $y = 1$ and $y = -1$.

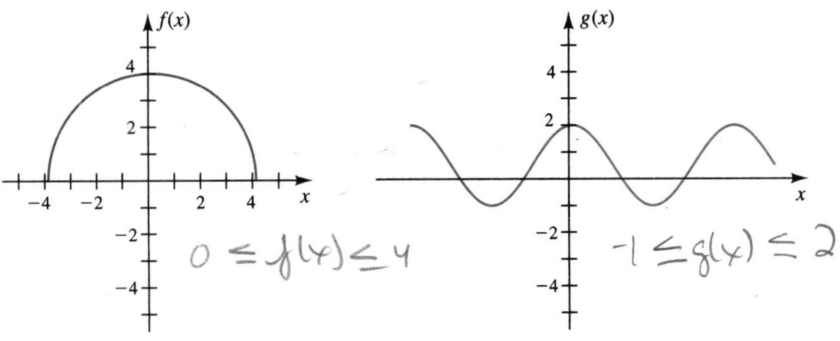

(a) Bounded functions

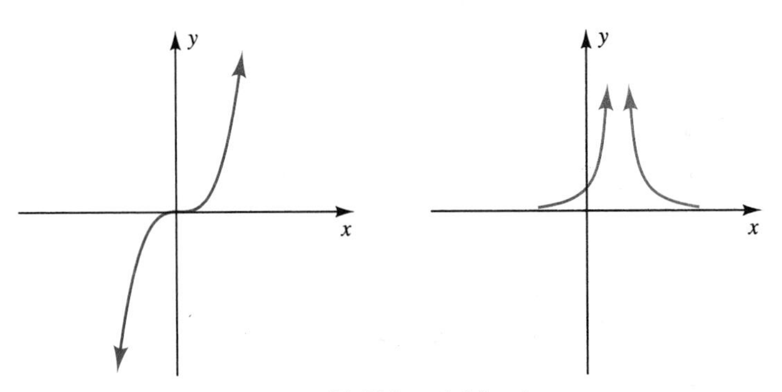

(b) Unbounded functions

Figure 5.7

Example 2

(a) The functional values of $y = x^2$ for $x \geq 0$ are *unbounded*, since the values x^2 get larger and larger as x takes on larger and larger values. The graph is shown in Figure 5.8(a).

(b) The functional values of $y = 1/(x - 3)$ are *unbounded* for x near 3, because the closer the value of x is to 3, the larger the value of y will be. See Figure 5.8(b).

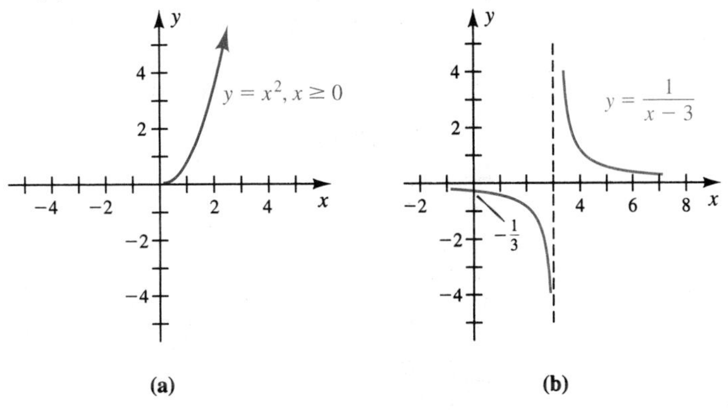

(a) **(b)**

Figure 5.8 ▲

Even and Odd Functions (See Sections 1.3 and 1.4)

Some functions have the property that their functional values do not change when the sign of the independent variable is changed. For example, if $f(x) = x^2$, then $f(-x) = (-x)^2 = x^2 = f(x)$. In general we denote this property by

$$f(x) = f(-x)$$

Functions that have this property are called *even* functions.

With some functions, when the sign of the independent variable changes, so does the sign of the dependent variable. For example, if $f(x) = x^3$, then $f(-x) = (-x)^3 = -x^3 = -f(x)$. This property is written in general as

$$f(-x) = -f(x)$$

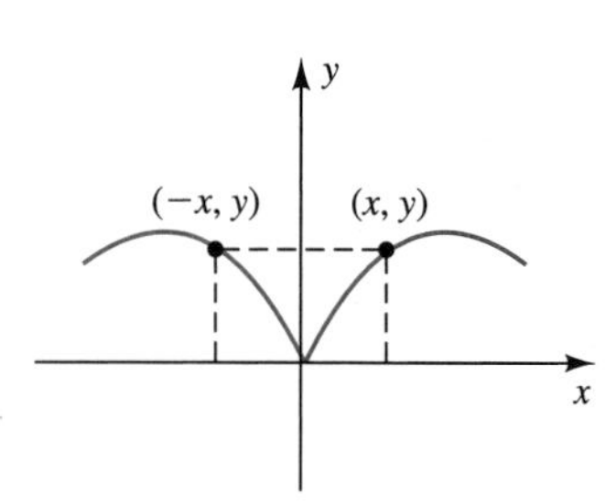

Figure 5.9
Symmetry with respect to y-axis

Functions that obey this rule are called *odd* functions.

The graphs of even and odd functions have an interesting kind of geometric symmetry.

1. For an even function $y = f(x)$, if the point (x, y) is on the graph of the function, so is the point $(-x, y)$. Graphs with this property are said to be *symmetric with respect to the y-axis*. (See Figure 5.9.)
2. For an odd function $y = f(x)$, if the point with coordinates (x, y) is on the graph of the function, so is the point $(-x, -y)$. Graphs with this property are said to be *symmetric with respect to the origin*. (See Figure 5.10.)

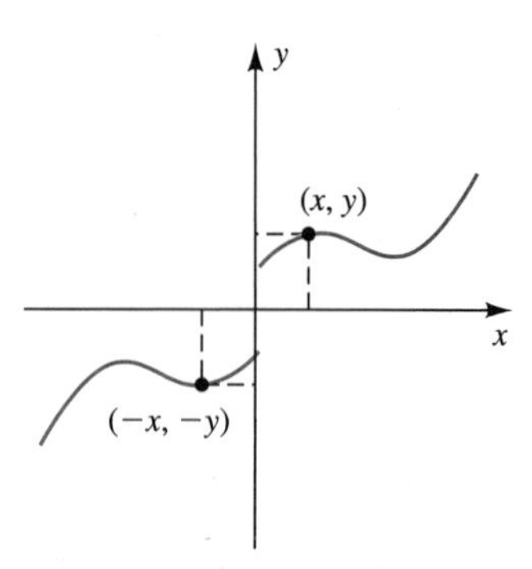

Figure 5.10
Symmetry with respect to origin

Usually, of course, a function is neither even nor odd. However, it can be proved that every function may be expressed as the sum of an even and an odd function.

By referring to Figure 5.4, we can demonstrate that $\sin x$ is an odd function; that is,

$$\sin(-x) = -\sin x$$

Similarly we can show that $\cos x$ is an even function; that is,

$$\cos(-x) = \cos x$$

Thus the graphs of these two functions exhibit the symmetry properties of odd and even functions, respectively. The calculator exercises at the end of the section will help you confirm these properties of $\sin x$ and $\cos x$.

Example 3 Determine whether the following functions are even, odd, or neither.

(a) $f(x) = x + \sin x$ **(b)** $g(x) = x + \cos x$ **(c)** $h(x) = x^2 + \cos x$

SOLUTION

(a) Since $f(x) = x + \sin x$, we have $f(-x) = (-x) + \sin(-x) = -x - \sin x = -(x + \sin x) = -f(x)$. Hence, since $f(-x) = -f(x)$, this function is odd.

(b) $g(-x) = (-x) + \cos(-x) = -x + \cos x$, which is equal to neither $g(x)$ nor $-g(-x)$. Hence this function is neither even nor odd.

(c) $h(-x) = (-x)^2 + \cos(-x) = x^2 + \cos x = h(x)$. Since $h(x) = h(-x)$, h is even. ▲

Variation of the Trigonometric Functions

As an angle increases from 0 to 2π, its terminal side rotates in a counterclockwise direction. During this interval all values of the trigonometric functions are included. Table 5.1 shows the variation in the values of $\sin u$ and $\cos u$ during one revolution of the terminal side. Because of the property of periodicity, the pattern in Table 5.1 will be repeated for $2\pi \leq u \leq 4\pi$, for $4\pi \leq u \leq 6\pi$, and so on.

Table 5.1 Variation of $\sin u$ and $\cos u$

When u increases	$\sin u$	$\cos u$
From 0 to $\frac{1}{2}\pi$	Increases from 0 to 1	Decreases from 1 to 0
From $\frac{1}{2}\pi$ to π	Decreases from 1 to 0	Decreases from 0 to -1
From π to $\frac{3}{2}\pi$	Decreases from 0 to -1	Increases from -1 to 0
From $\frac{3}{2}\pi$ to 2π	Increases from -1 to 0	Increases from 0 to 1

Graphs of the Sine and Cosine Functions

The graph of any function $y = f(x)$ is the set of points in the plane with the coordinates $(x, f(x))$. In general such a graph is obtained by tabulating values and plotting the corresponding points. But in particular cases the task can be significantly shortened by using some general properties of the given function.

We first examine the graphs of the sine and cosine functions considered as functions of *real numbers*. The procedures for graphing the sine and the cosine are almost identical.

Let us review some of the analytical properties previously discussed.

1. Both $\sin x$ and $\cos x$ are *bounded*, above by 1 and below by -1.

Thus the graph of each function lies between the lines $y = 1$ and $y = -1$.

2. Both $\sin x$ and $\cos x$ are *periodic* with period 2π. Thus only one interval of length 2π need be considered when the two functions under consideration are being graphed. Outside this interval the graph repeats itself.
3. The sine function is odd; that is, $\sin x = -\sin(-x)$. Thus the graph of $\sin x$ is symmetric about the origin. The cosine function is even; that is, $\cos x = \cos(-x)$. Thus its graph is symmetric about the y-axis.
4. $\sin x = 0$ for $x = 0, \pm\pi, \pm 2\pi, \pm 3\pi$, and so on.
   $\cos x = 0$ for $x = \pm\frac{1}{2}\pi, \pm\frac{3}{2}\pi, \pm\frac{5}{2}\pi$, and so on.
   In each case, these are the intercepts on the x-axis.
5. The numerical values of the sine and cosine functions for $0 \leq x \leq \frac{1}{2}\pi$ correspond to the values of $\sin x$ and $\cos x$ in the first quadrant. The other three quadrants yield values that are numerically the same (though there may be a difference in sign).

To obtain the specific graph, we need to calculate a reasonable number of points for $0 \leq x \leq \frac{1}{2}\pi$. (See Table 5.2.) Then we can draw a smooth curve through these points, after which we can use the general properties to obtain the remainder of the curve.

Table 5.2 Values of $\sin x$ and $\cos x$ from $x = 0$ to $x = 2\pi$ in increments of $\frac{1}{6}\pi$

x	0	$\frac{1}{6}\pi$	$\frac{1}{3}\pi$	$\frac{1}{2}\pi$	$\frac{2}{3}\pi$	$\frac{5}{6}\pi$	π	$\frac{7}{6}\pi$	$\frac{4}{3}\pi$	$\frac{3}{2}\pi$	$\frac{5}{3}\pi$	$\frac{11}{6}\pi$	2π
sin x	0	0.5	0.87	1	0.87	0.5	0	−0.5	−0.87	−1	−0.87	−0.5	0
cos x	1	0.87	0.5	0	−0.5	−0.87	−1	−0.87	−0.5	0	0.5	0.87	1

Figure 5.11 shows the graph of one period of the sine function. To the left of the graph is a circle of radius 1. The sine function has values numerically equal to the y-coordinates of points on this circle. This figure dis-

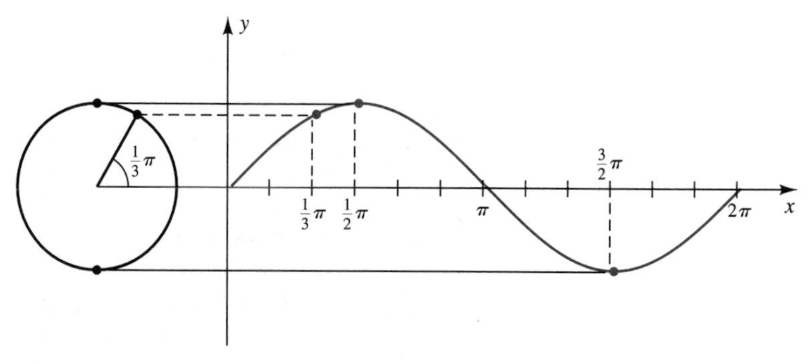

Figure 5.11

plays the relationship between points on the unit circle and points on the graph of the function.

Figure 5.12 is a graph of several periods of the function $y = \sin x$, and Figure 5.13 is a graph of several periods of the cosine function. Although in the figures the graphs terminate, in reality they continue indefinitely. A statement of the period is often included with the graph to emphasize this indefinite continuation.

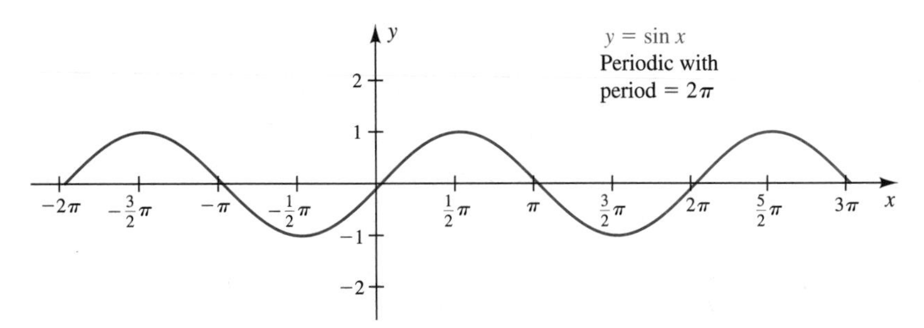

Figure 5.12

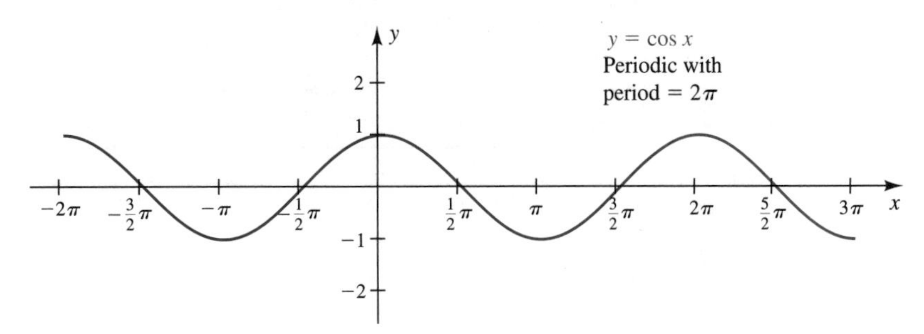

Figure 5.13

COMMENT Notice that in sketching the graphs in Figures 5.12 and 5.13 we used different units on the coordinate axes—we used integer units on the y-axis and units that are multiples of π on the x-axis. This approach causes some distortion in the shape of the curve, but it greatly simplifies the plotting process. Using different scale factors for the coordinate axes is acceptable as long as the basic units are clearly labeled.

The graph of the sine function is called a sine wave, or a **sinusoid**. The term *sinusoid* may be applied to any curve having the shape of the sine function. For example, the graph of the cosine function is properly called a sinusoid, since it is obtained simply by shifting the graph of the sine function $\frac{1}{2}\pi$ units to the left.

DEFINITION 5.2 A **cycle** is the shortest segment of the graph that includes one period. The **frequency** of the sinusoid is defined to be the reciprocal of the period.

Thus frequency represents the number of cycles of the function in each unit interval. The graph of any sinusoid should clearly demonstrate the boundedness, the periodicity, and the intercepts.

A sketch of a sinuoid can be obtained by connecting known points with a smooth curve. Although there is no rule stating how many points should be plotted for a given sinusoid, the idea is to choose just enough points to make the shape obvious. In this section we suggest you use six to ten points per period of the graph. Example 4 shows the point-plotting approach to graphing $y = \cos \frac{1}{2}x$, using multiples of $\frac{1}{2}\pi$ for x.

Example 4 Sketch the graph of $y = \cos \frac{1}{2}x$ on the interval $0 \leq x \leq 4\pi$, using multiples of $\frac{1}{2}\pi$ for x.

SOLUTION The table shows the various values for this interval. The graph (Figure 5.14) is then obtained by plotting these points. Notice that we have completed one period on the interval $[0, 4\pi]$. We conclude from this that the period of $y = \cos \frac{1}{2}x$ is 4π.

x	0	$\frac{1}{2}\pi$	π	$\frac{3}{2}\pi$	2π	$\frac{5}{2}\pi$	3π	$\frac{7}{2}\pi$	4π
y	1	0.71	0	−0.71	−1	−0.71	0	0.71	1

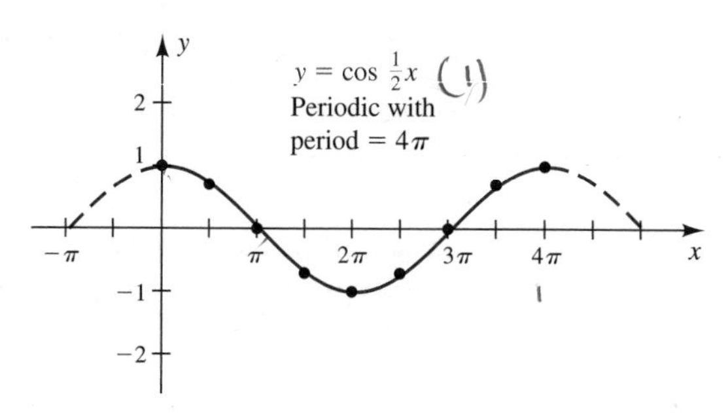

Figure 5.14 ▲

Example 5 Sketch the graph of $y = 3 \sin 2\pi x$ on the interval $0 \leq x \leq 1$, using multiples of $\frac{1}{6}$ for x. Multiples of π are frequently used for domain values, but as this example shows, that is not always the best choice.

SOLUTION We first construct a table of values, using a calculator to evaluate y. The corresponding graph is shown in Figure 5.15. Notice that the period is 1.

x	0	$\frac{1}{6}$	$\frac{1}{3}$	$\frac{1}{2}$	$\frac{2}{3}$	$\frac{5}{6}$	1
y	0	2.6	2.6	0	−2.6	−2.6	0

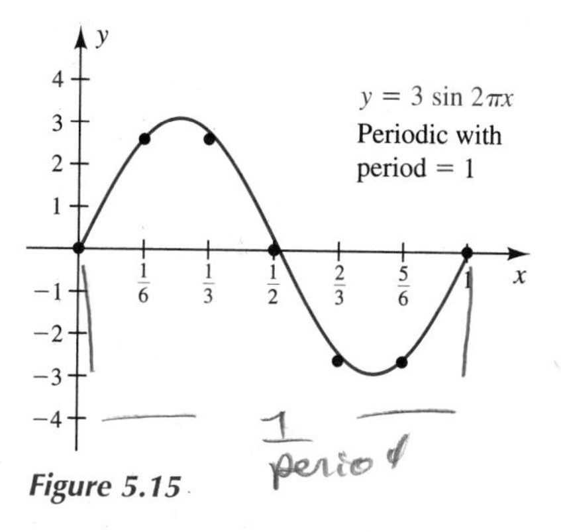

Figure 5.15

▲

Try graphing $y = \sin x$ and $y = \cos x$ on a graphing calculator. Initially use the default subdivisions for the x- and y-axes. Then use the stepping cursor in the trace mode to see how the wave gets generated by the points. Finally, change the subdivision scale on the axes and see what happens to the graph. Continue to use the cursor in the trace mode to step through the points on the graph. The calculator displays for $y = \sin x$ and $y = \cos x$ are shown in the figures below. The x-axis extends from -2π to 2π and the y-axis extends from -3 to 3.

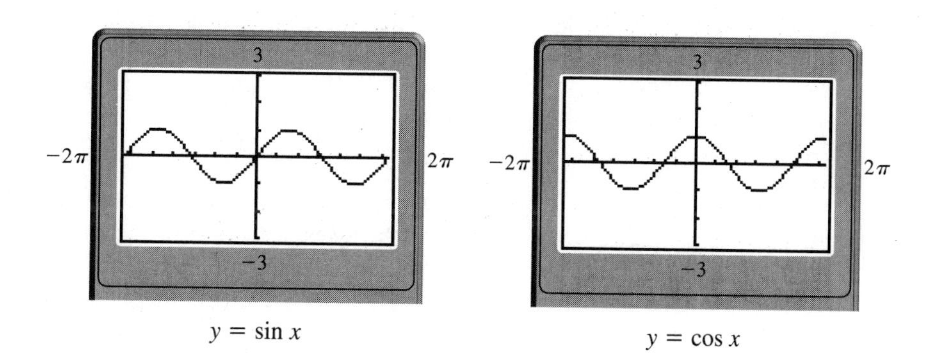

$y = \sin x$ $y = \cos x$

Exercises for Section 5.2

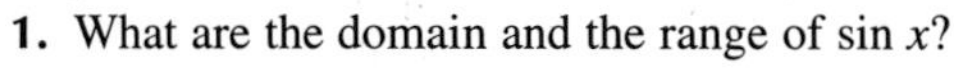

1. What are the domain and the range of $\sin x$?
2. What are the domain and the range of $\cos x$?

In Exercises 3–10 determine if the functions are bounded or unbounded on their interval of definition. If they are bounded, give the upper and lower extremes of their boundedness.

3. $y = x, x \geq 0$
4. $y = 2 \sin x$
5. $y = -3 \sin x$
6. $y = (x - 2)^2$
7. $y = 1/(x - 4)$
8. $y = 1/(x + 2)$
9. $y = 1/\sin x$
10. $y = 1/\cos x$
11. $y = \dfrac{\sin x}{\cos x}$
12. $y = x \sin x$
13. $y = x \cos x$
14. $y = \cos x + 2 \sin x$

In Exercises 15–24 determine whether the functions are even, odd, or neither.

15. $y = x \sin x$
16. $y = x \cos x$
17. $y = \sin x \cos x$
18. $y = \sin^2 x$
19. $y = \cos x - \sin x$
20. $y = (\cos x)/x$
21. $y = \cos x + 1$
22. $y = 3 \cos x - 2$
23. $1 - 2 \cos^2 x$
24. $y = \cos x + \sin x$

In Exercises 25–28 indicate the period of each sinusoid whose graph is shown, and tell whether it is a sine or cosine wave.

25.

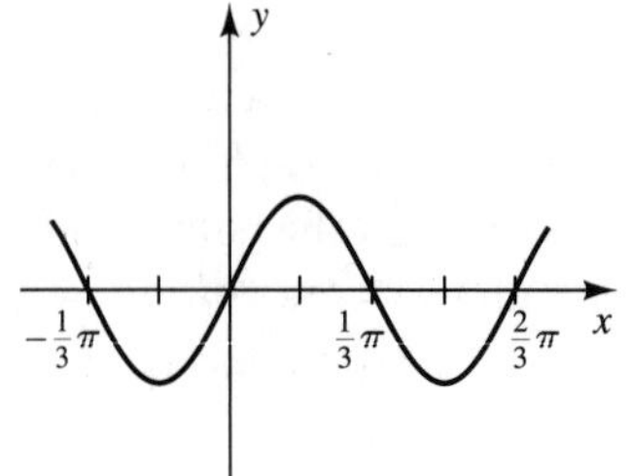

26.

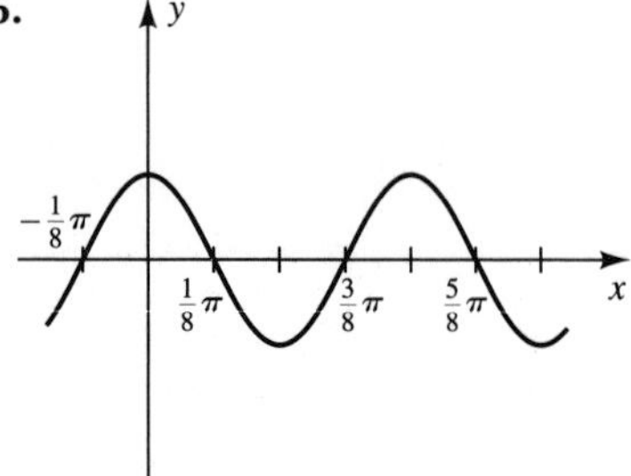

27.

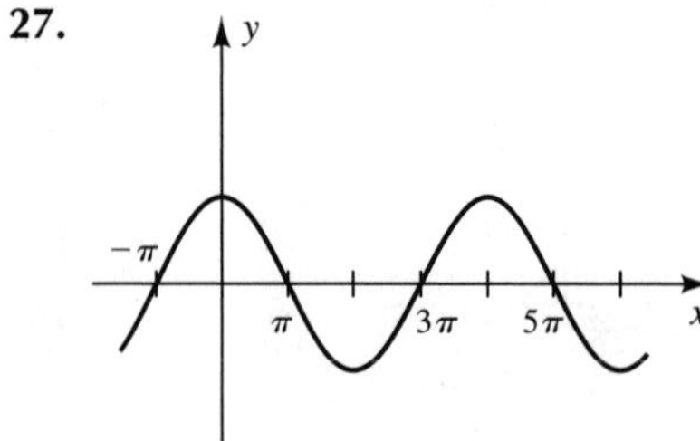

28.

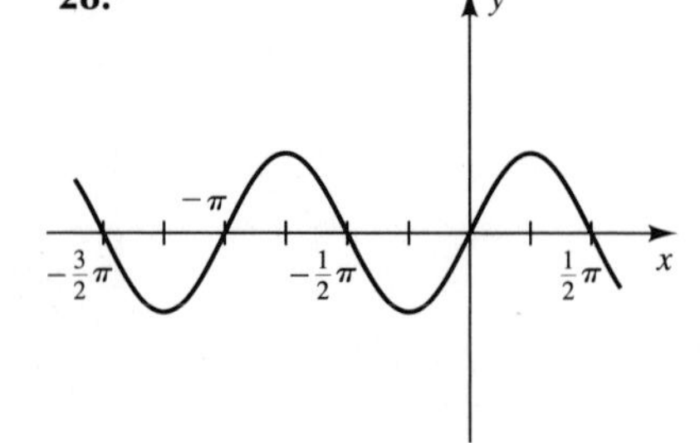

29. Sketch one period of $y = \sin x$, using multiples of $\frac{1}{4}\pi$ for x.
30. Sketch one period of $y = \cos x$, using multiples of $\frac{1}{4}\pi$ for x.

31. Sketch $y = 2 \sin x$ on the interval $0 \le x \le 2\pi$, using multiples of $\frac{1}{4}\pi$ for x.

32. Sketch $y = 3 \cos x$ on the interval $0 \le x \le 2\pi$, using multiples of $\frac{1}{4}\pi$ for x.

33. Sketch $y = \cos 2x$ on the interval $0 \le x \le \pi$, using multiples of $\frac{1}{8}\pi$. What is its period?

34. Sketch $y = \sin 2x$ on the interval $0 \le x \le \pi$, using multiples of $\frac{1}{8}\pi$. What is its period?

Sin x and cos x are defined for all real values of x, not just multiples of π. In Exercises 35–40 use a calculator to evaluate the given function at $x = 0, 1, 2, 3, 4, 5, 6, 7$. Sketch the graph of the given function on the interval $0 \le x \le 7$ by plotting these points and connecting them with a smooth curve.

35. $y = \cos x$ **36.** $y = \sin x$ **37.** $y = 1.5 \sin x$

38. $y = 3 \cos x$ **39.** $y = \cos \frac{1}{2}x$ **40.** $y = \sin \frac{1}{2}x$

41. The function $f(x) = |\sin x|$ is called the **full-wave rectified sine wave**. Sketch this function on the interval $0 \le x \le 2\pi$, using multiples of $\frac{1}{4}\pi$. What is its period?

42. The function defined by

$$f(x) = \begin{cases} \sin x, & 0 \le x \le \pi \\ 0, & \pi \le x \le 2\pi \end{cases}$$

and periodic with period 2π is called the **half-wave rectified sine wave**. Sketch this function on the interval $0 \le x \le 4\pi$, using multiples of $\frac{1}{4}\pi$.

43. Sketch $y = |\cos x|$ on the interval $-\frac{1}{2}\pi \le x \le \frac{3}{2}\pi$, using multiples of $\frac{1}{4}\pi$.

44. Make a sketch of $y = x$ and $y = \sin x$ on the same axes, and convince yourself that $\sin x < x$ for $x > 0$.

45. The velocity of a particle is given by the equation $v = \cos t$. Sketch the graph of velocity as a function of time. What is the velocity when $t = 0$? (This is called the *initial velocity*.) For which times is the velocity equal to zero? How would you describe these points graphically?

46. The vertical component of the motion of a particle that moves in a circular orbit is given by $v = R \sin t$, where R is the radius of the orbit. Sketch the variation in the value of the vertical component for $0 \le t \le \pi$ for a radius of 3 ft.

47. The output of an electrical circuit designed to produce triangular waves is given by

$$E = \begin{cases} \cos t, & 0 \le t \le \frac{1}{2}\pi \\ \sin t, & \frac{1}{2}\pi < t \le \pi \\ -\cos t, & \pi < t \le \frac{3}{2}\pi \\ -\sin t, & \frac{3}{2}\pi < t \le 2\pi \end{cases}$$

Sketch the graph of this output for the given interval.

Graphing Calculator Exercises

Exercises 1–6 are the same functions that you graphed by hand in Exercises 35–40. Use a graphing calculator to sketch the graph of the function. Set the RANGE of the viewing screen for $0 \le x \le 7$ and $-3 \le y \le 3$.

1. $y = \cos x$
2. $y = \sin x$
3. $y = 1.5 \sin x$
4. $y = 3 \cos x$
5. $y = \cos \frac{1}{2}x$
6. $y = \sin \frac{1}{2} x$

7. Using your calculator, fill in the values for $\sin x$. Write a sentence or two explaining how this table confirms the fact that $\sin x$ is an odd function.

x	$\sin x$
0.1	
−0.1	
1.2	
−1.2	
2.5	
−2.5	
3.7	
−3.7	

8. Using your calculator, fill in the values for $\cos x$. Write a sentence or two explaining how this table confirms the fact that $\cos x$ is an even function.

x	$\cos x$
0.1	
−0.1	
1.2	
−1.2	
2.5	
−2.5	
3.7	
−3.7	

9. Use a graphing calculator to plot the graph of $y = \sin x$ on the interval $[-4, 4]$. Using the TRACE feature, evaluate $\sin x$ for several values of $\pm x$. Do the results support the fact that $\sin(-x) = -\sin x$?
10. Use a graphing calculator to plot the graph of $y = \cos x$ on the interval $[-4, 4]$. Using the TRACE feature, evaluate $\cos x$ for several values of $\pm x$. Do the results support the fact that $\cos(-x) = \cos x$?

11. Use a graphing calculator to sketch the graphs of $y = \sin x$, $y = |\sin x|$, and $y = \sin |x|$ on the interval $[-7, 7]$. In a sentence or two describe the similarities and differences of the graphs of these three functions.

12. Use a graphing calculator to sketch the graphs of $y = \cos x$, $y = |\cos x|$, and $y = \cos |x|$ on the interval $[-7, 7]$. In a sentence or two describe the similarities and differences of the graphs of these three functions.

5.3 Expansion and Contraction of Sine and Cosine Graphs

In Section 5.2 you saw how to graph the sine and cosine functions by plotting points. Graphing modified functions such as $\sin 2x$, $4 \cos x$, and $\cos(x + 3)$ with the point-plotting process can be tedious. In this section you will learn how to simplify the procedure for graphing the sine and cosine functions.

Five-Point Plotting Procedure

The procedure for graphing one period of the sine function can be simplified by observing that the graph of $y = \sin x$ starts at the origin, reaches a maximum at $\frac{1}{2}\pi$, crosses the x-axis at π, reaches a minimum at $\frac{3}{2}\pi$, and completes the period by crossing the x-axis at 2π. In general the zeros and the maximum and minimum values of the sine function occur at multiples of $\frac{1}{4}$ of the period. Therefore the graph of one period of any sine function begins on the x-axis, reaches its maximum at $\frac{1}{4}$ period, crosses the x-axis at $\frac{1}{2}$ period, reaches a minimum of $\frac{3}{4}$ period, and completes the period at the x-axis. The sine curve in Figure 5.16 shows this pattern, which is the basis of the five-point plotting procedure.

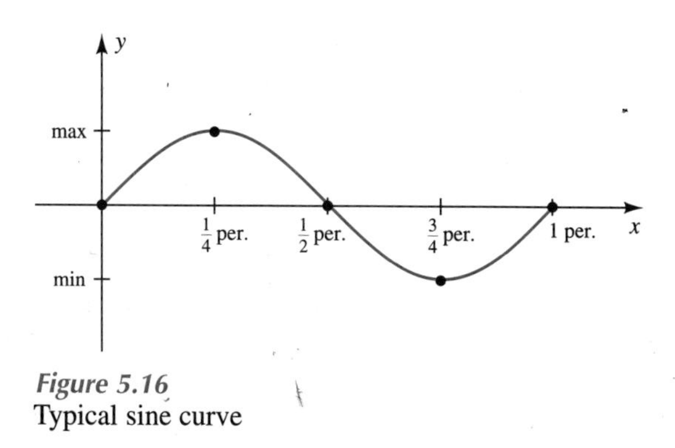

Figure 5.16
Typical sine curve

A similar five-point plotting procedure can be used for the cosine function. The graph of the cosine function over one period begins at a maximum, crosses the x-axis at $\frac{1}{4}$ period, reaches a minimum at $\frac{1}{2}$ period, crosses the x-axis at $\frac{3}{4}$ period, and reaches a maximum at the end of the period. (See Figure 5.17.)

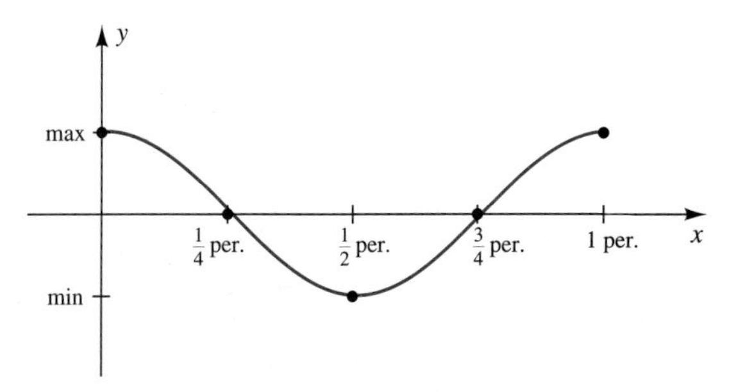

Figure 5.17
Typical cosine curve

The five-point plotting procedure is especially helpful in drawing graphs of sine and cosine functions that have been modified by the addition of or multiplication by a constant. In this section and the next we will discuss the following four modifications of sine and cosine functions:

1. Multiplication of the function by a constant
2. Multiplication of the argument by a constant
3. Addition of a constant to or subtraction of a constant from the argument
4. Addition of a constant to or subtraction of a constant from the functional value

Multiplication of the Function by a Constant

We have seen that the values of the sine function oscillate between $+1$ and -1. Consider the equation $y = A \sin x$. Since

$$-1 \leq \sin x \leq 1$$

it follows that

$$-|A| \leq A \sin x \leq |A|$$

The number A is called the amplitude coefficient, and $|A|$ is the **amplitude** of the sine wave. If $|A|$ is greater than 1, the amplitude of the sine wave is increased; if $|A|$ is less than 1, the amplitude is decreased. Sometimes $|A|$ is called the **maximum**, or **peak**, value of the function. Figure 5.18 shows the graph of $y = A \sin x$ for $A = 1, \frac{1}{2}$, and 2. The graphs of these three func-

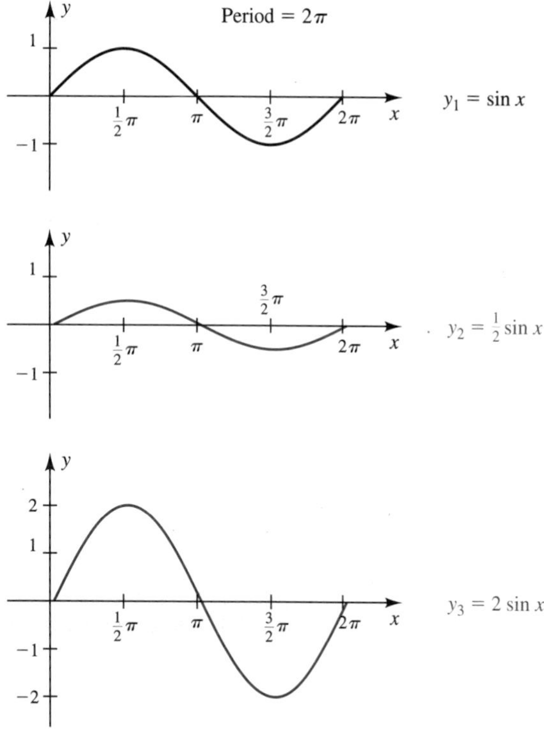

Figure 5.18

tions have exactly the same shape and cross the axes at the same places, but y_2 has an amplitude that is one-half that of y_1, and y_3 has an amplitude that is twice that of y_1. Similarly, $|A|$ is the amplitude of $y = A \cos x$.

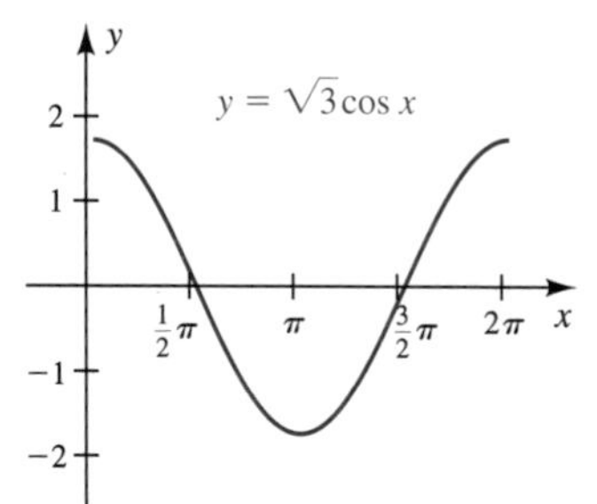

Figure 5.19

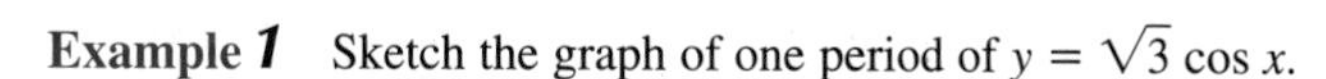

Example 1 Sketch the graph of one period of $y = \sqrt{3} \cos x$.

SOLUTION The function $y = \sqrt{3} \cos x$ has an amplitude of $\sqrt{3}$ and a period of 2π. The graph of $y = \cos x$ is at its maximum at $x = 0$, has a zero at $x = \frac{1}{2}\pi$, is at its minimum at $x = \pi$, has a zero at $x = \frac{3}{2}\pi$, and is at its maximum again at $x = 2\pi$. The graph of the given function is shown in Figure 5.19. ▲

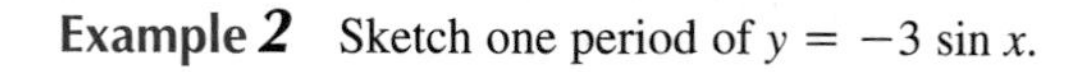

Example 2 Sketch one period of $y = -3 \sin x$.

SOLUTION The amplitude coefficient in this case is negative, which means that the sign of each functional value will be opposite what it is for $y = 3 \sin x$; $y < 0$ when $\sin x > 0$, and $y > 0$ when $\sin x < 0$. Thus the graph of $y = -3 \sin x$ is the reflection in the x-axis of the graph of $y = 3 \sin x$. Both graphs are shown in Figure 5.20 to emphasize the relationship between the two curves.

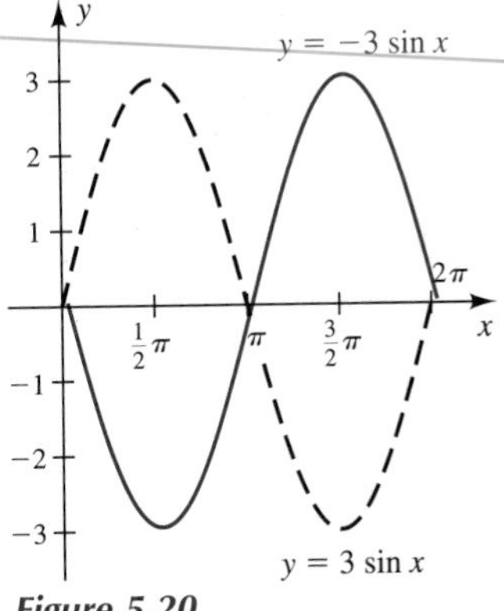

Figure 5.20

▲

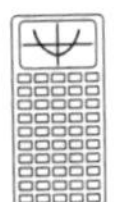

A graphing calculator can be used to display more than one graph at a time. The figure below shows the graphs of $y_1 = \sin x$, $y_2 = \frac{1}{2} \sin x$, and $y_3 = 2 \sin x$ as they might appear on a graphing calculator. The horizontal axis extends from -2π to 2π, and the vertical axis extends from -3 to 3. Can you tell which graph represents which function?

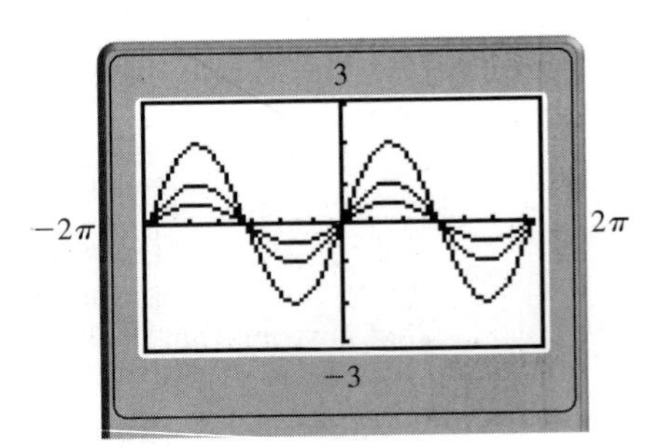

Multiplication of the Argument by a Constant

If we multiply the argument of $\sin x$ by a positive constant B, the function becomes $\sin Bx$. The graph of this function is sinusoidal, but since the argument is Bx, one period of $\sin Bx$ is contained in the interval

$$0 \le Bx \le 2\pi$$

or

$$0 \le x \le \frac{2\pi}{B}$$

Therefore the period of sin Bx is $2\pi/B$. For example, the period of sin $2x$ is π; the period of sin $\frac{1}{2}x$ is 4π. Graphically, increasing B has the effect of squeezing the sine curve together like an accordion, and decreasing B has the effect of pulling it apart. (See Figure 5.21.)

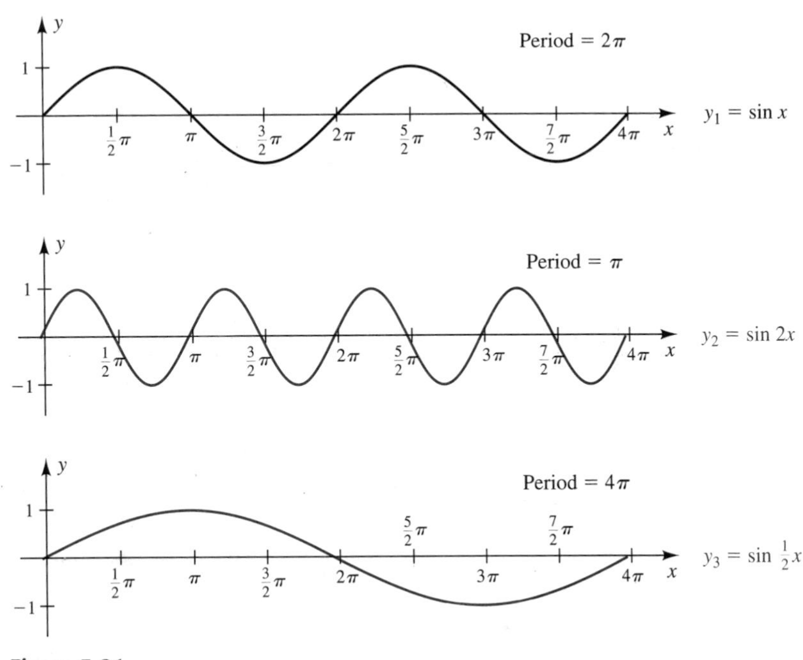

Figure 5.21

Similarly, the period of cos Bx is $2\pi/B$. Thus

> the period of both sin Bx and cos Bx is $\dfrac{2\pi}{B}$, $(B > 0)$.

Figure 5.21 shows the graphs of $y_1 = \sin x$, $y_2 = \sin 2x$, and $y_3 = \sin \frac{1}{2}x$. Notice that the graphs all have the same amplitude, but the period of each is different. The period of y_2 is one-half that of y_1, and the period of y_3 is twice that of y_1.

Example 3

(a) sin $3x$ has a period of $\dfrac{2}{3}\pi$.

(b) $\cos \dfrac{1}{4}x$ has a period of $\dfrac{2\pi}{1/4} = 8\pi$.

(c) $\cos 10x$ has a period of $\dfrac{2\pi}{10} = \dfrac{1}{5}\pi$.

(d) $\sin 4\pi x$ has a period of $\dfrac{2\pi}{4\pi} = \dfrac{1}{2}$. ▲

A general procedure for sketching sine and cosine functions when you are given the amplitude and the period is summarized below.

- For the argument of the function, choose units that are equal to $\frac{1}{4}$ of the period of the function.
- Remember that the zeros and the maximum and minimum values occur at multiples of $\frac{1}{4}$ of the period. So if the period is 2π, for example, use $\frac{1}{4}(2\pi) = \frac{1}{2}\pi$ as a basic unit on the x-axis; if the period is 5, use $\frac{1}{4}(5)$, or $\frac{5}{4}$.
- Remember that the graph of the sine function over one period crosses the x-axis at its initial point, reaches a maximum at $\frac{1}{4}$ period, crosses the x-axis at $\frac{1}{2}$ period, reaches a minimum at $\frac{3}{4}$ period, and crosses the x-axis at the end of the period.
- Remember that the graph of the cosine function over one period begins at a maximum, crosses the x-axis at $\frac{1}{4}$ period, reaches a minimum at $\frac{1}{2}$ period, crosses the x-axis at $\frac{3}{4}$ period, and reaches a maximum at the end of the period.

Example 4 Sketch the graph of one period of $y = 2\cos 5x$.

SOLUTION Here the amplitude is 2 and the period is $\frac{2}{5}\pi$. The graph is shown in Figure 5.22. Notice that the units along the x-axis are in multiples of $\frac{1}{4}$ of the period—that is, $\frac{1}{4}(\frac{2}{5}\pi) = \frac{1}{10}\pi$.

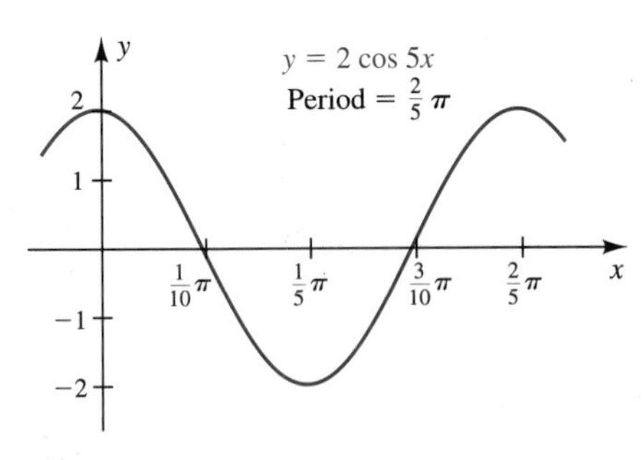

Figure 5.22 ▲

Example 5 Sketch one period of the graph of $s = 2.1 \sin 3\pi t$.

SOLUTION In this case the amplitude is 2.1 and the period is $2\pi/3\pi = 2/3$. The graph appears in Figure 5.23. The units along the x-axis are in multiples of $\frac{1}{4}$ of the period—that is, $\frac{1}{4}(\frac{2}{3}) = \frac{1}{6}$.

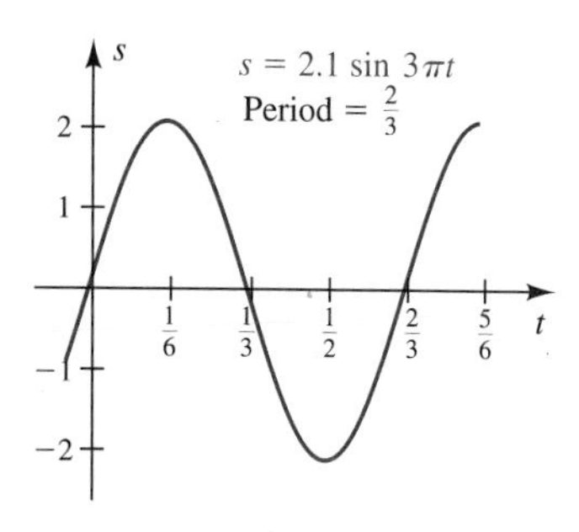

Figure 5.23 ▲

Example 6 Write the equation of the sine function whose amplitude is 3 and whose period is 7.

SOLUTION The general form of the desired sine function is $y = A \sin Bx$. The amplitude 3 yields $A = 3$. The period is given by $2\pi/B$, so we must have $2\pi/B = 7$, from which $B = 2\pi/7$. The desired function is $y = 3 \sin 2\pi/7x$. ▲

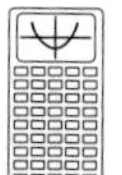

Try drawing $y_1 = \cos x$ and $y_2 = \cos 2x$ on a graphing calculator. Then try other arguments, such as $3x$, $\frac{1}{4}x$, and so on, and observe how they change the graph. The screen in the following figure shows the graphs of $y = 2 \sin \frac{1}{2}x$ and $y = \cos x$ on the same coordinate system. Can you tell which is which? If the x-axis extends from -2π to 2π, what is the period of each graph?

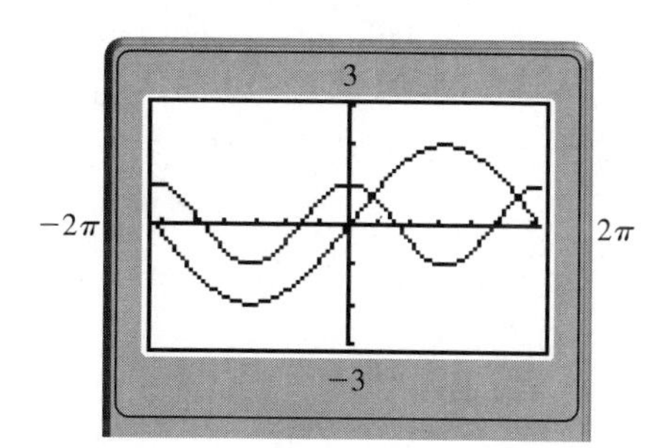

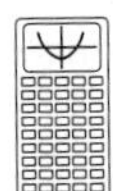

When using a calculator to graph sine or cosine functions, keep the amplitude and period of the function in mind when you choose a scale for the axes. For example, if we sketch $y = \sin 0.1x$ with the x-axis extending from -2π to 2π, as shown in the following figure, it is difficult to recognize that this is a sine function because we see only part of one period. Since the period of $\sin 0.1x$ is $2\pi/0.1 = 20\pi$, a better choice for the extent of the x-axis is from -20π to 20π, as shown on the right. Notice how a choice of reasonable limitations on the x values enhances our understanding of the graph.

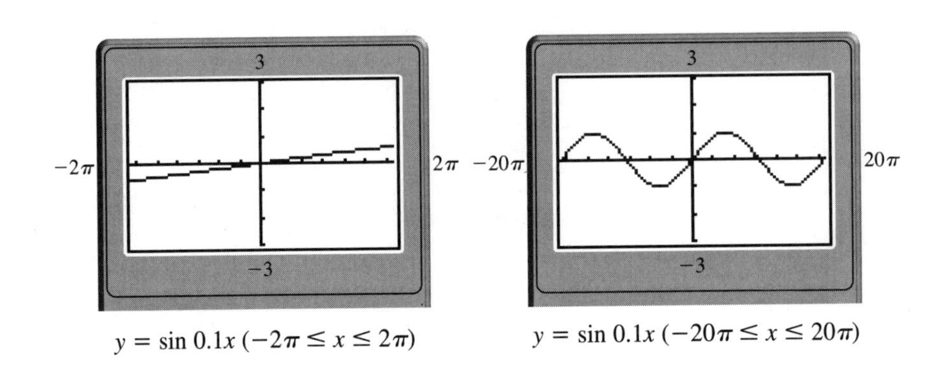

$y = \sin 0.1x \ (-2\pi \le x \le 2\pi)$ $\quad$ $y = \sin 0.1x \ (-20\pi \le x \le 20\pi)$

▲▼ Exercises for Section 5.3

In Exercises 1–20 sketch the graphs of the given functions. In each case give the amplitude and the period.

1. $y = 3 \sin x$
2. $y = \frac{1}{2} \sin x$
3. $y = 6 \cos x$
4. $y = \frac{1}{3} \sin x$
5. $y = \sin \frac{2}{3}x$
6. $y = 0.3 \sin 3x$
7. $y = \sin \pi x$
8. $y = \cos 0.1x$
9. $s = \frac{1}{2} \cos 2t$
10. $y = 100 \cos 3x$
11. $y = 8.2 \sin 0.4x$
12. $v = 3 \sin \frac{7}{6}t$
13. $y = -\cos \frac{2}{5}x$
14. $y = -5 \cos 7x$
15. $p = \pi \cos 100t$
16. $i = -0.02 \sin \pi t$
17. $y = -12 \sin 0.2x$
18. $P = 10^6 \cos \dfrac{\pi}{1000} x$
19. $v = \dfrac{\cos 1000\pi x}{50}$
20. $y = \dfrac{3 \sin 25\pi t}{200}$

In Exercises 21–28 write the equation of the sine function having the indicated amplitude and period.

21. Amplitude $= \frac{1}{3}$, period $= 12$
22. Amplitude $= \frac{1}{2}$, period $= 15$
23. Amplitude $= 20$, period $= \frac{3}{8}$
24. Amplitude $= \sqrt{5}$, period $= \frac{1}{3}$

25. Amplitude $= 2.4$, period $= \frac{1}{3}\pi$

26. Amplitude $= 0.94$, period $= \frac{1}{6}\pi$

27. Amplitude $= \pi$, period $= 3\pi$

28. Amplitude $= 2/\pi$, period $= 7\pi$

29. The motion of a pendulum can be represented by the equation $x = A \sin Bt$. Write the equation of motion of a pendulum oscillating with an amplitude of 3.2 ft and a period of 2.5 sec. (See Figure 5.24.)

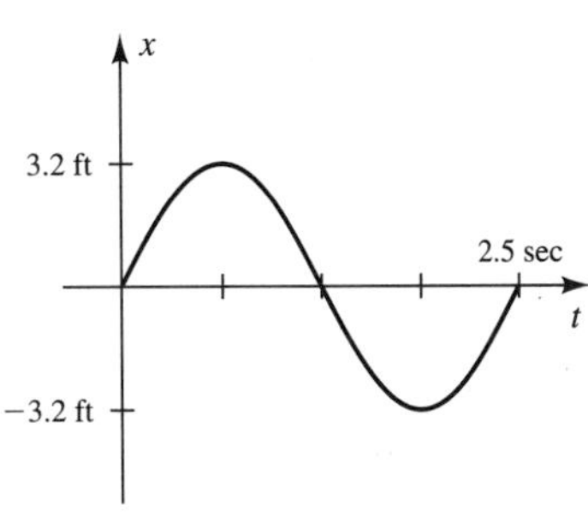

Figure 5.24

30. The equation for the voltage drop across the terminals of an ordinary electric outlet is given approximately by

$$E = 156 \sin(110\pi t)$$

Sketch the voltage curve for several cycles.

31. If B is small, the equation $y = \sin Bx$ approximates the shape of ocean waves. Sketch several cycles of an ocean wave described by

$$y = \sin \frac{1}{20}\pi x$$

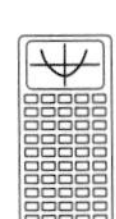

Graphing Calculator Exercises

1. Use a graphing calculator to display the graphs of the trigonometric functions in Exercises 1, 5, 7, 9, and 15 above. Use a scale on the x-axis that will show two periods of the graph. In each case, compare the calculator graph with the one you drew.

2. Use a graphing calculator to display the graphs of $y_1 = 3 \sin x$ and $y_2 = \sin 3x$ on the same coordinate axes for $0 \le x \le 2\pi$. On the basis of the graphs, would you say that $3 \sin x = \sin 3x$? Why?

3. Use a graphing calculator to display the graphs of $y_1 = \cos \frac{1}{3}x$ and $y_2 = \frac{1}{3} \cos x$ on the same coordinate axes for $0 \le x \le 6\pi$. On the basis of the graphs, would you say that $\cos \frac{1}{3}x = \frac{1}{3} \cos x$? Why?

4. Use a graphing calculator to display the graphs of $y_1 = \sin(-x)$ and $y_2 = -\sin x$ on the same coordinate axes for $-2\pi \le x \le 2\pi$. On the basis of the graphs, would you say that $\sin(-x) = -\sin x$? Why?

Many graphing calculators have the capability of sequentially graphing functions with a parameter such as $k \sin x$ or $\sin kx$ for a variety of (integer) values of k. In this manner, the impact of the constants on the graphs can be displayed. For example, the calculator will graph the function $\{1,2,3\}\sin x$ as $\sin x$, $2 \sin x$, and $3 \sin x$ consecutively. Similarly the function $\sin(\{1,2,3\}x)$ results in $\sin x$, $\sin 2x$, and $\sin 3x$.

This is done on the TI-85 using the GRAPH and the LIST keys. The key stroke sequence to graph $\{1,2,3\}\sin x$ is as follows:

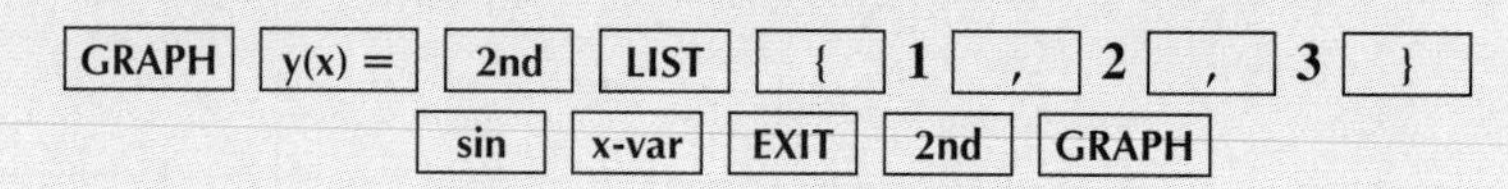

5. Using a graphing calculator and a LIST, graph $k \cos x$ for $k = 1,2,3$.
6. Repeat Exercise 5 for $\sin kx$.

5.4 Vertical and Horizontal Translation

A vertical or horizontal relocation of the graph of a function in which the graph's shape is not changed or distorted is called a **translation**. A function is translated vertically if a constant is added to the value of the function, and it is translated horizontally if a constant is added to the argument of the function. Both of these translations, which are important in applied work, are explained here in the context of sine and cosine functions.

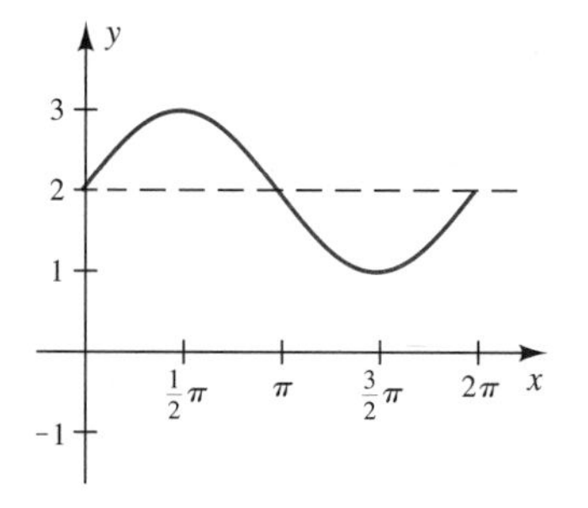

Figure 5.25

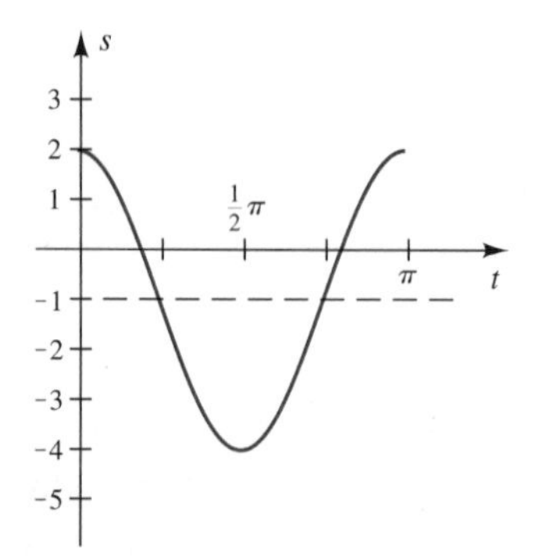

Figure 5.26

Addition of a Constant to the Value of the Function

Consider the graph of $y = D + \sin x$. Since D is added to $\sin x$ for each x, the graph of $D + \sin x$ is simply the graph of $\sin x$ displaced D units vertically. The graph is translated up if D is positive and down if D is negative. D is called the **mean value,** or **average value**, of the function.

Example 1 Sketch the graph of $y = 2 + \sin x$.

SOLUTION The graph is shown in Figure 5.25. The mean value is 2, the amplitude is 1, and the period is 2π. ▲

Example 2 Sketch the graph of $s = -1 + 3 \cos 2t$.

SOLUTION The graph is shown in Figure 5.26. The mean value is -1, the amplitude is 3, and the period is π. ▲

Addition of a Constant to the Argument

The addition of a constant to the argument of $\sin x$ is written $\sin(x - C)$. The constant C has the effect of shifting the graph of the sine function to the right or to the left. Notice that $\sin(x - C)$ is zero when $x - C = 0$ (that is, for $x = C$). The value of x for which the argument of the sine function is zero is called the **phase shift**. If C is positive the shift is to the right, and if C is negative the shift is to the left. Figure 5.27 shows three sine waves with phase shifts of 0, $-\frac{1}{4}\pi$, and $\frac{1}{4}\pi$.

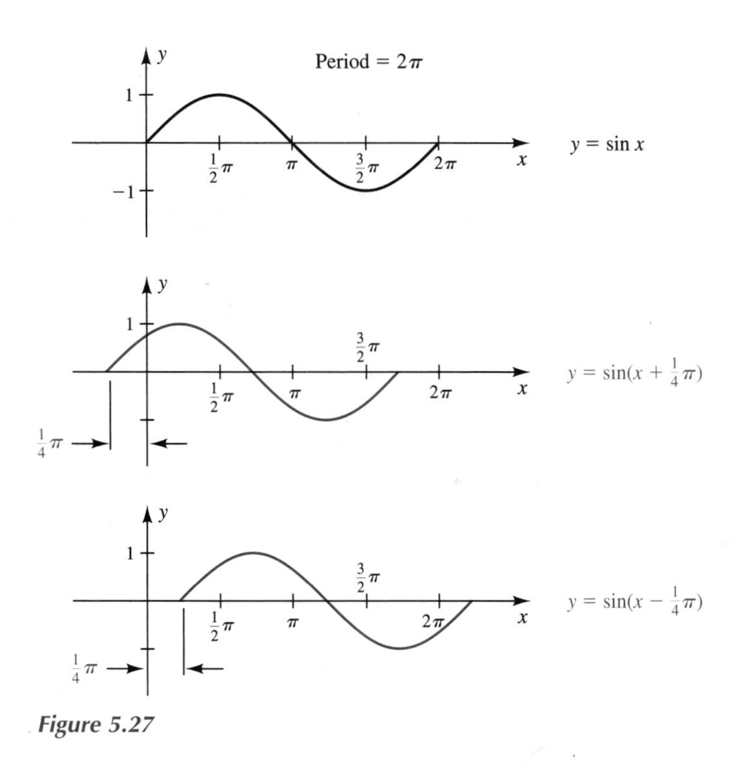

Figure 5.27

In the more general case the effects of changes in amplitude, period, and phase shift are all combined. The function

$$y = A \sin(Bx - C) = A \sin B\left(x - \frac{C}{B}\right)$$

has an amplitude of A, a period of $2\pi/B$, and a phase shift corresponding to the value of x given by $Bx - C = 0$ (that is, $x = C/B$). Figure 5.28 shows a graph of the basic sine curve and the graph of $y = 3 \sin(2x - \frac{1}{3}\pi)$.

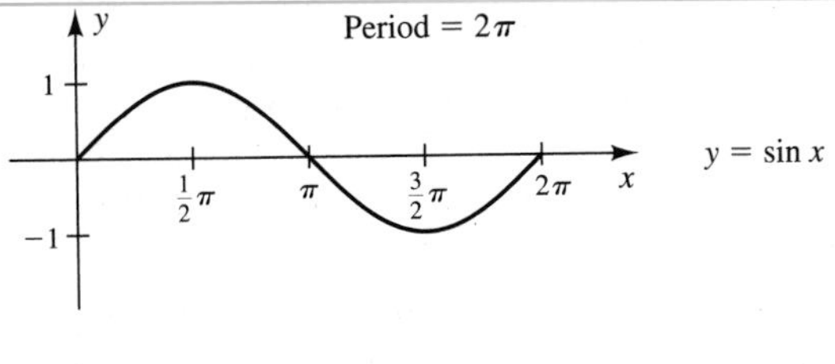

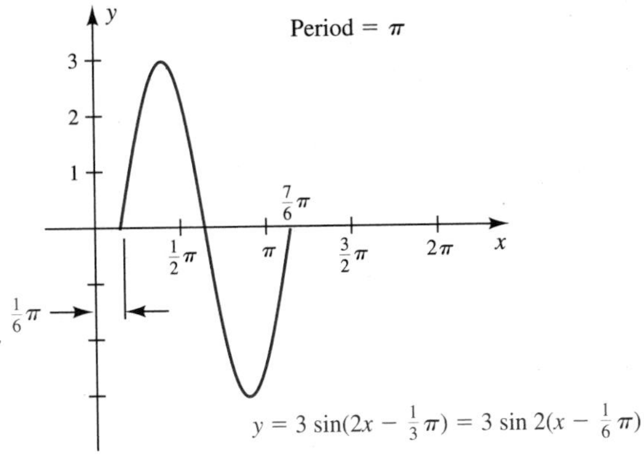

Figure 5.28

COMMENT Figure 5.28 shows the graph of $y = 3\sin(2x - \frac{1}{3}\pi)$ over a fundamental period. The following step-by-step approach can be used in making such a sketch of $y = A\sin(Bx - C)$.

- The amplitude is the coefficient of the sine function A. In this case $A = 3$. This gives the upper and lower values of the range of the graph.
- A fundamental period, which is $2\pi/B$, begins at the point where $Bx - C = 0$ and ends where $Bx - C = 2\pi$. In the particular function in Figure 5.28, we have

$$2x - \frac{1}{3}\pi = 0, \text{ which gives}$$

$$x = \frac{1}{6}\pi \text{ for the left-hand endpoint of the fundamental period}$$

$$2x - \frac{1}{3}\pi = 2\pi, \text{ which gives}$$

$$x = \frac{7}{6}\pi \text{ for the right-hand endpoint of a fundamental period}$$

Notice that the left-hand endpoint of the fundamental period also gives us the phase shift.

▶ The period of this function, which is $2\pi/2 = \pi$, can be computed as the difference between the right- and left-hand endpoints (in this case $\frac{7}{6}\pi - \frac{1}{6}\pi = \pi$). The latter computation is often used as an accuracy check.

Example 3 Sketch the graph of $y = 2\sin(\frac{1}{3}x + \frac{1}{9}\pi)$.

SOLUTION The amplitude of the graph is 2 and the fundamental period is $2\pi/(\frac{1}{3}) = 6\pi$. The phase shift is $-\frac{1}{3}\pi$, which is found by solving $\frac{1}{3}x + \frac{1}{9}\pi = 0$ for x. Hence the fundamental period starts $\frac{1}{3}\pi$ units to the left of the origin. The end of the fundamental period is found by solving $\frac{1}{3}x + \frac{1}{9}\pi = 2\pi$ for x; that is, $x = \frac{17}{3}\pi$. (See Figure 5.29.) For convenience, the x-axis is marked off in increments of $\frac{1}{3}\pi$.

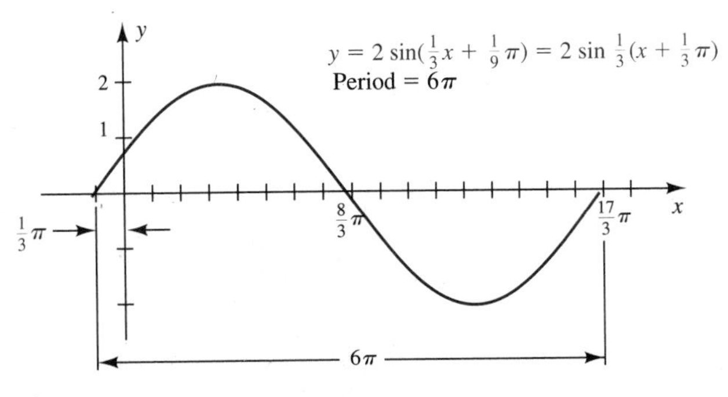

Figure 5.29 ▲

In all cases in which a sine curve is modified, the distinctive shape of the curve remains unaltered. This basic shape is expanded or contracted vertically because of multiplication by amplitude constant A, expanded or contracted horizontally by the constant B, and shifted to the right or left by the constant C/B.

A similar analysis could be made for the cosine function. We will not discuss in detail the function $y = D + A\cos(Bx - C)$, but the constants A, B, C, and D alter the basic cosine graph the same way they alter the sine graph.

Example 4 Sketch the graph of $y = 3\cos(\frac{1}{2}x + \frac{1}{4}\pi)$.

SOLUTION The amplitude is 3, since the basic cosine function is multiplied by 3. The period is $2\pi/(\frac{1}{2}) = 4\pi$. The phase shift is found by solving $\frac{1}{2}x + \frac{1}{4}\pi = 0$; that is, $x = -\frac{1}{2}\pi$. Hence the fundamental period starts $\frac{1}{2}\pi$ units to the left of the origin.

The fundamental period ends at $\frac{1}{2}x + \frac{1}{4}\pi = 2\pi$, which gives $x = 7\pi/2$. The graph is shown in Figure 5.30.

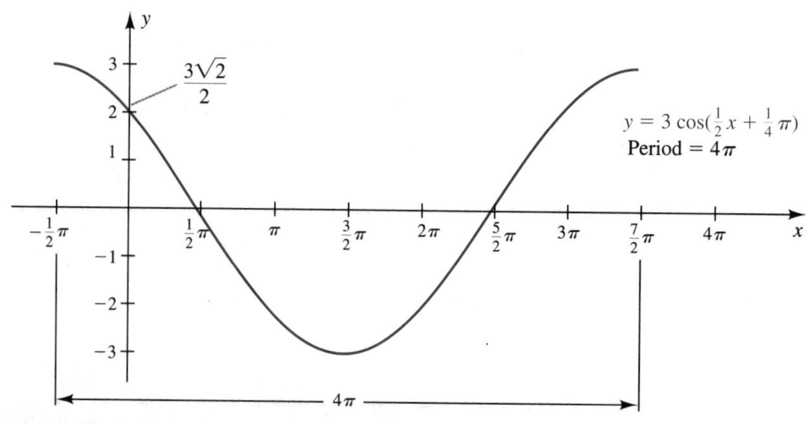

Figure 5.30 ▲

If a function is multiplied by a negative constant, you may use one of the relationships in Equation (2.5) of Section 2.4 (page 82) to put the expression into a form with a positive constant. See Example 5(b).

Example 5

(a) Sketch the graph of $y = -2\sin(3x + 1)$.

(b) Express the given function in the form $y = A\sin(Bx + C)$, where A and B are positive constants.

SOLUTION

(a) To graph this function, sketch the function $y = 2\sin(3x + 1) = 2\sin 3(x + \frac{1}{3})$ and then reflect this graph in the x-axis. See Figure 5.31.

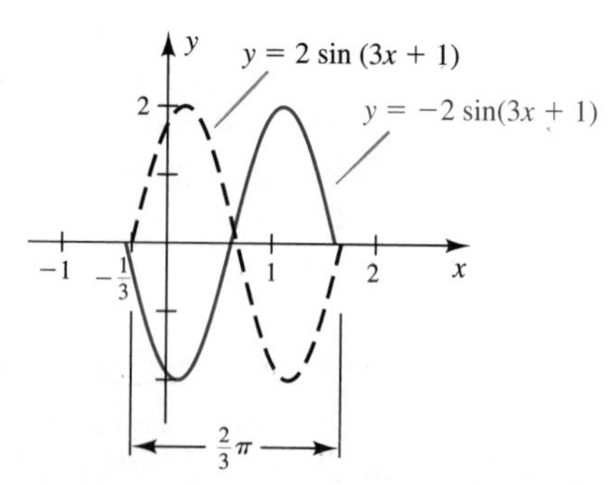

Figure 5.31

(b) From Section 2.4 we know that $\sin x = -\sin(-x)$, so the given function may be written

$$y = -2\sin(3x + 1) = 2\sin(-3x - 1)$$

Also from Section 2.4, $\sin(x + \pi) = \sin(-x)$, so

$$\begin{aligned} y &= 2\sin(-3x - 1) \\ &= 2\sin[-(3x + 1)] \\ &= 2\sin[(3x + 1) + \pi] \\ &= 2\sin[3x + (1 + \pi)] \end{aligned}$$

▲

The graph of $y = \sin(x + 0.5)$ is shown below as it might appear on the screen of a graphing calculator. We recognize that this function has a phase shift of 0.5 unit to the left, so its graph should cross the x-axis at -0.5. To obtain the phase shift from the graph, move the cursor to the point where the graph crosses the x-axis just to the left of the origin (see the figure on the left). The readout of this point is -0.47619, which is the calculator estimate of the phase shift. The calculator does not give the exact value because the x-axis is composed of a finite number of points, and -0.5 lies between two of these points. To obtain a closer estimate of the phase shift, use the zoom feature. The figure on the right shows $x = -0.49659$ as the phase shift once the zoom feature was used.

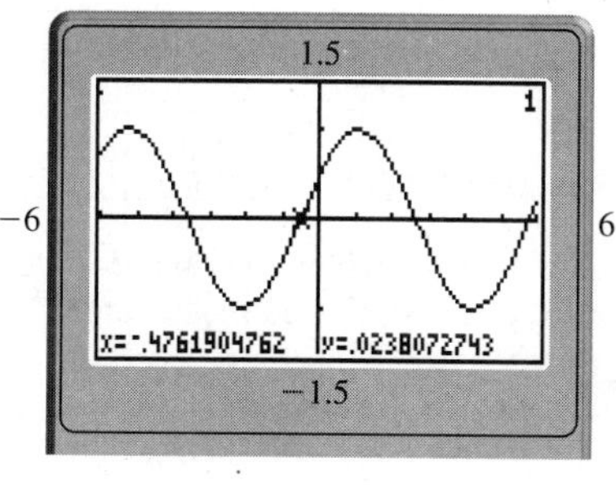

$y = \sin(x + 0.5)$

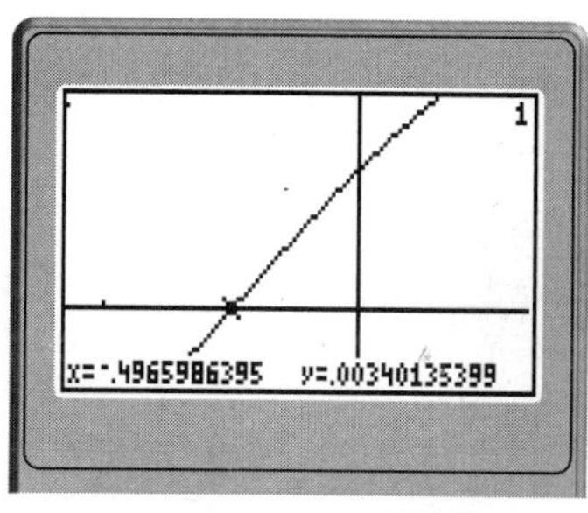

$y = \sin(x + 0.5)$ zoom

Example **6** Write the equation of the sinusoid whose graph over one period is shown below.

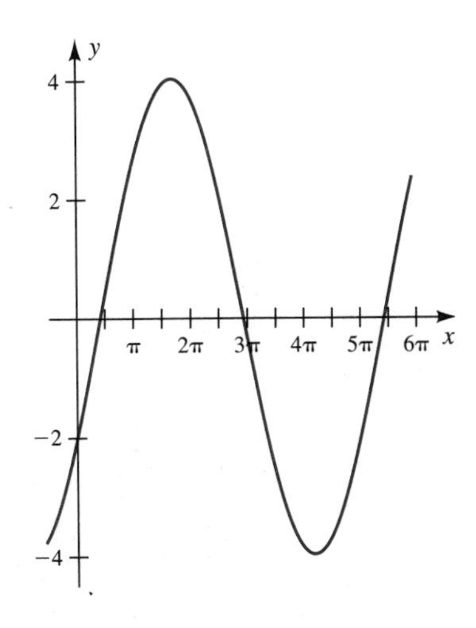

SOLUTION We wish to find a function of the form $y = A\sin(Bx - C)$. The value of A is the amplitude, which in this case is seen to be four. The period is always given by $2\pi/B$. In this case the period is 5π, so we have the equality $2\pi/B = 5\pi$, which gives $B = 2/5$. The phase shift is C/B. In this case the phase shift is $\pi/2$, so $C/(2/5) = \pi/2$, which gives $C = \pi/5$. Using these values we have

$$y = 4\sin\left(\frac{2}{5}x - \frac{\pi}{5}\right)$$

Notice that other correct answers are possible here. Since the given sinusoid has period 5π, any multiple of 5π may be added to the argument and the answer will still be correct. ▲

▲▼ Exercises for Section *5.4*

In Exercises 1–20 sketch one period of the graph of the function. Give the amplitude, period, phase shift, and mean value.

1. $y = 3 + \cos 2x$

2. $s = 4 + \sin 3x$

3. $v = 6 + 8\sin t$

4. $y = -2 + 3\cos 6x$

5. $y = -2 + \sin \frac{1}{2}x$

6. $i = 0.2 + 1.3\cos 0.2t$

7. $M = 3 - 3\sin 3x$

8. $y = -2 - \sin \pi x$

9. $y = \cos(x + \frac{1}{3}\pi)$

10. $y = 2\sin \frac{1}{3}x$

11. $y = 2\cos(\frac{1}{2}x - \frac{1}{2}\pi)$

12. $y = \sin 2(x + \frac{1}{6}\pi)$

13. $y = \cos(2x + \pi)$

14. $y = 3\cos(3x - \pi)$

15. $y = 4\sin(\frac{1}{3}x + \frac{1}{3}\pi)$

16. $y = 0.2\sin(0.25x - \pi)$

17. $y = \cos(\pi x - \frac{1}{4}\pi)$

18. $y = \sqrt{3}\cos(\pi x + \pi)$

19. $y = 4 + \sqrt{3}\sin(4x - \pi)$

20. $y = 3 - 2\cos(\frac{1}{2}x + \frac{1}{8}\pi)$

In Exercises 21–25 sketch the graph and write each expression in the form $A\sin(Bx + C)$, where A and B are positive. [See Example 5(b) on page 200.]

21. $-\sin(x + 1)$

22. $-\sin(-2x + 3)$

23. $-\sin(2\pi x + \frac{1}{2})$

24. $3\cos(2\pi x + \pi)$

25. $-\cos(\pi x + 1)$

In Exercises 26–29 write the equation of the sinusoid whose graph over one period is shown.

26.

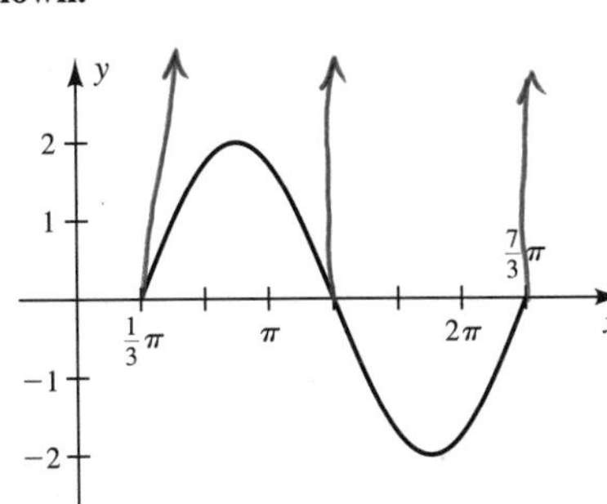

27.

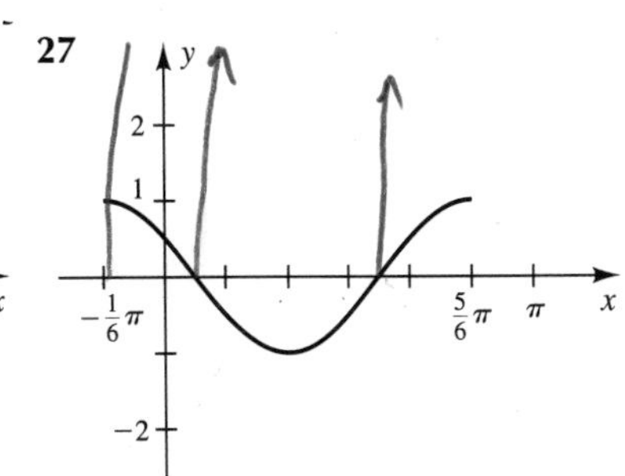

28.

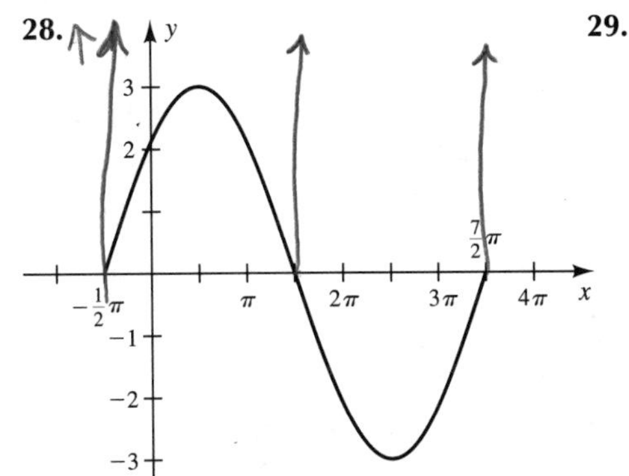

29.

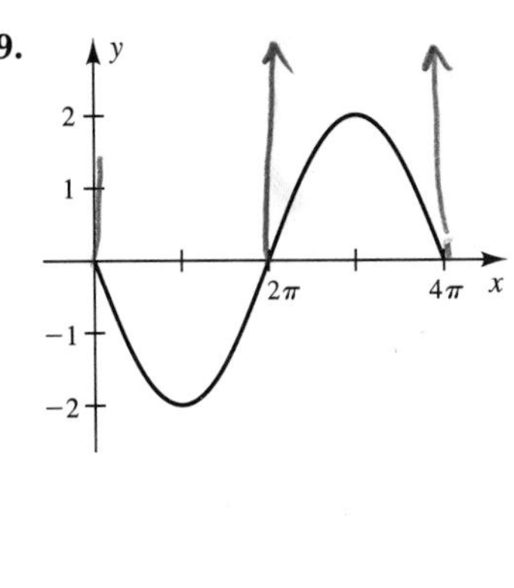

30. It is always possible to express functions of the type $A\sin(Bx + C)$ or $A\cos(Bx + C)$ in the form $A'\sin(B'x + C')$ or $A'\cos(B'x + C')$, where A', B', and C' are positive. Prove this.

31. How are the graphs of $y = \sin(-t)$ and $y = \sin(t)$ related?

32. How are the graphs of $\cos(-t)$ and $\cos(t)$ related?

33. How are the graphs of $\sin(t)$ and $\cos(\frac{1}{2}\pi - t)$ related?

34. A block is attached to a spring as shown in Figure 5.32. If the block is pulled to the right and then released, the spring will cause it to oscillate from side to side. The mathematical description of the motion of the block is

$$s = \begin{cases} \frac{7}{8}\cos 2t + \frac{1}{8}, & 0 \le t \le \frac{1}{2}\pi \\ \frac{5}{8}\cos 2t - \frac{1}{8}, & \frac{1}{2}\pi \le t \le \pi \end{cases}$$

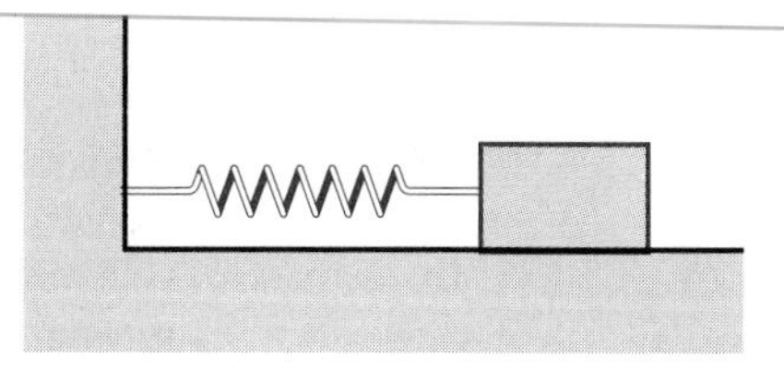

Figure 5.32

The block stops at the position given by $s(\pi)$. Sketch a graph of the block's motion.

35. Repeat Exercise 34, given that the motion is described by

$$s = \begin{cases} \frac{5}{2}\cos 3t + \frac{1}{2}, & 0 \le t \le \frac{1}{3}\pi \\ \frac{3}{2}\cos 3t - \frac{1}{2}, & \frac{1}{3}\pi \le t \le \frac{2}{3}\pi \\ \frac{1}{2}\cos 3t + \frac{1}{2}, & \frac{2}{3}\pi \le t \le \pi \end{cases}$$

The block stops at the position given by $s(\pi)$.

36. The transverse (vertical) displacement of a traveling wave on a plucked violin string is represented by the equation

$$y = A\cos\frac{2\pi}{\lambda}(x - Vt)$$

where A is the amplitude, λ is the wavelength, x is the distance along the string, V is the velocity of the wave, and t is the time in seconds. Graph this equation on the interval $0 \le x \le 4$ in., given that $A = 2$, $\lambda = 2$ in., $V = \frac{1}{2}$ in./sec, and $t = 1$ sec.

37. The output of an alternating current generator is given by

$$I = I_{max}\sin(2\pi ft + \phi)$$

where f is the frequency and ϕ is a constant. Sketch the output current for two periods, given that $f = 60$ Hz, $I_{max} = 157$, and $\phi = \pi/4$.

38. A sinusoid of amplitude .01 and frequency $1/(880\pi)$ cycles/sec can be used to approximate the musical note A above middle C. Write the expression for this sinusoid.

39. If a person's blood pressure is approximated by $P = 110 + 25\sin 5\pi/3\ t$, where P is the blood pressure in millimeters of mercury and t is the time in seconds:

(a) Find the period of this function.

(b) Sketch the graph of this blood pressure function.

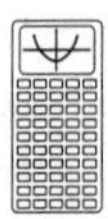

Graphing Calculator Exercises

Use a graphing calculator to sketch the graphs of the following functions. Choose a scale for the y-axis that will clearly show the mean value and amplitude of the graph. Choose a scale for the x-axis that will display two periods of the function. If there is a phase shift, compare the calculator estimate of the phase shift with the known value.

1. $y = 2 + \cos x$
2. $y = 5 + 2\cos x$
3. $y = \sin(x - 0.4)$
4. $y = \sin(x - 1)$
5. $y = 2\cos(x + \frac{1}{4}\pi)$
6. $y = 3\cos(x + \frac{1}{8}\pi)$
7. $y = \sin(2x + 1)$
8. $y = \sin(3x + 2)$
9. Display $y_1 = \sin x + 2$ and $y_2 = \sin(x + 2)$ on the screen of a graphing calculator. On the basis of the graphs, do you think $\sin x + 2 = \sin(x + 2)$? Explain your answer.
10. Display $y_1 = \sin x$ and $y_2 = \cos(x - \pi/2)$ on the screen of a graphing calculator. On the basis of the graphs, do you think $\sin x = \cos(x - \pi/2)$? Explain your answer.
11. Display $y_1 = -\sin x$ and $y_2 = \sin(x - \pi)$ on the screen of a graphing calculator. On the basis of the graphs, do you think $-\sin x = \sin(x - \pi)$? Explain your answer.

5.5 Sinusoidal Modeling

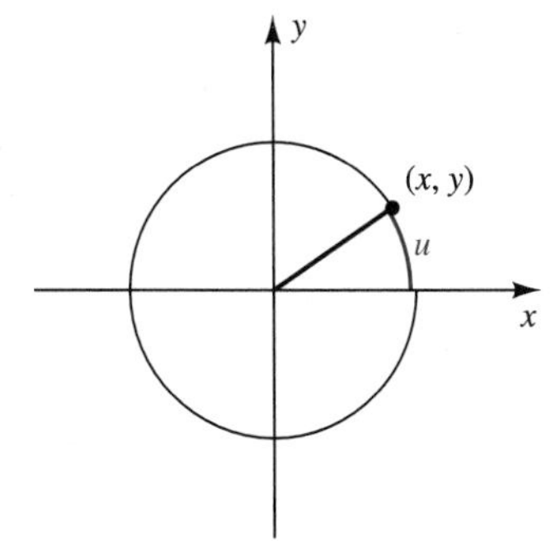

Figure 5.33

In Section 5.1 we established a correspondence between the x- and y-coordinates of a point on a unit circle and the sine and cosine of a real number u mapped onto the unit circle. (See Figure 5.33.) Since $\cos u = x$ and $\sin u = y$, the circular functions are excellent mathematical models of physical problems that involve circular motion. In this section we will describe some typical models.

Imagine an object hanging on a spring as shown in Figure 5.34. If the object is pulled down and released, it will oscillate up and down about its rest point. Assuming there is no frictional force, this oscillation will continue indefinitely. Vibratory motion of this type is called **simple harmonic motion** and can be described mathematically using sine or cosine functions or combinations of the two. In this section we shall restrict ourselves to the use of cosine functions to model harmonic motion.

Simple harmonic motion describes other phenomena besides oscillation of an object on a spring—such as the motion of a point on a guitar string that has been plucked and the motion of air brought about by certain sound waves and some radio and television devices.

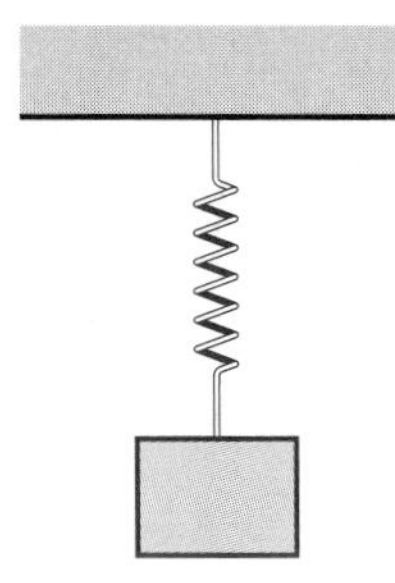
Figure 5.34

Simple Harmonic Motion

Consider a point Q moving at a constant angular velocity around the circumference of a circle, as depicted in Figure 5.35. Assume that our first observation is made when Q is at position (x, y). The point P directly below Q is called the projection of Q on the x-axis. As Q revolves, this projection moves back and forth along the x-axis between the extremes of the diameter of the circle. Note that the horizontal motion of P *is the same as the vertical motion of the weight hanging on a spring; that is,* P is in simple harmonic motion. The method used to describe the motion of P can also be used to describe any other simple harmonic motion.

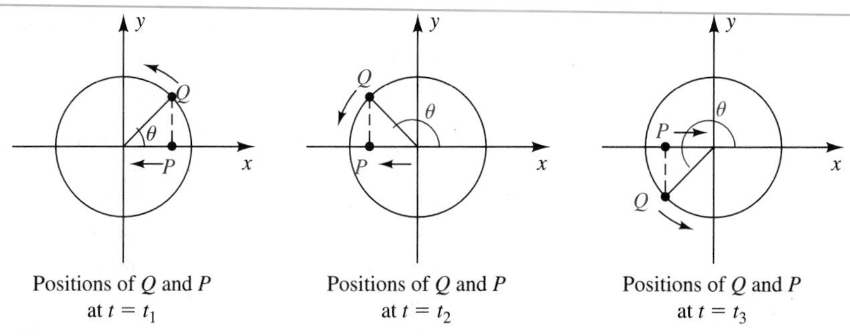

Figure 5.35

To construct a mathematical model of simple harmonic motion, we proceed as follows. The displacement of P from the origin is labeled $x = \overline{OP}$, the angle made by $\overline{OQ}$ and the positive x-axis is called θ, and the radius is $r = \overline{OQ}$. Then $x = r\cos\theta$. More generally, if Q has a constant angular velocity ω, we can write the model as a function of time by letting $\theta = \omega(t - t_0)$, where t_0 is some initial time for the motion, sometimes called the phase shift. Then we have

$$x = r\cos\omega(t - t_0) \qquad \text{Mathematical model of simple harmonic motion} \qquad \textbf{(5.1)}$$

In this context the amplitude of the motion is r, the period is $2\pi/\omega$ time units, and the phase shift is t_0 time units.

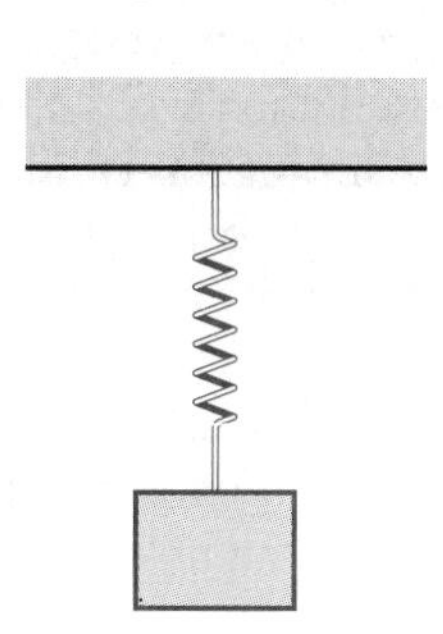

Figure 5.36

Example 1 The motion in the spring-mass system shown in Figure 5.36 is modeled by the equation $y = 4\cos 0.7t$, where y is the displacement in meters from the rest point and t is the elapsed time. How long does it take for one oscillation of the mass? Where is the mass relative to the rest point when **(a)** $t = 0.5$ sec, **(b)** $t = 2.0$ sec, and **(c)** $t = 3.0$ sec? Notice that when $t = 0$, the object is 4 m above the rest point.

SOLUTION The angular velocity is $\omega = 0.7$ rad/sec. Therefore the time required to complete one oscillation is $2\pi/0.7 \approx 9$ sec. To find the location of the weight at the indicated times, we evaluate $y = 4\cos 0.7t$.

(a) At $t = 0.5$,

$$y = 4\cos[0.7(0.5)] \approx 3.8$$

This means that the object is 3.8 m above the rest position.

(b) At $t = 2.0$,

$$y = 4\cos[0.7(2.0)] \approx 0.68$$

The object is now 0.68 m above the rest position.

(c) At $t = 3.0$,

$$y = 4\cos[0.7(3.0)] \approx -2.0$$

The object is approximately 2 m below the rest position at this time. Figure 5.37 shows the position of the object at the various times.

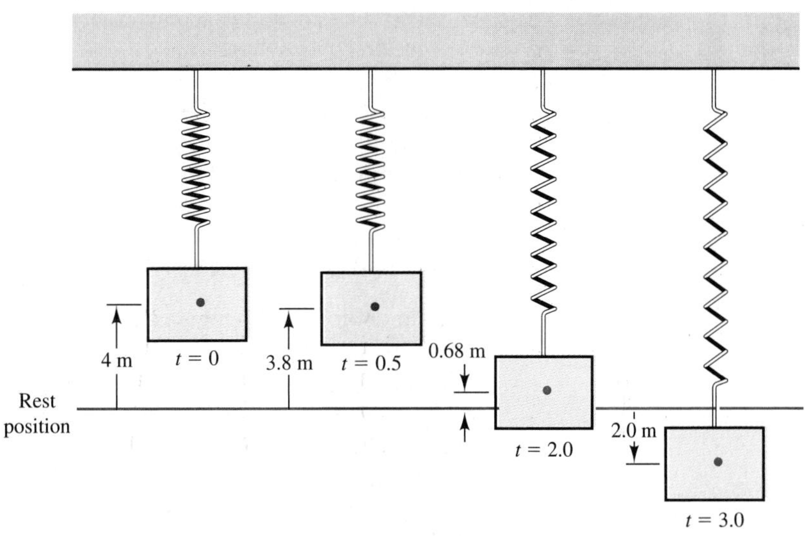

Figure 5.37 ▲

Example 2 The water wheel shown in Figure 5.38 rotates at a rate of 3 rpm. Twelve seconds after we make our first observation, point Q is at its greatest height.

(a) Model the distance h of point Q from the surface of the water in terms of the elapsed time.

(b) Determine h when $t = 15$ sec.

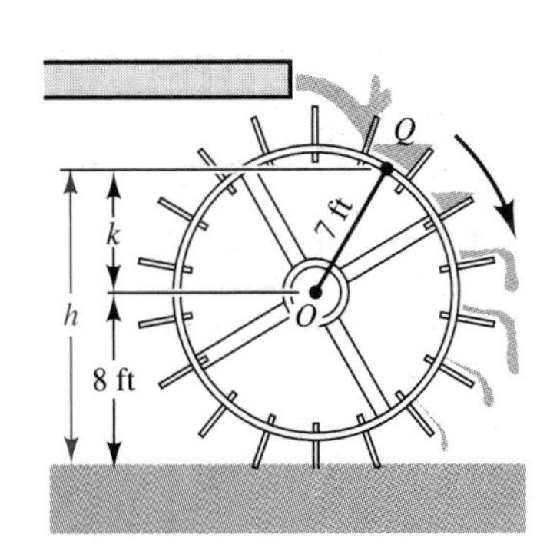

Figure 5.38

SOLUTION

(a) We begin by expressing the angular velocity of the wheel in radians per second:

$$\omega = 3\,\frac{\text{rev}}{\text{min}} \cdot 2\pi\,\frac{\text{rad}}{\text{rev}} \cdot \frac{1}{60}\,\frac{\text{min}}{\text{sec}} = \frac{1}{10}\pi\,\frac{\text{rad}}{\text{sec}}$$

Figure 5.38 shows that $r = 7$ ft. The phase shift is $t_0 = 12$ sec, the time required for Q to reach its maximum height. Using Equation (5.1), we find the distance k to be

$$k = 7\cos\frac{1}{10}\pi(t - 12)$$

Since the center of the wheel is 8 ft above the water level, our model of the distance h is

$$\begin{aligned} h &= 8 + k \\ &= 8 + 7\cos\frac{1}{10}\pi(t - 12)\text{ ft} \end{aligned}$$

(b) The value of h for $t = 15$ sec is

$$h = 8 + 7\cos\frac{1}{10}\pi(3) \approx \boxed{12.11449677} \approx 12.1$$

Therefore point Q is approximately 12 ft above the water level 15 sec after our first observation. ▲

Sinusoids are also used to model physical phenomena other than circular motion. The following examples will give you some idea of the variety of applications involving sinusoids.

The Predator-Prey Problem

Certain ecological systems can be represented by periodic functions. For instance, the sine function is frequently used to describe the predator-prey relationship in a balanced ecological system. If the number of predators in a region is relatively small, the number of prey increases. But as the prey becomes more plentiful, the number of predators increases because food is easy to find. As the number of predators continues to increase, the number of prey eventually begins to decrease; thus food for the predators eventually becomes scarce, which causes the predator population to decrease, which in turn allows more prey to survive, and so on. This cycle is repeated over and over again, with the two populations oscillating about their respective mean values.

Example 3 The population of rabbits in a certain region is given by

$$N = 500 + 150\sin\frac{1}{2}\pi t$$

where t is time in years. Discuss the variation in the rabbit population.

SOLUTION The constant term represents the mean population of rabbits, and the coefficient of the sine function represents the variation in the population. Thus the mean population is 500, and it varies from a high of 650 to a low of 350. The period of variation is $2\pi/\frac{1}{2}\pi = 4$ years. The population is graphed in Figure 5.39.

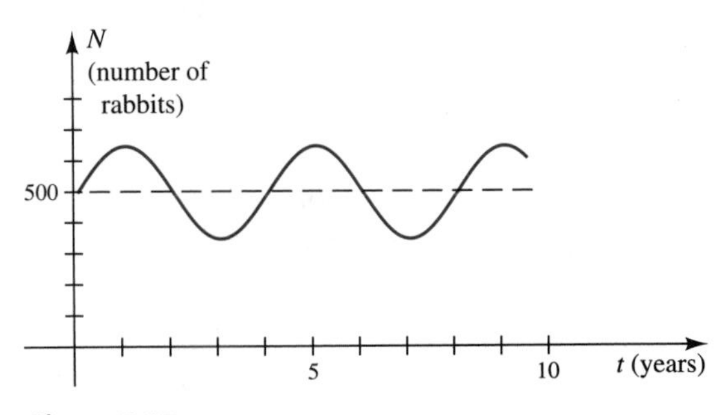

Figure 5.39

▲

Seasonal Temperature Variation

Sine and cosine curves can be used to describe physical conditions represented by meteorological data. Figure 5.40 shows a representation of the daily mean temperatures for Dayton, Ohio, from January 1993 to September 1994; the sinusoidal shape is unmistakable. The temperature variation can be approximated by the equation

$$T = T_m + A \sin\left[\frac{2\pi}{365}(t - C)\right]$$

where T_m is the mean annual temperature, A is the maximum temperature deviation from T_m, and C is the phase shift found by counting the number of days from January 1 to the point at which the temperature curve crosses the mean annual temperature line.

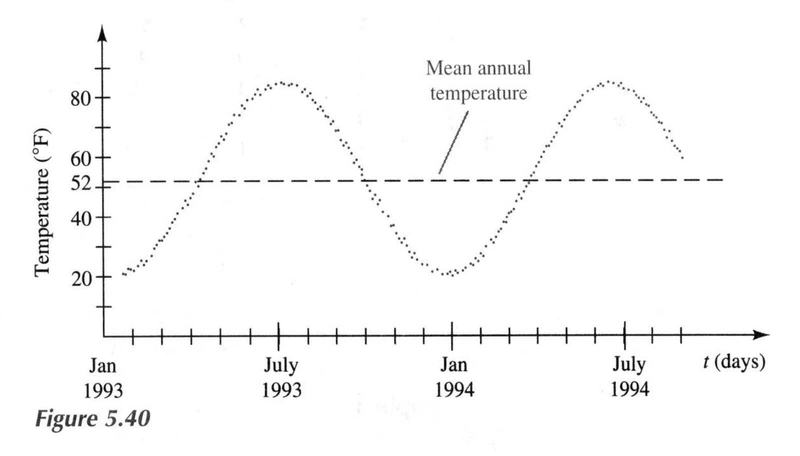

Figure 5.40

Example 4 Write an equation to approximate the temperature variations shown in Figure 5.40.

SOLUTION From Figure 5.40 we estimate that $T_m = 52°\text{F}$, $A = 32°\text{F}$, and $C = 102$ days. Figure 5.41 shows both the actual data and the graph of the equation

$$T = 52 + 32 \sin\left[\frac{2\pi}{365}(t - 102)\right]$$

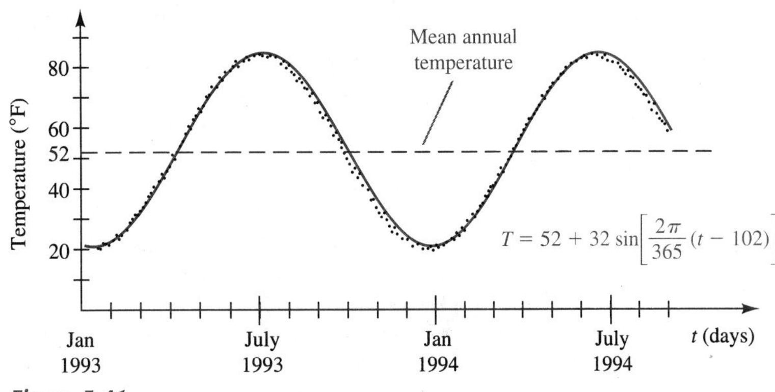

Figure 5.41

▲

Exercises for Section 5.5

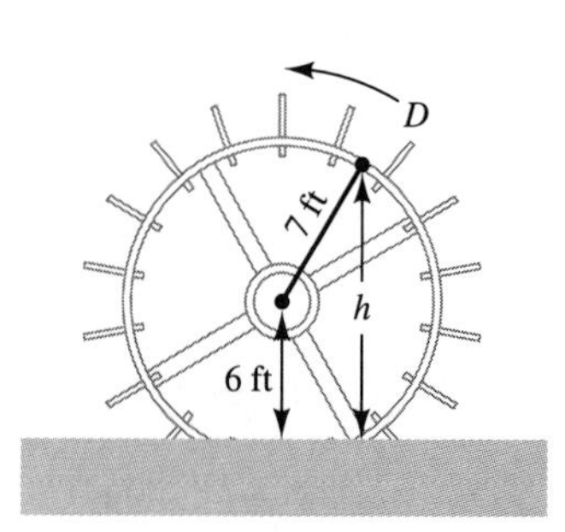

Figure 5.42

1. The motion of a spring-mass system similar to the one shown in Figure 5.36 is modeled by the equation $x = 15 \cos 2t$, where x is the displacement in centimeters from the rest position.
   (a) Determine the time required for one oscillation.
   (b) Determine the position of the weight when $t = 1$ sec.
2. The motion in a spring-mass system is described by $y = 2 \cos 3.2t$, where y is the displacement in meters from the rest position and t is the elapsed time in seconds.
   (a) Determine the time required for one oscillation of the mass.
   (b) Determine the location of the weight after 5 sec.
3. Suppose the water wheel illustrated in Figure 5.42 rotates at 6 rpm. Two seconds after a stopwatch is started, point D on the rim of the wheel is at its greatest height.
   (a) Model the distance h of point D from the surface of the water in terms of the time t in seconds on the stopwatch.
   (b) Sketch the graph of the sinusoidal model.
   (c) Find the time when D first emerges from below the water.
4. A space shuttle is fired into a circular orbit from its launch pad in Florida. Ten minutes after it leaves, it reaches its farthest distance north of the equator, which is 4000 km. A half-cycle later it reaches its farthest distance south of the equator (also 4000 km).
   (a) Write a sinusoidal model describing the relationship between distance from the equator, D, and time from launch, t, in minutes. Consider distance south of the equator to be negative distance.
   (b) How much time will elapse before the shuttle crosses the equator for the first time?
   (c) How far north of the equator is the launch site? Assume a 90-min orbit.
5. Tarzan is swinging back and forth on his grapevine, alternately going over land and water. (See Figure 5.43.) Consider values of y, the distance from the river bank, to be positive if Tarzan is over water and negative if he is over land. Write a sinusoidal model to describe Tarzan's distance from the river bank. Assume that Jane is measuring Tarzan's motion and finds that 2 sec after she starts her stopwatch, Tarzan is at one end of his swing, where $y = -23$ ft. When $t = 5$ sec, he is at the other end of his swing, where $y = 17$ ft.
   (a) Sketch a graph of this sinusoidal function.
   (b) Predict y when $t = 2.8$, 6.3, and 15 sec.
   (c) Where was Tarzan when Jane started her stopwatch?
   (d) Determine the time indicated on the stopwatch when Tarzan first reached a point directly over the bank.

Note: Exercises 3, 4, 5, 6, 11, and 12 are reprinted by permission of the author Paul A. Foerster from *Trigonometry: Functions and Applications*, 1977, Addison–Wesley Publishing Company, Inc., Reading, MA.

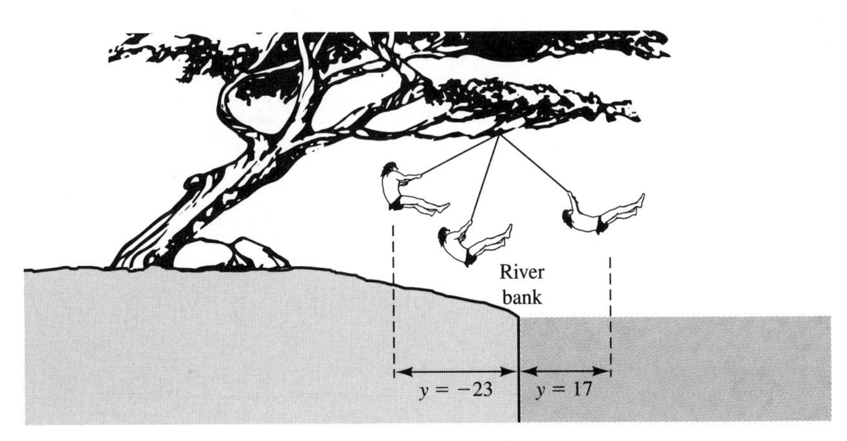

Figure 5.43

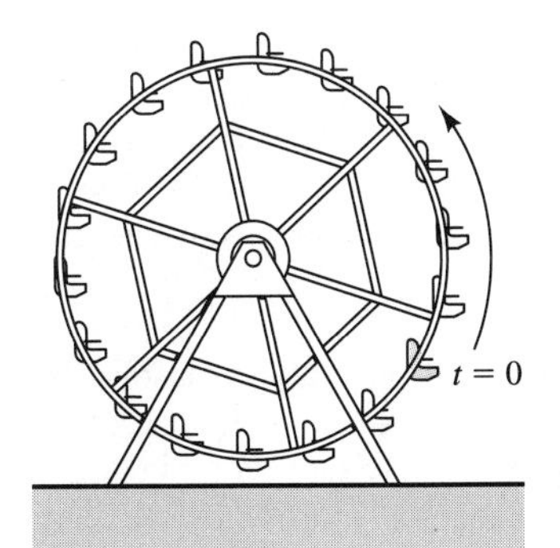

Figure 5.44

6. As you ride a ferris wheel, your distance from the ground varies sinusoidally with time. Suppose that when the last seat is filled and the ferris wheel ride begins, your seat is at the position shown in Figure 5.44. It takes you 3 sec to reach the top, which is 43 ft above the ground. The wheel, which travels at 7.5 rpm, has a diameter of 40 ft.
   (a) Sketch a graph of this sinusoid.
   (b) Write an equation for the sinusoid.
   (c) What is the lowest point you reach as the wheel turns?
   (d) How high above the ground are you when $t = 0$? when $t = 9$?
   (e) What is the value of t when, in your first descent, you reach a point 23 ft above the ground?
7. The number of deer in a region is given by $D = 1500 + 400 \sin 0.4t$, and the number of pumas in the region by $P = 500 + 200 \sin(0.4t - 0.8)$, where t is the time in years. Sketch the variation in these two populations on the same set of coordinates.
8. The rabbit population in an ecological region is given by the expression $R = 1000 + 200 \sin 4t$, and the fox population by $F = 100 + 10 \sin(4t - 0.8)$, where t is the time in years. Discuss the variation in each population and sketch the graphs of the populations on the same coordinate system.
9. Draw the graph of the normal mean temperature variation for a two-year period if the approximating equation is

$$T = 55 + 38 \sin\left[\frac{2\pi}{365}(t - 100)\right]$$

   Assume that T is in °F, that t is the number of days from January 1, and that neither year is a leap year. What is the mean annual temperature of the area described by this equation?
10. Write the equation for the temperature variation shown in Figure 5.45.

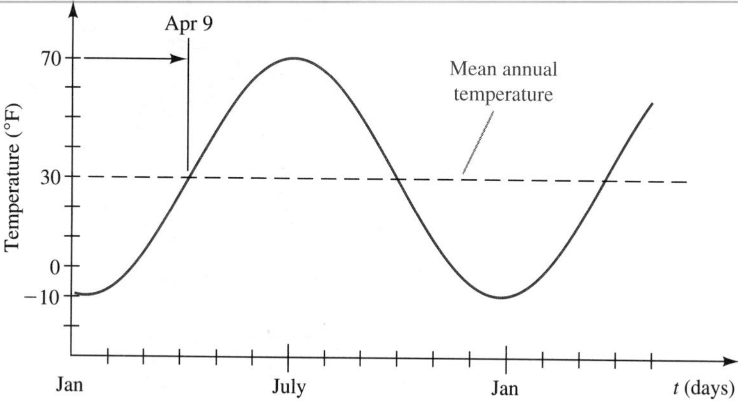

Figure 5.45

11. For several hundred years astronomers have kept track of the number of solar flares, or sunspots, that occur on the surface of the sun. The number of sunspots counted in a given year varies periodically from a minimum of about 10 per year to a maximum of about 110 per year. Between the maximums that occurred in the years 1750 and 1948, there were 18 complete cycles. (Check the September 1975 issue of *Scientific American*, page 166, to see how closely the sunspot cycle resembles a sinusoid.)

(a) What is the period of the sunspot cycle?

(b) Assume that the number of sunspots counted in a year varies sinusoidally with the year. Sketch a graph of two sunspot cycles, starting with 1948.

(c) Write the equation expressing the number of sunspots per year (y) in terms of the year (t).

(d) How many sunspots should you expect this year? in the year 2000?

(e) What is the first year after 2000 in which the maximum number of sunspots will occur?

12. A portion of a roller coaster track is to be built in the shape of a sinusoid. (See Figure 5.46.)

(a) The high and low points on the track are separated horizontally by 50 m and vertically by 30 m. The low point is 3 m below the ground. Letting y be the number of meters the track is above or below the ground and x the number of meters horizontally from the high point, write an equation expressing y in terms of x.

(b) How long is the vertical timber at the high point? the one at $x = 4$ m? the one at $x = 32$ m?

(c) How long is the horizontal timber that is 25 m above the ground and the one that is 5 m above the ground?

(d) Where does the track first go below the ground?

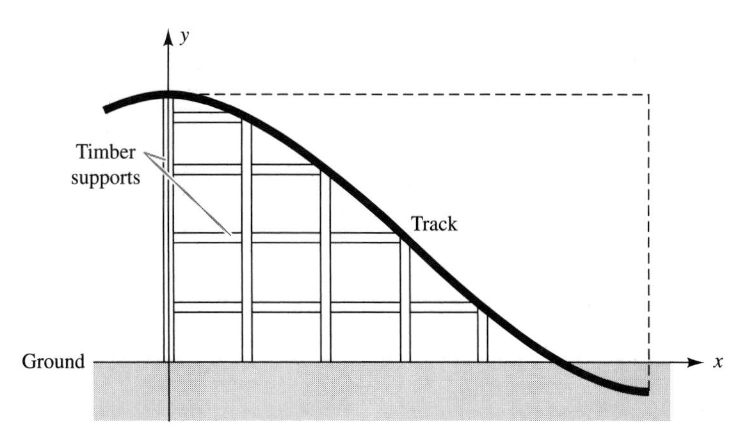

Figure 5.46

(e) The vertical timbers are spaced every 2 m, starting at $x = 0$ and ending where the track goes below the ground. What is the total length of timber needed to make the vertical supports? (If you have access to a computer, do this exercise by writing a program that prints out the length of each timber and then totals these lengths.)

5.6 Addition of Ordinates

Functions that are the sum of two or more elementary functions, such as

$$y_1 = \sin x + \cos x \qquad \text{and} \qquad y_2 = x + \sin x$$

occur frequently. It can be a very tedious process to graph such functions if you use the method of substituting values of x and determining corresponding ordinates. Sometimes a technique called **addition of ordinates** can be useful in plotting such functions. Suppose $h(x) = f(x) + g(x)$. We sketch the graphs of $f(x)$ and $g(x)$ on the same coordinate system, as in Figure 5.47. Then for particular values of x, such as x_1, we find $h(x_1)$ as the sum of $f(x_1)$ and $g(x_1)$. A common way to do this is to draw a vertical line at the point $(x_1, 0)$, and then to add ordinates $f(x_1)$ and $g(x_1)$ by using dividers or markings on the edge of a strip of paper. This process is repeated as often as necessary to get a representation of the desired graph.

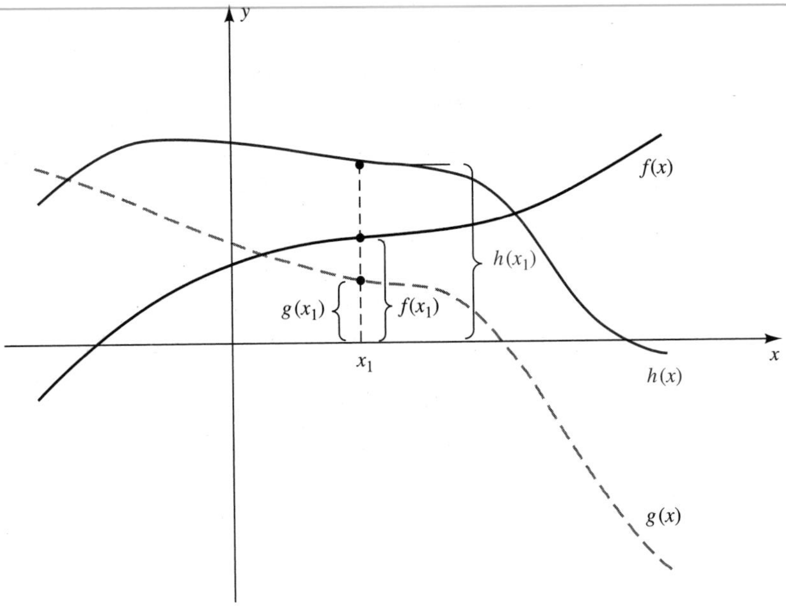

Figure 5.47

Example 1 Use the method of addition of ordinates to sketch the function $y = \sin x + \cos 2x$.

SOLUTION Both $\sin x$ and $\cos 2x$ are sketched in Figure 5.48, along with their sum. The period of the given function is 2π, even though that of $\cos 2x$ is π.

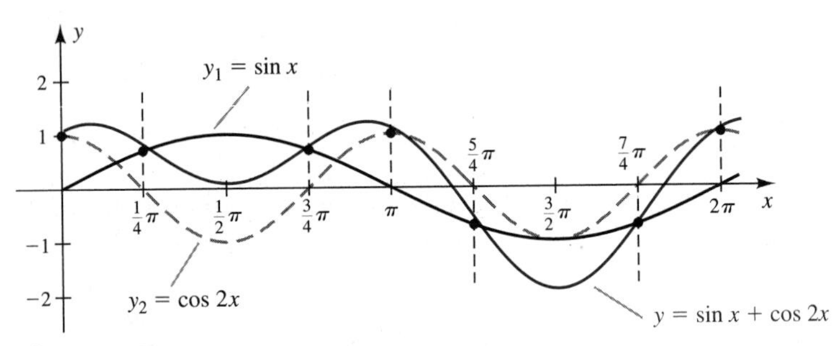

Figure 5.48

▲

When you are using addition of ordinates to graph $h(x) = f(x) + g(x)$, you will find that certain values of x are more useful than others. The following observations may be helpful.

- When the graph of $y_1 = f(x)$ crosses the x-axis, the ordinate of $y = h(x)$ is the same as that of $y_2 = g(x)$ and vice versa. For instance, in Figure 5.48 notice that the graph of $y_1 = \sin x$ crosses the x-axis at $x = 0$, π, and 2π and that the graph of $y = \sin x + \cos 2x$ crosses the graph of $y_2 = \cos 2x$ at these same values. Also note that $y_2 = \cos 2x$ crosses the x-axis at $x = \pi/4$, $3\pi/4$, $5\pi/4$, and $7\pi/4$ and that the graph of $y = \sin x + \cos 2x$ crosses the graph of $y_1 = \sin x$ at these values.

- When the maximum or minimum amplitude values of $y_1 = f(x)$ and $y_2 = g(x)$ occur at the same value of x, the graph of $y = h(x)$ may also have a maximum or minimum amplitude at this point. For instance, in Figure 5.48 the graphs of $y_1 = \sin x$ and $y_2 = \cos 2x$ both have minimum values at $x = 3\pi/2$, and the graph of $y = \sin x + \cos 2x$ also has a minimum at this value.

Example 2 Sketch the graph of $y = x + \sin x$.

SOLUTION See Figure 5.49. In this case the basic sine curve oscillates about the curve $y = x$. Note that the given function is *not* periodic and is *not* bounded.

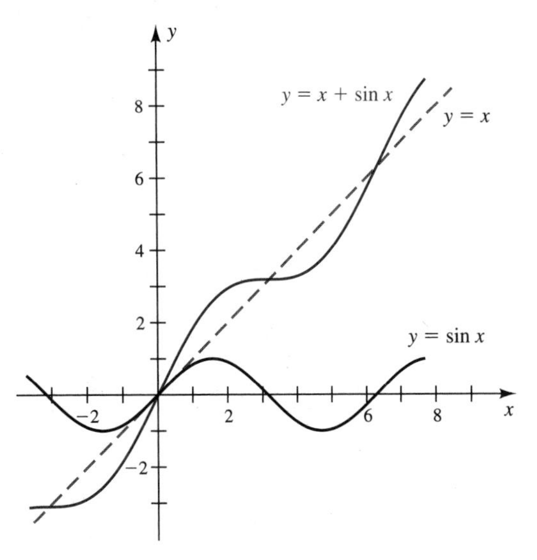

Figure 5.49

▲

Try using a graphing calculator to graph a function that is the sum of two elementary functions. Have the calculator sketch consecutively y_1, y_2, and then the sum. Take particular note of the relationship between a point on the graph and the points on y_1 and y_2 when the graph of y_1 or y_2 crosses the x-axis. The figure below shows the graph of $y = x + \sin x$ as it appears on a graphing calculator, the same function sketched in Example 2.

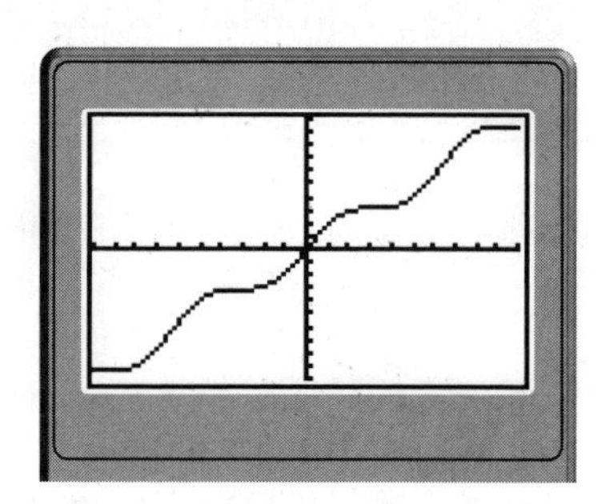

Exercises for Section 5.6

In Exercises 1–20 use the method of addition of ordinates to sketch the graph of each function. In each case tell whether the function is periodic.

1. $y = 3 + \sin x$

2. $y = 1 + 2\cos x$

3. $y = -1 + \cos x$

4. $y = -0.5 + \sin x$

5. $y = \sin x + 2$

6. $y = x - \sin x$

7. $y = \frac{1}{2}x + \cos x$

8. $y = \cos x - x$

9. $y = 2 - x + \sin x$

10. $y = 1 + x - \cos x$

11. $y = 0.1x^2 + \sin 2x$

12. $y = 0.1x^2 + \cos x$

13. $y = \sin x + \cos x$

14. $y = \sin 2x + 2\sin x$

15. $y = \sin \frac{1}{2}x - 2\cos x$

16. $y = \sin \frac{1}{2}x - \sin x$

17. $y = \sin x + \sin(x - \frac{1}{4}\pi)$

18. $y = \cos x + \sin(x - \frac{1}{4}\pi)$

19. $y = 2\sin \pi x + \sin x$

20. $y = \sin \pi x - \cos 2x$

21. Compare the graphs of the following functions.

$$y = \sqrt{2}\sin(x + \tfrac{1}{4}\pi) \quad \text{and} \quad y = \cos x + \sin x$$

22. Compare the graphs of the following functions.

$$y = \cos^2 x \quad \text{and} \quad y = \frac{1}{2} + \frac{1}{2}\cos 2x$$

23. An object oscillating on a spring has a motion described by

$$y(t) = y_0 \cos 2t + \left(\frac{v_0}{2}\right)\sin 2t$$

where y_0 is the initial displacement from equilibrium and v_0 is the initial velocity. Given that $y_0 = 5$ cm and $v_0 = 24$ cm/sec, make a graph of the motion for $0 \le t \le 2\pi$.

24. An electronic tuner/amplifier accepts signals of the form $I = I_m \sin 2\pi ft$. Graph the input to the tuner if two waves arrive simultaneously at the terminals, one with $I_m = 3$ mA, $f = 1000$ Hz, and the other with $I_m = 5$ mA, $f = 2000$ Hz. The tuner sums signals that are received simultaneously.

25. The current in a transmission line may be considered as the sum of an outgoing wave of the form $\sin(x + ct)$ and a reflected wave of the form $\sin(x - ct)$. Graph the sum of these two waveforms for $c = 1$ when $t = 1$ for $0 \le x \le 2\pi$.

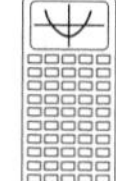

Graphing Calculator Exercises

Use a graphing calculator to sketch the graphs of the following functions. Compare the graph on the screen to the one you obtained previously by hand.

1. $y = -1 + \cos x$ (Exercise 3)

2. $y = \sin x + 2$ (Exercise 5)

3. $y = x - \sin x$ (Exercise 6)

4. $y = 2 - x + \sin x$ (Exercise 9)

5. $y = \sin 2x + 2 \sin x$ (Exercise 14)

6. $y = \sin \frac{1}{2}x - 2 \cos x$ (Exercise 15)

7. $y = \cos x + \sin(x - \frac{1}{4}\pi)$(Exercise 18)

8. $y = 2 \sin \pi x + \sin x$ (Exercise 19)

5.7 Graphs of the Tangent and Cotangent Functions

The properties of the tangent and cotangent functions are summarized below. Each property affects the nature of the graph in a very important manner.

1. Both $\tan x$ and $\cot x$ are periodic with period π. Thus only a one-period interval need be analyzed, such as $-\frac{1}{2}\pi < x < \frac{1}{2}\pi$ or $0 < x < \pi$.

2. Both $\tan x$ and $\cot x$ are **unbounded**, which means that their values become arbitrarily large. Tan x becomes unbounded near odd multiples of $\frac{1}{2}\pi$, whereas $\cot x$ becomes unbounded near multiples of π. The lines $x = \frac{1}{2}\pi + n\pi$ are called **vertical asymptotes** of the graph of $y = \tan x$; the lines $x = n\pi$ are vertical asymptotes of the graph of $y = \cot x$.

3. Tan x is zero for $x = 0, \pm\pi, \pm 2\pi$, and so on. Cot x is zero at $x = \pm\frac{1}{2}\pi, \pm\frac{3}{2}\pi, \pm\frac{5}{2}\pi$, and so on. The graphs cross the x-axis at these places.

4. Numerically (ignoring sign) the values of both functions are completely determined in the first quadrant—that is, for $0 < x < \frac{1}{2}\pi$.

Table 5.3 shows the variation in the values of $\tan x$ and $\cot x$ during one revolution of the terminal side of an angle.

Table 5.3 Variation of tan x and cot x

When x increases	tan x increases	cot x decreases
From 0 to $\frac{1}{2}\pi$	From 0 to become unbounded	From large values to 0
From $\frac{1}{2}\pi$ to π	From an unbounded negative to 0	From 0 to an unbounded negative
From π to $\frac{3}{2}\pi$	From 0 to become unbounded	From large values to 0
From $\frac{3}{2}\pi$ to 2π	From an unbounded negative to 0	From 0 to an unbounded negative

Figure 5.50 shows two graphs—one of several periods of the tangent function and one of several periods of the cotangent function. The x-intercepts and the asymptotes are indicated. The asymptotes for tan x are $x = \pm\frac{1}{2}\pi, \pm\frac{3}{2}\pi, \pm\frac{5}{2}\pi$, and so on. The asymptotes for cot x are $x = 0, \pm\pi, \pm 2\pi, \pm 3\pi$, and so on.

x	0	$\frac{1}{6}\pi$	$\frac{1}{4}\pi$	$\frac{1}{3}\pi$	$\frac{1}{2}\pi$	$\frac{3}{4}\pi$	π
tan x	0	0.58	1	1.73	undef.	−1	0
cot x	undef.	1.73	1	0.58	0	−1	undef.

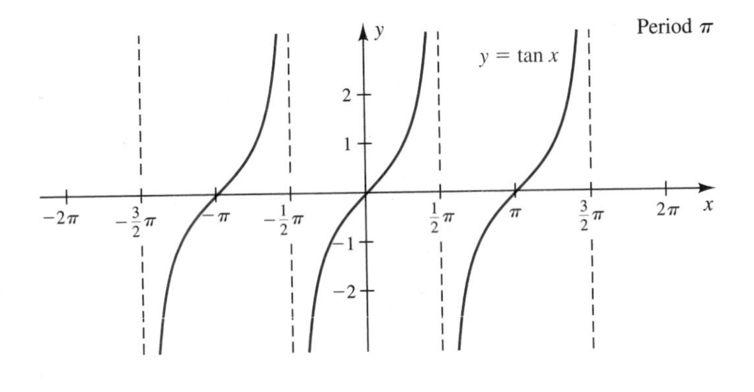

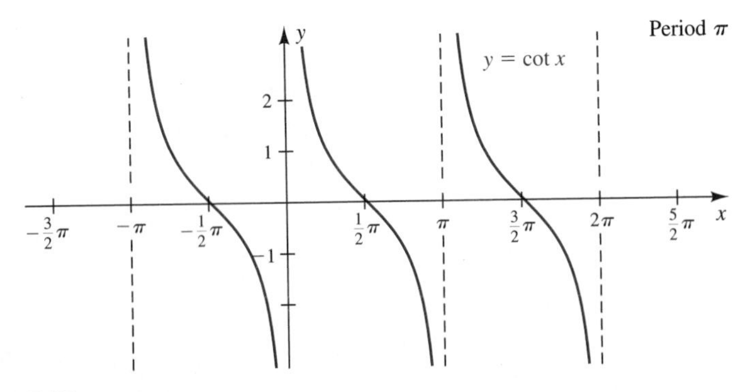

Figure 5.50

The graphs of the more general functions $y = A\tan(Bx - C)$ and $y = A\cot(Bx - C)$ can be analyzed in a manner similar to that described in Section 5.3. In the case of $y = A\tan x$, we do not call A the amplitude because this would imply that the function was bounded. The constant A multiplies each functional value but has no other graphical significance.

The period of $\tan Bx$ is π/B. Thus if $B > 1$, the period is shorter than that of the basic tangent function; if $B < 1$, the period is larger. The constant C is a phase shift constant and translates the basic function to the right or to the left.

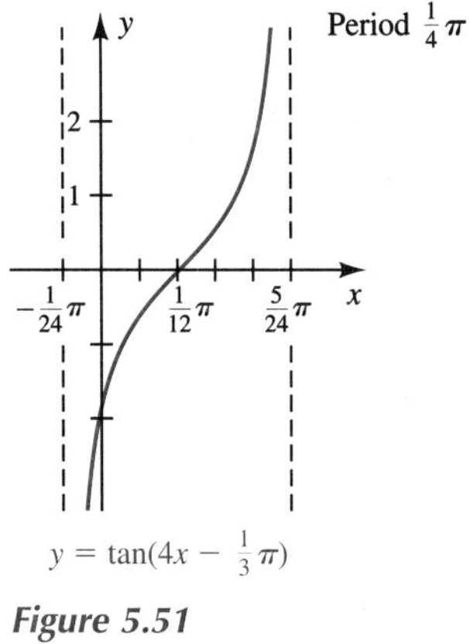

$y = \tan(4x - \frac{1}{3}\pi)$

Figure 5.51

Example 1 Sketch the function $y = \tan(4x - \frac{1}{3}\pi) = \tan 4(x - \frac{1}{12}\pi)$.

SOLUTION The period of this function is $\frac{1}{4}\pi$. The phase shift is located by determining where the argument $4x - \frac{1}{3}\pi$ is equal to zero. Thus the phase shift is $\frac{1}{12}\pi$. The graph is shown in Figure 5.51. If there were no phase shift, the left-hand asymptote of this graph would be located at $x = -\frac{1}{8}\pi$. Since this graph must be moved $\frac{1}{12}\pi$ units to the right, the left-hand asymptote occurs at $x = -\frac{1}{8}\pi + \frac{1}{12}\pi = -\frac{1}{24}\pi$. The right-hand asymptote is then $\frac{1}{4}\pi$ units to the right of $-\frac{1}{24}\pi$; that is, it is at $x = -\frac{1}{24}\pi + \frac{1}{4}\pi = \frac{5}{24}\pi$. ▲

Example 2 Draw the graph of $y = 3\cot\frac{1}{2}x$.

SOLUTION The period of this graph is $\pi/\frac{1}{2} = 2\pi$, and the phase shift is zero. The graph is shown in Figure 5.52.

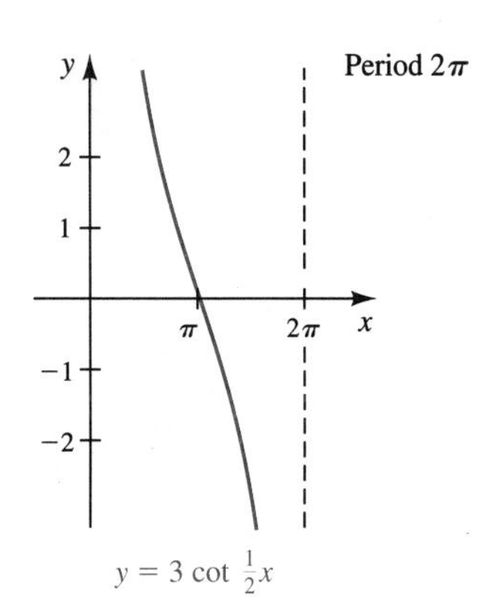

$y = 3\cot\frac{1}{2}x$

Figure 5.52

▲

Some graphing calculators include the asymptotes when graphing the tangent and cotangent functions. The graph of tan x is shown in the figure. The asymptotes of the graph can be used to determine the period and any phase shift for the function. The period is obtained by subtracting the readouts for any two successive asymptotes.

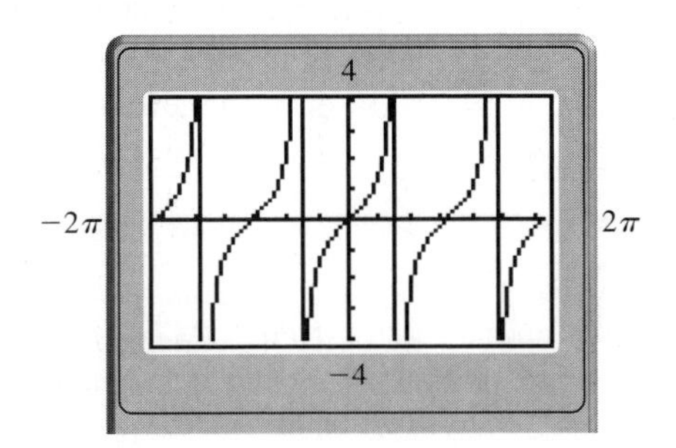

5.8 Graphs of the Secant and Cosecant Functions

The graphs of sec x and csc x can be sketched directly from the graphs of cos x and sin x, since they are reciprocals of those respective functions. These functions have several important general properties.

1. Both functions are unbounded. In fact, since both sin x and cos x are bounded by ± 1, the graphs of csc x and sec x lie above $y = 1$ and below $y = -1$.
2. Both sec x and csc x are periodic with period 2π.
3. Sec x is an even function; csc x is an odd function.
4. Both sec x and csc x are never 0.
5. Numerically, the functional values are determined for $0 < x < \frac{1}{2}\pi$.

Table 5.4 shows the variation in the values of sec x and csc x during one revolution of the terminal side of an angle.

Table 5.4 Variation of sec x and csc x

When x increases	sec x	csc x
From 0 to $\frac{1}{2}\pi$	Increases from 1 to become unbounded	Decreases from large positive values to 1
From $\frac{1}{2}\pi$ to π	Increases from large negative numbers to -1	Increases from 1 to large positive values
From π to $\frac{3}{2}\pi$	Decreases from -1 to large negative numbers	Increases from large negative values to -1
From $\frac{3}{2}\pi$ to 2π	Decreases from large positive numbers to 1	Decreases from -1 to large negative values

The graphs of both functions are sketched in Figure 5.53. In each case the reciprocal function is shown with a dashed line on the same coordinate system to show the relation between the two.

x	0	$\frac{1}{6}\pi$	$\frac{1}{4}\pi$	$\frac{1}{3}\pi$	$\frac{1}{2}\pi$	π	$\frac{3}{2}\pi$	2π
sec x	1	1.2	1.4	2	undef.	-1	undef.	1
csc x	undef.	2	1.4	1.2	1	undef.	-1	undef.

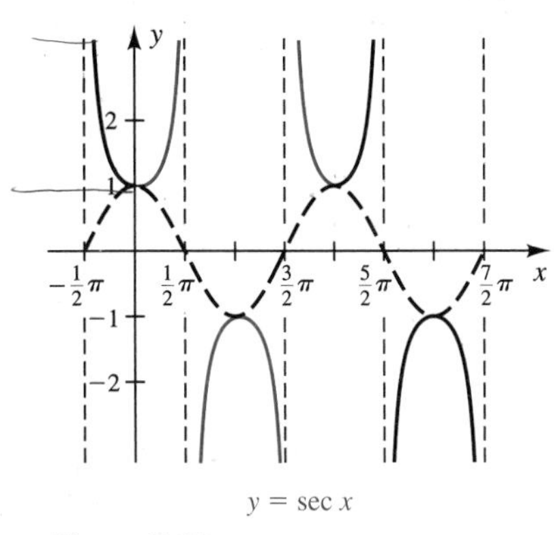

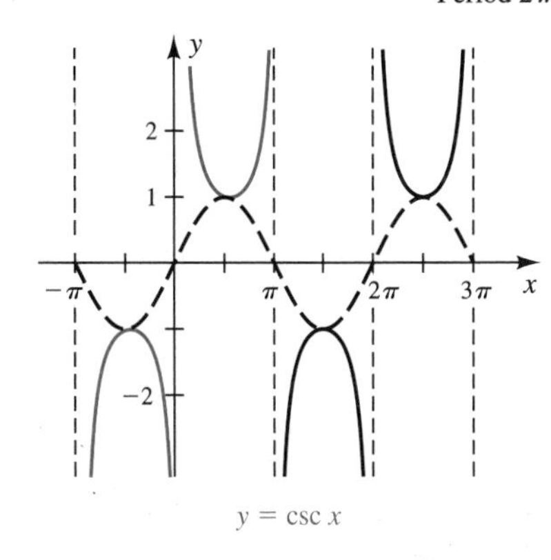

Figure 5.53

The more general functions

$$y = A \sec(Bx - C) \qquad \text{and} \qquad y = A \csc(Bx - C)$$

can be sketched using techniques similar to those discussed in Section 5.3. The basic waveforms remain the same, but the constants A and B cause vertical and horizontal stretching, and C effects a horizontal translation.

Example 1 Sketch the graph of $y = 0.3 \sec(x + \frac{1}{4}\pi)$.

SOLUTION Since 0.3 is the coefficient of the secant, the range is outside the interval $-0.3 < y < 0.3$. The period is 2π, and the phase shift is $\frac{1}{4}\pi$ unit to the left. The graph is shown in Figure 5.54.

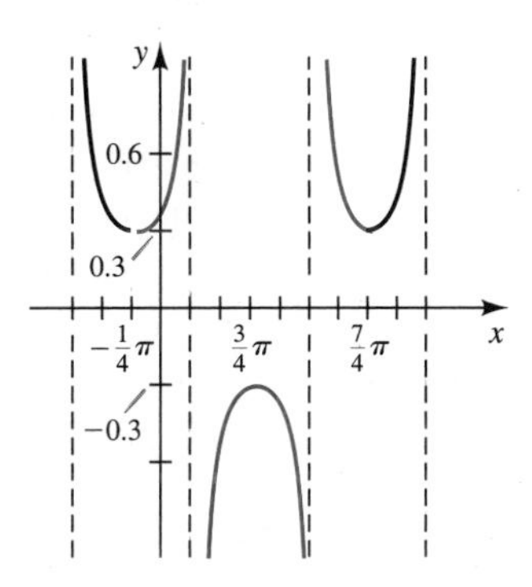

Figure 5.54

▲

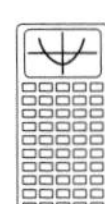

A graphing calculator will display the graphs of the secant and cosecant functions, but of course they will have to be input as 1/cos x and 1/sin x. Notice that the calculator inserts the vertical asymptotes.

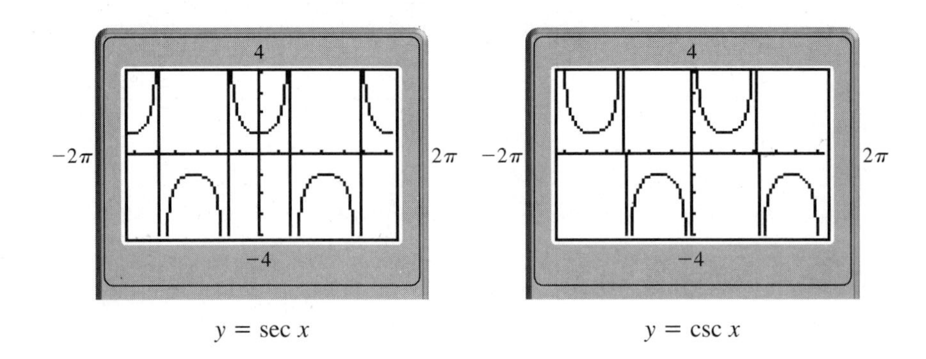

$y = \sec x$ $y = \csc x$

Exercises for Sections *5.7* and *5.8*

In Exercises 1–8 find the fundamental period of each function. [Recall from the definition of a period in Section 5.2 that the fundamental period is the *smallest* number p for which $f(x) = f(x + p)$.]

1. $\cot \frac{1}{2}x$ **2.** $\csc 3x$ **3.** $\sec \pi x$ **4.** $\tan 3\pi x$

5. $\tan \frac{1}{2}\pi x$ **6.** $\sec \frac{1}{3}x$ **7.** $\cot \frac{5}{6}x$ **8.** $\sec \dfrac{1}{\pi}x$

In Exercises 9–17 sketch the graph of each function over at least two periods. Give the period, the phase shift, and the asymptotes.

9. $y = \tan 2x$ **10.** $y = \tan(x + \frac{1}{2}\pi)$ **11.** $y = \cot(\frac{1}{4}\pi - x)$

12. $y = 2 \sec(x - \frac{1}{2}\pi)$ **13.** $y = \tan(2x + \frac{1}{3}\pi)$ **14.** $y = 2 \csc 2x$

15. $y = \csc(2x - 3\pi)$ **16.** $y = \sec(x + \frac{1}{3}\pi)$ **17.** $y = -\tan(x - \frac{1}{4}\pi)$

18. Does $\tan x$ exist at its asymptote?

19. How are the graphs of $\tan x$ and $\cot x$ related?

20. How are the graphs of $\sec x$ and $\csc x$ related?

21. How are the zeros of the tangent function and the asymptotes of the cotangent function related?

22. How are the zeros of the sine function and the asymptotes of the cosecant function related?

23. How are the graphs of $y = \tan x$ and $y = \tan(-x)$ related?

24. How are the graphs of $y = \tan x$ and $y = -\tan x$ related?

25. Show that the distance x from an observer to the base of the antenna in Figure 5.55 is given by $x = 20 \cot \theta$. Make a sketch of x as a function of θ.

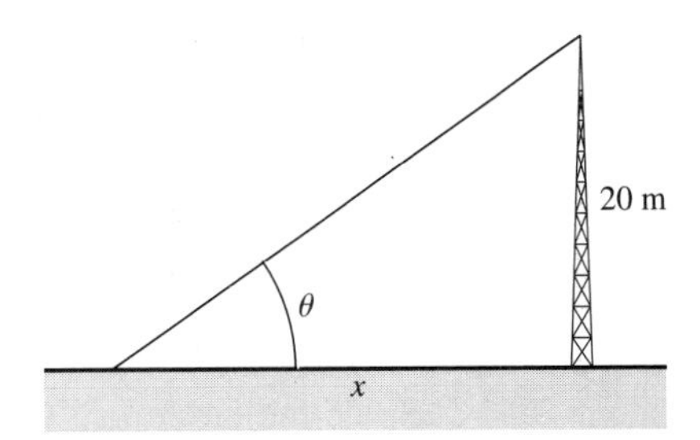

Figure 5.55

26. The force in a cable is given by $T = T_x \sec\theta$, where T_x is a measurable horizontal component of T. Assume that T_x is kept constant at 10 lb and sketch the graph of T as a function of θ for $0 \leq \theta \leq 60°$.
27. To determine the coefficient of sliding friction, μ, physicists perform a simple experiment in which the angle of inclination θ of an inclined plane is increased until a block on the inclined plane starts to slide. Then $\mu = \tan\theta$. Graph μ as a function of θ for $0 \leq \theta \leq 80°$.
28. A revolving light 1 mi from a straight shoreline makes 3 revolutions per minute. (See Figure 5.56.) If the light is initially directed parallel to the shore, show that the distance x is given by $x = \cot \frac{1}{10}\pi t$, where t is the time in seconds, $0 \leq t \leq 5$ sec. Make a sketch of x as a function of t for this time interval.

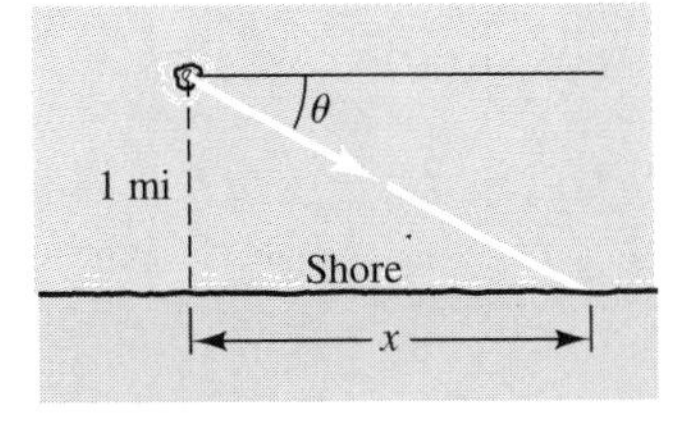

Figure 5.56

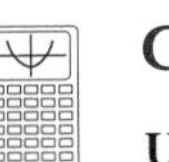

Graphing Calculator Exercises

Use a graphing calculator to sketch the graphs of the following functions. Choose a scale for the x-axis that will display two periods of the graph. If there is a phase shift, use the cursor to obtain an estimate of the phase shift and compare it to the known value.

1. $y = \tan 2x$
2. $y = \cot 3x$
3. $y = \sec 2x$
4. $y = \csc 3x$
5. $y = 0.5 \csc(x - 1)$
6. $y = 2 \sec(x + 1.2)$
7. $y = \cot(x + 0.8)$
8. $y = \tan(x - 0.5)$
9. $y = \tan(0.5x - 0.5)$
10. $y = 0.5 \sec(2x - 3)$

5.9 A Fundamental Inequality

The inequality

$$\sin x < x < \tan x; \qquad 0 < x < \tfrac{1}{2}\pi \tag{5.2}$$

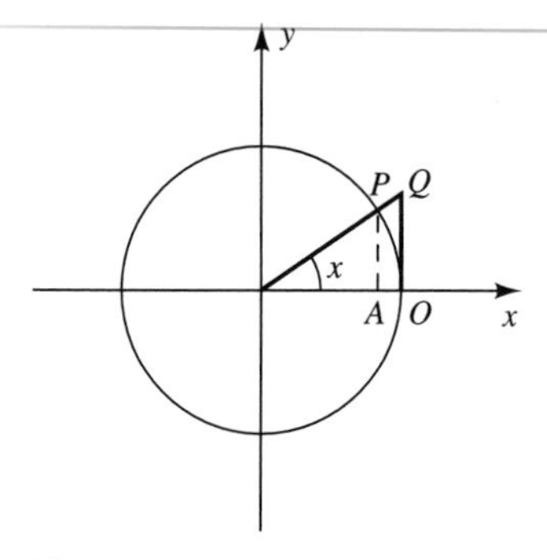

Figure 5.57

is fundamental to trigonometric analysis because it relates the arc length on the unit circle to two of the trigonometric functions. To demonstrate the validity of this inequality, we construct a unit circle as shown in Figure 5.57. From the figure it is obvious that the length of $\overline{AP}$ is less than the length of the arc OP, which in turn is less than $\overline{OQ}$.

$$\overline{AP} < \widehat{OP} < \overline{OQ}$$

Since the circle has radius 1, $\sin x = \overline{AP}$, $\tan x = \overline{OQ}$, and the $\widehat{OP} = x$. Hence

$$\sin x < x < \tan x$$

Figure 5.58 shows the three functions $y = \sin x$, $y = x$, and $y = \tan x$ and at the same time exhibits the fact that *the inequality is true only for* $0 < x < \frac{1}{2}\pi$.

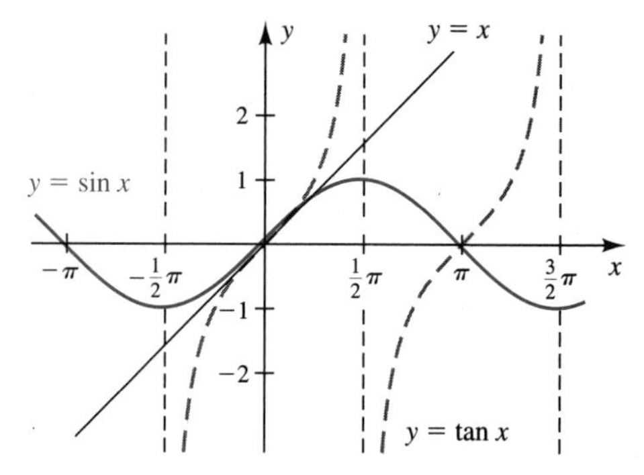

Figure 5.58

The ratio $(\sin x)/x$ is important in calculus. In the exercises for Section 5.1 you were asked to compute the values of this ratio for a few values of x near zero. You should have found that this ratio is just slightly less than 1 when x is a very small number. Using the fundamental inequality, (5.2), we can show this a bit more rigorously. Divide the members of the fundamental inequality by $\sin x$ to obtain

$$1 < \frac{x}{\sin x} < \frac{1}{\cos x}$$

(We are assuming x is positive.) Inverting and reversing the sense of the inequalities, we obtain the equivalent inequality

$$\cos x < \frac{\sin x}{x} < 1$$

Since $\cos x$ is near 1 when x is small and since the ratio is trapped between two functions close to 1, it follows that it also has values close to 1. (A similar argument for negative x leads to the same conclusion.)

The ratio $(\sin x)/x$ is undefined for $x = 0$, but if you look at a sketch of its graph (Figure 5.59), it looks as though its value "should be" 1.

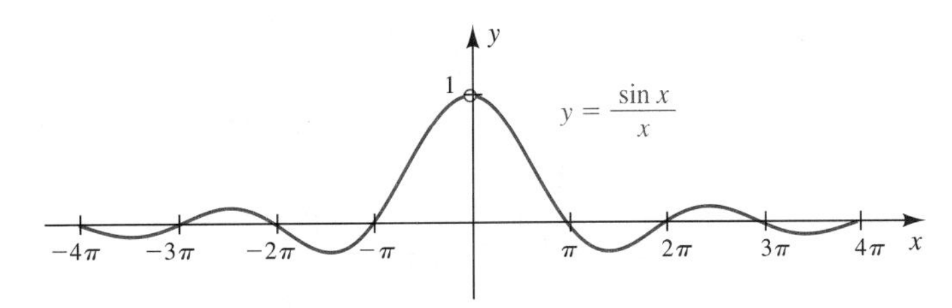

Figure 5.59

One of the side results of the fact that $(\sin x)/x$ is close to 1 for x close to zero is that for small values of x, $\sin x \approx x$. Thus in some applications if the absolute value of x is small, $\sin x$ is replaced by x.

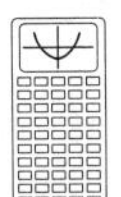

Use a graphing calculator to obtain the graph of $y = (\sin x)/x$. Note that the graph will look just like Figure 5.59, *with one exception*. The open dot at (0, 1) is not open on the calculator graph. Try computing $(\sin x)/x$ when $x = 0$ on your calculator and see what happens. What conclusion can you draw about the nature of the computations made by the calculator when graphing $y = (\sin x)/x$?

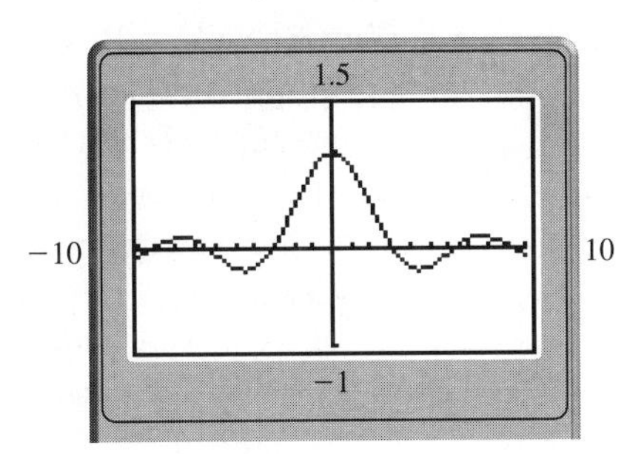

Exercises for Section 5.9

1. By carefully examining the graphs of $y = 2x/\pi$ and $y = \sin x$, convince yourself that $\frac{2x}{\pi} \leq \sin x \leq 1$ for $0 < x < \frac{\pi}{2}$.

2. What are the zeros of $\frac{\sin x}{x}$?

3. Sketch the function $f(x) = \frac{\sin 2x}{x}$ for $x \neq 0$. What do you think the value of the function approaches when x is near 0? (*Hint*: Use a calculator to calculate values of the function for $x = 0.1, 0.01, 0.001$, and so on. Then use a graphing calculator to obtain the graph of $f(x)$.) What is $f(0)$?

Key Topics for Chapter 5

Define and/or discuss each of the following.

Periodicity
Graphs of Trigonometric Functions
Bounded and Unbounded Trigonometric Functions
Even and Odd Functions
Cycle
Amplitude
Phase Shift
Harmonic Motion
Predator-Prey Problem
Addition of Ordinates
Asymptotes
Fundamental Inequality

Review Exercises for Chapter 5

1. Let $f(x) = 2\sin(3x - 1)$. Compute
   (a) $f(0)$ (b) $f(\frac{1}{3})$ (c) $f(\frac{2}{3}\pi)$
2. Determine where the function in Exercise 1 is equal to 0.
3. Determine where the function in Exercise 1 is equal to 1.

In Exercises 4–10 sketch the graph of each function and give its period and phase shift. Also give amplitude, mean value (avg), asymptotes, and intercepts where applicable.

4. $y = -\sin(x + \frac{1}{6}\pi)$
5. $y = 2 + \sin(x - \frac{1}{3}\pi)$
6. $y = \tan(\pi x + \pi) - 1$
7. $y = 4\cot(\frac{1}{2}x + \frac{1}{8}\pi)$
8. $y = 2\cos(2 - x) + \frac{1}{2}$
9. $y = \sec\frac{1}{3}(\pi - x) + 2$
10. $y = \csc(2x + 1)$
11. Sketch the graph of the harmonic motion described by $x = 9\cos 3t$ and give the amplitude and period.
12. Use the method of addition of ordinates to sketch $y = x + \cos x$.
13. Locate the asymptotes for the graph of $y = 2\tan(x + 3)$.
14. Completely describe the impact of the constants A, B, C, and D on the graph of $y = D + A\sin(Bx + C)$.
15. A weight at the end of a spring oscillates according to the formula

$$y(t) = 10\cos(3t - \tfrac{1}{2}\pi)$$

   Make a sketch of the motion of the weight.
16. The current in a certain coil is given by

$$i(t) = 10\sin(120\pi t - \tfrac{1}{6}\pi)$$

   Make a sketch of current versus time.
17. A block attached to a spring oscillates with decreasing amplitude according to the following formula:

$$x(t) = \begin{cases} \frac{5}{2}\cos\frac{3}{2}t + \frac{1}{2}, & 0 \le t \le \frac{2}{3}\pi \\ \frac{3}{2}\cos\frac{3}{2}t - \frac{1}{2}, & \frac{2}{3}\pi \le t \le \frac{4}{3}\pi \\ \frac{1}{2}\cos\frac{3}{2}t + \frac{1}{2}, & \frac{4}{3}\pi \le t \le 2\pi \\ 0, & 2\pi \le t \end{cases}$$

   Make a sketch of the motion of the block.

18. The current in a particular circuit is given by $i = 0.5 \sin 1$. Find the value of i.

19. The horizontal displacement of a simple oscillator is found to be

$$x = A \cos 2\pi ft \text{ cm}$$

What is x if $A = 1.4, f = 0.1$, and $t = 4$?

20. Compare the values of $\sin x$ for $x = 1$ radian, 1 degree, and the real number 1.

21. For which values of x is $\sin x = \tan x = 0$?

22. Why are the trigonometric functions called the circular functions?

23. A piston is connected to the rim of a wheel, as shown in Figure 5.60. The radius of the wheel is 2 ft, and the length of the connecting rod ST is 5 ft. The wheel rotates counterclockwise at the rate of 1 revolution per second. Find a formula for the position of the point S, t seconds after it has coordinates (2, 0). Find the position of the point S when $t = \frac{1}{2}, \frac{3}{4}$, and 2.

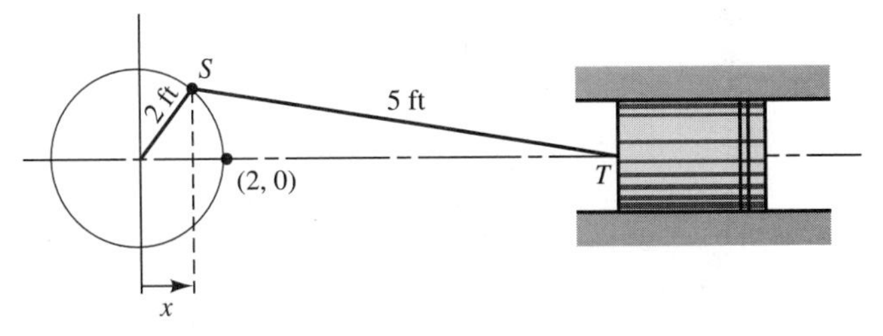

Figure 5.60

24. One end of a shaft is fastened to a piston that moves vertically. The other end is connected to the rim of a wheel by means of prongs, as shown in Figure 5.61. If the radius of the wheel is 2 ft and the shaft is 5 ft long, find a formula for the distance d, in feet, between the bottom of the piston and the x-axis, t seconds after P is at (2, 0). Assume the wheel rotates at 2 revolutions per second.

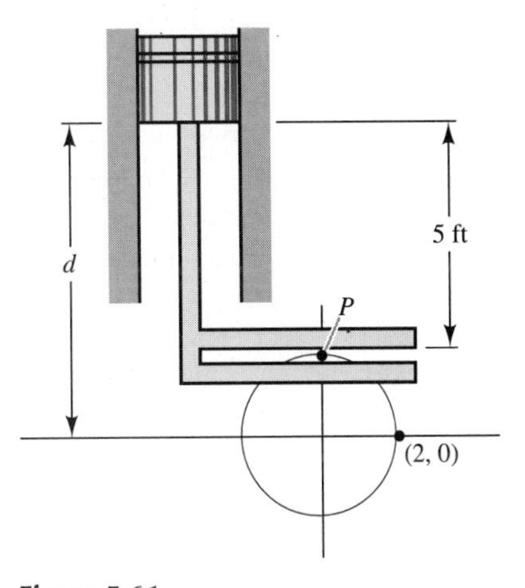

Figure 5.61

Practice Test Questions for Chapter 5

In Exercises 1–10 answer *true* or *false*.

1. A function f is even if $f(x) = f(-x)$ for all x in the domain of f.
2. The x-coordinate of the point in the plane associated with the real number u is equal to $\cos u$.
3. The period of $y = \cos 2x$ is 4π.
4. The graph of an odd function is symmetric with respect to the origin.
5. The phase shift of $\sin(3x - \pi/2)$ is $\pi/2$ units.
6. Tan x is an odd function.
7. The zeros of $\tan x$ and $\cot x$ coincide.
8. $y = \sin x$ and $y = \cos(x - \pi/2)$ have identical graphs.
9. A cycle is the shortest segment of a graph that includes one period.
10. An angle of u radians corresponds to the real number u.

In Exercises 11–20 fill in the blank to make the statement true.

11. A function f is ____________ if there exists a number $p > 0$ such that $f(x) = f(x + p)$.
12. A function f is ____________ if there exists a number $M > 0$ such that $|f(x)| \leq M$ for all x in the domain of f.
13. The amplitude of $y = 3 \cos 5x$ is ____________.
14. The period of $y = 2 \sin(7x + \pi)$ is ____________.
15. $\tan(\pi/4) =$ ____________.
16. The line $x = \pi/2$ is ____________ of the graph of $y = \tan x$.
17. The period of $\sec x$ is ____________.
18. When x increases from 0 to $\pi/2$, $\cos x$ decreases from ____________ to ____________.
19. The ____________ of a sinusoid is the reciprocal of the period.
20. The graph of $y = \tan(x - \pi/4)$ crosses the y-axis at ____________.

Solve the stated problem in the following exercises. Show all your work.

21. Sketch one period of the graph of $y = 1 + 5 \cos \frac{1}{2}x$. Indicate amplitude, period, mean value, and phase shift.
22. Sketch one period of the graph of $y = \sqrt{13} \sin(x - \pi/4)$. Indicate amplitude, period, mean value, and phase shift.
23. Sketch the graph of $y = -\tan 2x$ on the interval $-\pi/4 \leq x \leq 3\pi/4$. Indicate the period.
24. Sketch the graph of $y = -3 \sin(\pi x/2)$ on the interval $-4 \leq x \leq 4$. Indicate the period.
25. The motion of a spring-mass system is described by $y = 2.5 \cos(1.7t)$, where y is the distance in feet from the rest position and t is time in seconds.
    (a) Determine the time required for the mass to complete one oscillation.
    (b) Calculate the location of the mass when $t = 3$ sec.
26. Use addition of ordinates to obtain the graph of $y = x + \cos 2x$ for one period of the $\cos x$.

Identities, Equations, and Inequalities

*T*he values of the various trigonometric functions are ultimately related to the *x*- and *y*-coordinates of a point on the unit circle. Consequently the values of all six functions have a vast and seemingly endless variety of relationships to one another. We introduced some of these basic identities in Section 2.2. In Sections 6.1 and 6.2 we show how to derive and prove identities involving trigonometric functions. Then in Section 6.3 we show how to solve equations involving trigonometric functions. For perhaps the first time in your mathematics studies, you will see equations that have infinitely many solutions. The study of parametric equations in Section 6.4 is not limited to, but definitely emphasizes, the use of trigonometric functions in such equations. The Lissajous figure is a nice application of this concept. Here again the graphing calculator is an invaluable aid.

6.1 Fundamental Trigonometric Relations

Any combination of trigonometric functions such as $3 \sin x + \cos x$ or $\sec^2 x + \tan^2 x + 2 \sin x$ is called a **trigonometric expression**. One of the important skills you will learn in this chapter is how to simplify or alter the form of trigonometric expressions using certain fundamental trigonometric relations.

There are eight *fundamental relations*, or identities, that you must know if you are to work the problems in the remainder of this book efficiently. You are already familiar with most of these relations, but we will review them here. The fundamental relations fall into three groups: the reciprocal relations, the quotient relations, and the Pythagorean relations.

The Reciprocal Relations

$$\sin\theta = \frac{1}{\csc\theta} \quad \textbf{(6.1)}$$

$$\cos\theta = \frac{1}{\sec\theta} \quad \textbf{(6.2)}$$

$$\tan\theta = \frac{1}{\cot\theta} \quad \textbf{(6.3)}$$

We establish Equation (6.1) by observing that for any angle θ in standard position and (x, y) on the terminal side with length r,

$$\sin\theta = \frac{y}{r} = \frac{1}{r/y} = \frac{1}{\csc\theta}$$

The other two relations are established in a similar manner.

The Quotient Relations

$$\tan\theta = \frac{\sin\theta}{\cos\theta} \quad \textbf{(6.4)}$$

$$\cot\theta = \frac{\cos\theta}{\sin\theta} \quad \textbf{(6.5)}$$

To establish Equation (6.4) we note that

$$\tan\theta = \frac{y}{x} = \frac{y/r}{x/r} = \frac{\sin\theta}{\cos\theta}$$

Therefore

$$\tan\theta = \frac{\sin\theta}{\cos\theta}$$

The Pythagorean Relations

$$\sin^2\theta + \cos^2\theta = 1 \quad \textbf{(6.6)}$$

$$\tan^2\theta + 1 = \sec^2\theta \quad \textbf{(6.7)}$$

$$\cot^2\theta + 1 = \csc^2\theta \quad \textbf{(6.8)}$$

We prove Equation (6.6) by dividing $x^2 + y^2 = r^2$ by r^2 to get

$$\frac{x^2}{r^2} + \frac{y^2}{r^2} = 1$$

Then since

$$\cos \theta = \frac{x}{r} \qquad \text{and} \qquad \sin \theta = \frac{y}{r}$$

we have

$$\cos^2\theta + \sin^2\theta = 1$$

Note that Equation (6.7) is derived from Equation (6.6). Dividing both sides of Equation (6.6) by $\cos^2\theta$ gives

$$\frac{\sin^2\theta}{\cos^2\theta} + 1 = \frac{1}{\cos^2\theta}$$

Then using the fact that

$$\frac{\sin \theta}{\cos \theta} = \tan \theta \text{ and } \frac{1}{\cos \theta} = \sec \theta$$

we can convert the equation above to

$$\tan^2\theta + 1 = \sec^2\theta$$

Similarly, Equation (6.8) is derived from Equation (6.6) by first dividing both sides by $\sin^2\theta$ and then applying Equations (6.1) and (6.5).

These eight **fundamental identities** of trigonometry are valid for all values of the argument for which the functions in the expression have meaning and for which the denominators are nonzero. As before, the variable (which is often the letter x rather than θ) may be regarded as either a real number or an angle, the interpretation depending on the context. Using the fundamental identities, you can manipulate trigonometric expressions into alternative forms.

Example 1 Write the following expression as a single trigonometric term:

$$\frac{\tan x \csc^2 x}{1 + \tan^2 x}$$

SOLUTION Equation (6.7) shows that the denominator may be written as $\sec^2 x$. Thus

$$\frac{\tan x \csc^2 x}{1 + \tan^2 x} = \frac{\tan x \csc^2 x}{\sec^2 x}$$

We can now express $\tan x$, $\csc x$, and $\sec x$ in terms of the sine and cosine functions.

$$\frac{\tan x \csc^2 x}{1 + \tan^2 x} = \frac{\dfrac{\sin x}{\cos x} \cdot \dfrac{1}{\sin^2 x}}{\dfrac{1}{\cos^2 x}}$$

$$= \frac{\cos^2 x \sin x}{\sin^2 x \cos x} \qquad \text{Invert } \frac{1}{\cos^2 x} \text{ and multiply}$$

$$= \frac{\cos x}{\sin x} \qquad \text{Apply the cancellation law}$$

$$= \cot x \qquad \frac{\cos x}{\sin x} = \cot x$$

Therefore

$$\frac{\tan x \csc^2 x}{1 + \tan^2 x} = \cot x$$ ▲

As the preceding example shows, a large part of the process is algebraic. The steps used in the example are not the only way to simplify the expression. For example, we could have initially written the complete expression in terms of the sine and cosine functions. However, writing the entire expression in terms of sine and cosine functions is not necessarily the shortest or easiest method.

Example 2 Simplify the expression $(\sec x + \tan x)(1 - \sin x)$.

SOLUTION We write each of the functions in terms of the sine and cosine functions.

$$(\sec x + \tan x)(1 - \sin x) = \left(\frac{1}{\cos x} + \frac{\sin x}{\cos x}\right)(1 - \sin x)$$

$$= \frac{(1 + \sin x)(1 - \sin x)}{\cos x} \qquad \text{Add fractions}$$

$$= \frac{(1 - \sin^2 x)}{\cos x} \qquad (a + b)(a - b) = a^2 - b^2$$

$$= \frac{\cos^2 x}{\cos x} \qquad 1 - \sin^2 x = \cos^2 x$$

$$= \cos x \qquad \text{Apply the cancellation law}$$

Therefore $(\sec x + \tan x)(1 - \sin x) = \cos x$. ▲

Example 3 Expand and simplify the expression $(\sin x + \cos x)^2$.

SOLUTION Note that this is *not* the same expression as $\sin^2 x + \cos^2 x$. By squaring the expression we obtain

$$(\sin x + \cos x)^2 = \sin^2 x + 2 \sin x \cos x + \cos^2 x \qquad \text{Expand the binomial}$$

$$= 1 + 2 \sin x \cos x \qquad \sin^2 x + \cos^2 x = 1$$ ▲

Example 4 Simplify the expression $\sin^4 x - \cos^4 x + \cos^2 x$.

SOLUTION We write the expression in a form involving only the cosine function. To do so we note that $\sin^4 x = (\sin^2 x)^2 = (1 - \cos^2 x)^2$. Thus

$$\begin{aligned}\sin^4 x - \cos^4 x + \cos^2 x &= (1 - \cos^2 x)^2 - \cos^4 x + \cos^2 x \\ &= 1 - 2\cos^2 x + \cos^4 x - \cos^4 x + \cos^2 x \\ &= 1 - \cos^2 x \\ &= \sin^2 x\end{aligned}$$

Therefore $\sin^4 x - \cos^4 x + \cos^2 x = \sin^2 x$. ▲

Certain algebraic expressions encountered in calculus are often transformed into trigonometric expressions in order to make hard-to-handle terms such as radicals disappear.

Example 5 Using the substitution $x = 2\sin\theta$, simplify the expression $\sqrt{4 - x^2}$ and determine an interval for the variable θ that corresponds to $0 \le x \le 2$ in a one-to-one manner. What is $\tan\theta$?

SOLUTION Substituting $x = 2\sin\theta$ into the radical gives

$$\begin{aligned}\sqrt{4 - x^2} &= \sqrt{4 - 4\sin^2\theta} \\ &= \sqrt{4(1 - \sin^2\theta)} \\ &= |2\cos\theta| = 2|\cos\theta|\end{aligned}$$

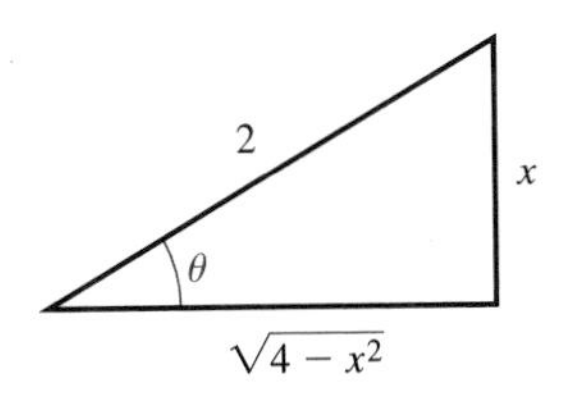

Figure 6.1

When $x = 0$, $\theta = 0$, and when $x = 2$, $\theta = \frac{1}{2}\pi$, so the interval $0 \le x \le 2$ corresponds to $0 \le \theta \le \frac{1}{2}\pi$. On this interval $\cos\theta \ge 0$, so $|\cos\theta| = \cos\theta$. Hence

$$\sqrt{4 - x^2} = 2\cos\theta \qquad \text{for } 0 \le x \le 2 \text{ and } 0 \le \theta \le \tfrac{1}{2}\pi$$

Since $\sin\theta = x/2$, the right triangle in Figure 6.1 shows the relations necessary to establish that

$$\tan\theta = \frac{x}{\sqrt{4 - x^2}}$$

▲

▲▼ Exercises for Section 6.1

In Exercises 1–20 reduce each expression to a single trigonometric function.

1. $\cos\theta + \tan\theta\sin\theta$

2. $\csc\theta - \cot\theta\cos\theta$

3. $(\tan x + \cot x)\sin x$

4. $\dfrac{1 + \cos x}{1 + \sec x}$

5. $\dfrac{(\tan x)(1 + \cot^2 x)}{1 + \tan^2 x}$

6. $\sec x - \sin x\tan x$

7. $\cos x\csc x$

8. $\cos x(\tan x + \cot x)$

9. $(\cos^2 x - 1)(\tan^2 x + 1)$

10. $\dfrac{\sec^2 x - 1}{\sec^2 x}$

11. $\dfrac{\sec x - \cos x}{\tan x}$

12. $\dfrac{1 + \tan^2 x}{\tan^2 x}$

13. $(\sin^2 x + \cos^2 x)^3$

14. $\dfrac{1 + \sec x}{\tan x + \sin x}$

15. $(\csc x - \cot x)^4(\csc x + \cot x)^4$

16. $\dfrac{\sec x}{\tan x + \cot x}$

17. $(\tan x)(\sin x + \cot x \cos x)$

18. $1 + \dfrac{\tan^2 x}{1 + \sec x}$

19. $\dfrac{\tan x \sin x}{\sec^2 x - 1}$

20. $(\cos x)(1 + \tan^2 x)$

In Exercises 21–26 use substitutions to reduce each given expression to one involving only trigonometric functions. Assume $0 \leq \theta < \frac{1}{2}\pi$.

21. $\sqrt{a^2 + x^2}$, let $x = a \tan \theta$. What is $\sin \theta$?

22. $\sqrt{36 + 16x^2}$, let $x = \frac{3}{2} \tan \theta$. What is $\sin \theta$?

23. $\dfrac{\sqrt{x^2 - 4}}{x}$, let $x = 2 \sec \theta$.

24. $x^2\sqrt{4 + 9x^2}$, let $x = \frac{2}{3} \tan \theta$.

25. $\sqrt{3 - 5x^2}$, let $x = \sqrt{\frac{3}{5}} \sin \theta$.

26. $\sqrt{2 + x^2}$, let $x = \sqrt{2} \tan \theta$.

27. In calculus class a dispute arises over the answer to a problem. One group says the answer is $\tan x$, and the other group says the answer is

$$\frac{\cos x - \cos^3 x}{\sin x - \sin^3 x}$$

Can you show that both groups are correct?

28. The polar equation of a parabola can be written as

$$r = \frac{k}{1 - \cos \theta}$$

Show that this is equivalent to

$$r = k \csc \theta(\csc \theta + \cot \theta)$$

29. Working a problem in statics, you get an answer of $\sec x + \tan x$. The answers section of the book has $(\sec x - \tan x)^{-1}$. Is there a printing error in the book?

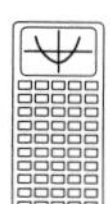

Graphing Calculator Exercises

1. Use a graphing calculator to show that the two functions in Exercise 29 have the same graph. If two functions have the same graph, what can you say about the functions?

2. Use a graphing calculator to display the graph of

$$y = \cos x + \tan x \sin x$$

Does the graph look familiar? On the basis of its graph, what can you say about the function? Determine the period of the function from the graph.

6.2 Trigonometric Identities

When both sides of an equation are equal for all values of the variable for which the equation is defined, the equation is called an **identity**. The eight fundamental relations (6.1)–(6.8) are trigonometric identities. In addition to the fundamental identities, there are many other trigonometric identities that arise in applications of mathematics. In most cases you are required to verify, or **prove**, that the given relation is an identity. There are several techniques that can be used to prove an identity, but the most common one involves using other known identities to transform one side of the equation into precisely the same form as the other. In this section you will learn to use the fundamental identities to prove other identities. Although there is no general approach to proving identities, you may find it desirable to write the given expressions in terms of sines and cosines only. Often doing so allows you to see the manipulations necessary to complete the verification.

COMMENT It is incorrect to verify an identity by beginning with the assumption that it *is* an identity. Thus the expressions on either side of the equal sign are not equated initially.

Example 1 Verify the identity $\cot x + \tan x = \csc x \sec x$.

SOLUTION Here we express the left-hand side in terms of sines and cosines. Thus

$$\begin{aligned}
\cot x + \tan x &= \frac{\cos x}{\sin x} + \frac{\sin x}{\cos x} && \text{Change to sine and cosine} \\
&= \frac{\cos^2 x + \sin^2 x}{\sin x \cos x} && \text{Add fractions} \\
&= \frac{1}{\sin x \cos x} && \cos^2 x + \sin^2 x = 1 \\
&= \csc x \sec x && \frac{1}{\sin x} = \csc x;\ \frac{1}{\cos x} = \sec x
\end{aligned}$$

Therefore $\cot x + \tan x = \csc x \sec x$. ▲

Example 2 Verify the identity

$$\frac{\sec^4 x - \tan^4 x}{\sec^2 x} = 1 + \sin^2 x$$

SOLUTION We begin by factoring the numerator as the difference of two squares.

$$\frac{\sec^4 x - \tan^4 x}{\sec^2 x} = \frac{(\sec^2 x - \tan^2 x)(\sec^2 x + \tan^2 x)}{\sec^2 x} \qquad \text{Factor } \sec^4 x - \tan^4 x$$

$$= \frac{\sec^2 x + \tan^2 x}{\sec^2 x} \qquad \text{Replace } \sec^2 x - \tan^2 x \text{ with 1 (recall that } \tan^2 x + 1 = \sec^2 x)$$

$$= \frac{\dfrac{1}{\cos^2 x} + \dfrac{\sin^2 x}{\cos^2 x}}{\dfrac{1}{\cos^2 x}} \qquad \text{Change to } \sin x \text{ and } \cos x$$

$$= 1 + \sin^2 x \qquad \text{Multiply numerator and denominator by } \cos^2 x$$

Therefore

$$\frac{\sec^4 x - \tan^4 x}{\sec^2 x} = 1 + \sin^2 x$$ ▲

Example 3 Show that $\sin\theta(\csc\theta - \sin\theta) = \cos^2\theta$ is an identity.

SOLUTION Here the most expedient approach is to expand the left-hand side. Thus

$$\sin\theta(\csc\theta - \sin\theta) = \sin\theta\csc\theta - \sin^2\theta \qquad \text{Expand}$$

$$= \sin\theta\frac{1}{\sin\theta} - \sin^2\theta \qquad \csc\theta = \frac{1}{\sin\theta}$$

$$= 1 - \sin^2\theta \qquad \text{Apply the cancellation law}$$

$$= \cos^2\theta \qquad 1 - \sin^2\theta = \cos^2\theta$$

Therefore

$$\sin\theta(\csc\theta - \sin\theta) = \cos^2\theta$$ ▲

Example 4 Verify the identity

$$\frac{\cos x}{1 - \sin x} = \frac{1 + \sin x}{\cos x}$$

SOLUTION Here we start on the left-hand side. One way to get $1 + \sin x$ into the numerator of the left side is to multiply the left side by $(1 + \sin x)/(1 + \sin x)$, as follows.

$$\frac{\cos x}{1 - \sin x} = \frac{\cos x}{1 - \sin x} \cdot \frac{1 + \sin x}{1 + \sin x} \qquad \text{Multiply by } \frac{1 + \sin x}{1 + \sin x}$$

$$= \frac{\cos x(1 + \sin x)}{1 - \sin^2 x} \qquad \text{Complete the multiplication}$$

$$= \frac{\cos x(1 + \sin x)}{\cos^2 x} \qquad 1 - \sin^2 x = \cos^2 x$$

$$= \frac{1 + \sin x}{\cos x} \qquad \text{Apply the cancellation law}$$

Therefore

$$\frac{\cos x}{1 - \sin x} = \frac{1 + \sin x}{\cos x}$$ ▲

Example 5 Verify the identity

$$(\csc x + \cot x)^2 = \frac{1 + \cos x}{1 - \cos x}$$

SOLUTION We start on the left-hand side by squaring the binomial.

$$\begin{aligned}
(\csc x + \cot x)^2 &= \csc^2 x + 2 \csc x \cot x + \cot^2 x && \text{Expand} \\
&= \frac{1}{\sin^2 x} + \frac{2 \cos x}{\sin^2 x} + \frac{\cos^2 x}{\sin^2 x} && \text{Change to } \sin x \text{ and } \cos x \\
&= \frac{1 + 2 \cos x + \cos^2 x}{\sin^2 x} && \text{Add fractions} \\
&= \frac{(1 + \cos x)^2}{\sin^2 x} && \text{Factor the numerator} \\
&= \frac{(1 + \cos x)^2}{1 - \cos^2 x} && \sin^2 x = 1 - \cos^2 x \\
&= \frac{(1 + \cos x)^2}{(1 + \cos x)(1 - \cos x)} && \text{Factor the denominator} \\
&= \frac{1 + \cos x}{1 - \cos x} && \text{Apply the cancellation law}
\end{aligned}$$

Therefore

$$(\csc x + \cot x)^2 = \frac{1 + \cos x}{1 - \cos x}$$ ▲

Sometimes you can verify an identity by manipulating the left-hand side and the right-hand side into forms that are precisely the same.

Example 6 Verify the identity

$$\cos^2 x \tan^2 x + 1 = \sec^2 x + \sin^2 x - \sin^2 x \sec^2 x$$

SOLUTION We transform the two sides of the given expression into precisely the same form. The left-hand side becomes

$$\begin{aligned}
\cos^2 x \tan^2 x + 1 &= \cos^2 x \left(\frac{\sin^2 x}{\cos^2 x}\right) + 1 && \text{Change to sine and cosine} \\
&= \sin^2 x + 1 && \text{Apply the cancellation law}
\end{aligned}$$

The right-hand side may be transformed as follows:

$$\begin{aligned}
\sec^2 x + \sin^2 x - \sin^2 x \sec^2 x &= \frac{1}{\cos^2 x} + \sin^2 x - \frac{\sin^2 x}{\cos^2 x} && \text{Change to sine and cosine} \\
&= \frac{1 - \sin^2 x}{\cos^2 x} + \sin^2 x && \text{Combine fractions} \\
&= \frac{\cos^2 x}{\cos^2 x} + \sin^2 x && 1 - \sin^2 x = \cos^2 x \\
&= 1 + \sin^2 x && \text{Apply the cancellation law}
\end{aligned}$$

Since the right and left sides of the identity have been transformed into the same expression, the identity is verified. Therefore $\cos^2 x \tan^2 x + 1 = \sec^2 x + \sin^2 x - \sin^2 x \sec^2 x$. ▲

To show that a relation is not an identity, we need only show that the two expressions are unequal for a particular value of x.

Example 7 Show that the expression $\sin x = \sqrt{\sin^2 x}$ is not an identity.

SOLUTION Note that there are many values of x for which the two expressions $\sin x$ and $\sqrt{\sin^2 x}$ *are* equal. But consider $x = -\pi/4$. The value of $\sin(-\pi/4)$ is $-\sqrt{2}/2$, whereas

$$\sqrt{\sin^2\left(\frac{-\pi}{4}\right)} = \sqrt{\left(\frac{-\sqrt{2}}{2}\right)^2} = \sqrt{\frac{1}{2}} = \frac{\sqrt{2}}{2}$$

Hence the two expressions are not equal for this value of x, and the given expression is not an identity. (There may also be other values of x for which the expressions are not equal.) ▲

Following are some guidelines for verifying trigonometric identities.

- Know your fundamental identities well. For example, you should know not only $\sin^2 x + \cos^2 x = 1$ but also all the variations of this identity, such as $\sin^2 x = 1 - \cos^2 x$. Quick recall of the fundamental identities is the single most important ability you can develop.
- Start with the most complicated side. If both sides appear fairly complicated, you might try manipulating the sides *independently* until they are in the same form.
- Sometimes it helps to change everything into sines and cosines or, in some instances, into the same function.
- Perform algebraic simplifications whenever possible, but be aware that there may be two (or more) possible ways to modify an expression. For example,

$$\frac{1}{\sin x} - \sin x = \frac{1 - \sin^2 x}{\sin x} = \frac{\cos^2 x}{\sin x}$$

or

$$\frac{1}{\sin x} - \sin x = \frac{1 - \sin^2 x}{\sin x} = \frac{(1 - \sin x)(1 + \sin x)}{\sin x}$$

- As you simplify expressions, be guided by what comes next, or by the result you wish to achieve. Consider the expressions

$$\frac{1 - \sin^2 x}{1 + \sin x} \quad \text{and} \quad \frac{1 - \sin^2 x}{\cos x}$$

which have the same numerator. With the expression on the left, the logical next step is to factor the numerator and simplify as follows:

$$\frac{(1 - \sin x)(1 + \sin x)}{1 + \sin x} = 1 - \sin x$$

With the expression on the right, the best thing to do is to recall the identity $1 - \sin^2 x = \cos^2 x$ and simplify as follows:

$$\frac{1 - \sin^2 x}{\cos x} = \frac{\cos^2 x}{\cos x} = \cos x$$

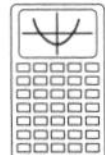

One way to check to see whether a trigonometric expression is an identity is to graph the left-hand side function and the right-hand side function separately. If the two graphs are not identical, we conclude that the expression is not an identity. If the two graphs appear to be the same, then we can say at least that the expression is *possibly an identity*. This approach to verifying trigonometric identities is especially appealing when using a graphing calculator. But be careful when analyzing graphs in the display; it is sometimes difficult to discern from the display whether two graphs are identical.

Example 8 Use a graphing calculator to show that

$$\sin x = \sqrt{\sin^2 x}$$

is not an identity. (This was verified in Example 7, where we showed that $\sin(-\pi/4) \neq \sqrt{\sin^2(-\pi/4)}$.)

SOLUTION Let $y_1 = \sin x$ and $y_2 = \sqrt{\sin^2 x}$. The graph of y_1 is shown on the left below; the graph of y_2 is shown on the right. Clearly, the two graphs are not identical, so we conclude that the given expression is not an identity; that is, $\sin x \neq \sqrt{\sin^2 x}$.

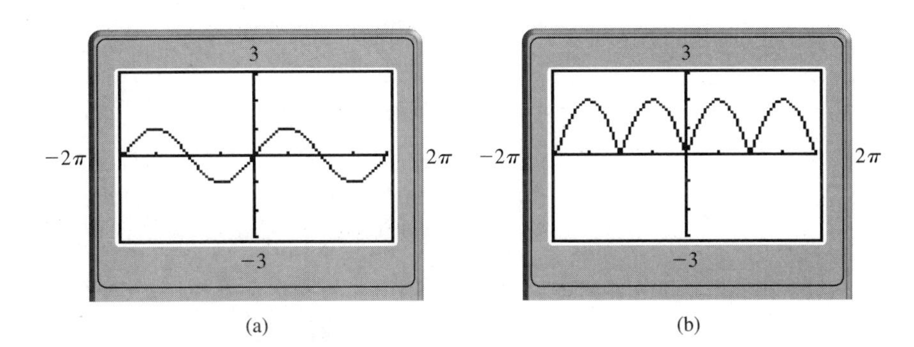

(a) (b)

(We have shown the graphs of y_1 and y_2 on two separate screens to avoid any confusion as to which graph is which. In practice, both graphs can be displayed on the same screen.) ▲

CAUTION To reiterate: Just because two functions *appear* to have the same graph does not *prove* that the two functions are the same. Analytical methods must be used to show that two functions are identical.

Exercises for Section 6.2

In Exercises 1–82 verify the identities.

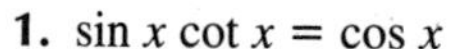

1. $\sin x \cot x = \cos x$
2. $\cos x \tan x = \sin x$
3. $\sec x \cot x = \csc x$
4. $(1 + \tan^2 x)\sin^2 x = \tan^2 x$
5. $\sin^2 x(1 + \cot^2 x) = 1$
6. $\cot^2 x - \cos^2 x = \cot^2 x \cos^2 x$
7. $\csc x - \sin x = \cot x \cos x$
8. $\sec^2 x \csc^2 x = \sec^2 x + \csc^2 x$
9. $(\sin^2 x - 1)(\cot^2 x + 1) = 1 - \csc^2 x$
10. $\dfrac{\sin^2 x + \cos^2 x}{\cos^2 x} = \sec^2 x$
11. $\dfrac{2 + \sec x}{\csc x} - 2 \sin x = \tan x$
12. $\dfrac{\sin^4 x - \cos^4 x}{\sin x - \cos x} = \sin x + \cos x$
13. $\dfrac{\sin x}{1 - \cos x} = \csc x + \cot x$
14. $\dfrac{\tan x - 1}{\tan x + 1} = \dfrac{1 - \cot x}{1 + \cot x}$
15. $\dfrac{\cot x + 1}{\cot x - 1} = -\dfrac{\tan x + 1}{\tan x - 1}$
16. $\dfrac{\cos x}{\sec x} + \dfrac{\sin x}{\csc x} = \sec^2 x - \tan^2 x$
17. $\dfrac{1 + \sec x}{\sin x + \tan x} = \csc x$
18. $\sec^2 x - \csc^2 x = \tan^2 x - \cot^2 x$
19. $\dfrac{1 - \sin x}{1 + \sin x} = (\sec x - \tan x)^2$
20. $\cos^2 x - \sin^2 x = 2 \cos^2 x - 1$
21. $(\sin^2 x + \cos^2 x)^4 = 1$
22. $\dfrac{\csc^2 x - \cot^2 x}{\sec^2 x} = \cos^2 x$
23. $\dfrac{\tan x + \cot x}{\tan x - \cot x} = \dfrac{\sec^2 x}{\tan^2 x - 1}$
24. $\sin^2 x \sec^2 x + 1 = \sec^2 x$
25. $\dfrac{\sin x}{\csc x(1 + \cot^2 x)} = \sin^4 x$
26. $\sec^2 x - (\cos^2 x + \tan^2 x) = \sin^2 x$
27. $1 - \tan^4 x = 2 \sec^2 x - \sec^4 x$
28. $\sec x - \cos x = \sin x \tan x$
29. $(\cot x + \csc x)^2 = \dfrac{1 + \cos x}{1 - \cos x}$
30. $\sin^2 x(\csc^2 x - 1) = \cos^2 x$
31. $\sec x \csc x - 2 \cos x \csc x = \tan x - \cot x$
32. $\sec^4 x + \tan^4 x = 1 + 2 \sec^2 x \tan^2 x$
33. $\tan^2 x - \cot^2 x = \sec^2 x - \csc^2 x$
34. $\dfrac{1 - \tan^2 x}{1 - \cot^2 x} = 1 - \sec^2 x$
35. $(1 - \sin^2 x)(1 + \tan^2 x) = 1$
36. $\dfrac{\tan x + \sin x}{\tan x - \sin x} = \dfrac{\sec x + 1}{\sec x - 1}$
37. $\dfrac{\cos x + \tan x}{\sin x \cos x} = \csc x + \sec^2 x$

38. $\cos^2 x \tan x = \dfrac{2 \sin x}{\cos x + \sec x + \sin^2 x \sec x}$

39. $(\sin x - \cos x)^2 = 1 - 2 \sin x \cos x$

40. $\dfrac{\cos x}{\cos x - \sin x} = \dfrac{1}{1 - \tan x}$

41. $\cos^2 x - \sin x \tan x = \cos x \cot x \sin x - \tan x \sin x$

42. $\dfrac{\tan^2 x}{\sin^4 x} = \dfrac{1 + \tan^2 x}{1 - \cos^2 x}$

43. $\tan x - \cot x = -\dfrac{\cos x - \sin^2 x \sec x}{\sin x}$

44. $1 - \sin x = \dfrac{\cot x - \cos x}{\cot x}$

45. $\csc^4 x + \cot^4 x = 1 + 2 \csc^2 x \cot^2 x$

46. $\dfrac{\sec^2 x + 2 \tan x}{1 + \tan x} = 1 + \tan x$

47. $(1 + \cos x)^2 = \sin^2 x \dfrac{\sec x + 1}{\sec x - 1}$

48. $(\sec x - \tan x)^2 = \dfrac{1 - \sin x}{1 + \sin x}$

49. $(\sec x + \tan x)^2 = \dfrac{\sec x + \tan x}{\sec x - \tan x}$

50. $(\csc x - \cot x)^2 = \dfrac{\csc x - \cot x}{\csc x + \cot x}$

51. $\dfrac{\csc x}{\csc x - \tan x} = \dfrac{\cos x}{\cos x - \sin^2 x}$

52. $(\tan x - 1) \cos x = \sin x - \cos x$

53. $\dfrac{1}{1 - \sin x} - \dfrac{1}{1 + \sin x} = 2 \tan x \sec x$

54. $\dfrac{\tan x - \csc x}{\tan x + \csc x} = \dfrac{\sin^2 x - \cos x}{\sin^2 x + \cos x}$

55. $\sec^4 x - \tan^4 x = \dfrac{1 + \sin^2 x}{\cos^2 x}$

56. $\dfrac{\tan x}{\sec x - \cos x} = \csc x$

57. $(\csc x - \cot x)(\sec x + 1) = \tan x$

58. $\dfrac{\cos^2 x}{1 + \sin x} = 1 - \sin x$

59. $(1 + \sin x)(\sec x - \tan x) = \cos x$

60. $\cos^4 x - \sin^4 x = 1 - 2 \sin^2 x$

61. $\sec x \csc x - 2 \cos x \csc x = \tan x - \cot x$

62. $\dfrac{\sin x}{\sin x + \cos x} = \dfrac{\tan x}{1 + \tan x}$

63. $\dfrac{\sin x}{1 + \cos x} + \dfrac{1 + \cos x}{\sin x} = 2 \csc x$

64. $2 \sin^4 x - 3 \sin^2 x + 1 = \cos^2 x(1 - 2 \sin^2 x)$

65. $\dfrac{\csc x}{\tan x + \cot x} = \cos x$

66. $\dfrac{1 - \sin x}{1 + \sin x} = \left(\dfrac{\cos x}{1 + \sin x}\right)^2$

67. $\dfrac{1 + \cos x}{1 - \cos x} = (\csc x + \cot x)^2$

68. $\dfrac{\sec^3 x - \cos^3 x}{\sec x - \cos x} = 1 + \cos^2 x + \sec^2 x$

69. $\dfrac{\cos^2 x}{1 - \sin x + \cos^2 x} = \dfrac{1 + \sin x}{2 + \sin x}$

70. $(1 + \tan x)^2 = \sec^2 x(1 + 2 \cos x \sin x)$

71. $(1 + \cot x)^2 = \csc^2 x(1 + 2 \cos x \sin x)$

72. $(\sin x + \cos x)^2 = \dfrac{\sec x \csc x + 2}{\sec x \csc x}$

73. $(\cos x + \sin x + \tan x)^2 = \sec^2 x(1 + 2 \sin^2 x \cos x) + 2 \sin x(1 + \cos x)$

74. $\dfrac{1 + \cos x}{2 - \cos x} = \dfrac{\sin^2 x}{2 - 3 \cos x + \cos^2 x}$

75. $(1 + \sin^2 x)^4 = 16 - 32 \cos^2 x + 24 \cos^4 x - 8 \cos^6 x + \cos^8 x$

76. $\dfrac{\tan x - \cot x}{\sec^2 x - \csc^2 x} = \sin x \cos x$

77. $\dfrac{\tan x - \tan y}{\cot x - \cot y} = \dfrac{1 - \tan x \tan y}{1 - \cot x \cot y}$

78. $\dfrac{\cos x \cos y - \sin x \sin y}{\cos x \sin y + \cos y \sin x} = \dfrac{1 - \tan x \tan y}{\tan x + \tan y}$

79. $\tan x + \cot x = \sec x \csc x$

80. $\dfrac{\tan x}{1 - \cot x} + \dfrac{\cot x}{1 - \tan x} - 1 = \sec x \csc x$

81. $(2 \cos x - \sin x)^2 + (2 \sin x + \cos x)^2 = 5$

82. $(a \cos x - b \sin x)^2 + (a \sin x + b \cos x)^2 = a^2 + b^2$

83. A student finds that two calculus books give different formulas for the derivative of $\sec x$. One book gives $\sec x \tan x$ and the other one gives $\sin x/\cos^2 x$. Show that the two forms are equivalent.

84. In calculating the slope of the path of a projectile, an aerospace engineer is confronted with the expression

$$\frac{-2v_0 \sin \theta + v_0 \sin \theta}{v_0 \cos \theta}$$

where v_0 is the muzzle velocity and θ is the initial angle of elevation. Show that this expression is equal to $-\tan \theta$.

85. In finding the point of intersection of

$$r = \frac{1}{1 - \cos \theta} \quad \text{and} \quad r = \frac{3}{1 + \cos \theta}$$

you must simplify the equality

$$\frac{1 + \cos \theta}{1 - \cos \theta} = 3$$

Show that the left-hand side may be written as

$$\frac{(1 + \cos \theta)^2}{\sin^2 \theta}$$

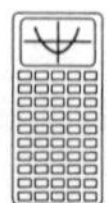

Graphing Calculator Exercises

In Exercises 1–12 use a graphing calculator to show that the given expressions are not identities. Then explain why not.

1. $\cos t = \sqrt{\cos^2 t}$

2. $1 = \tan(\cot x)$

3. $1 = \sec(\cos x)$

4. $\sin x = (1 - \cos x)^2$

5. $\cos^2 x = \dfrac{1 - \sin x}{2}$

6. $\sin x + \cos x = \sqrt{\sin^2 x + \cos^2 x}$

7. $\sin \frac{1}{2}x = \frac{1}{2} \sin x$

8. $\tan 2x = 2 \tan x$

9. $\sin(x + \pi) = \sin x$

10. $\cos(x + \pi) = \cos x - 1$

11. $\cos x^2 = \cos^2 x$

12. $\sin x^2 = \cos(1 - x^2)$

In Exercises 13–22 use a graphing calculator to determine which of the expressions are not identities. If the two graphs appear to be identical, prove the identity.

13. $(\cos x - \sin x)(\cos x + \sin x) = 2 \cos^2 x - 1$

14. $\sin x \sec x = \tan x$

15. $\cos x = \cot x$

16. $1 - \cot x = \cot x \tan x - \cot x$

17. $1 - \dfrac{2}{\sec^2 x} = \sin^2 x - \cos^2 x$

18. $\cos x + 1 = \sin x$

19. $\dfrac{\cos x}{1 - \sin x} = \dfrac{1 + \sin x}{\cos x}$

20. $\sin x \cot x \tan^2 x = \sec x - \sin x \cot x$

21. $\sin x \tan x + \cos x = \sec x$

22. $\sin x \tan^2 x = \sin x$

6.3 Trigonometric Equations

A **trigonometric equation** is any statement involving a conditional equality of two trigonometric expressions. A **solution** to the trigonometric equation is a value of the variable (within the domain of the function) that makes the statement true. The **solution set** is the set of all values of the variable that are solutions. Solving a trigonometric equation means finding the solution set for some indicated domain. If no domain is specifically mentioned, the domain is assumed to be all values of the independent variable for which the terms of the equation have meaning.

To solve trigonometric equations, we proceed in a series of algebraic steps until we reach a point at which an explicit determination of the solution set can be made. Usually some specific knowledge about certain values of the trigonometric functions is necessary before we can make this determination.

We say that two trigonometric equations are **equivalent** if they have the same solution sets. Any operation on a given equation is **allowable** if the consequence of the operation is an equivalent equation. The permissible operations are (1) adding the same expression to, or subtracting it from, both sides of an equality and (2) multiplying or dividing both sides by the same nonzero expression.

COMMENT Generally trigonometric equations have infinitely many solutions. But obtaining the roots over some fundamental interval is the key to writing the complete solution set. All roots can be obtained from those on the fundamental interval simply by adding (or subtracting) multiples of the period.

Example 1 Solve the equation $\cos x = \frac{1}{2}$.

SOLUTION The period of $\cos x$ is 2π. Since $\cos \frac{1}{3}\pi = \frac{1}{2}$, the only solutions to this equation on the fundamental interval $0 \le x < 2\pi$ are $x = \frac{1}{3}\pi$ and $\frac{5}{3}\pi$. The complete solution set is composed of the values that can be written in the form $\frac{1}{3}\pi + 2n\pi$ and $\frac{5}{3}\pi + 2n\pi$, where n is an integer. Figure 6.2 illustrates that the solution set consists of the points of intersection of the curve $y = \cos x$ with the line $y = \frac{1}{2}$.

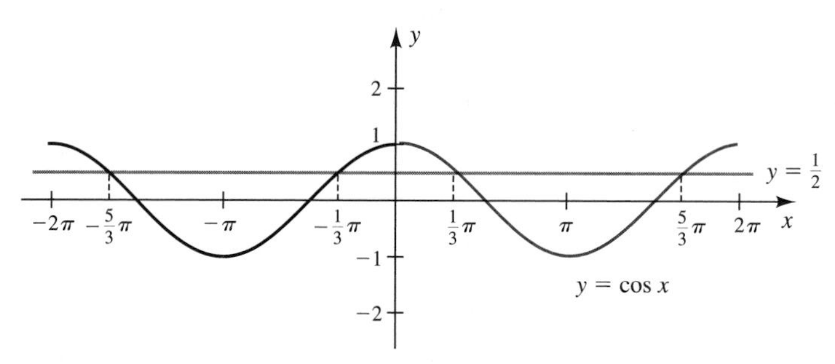

Figure 6.2 ▲

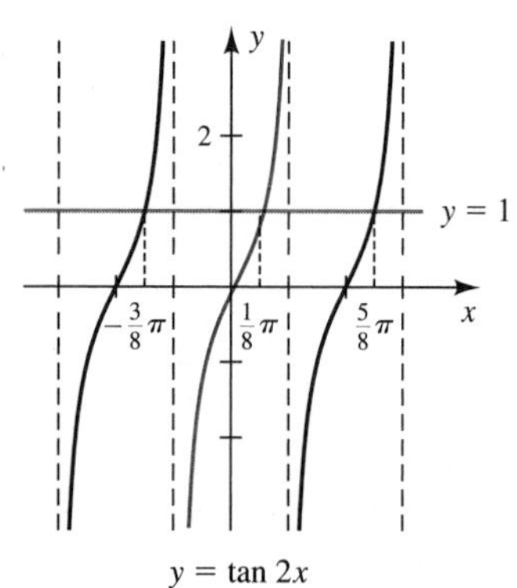

Figure 6.3

Example 2 Solve the trigonometric equation $\tan 2x = 1$.

SOLUTION Since the period of this function is $\frac{1}{2}\pi$, we first look for roots on the fundamental interval $-\frac{1}{4}\pi \le x < \frac{1}{4}\pi$. The value of θ at which $\tan \theta = 1$ is $\theta = \frac{1}{4}\pi$. Hence $\tan 2x = 1$ has the solution $x = \frac{1}{8}\pi$. Figure 6.3 shows the solution set as the intersection of the curve $y = \tan 2x$ with the line $y = 1$. Then all solutions are given by $x = \frac{1}{8}\pi + \frac{1}{2}n\pi$. ▲

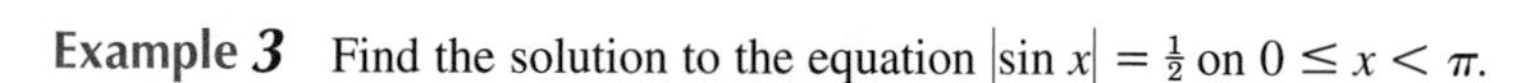

Example 3 Find the solution to the equation $|\sin x| = \frac{1}{2}$ on $0 \le x < \pi$.

SOLUTION Figure 6.4 shows the points of intersection of the curve $y = |\sin x|$ and the line $y = \frac{1}{2}$. Since $|\sin x|$ is periodic with period π, we need find only those values of x on the interval $0 \le x < \pi$ for which the equation is true. From our knowledge of the sine function, we know that these values are $x = \frac{1}{6}\pi$ and $x = \frac{5}{6}\pi$.

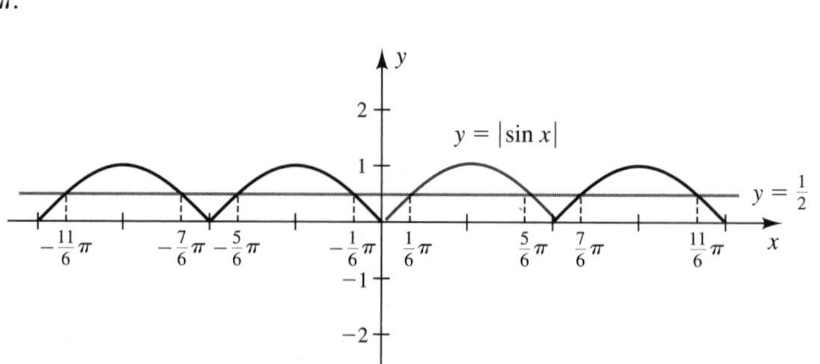

Figure 6.4 ▲

Some trigonometric equations that are quadratic in one of the functions can be factored into a product of linear factors. (You may need the quadratic formula.) The total solution set is found by finding the solution to each of the resulting linear equations.

Example 4 Solve the equation $2\cos^2 x + 3\cos x + 1 = 0$ on $0 \le x < 2\pi$.

SOLUTION This is a quadratic equation in $\cos x$ and may be factored as

$$(2\cos x + 1)(\cos x + 1) = 0$$

Setting each factor equal to zero and solving for x, we get

$$\begin{array}{r|r} 2\cos x + 1 = 0 & \cos x + 1 = 0 \\ \cos x = -\frac{1}{2} & \cos x = -1 \\ x = \frac{2}{3}\pi, \frac{4}{3}\pi & x = \pi \end{array}$$

Hence the solutions are $x = \frac{2}{3}\pi$, π, and $\frac{4}{3}\pi$. ▲

If more than one trigonometric function occurs in an equation, use trigonometric identities to write an equivalent equation involving only one function.

Example 5 Solve the equation $2\cos^2 x - \sin x - 1 = 0$ on $0 \le x < 2\pi$.

SOLUTION Since $\cos^2 x = 1 - \sin^2 x$,

$$\begin{aligned} 2(1 - \sin^2 x) - \sin x - 1 &= 0 \\ 2 - 2\sin^2 x - \sin x - 1 &= 0 \\ 2\sin^2 x + \sin x - 1 &= 0 \end{aligned}$$

Factoring yields

$$(2\sin x - 1)(\sin x + 1) = 0$$

Equating each factor to zero and solving for x, we have

$$\begin{array}{r|r} 2\sin x - 1 = 0 & \sin x + 1 = 0 \\ \sin x = \frac{1}{2} & \sin x = -1 \\ x = \frac{1}{6}\pi, \frac{5}{6}\pi & x = \frac{3}{2}\pi \end{array}$$

Hence the solutions are $x = \frac{1}{6}\pi$, $\frac{5}{6}\pi$, and $\frac{3}{2}\pi$. ▲

The following is normally a good procedure for solving trigonometric equations.

1. Gather the entire expression on one side of the equal sign.
2. Use the fundamental identities to express the conditional equality in terms of one function or, failing this, as a product of two expressions, each involving one function.

3. Use some algebraic technique, such as substitution into the quadratic formula or techniques of factoring, to write the expression as a product of linear factors.
4. Determine the zeros, if any, of each of the linear factors. The solution set consists of all zeros of these linear factors.

Example 6 Solve the equation $2 \tan \theta \sec \theta - \tan \theta = 0$ on $0° \leq \theta < 360°$.

SOLUTION The given equation can be factored as

$$\tan \theta(2 \sec \theta - 1) = 0$$

Equating each factor to zero, we get

$\tan \theta = 0$	$2 \sec \theta - 1 = 0$
$\theta = 0°$ and $180°$	$\sec \theta = \frac{1}{2}$
	No solution is possible, because $\sec \theta$ cannot be less than 1.

Thus the solutions are $\theta = 0°$ and $180°$. ▲

WARNING Squaring both sides of an equality is not an allowable operation because it does not necessarily yield an equivalent equation. In practice you need not restrict yourself to allowable operations, but when you use nonallowable operations to solve an equation, you must check each apparent solution for validity. (Of course, it is *always* a good idea to check your work.)

Example 7 Solve the equation $\sin x + \cos x = 1$ on $0 \leq x < 2\pi$.

SOLUTION Squaring both sides of this equation, we get

$$\sin^2 x + 2 \sin x \cos x + \cos^2 x = 1$$

Since $\sin^2 x + \cos^2 x = 1$,

$$\sin x \cos x = 0$$

The solution to this equation consists of the values of x for which $\sin x = 0$ (that is, $x = 0$ and π) and the values of x for which $\cos x = 0$ (that is, $x = \frac{1}{2}\pi$ and $\frac{3}{2}\pi$). Hence the possible solutions are $x = 0, \frac{1}{2}\pi, \pi$, and $\frac{3}{2}\pi$.

Since squaring does not yield an equivalent equation, you must check these values to determine whether they are solutions to the original equation. It is easy to show that only $x = 0$ and $x = \frac{1}{2}\pi$ are valid solutions. ▲

Example 8 Solve the equation $\sin^2 x + 3 \sin x - 2 = 0$ on $0 \leq x \leq 2\pi$.

SOLUTION Since we cannot factor the given quadratic by inspection, we use the quadratic formula to obtain

$$\sin x = \frac{-3 \pm \sqrt{3^2 - 4(1)(-2)}}{2(1)} = \frac{-3 \pm \sqrt{17}}{2}$$

which can be evaluated as follows:

$$\sin x = \frac{-3 + \sqrt{17}}{2} \approx 0.5616 \qquad \text{and} \qquad \sin x = \frac{-3 - \sqrt{17}}{2} \approx -3.5616$$

Since $\sin x$ cannot be less than -1, $\sin x = -3.5616$ has no solution. Using a calculator in the radian mode to solve $\sin x = 0.5616$, we find that $x \approx 0.596$. Since $\sin x$ is also positive in the second quadrant, another solution is $\pi - 0.596 \approx 2.546$. Therefore the desired solutions are $x \approx 0.596$ and 2.546. (*Note:* In Section 6.5 we will show how to solve this equation using a graphing calculator.) ▲

Equations involving trigonometric functions with phase shifts can also be solved. As in the previous problems, the period of the function and its solutions on the fundamental period are obtained. The solution set then consists of the solutions on the fundamental period plus integer multiples of the period.

Example 9 Solve the equation $2\cos(3x - 1) + 1 = 0$.

SOLUTION The period of $2\cos(3x - 1)$ is $2\pi/3$. To find the set of all solutions of the given equation, we first find its solutions on the fundamental period and then add integer multiples of $2\pi/3$ to each of these solutions. Beginning with $2\cos(3x - 1) + 1 = 0$, we solve for $\cos(3x - 1)$ to obtain

$$\cos(3x - 1) = -\frac{1}{2}$$

The fundamental period for $\cos(3x - 1)$ is $0 \le 3x - 1 \le 2\pi$ or $1/3 \le x \le (2\pi + 1)/3$. Since the solutions of $\cos x = -1/2$ on $0 \le x \le 2\pi$ are $x = 2\pi/3$ and $4\pi/3$, it follows that the solutions of $\cos(3x - 1) = -1/2$ on $1/3 \le x \le (2\pi + 1)/3$ are given by

$$3x - 1 = \frac{2\pi}{3} \qquad \text{and} \qquad 3x - 1 = \frac{4\pi}{3}$$

from which

$$x = \frac{2\pi}{9} + \frac{1}{3} \qquad \text{and} \qquad \frac{4\pi}{9} + \frac{1}{3}$$

The solutions of the given equation are then

$$x = \frac{2\pi}{9} + \frac{1}{3} + \frac{2n\pi}{3} \qquad \text{and} \qquad \frac{4\pi}{9} + \frac{1}{3} + \frac{2n\pi}{3}$$

where n is an integer. ▲

Exercises for Section 6.3

In Exercises 1–10 solve each equation. Make a sketch showing the solution set as the intersection of a line with the graph of the trigonometric function.

1. $\sin x = \frac{1}{2}$

2. $\cos 2x = \frac{1}{2}\sqrt{2}$

3. $\tan x = \sqrt{3}$

4. $\cos x = 1$

5. $\sin x = \frac{1}{2}\sqrt{3}$
6. $\cos x = -\frac{1}{2}$
7. $\sin 2x = -\frac{1}{2}$
8. $\tan \frac{1}{2}x = 1$
9. $|\cos x| = \frac{1}{2}\sqrt{2}$
10. $|\sin x| = \frac{1}{2}\sqrt{3}$

In Exercises 11–37 solve each trigonometric equation over the interval $0 \le x < 2\pi$ unless another interval is indicated.

11. $2 \sin x + 1 = 0$
12. $\sin 2x + 1 = 0,\ [0, \pi)$
13. $\cos 3x = 1,\ [0, \frac{2}{3}\pi)$
14. $\tan 2x + 1 = 0,\ [0, \frac{1}{2}\pi)$
15. $\cos^2 x + 2 \cos x + 1 = 0$
16. $\tan^2 x - 1 = 0,\ [0, \pi)$
17. $2 \sin^2 x = \sin x$
18. $\sec^2 2x = 1,\ [0, \pi)$
19. $\sec^2 x + 1 = 0$
20. $\cos^2 x = 2$
21. $\cos x = \sin x$
22. $2 \sec x \tan x + \sec^2 x = 0$
23. $\sec^2 x - 2 = \tan^2 x$
24. $4 \sin^2 x - 1 = 0$
25. $2 \cos^2 x - \sin x = 1$
26. $2 \sec x + 4 = 0$
27. $\sin^2 x - 2 \sin x + 1 = 0$
28. $\cot^2 x - 5 \cot x + 4 = 0,\ [0, \pi)$
29. $\tan^2 x - \tan x = 0$
30. $\cos 2x + \sin 2x = 0,\ [0, \pi)$
31. $\sin x \tan^2 x = \sin x$
32. $\cos x + 2 \sin^2 x = 1$
33. $\tan x + \sec x = 1$
34. $\tan x + \cot x = \sec x \csc x$
35. $\cos x + 1 = \sin x$
36. $2 \tan x - \sec^2 x = 0,\ [0, \frac{1}{2}\pi)$
37. $\csc^5 x - 4 \csc x = 0$

In Exercises 38–51 solve each equation for all values of θ over the interval $0 \le \theta < 360°$.

38. $4 \sin^2\theta = 3$
39. $2 \cos^2\theta = 1$
40. $\tan^2\theta - 3 = 0$
41. $\csc^2\theta = 1$
42. $(2 \cos \theta - \sqrt{3})(\sqrt{2} \sin \theta + 1) = 0$
43. $(\sin \theta - 1)(2 \cos \theta + \sqrt{3}) = 0$
44. $\csc \theta = 1 + \cot \theta$
45. $2 \sin \theta \tan \theta + \sqrt{3} \tan \theta = 0$
46. $2 \cos^3\theta = \cos \theta$
47. $\tan^2\theta - \tan \theta = 0$
48. $2 \sin^2\theta = 21 \sin \theta + 11$
49. $2 \cos^2\theta + 7 \sin \theta = 5$
50. $5 \tan^2\theta - 11 \sec \theta + 7 = 0$
51. $\sin \theta + \sqrt{3} \cos \theta = 0$

In Exercises 52–61 solve the given equation. Some will require a calculator.

52. $\sin(x + \frac{1}{3}\pi) = \frac{1}{2}$
53. $\cos(x - \frac{1}{4}\pi) = \dfrac{\sqrt{3}}{2}$
54. $3 \tan(x + \frac{1}{3}\pi) = \sqrt{3}$
55. $\sin(3x + \pi) = 1$
56. $\cos(\frac{1}{3}x + \frac{1}{6}\pi) = 0.2$
57. $2 \sin(2x + 1) = \sqrt{2}$
58. $5 \sin(2x + 1) = 1$
59. $4 \cos^2(3x + 2) = 3$
60. $\tan^2(2x + 5) = \frac{1}{3}$
61. $5 \sin^2(x + 2) = 3$

In Exercises 62–70 solve each equation for x, $0 \le x < 2\pi$. Note: The solutions to these problems are not necessarily multiples of π.

62. $\tan^2 x = 3.2$
63. $\sin^2 x + 2 \cos^2 x = 1.7$
64. $\cos x + 2 \sin x \tan x = 1$
65. $3 \sin^2 x + \sin x = 0$
66. $\tan^2 x + 2 \sec^2 x = 1$
67. $2 \sin x - \cot x \cos x = 1$

68. $\csc^2 x + 2\cot^2 x - 1 = 0$

69. $\sin x - \csc x + 1 = 0$

70. $2\tan^2 x + 3\sec x - 5 = 0$

71. To solve a vibrations problem, a physics student must find the values of x for which $\sqrt{3}\cos \pi x + \sin \pi x = 0$. What are those values?

72. The undamped motion of an object at the end of a spring is governed by the formula $y = \cos t + \sqrt{3}\sin t$. For which values of t is $y = 0$?

73. In radio engineering, a carrier wave is described by $V\sin(\omega_c t + \phi)$, where V is the amplitude and ϕ is the phase angle of the carrier. If V is varied, the wave is said to be amplitude modulated. If $V \neq 0$, for what values of t is the carrier wave equal to 0?

74. The approximate solution to Bessel's equation, which is used in solving vibration problems, is given by

$$\frac{A}{\sqrt{x}}\cos(x + \beta)$$

Find the values of x for which this function is zero.

6.4 Parametric Equations

When a quarterback throws a football, the path of the ball is approximately parabolic. (See Figure 6.5.) Using the laws of physics, we can show that if the ball is thrown with a velocity v_0 ft/sec at an angle θ with the horizontal, then the horizontal component of velocity is constant and is given by

$$v_x = v_0\cos\theta$$

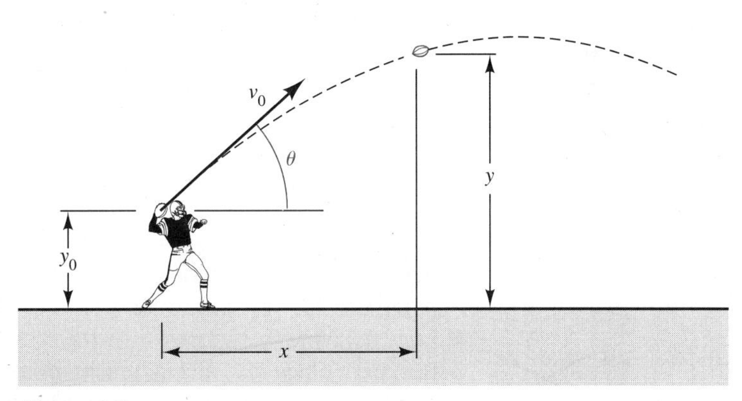

Figure 6.5

The horizontal component of the displacement of the ball after t seconds is then

$$x = v_x t = (v_0\cos\theta)t$$

It can also be shown that the vertical component of velocity is independent of v_x and varies with time, so that the height (the vertical component of the position) of the ball after t seconds is

$$y = y_0 + (v_0 \sin \theta)t - 16t^2$$

Thus for a given y_0, v_0, and θ, the path of the football is described as a function of time. Each value of t gives the coordinates (x, y) of the football at that time. (See Exercise 9, page 254.)

A convenient way to define the locus* of a point in the plane is to create two equations, one for x and one for y, in terms of some variable, say t. Thus two functions of t,

$$x = f(t) \qquad \text{and} \qquad y = g(t)$$

determine the location of a set of points. As t increases, the point describes a curve in the plane—representing, for example, the path of a football being thrown by a quarterback. The equations are called **parametric equations** of the curve; the variable t is called the **parameter**. Each value of t gives a pair of values x and y that represents a point on the curve. For any interval of the parameter, a smooth curve can be drawn connecting these points. Such a curve is said to be defined **parametrically** in terms of the parameter t. (The parameter t may be thought of as representing time.)

Example 1 Determine the values of (x, y) for values of t between 0 and 2π if $x = 2 \cos t$, $y = 2 \sin t$. Then sketch the curve given by these points.

SOLUTION The following table shows the values for $0 \leq t \leq \pi$. The values for $\pi \leq t \leq 2\pi$ are obtained in the same way.

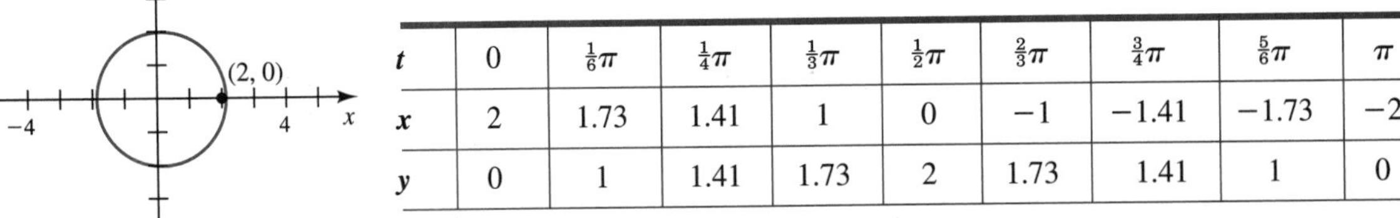

t	0	$\frac{1}{6}\pi$	$\frac{1}{4}\pi$	$\frac{1}{3}\pi$	$\frac{1}{2}\pi$	$\frac{2}{3}\pi$	$\frac{3}{4}\pi$	$\frac{5}{6}\pi$	π
x	2	1.73	1.41	1	0	−1	−1.41	−1.73	−2
y	0	1	1.41	1.73	2	1.73	1.41	1	0

Figure 6.6

Figure 6.6 shows the curve. It is a circle of radius 2 centered at the origin. ▲

A major advantage of the parametric method is that curves are **oriented** by the parameter. For example, consider the curve described parametrically by $x = 2 \sin t$, $y = 3 \cos t$. (The graph is shown in Figure 6.7.) Note that when $t = 0$, the point is at $(0, 3)$. The point goes around the curve clockwise as t increases, completing a revolution every 2π units.

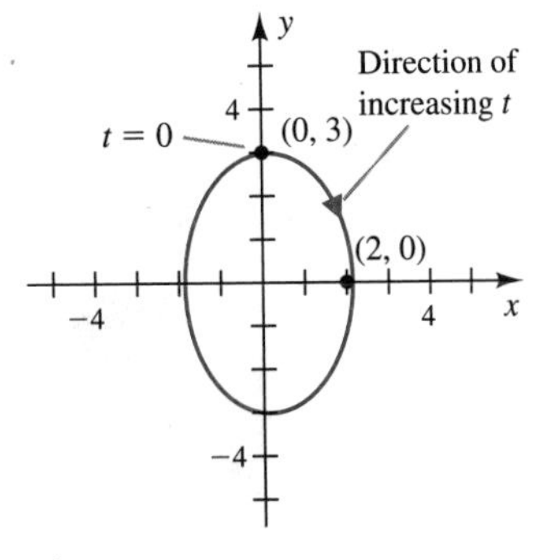

Figure 6.7

Sometimes a curve defined parametrically can also be expressed in the form of an equation involving only x and y. The basic idea in solving parametric equations is to *eliminate the parameter t* from the two given parametric equations. Two approaches are possible.

- Use trigonometric identities so that certain algebraic combinations of x and y will cause the parameter to disappear.

* The **locus** of a point in the plane is the path taken by the point in satisfying a given condition, such as a mathematical formula.

▶ Solve for t in terms of either x or y and substitute into the remaining equation.

Example 2 Eliminate the parameter from the following parametrically defined curves:

(a) $x = 2 \cos t$, $y = 2 \sin t$

(b) $x = 1 + 2t$, $y = 2 - t$

SOLUTION

(a) Square both sides of equations $x = 2 \cos t$, $y = 2 \sin t$, to obtain $x^2 = 4 \cos^2 t$, $y^2 = 4 \sin^2 t$. Adding these two equations gives

$$\begin{aligned} x^2 + y^2 &= 4 \cos^2 t + 4 \sin^2 t \\ &= 4(\cos^2 t + \sin^2 t) \\ &= 4 \end{aligned}$$

We note that $x^2 + y^2 = 4$ is the equation of a circle centered at the origin with a radius of 2. Hence $x = 2 \cos t$ and $y = 2 \sin t$ are parametric equations of a circle.

(b) For the equations $x = 1 + 2t$, $y = 2 - t$, solve the second equation for t and substitute into the first one. Thus $t = 2 - y$, and

$$\begin{aligned} x &= 1 + 2t = 1 + 2(2 - y) = 1 + 4 - 2y \\ &= 5 - 2y \end{aligned}$$

Rearranging yields $x + 2y = 5$, which is the equation of a straight line. Hence $x = 1 + 2t$, $y = 2 - t$ are parametric equations of a straight line. ▲

When you are eliminating the parameter, be sure that the answer you arrive at through the process of elimination neither includes more points than the original parametric equations allow nor excludes points that are solutions to the original equations. For example, the process of squaring may produce extraneous roots.

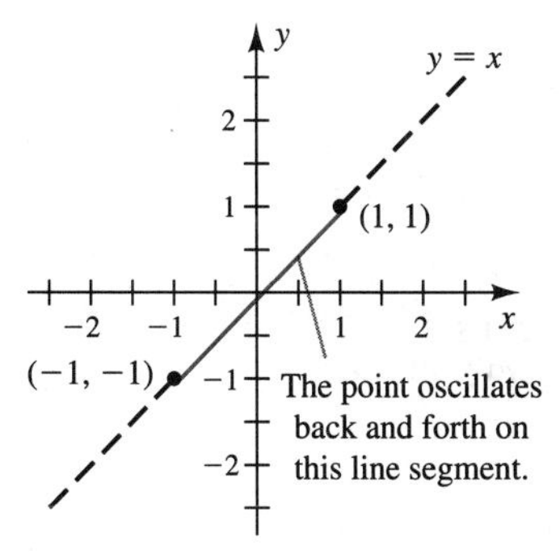

Figure 6.8

Example 3 Sketch the graph of the curve described parametrically by $x = \sin t$, $y = \sin t$.

SOLUTION The parameter can be eliminated because x and y represent the same function. Hence the nonparametric relation is $y = x$. The graph of $y = x$ is the line that splits the first and third quadrants. (See Figure 6.8.) However, by the nature of the parametric equations, the values of x and y are limited between 1 and -1. Thus the point moves on the line $y = x$, but it oscillates between the points $(1, 1)$ and $(-1, -1)$. ▲

Example 4 Eliminate t from the parametric equations and sketch

$$x = 2 \sec t, \ y = \tan t$$

SOLUTION Rewrite the first equation as $x/2 = \sec t$. Then squaring and subtracting gives $x^2/4 - y^2 = \sec^2 t - \tan^2 t$. From a variation of the Pythagorean

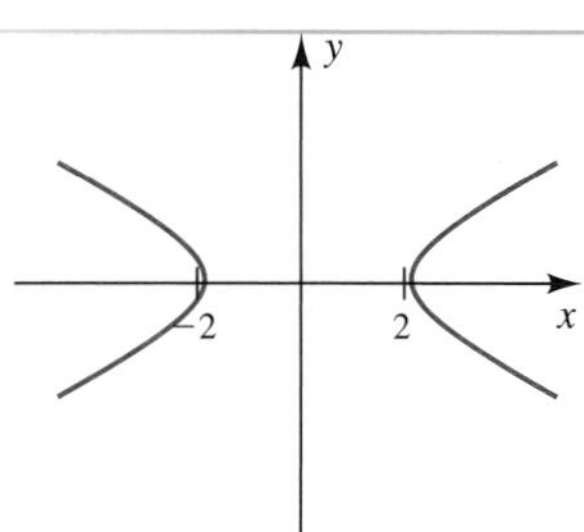

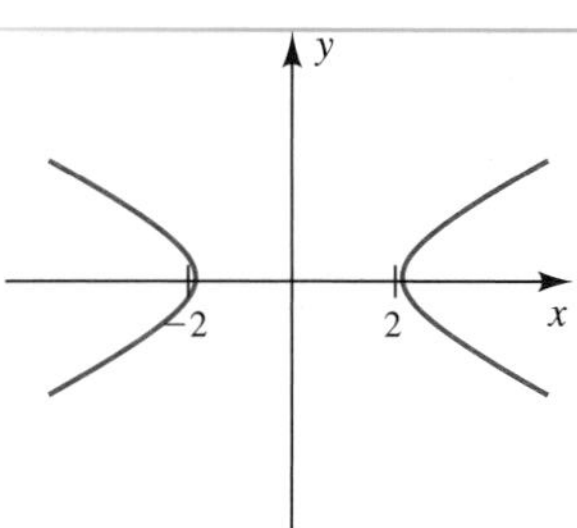

Figure 6.9

identity, the right-hand side of this equation is 1. Thus the given curve is described as a relation in x and y by

$$\frac{x^2}{4} - y^2 = 1$$

This curve is studied in more detail in analytic geometry. It is called a hyperbola. (See Figure 6.9.) ▲

When the parameters are sinusoids, the wave forms can be displayed on an oscilloscope. Voltages corresponding to the x and y functions are applied to the horizontal and vertical input terminals of the scope, respectively. When the parametric equations are displayed in this manner, the resulting figure is called a **Lissajous figure**.

Example 5 Plot the Lissajous figure corresponding to input voltages of $x = \sin 2t$, $y = \cos t$.

SOLUTION Elimination of the parameter is not convenient in this case, so we generate a table of values for $0 \le t \le 2\pi$.

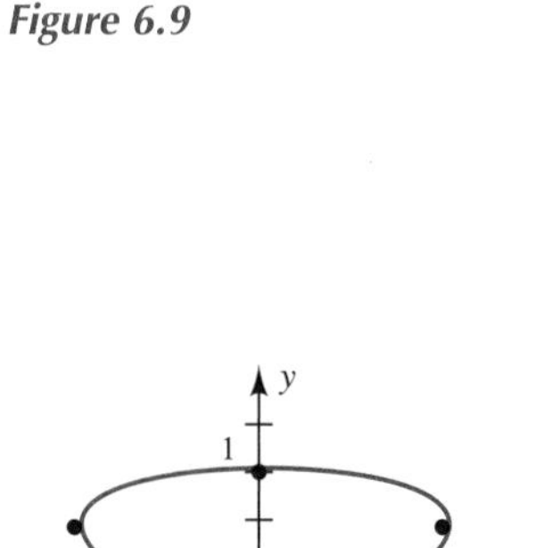

Figure 6.10

t	0	$\frac{1}{4}\pi$	$\frac{1}{2}\pi$	$\frac{3}{4}\pi$	π	$\frac{5}{4}\pi$	$\frac{3}{2}\pi$	$\frac{7}{4}\pi$	2π
$x = \sin 2t$	0	1	0	−1	0	1	0	−1	0
$y = \cos t$	1	0.71	0	−0.71	−1	−0.71	0	0.71	1

Using these points we obtain Figure 6.10. ▲

Another way of displaying Lissajous figures and other parametric curves is to use a graphing calculator in the parametric mode. In this mode you input the $x(t)$ and the $y(t)$, along with the range values for x, y, and t. The figure below shows a graphing calculator display of the Lissajous figure discussed in Example 5. Notice the orientation of the graph as it is sketched on the screen of the calculator.

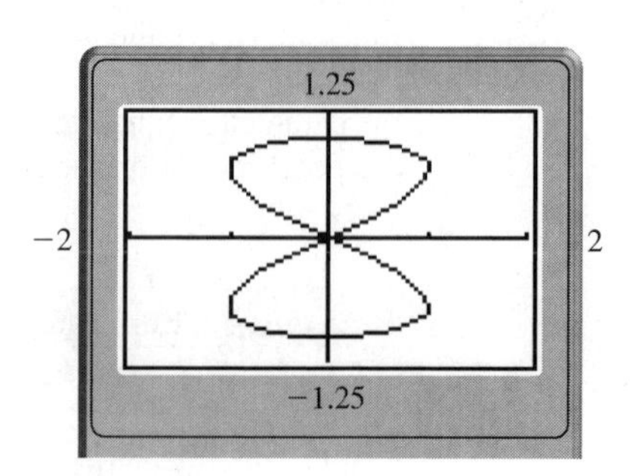

Exercises for Section 6.4

In Exercises 1–10 sketch the curve described by each parametric equation. In each case indicate the orientation of the curve.

1. $x = \sin t$, $y = \cos t$
2. $x = 2\cos t$, $y = \sin t$
3. $x = \cos^2 t$, $y = 1$
4. $x = 3t$, $y = 2$
5. $x = t^2$, $y = t^3$
6. $x = \cos t$, $y = \sin 2t$
7. $x = \sin t$, $y = \cos 2t$
8. $x = \sin t + \cos t$, $y = \sin t - \cos t$
9. $x = 2\sin t + \cos t$, $y = 2\cos t - \sin t$
10. $x = \sin t$, $y = \sin 2t$

In Exercises 11–20 eliminate *t* from each parametric equation to obtain one equation in *x* and *y*. Sketch the curve, being careful to observe any limitations imposed by the parameterization.

11. $x = \sin t$, $y = \cos t$
12. $x = 2\cos t$, $y = \sin t$
13. $x = t$, $y = 1 + t$
14. $x = -\cos t$, $y = \sin t$
15. $x = \cos t$, $y = \cos t$
16. $x = 1 - t^2$, $y = t$
17. $x = \sin^2 t$, $y = t$
18. $x = \cos^2 t$, $y = \cos^2 t$
19. $x = t - 3$, $y = t^2 + 1$
20. $x = 2\sin t + \cos t$, $y = 2\cos t - \sin t$
21. The approximate path of the earth about the sun is given by $x = aR\cos 2\pi t$, $y = bR\sin 2\pi t$, where t is the time in years, R is the radius of the earth, and b and a are constants that are very nearly equal. Sketch this path for $a = 1$ and $b = 1.1$.
22. A piston is connected to the rim of a wheel as shown in Figure 6.11. At a time t seconds after it has coordinates (2, 0), the point S has coordinates given by $x = 2\cos 2\pi t$, $y = 2\sin 2\pi t$. Find the position of the point S when $t = \frac{1}{2}, \frac{3}{4}$, and 2. Eliminate the parameter to find an equation relating x and y.

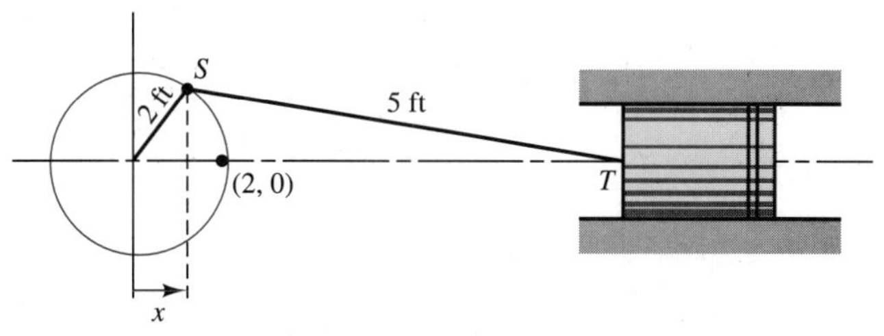

Figure 6.11

23. Lissajous figures are sometimes described by saying that the plotting is at right angles. The right-angle plot idea comes from the fact that the voltages are applied at right angles on the scope. Sketch the Lissajous figure for the equations $x = \sin t$, $y = \cos(t - \frac{1}{4}\pi)$.
24. Sketch the Lissajous figure for $x = 2\sin 3t$, $y = \cos t$.

25. Sketch the Lissajous figure for $x = 2 \cos 3t$, $y = \sin 2t$.

26. The curve traced by a point on the circumference of a bicycle wheel is given by the set of parametric equations $x = t - \sin t$, $y = 1 - \cos t$. Sketch this curve. (The curve is called a **cycloid**.)

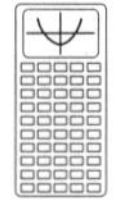

Graphing Calculator Exercises

In Exercises 1–8 use a graphing calculator to sketch the graph of the parametric equations. How does the calculator graph compare with the freehand sketch you made in the indicated exercise?

1. $x = \sin t$, $y = \cos t$ (See Exercise 1)
2. $x = 2 \cos t$, $y = \sin t$ (Exercise 2)
3. $x = \cos^2 t$, $y = 1$ (Exercise 3)
4. $x = 3t$, $y = 2$ (Exercise 4)
5. $x = \sin t$, $y = \cos 2t$ (Exercise 7)
6. $x = \sin t$, $y = \sin 2t$ (Exercise 10)
7. $x = \sin t$, $y = \cos(t - \frac{1}{4}\pi)$ (Exercise 23)
8. $x = 2 \sin 3t$, $y = \cos t$ (Exercise 24)

9. Sketch the path of a football thrown by a 6-ft tall quarterback with an initial velocity $v_0 = 60$ ft/sec at $\theta = 30°$. By how much would he have to increase the angle of his throw in order for the ball to go 5 ft farther? (*Hint:* Vary θ until x is increased by 5.)

6.5 Graphical Solutions of Trigonometric Equations and Inequalities

Trigonometric Equations

Equations containing a mixture of trigonometric and other functions may be quite difficult to solve by analytic methods. A graphical analysis usually yields at least an approximation of the roots and often gives helpful information even when a problem can be solved analytically.

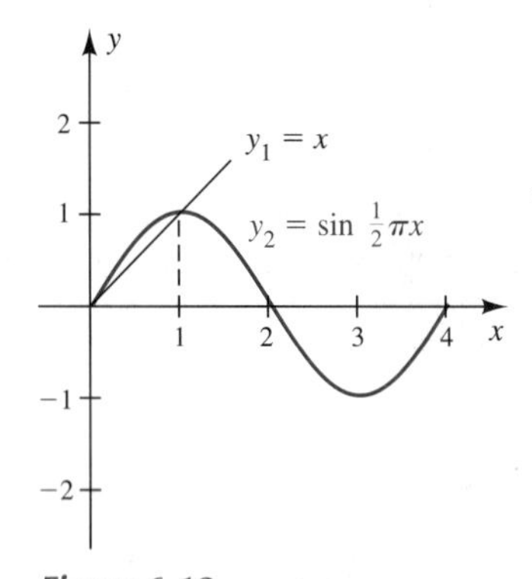

Figure 6.12

Example 1 Solve the equation $x = \sin \frac{1}{2}\pi x$ graphically.

SOLUTION The functions $y_1 = x$ and $y_2 = \sin \frac{1}{2}\pi x$ are sketched in Figure 6.12. Since both functions are odd (that is, their graphs are symmetric with respect to the origin), the graph is drawn for positive values of x only. The solution set to the equation is the set of x-coordinates of the points of intersection of the two curves. The figure shows that the values are $x = 0$ and $x = 1$; hence because the graph is symmetrical, the solution set is $\{-1, 0, 1\}$. ▲

Example 2 Solve the equation $x \tan x = 1$ graphically.

SOLUTION We write the equation in the form $\tan x = 1/x$ and then graph the two functions $y_1 = \tan x$ and $y_2 = 1/x$, as shown in Figure 6.13. As in the previous example, we may without loss of generality sketch only the part of the graphs for $x \geq 0$. Figure 6.13 shows that there are infinitely many solutions, of which the first positive one is approximately 0.86.

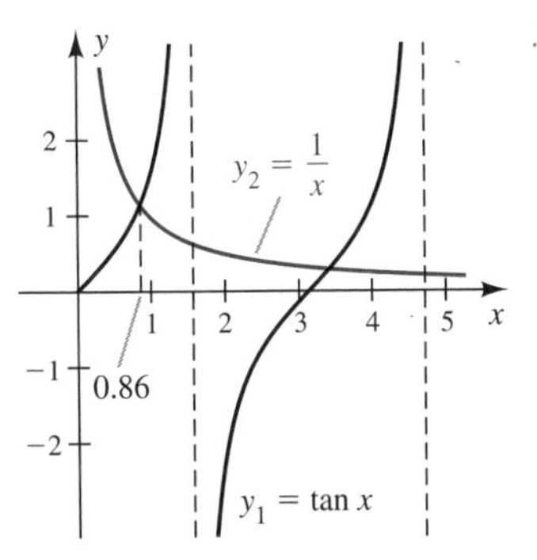

Figure 6.13 ▲

A graphing calculator can be used to estimate solutions. For example, to solve $x \tan x = 1$ on $(0, 5)$ we display both $y_1 = \frac{1}{x}$ and $y_2 = \tan x$ on the screen for $0 < x < 5$ (set the range for approximately $-2 < y < 3$). The intersection of the two curves gives the solutions. To obtain a numerical estimate, move the cursor to the point of intersection and read the coordinate values. The x-values of the first two points of intersection are found to be 0.86 and 3.5. The accuracy of such a graphical estimate may be increased by using the zoom feature.

Another way of graphically solving a trigonometric equation is to equate all terms in the equation to zero and then graph the nonzero terms. For example, write the equation in Example 2 in the form $x \tan x - 1 = 0$. Now use the calculator to draw the graph of $y = x \tan x - 1$ on the desired interval. In this case the solutions of $x \tan x = 1$ correspond to the x-intercepts of the graph of $y = x \tan x - 1$.

A further refinement can be obtained using the SOLVER menu. For this routine, a rough estimate of a zero is required. We used as a first estimate $x = 0.1$ and obtained 0.8603335890194 **. For the next zero, we estimated $x = 3.14$ and obtained** 3.4256184594817 **. Continuing by making estimates of approximately $n\pi$, we can obtain as many zeros as desired.**

Example 3 Use a graphing calculator to solve $\sin \frac{1}{2}\pi x = x$.

SOLUTION The approach used in Examples 1 and 2 to illustrate graphical solutions to trigonometric equations is the basis for solving such equations on a calculator. For example, to solve the equation $x = \sin \frac{1}{2}\pi x$ on one period (that is, $0 \le x \le 4$), we display both

$$y_1 = \sin \tfrac{1}{2}\pi x \qquad \text{and} \qquad y_2 = x$$

on the screen with the x-axis set for $-2 \le x \le 4$, and the y-axis for $-2 \le y \le 2$. The display is shown in the figure.

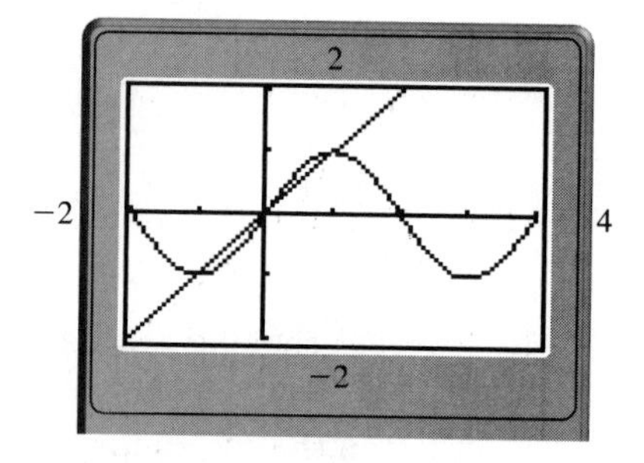

The solutions are represented by the intersection of the two curves. We obtain a numerical estimate of the nonzero solution by moving the cursor to the point of intersection. In this case we get $x = y = 1.01$. How does this compare with he answer obtained in Example 1? The accuracy of the calculator approximation can be increased dramatically by using a zoom feature at the point of intersection. By using the zoom feature, the approximation of x is quickly improved to $x = 1.0016621$. ▲

Although a trigonometric equation like $\sin^2 x + 3 \sin x - 2 = 0$ can be solved by analytic methods, a graphical solution provides some additional insight into the roots of the equation. The next example shows the graphical solution on the interval $0 \le x \le 2\pi$.

Example 4 Use a graphing calculator to solve the trigonometric equations

$$\sin^2 x + 3 \sin x - 2 = 0.$$

SOLUTION We display the graph of $y = \sin^2 x + 3 \sin x - 2$ on $0 \leq x < 2\pi$. The solutions correspond to the points at which this graph crosses the x-axis on the indicated interval. Estimates of the value of x at these points, obtained using the cursor and the zoom feature, are $x \approx 0.597$ and $x \approx 2.544$. Compare these values to those given in Example 8 of Section 6.3 (page 246). As mentioned in Example 2, the SOLVER menu can be used to improve the accuracy of the estimates.

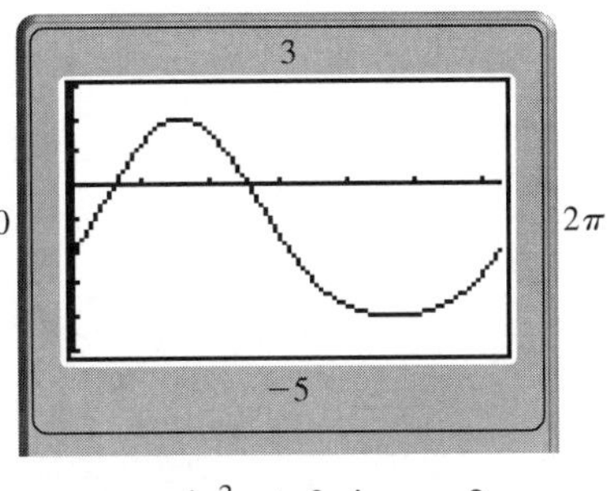

$y = \sin^2 x + 3 \sin x - 2$ ▲

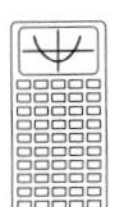

Bear in mind that the graphical approach is just another way to solve this equation, *not necessarily* the best method. Since this equation is a quadratic in sin x, the method used in Section 6.3 might arguably be better. However, the graphical approach has the appeal of being applicable to a wider variety of equations. For example, $\sin^3 x + 3 \sin x - 2 = 0$ or $\sin^4 x + 3 \sin x - 2 = 0$ are not readily factorable, and hence a graphical solution would be not only the best approach but the only reasonable approach.

Trigonometric Inequalities

A **trigonometric inequality** is any conditional statement of inequality involving trigonometric expressions. Like the solution set of a trigonometric equation, the solution set of a trigonometric inequality is defined as the set of values for which the conditional statement is true. Other related terminology is consistent with that for trigonometric equations. The allowable operations are slightly more restricted for inequalities in that multiplication or division of both sides is permitted only if the expressions are positive. Multiplication of both sides by a negative quantity results in a reversal of the inequality.

Aside from some of the very basic kinds of inequalities such as $\sin x < 1$ and $\cos x > 0$, most trigonometric inequalities are best solved by some combination of graphical and analytical methods.

Example 5 Solve the inequality $\sin x > \cos x$.

SOLUTION In Figure 6.14, both the sine and the cosine function are sketched. The points of intersection occur at the real numbers that can be written in the form $\frac{1}{4}\pi + n\pi$. The graph shows that the sine function is greater than the cosine function over those intervals whose left-hand end points are even values of n. In Figure 6.14 the solution set for one period is shown as the interval $\frac{1}{4}\pi < x < \frac{5}{4}\pi$.

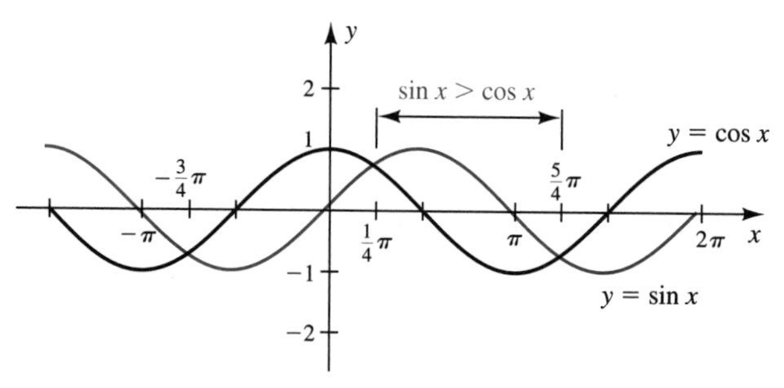

Figure 6.14 ▲

Inequalities that are not strictly trigonometric but that include other functions are also best analyzed graphically.

Example 6 Solve the inequality $x > \cos x$.

SOLUTION From Figure 6.15 we see that $x = \cos x$ at approximately $x = 0.74$. Hence we determine from the graph that the solution is $x > 0.74$. ▲

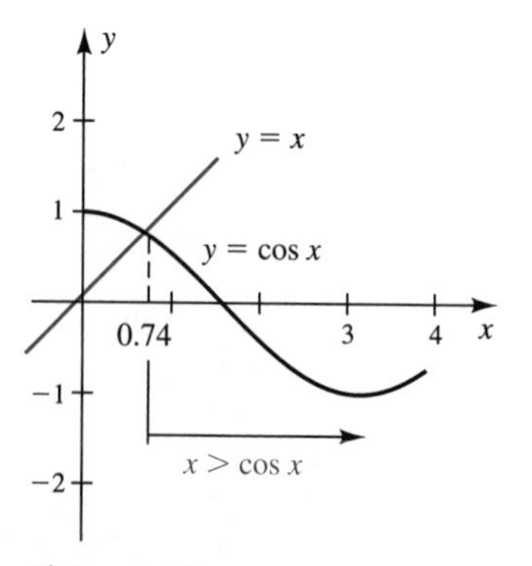

Figure 6.15

To solve a trigonometric inequality such as $\sin x > \cos x$ on a graphing calculator, first write it in the form $\sin x - \cos x > 0$. Then graph the function $y = \sin x - \cos x$ on $0 \leq x \leq 2\pi$. The desired solutions are those intervals on the x-axis on $[0, 2\pi]$ for which the graph is above the x-axis. The endpoints of these intervals are the points at which the graph of y crosses the x-axis; these can be obtained by using the trace capability and moving the cursor to each point where the x-coordinate may be read directly from the screen. From the screen below, we estimate the solution interval to be $0.79 \leq x \leq 3.90$. How does this compare with the exact solution found in Example 5?

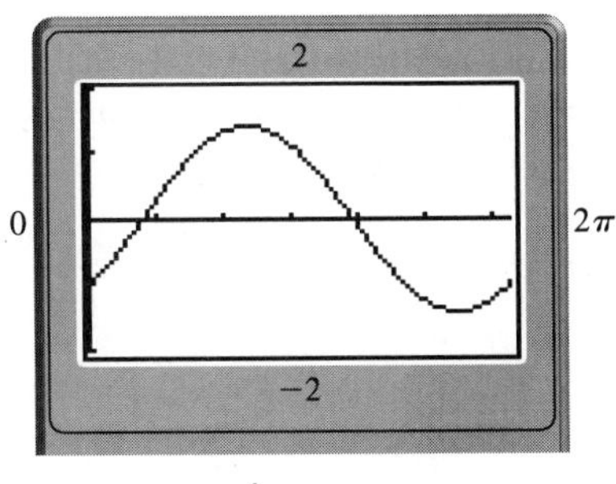

$y = \sin x - \cos x$

Exercises for Section 6.5

In Exercises 1–10 use graphical methods to approximate the solutions to each equation on one fundamental period.

1. $x = \sin x$
2. $x = \sin 2x$
3. $x \sin x = 1$
4. $x = \tan x$
5. $x = \cos x$
6. $x^2 = \cos 2\pi x$
7. $x^2 - \tan x = 0$
8. $\sin x = \cos x$
9. $\tan x = \cos x$
10. $\tan x = \sin x$

In Exercises 11–20 use graphical methods to solve each trigonometric inequality on one fundamental period.

11. $\sin x \geq \frac{1}{2}$
12. $\frac{1}{2} > \cos x$
13. $2 \cos^2 x \leq 1$
14. $\tan^2 x \leq 3$
15. $\sec x > \csc x$
16. $x > \sin x$
17. $x > \sin 2\pi x$
18. $x \sin x > 1$
19. $|\sin x| < |\cos x|$
20. $\sec^2 x < 1$

21. Two ramps are built, one in the shape of the straight line $y = 2x$ and the other in the shape of the curve $y = \tan x$. For which values of x is the straight-line ramp above the curved ramp?

22. To solve a problem involving phase modulation, a communications engineer must find the zeros of

$$(\sin t)(\tfrac{1}{2}t - \sin t) = 0$$

Find the zeros of this function corresponding to $\frac{1}{2}t - \sin t = 0$.

Graphing Calculator Exercises

In Exercises 1–4 use a graphing calculator to solve the given trigonometric equation on $0 \le x < 2\pi$ and compare your result with that obtained with an analytic approach.

1. $\sin^2 x + 2\cos x = 0$

2. $\cos^2 x - \sin x = 0$

3. $\cos^2 x + 3\cos x - 1 = 0$

4. $\sin^2 x - 2\sin x + 1 = 0$

In Exercises 5–16 use a graphing calculator to solve the given trigonometric equation or inequality on $0 \le x < 2\pi$.

5. $\cos^3 x + 3\cos x - 1 = 0$

6. $\sin^3 x - 2\sin x + 1 = 0$

7. $\cos^4 x + 3\cos x - 1 = 0$

8. $\sin^4 x - 2\sin x - 5 = 0$

9. $x = \cos x$

10. $\sin x = 1/x$

11. $\sin x = x^2$

12. $4 - x^2 = \cos x$

13. $\cos x > \sin x$

14. $\cos^2 x < x$

15. $x\sin x > 1$

16. $x\cos x < 1$

Key Topics for Chapter 6

Define and/or discuss each of the following.

Fundamental Identities
Verifying and Proving Identities
Trigonometric Equations
Parametric Equations
Graphical Solutions of Trigonometric Equations
Trigonometric Inequalities

Review Exercises for Chapter 6

In Exercises 1–10 prove each of the given identities.

1. $1 + \sec x = \csc x(\tan x + \sin x)$

2. $\cos^2 x = \dfrac{\csc^2 x - 1}{\csc^2 x}$

3. $\sin x = \dfrac{1 - \cos^2 x}{\sin x}$

4. $\cos x = \sec x - \sin x \tan x$

5. $\cos x + \sin x = \dfrac{2\cos^2 x - 1}{\cos x - \sin x}$

6. $\sec x + \tan x = \dfrac{1}{\sec x - \tan x}$

7. $1 - \sin x = \dfrac{\cos x}{\sec x + \tan x}$

8. $\dfrac{\csc x + 1}{\cot x} = \dfrac{\cot x}{\csc x - 1}$

9. $\sec x - \cos x = \dfrac{\tan^2 x}{\sec x}$

10. $2\sin^2 x = \sin^4 x - \cos^4 x + 1$

In Exercises 11–15 solve each trigonometric equation over one fundamental interval.

11. $\sin x = -\sqrt{3}/2$

12. $\cos 2x = 0.6789$

13. $\sin x + 2\cos^2 x = 1$

14. $\sin^2 x - \sin x = 0$

15. $\sec^5 x - 4\sec x = 0$

In Exercises 16–20 eliminate the parameter and sketch the graph of each parametric equation.

16. $x = 3\sin t,\ y = \cos t$

17. $x = \tan t,\ y = 2\sec t$

18. $x = 1 + \sin t,\ y = 2\cos t$

19. $x = \tan t,\ y = \tan t,\ 0 \leq t \leq \pi$

20. $x = 1,\ y = \sin t$

21. Using graphical methods, find approximate solutions to $x^2 = \sin x$ on $(-\pi, \pi)$.

22. Find approximate solutions to $x \cos x = 1$ on $[0, 2\pi]$.

23. Solve the inequality $\cos x > \sin x$ on $[0, 2\pi]$.

24. Solve the inequality $\sec x > 2$ on $[-\pi, \pi]$.

25. Plot the Lissajous figure given by $x = \cos t,\ y = 2\cos t$.

26. Plot the Lissajous figure given by $x = 2\sin t,\ y = \cos 3t$.

27. In analyzing conditions corresponding to overthrust faulting, a geologist is confronted with the expression

$$\frac{1}{2}(\sigma_x - \sigma_z) = \left[\frac{1}{2}(\sigma_x + \sigma_z) + \tau_0 \cot\phi\right]\sin\phi$$

Show that this simplifies to

$$\sigma_x = a + b\sigma_z$$

where

$$a = 2\tau_0\sqrt{b} \qquad \text{and} \qquad b = \frac{1 + \sin\phi}{1 - \sin\phi}$$

28. The values of time t for which the function $y(t) = 6\cos t - 8\sin t$ is equal to 0 represent times at which a mass at the end of a spring attains its extreme positions. Find these times.

29. In simplifying a calculus problem, a student substitutes $x = 2\tan\theta$, where $-\frac{1}{2}\pi < \theta < \frac{1}{2}\pi$, into the expression $x\sqrt{x^2 + 4}$. Carry out this substitution and simplify.

30. Using the substitution $x = 3\sin\theta$, where $-\frac{1}{2}\pi < \theta < \frac{1}{2}\pi$, reduce the expression $\sqrt{9 - x^2}$ to one with one trigonometric function. What is $\tan\theta$?

31. A civil engineer designs two ramps for a parking garage, the curves of which approximate those of $y = x^2$ and $y = \sin 3x$ for small values of x. Determine graphically where these curves meet.

32. The function

$$y = \begin{cases} \cos x & 0 \leq x \leq x_1 \\ \frac{3}{4} - \frac{3}{8}x, & x_1 \leq x \leq 2 \end{cases}$$

can be used to approximate a temperature distribution on a wire. Sketch a graph to find the value of x_1 such that $\cos x_1 = \frac{3}{4} - \frac{3}{8}x_1$.

33. Voltages of $3 \sin t$ and $2 \cos t$ are applied to the horizontal and vertical terminals of an oscilloscope, respectively. Make a sketch of the waveform that appears on the screen.

Practice Test Questions for Chapter 6

In Exercises 1–6 answer *true* or *false*.

1. To verify an identity, we begin with the assumption that it is an identity.

2. $\sin x = \dfrac{1}{\csc x}$, $0 < x < \pi$.

3. Every trigonometric equation is an identity.

4. Squaring both sides of an equality does not necessarily yield an equivalent equality.

5. Two functions are not equal if they have different graphs.

6. $\sec^2\theta = 1 + \cot^2\theta$.

In Exercises 7–12 fill in the blank to make the statement true.

7. $\sin^2 x +$ ________ $= 1$.

8. $\sec^2 x - 1 =$ ________.

9. The value of $\tan x$ is equal to the reciprocal of ________.

10. The curve $x^2 + y^2 = 1$ is said to be defined ________ by $x = \cos t$, $y = \sin t$.

11. An equation that is satisfied by all values of the variable for which it is defined is a(n) ________.

12. An operation on an equation is ________ if the consequence of the operation is an equivalent equation.

In the following exercises solve the stated problem. Show all your work.

13. Verify that $\cos x \tan x \cot^2 x = \csc x - \cos x \tan x$.

14. Verify that $\cos^2 x\,(1 + \tan^2 x) = 1$.

15. Verify that $\dfrac{\cot^2 x - 1}{1 + \cot^2 x} = 2\cos^2 x - 1$.

16. Verify that $\dfrac{\cos x}{\sec x\,(1 + \tan^2 x)} = \cos^4 x$.

17. Solve the equation $\tan x + 1 = 0$, on the interval $0 \le x \le 2\pi$.

18. Solve the equation $2\sin^2 x + 5\sin x + 3 = 0$, on the interval $0 \le x \le 2\pi$.

19. Solve the equation $\sin x + 2\sin x \cos x = 0$, on the interval $0 \le x \le 2\pi$.

20. Solve the equation $\cos^2\theta - 3\sin^2\theta = 0$, on the interval $0 \le x \le 2\pi$.

21. Eliminate the parameter and sketch the graph of $x = \sin t$, $y = 2\cos t$.

22. Eliminate the parameter and sketch the graph of $x = 1 - \cos t$, $y = 2 + \sin t$.

23. Graphically solve the equation $\sin x = x^2$ on $0 \le x < \pi/2$. Give answers to two decimal places.

24. Graphically solve the inequality $x \cos x > 1$, on $0 \le x < 2\pi$.

Composite Angle Identities

*T*he argument of a trigonometric function can be varied by multiplying it by or adding it to a constant. In Chapter 5 we saw how such changes in the argument affect the graphs of the trigonometric functions. In this chapter we develop identities that relate trigonometric functions with such changes in their arguments to trigonometric functions of unaltered arguments. We begin by deriving very important formulas for the cosine and sine of a sum of two numbers in terms of the cosine and sine of the individual numbers. After the fundamental identities introduced in Section 2.2, these two identities are the most important in all of analytic trigonometry. We use these two identities in Sections 7.3 to derive identities for doubling or halving the argument, the so called double- and half-angle formulas We conclude the chapter with some formulas for the sum and product of trigonometric functions which become useful in advanced mathematics.

7.1 The Cosine of the Difference or Sum of Two Angles

The cosine of the difference of A and B is denoted by $\cos(A - B)$. Beginning trigonometry students tend to think that $\cos(A - B)$ should equal $\cos A - \cos B$. However, as Example 1 shows, these two expressions, are not equivalent.

Example 1 Show that $\cos(A - B) = \cos A - \cos B$ is not an identity.

SOLUTION Suppose we let $A = \frac{1}{2}\pi$ and $B = \frac{1}{6}\pi$. Then

$$\cos(A - B) = \cos\left(\frac{1}{2}\pi - \frac{1}{6}\pi\right) = \cos\frac{1}{3}\pi = \frac{1}{2}$$

But

$$\cos A - \cos B = \cos\frac{1}{2}\pi - \cos\frac{1}{6}\pi = 0 - \frac{1}{2}\sqrt{3}$$

Since the two expressions yield different values for the chosen numbers, we conclude that $\cos(A - B) = \cos A - \cos B$ is not an identity. ▲

The purpose of this section is to show precisely which formulas the trigonometric functions of sums and differences *do* obey. The formulas will be derived using angles, but they are also valid for real numbers. The most basic of the **addition formulas** is the formula for $\cos(A - B)$. This formula is derived directly from the definitions of the trigonometric functions.

Let A and B represent angles in standard position superimposed on a circle of radius 1. Figure 7.1(a) is a picture of the general situation. The terminal side of A intersects the unit circle at the point $(x_A, y_A) = (\cos A, \sin A)$. Similarly, the terminal side of B intersects the circle at $(x_B, y_B) = (\cos B, \sin B)$. The distance D between these two points is given by

$$\begin{aligned} D^2 &= (x_A - x_B)^2 + (y_A - y_B)^2 \\ &= (\cos A - \cos B)^2 + (\sin A - \sin B)^2 \\ &= (\cos^2 A - 2\cos A\cos B + \cos^2 B) \\ &\quad + (\sin^2 A - 2\sin A\sin B + \sin^2 B) \end{aligned}$$

Since $\cos^2 A + \sin^2 A = 1$ and $\cos^2 B + \sin^2 B = 1$, we have

$$D^2 = 2(1 - \cos A\cos B - \sin A\sin B)$$

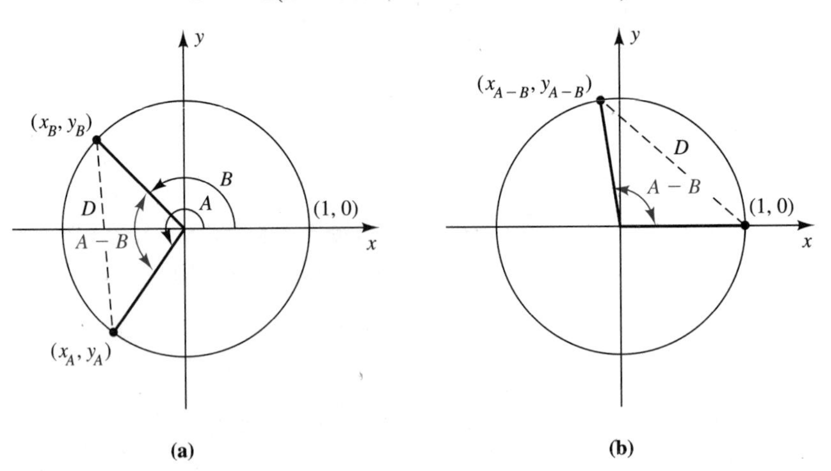

Figure 7.1

Now we rotate the angle $A - B$ until it is in standard position, as shown in Figure 7.1(b). The coordinates of the point of intersection of the terminal side of the angle $A - B$ and the unit circle are

-VIMMY-

$$(x_{A-B}, y_{A-B}) = [\cos(A - B), \sin(A - B)]$$

D is now the distance connecting (1, 0) to this point. Using the distance formula, we have

$$\begin{aligned} D^2 &= (x_{A-B} - 1)^2 + (y_{A-B})^2 \\ &= [\cos(A - B) - 1]^2 + \sin^2(A - B) \\ &= \cos^2(A - B) - 2\cos(A - B) + 1 + \sin^2(A - B) \end{aligned}$$

Since $\cos^2(A - B) + \sin^2(A - B) = 1$, we can write D^2 as

$$D^2 = 2[1 - \cos(A - \mathrm{B})]$$

We can equate this result to the first expression we derived for D^2:

$$1 - \cos A \cos B - \sin A \sin B = 1 - \cos(A - B)$$

Simplifying this expression yields the fundamental formula

$$\cos(A - B) = \cos A \cos B + \sin A \sin B \tag{7.1}$$

Formula (7.1) was derived under the conditions that $A > B$ and that A and B are between 0 and 2π. However, since

$$\cos(A - B) = \cos(B - A) = \cos(A - B + 2n\pi)$$

it follows that the formula is general. Since it is true for all angles A and B and consequently for all real numbers, it is an **identity**.

The principal use of this formula is to derive other important relations. However, it can also be used to obtain the value of the cosine function at a particular angle (or real number) if that angle can be expressed as the difference of two angles for which the exact value of the cosine is known.

Example 2 Without using a calculator, find the exact value of $\cos \frac{1}{12}\pi$.

SOLUTION Since $\frac{1}{12}\pi = \frac{4}{12}\pi - \frac{3}{12}\pi = \frac{1}{3}\pi - \frac{1}{4}\pi$,

$$\begin{aligned} \cos \tfrac{1}{12}\pi &= \cos(\tfrac{1}{3}\pi - \tfrac{1}{4}\pi) \\ &= \cos \tfrac{1}{3}\pi \cos \tfrac{1}{4}\pi + \sin \tfrac{1}{3}\pi \sin \tfrac{1}{4}\pi \\ &= \frac{1}{2}\left(\frac{\sqrt{2}}{2}\right) + \left(\frac{\sqrt{3}}{2}\right)\left(\frac{\sqrt{2}}{2}\right) \\ &= \frac{\sqrt{2} + \sqrt{6}}{4} \end{aligned}$$

▲

If we replace B with $-B$ in Formula (7.1), we obtain

$$\cos(A - (-B)) = \cos A \cos(-B) + \sin A \sin(-B)$$

Since the cosine function is even, $\cos(-B) = \cos B$; since the sine function is odd, $\sin(-B) = -\sin B$. Hence the formula for $\cos(A + B)$ is

$$\cos(A + B) = \cos A \cos B - \sin A \sin B \tag{7.2}$$

Example 3

(a) $-\sin 2\alpha \sin 3\alpha + \cos 2\alpha \cos 3\alpha = \cos(2\alpha + 3\alpha) = \cos 5\alpha$

(b) $7 \cos(2x + 4y) = 7 \cos 2x \cos 4y - 7 \sin 2x \sin 4y$ ▲

Example 4 Without using a calculator, find the exact value of cos 75°.

SOLUTION Since $75° = 30° + 45°$,

$$\begin{aligned}\cos 75° &= \cos(30° + 45°)\\ &= \cos 30° \cos 45° - \sin 30° \sin 45°\\ &= \frac{\sqrt{3}}{2} \cdot \frac{\sqrt{2}}{2} - \frac{1}{2} \cdot \frac{\sqrt{2}}{2}\\ &= \frac{\sqrt{6} - \sqrt{2}}{4}\end{aligned}$$ ▲

Example 5 Without using a calculator, find the exact value of $\cos(A - B)$, given that $\sin A = \frac{3}{5}$ in quadrant II and $\tan B = \frac{1}{2}$ in quadrant I.

SOLUTION From Figure 7.2 we see that $\cos A = -4/5$, $\sin B = 1/\sqrt{5}$, and $\cos B = 2/\sqrt{5}$. Hence

$$\begin{aligned}\cos(A - B) &= \cos A \cos B + \sin A \sin B\\ &= \left(-\frac{4}{5}\right)\left(\frac{2}{\sqrt{5}}\right) + \left(\frac{3}{5}\right)\left(\frac{1}{\sqrt{5}}\right)\\ &= -\frac{5}{5\sqrt{5}} = -\frac{\sqrt{5}}{5}\end{aligned}$$

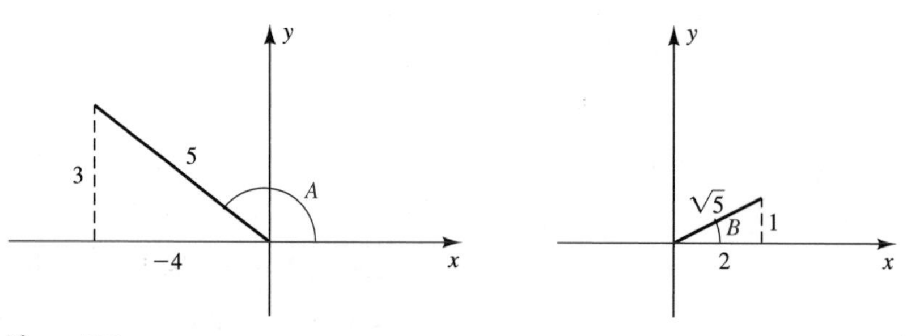

Figure 7.2 ▲

In certain applied problems you will encounter trigonometric functions of the form $c_1\cos Bx + c_2\sin Bx$. It is easy to see that this function is periodic with a period of $2\pi/B$; however, it is not easy to recognize the amplitude and phase shift of the oscillation of the function in its present form. To obtain these properties, as well as the period, we make use of the identity

$$c_1\cos Bx + c_2\sin Bx = A\cos(Bx - C)$$

where $A = \sqrt{c_1^2 + c_2^2}$ and $\tan C = c_2/c_1$.

To verify this identity we note that by Equation (7.2),

$$A\cos(Bx - C) = A\cos C\cos Bx + A\sin C\sin Bx$$

Therefore

$$c_1\cos Bx + c_2\sin Bx = A\cos C\cos Bx + A\sin C\sin Bx$$

if and only if $c_1 = A\cos C$ and $c_2 = A\sin C$. Squaring c_1 and c_2 and adding, we get

$$c_1^2 + c_2^2 = A^2\cos^2 C + A^2\sin^2 C = A^2(\cos^2 C + \sin^2 C)$$

or

$$A = \sqrt{c_1^2 + c_2^2}$$

Also, the ratio of c_2 to c_1 yields

$$\frac{c_2}{c_1} = \frac{A\sin C}{A\cos C} = \tan C$$

Note that C cannot be just *any* angle for which $\tan C = c_2/c_2$. The angle must be chosen so that its terminal side passes through the point (c_1, c_2).

Example 6 Express $f(x) = \sin x + \cos x$ as a cosine function and sketch.

SOLUTION If we refer to the formulas we just discussed, we see that

$$f(x) = A\cos(x - C)$$

where $A = \sqrt{1^2 + 1^2} = \sqrt{2}$ and $\tan C = 1$. Since $c_1 = c_2 = 1$ and $(1, 1)$ is in the first quadrant, we let $C = \frac{1}{4}\pi$. Therefore

$$f(x) = \sqrt{2}\cos(x - \tfrac{1}{4}\pi)$$

This is a function with amplitude $\sqrt{2}$, period 2π, and phase shift $\frac{1}{4}\pi$. The graph is shown in Figure 7.3 on page 268, along with the graphs of $\sin x$ and $\cos x$.

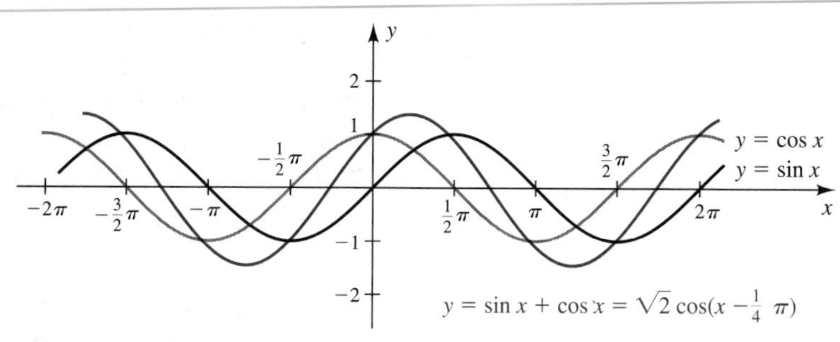

Figure 7.3

Exercises for Section 7.1

1. Show that $\sin(x + y) \neq \sin x + \sin y$. (Let $x = \frac{1}{3}\pi$ and $y = \frac{1}{6}\pi$.)
2. Show that $\tan(x + y) \neq \tan x + \tan y$. (Let $x = \frac{1}{3}\pi$ and $y = \frac{1}{6}\pi$.)
3. Show that $\sin 2x \neq 2 \sin x$. (Let $x = \frac{1}{4}\pi$.)
4. Show that $\cos 2x \neq 2 \cos x$. (Let $x = \frac{1}{4}\pi$.)
5. Show that $\tan 2x \neq 2 \tan x$. (Let $x = \frac{1}{4}\pi$.)

In Exercises 6–10 find the *exact* value of each trigonometric function without using a calculator.

6. $\cos 105°$
7. $\cos \frac{1}{12}\pi$
8. $\cos \frac{11}{12}\pi$
9. $\cos 195°$
10. $\cos 345°$

In Exercises 11–13 use Equation (7.1) to show that the expressions are true.

11. $\cos(\pi - \theta) = -\cos\theta$
12. $\cos(\frac{1}{2}\pi - \theta) = \sin\theta$
13. $\cos(\frac{1}{2}\pi + \theta) = -\sin\theta$

In Exercises 14–17 reduce each expression to a single term.

14. $\cos 2x \cos 3x + \sin 2x \sin 3x$
15. $\cos 7x \cos x - \sin 7x \sin x$
16. $\cos \frac{1}{6}x \cos \frac{5}{6}x - \sin \frac{1}{6}x \sin \frac{5}{6}x$
17. $\cos 5\theta \cos 3\theta + \sin 5\theta \sin 3\theta$
18. Using Equation (7.1), give a proof that the cosine is an even function.

In Exercises 19–28 verify each of the given identities.

19. $\cos(\frac{1}{3}\pi - x) = \dfrac{\cos x + \sqrt{3}\sin x}{2}$
20. $\cos(\frac{1}{4} + \theta) = \dfrac{\cos\theta - \sin\theta}{\sqrt{2}}$
21. $\cos(\frac{3}{2}\pi + x) = \sin x$
22. $\cos 2x = \cos^2 x - \sin^2 x$
23. $\cos 5x + \sin 3x \sin 2x = \cos 3x \cos 2x.$
24. $\cos 3x - \cos 4x \cos x = \sin 4x \sin x$
25. $\dfrac{\cos 3x}{\sin 3x} - \dfrac{\sin x}{\cos x} = \dfrac{\cos 4x}{\sin 3x \cos x}$

26. $\cot x + \tan 2x = \dfrac{\cos x}{\sin x \cos 2x}$

27. $\cos(x + y)\cos(x - y) = \cos^2 x - \sin^2 y$

28. $\cos(x + y) + \cos(x - y) = 2 \cos x \cos y$

In Exercises 29–32 find the exact value of $\cos(A + B)$ for each of the given conditions, without using a calculator.

on test

29. $\cos A = \frac{1}{3}$, $\sin B = -\frac{1}{2}$, A in quadrant I, B in quadrant IV

30. $\cos A = \frac{3}{5}$, $\tan B = \frac{12}{5}$, both A and B acute

31. $\tan A = \frac{24}{7}$, $\sec B = \frac{5}{3}$, A in quadrant III, B in quadrant I

32. $\sin A = \frac{1}{4}$, $\cos B = \frac{1}{2}$, both A and B in quadrant I

In Exercises 33 and 34 let both A and B be acute angles. Do not use a calculator.

33. Find the exact value of $\cos A$ if $\cos(A + B) = \frac{5}{6}$ and $\sin B = \frac{1}{3}$.

34. Find the exact value of $\cos A$ if $\cos(A - B) = \frac{3}{4}$ and $\cos B = \frac{2}{3}$.

In Exercises 35–38 express each of the given functions as a single cosine function. Give the amplitude and phase shift and sketch the graph.

35. $f(x) = \cos x - \sin x$

36. $f(x) = 2 \cos x + 2 \sin x$

37. $f(x) = \cos 2x + \sqrt{3} \sin 2x$

38. $f(x) = -\cos 2x + \sqrt{3} \sin 2x$

In Exercises 39–44 show that $\cos(A + B) \neq \cos A + B$. Then verify that $\cos(A + B) = \cos A \cos B - \sin A \sin B$. Use a calculator when necessary.

39. $A = 0$, $B = 0$

40. $A = 30°$, $B = 30°$

41. $A = 0.987$, $B = 0.111$

42. $A = -0.912$, $B = 0.912$

43. $A = 23.1$, $B = 14.14$

44. $A = 1.57$, $B = -1.57$

45. An electrical engineer knows that the output voltage of an a.c. generator is approximated by

$$v = 156 \cos(2\pi ft - \tfrac{1}{3}\pi)$$

where $f = 60$ Hz. Express this output as the sum of the sine and a cosine.

46. Suppose you are given that $\tan \phi = \frac{2}{3}$ and $\tan \theta = \frac{3}{2}$, both in quadrant I. Without using a calculator, find the exact value of $\cos(\theta - \phi)$.

47. Ocean waves (Figure 7.4) can be represented by

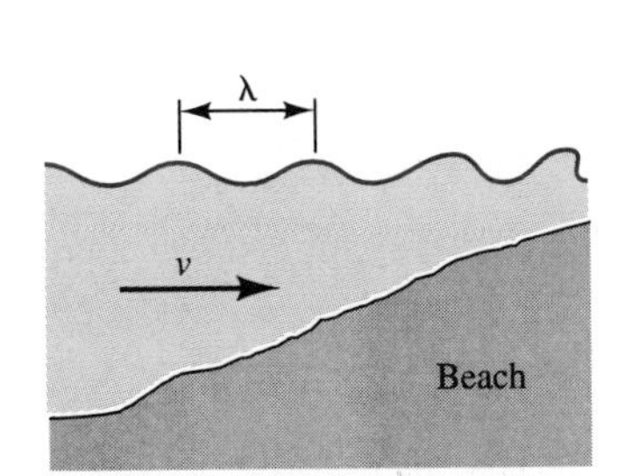

Figure 7.4

$$y = A \cos \frac{2\pi}{\lambda}(x - vt)$$

where λ is the wavelength, v is the wave velocity, and t is time. If $t = 1$ and λ and v are constant, then the wave can be represented by

$$y = c_1 \cos \frac{2\pi}{\lambda}x + c_2 \sin \frac{2\pi}{\lambda}x$$

Show that this is true and find expressions for c_1 and c_2.

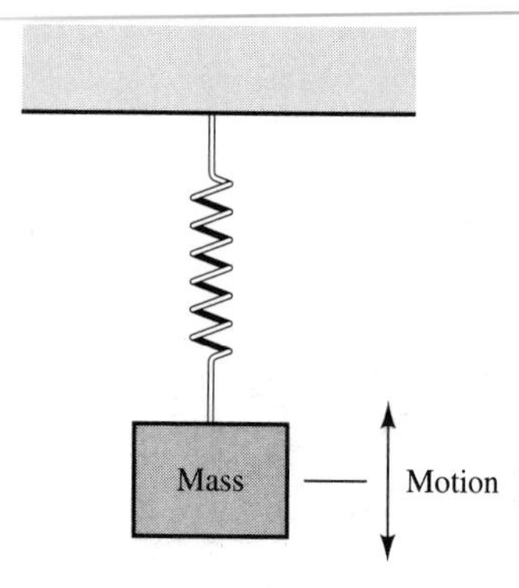

Figure 7.5

48. If a mass is attached to a spring, as in Figure 7.5, its motion may be approximated by

$$y = A \cos \omega t + \frac{F}{\omega Z} \sin \omega t$$

Write this expression as one cosine function of time, t. Assume that A, F, Z, and ω are constants.

49. A vibrating string has a vertical deflection that can be approximated by

$$y = 3(\cos 2\pi x \cos ct + \sin 2\pi x \sin ct)$$

Write this expression as one cosine function of x. Assume that c and t are constants.

50. In studying the theory of phase modulation, a communications engineer is confronted with the sum

$$\cos k\theta \cos(x \sin \theta) + \sin k\theta \sin(x \sin \theta)$$

Show that this sum is equal to $\cos(k\theta - x \sin \theta)$.

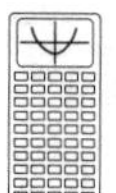

Graphing Calculator Exercise

1. Use a graphing calculator to display the graphs of the functions on each side of the following expressions. Display both graphs on the same coordinate axes. On the basis of the graphs, indicate which of the given expressions are identities.

(a) $\cos(x + 3\pi/2) = \sin x$
(b) $\sin(\pi/4 - x) = -\cos x$
(c) $\sin(x + \frac{1}{2}\pi) = \sin x + \sin \frac{1}{2}\pi$
(d) $\sin x + \cos x = \sqrt{2} \cos(x - \frac{1}{4}\pi)$
(e) $\sin x - \cos x = \sqrt{2} \sin(x - \frac{1}{4}\pi)$
(f) $\sin 2x = 2 \sin x$
(g) $\cos 2x = \cos^2 x - \sin^2 x$
(h) $\cos 2x = 1 - 2 \sin^2 x$

7.2 Other Addition Formulas

From the formula for the cosine of the difference of two angles, it is easy to establish that

$$\cos(\tfrac{1}{2}\pi - \theta) = \sin \theta \quad \text{and} \quad \sin(\tfrac{1}{2}\pi - \theta) = \cos \theta \qquad (7.3)$$

Note that this is a general statement of the identity relating cofunctions of complementary angles.

Example 1

(a) $\cos(-10°) = \cos(90° - 100°) = \sin 100°$

(b) $\sin(\frac{5}{6}\pi) = \sin[\frac{1}{2}\pi - (-\frac{1}{3}\pi)] = \cos(-\frac{1}{3}\pi) = \cos \frac{1}{3}\pi$ ▲

The fundamental Pythagorean relationship is $\sin^2 x + \cos^2 x = 1$. A similar formula involving two numbers x and y is true if x and y have a sum of $\frac{1}{2}\pi$.

Example 2 Show that if x and y are complementary numbers (that is, their sum is $\frac{1}{2}\pi$), then $\sin^2 x + \sin^2 y = 1$.

SOLUTION Since $x + y = \frac{1}{2}\pi$, $y = \frac{1}{2}\pi - x$. Therefore $\sin y = \sin(\frac{1}{2}\pi - x) = \cos x$. Since $\cos x = \sin y$, the Pythagorean relation $\sin^2 x + \cos^2 x = 1$ can be written

$$\sin^2 x + \sin^2 y = 1$$

This identity is valid for $x + y = \frac{1}{2}\pi$. ▲

If we let $\theta = A + B$ in Equation (7.3), we can write

$$\begin{aligned}\sin(A + B) &= \cos[\tfrac{1}{2}\pi - (A + B)] \\ &= \cos[(\tfrac{1}{2}\pi - A) - B]\end{aligned}$$

Using Equation (7.1) for the cosine of a difference, we obtain

$$\sin(A + B) = \cos(\tfrac{1}{2}\pi - A)\cos B + \sin(\tfrac{1}{2}\pi - A)\sin B$$

Then applying Equaltion (7.3) again yields the following identity for $\sin(A + B)$.

$$\sin(A + B) = \sin A \cos B + \cos A \sin B \qquad \textbf{(7.4)}$$

Similarly, if $\theta = A - B$, in Equation (7.3) we get

$$\sin(A - B) = \sin A \cos(-B) + \cos A \sin(-B)$$

Since the cosine function is even and the sine function is odd, this becomes

$$\sin(A - B) = \sin A \cos B - \cos A \sin B \qquad \textbf{(7.5)}$$

Example 3 Without using a calculator, find the exact value of $\sin(\frac{1}{12}\pi)$.

SOLUTION Since $\frac{1}{12}\pi = \frac{4}{12}\pi - \frac{3}{12}\pi = \frac{1}{3}\pi - \frac{1}{4}\pi$,

$$\begin{aligned}\sin(\tfrac{1}{12}\pi) &= \sin(\tfrac{1}{3}\pi)\cos(\tfrac{1}{4}\pi) - \cos(\tfrac{1}{3}\pi)\sin(\tfrac{1}{4}\pi) \\ &= \frac{\sqrt{3}}{2}\frac{\sqrt{2}}{2} - \frac{1}{2}\frac{\sqrt{2}}{2} \\ &= \frac{1}{4}(\sqrt{6} - \sqrt{2})\end{aligned}$$

▲

Example 4 Show that $\sin x + \cos x = \sqrt{2}\sin(x + \frac{1}{4}\pi)$.

SOLUTION If $\sin x + \cos x = A \sin(x + C) = A \sin x \cos C + A \cos x \sin C$, then $A \cos C = 1$ and $A \sin C = 1$. If we square these two equations and add the results, we get

$$A^2\cos^2C + A^2\sin^2C = 1^2 + 1^2 = 2$$

or $A = \sqrt{2}$.

Also,

$$\frac{A \sin C}{A \cos C} = \frac{1}{1}$$

so $\tan C = 1$.

We choose C to be a first-quadrant angle, since its terminal side must pass through (1, 1). Thus $C = \frac{1}{4}\pi$, and hence

$$\sin x + \cos x = \sqrt{2} \sin(x + \tfrac{1}{4}\pi)$$

The sketch of this function is precisely the one given in Figure 7.3. Can you verify this? ▲

Sum and difference formulas for the tangent follow directly from those for the sine and cosine.

$$\tan(A + B) = \frac{\sin(A + B)}{\cos(A + B)}$$

$$= \frac{\sin A \cos B + \cos A \sin B}{\cos A \cos B - \sin A \sin B}$$

Now we divide both the numerator and denominator by $\cos A \cos B$.

$$\tan(A + B) = \frac{\dfrac{\sin A \cos B}{\cos A \cos B} + \dfrac{\cos A \sin B}{\cos A \cos B}}{\dfrac{\cos A \cos B}{\cos A \cos B} - \dfrac{\sin A \sin B}{\cos A \cos B}}$$

Simplifying, we get

$$\tan(A + B) = \frac{\tan A + \tan B}{1 - \tan A \tan B} \qquad \textbf{(7.6)}$$

Similarly,

$$\tan(A - B) = \frac{\tan A - \tan B}{1 + \tan A \tan B} \qquad \textbf{(7.7)}$$

Example 5 Verify that $\tan(\theta + \pi) = \tan \theta$.

SOLUTION Using Equation (7.6) and the fact that $\tan \pi = 0$, we have

$$\tan(\theta + \pi) = \frac{\tan \theta + \tan \pi}{1 - \tan \theta \tan \pi} = \tan \theta$$ ▲

Example 6 Reduce $\dfrac{\tan(x + y) - \tan(x - y)}{1 + \tan(x + y)\tan(x - y)}$ to a single term.

SOLUTION We recognize this expression as the right-hand side of Equation (7.7), with $A = x + y$ and $B = x - y$. Thus,

$$\frac{\tan(x + y) - \tan(x - y)}{1 + \tan(x + y)\tan(x - y)} = \tan[(x + y) - (x - y)] = \tan 2y$$ ▲

The sum and difference formulas that we have derived can be summarized as follows:

$$\sin(A \pm B) = \sin A \cos B \pm \cos A \sin B$$

$$\cos(A \pm B) = \cos A \cos B \mp \sin A \sin B$$

$$\tan(A \pm B) = \frac{\tan A \pm \tan B}{1 \mp \tan A \tan B}$$

By convention, when the symbols $\pm$ and $\mp$ are used in the same formula, the top signs go together and the bottom signs go together.

Exercises for Section 7.2

In Exercises 1–6 find the exact value of each expression without using a calculator.

1-29 odd

28, 30

1. $\sin(\frac{5}{12}\pi)$
2. $\tan 15°$
3. $\sin(\frac{7}{12}\pi)$
4. $\sin(345°)$
5. $\cot(\frac{5}{12}\pi)$
6. $\tan(\frac{1}{12}\pi)$

In Exercises 7–23 verify each identity.

7. $\sin(x + \pi) = -\sin x$
8. $\sin(x + \frac{1}{4}\pi) = \dfrac{\sqrt{2}}{2}(\sin x + \cos x)$
9. $\tan(x + \frac{1}{2}\pi) = -\cot x$
10. $\tan(x + \frac{1}{4}\pi) = \dfrac{1 + \tan x}{1 - \tan x}$
11. $\dfrac{\sin 2x \cos x + \cos 2x \sin x}{\cos 2x \cos x - \sin 2x \sin x} = \tan 3x$
12. $\dfrac{1}{\sin 3x \cos 2x - \cos 3x \sin 2x} = \csc x$
13. $\cot(A + B) = \dfrac{\cot A \cot B - 1}{\cot A + \cot B}$
14. $\dfrac{\sin(A + B)}{\sin(A - B)} = \dfrac{\tan A + \tan B}{\tan A - \tan B}$

15. $\sin(A + B)\sin(A - B) = \sin^2 A - \sin^2 B$

16. $\sin(A + B) + \sin(A - B) = 2 \sin A \cos B$

17. $\tan A + \tan B = \dfrac{\sin(A + B)}{\cos A \cos B}$

18. $\dfrac{\sin(x + y) + \sin(x - y)}{\cos(x + y) + \cos(x - y)} = \tan x$

19. $\dfrac{\cos(x - y) - \cos(x + y)}{\sin(x + y) + \sin(x - y)} = \tan y$

20. $\dfrac{\cos 3x}{\cos x} + \dfrac{\sin 3x}{\sin x} = \sin 4x \csc x \sec x$

21. $\dfrac{\sin 5x}{\sin 2x} - \dfrac{\cos 5x}{\cos 2x} = \sin 3x \csc 2x \sec 2x$

22. $\sec(x + y) = \dfrac{\sec x \sec y \csc x \csc y}{\csc x \csc y - \sec x \sec y}$

23. $\csc(x + y) = \dfrac{\sec x \sec y \csc x \csc y}{\sec x \csc y + \csc x \sec y}$

In Exercises 24–27 reduce each expression to a single term.

24. $\sin 2x \cos 3x + \sin 3x \cos 2x$

25. $\dfrac{\tan 3x - \tan 2x}{1 + \tan 3x \tan 2x}$

26. $\sin \frac{1}{3}x \cos \frac{2}{3}x + \sin \frac{2}{3}x \cos \frac{1}{3}x$

27. $\dfrac{\tan(x + y) + \tan z}{1 - \tan(x + y)\tan z}$

In Exercises 28–30 find the exact values of $\sin(A + B)$ and $\tan(A + B)$. Do not use a calculator.

28. $\sin A = \frac{3}{5}$, $\cos B = \frac{4}{5}$, both A and B in quadrant I

29. $\tan A = -\frac{7}{24}$, $\tan B = \frac{5}{12}$, A in quadrant II, B in quadrant III

30. $\cos A = \frac{1}{3}$, $\cos B = -\frac{1}{3}$, A in quadrant IV, B in quadrant III

In Exercises 31–34 express each function as a sine function; state the phase shift and sketch the function.

31. $\sin 2x + \cos 2x$

32. $\cos x$

33. $\sqrt{3} \sin \pi x + \cos \pi x$

34. $7 \sin 2x - 24 \cos 2x$

In Exercises 35–40 use a calculator to show that in each case the left-hand side is equal to the right-hand side.

35. $\sin(27° + 96°) = \sin 27° \cos 96° + \cos 27° \sin 96°$

36. $\cos(12.6° + 8.7°) = \cos 12.6° \cos 8.7° - \sin 12.6° \sin 8.7°$

37. $\cos(1.1 - 0.3) = \cos 1.1 \cos 0.3 + \sin 1.1 \sin 0.3$

38. $\sin(2.8 - 1.6) = \sin 2.8 \cos 1.6 - \cos 2.8 \sin 1.6$

39. $\tan(0.4 + 0.3) = \dfrac{\tan 0.4 + \tan 0.3}{1 - \tan 0.4 \tan 0.3}$

40. $\tan(2.9 - 1.2) = \dfrac{\tan 2.9 - \tan 1.2}{1 + \tan 2.9 \tan 1.2}$

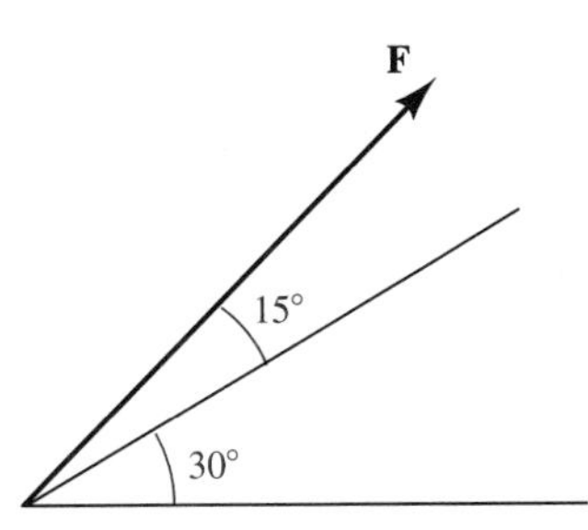

Figure 7.6

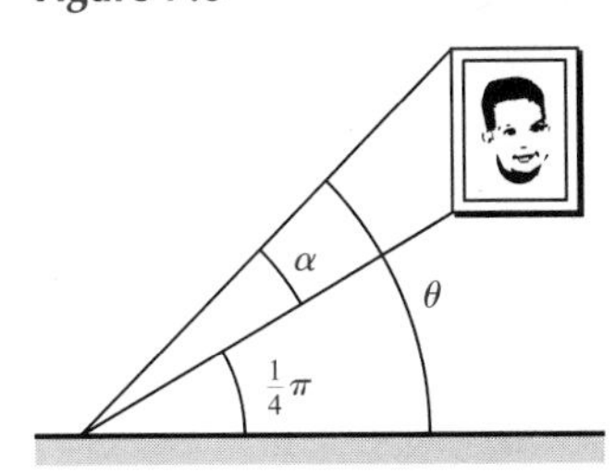

Figure 7.7

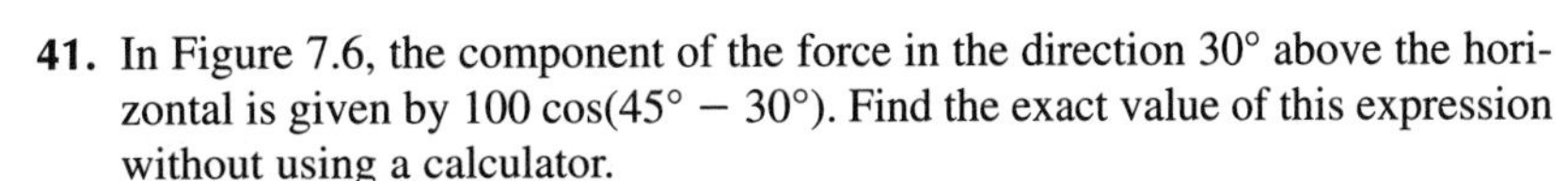

41. In Figure 7.6, the component of the force in the direction 30° above the horizontal is given by $100 \cos(45° - 30°)$. Find the exact value of this expression without using a calculator.
42. The angle α in Figure 7.7 subtends the picture hanging on the wall. This angle can be changed by raising or lowering the picture or by moving the origin of the angle closer to the wall. Find $\tan \alpha$ given that $\tan \theta = 2$.
43. The output of a 60-volt signal generator is $v(t) = 60 \sin(2\pi ft + \frac{1}{3}\pi)$. Express $v(t)$ as a sum of sine and cosine terms.
44. In analyzing the luminous intensity from a point source, an architect must simplify the expression $2.1 \cos \alpha t - 3.2 \sin \alpha t$. Write this expression as one sine function.
45. In measuring the angle β between two curves, a surveyor derives the expression $\tan(\theta_2 - \theta_1)$, where θ_2 and θ_1 are the angles with respect to the horizontal of each of the curves. Express $\tan \beta = \tan(\theta_2 - \theta_1)$ in terms of $m_2 = \tan \theta_2$ and $m_1 = \tan \theta_1$.

Graphing Calculator Exercises

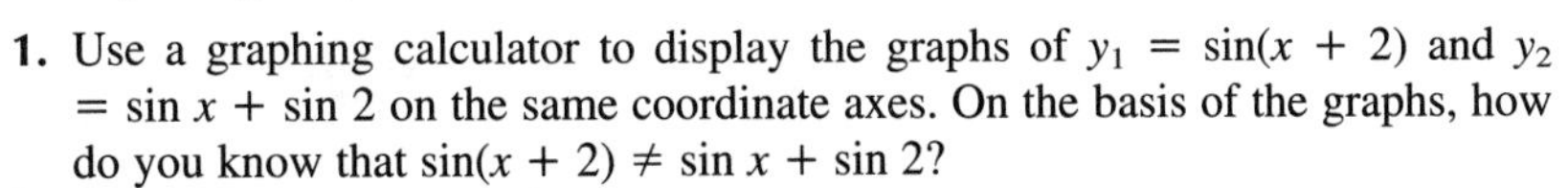

1. Use a graphing calculator to display the graphs of $y_1 = \sin(x + 2)$ and $y_2 = \sin x + \sin 2$ on the same coordinate axes. On the basis of the graphs, how do you know that $\sin(x + 2) \neq \sin x + \sin 2$?
2. Use a graphing calculator to display the graphs of $y_1 = \sin(x + 2)$ and $y_2 = \sin x \cos 2 + \sin 2 \cos x$ on the same coordinate axes. On the basis of the graphs, is $\sin(x + 2) = \sin x \cos 2 + \sin 2 \cos x$ an identity? How does this expression relate to Equation (7.4)?

▲▼ 7.3 Double- and Half-Angle Formulas

In the previous two sections, we were primarily interested in expanding trigonometric functions whose arguments are $A + B$. Now we will derive formulas for functions of $2A$ and $\frac{1}{2}A$. If A represents an angle, the formulas are called the **double-** and **half-angle formulas,** respectively.

The double-angle formulas are easily proved by choosing $B = A$ in the formulas for the sum of two angles. Thus

$$\sin 2A = \sin(A + A)$$
$$= \sin A \cos A + \sin A \cos A$$

or

$$\sin 2A = 2 \sin A \cos A \tag{7.8}$$

The proof of the double-angle formula for cosines is similar:

$$\cos 2A = \cos(A + A)$$
$$= \cos A \cos A - \sin A \sin A$$

or

$$\cos 2A = \cos^2 A - \sin^2 A \qquad \textbf{(7.9)}$$

Using the Pythagorean relation $\sin^2 A + \cos^2 A = 1$, we can express this last formula in the equivalent forms

$$\cos 2A = 2\cos^2 A - 1 \qquad \textbf{(7.9a)}$$

and

$$\cos 2A = 1 - 2\sin^2 A \qquad \textbf{(7.9b)}$$

Similarly, letting $B = A$ in Equation (7.6), we have

$$\tan 2A = \frac{\tan A + \tan A}{1 - \tan A \tan A}$$

$$\tan 2A = \frac{2\tan A}{1 - \tan^2 A} \qquad \textbf{(7.10)}$$

Example 1 Without using a calculator, find the exact value of $\sin 2A$ if $\sin A = \frac{1}{3}$ and A is in quadrant II.

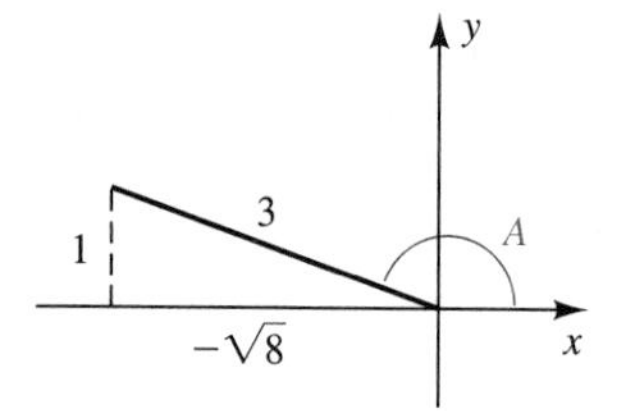

Figure 7.8

SOLUTION Since $\sin A = \frac{1}{3}$, Figure 7.8 shows that $\cos A = -\frac{1}{3}\sqrt{8}$. Thus,

$$\sin 2A = 2 \sin A \cos A$$
$$= 2\left(\frac{1}{3}\right)\left(\frac{-\sqrt{8}}{3}\right) = \frac{-2\sqrt{8}}{9}$$

Note that $\sin 2A \neq 2 \sin A$. ▲

Example 2 Sketch the graph of $y = \sin x \cos x$. Where does the maximum value of this function occur?

SOLUTION By multiplying and dividing the right-hand side of this function by 2, we obtain

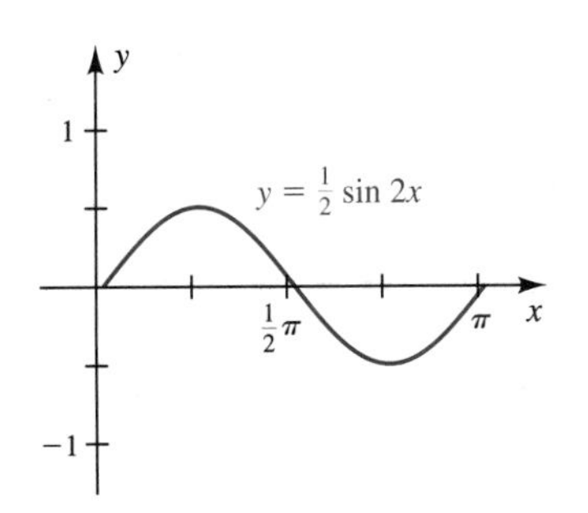

Figure 7.9

$$y = \frac{2\sin x\cos x}{2} = \frac{1}{2}\sin 2x$$

Thus the graph of this function is a sine wave with amplitude $\frac{1}{2}$ and period π. The maximum value of $\frac{1}{2}$ occurs at $x = \frac{1}{4}\pi + n\pi$. (See Figure 7.9.) ▲

The half-angle formulas are directly related to the formulas for the cosine of a double angle. Since

$$\cos 2A = 2\cos^2 A - 1$$

solving for $\cos A$ yields

$$\cos A = \pm\sqrt{\frac{1 + \cos 2A}{2}}$$

The positive sign is used when angle A is in either quadrant I or quadrant IV, since $\cos A$ is positive in those quadrants. The negative sign is used when angle A is in either quadrant II or quadrant III.

Similarly, if we solve the formula $\cos 2A = 1 - 2\sin^2 A$ for $\sin A$, we get

$$\sin A = \pm\sqrt{\frac{1 - \cos 2A}{2}}$$

Letting $A = \frac{1}{2}x$ in both of these formulas, we have

$$\cos\frac{1}{2}x = \pm\sqrt{\frac{1 + \cos x}{2}} \qquad \textbf{(7.11)}$$

and

$$\sin\frac{1}{2}x = \pm\sqrt{\frac{1 - \cos x}{2}} \qquad \textbf{(7.12)}$$

In Equations (7.11) and (7.12), whether the plus or the minus sign is used depends on the quadrant in which the angle $\frac{1}{2}x$ lies.

To get the formula for $\tan\frac{1}{2}x$, we write

$$\left|\tan\frac{1}{2}x\right| = \frac{\left|\sin\frac{1}{2}x\right|}{\left|\cos\frac{1}{2}x\right|} = \frac{\sqrt{(1 - \cos x)/2}}{\sqrt{(1 + \cos x)/2}} = \sqrt{\frac{1 - \cos x}{1 + \cos x}}$$

Multiplying the numerator and denominator of this expression by $(1 + \cos x)$ yields

$$\left|\tan \frac{1}{2}x\right| = \sqrt{\frac{1 - \cos^2 x}{(1 + \cos x)^2}}$$

$$= \sqrt{\frac{\sin^2 x}{(1 + \cos x)^2}}$$

$$= \frac{|\sin x|}{|1 + \cos x|}$$

The expression $1 + \cos x$ is nonnegative because $\cos x$ is never less than -1. To show that $\tan \frac{1}{2}x$ and $\sin x$ have the same sign for all x for which $\tan \frac{1}{2}x$ is defined, we note that

$$\tan \frac{1}{2}x \sin x = \tan \frac{1}{2}x(2 \sin \frac{1}{2}x \cos \frac{1}{2}x) = 2 \sin^2 \frac{1}{2}x$$

Since $\sin^2 \frac{1}{2}x$ is always positive, $\tan \frac{1}{2}x$ and $\sin x$ must have the same sign. Thus we can drop the absolute-value signs and write

$$\tan \frac{1}{2}x = \frac{\sin x}{1 + \cos x} \tag{7.13}$$

Example 3 If $\tan \theta = -\frac{4}{3}$ and $-\frac{1}{2}\pi < \theta < 0$, find the exact values of $\sin \frac{1}{2}\theta$ and $\cos \frac{1}{2}\theta$. Do not use a calculator.

SOLUTION Figure 7.10 shows the angle θ in standard position. From this figure we can see that $\cos \theta = \frac{3}{5}$, and hence

$$\sin \frac{1}{2}\theta = -\sqrt{\frac{1 - (\frac{3}{5})}{2}} = -\frac{\sqrt{5}}{5}$$

(A minus sign is chosen because $\sin \frac{1}{2}\theta$ is negative for θ in $-\frac{1}{2}\pi < \theta < 0$.) Similarly,

$$\cos \frac{1}{2}\theta = \sqrt{\frac{1 + (\frac{3}{5})}{2}} = \sqrt{\frac{4}{5}} = \frac{2\sqrt{5}}{5}$$

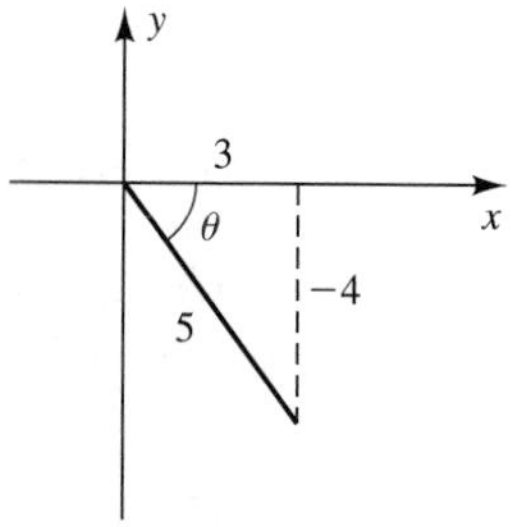

Figure 7.10

▲

Example 4 Find an expression for $\sin 3\theta$ in terms of $\sin\theta$.

SOLUTION

$\sin 3\theta = \sin(2\theta + \theta)$	
$= \sin 2\theta \cos\theta + \sin\theta \cos 2\theta$	Use $\sin(A + B) = \sin A \cos B + \sin B \cos A$
$= 2 \sin\theta \cos^2\theta + \sin\theta(1 - 2\sin^2\theta)$	Use $\sin 2A = 2 \sin A \cos A$ and $\cos 2A = 1 - 2\sin^2 A$
$= 2\sin\theta(1 - \sin^2\theta) + \sin\theta(1 - 2\sin^2\theta)$	Use $\cos^2 A = 1 - \sin^2 A$
$= 3\sin\theta - 4\sin^3\theta$	Expand and collect like terms

This proves that $\sin 3\theta = 3\sin\theta - 4\sin^3\theta$. ▲

Example 5 Solve the equation $\cos 2x = \cos x$ on the interval $0 \le x \le 2\pi$.

SOLUTION We use the identity for $\cos 2x$ to transform the equation into one involving $\cos x$.

$$2\cos^2 x - 1 = \cos x$$

Subtracting $\cos x$ from both sides gives

$$2\cos^2 x - \cos x - 1 = 0$$

Factoring, we get

$$(2\cos x + 1)(\cos x - 1) = 0$$

We now have two separate equations,

$$2\cos x + 1 = 0 \quad \text{and} \quad \cos x - 1 = 0$$

The solution to the first equation on $0 \le x \le 2\pi$ is $x = \frac{2}{3}\pi$ and $\frac{4}{3}\pi$; the solution to the second is $x = 0$ and 2π. Hence the complete solution is $0, \frac{2}{3}\pi, \frac{4}{3}\pi$, and 2π. ▲

Example 6 If $\cot 2\theta = -\frac{7}{24}$ and θ is in quadrant I, find $\sin\theta$ and $\cos\theta$. Do not use a calculator.

SOLUTION A sketch such as the one shown in Figure 7.11 is helpful for this kind of problem. From this figure we can see immediately that $\cos 2\theta = -\frac{7}{25}$.

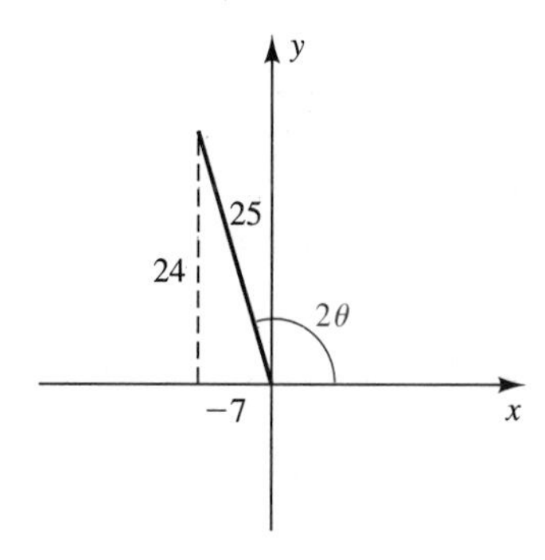

Figure 7.11

Hence

$$\sin\theta = \left(\frac{1-\cos 2\theta}{2}\right)^{1/2}$$
$$= \left(\frac{1+\frac{7}{25}}{2}\right)^{1/2}$$
$$= \left(\frac{25+7}{50}\right)^{1/2} = \frac{4}{5}$$

Similarly, $\cos\theta = \frac{3}{5}$. ▲

The double- and half-angle formulas are summarized below.

$$\sin 2A = 2\sin A\cos A$$
$$\cos 2A = \cos^2 A - \sin^2 A$$
$$= 2\cos^2 A - 1$$
$$= 1 - 2\sin^2 A$$
$$\tan 2A = \frac{2\tan A}{1-\tan^2 A}$$
$$\sin\frac{1}{2}A = \pm\sqrt{\frac{1-\cos A}{2}}$$
$$\cos\frac{1}{2}A = \pm\sqrt{\frac{1+\cos A}{2}}$$
$$\tan\frac{1}{2}A = \frac{\sin A}{1+\cos A}$$

▲▼ Exercises for Section 7.3

In Exercises 1–6 rewrite each expression as a single trigonometric function.

1. $2\sin 3x\cos 3x$ **2.** $6\sin\frac{1}{2}x\cos\frac{1}{2}x$ **3.** $\sin^2 4x - \cos^2 4x$

4. $4\sin^2 x\cos^2 x$ **5.** $\dfrac{2\tan\frac{1}{6}x}{1-\tan^2\frac{1}{6}x}$ **6.** $\dfrac{\sin 6x}{1+\cos 6x}$

7. Sketch the graph of the function $f(x) = \sin 2x\cos 2x$. What is the maximum value of the function? Where does it occur?

8. Sketch the graph of the two functions $f(x) = \cos^2 x - \sin^2 x$ and $g(x) = \cos^2 x + \sin^2 x$. Tell what the maximum values are for each.

9. Sketch the graph of the function $f(x) = \sec x\csc x$. Where is this function undefined?

10. Sketch the graph of the function

$$f(x) = \frac{2\tan x}{1+\tan^2 x}$$

What is the maximum value of this function? What is the period?

11. Sketch the graph of the function $f(x) = \tan x + \cot x$. Where is this function undefined? What is the period?
12. Sketch the graph of the function $f(x) = \cot x - \tan x$. What are the zeros of this function? Is it bounded or unbounded? What is the period?

13–15 all * 4

In Exercises 13–16 find $\sin 2A$, $\cos 2A$, and $\tan 2A$ without using a calculator.

13. $\sin A = \frac{3}{5}$, A in quadrant I
14. $\cos A = -\frac{12}{13}$, A in quadrant III
15. $\tan A = \frac{7}{24}$, A in quadrant III
16. $\sec A = -\frac{13}{5}$, A in quadrant II

In Exercises 17–21 use the half-angle identities to determine the exact value of each expression. Do not use a calculator.

17. $\sin \frac{1}{8}\pi$
18. $\cos \frac{5}{8}\pi$
19. $\tan 157.5°$
20. $\sin 67.5°$
21. $\cos \frac{1}{12}\pi$

In Exercises 22–42 verify each of the identities.

22. $(\sin x + \cos x)^2 = 1 + \sin 2x$
23. $\cos 3x = 4\cos^3 x - 3\cos x$
24. $\sin 4x = 4\cos x \sin x(1 - 2\sin^2 x)$
25. $\tan x + \cot x = 2\csc 2x$
26. $\cos 4x = 8\cos^4 x - 8\cos^2 x + 1$
27. $\cot^2 \frac{1}{2}x = \dfrac{\sec x + 1}{\sec x - 1}$
28. $\cos^4 x = \frac{3}{8} + \frac{1}{2}\cos 2x + \frac{1}{8}\cos 4x$
29. $\sec^2 x = \dfrac{4\sin^2 x}{\sin^2 2x}$
30. $\tan 2x = \dfrac{2}{\cot x - \tan x}$
31. $2\cot 2x \cos x = \csc x - 2\sin x$
32. $\dfrac{1 - \tan x}{1 + \tan x} = \dfrac{1 - \sin 2x}{\cos 2x}$
33. $\tan^2 x + \cos 2x = 1 - \cos 2x \tan^2 x$
34. $\dfrac{2\tan x}{\tan 2x} = 1 - \tan^2 x$
35. $\dfrac{\cos 3x}{\sec x} - \dfrac{\sin x}{\csc 3x} = \cos^2 2x - \sin^2 2x$
36. $\dfrac{\csc x - \sec x}{\csc x + \sec x} = \dfrac{\cos 2x}{1 + \sin 2x}$
37. $1 + \cot x \cot 3x = \dfrac{2\cos 2x}{\cos 2x - \cos 4x}$
38. $-4\sin^3 x + 3\sin x = \sin 3x$
39. $\sin 6x \tan 3x = 2\sin^2 3x$
40. $\dfrac{\cos^3 x + \sin^3 x}{2 - \sin 2x} = \dfrac{1}{2}(\sin x + \cos x)$
41. $\cos 4x \sec^2 2x = 1 - \tan^2 2x$
42. $16\cos^5 x = 5\cos 3x + 10\cos x + \cos 5x$

In Exercises 43–45 find the indicated exact functional value. (Assume that 2θ is in quadrant I.) Do not use a calculator.

43. $\tan\theta$ if $\sin 2\theta = \frac{5}{13}$
44. $\sin\theta$ if $\sin 2\theta = \frac{3}{5}$
45. $\cos\theta$ if $\cos 2\theta = \frac{24}{25}$

In Exercises 46–54 solve each equation on the interval $0 \le x \le 2\pi$. Give the answer in multiples of π.

46. $\sin 2x = \sin x$

47. $\sin x = \cos x$

48. $\sin 2x \sin x + \cos x = 0$

49. $\tan 2x = \tan x$

50. $\cos x - \sin 2x = 0$

51. $\sin 2x + \cos 2x = 0$

52. $\sin 2x - 2 \cos x + \sin x - 1 = 0$

53. $2(\sin^2 2x - \cos^2 2x) = 1$

54. $\sin 2x \cos x - \frac{1}{2} \sin 3x = \frac{1}{2} \sin x$

In Exercises 55–59 show that each expression reduces to 1.

55. $(\sin x + \cos x)^2 - \sin 2x$

56. $\dfrac{\sin 2x \sin x}{2 \cos x} + \cos^2 x$

57. $\sec^4 x - \tan^4 x - 2 \tan^2 x$

58. $\left[\dfrac{\sin 2x}{\sin x} - \dfrac{\cos 2x}{\cos x}\right] \sec x - \tan^2 x$

59. $\cos 2x + \sin 2x \tan x$

In Exercises 60–64 use a calculator to show that the left-hand side of each equation is equal to the right-hand side.

60. $\sin 2(1.05) = 2 \sin 1.05 \cos 1.05$

61. $\cos^2 0.47 = \frac{1}{2}[1 + \cos 2(0.47)]$

62. $\sin^2 1.3 = \frac{1}{2}[1 - \cos 2(1.3)]$

63. $\cos 2(0.22) = \cos^2 0.22 - \sin^2 0.22$

64. $\tan \dfrac{1}{2}(3.0) = \dfrac{\sin 3.0}{1 + \cos 3.0}$

65. The angle of rotation, θ, of the coordinate axes used by a numerically controlled laser saw is given in terms of $\tan 2\theta$. Find $\cos \theta$ and $\sin \theta$ (θ acute) when $\tan 2\theta = 2$.

66. Find the cosine and sine of the acute rotation angle θ in Exercise 65 when $\tan 2\theta = -2$.

67. The horizontal displacement of a projectile fired at an angle θ with initial velocity v_0 is given by

$$R = 32v_0^2 \sin \theta \cos \theta$$

Express R as a function of 2θ. (See Figure 7.12.)

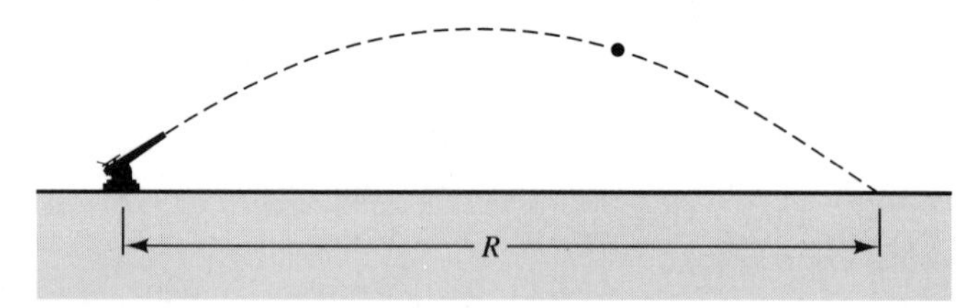

Figure 7.12

68. In the analysis of an electron microscope image, a biologist encounters the expression

$$\sin \frac{\alpha}{2} = \frac{\lambda}{a}$$

Find $\sin \alpha$ for the biologist.

7.4 Sum and Product Formulas

Sum Formulas

Sometimes you will need to write a sum of sines and cosines as a product. Example 1 illustrates an approach based on the addition formulas.

Example 1 Write $\sin 7x + \sin 3x$ as a product of sine and cosine functions.

SOLUTION

$$\begin{aligned}\sin 7x + \sin 3x &= \sin(5x + 2x) + \sin(5x - 2x)\\ &= \sin 5x \cos 2x + \sin 2x \cos 5x + \sin 5x \cos 2x - \sin 2x \cos 5x\\ &= 2 \sin 5x \cos 2x\end{aligned}$$

▲

The method illustrated in Example 1 is called the **average angle method**. The procedure for the general case is as follows:

$$\begin{aligned}\sin A + \sin B &= \sin\left(\frac{A+B}{2} + \frac{A-B}{2}\right) + \sin\left(\frac{A+B}{2} - \frac{A-B}{2}\right)\\ &= \sin\frac{A+B}{2}\cos\frac{A-B}{2} + \cos\frac{A+B}{2}\sin\frac{A-B}{2}\\ &\quad + \sin\frac{A+B}{2}\cos\frac{A-B}{2} - \cos\frac{A+B}{2}\sin\frac{A-B}{2}\end{aligned}$$

Combining terms yields

$$\sin A + \sin B = 2 \sin\frac{A+B}{2}\cos\frac{A-B}{2} \tag{7.14}$$

The following formulas are derived analogously.

$$\sin A - \sin B = 2 \cos\frac{A+B}{2}\sin\frac{A-B}{2} \tag{7.15}$$

$$\cos A + \cos B = 2 \cos\frac{A+B}{2}\cos\frac{A-B}{2} \tag{7.16}$$

$$\cos A - \cos B = -2 \sin\frac{A+B}{2}\sin\frac{A-B}{2} \tag{7.17}$$

Example 2 In calculus the difference quotient of a function f is denoted Δf and is defined by

$$\Delta f = \frac{f(x+h) - f(x)}{h}$$

Find the difference quotient for the sine function and express it as the product of a sine and a cosine.

SOLUTION First compute $\sin(x + h) - \sin x$ using Equation (7.15).

$$\sin(x + h) - \sin x = 2 \cos \frac{(x + h) + x}{2} \sin \frac{(x + h) - x}{2}$$
$$= 2 \sin \frac{h}{2} \cos \frac{2x + h}{2}$$

Thus the difference quotient for the sine function is

$$\frac{\sin(h/2)\cos[x + (h/2)]}{h/2}$$ ▲

Example 3 Prove the identity

$$\frac{\sin 6x - \sin 4x}{\cos 6x + \cos 4x} = \tan x$$

SOLUTION
$$\frac{\sin 6x - \sin 4x}{\cos 6x + \cos 4x} = \frac{2 \cos \frac{1}{2}(6x + 4x)\sin \frac{1}{2}(6x - 4x)}{2 \cos \frac{1}{2}(6x + 4x)\cos \frac{1}{2}(6x - 4x)}$$
$$= \frac{\cos 5x \sin x}{\cos 5x \cos x}$$
$$= \tan x$$ ▲

Example 4 Solve the equation $\sin 3x + \sin x = 0$ on the interval $0 \le x \le \pi$.

SOLUTION To put the given equation in factored form, use Equation (7.14) with $A = 3x$ and $B = x$:

$$\sin 3x + \sin x = 2 \sin 2x \cos x$$

The given equation is then

$$\sin 3x + \sin x = 2 \sin 2x \cos x = 0$$

Now, since $\sin 2A = 2 \sin A \cos A$, we can write this equation as

$$2(2 \sin x \cos x)\cos x = 0$$
$$4 \sin x \cos^2 x = 0$$

Equating each factor to zero, we have

$\sin x = 0$	$\cos^2 x = 0$
$x = 0, \pi$	$x = \frac{1}{2}\pi$

The desired solution set is $x = 0, \frac{1}{2}\pi, \pi$. ▲

Example 5 In the analysis of some types of harmonic motion, the governing equation is $y(t) = A(\cos \omega t - \cos \omega_0 t)$, where the difference between ω and ω_0 is considered to be very small. Make a sketch of the graph of this function.

SOLUTION Using Equation (7.17), we write

$$y(t) = 2A \sin\left(\frac{\omega_0 - \omega}{2}t\right) \sin\left(\frac{\omega_0 + \omega}{2}t\right)$$

If ω is close to ω_0, the oscillation can be interpreted as having a frequency close to $\omega_0/2\pi$ (and close to $\omega/2\pi$) with variable amplitude given by

$$2A \sin \frac{\omega_0 - \omega}{2} t$$

which fluctuates with frequency $(\omega - \omega_0)/\pi$. Oscillations of this type are called **beats**. (See Figure 7.13.)

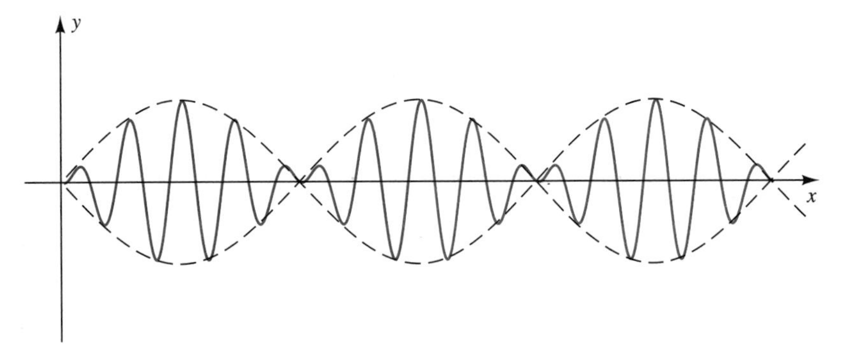

Figure 7.13 ▲

Product Formulas

Sometimes you may need to express products of trigonometric functions as sums of trigonometric functions. If we add the formulas for $\sin(A + B)$ and for $\sin(A - B)$, we obtain

$$\sin A \cos B = \tfrac{1}{2}[\sin(A + B) + \sin(A - B)] \tag{7.18}$$

Subtracting $\sin(A - B)$ from $\sin(A + B)$ and simplifying yields

$$\cos A \sin B = \tfrac{1}{2}[\sin(A + B) - \sin(A - B)] \tag{7.19}$$

In like manner, by first adding and then subtracting the formulas for $\cos(A + B)$ and $\cos(A - B)$, we obtain

$$\cos A \cos B = \tfrac{1}{2}[\cos(A + B) + \cos(A - B)] \tag{7.20}$$

and

$$\sin A \sin B = \tfrac{1}{2}[\cos(A - B) - \cos(A + B)] \qquad (7.21)$$

Example 6 Express $\sin mx \cos nx$ as a sum of functions.

SOLUTION Using Equation (7.18) with $A = mx$ and $B = nx$, we have

$$\begin{aligned}\sin mx \cos nx &= \tfrac{1}{2}[\sin(mx + nx) + \sin(mx - nx)]\\ &= \tfrac{1}{2}[\sin(m + n)x + \sin(m - n)x]\end{aligned}$$

▲

Example 7 Prove that

$$\frac{\sin(x + y) + \sin(x - y)}{\sin(x + y) - \sin(x - y)} = \tan x \cot y$$

SOLUTION

$$\begin{aligned}\frac{\sin(x + y) + \sin(x - y)}{\sin(x + y) - \sin(x - y)} &= \frac{\frac{1}{2}[\sin(x + y) + \sin(x - y)]}{\frac{1}{2}[\sin(x + y) - \sin(x - y)]}\\ &= \frac{\sin x \cos y}{\cos x \sin y}\\ &= \tan x \cot y\end{aligned}$$

▲

Exercises for Section 7.4

In Exercises 1–7 express each sum or difference as a product.

1. $\sin 3\theta + \sin \theta$
2. $\cos 3\alpha - \cos 8\alpha$
3. $\sin 8x + \sin 2x$
4. $\sin \frac{1}{2}x - \sin \frac{1}{4}x$
5. $\cos 50° - \cos 30°$
6. $\sin \frac{3}{4}\pi - \sin \frac{1}{4}\pi$
7. $\sin \frac{3}{4} - \sin \frac{1}{4}$

In Exercises 8–12 express each product as a sum or difference.

8. $\sin 3x \cos x$
9. $\cos x \sin \frac{1}{2}x$
10. $\cos \frac{1}{3}\pi \sin \frac{2}{3}\pi$
11. $\cos 6x \cos 2x$
12. $\sin \frac{1}{4}\pi \sin \frac{1}{12}\pi$

In Exercises 13–18 verify each identity.

13. $\dfrac{\sin x + \sin y}{\cos x + \cos y} = \tan \dfrac{1}{2}(x + y)$
14. $\dfrac{\sin x + \sin y}{\sin x - \sin y} = \dfrac{\tan \frac{1}{2}(x + y)}{\tan \frac{1}{2}(x - y)}$
15. $\dfrac{\cos 3x + \cos x}{\sin 3x + \sin x} = \cot 2x$
16. $\cos 7x + \cos 5x + 2 \cos x \cos 2x = 4 \cos 4x \cos 2x \cos x$
17. $\dfrac{\sin 2x + \sin 2y}{\cos 2x + \cos 2y} = \tan(x + y)$
18. $\dfrac{\sin 9x - \sin 5x}{\sin 14x} = \dfrac{\sin 2x}{\sin 7x}$
19. Find the difference quotient for $\cos x$. (See Example 2, page 283.)

In Exercises 20–23 solve each of the given equations on $0 \le x \le \pi$.

20. $\sin 3x + \sin 5x = 0$

21. $\sin x - \sin 5x = 0$

22. $\cos 3x - \cos x = 0$

23. $\cos 2x - \cos 3x = 0$

24. Let $f(x) = \sin(2x + 1) + \sin(2x - 1)$. Make a sketch of the graph of the function, and give the period and the amplitude.

25. Let $f(x) = \cos(3x + 1) + \cos(3x - 1)$. Make a sketch of the graph of this function, and give the period and the amplitude.

26. Make a sketch of the graph of the function $f(x) = \cos 99x - \cos 101x$.

In Exercises 27–30 use a calculator to show that the left-hand side of each equation is equal to the right-hand side.

27. $\sin 0.3 + \sin 1.2 = 2 \sin \frac{1}{2}(0.3 + 1.2)\cos \frac{1}{2}(0.3 - 1.2)$

28. $\cos 2.05 + \cos 0.72 = 2 \cos \frac{1}{2}(2.05 + 0.72)\cos \frac{1}{2}(2.05 - 0.72)$

29. $\cos 200° - \cos 76° = -2 \sin \frac{1}{2}(200° + 76°)\sin \frac{1}{2}(200° - 76°)$

30. $\sin 15° \cos 29° = \frac{1}{2}[\sin(15° + 29°) + \sin(15° - 29°)]$

31. The Fourier analysis of a vibrating string with a particular initial condition yields an expression whose first two terms are

$$\cos \tfrac{1}{2}x + \cos \tfrac{3}{2}x$$

Using trigonometric identities, express this sum as a product.

32. In the analysis of a half-wave rectified sine wave, a communications engineer encounters the product

$$\sin \tfrac{1}{3}\pi x \sin \tfrac{1}{3}n\pi x$$

Show that this product can be written as a difference of cosines.

Key Topics for Chapter 7

Define and/or discuss each of the following.

Cosine of Sum and Difference
Sine of Sum and Difference
Tangent of Sum and Difference
Double-Angle Formulas
Half-Angle Formulas
Sum and Difference Formulas
Product Formulas

Review Exercises for Chapter 7

In Exercises 1–4 do not use a calculator.

1. Find $\sin(A + B)$ if $\cos A = \frac{3}{5}$, $\sin B = \frac{2}{3}$, A in quadrant I, B in quadrant II.

2. Find $\cos(A + B)$ if $\sin A = \frac{1}{4}$, $\tan B = 1$, A and B in quadrant I.

3. Find $\tan 2A$ if $\cos A = -\frac{4}{5}$ in quadrant II.

4. Find $\sin 2A$ if $\tan A = \frac{1}{2}$ in quadrant III.

In Exercises 5–20 verify each identity.

5. $\cos(x - \frac{1}{3}\pi) = \frac{1}{2}(\cos x + \sqrt{3}\sin x)$
6. $\sin(2x + 1) = \sin 2x \cos 1 + \cos 2x \sin 1$
7. $\tan\left(2x + \frac{1}{4}\pi\right) = \frac{1 + \tan 2x}{1 - \tan 2x}$
8. $2\cot 2x = \cot x - \tan x$
9. $\csc x = \frac{1}{2}\csc \frac{1}{2}x \sec \frac{1}{2}x$
10. $\frac{\cos 2x}{\cos x} = \cos x - \tan x \sin x$
11. $\cos(1 + h) - \cos 1 = -2\sin \frac{1}{2}(2 + h)\sin \frac{1}{2}h$
12. $\sin 4x - \sin 2x = 2\sin x \cos 3x$
13. $\frac{\cos 2x - \cos 4x}{\sin 2x + \sin 4x} = \tan x$
14. $\frac{\cos 3t}{\cos t} = 1 - 4\sin^2 t$
15. $\frac{\sin 3\theta}{\sin\theta} = 3 - 4\sin^2\theta$
16. $\frac{\sin x + \sin 3x}{\cos x + \cos 3x} = \tan 2x$
17. $\sin 4a \sin 2a + \sin^2 a = \frac{1}{2}(1 - \cos 6a)$
18. $\frac{\sin^3 x - \cos^3 x}{\sin x - \cos x} = 1 + \frac{1}{2}\sin 2x$
19. $\cos 4x = 4\cos 2x - 3 + 8\sin^4 x$
20. $\sin\frac{1}{2}A = \frac{\sec A - 1}{2\sin\frac{1}{2}A\sec A}$

In Exercises 21–25 write each of the functions as a cosine and indicate the amplitude, the period, and the phase shift. Sketch the graph.

21. $f(x) = 5\cos 2x - 12\sin 2x$
22. $f(x) = 3\sin x - 4\cos x$
23. $f(x) = \cos 3x - \sin 3x$
24. $f(x) = \sin 3x + \cos 3x$
25. $f(x) = \sqrt{3}\sin x + \cos x$

In Exercises 26–29 find the exact value for sin *θ*. Do not use a calculator.

26. $\sin 2\theta = \frac{1}{4}, 0 \le \theta \le \frac{1}{2}\pi$
27. $\sin 2\theta = -\frac{1}{3}, -\frac{1}{2}\pi \le \theta \le 0$
28. $\tan 2\theta = -0.7, 0 \le \theta \le \frac{1}{2}\pi$
29. $\sin\frac{1}{2}\theta = \frac{4}{5}, \frac{1}{2}\pi \le \theta \le \pi$
30. In the study of Mohr's stress circle in geology, a geologist must solve the two equations

$$\sigma_x \cos\alpha = \tau\sin\alpha + \sigma\cos\alpha$$
$$\sigma_z \sin\alpha = \tau\cos\alpha + \sigma\sin\alpha$$

Show that by solving for σ and τ and simplifying, the geologist ends up with

$$\sigma = \tfrac{1}{2}(\sigma_x + \sigma_z) + \tfrac{1}{2}(\sigma_x - \sigma_z)\cos 2\alpha$$
$$\tau = \tfrac{1}{2}(\sigma_x - \sigma_z)\sin 2\alpha$$

31. Rotation of the coordinate axes through an angle θ can be used to simplify some expressions in analytic geometry and calculus. Given that $\tan 2\theta = -\frac{4}{3}$, find $\sin\theta$ and $\cos\theta$. Find the exact values; do not use your calculator. (Assume that θ is acute.)
32. The voltage across the output terminals of a signal generator is given by $V = 3\cos 2\pi ft - 4\sin 2\pi ft$. Express V as one cosine function and sketch its graph, given that $f = 60$ Hz. Then express V as one sine function.
33. The expression $\sin x + \sin 2x$ represents an approximation to the solution of a problem in harmonic analysis. Write this sum as a product of functions.

▲▼ Practice Test Questions for Chapter 7

In Exercises 1–6 answer *true* or *false*.

1. $\cos 2x = 2 \cos x$.
2. $\sin x = \cos(x - \frac{1}{2}\pi)$.
3. $\cos^2\theta - \sin^2\theta = \cos 2\theta$.
4. $\sin t \cos t = \frac{1}{2} \sin 2t$.
5. $\tan(x + y) = \tan x + \tan y$.
6. The graph of $y = \sin x \cos x$ is bounded by $y = -\frac{1}{2}$ and $\frac{1}{2}$.

In Exercise 7–12 fill in the blank to make the statement true.

7. The graph of $y = 5 \sin x \cos x$ has a period of ____________.
8. If $f(x) = \cos x$ then $f(x + y) =$ ____________.
9. $\sin x \cos y + \cos x \sin y =$ ____________.
10. $\sin A + \sin B = 2 \sin($____________$) \cos ($____________$)$.
11. The amplitude of the graph of $y = 2 \cos 3t + 5 \sin 3t$ is ____________.
12. If $\sin \alpha = \frac{3}{5}$ and $\cos \alpha = \frac{4}{5}$, then $\sin 2\alpha =$ ____________.

In the following exercises solve the stated problem. Show all your work.

13. Verify the identity $\sin 3x = \sin x(3 \cos^2 x - \sin^2 x)$.
14. Express as a cosine function with a phase shift: $y = \sin \frac{1}{2}x + 2 \cos \frac{1}{2}x$.
15. If $\sin 2\theta = \frac{3}{5}$ and θ is in quadrant I, find $\sin \theta$.
16. Determine the exact value of $\cos \frac{1}{8}\pi$.
17. Verify the identity $\cos 2x = \dfrac{1 - \tan^2 x}{1 + \tan^2 x}$
18. Compute the exact value of $\sin 195°$ from the functions of $60°$ and $135°$.
19. Solve $\sin 2x = -2 \sin x$, $0 \leq x < 2\pi$.
20. Solve $4 \sin x - \cos 2x - 5 = 0$, $0 \leq x < 2\pi$.

In Exercises 21–24 verify the identities.

21. $\tan 3x \csc 3x = \sec 3x$
22. $\sec 2x = \dfrac{\sec^2 x}{1 - \tan^2 x}$
23. $\cot 2\phi + \csc 2\phi = \cot \phi$
24. $\tan \frac{1}{2}\theta = \csc \theta - \cot \theta$
25. Sketch one period of the graph of $y = \sqrt{3} \cos 2x + \sin 2x$. What is the amplitude, period and phase shift?
26. Find the exact value of $\sin (A + B)$ if $\cos A = \dfrac{5}{13}$, $\sin B = \dfrac{-\sqrt{3}}{2}$, A and B in quadrant IV.
27. Find the exact value of $\sin 2\theta$ if $\tan \theta = -\frac{3}{4}$ in quadrant II.

The Inverse Trigonometric Functions

The statement $\frac{1}{2} = \sin 30°$ means that $\frac{1}{2}$ is the sine of 30°. If we write $\sin x = \frac{1}{2}$, we understand that we wish to find the angle x whose sine is $\frac{1}{2}$. In this chapter we will introduce and discuss the inverse trigonometric functions as a way of writing $\sin x = \frac{1}{2}$, which will emphasize the fact that we know the sine of an angle x is $\frac{1}{2}$ and we wish to find the angle x. The difficulty with solving $\sin x = \frac{1}{2}$ for x is that there are infinitely many solutions; for instance, 30°, 150°, −210°, and 390° are four of the solutions. To overcome this nonuniqueness of the solution of trigonometric equations, we agree to limit the domain of the trigonometric functions so that there will be only one solution of $\sin x = \frac{1}{2}$. Section 8.1 describes this process for functions in general, and Section 8.2 shows how it is done for the trigonometric functions to generate the inverse trigonometric functions. The inverse trigonometric functions are discussed from a view of reversing the role of the range and the domain of trigonometric functions. Scientific calculators have the inverse sine, inverse cosine, and inverse tangent as a part of the accessible functions.

8.1 Relations, Functions, and Inverses

Functions and Relations

The idea of a function, which was introduced in Chapter 1, is the basis for our discussion of inverse trigonometric functions, so we will review some of the main points before we introduce the main topic. Recall that a function f is a pairing of numbers from a set X called the **domain** to a set Y called the **range**. Central to the concept of function is the rule of correspondence which specifies that the pairing must be unique; that is, for each x in X there is exactly one value y in Y. The general pairing indicated by f is written

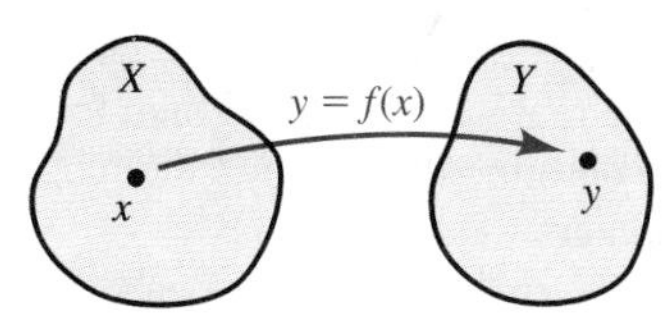

Figure 8.1

$$y = f(x)$$

and is depicted as shown in Figure 8.1. The figure emphasizes the idea that the pairings are ordered; that is, each y is obtained from an x. The figure also shows that the functional pairing is unique; that is, only one value of y is paired with each x.

Functions are frequently expressed by a formula such as $y = 3x^2$, but other means may be used. For instance, a set of ordered pairs $\{(x, y)\}$ is a function if no two distinct pairs have the same first element. This follows from the fact that a function assigns to each x in X a unique y in Y. Thus the set $\{(2, 3), (-1, 4), (0, -5), (3, 4)\}$ is a function, since each first element is paired with a unique second element. The following illustrations clarify the important points of the definition of a function.

- A formula such as $y = \pm\sqrt{x}$ does *not* define y as a function of x because it assigns two values to each nonzero value of x. For example, $+2$ and -2 are both images of $x = 4$. However, the formulas $y = \sqrt{x}$ and $y = -\sqrt{x}$ taken separately do define functions.
- The expression $y = 8$ defines a function because y has the value 8 for every value of x. The definition does not require that y have a different value for each x, only that the pairing be unique.
- The expression $x = 5$ is not a function because many values of y correspond to $x = 5$. For example, $(5, -1)$, $(5, 0)$, and $(5, 3)$ are some of the ordered pairs that satisfy the expression $x = 5$.
- The set $\{(-3, 4), (2, 5), (2, -6), (9, 7)\}$ is not a function because the distinct pairs $(2, 5)$ and $(2, -6)$ have the same first element. Therefore two different values of y are assigned to the same x.

COMMENT Rules of correspondence that permit more than one value of y to be paired with a value of x are called *relations*. Hence $y = \pm\sqrt{x}$, $x = 5$, and $\{(-3, 4), (2, 5), (2, -6), (9, 7)\}$ are relations. Note that every function is a relation, but a relation is not necessarily a function.

The domain of a function can be quite arbitrary. *If the domain is not specified, we assume that it consists of all real numbers for which the rule of correspondence will yield a real number.* Examples 2 and 3 show functions with restricted domains.

Example 1 The equation $y = x^2 + 5$ defines a function since each value of x determines only one value of y. The domain consists of all real numbers, and the range consists only of those real numbers greater than or equal to 5 (since the smallest possible value of x^2 is 0). ▲

Example 2 Find the domain and range of the function $y = \sqrt{x}$.

SOLUTION If we substitute a negative real number for x in $y = \sqrt{x}$, we do not get a real number for y. However, each nonnegative real number substituted for x yields a nonnegative real number for y. Therefore both the domain and the range of this function consist of all nonnegative real numbers. ▲

Example 3 Find the domain and range of the function $y = \dfrac{4}{x-3}$.

SOLUTION Since division by zero is not allowed, we must exclude 3 as a domain element; we conclude that the domain consists of all real numbers except 3.

To find the range of this function, we solve for x and note any restrictions on y. Thus

$$y = \frac{4}{x-3}$$

$$xy - 3y = 4 \qquad \text{Multiply both sides by } x - 3$$

$$xy = 3y + 4 \qquad \text{Add } 3y \text{ to both sides}$$

$$x = \frac{3y+4}{y} \qquad \text{Divide both sides by } y$$

The only limitation on y is that it cannot equal 0. Therefore the range consists of all real numbers except 0. ▲

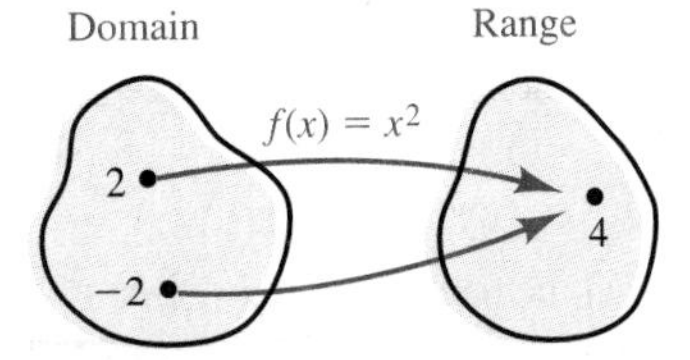

Figure 8.2

Inverse Functions

A number in the range of a function may correspond to more than one number in its domain. For example, if $f(x) = x^2$, then both $f(2) = 4$ and $f(-2) = 4$. This is illustrated in Figure 8.2.

If every number in the domain is assigned a different number in the range, the function is called a **one-to-one function**.

DEFINITON 8.1 A function f is one-to-one if, for a and b in the domain of f, $f(a) = f(b)$ implies that $a = b$.

The function $y = x^2$ is *not* one-to-one, because $f(2) = 4 = f(-2)$, but $2 \neq -2$.

Recall from Section 1.4 that if f is a function, a vertical line will intersect its graph in at most one point. A one-to-one function is one for which *both* horizontal *and* vertical lines intersect the graph in, at most, one point. In Figure 8.3 the function on the left is one-to-one and the function on the right is not.

If f is a one-to-one function, we know that each element in the range corresponds to a unique element in the domain. Under this condition it is possible to define a function f^{-1} as follows.

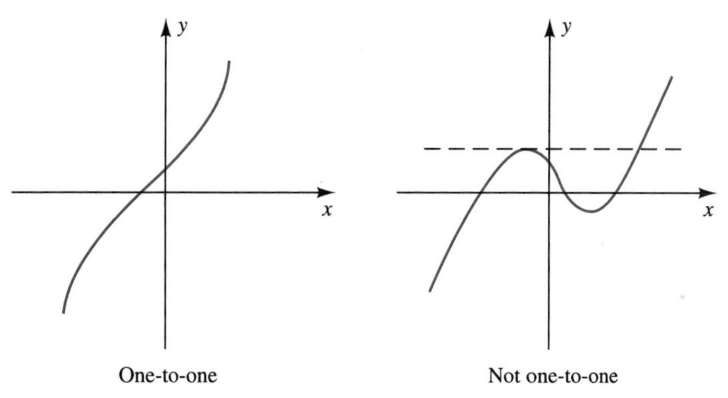

Figure 8.3

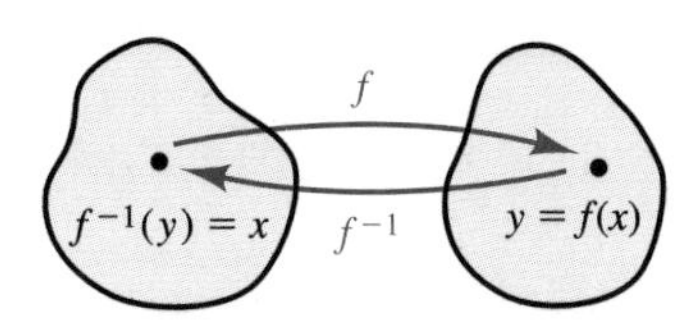

DEFINITON 8.2 If f is one-to-one, then f^{-1} is the function obtained by interchanging the numbers in the functional pairing (x, y). Under this transformation the domain of f^{-1} is the range of f and the range of f^{-1} is the domain of f. Thus $f^{-1}(x) = y$ if and only if $f(y) = x$. The function f^{-1} is called the inverse function of f. The relationship between f and f^{-1} is shown in Figure 8.4.

WARNING The notation f^{-1} does not mean f to a power of -1; that is,

$$f^{-1}(x) \neq \frac{1}{f(x)}$$

Example 4 The one-to-one function $f = \{(2, 1), (-3, 2), (0, 5)\}$ has the inverse $f^{-1} = \{(1, 2), (2, -3), (5, 0)\}$. ▲

The algebraic procedure for finding the inverse of a function given by the formula $y = f(x)$ is summarized as follows:

1. Interchange the x and y variables.
2. Solve the new equation for the y variable. The resulting expression is $f^{-1}(x)$.

Example 5 If $f(x) = 2x + 4$, find $f^{-1}(x)$.

SOLUTION Let $y = 2x + 4$, and then interchange x and y to get

$$x = 2y + 4$$

Solving this equation for y yields

$$2y = x - 4$$
$$y = \tfrac{1}{2}x - 2$$

Substituting $f^{-1}(x)$ for y, we find that the inverse of $f(x) = 2x + 4$ is

$$f^{-1}(x) = \tfrac{1}{2}x - 2.$$

▲

Example 6 Let

$$f(x) = \frac{2x + 1}{x + 3}$$

Find the inverse of f.

SOLUTION Let

$$y = \frac{2x + 1}{x + 3}$$

Then interchange the x and y variables to get

$$x = \frac{2y + 1}{y + 3}$$

Solve for y:

$$xy + 3x = 2y + 1$$
$$xy - 2y = 1 - 3x$$
$$y(x - 2) = 1 - 3x$$
$$y = \frac{1 - 3x}{x - 2}$$

Hence the inverse of f is

$$f^{-1}(x) = \frac{1 - 3x}{x - 2}$$

▲

Notice that the domain of the inverse function in Example 6 is all real numbers except $x = 2$. Restrictions of this type should always be specified when inverse functions are computed. Furthermore, the domain of f is the range of f^{-1}, and the range of f is the domain of f^{-1}. Thus the function and its inverse are obtained by interchanging the domain and range and keeping the same correspondence.

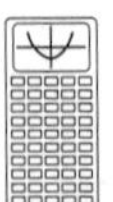

Some calculators have an inv button, but this button can be used only to perform a few selected preprogramed functions and not the inverse operation in general.

A function that is not one-to-one does not have an inverse. Sometimes we restrict the domain of such a function to a special interval in order to make it one-to-one, thereby allowing the possibility of an inverse function.

Example 7 The function $y = x^2 - 1$ is not a one-to-one function because a horizontal line will intersect its graph in two places for all $y > -1$. See Figure 8.5(a). However, if we restrict the domain to the interval $x \geq 0$, the function is one-to-one. We can compute the inverse function on this interval by interchanging x and y in the original function to get $x = y^2 - 1$. Solving for y yields the inverse function $y = \sqrt{x + 1}$. Figure 8.5(b) shows the graph of the inverse function as a dashed line.

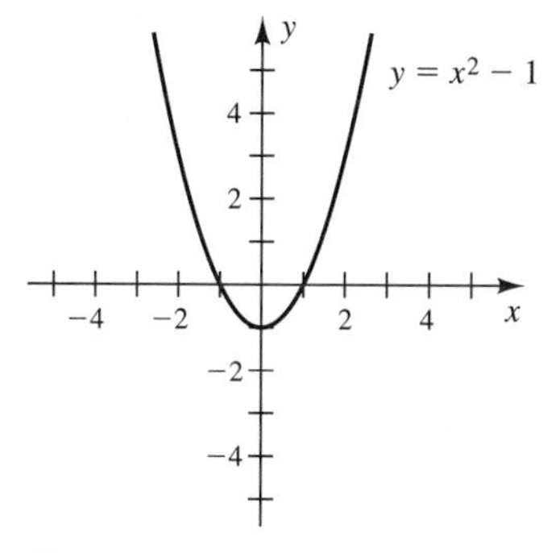

Figure 8.5 ▲

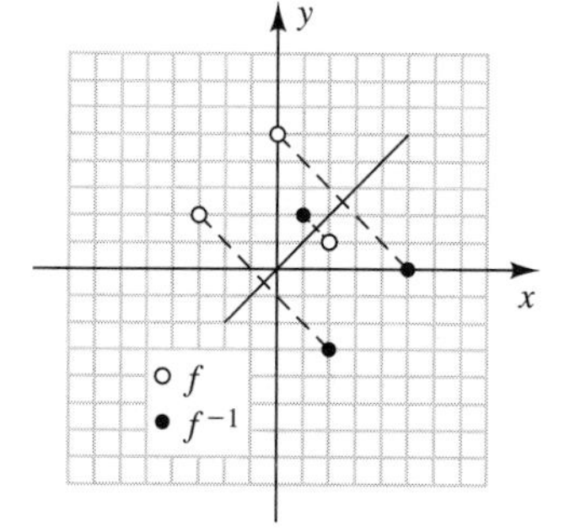

Figure 8.6

The graphs of f and f^{-1} have an interesting relationship. As the next two examples show, the graphs of f and f^{-1} are mirror reflections of each other in the line that bisects the first and third quadrants. This is because of the fact that if (x, y) is in f, then (y, x) is in f^{-1}.

Example 8 Draw the graphs of $f = \{(2, 1), (-3, 2), (0, 5)\}$ and its inverse function $f^{-1} = \{(1, 2), (2, -3), (5, 0)\}$ on the same coordinate axes.

SOLUTION The second set of ordered pairs is the inverse of the first. The graphs of f and f^{-1} are shown in Figure 8.6. You can see that the points in the graphs are mirror reflections in the 45° line. ▲

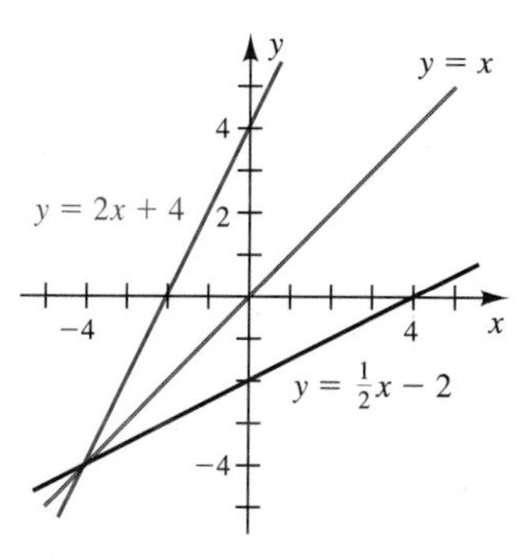

Figure 8.7

Example 9 Draw the graphs of $y = 2x + 4$ and $y = \frac{1}{2}x - 2$ on the same coordinate axes.

SOLUTION The inverse nature of these functions was established in Example 5. Several solution pairs for each equation are given in the tables below, and the graphs are plotted in Figure 8.7. Notice that the graphs are mirror reflections in the 45° line.

$y = 2x + 4$

x	−6	−4	−2	0	2
y	−8	−4	0	4	8

$y = \frac{1}{2}x - 2$

x	−8	−4	0	4	8
y	−6	−4	−2	0	2

▲

Exercises for Section 8.1

1. If $f(x) = 2x + 3$, for which x is $f(x) = 6$?
2. If $f(x) = \dfrac{4}{x}$, for which x is $f(x) = 3$?
3. If $v = 10 + 4t$, for which t is $v = 20$?
4. If $f = \{(3, 0), (2, 6), (1, 5)\}$, for which x is $f(x) = 5$? For which x is $f(x) = 0$?
5. If $f = \{(0, 1), (2, 5), (3, 8), (5, 6)\}$, for which x is $f(x) = 8$? For which x is $f(x) = 3$?

In Exercises 6–21 determine the inverse function (if one exists).

6. $\{(2, 6), (3, 5), (0, 4)\}$
7. $\{(3, 7), (5, 9), (7, 3), (9, 5)\}$
8. $\{(-1, 2), (2, 3), (6, -2)\}$
9. $\{(1, 2), (2, 2)\}$
10. $\{(3, 2), (5, 4), (7, 2), (9, 8)\}$
11. $\{(-2, 3), (-1, 4), (0, 0)\}$
12. $\{(0, 1), (1, 0)\}$
13. $\{(0, 3), (1, 5), (2, 3), (6, 7)\}$
14. $f(x) = 3x + 2$
15. $f(x) = x - 3$
16. $f(x) = \frac{1}{2}x + 5$
17. $f(x) = \dfrac{1}{x + 1}$
18. $f(x) = x^2 + 2x$
19. $f(x) = 2 - 3x^2$
20. $f(x) = \dfrac{2x - 1}{3x + 5}$
21. $f(x) = \dfrac{x - 1}{x + 1}$
22. The function $f(x) = x^2 - 4x$ is not one-to-one and hence does not have an inverse. Show how to define an inverse on a restricted domain.

In Exercises 23–26 determine which functions whose graphs are shown have inverses that are functions. Sketch the graph of the inverse where one exists.

23.

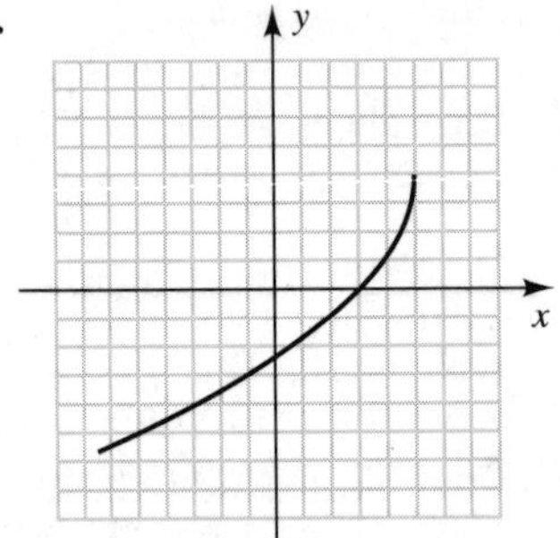

24.

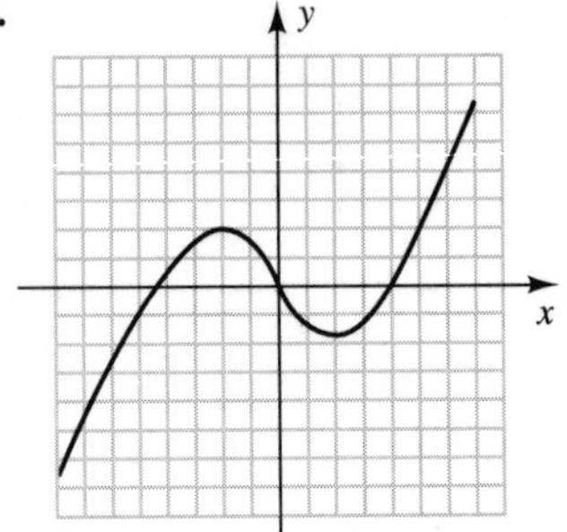

25.

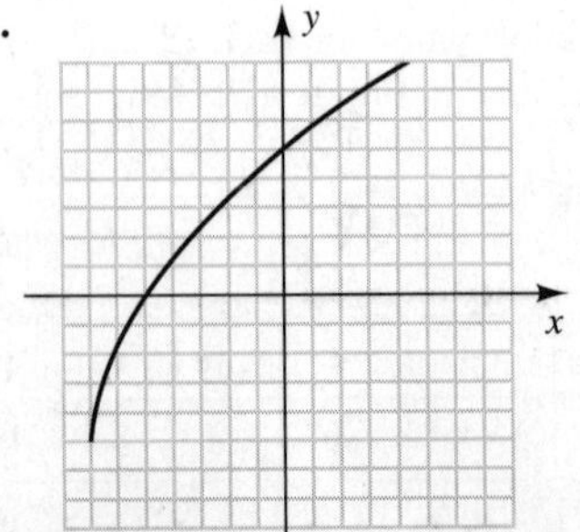

26.

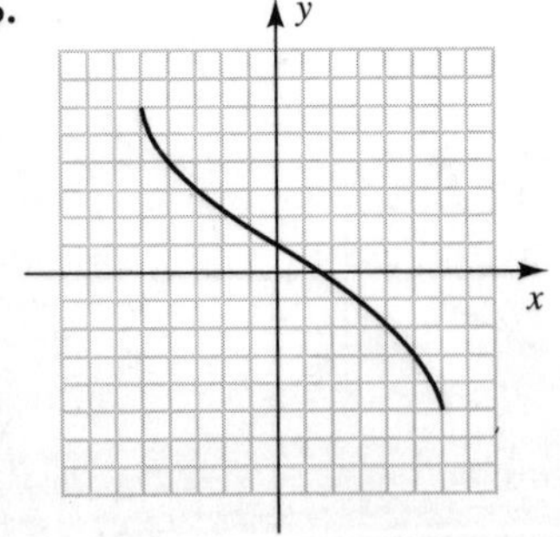

In Exercises 27–34 determine graphically which pairs of functions are inverses.

27. $y = -2x$, $y = -\frac{1}{2}x$

28. $y = 3x$, $y = \frac{1}{3}x$

29. $y = x + 1$, $y = x - 2$

30. $y = 6 - 3x$, $y = -\frac{1}{3}x + 2$

31. $y = 5(x + 1)$, $y = \frac{1}{5}x - 1$

32. $y = \frac{1}{2}x + 1$, $y = 2x - 1$

33. $y = x^3$, $y = \sqrt[3]{x}$

34. $y = x^2$, $y = \sqrt{x}$

8.2 The Inverse Trigonometric Functions

You learned in Section 5.1 that $y = \sin x$ can be treated as a function whose domain is the set of all real numbers and whose range is the interval $[-1, 1]$. For instance, $y = \sin \frac{1}{6}\pi = \frac{1}{2}$. Sometimes instead of computing the value of the function for a given domain element, we wish to determine the domain element that corresponds to a given functional value. For example, we might want the value of x for which $\sin x = \frac{1}{2}$. Unfortunately, since $\sin x$ is not a one-to-one function, $x = \frac{1}{6}\pi$ is not the only solution. In fact, there are infinitely many solutions, some of which are shown in Figure 8.8.

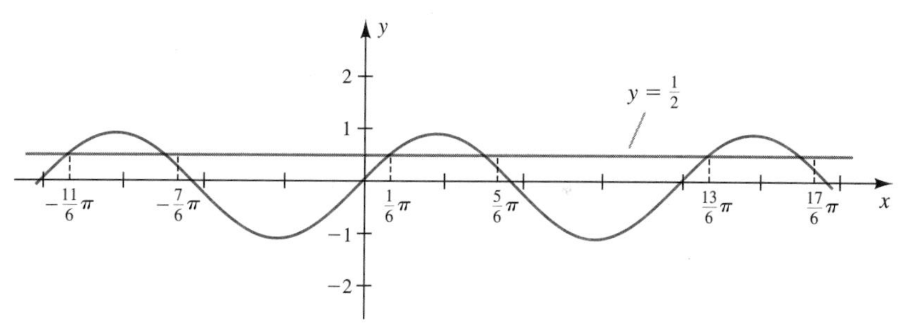

Figure 8.8
Solutions to $\sin x = \frac{1}{2}$

The fact that $-\frac{7}{6}\pi$, $\frac{1}{6}\pi$, $\frac{5}{6}\pi$, $\frac{13}{6}\pi$, and so on all satisfy $\sin x = \frac{1}{2}$ causes certain difficulties. For example, if you were asked to design a calculator to solve $\sin x = \frac{1}{2}$, what value would you have the calculator display as the solution? To avoid ambiguity, it is customary to restrict the domain of $\sin x$ to the interval $[-\frac{1}{2}\pi, \frac{1}{2}\pi]$ so that $\sin x$ is a one-to-one function. With this added restriction, $x = \frac{1}{6}\pi$ is the only value for which $\sin x = \frac{1}{2}$.

The situation described for the sine function is true for each trigonometric function. Hence we restrict the domain so that each of the six functions has one and only one domain value for each range value. The interval to which the domain of the trigonometric function is restricted in order to make the function one-to-one is called the **principal value interval** of the function. The principal value intervals for each of the six trigonometric functions are given in Table 8.1.

Table 8.1 Table of principal value intervals of the trigonometric functions

sin x	$-\frac{1}{2}\pi \le x \le \frac{1}{2}\pi$	cot x	$0 < x < \pi$
cos x	$0 \le x \le \pi$	csc x	$-\frac{1}{2}\pi \le x \le \frac{1}{2}\pi,\ x \ne 0$
tan x	$-\frac{1}{2}\pi < x < \frac{1}{2}\pi$	sec x	$0 \le x \le \pi,\ x \ne \frac{1}{2}\pi$

Example 1 Find the principal value of x for which **(a)** $\cos x = -\frac{1}{2}$ and **(b)** $\sin x = -\frac{1}{2}$.

SOLUTION Figure 8.9(a) shows the cosine function intersecting the line $y = -\frac{1}{2}$ and Figure 8.9(b) shows the sine function intersecting the line $y = -\frac{1}{2}$. In each case the principal value interval is printed in color so that you can see that only one value of x on that interval satisfies the equation. Thus the principal value of x for which $\cos x = -\frac{1}{2}$ is $x = \frac{2}{3}\pi$. The principal value of x for which $\sin x = -\frac{1}{2}$ is $x = -\frac{1}{6}\pi$.

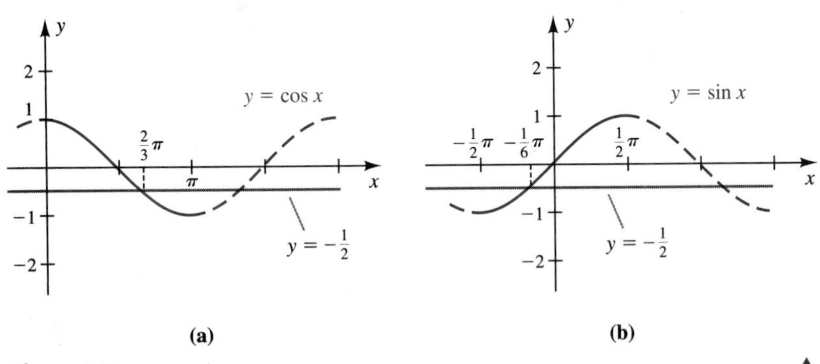

Figure 8.9 ▲

Since $\sin x$ is a one-to-one function on its principal value interval, it has an inverse function on that interval.

DEFINITION 8.3 The inverse sine function, denoted by **arcsin** or **$\sin^{-1}$**, is defined by

$$y = \sin^{-1}x \qquad \text{if and only if} \qquad \sin y = x$$

where $-1 \le x \le 1$ and $-\frac{1}{2}\pi \le y \le \frac{1}{2}\pi$.

The notation $y = \arcsin x$ or $y = \sin^{-1}x$ means "y is the arc length whose sine is x" and is usually read "the inverse sine of x." We will use $\arcsin x$ and $\sin^{-1}x$ interchangeably throughout this chapter.

WARNING

$$y = \sin^{-1}x \text{ does } NOT \text{ mean } y = \frac{1}{\sin x}.$$

Example 2 Evaluate **(a)** $y = \arcsin \dfrac{\sqrt{3}}{2}$ and **(b)** $y = \sin^{-1}(\cos \frac{1}{3}\pi)$.

SOLUTION

(a) Finding y such that $y = \arcsin \sqrt{3}/2$ is the same as finding the principal value of the angle y for which $\sin y = \sqrt{3}/2$. Recalling the special-angle values, we see that $y = \arcsin \sqrt{3}/2 = \frac{1}{3}\pi$. The equation $y = \arcsin \sqrt{3}/2$ is equivalent to $\sin y = \sqrt{3}/2$ if y is restricted to its principal value interval.

(b) Since $\cos \frac{1}{3}\pi = \frac{1}{2}$, we wish to find y such that $\sin y = \frac{1}{2}$ for $-\frac{1}{2}\pi \le y \le \frac{1}{2}\pi$. Hence

$$y = \sin^{-1}(\cos \tfrac{1}{3}\pi) = \sin^{-1}(\tfrac{1}{2}) = \tfrac{1}{6}\pi$$ ▲

The definition of each of the other five inverse trigonometric functions parallels that of the inverse sine function. Table 8.2 lists the domain and principal value interval for each of the inverse trigonometric functions.

Table 8.2 The inverse trigonometric functions

Function	Domain	Principal Value Interval
$y = \sin^{-1}x$ $(x = \sin y)$	$-1 \le x \le 1$	$-\frac{1}{2}\pi \le \sin^{-1}x \le \frac{1}{2}\pi$
$y = \cos^{-1}x$ $(x = \cos y)$	$-1 \le x \le 1$	$0 \le \cos^{-1}x \le \pi$
$y = \tan^{-1}x$ $(x = \tan y)$	$-\infty < x < \infty$	$-\frac{1}{2}\pi < \tan^{-1}x < \frac{1}{2}\pi$
$y = \cot^{-1}x$ $(x = \cot y)$	$-\infty < x < \infty$	$0 < \cot^{-1}x < \pi$
$y = \sec^{-1}x$ $(x = \sec y)$	$x \le -1$ or $x \ge 1$	$0 \le \sec^{-1}x \le \pi$, $\sec^{-1}x \ne \frac{1}{2}\pi$
$y = \csc^{-1}x$ $(x = \csc y)$	$x \le -1$ or $x \ge 1$	$-\frac{1}{2}\pi \le \csc^{-1}x \le \frac{1}{2}\pi$, $\csc^{-1}x \ne 0$

Study Table 8.2 carefully so that you will know what limitations apply to the various inverse trigonometric functions. The information in this table is fundamental to the examples and exercises that follow.

COMMENT The choice of principal values is a matter of convention, so other principal value intervals are possible. For example, some authors use $(0, \frac{1}{2}\pi] \cup (\pi, \frac{3}{2}\pi]$ for $\csc^{-1}x$ and $[0, \frac{1}{2}\pi) \cup [\pi, \frac{3}{2}\pi)$ for $\sec^{-1}x$.

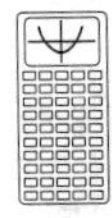

Calculators display the principal values of the inverse trigonometric functions. This is why your calculator, when in the degree mode, gives

$$\cos^{-1}(-0.5) = \boxed{\text{2nd}}\ \boxed{\cos}\ -0.5\ \boxed{=}\ 120^\circ \qquad \textbf{Second quadrant}$$

and

$$\sin^{-1}(-0.5) = \boxed{\text{2nd}}\ \boxed{\sin}\ -0.5\ \boxed{=}\ -30^\circ \qquad \textbf{Fourth quadrant}$$

Example 3 Evaluate the following without using a calculator.
(a) $\cos^{-1}(0.5)$ **(b)** $\cos^{-1}(-0.5)$ **(c)** $\arcsin(-2)$

SOLUTION

(a) Let $y = \cos^{-1}(0.5)$. Then

$$0.5 = \cos y \qquad \text{where } 0 \le y \le \pi$$

The angle or number whose cosine is 0.5 is $\frac{1}{3}\pi$. Thus $\cos^{-1}(0.5) = \frac{1}{3}\pi$.

(b) Let $y = \cos^{-1}(-0.5)$. Then

$$-0.5 = \cos y \qquad \text{where } 0 \le y \le \pi$$

The angle or number whose cosine is -0.5 is $\frac{2}{3}\pi$. Thus $\cos^{-1}(-0.5) = \frac{2}{3}\pi$.

(c) Let $y = \arcsin(-2)$. Then y is undefined because -2 is not in the domain of $\arcsin x$. ▲

Example 4 Find $\arctan(-1)$ and $\text{arccot}(-1)$.

SOLUTION Let $y = \arctan(-1)$ and $u = \text{arccot}(-1)$. Then

$$-1 = \tan y, \text{ where } -\tfrac{1}{2}\pi < y < \tfrac{1}{2}\pi \qquad \text{and} \qquad -1 = \cot u, \text{ where } 0 < u < \pi$$

Thus,

$$y = \arctan(-1) = -\tfrac{1}{4}\pi \qquad \text{and} \qquad u = \text{arccot}(-1) = \tfrac{3}{4}\pi$$ ▲

COMMENT Since the tangent and cotangent are reciprocal functions, you might have expected that $\arctan(-1)$ and $\text{arccot}(-1)$ would yield the same value. The previous example shows the necessity of strictly adhering to the definitions of the inverse functions, giving close attention to the principal value interval.

Most calculators are capable of displaying values of the inverse sine, inverse cosine, and inverse tangent functions. These functions are usually activated by pressing an $\boxed{\text{inv}}$ or $\boxed{\text{2nd}}$ key, followed by the appropriate trigonometric function key. The values of arccot, arcsec, and arccsc are obtained from these by first using the $\boxed{1/x}$ or $\boxed{x^{-1}}$ key.

Example 5 Find $\sec^{-1}2.53$ to the nearest tenth of a degree.

SOLUTION Let $\theta = \sec^{-1}2.53$. Then

$$\theta \approx \cos^{-1}\frac{1}{2.53} \approx 66.7^\circ$$

The keystroke sequence for θ is

[2nd] [cos] 2.53 [1/x] [=] [66.71800172] ▲

The following examples concern trigonometric functions of some inverse trigonometric functions. In evaluating such functions it helps to show the functional value by drawing a right triangle, always keeping track of the principal value interval.

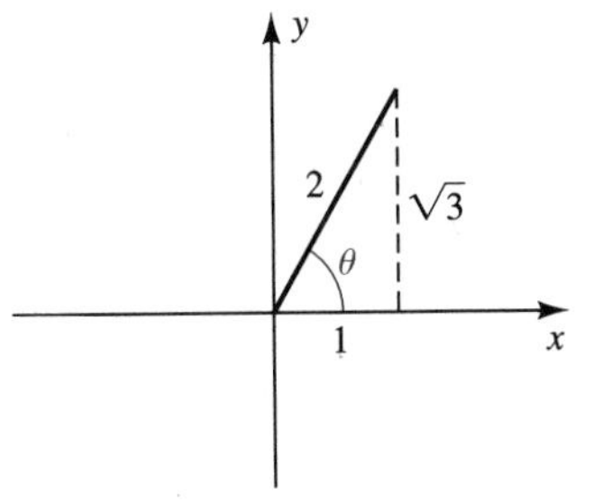

Figure 8.10

Example 6 Find $\sin(\cos^{-1}\frac{1}{2})$ without using a calculator.

SOLUTION We first let $\theta = \cos^{-1}\frac{1}{2}$. Then θ is the angle shown in Figure 8.10, from which it is easy to see that

$$\sin(\cos^{-1}\tfrac{1}{2}) = \sin\theta = \frac{\sqrt{3}}{2}$$ ▲

Example 7 Find $\cos(\sin^{-1}x)$ in terms of x.

SOLUTION We want to find $\cos\theta$, where $\theta = \sin^{-1}x$. Since the range values of $\sin^{-1}x$ are $[-\frac{1}{2}\pi, \frac{1}{2}\pi]$, the angle θ will be one of the angles shown in Figure 8.11. In either case, $\cos\theta = \sqrt{1 - x^2}$.

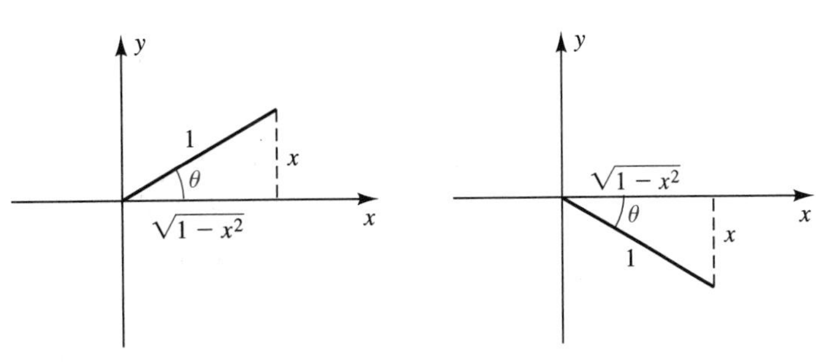

Figure 8.11 ▲

Example 8 Find $\sin[\arccos\frac{1}{3} + \arctan(-2)]$ without using a calculator.

SOLUTION We let $\theta = \arccos\frac{1}{3}$ and $\phi = \arctan(-2)$. See Figure 8.12. Can you tell why θ is drawn as a first-quadrant angle and ϕ as a fourth-quadrant angle?

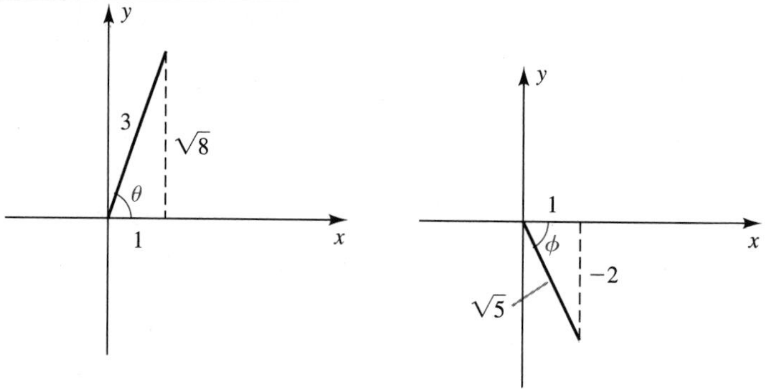

Figure 8.12

Since $\sin(\theta + \phi) = \sin\theta\cos\phi + \sin\phi\cos\theta$,

$$\sin\left(\arccos\frac{1}{3} + \arctan(-2)\right) = \frac{\sqrt{8}}{3}\left(\frac{1}{\sqrt{5}}\right) + \frac{-2}{\sqrt{5}}\left(\frac{1}{3}\right)$$
$$= \frac{-2 + \sqrt{8}}{3\sqrt{5}}$$

▲

Example 9 Verify the identity $\cos(2\cos^{-1}x) = 2x^2 - 1$.

SOLUTION Letting $\theta = \cos^{-1}x$, we have

$$\cos(2\cos^{-1}x) = \cos 2\theta = \cos^2\theta - \sin^2\theta$$

From Figure 8.13 we see that the value of $\cos\theta$ is x and the value of $\sin\theta$ is $\sqrt{1 - x^2}$. Hence

$$\cos(2\cos^{-1}x) = x^2 - (1 - x^2) = 2x^2 - 1$$

▲

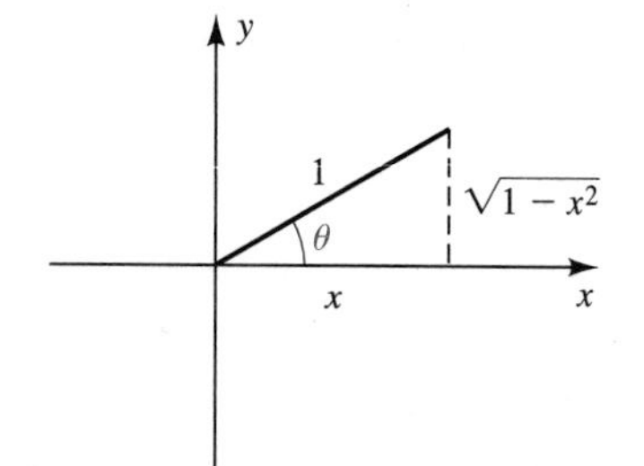

Figure 8.13

Example 10 Find x if

(a) $\cos^{-1}x = 0.578$ **(b)** $\cos^{-1}x = 2\pi$

SOLUTION

(a) If $\cos^{-1}x = 0.578$, then $x = \cos 0.578$, if 0.578 is within the principal value interval $[0, \pi]$. Since it is within this interval,

$$x = \cos 0.578 \approx 0.8376$$

(b) Since 2π is not in the principal value interval for the cosine function, the equation $\cos^{-1}x = 2\pi$ has no solution. If you were not on the lookout for the principal value interval, you might conclude that since $\cos^{-1}x = 2\pi$, $x = \cos 2\pi = 1$, which is incorrect.

▲

In passing we note that

$$\sin(\sin^{-1}x) = x$$
$$\cos(\cos^{-1}x) = x$$

and so forth, for all six trigonometric functions. This result is a direct consequence of the inverse nature of the functions involved. Also, *if x is limited to the principal value interval* of the function, then

$$\sin^{-1}(\sin x) = x$$
$$\cos^{-1}(\cos x) = x$$

and so forth. For example, $\sin^{-1}(\sin \frac{1}{4}\pi) = \frac{1}{4}\pi$, since $\frac{1}{4}\pi$ is within the principal value interval of the $\sin^{-1}$ function. But $\sin^{-1}(\sin \frac{5}{6}\pi) = \frac{1}{6}\pi$, since $\frac{5}{6}\pi$ is not within the principal value interval.

Exercises for Section 8.2

In Exercises 1–25 find the exact value of each expression without using a calculator. Express the value as a real number (not in degrees).

1. $\sin^{-1}\frac{1}{2}$

2. $\sin^{-1}1$

3. $\tan^{-1}1$

4. $\cos^{-1}\dfrac{\sqrt{3}}{2}$

5. $\arccos\dfrac{-\sqrt{3}}{2}$

6. $\sin^{-1}\dfrac{-\sqrt{2}}{2}$

7. $\sec^{-1}(-2)$

8. $\cot^{-1}(-\sqrt{3})$

9. $\sec^{-1}1$

10. $\operatorname{arccsc}\sqrt{2}$

11. $\tan^{-1}(-\sqrt{3})$

12. $\sin[\cos^{-1}(-\frac{3}{5})]$

13. $\cos[\sin^{-1}(-\frac{5}{13})]$

14. $\sin(\sin^{-1}1)$

15. $\sin(\tan^{-1}2)$

16. $\sec(\cos^{-1}\frac{1}{3})$

17. $\cos(\arcsin\frac{1}{4})$

18. $\sin(2\arcsin\frac{1}{3})$

19. $\cos(2\sin^{-1}\frac{1}{4})$

20. $\cos[\cos^{-1}(-\frac{1}{3}) - \sin^{-1}(-\frac{1}{3})]$

21. $\sin(\tan^{-1}1 - \tan^{-1}0.8)$

22. $\cos[\sin^{-1}\frac{1}{4} + \cos^{-1}(-\frac{1}{3})]$

23. $\tan[\sin^{-1}(-\frac{1}{2}) - \tan^{-1}(-2)]$

24. $\cos(\tan^{-1}2 - \cos^{-1}\frac{1}{2})$

25. $\tan(2\tan^{-1}2)$

In Exercises 26–30 simplify the given expression.

26. $\sin(\cos^{-1}x^2)$

27. $\tan(\sin^{-1}x)$

28. $\tan(2\cos^{-1}x)$

29. $\sin(\arcsin y + \arcsin x)$

30. $\cos(\arccos x + \arcsin y)$

In Exercises 31–40 use a calculator to solve for x or tell why there is no solution.

31. $\cos^{-1}x = 0.241$

32. $\sin^{-1}x = -0.314$

33. $\cos^{-1}x = -0.5$

34. $\operatorname{arccsc} x = -\pi$

35. $\arctan x = 1.2$

36. $\tan^{-1}x = 2.43$

37. $\cot^{-1}x = 1.34$

38. $\cos^{-1}x = \frac{3}{4}\pi$

39. $\operatorname{arccsc} x = 2.8947$

40. $\sec^{-1}x = 2.815$

In Exercises 41–45 verify the given identity.

41. $\cos(2\sin^{-1}x) = 1 - 2x^2$

42. $\tan(\tan^{-1}x + \tan^{-1}1) = \dfrac{1+x}{1-x}$

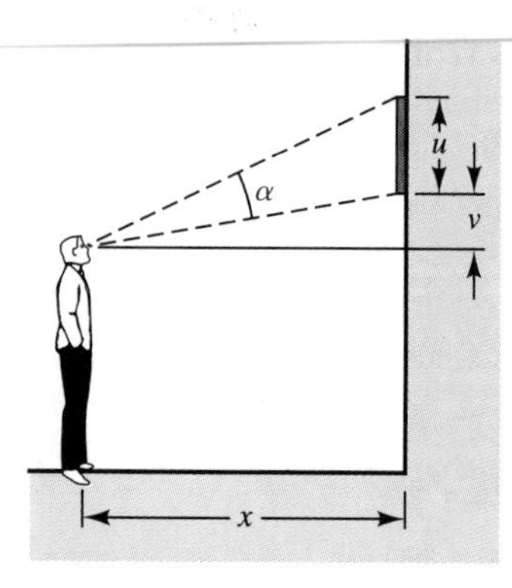

Figure 8.14

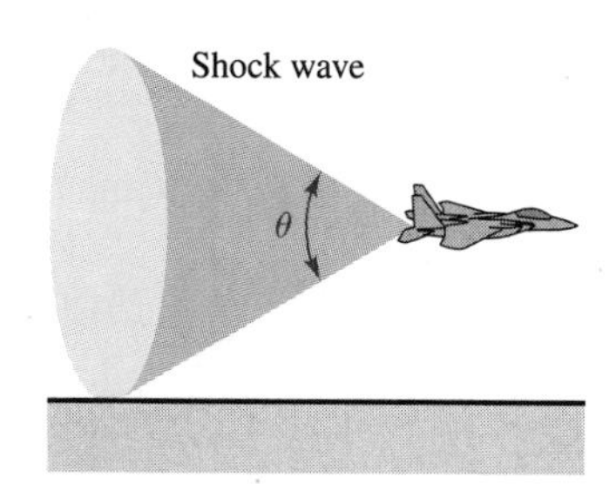

Figure 8.15

43. $\tan(2\tan^{-1}x) = \dfrac{2x}{1-x^2}$

44. $\sin(3\sin^{-1}\theta) = 3\theta - 4\theta^3$

45. $\cos(2\arccos y) = 2y^2 - 1$

46. Verify that $\sin^{-1}2x \neq 2\sin^{-1}x$.

47. Verify that $\cos^{-1}x + \cos^{-1}y \neq \cos^{-1}(x+y)$.

48. Sketch the graph of $y = \sin^{-1}(\sin x)$.

49. Sketch the graph of $y = \sin^{-1}(\cos x)$.

50. A picture u ft high is placed on a wall with its base v ft above the level of the observer's eye. (See Figure 8.14.) If the observer stands x ft from the wall, show that the angle of vision α subtended by the picture is given by

$$\alpha = \operatorname{arccot}\frac{x}{u+v} - \operatorname{arccot}\frac{x}{v}$$

51. Assuming that you can find $\sin^{-1}x$, $\cos^{-1}x$, and $\tan^{-1}x$ on your calculator, explain how to use it to find the values of $\cot^{-1}x$, $\sec^{-1}x$, and $\csc^{-1}x$.

52. A supersonic airplane such as the F-16 produces a conical shockwave when it flies faster than the speed of sound. (See Figure 8.15.) Aeronautical engineers can show that the angle θ between the sides of the cone obeys the law

$$\sin\frac{1}{2}\theta = \frac{v_s}{v_p}$$

where v_s is the speed of sound and v_p is the speed of the airplane. Solve this expression for θ.

53. The range R of a projectile with muzzle velocity v_0, projected at an angle θ with the horizontal, is given approximately by $R = 16v_0^2 \sin 2\theta$. Solve for the angle θ.

54. In using an equal-arm analytical balance, a technician must determine the equilibrium angle θ from the equation

$$Mgl\sin\theta = (m_2 - m_1)gL\cos\theta$$

Solve for the angle θ.

In Exercises 55–66 use a calculator to determine the given values. Give the answer in radians or real numbers.

55. $\tan^{-1}2.659$ **56.** $\sin^{-1}0.7863$ **57.** $\sec^{-1}5.78$
58. arccos 0.3547 **59.** $\sin^{-1}0.9866$ **60.** $\tan^{-1}2.76$
61. arccot 0.8966 **62.** $\cos^{-1}0.9034$ **63.** $\csc^{-1}1.3751$
64. $\csc^{-1}0.5$ **65.** $\cot^{-1}2$ **66.** arccot 0.32

8.3 Graphs of the Inverse Trigonometric Functions

The graphs of the six inverse trigonometric functions can be determined from their definitions and the graphs of the trigonometric functions. For example, $y = \sin^{-1}x$ if and only if $x = \sin y$, where $-\frac{1}{2} \leq y \leq \frac{1}{2}\pi$. It follows that $y = \sin^{-1}x$ has the same shape as a portion of the relation $x = \sin y$. [See Figure 8.16(a).]

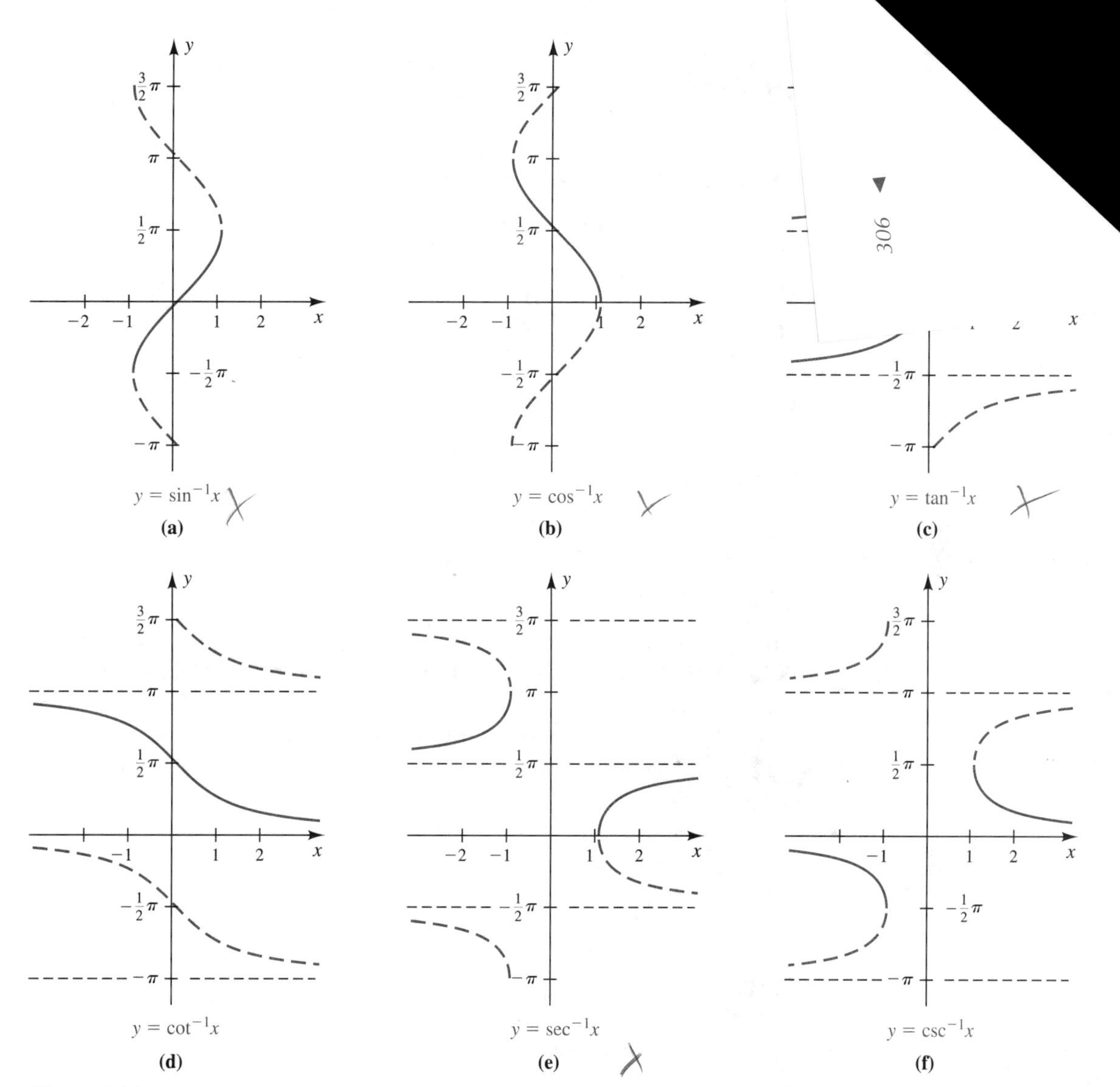

Figure 8.16
The inverse trigonometric functions

The other parts of Figure 8.16 show the graphs of the remaining inverse functions. In each case first think of the graph of the original function wrapped around the y-axis and then consider only that portion corresponding to the principal value interval.

Adding or multiplying by certain constants modifies the graphs of the inverse trigonometric functions in much the same way that it modifies the graphs of the trigonometric functions. Observing the effects of constants on the shape and location of the graphs can facilitate the process of graphing the inverse trigonometric functions. We will explain the modifications in terms of $\sin^{-1}x$, but the results apply to the other inverse trigonometric functions as well.

Multiplication of the Argument by a Constant

The function $y = \sin^{-1}Ax$ is equivalent to $\sin y = Ax$ or $(1/A)\sin y = x$. We interpret this to mean that the domain of $y = \sin^{-1}Ax$ is $1/A$ times the domain of $y = \sin^{-1}x$. Thus **the constant A expands or contracts the domain of the inverse function**. The principal value interval remains unaltered. The domain of $\sin^{-1}Ax$ is $-1/A \leq x \leq 1/A$.

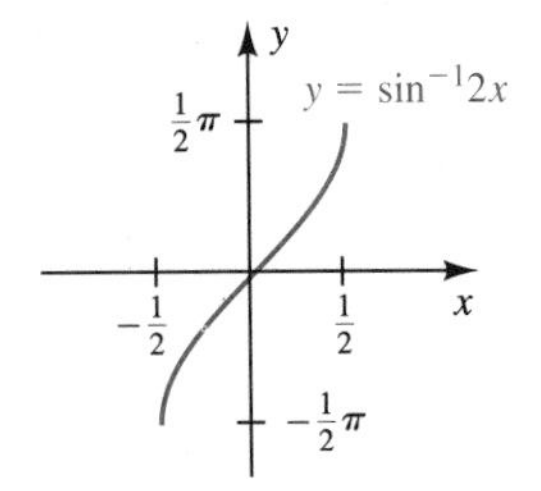

Figure 8.17

Example 1 Sketch the graph of $y = \sin^{-1}2x$.

SOLUTION In this case $A = 2$, so we multiply the domain elements by $\frac{1}{2}$. Thus the domain of $\sin^{-1}2x$ is $-\frac{1}{2} \leq x \leq \frac{1}{2}$. The graph for $y = \sin^{-1}2x$ is shown in Figure 8.17. ▲

Multiplication of the Function by a Constant

Consider the multiplication of $\sin^{-1}x$ by a constant B. In this case the function $y = B\sin^{-1}x$ is equivalent to $\sin(y/B) = x$. The principal values for $\sin(y/B)$ can be written $-\frac{1}{2}\pi \leq y/B \leq \frac{1}{2}\pi$, or $-\frac{1}{2}\pi B \leq y \leq \frac{1}{2}\pi B$. We conclude from this that **the constant B alters the principal value interval**, leaving the domain unchanged. To find the principal value interval of $\sin^{-1}Bx$, simply multiply the endpoints by B.

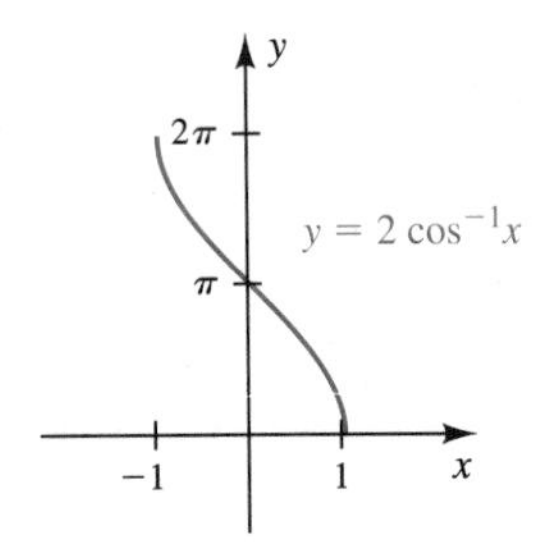

Figure 8.18

Example 2 Sketch the graph of $2\cos^{-1}x$.

SOLUTION Let $y = 2\cos^{-1}x$. Since the principal value interval for $y = \cos^{-1}x$ is $0 \leq y \leq \pi$, it follows that the principal value interval for $y = 2\cos^{-1}x$ is $0 \leq y \leq 2\pi$. (See Figure 8.18.) ▲

Addition of a Constant to the Function

The addition of a constant C to $\sin^{-1}x$ can be represented by $y = C + \sin^{-1}x$ or $y - C = \sin^{-1}x$. The principal value interval for $C + \sin^{-1}x$ is $-\frac{1}{2}\pi \leq y - C \leq \frac{1}{2}\pi$, or $-\frac{1}{2}\pi + C \leq y \leq \frac{1}{2}\pi + C$. Thus the graph of $y = C + \sin^{-1}x$ is that of $y = \sin^{-1}x$ translated C units up or down. **If C is positive, the graph moves up C units; if it is negative, the graph moves down C units.**

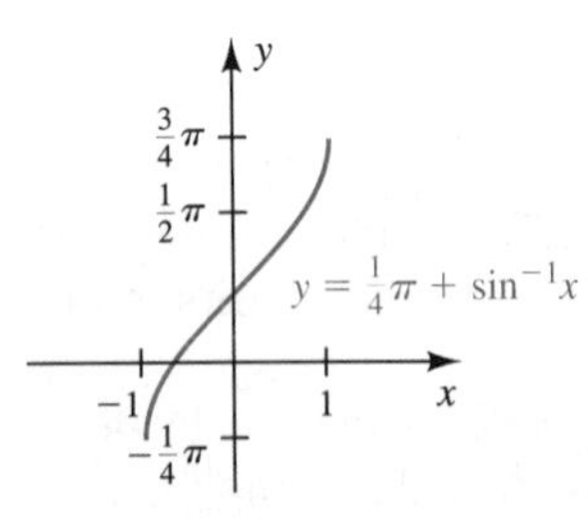

Figure 8.19

Example 3 Sketch the graph of $y = \frac{1}{4}\pi + \sin^{-1}x$.

SOLUTION The graph of this function is just the graph of $y = \sin^{-1}x$ translated up $\frac{1}{4}\pi$ units. Therefore the principal value interval for $y = \frac{1}{4}\pi + \sin^{-1}x$ is $-\frac{1}{4}\pi \leq y \leq \frac{3}{4}\pi$. (See Figure 8.19.) ▲

Addition of a Constant to the Argument

The function $y = \sin^{-1}(x + D)$ is equivalent to $\sin y = x + D$; thus the graph will be that of $\sin^{-1}x$ moved left or right D units. The translation will be D units to the left if D is positive and D units to the right if D is negative. The graph of $y = \sin^{-1}(Ax + D)$ is translated D/A units left or right.

Example 4 Sketch the graph of $y = \cos^{-1}(x - 1)$.

SOLUTION Note that the graph is that of $y = \cos^{-1}x$ translated 1 unit to the right. The graph is shown in Figure 8.20.

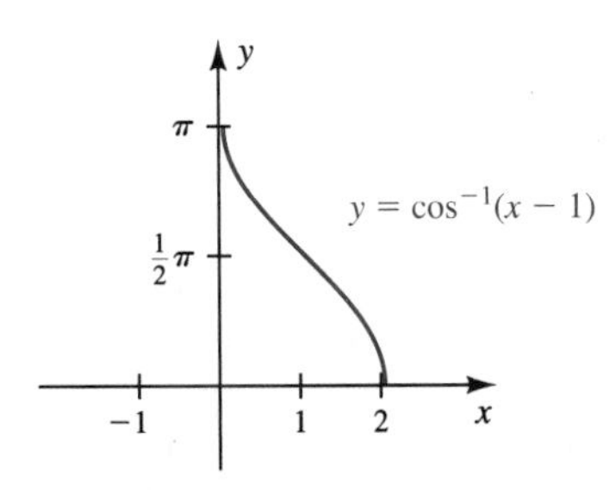

Figure 8.20

▲

The following example considers the different translations, contractions, and expansions simultaneously.

Example 5 Sketch the graph of $y = \frac{1}{2}\pi + 3\arcsin(2x - 1)$.

SOLUTION The constant $\frac{1}{2}\pi$ causes a vertical upward translation of the function $f(x) = 3\arcsin 2x$, and the constant -1 causes a translation $\frac{1}{2}$ unit to the right. The range of $3\arcsin(2x - 1)$ is

$$3(-\tfrac{1}{2}\pi) \le y \le 3(\tfrac{1}{2}\pi)$$

and its domain is

$$-1 \le 2x - 1 \le 1$$

or $0 \le x \le 1$. Figure 8.21 shows a sketch of $f(x)$ and a sketch of the function in its translated form. Notice that the graph of $f(x)$ crosses the x-axis where $y = 0$ [that is, where $\frac{1}{2}\pi + 3\arcsin(2x - 1) = 0$]. Solving this equation, we have

$$\arcsin(2x - 1) = -\tfrac{1}{6}\pi$$

which gives

$$2x - 1 = \sin(-\tfrac{1}{6}\pi) = -0.5$$
$$x = 0.25$$

Figure 8.21

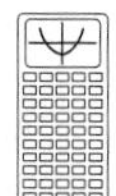

A sketch of the graph of $y = \frac{1}{2}\pi + 3\sin^{-1}(2x - 1)$ is shown in the following figure. Compare this to Figure 8.21. The range on the x-axis is from -0.5 to 1.5 in increments of 0.25; the range on the y-axis extends from $-\pi$ to 2π in increments of $\frac{1}{2}\pi$. You must know the range and increments in order to properly interpret a graph on the screen of a graphing calculator!

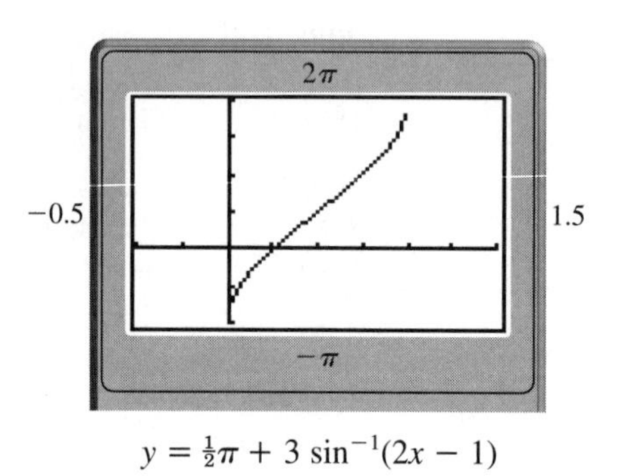

$y = \frac{1}{2}\pi + 3\sin^{-1}(2x - 1)$

Exercises for Section *8.3*

In Exercises 1–22 sketch the graph of each function.

1. $y = \sin^{-1}2x$

2. $y = \cos^{-1}2x$

3. $y = 2\cos^{-1}\frac{1}{2}x$

4. $y = \frac{1}{2}\sin^{-1}\frac{1}{2}x$

5. $y = \tan^{-1}3x$

6. $y = \tan^{-1}x + \frac{1}{2}\pi$

7. $y = 2\sin^{-1}x$

8. $y = \frac{1}{4}\cos^{-1}x$

9. $y = \sin^{-1}x + \pi$

10. $y = 3\tan^{-1}2x$

11. $y = \frac{1}{2}\sin^{-1}x + \frac{1}{2}\pi$

12. $y = \pi + 2\sin^{-1}\frac{1}{2}x$

13. $y = -\arcsin x$

14. $y = -2\arccos x$

15. $y = \sin^{-1}(x + 1)$

16. $y = \cos^{-1}(x + 1)$

17. $y = \tan^{-1}(x - 1)$

18. $y = \tan^{-1}(x + 1)$

19. $y = \sin^{-1}(x + \frac{1}{2})$

20. $y = \cos^{-1}(x - \frac{1}{2}\sqrt{3})$

21. $y = \sin^{-1}(2x - 1)$

22. $y = \arccos(2x - \frac{1}{2}\sqrt{3})$

23. The induced emf (electromotive force) in a dynamo is given by the equation $E = E_m \sin 2\pi ft$, where f is the frequency and t is the time in seconds. Graph t as a function of E/E_m for $0 \le E/E_m \le 1$. Assume f to be 60 Hz.

24. The angle of reflection θ of a light ray from air to a medium of index n is given by Snell's law as

$$\sin\theta = \frac{1}{n}$$

Solve for θ and plot θ as a function of n for $1 \le n \le 3$ in increments of 0.1.

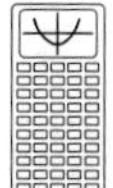

Graphing Calculator Exercises

Use a graphing calculator to sketch the graph of the given function.

1. $y = \sin^{-1}2x$

2. $y = \cos^{-1}3x$

3. $y = 3\cos^{-1}x$

4. $y = 4\sin^{-1}x$

5. $y = \tan^{-1}x + \pi$

6. $y = \pi/2 - \sin^{-1}x$

7. $y = \pi + 2\cos^{-1}3x$

8. $y = \pi/4 + 3\tan^{-1}4x$

9. Use a graphing calculator to draw the graph of $y = \arcsin x$ on $-1 \le x \le 1$. Using the **trace** key; estimate the value of $\arcsin\frac{1}{2}$ from the graph.

10. Use a graphing calculator to draw the graph of $y = \arccos x$ on $-1 \le x \le 1$. Using the **trace** key, estimate the value of $\arccos -\frac{1}{2}$ from the graph.

▲▼ Key Topics for Chapter 8

Define and/or discuss each of the following.

Inverse Functions
Principal Value Intervals of Inverse Trigonometric Functions
Definitions of Inverse Trigonometric Functions
Trigonometric Functions of Inverse Functions
Graphs of Inverse Trigonometric Functions

Review Exercises for Chapter 8

1. If $y = 3x - 2$, for which value of x is $y = 10$?

2. If $y = \dfrac{2x + 1}{x - 1}$, for which value of x is $y = 3$?

3. Determine the function that is the inverse of $y = 2x + 5$.

4. Determine the function that is the inverse of $y = \dfrac{2x + 5}{3x - 1}$.

5. Determine the function that is the inverse of $y = \sin(2x - 5) + 3$.

In Exercises 6–15 find the exact value of the given expression.

6. $\sin^{-1}(-\frac{1}{2})$
7. $\tan^{-1}\sqrt{3}$
8. $\sin(\arccos \frac{1}{3})$
9. $\tan[\sin^{-1}(-0.4)]$
10. $\cos[2 \cos^{-1}(0.3)]$
11. $\sin[2 \arcsin(-0.6)]$
12. $\sin(\tan^{-1}2 - \sin^{-1}\frac{1}{3})$
13. $\cos[\cos^{-1}\frac{1}{4} + \cos^{-1}(-\frac{1}{3})]$
14. $\tan[\tan^{-1}1 - \cot^{-1}(-2)]$
15. $\sin[\cos^{-1}(-\frac{1}{3}) + \sin^{-1}(-\frac{1}{3})]$

In Exercises 16–25 sketch the graph of the given function.

16. $y = \sin^{-1}3x$
17. $y = \frac{1}{2}\pi + \sin^{-1}x$
18. $y = \arccos 2x - \frac{1}{4}\pi$
19. $y = \cos^{-1}(2x - 1)$
20. $y = 2 \tan^{-1}x$
21. $y = \tan^{-1}2x + \pi$
22. $y = \sin^{-1}(x - \frac{1}{2}\sqrt{2})$
23. $y = -\tan^{-1}2x$
24. $y = 2 \tan^{-1}(-x + 1)$
25. $y = -3 \arccos(-2x + 1)$

In Exercises 26–27 verify the given equality for x and y in the domain of the arcsin function.

26. $\sin(2 \arcsin x) = 2x(1 - x^2)^{1/2}$
27. $\sin(\sin^{-1}x + \sin^{-1}y) = x(1 - y^2)^{1/2} + y(1 - x^2)^{1/2}$
28. The electrostatic distribution in a plate is expressed in the equation

$$\frac{y}{x} = 5 \tan(v + \tfrac{1}{4}\pi) - 10$$

Solve the equation for v.

Practice Test Questions for Chapter 8

In Exercises 1–10 answer *true* or *false*.

1. The inverse function of a function f, if it exists, is the reciprocal of f.
2. The principal value interval of arcsin x is $-1 \leq x \leq 1$.
3. $y = \tan^{-1}Ax$ is equivalent to $x = \dfrac{1}{A} \tan y$.
4. A function f is one-to-one if for a and b in the domain of f, $f(a) = f(b)$ implies that $a = b$.
5. If $y = x^3$, then its inverse function is $y = x^{1/3}$.
6. The function defined by $f = \{(2, 5), (3, 7), (0, -2), (-1, 7)\}$ is one-to-one.
7. $\sin^{-1}x = 1/\sin x$.
8. $\tan x(\arctan x) = x$.
9. $\sin^{-1}x$ and $\cos^{-1}x$ have the same domain.
10. $y = \arccos -\frac{1}{2} = 30°$.

In Exercises 11–20 fill in the blank to make the statement true.

11. The inverse function of a function g is denoted by ________.
12. A function f is ________ if for a and b in the domain of f, $f(a) = f(b)$ implies $a = b$.

13. If a function $f = \{(1, 2), (3, 7), (8, 1)\}$, then its inverse is ________.
14. The principal value interval for $y = \sin^{-1}x$ is ________.
15. The principal value interval for $y = \cos^{-1}x$ is ________.
16. The domain of $y = \arctan x$ is ________.
17. $\sin(\arcsin u) =$ ________.
18. Given that $y = \cos^{-1}3x$, then $x =$ ________.
19. The graph of $y = \tan^{-1}(x + 2)$ is the graph of ________ translated two units to the left.
20. The domain of $y = 3 \arcsin 2x$ is ________.

In the following exercises solve the stated problem. Show all your work.

21. Find the inverse function of $y = 3x + 7$.
22. Find the inverse function of $y = x^3 + 1$.
23. Evaluate $\cos(2 \arctan 2)$.
24. Evaluate $\sin(\tan^{-1}3 - \cos^{-1}\frac{1}{3})$.
25. Evaluate $\sin(2 \arcsin \frac{1}{2})$.
26. Evaluate $\arcsin(\sin \frac{5}{4}\pi)$.
27. Sketch the graph of $y = \tan^{-1}2x$.
28. Sketch the graph of $y = \frac{1}{3}\cos^{-1}(2x)$.
29. Sketch the graph of $y = 3 \cos^{-1}(x + 2)$.
30. Sketch the graph of $y = \frac{1}{2}\pi + \arcsin x$.

Complex Numbers and Polar Equations

A solution of the equation $x^2 - 2x + 5 = 0$ is $x = 1 + 2i$, where $i = \sqrt{-1}$. Numbers of this kind are called *complex numbers* and have been known to mathematicians for centuries. The description of complex numbers in terms of the trigonometric functions is one of the important events in the history of mathematics because it permits the understanding of these numbers at a very elementary level. The trigonometric representation of complex numbers, which is also called the polar form of a complex number, is discussed in Section 9.2. In Section 9.3 the trigonometric form of complex numbers is used to give new understanding to what is meant by the root of a number and to find the *n*th roots of a number. The chapter concludes with a discussion of the graphs of equations that are in polar form. You will find a graphing calculator helpful in constructing and understanding these graphs.

9.1 Complex Numbers

We sometimes think of numbers as an invention of the human mind because they were developed to obtain solutions to certain types of equations. For example, the negative integers were invented so that an equation like $x + 7 = 4$ could be solved. Similarly, the set of rational numbers was invented so that linear equations such as $2x = 3$ would have a solution. To solve $x^2 = 2$, it was necessary to invent the irrational numbers, such as $\pm\sqrt{2}$. The irrational numbers and the rational numbers together compose the set of real numbers.

For most applications the set of real numbers is sufficient, but there are cases in which this set is inadequate. For instance, in solving the equations $x^2 + 1 = 0$, we obtain the root $x = \sqrt{-1}$. Since the square of every real number is nonnegative, it is apparent that $\sqrt{-1}$ is not a real number. If we use i to represent $\sqrt{-1}$, with the understanding that i is a number such that

$i^2 = -1$, we can write the roots of $x^2 + 1 = 0$ as $x = \pm i$. The number i is called a **pure imaginary*** number and in general is the square root of -1. Thus for $a > 0$

$$\sqrt{-a} = i\sqrt{a}$$

Specifically,

$$\sqrt{-7} = i\sqrt{7} \text{ and } \sqrt{-4} = 2i$$

Numbers of the form bi, where b is a real number, make up the set of imaginary numbers.

Example 1 Solve the equation $x^2 + 9 = 0$.

SOLUTION

$$x^2 + 9 = 0$$
$$x^2 = -9$$
$$x = \pm\sqrt{-9} = \pm 3i$$

▲

In solving the equation $x^2 - 2x + 5 = 0$, we find that $x = 1 \pm \sqrt{-4}$. Using the concept of an imaginary number, we can write this expression as $x = 1 \pm 2i$. Thus we have a number that is a combination of a real number and an imaginary number; such numbers are called **complex numbers.**

DEFINITION 9.1 A complex number z is a number of the form $z = a + bi$, where a and b are real numbers and $i = \sqrt{-1}$.

The real number a is called the **real part** of z, and the real number b is called the **imaginary part** of z. By convention, if $b = 1$, the number is written $a + i$. Furthermore, if $b = 0$, the imaginary part is customarily omitted and the number is said to be a pure real number. If $a = 0$ and $b \neq 0$, the real part is omitted and the number is said to be a pure imaginary number.

Two complex numbers are equal if and only if their real parts are equal and their imaginary parts are equal. Thus $a + bi$ and $c + di$ are equal if and only if $a = c$ and $b = d$.

Example 2

(a) 3 is a real number.

(b) $2i$ is an imaginary number.

(c) $-1 + 5i$ is a complex number.

(d) If $x + yi = 3 - 2i$, then $x = 3$ and $y = -2$. ▲

* The word "imaginary" is an unfortunate choice, since it could lead you to believe that imaginary numbers are less important than the real numbers.

COMMENT Real numbers are complex numbers in which the imaginary part is 0, and imaginary numbers are complex numbers in which the real part is 0. Thus in Example 2, $3 = 3 + 0i$ and $2i = 0 + 2i$.

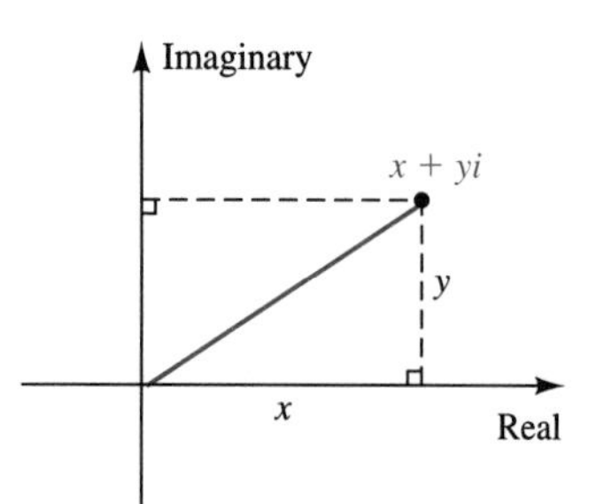

Figure 9.1

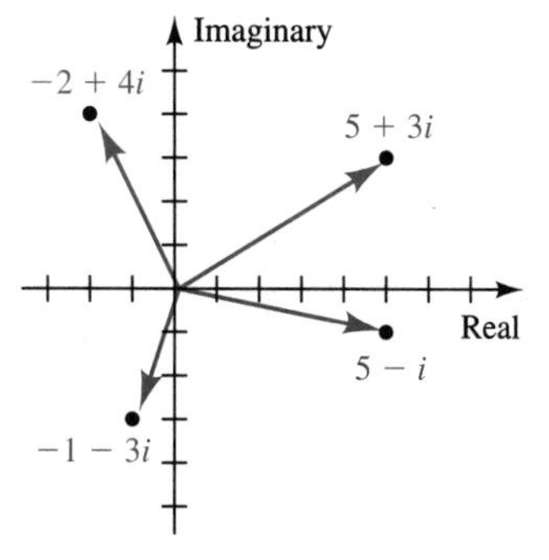

Figure 9.2

Graphical Representation of Complex Numbers

Since complex numbers are ordered pairs of real numbers, some two-dimensional configuration is needed to represent them graphically. The Cartesian coordinate system is often used for this purpose, in which case it is called the **complex plane**. The x-axis is used to represent the real part of the complex number, and the y-axis is used to represent the imaginary part. Hence the two axes are called the real axis and the imaginary axis. The complex number $x + iy$ is represented by the point whose coordinates are (x, y), as shown in Figure 9.1. Because x and y correspond to rectangular coordinates, the complex number $z = x + iy$ is said to be written in **rectangular form** or **standard form**.

It is often convenient to think of a complex number $x + iy$ as representing a vector. The complex number $x + iy$ in Figure 9.1 can be represented by the vector drawn from the origin to the point $x + iy$ with coordinates (x, y).

Example 3 The complex numbers $5 + 3i$, $-2 + 4i$, $-1 - 3i$, and $5 - i$ can be represented in the complex plane as shown in Figure 9.2. ▲

Operations on Complex Numbers

Combinations of complex numbers obey the ordinary algebraic rules for real numbers. Thus we find the sum, difference, product, and quotient of two complex numbers the same way we find the sum, difference, product, and quotient of two real binomials—with the understanding that $i^2 = -1$.

Example 4 Find the sum and difference of $3 + 5i$ and $-9 + 2i$.

SOLUTION

$$(3 + 5i) + (-9 + 2i) = (3 - 9) + (5 + 2)i$$
$$= -6 + 7i$$
$$(3 + 5i) - (-9 + 2i) = (3 + 9) + (5 - 2)i$$
$$= 12 + 3i$$
▲

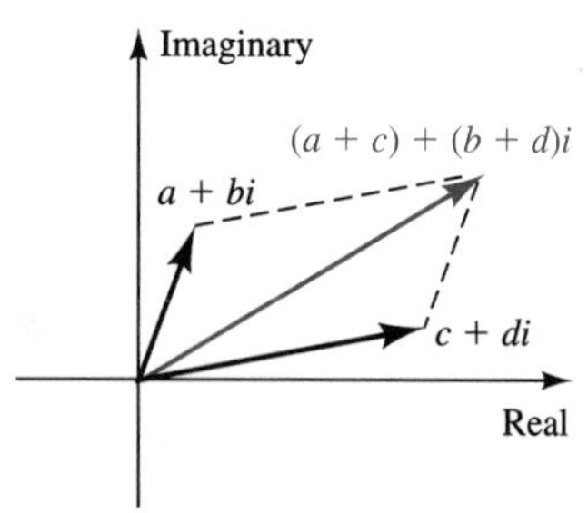

Figure 9.3
Addition of complex numbers

It is helpful to consider the graphical representation of the sum of two complex numbers. The numbers $a + bi$, $c + di$, and $(a + c) + (b + d)i$ are represented in Figure 9.3. The result is the same as if we had applied the parallelogram law to the vectors representing $a + bi$ and $c + di$. Note that we subtract $c + di$ from $a + bi$ by plotting $a + bi$ and $-c - di$ and then use the parallelogram law.

The product of two complex numbers is obtained by expanding the product of two binomials and replacing i^2 with -1.

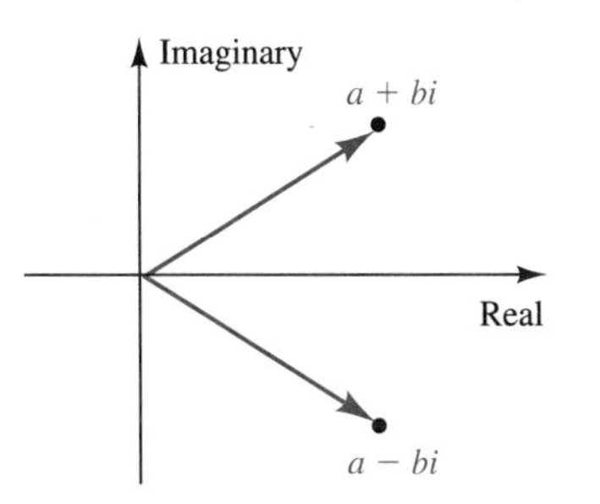

Figure 9.4

Example 5 Find the product $(3 - 2i)(4 + i)$.

SOLUTION

$$\begin{aligned}(3 - 2i)(4 + i) &= 12 + 3i - 8i - 2i^2 \\ &= 12 - 5i + 2 \\ &= 14 - 5i\end{aligned}$$

▲

The number $2 - 3i$ is called the **conjugate** of $2 + 3i$. The conjugate of a complex number $a + bi$ is denoted $\overline{a + bi}$ and is obtained by changing the sign of the imaginary part. In general $a - bi$ is the conjugate of $a + bi$. The graphs of $a - bi$ and $a + bi$ are shown in Figure 9.4. Notice that $a - bi$ is the mirror image of $a + bi$ in the x-axis.

The product of a complex number and its conjugate is a nonnegative real number. To prove this, we note that

$$(a + bi)(a - bi) = a^2 - abi + abi - b^2i^2 = a^2 + b^2$$

The expression $a^2 + b^2$ is a nonnegative real number.

Example 6

(a) $(-3 + 2i)(-3 - 2i) = (-3)^2 + 2^2 = 13$

(b) The conjugate of $4i$ is $-4i$; the product of $4i$ and its conjugate is 16. ▲

We use the conjugate to find the quotient of two complex numbers. The quotient of two complex numbers is found by multiplying the numerator and the denominator of the quotient by the conjugate of the denominator. This operation makes the denominator a real number. The technique is illustrated in the next example.

Example 7 Find the quotient $\dfrac{2 + 3i}{4 - 5i}$.

SOLUTION

$$\begin{aligned}\frac{2 + 3i}{4 - 5i} &= \frac{(2 + 3i)(4 + 5i)}{(4 - 5i)(4 + 5i)} \\ &= \frac{8 + (12 + 10)i + 15i^2}{16 - 25i^2} \\ &= \frac{-7 + 22i}{16 + 25} \\ &= \frac{-7 + 22i}{41} \\ &= -\frac{7}{41} + \frac{22}{41}i\end{aligned}$$

▲

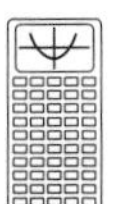

Some calculators can perform elementary arithmetic operations on complex numbers. Check to see if yours does!

Exercises for Section 9.1

In Exercises 1–10 plot the number, its negative, and its conjugate on the same coordinate system.

1. $-3 + 2i$
2. $4 - 3i$
3. $-2i$
4. $5 + i$
5. $-1 - i$
6. $3 + 5i$
7. -2
8. $3i$
9. $1 - i\sqrt{2}$
10. $\sqrt{3} + i\sqrt{5}$

In Exercises 11–34 perform the operations indicated, expressing all answers in the form $a + bi$. Check your answers to Exercises 11–14 graphically.

11. $(3 + 2i) + (4 + 3i)$
12. $(6 + 3i) + (5 - i)$
13. $(5 - 2i) + (-7 + 5i)$
14. $(-1 + i) + (2 - i)$
15. $(1 + i) + (3 - i)$
16. $7 - (5 + 3i)$
17. $(3 + 5i) - 4i$
18. $(3 + 2i) + (3 - 2i)$
19. $(2 + 3i)(4 + 5i)$
20. $(7 + 2i)(-1 - i)$
21. $(5 - i)(5 + i)$
22. $(6 - 3i)(6 + 3i)$
23. $(4 + \sqrt{3}i)^2$
24. $(5 - 2i)^2$
25. $6i(4 - 3i)$
26. $3i(-2 - i)$
27. $\dfrac{3 + 2i}{1 + i}$
28. $\dfrac{4i}{2 + i}$
29. $\dfrac{3}{2 - 3i}$
30. $\dfrac{7 - 2i}{6 - 5i}$
31. $\dfrac{1}{5i}$
32. $\dfrac{-3 + i}{-2 - i}$
33. $\dfrac{-1 - 3i}{4 - \sqrt{2}i}$
34. $\dfrac{i}{2 + \sqrt{5}i}$
35. Show that the sum of a complex number and its conjugate is a real number.
36. Show that the product of a complex number and its conjugate is a real number.
37. In the science of optics, it is often helpful to be able to locate a point in the complex plane, for then its mirror image is its conjugate. Find the mirror image of the point whose complex representation is given by $5 + 2i$.
38. An electrical impedance is expressed as the complex number $3 - 5i$. Plot this impedance on the complex plane.
39. The complex expression $x(t) + iy(t)$ is used to represent curves in the plane, where $x = x(t)$ and $y = y(t)$ and t is the parameter. Sketch the curve $z = \cos t + i(2 \sin t)$.
40. How is the curve $z(t)$ in Exercise 39 related to the curve $\bar{z}(t)$? Sketch the curve $\bar{z} = \cos t - i(2 \sin t)$. [Note that $\bar{z}(t)$ is the conjugate of $z(t)$.]

9.2 Polar Representation of Complex Numbers

The rectangular coordinate system was used exclusively in the first eight chapters of this book. Another coordinate system widely used in science and mathematics is the **polar coordinate system**. In this system, the position of a point is determined by specifying a distance from a given point and the direction from a given line. Actually this concept is not new; we frequently use this system to describe the relative locations of geographic points. When we say that Cincinnati is about 300 miles southeast of Chicago, we are using polar coordinates.

Figure 9.5

To establish a frame of reference for the polar coordinate system, we begin by choosing a point O and extending a line from this point. The point O is called the **pole**, and the extended line OA is called the **polar axis**. We can then determine the position of any point P in the plane if we know the distance OP and the angle AOP, as indicated in Figure 9.5. The directed distance OP is called the **radius vector** of P and is denoted by r. The angle AOP is called the **vectorial angle** and is denoted by θ. The coordinates of a point P are then written as the ordered pair (r, θ). Notice that the radius vector is the first element and the vectorial angle is the second.

Polar coordinates, like rectangular coordinates, are regarded as signed quantities. When we state the polar coordinates of a point, it is customary to use the following sign conventions.

1. The radius vector is positive when it coincides with the terminal side of the vectorial angle and negative when it extends in the opposite direction from the terminal side of the vectorial angle.
2. The vectorial angle is positive when generated by a counterclockwise rotation from the polar axis and negative when generated by a clockwise rotation.

The polar coordinates of a point uniquely determine the location of the point. However, the converse is not true. That is, the location of a point can determine several pairs of polar coordinates, as Figure 9.6 shows. Ignoring vectorial angles that are numerically greater than 360°, we have four pairs of coordinates that yield the same point: (5, 60°), (5, −300°), (−5, 240°), and (−5, −120°).

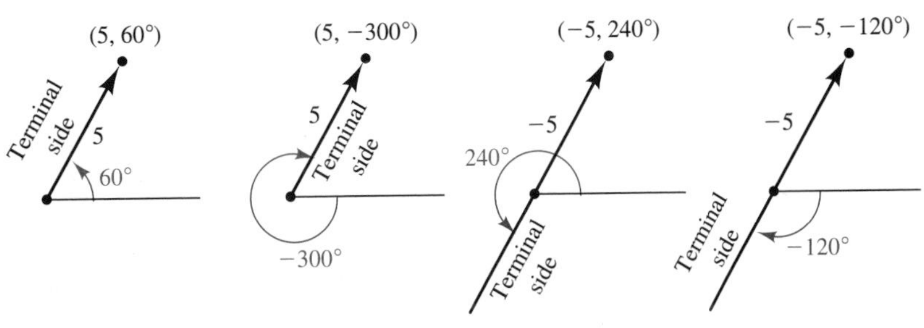

Figure 9.6

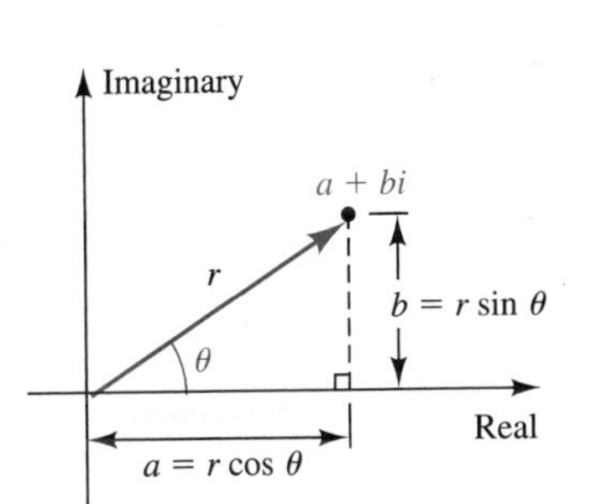

Figure 9.7
Polar form

The graphical representation of a complex number as a point in the complex plane is usually thought of as a vector drawn from the origin to the point. We may thus use polar coordiantes to describe a complex number. Figure 9.7 shows that a complex number $a + bi$ can be located with the polar coordinates (r, θ), where

$$r = \sqrt{a^2 + b^2} \quad \text{and} \quad \tan\theta = \frac{b}{a} \tag{9.1}$$

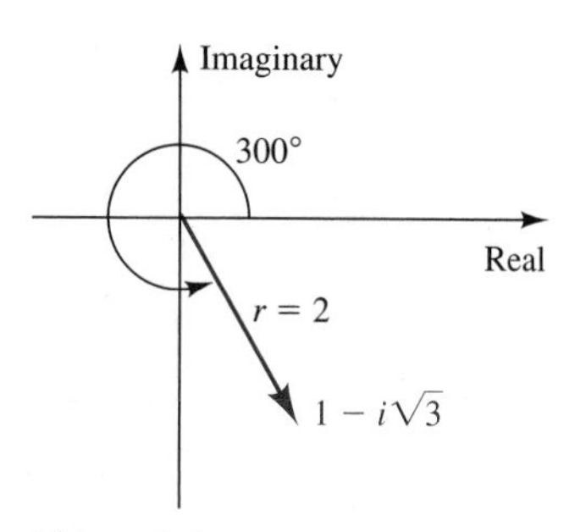

Figure 9.8

Example 1 Determine the polar coordinates for the complex number $1 - i\sqrt{3}$.

SOLUTION Using Equation (9.1) with $a = 1$ and $b = -\sqrt{3}$, we get

$$r = \sqrt{(1)^2 + (-\sqrt{3})^2} = 2 \quad \text{and} \quad \tan\theta = \frac{-\sqrt{3}}{1} = -\sqrt{3}$$

The vectorial angle θ is found by observing that $1 - i\sqrt{3}$ is in the fourth quadrant. (See Figure 9.8.) The reference angle for θ is $\alpha = \arctan\sqrt{3} = 60°$, so

$$\theta = 360° - 60° = 300°$$

▲

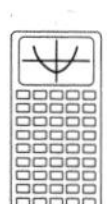

Many calculators are capable of making conversions from rectangular to polar coordinates. For example, on the TI-81 under the "math" menu, enter R → P(x, y) to obtain the value of r; then you can obtain the value of θ simply by entering the letter θ. Conversely, entering P → R(r, θ) yields x; then the value of y is obtained by entering the letter y.

Complex numbers are frequently expressed in terms of r and θ. Referring again to Figure 9.7, we see that

$$a = r\cos\theta$$

and

$$b = r\sin\theta$$

Therefore, the complex number $a + bi$ can be written $r\cos\theta + ir\sin\theta$, or in factored form,

$$z = a + bi = r(\cos\theta + i\sin\theta) \tag{9.2}$$

The right-hand side of Equation 9.2 is called the **polar form** of the complex number $a + bi$. The quantity $\cos\theta + i\sin\theta$ is sometimes written cis θ, in which case the polar form of the complex number is written

$$z = r \operatorname{cis}\theta \tag{9.3}$$

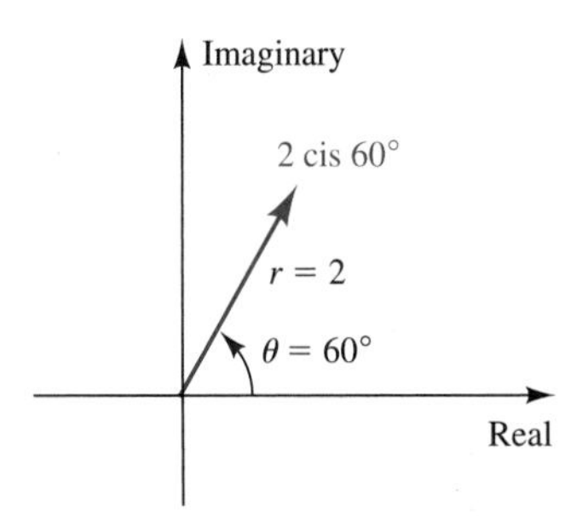

Figure 9.9

The number r is called the **modulus**, or magnitude, of z, and θ is called the **argument**. Note that a given complex number has many arguments, all differing by multiples of 2π. Sometimes we limit the argument to some interval of length 2π and thus obtain the **principal value**. In this book, unless we state otherwise, the principal values will be between $-\pi$ and π; that is, they will be between $-180°$ and $180°$.

Example 2 Represent $z = 1 + \sqrt{3}i$ in polar form. (See Figure 9.9.)

SOLUTION Since

$$r = \sqrt{1^2 + (\sqrt{3})^2} = \sqrt{4} = 2$$

and

$$\theta = \arctan\sqrt{3} = 60°$$

we have

$$1 + \sqrt{3}i = 2(\cos 60° + i \sin 60°) = 2 \text{ cis } 60°$$

▲

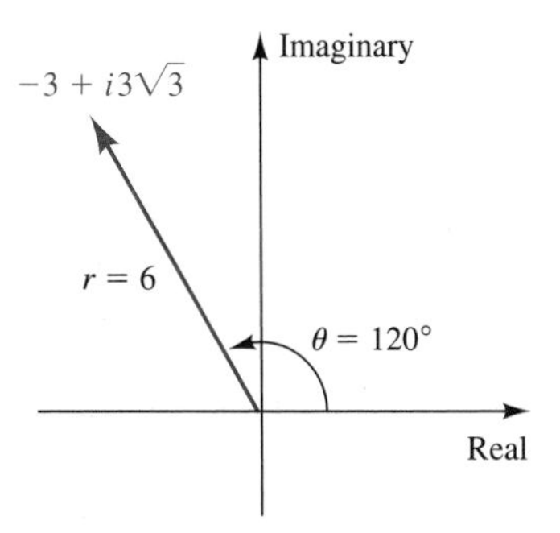

Figure 9.10

Example 3 Express $z = 6(\cos 120° + i \sin 120°)$ in rectangular form. (See Figure 9.10.)

SOLUTION Using the fact that $a = r \cos \theta$ and $b = r \sin \theta$, we have

$$a = 6 \cos 120° = 6\left(-\frac{1}{2}\right) = -3$$

$$b = 6 \sin 120° = 6\left(\frac{\sqrt{3}}{2}\right) = 3\sqrt{3}$$

Therefore

$$z = a + bi = -3 + 3\sqrt{3}i$$

▲

It is easy to give a geometric interpretation to the product of two complex numbers in polar form. If $z_1 = r_1 \text{ cis } \theta_1$ and $z_2 = r_2 \text{ cis } \theta_2$, the product $z_1 z_2$ may be written

$$\begin{aligned} z_1 z_2 &= r_1(\cos \theta_1 + i \sin \theta_1) \cdot r_2(\cos \theta_2 + i \sin \theta_2) \\ &= r_1 r_2[\cos \theta_1 \cos \theta_2 + i \cos \theta_1 \sin \theta_2 \\ &\quad + i \sin \theta_1 \cos \theta_2 + i^2 \sin \theta_1 \sin \theta_2] \\ &= r_1 r_2[(\cos \theta_1 \cos \theta_2 - \sin \theta_1 \sin \theta_2) \\ &\quad + i(\cos \theta_1 \sin \theta_2 + \sin \theta_1 \cos \theta_2)] \end{aligned}$$

Now using the identities for the sine and cosine of the sum of two angles, we have

$$z_1 z_2 = r_1 r_2[\cos(\theta_1 + \theta_2) + i \sin(\theta_1 + \theta_2)] = r_1 r_2 \text{cis}(\theta_1 + \theta_2) \tag{9.4}$$

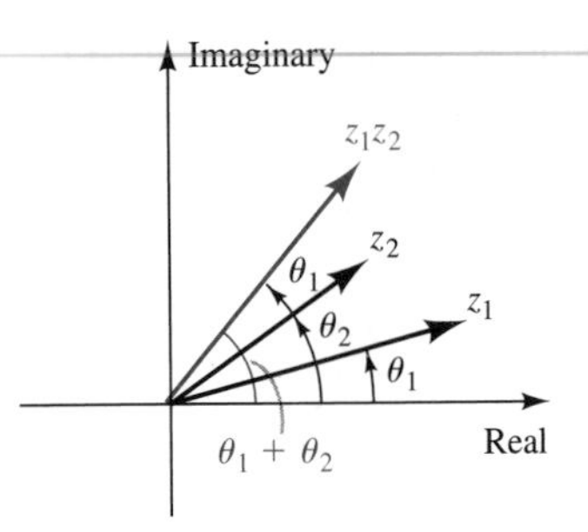

Figure 9.11
Multiplication of complex numbers

Therefore the modulus of the product of two complex numbers is the product of the individual moduli, and the argument of the product is the sum of the individual arguments. Graphically, multiplication of z_1 by z_2 results in a rotation of the vector through z_1 by an angle equal to the argument of z_2, and it causes an expansion or contraction of the modulus, depending on whether $|z_2| > 1$ or $|z_2| < 1$. (See Figure 9.11.)

Example 4 Multiply $z_1 = -1 + \sqrt{3}i$ and $z_2 = 1 + i$, using the polar form of each.

SOLUTION Computing the modulus and argument of each complex number yields

$$r_1 = \sqrt{(-1)^2 + (\sqrt{3})^2} = 2 \qquad \tan\theta_1 = \frac{\sqrt{3}}{-1},\ \theta_1 = 120^\circ$$

$$r_2 = \sqrt{1^2 + 1^2} = \sqrt{2} \qquad \tan\theta_2 = \frac{1}{1},\ \theta_2 = 45^\circ$$

Therefore

$$\begin{aligned} z_1z_2 &= (2 \text{ cis } 120^\circ)(\sqrt{2} \text{ cis } 45^\circ) = 2\sqrt{2}\,\text{cis}(120^\circ + 45^\circ) \\ &= 2\sqrt{2} \text{ cis } 165^\circ \end{aligned}$$

▲

Using the same procedure we employed to derive Equation (9.4), we can show that if $z_1 = r_1\text{cis } \theta_1$ and $z_2 = r_2\text{cis } \theta_2$, then

$$\frac{z_1}{z_2} = \frac{r_1}{r_2}[\cos(\theta_1 - \theta_2) + i\sin(\theta_1 - \theta_2)] = \frac{r_1}{r_2}\text{cis}(\theta_1 - \theta_2) \qquad \textbf{(9.5)}$$

In words, the modulus of the quotient of two complex numbers is the quotient of the individual moduli, and the argument is the difference of the individual arguments.

Example 5 Divide $z_1 = 2 \text{ cis } 120^\circ$ by $z_2 = \sqrt{2} \text{ cis } 45^\circ$.

SOLUTION $$\frac{z_1}{z_2} = \frac{2 \text{ cis } 120^\circ}{\sqrt{2} \text{ cis } 45^\circ} = \frac{2}{\sqrt{2}} \text{cis}(120^\circ - 45^\circ) = \sqrt{2} \text{ cis } 75^\circ$$ ▲

▲▼ Exercises for Section 9.2

In Exercises 1–10 express each complex number in polar form.

1. $1 - \sqrt{3}i$
2. $3 + 4i$
3. $\sqrt{5} + 2i$
4. $\sqrt{3} - i$
5. 9
6. $5i$
7. $3 - 4i$
8. $-1 + i$
9. $5 - 6i$
10. $-3 - 4i$

In Exercises 11–20 express each complex number in rectangular form.

11. $2 \text{ cis } 30°$
12. $4 \text{ cis } 60°$
13. $5 \text{ cis } 135°$
14. $10 \text{ cis } 90°$
15. $\sqrt{3} \text{ cis } 210°$
16. $\sqrt{5} \text{ cis } 180°$
17. $3 \text{ cis } 300°$
18. $7 \text{ cis } 0°$
19. $10 \text{ cis } 20°$
20. $2 \text{ cis } 100°$

In Exercises 21–36 perform the indicated operations. If the complex numbers are not already in polar form, express them in polar form before proceeding.

21. $(4 \text{ cis } 30°)(3 \text{ cis } 60°)$
22. $(2 \text{ cis } 120°)(\sqrt{5} \text{ cis } 180°)$
23. $(\sqrt{2} \text{ cis } 90°)(\sqrt{2} \text{ cis } 240°)$
24. $(5 \text{ cis } 180°)(3 \text{ cis } 90°)$
25. $(10 \text{ cis } 35°)(2 \text{ cis } 100°)$
26. $(3 \text{ cis } 45°)(2 \text{ cis } 120°)$
27. $(3 + 4i)(\sqrt{3} - i)$
28. $3i(2 - i)$
29. $\dfrac{10 \text{ cis } 30°}{2 \text{ cis } 90°}$
30. $\dfrac{5 \text{ cis } 29°}{3 \text{ cis } 4°}$
31. $\dfrac{4 \text{ cis } 26°40'}{2 \text{ cis } 19°10'}$
32. $\dfrac{12 \text{ cis } 100°}{3 \text{ cis } 23°}$
33. $\dfrac{1 - i}{\sqrt{3} + i}$
34. $\dfrac{\sqrt{3} + i}{\sqrt{3} - i}$
35. $\dfrac{4i}{-1 + i}$
36. $\dfrac{5}{1 + i}$

37. Prove *Euler's identities:*

$$\cos \theta = \frac{1}{2}[\text{cis } \theta + \text{cis}(-\theta)]$$

$$\sin \theta = \frac{1}{2i}[\text{cis } \theta - \text{cis}(-\theta)]$$

38. A force is represented by the complex number 20 cis 2.54. What is the magnitude of the force? What angle, in degrees, does the force make with the positive x-axis? What are the horizontal and vertical components of the force?

39. The power in a circuit is the product of the current, i, and the voltage, v. Write the equation for power, p, if $i = 50 \text{ cis}(-1.7t)$ and $v = 20 \text{ cis}(0.5t)$. What is the power when $t = 1$? Give the horizontal (resistive) and vertical (reactive) components of power.

9.3 DeMoivre's Theorem

The square of the complex number $z = r \text{ cis } \theta$ is given by

$$\begin{aligned} z^2 &= (r \text{ cis } \theta)(r \text{ cis } \theta) = r^2 \text{cis}(\theta + \theta) \\ &= r^2 \text{cis } 2\theta \end{aligned}$$

Likewise,

$$z^3 = z^2 \cdot z = (r^2\text{cis } 2\theta) \cdot (r \text{ cis } \theta)$$
$$= r^3\text{cis } 3\theta$$

We expect the pattern exhibited by z^2 and z^3 to apply as well to z^4, z^5, z^6, and so on. As a matter of fact, if $z = r \text{ cis } \theta$, then

$$z^n = r^n\text{cis } n\theta \tag{9.6}$$

This result is known as **DeMoivre's theorem**. The theorem is true for all real values of n, a fact that we shall accept without proof.

Example 1 Use DeMoivre's theorem to find $(-2 + 2i)^4$.

SOLUTION Here we have

$$r = \sqrt{2^2 + (-2)^2} = \sqrt{8} \qquad \text{and} \qquad \theta = 135°$$

Therefore

$$\begin{aligned}(-2 + 2i)^4 &= (\sqrt{8})^4\text{cis } 4(135°)\\ &= (\sqrt{8})^4[\cos 4(135°) + i \sin 4(135°)]\\ &= 64[\cos 540° + i \sin 540°]\\ &= 64[\cos 180° + i \sin 180°]\\ &= -64\end{aligned}$$

▲

In the system of real numbers, there is no square root of -1, no fourth root of -81, and so on. However, if we use complex numbers, we can find the nth roots of any number by using DeMoivre's theorem.

Since DeMoivre's theorem is valid for all real n, it is possible to evaluate $[r \text{ cis } \theta]^{1/n}$ as

$$[r \text{ cis } \theta]^{1/n} = r^{1/n}\text{cis } \frac{\theta}{n} = \sqrt[n]{r} \text{ cis } \frac{\theta}{n} \tag{9.7}$$

Since $\cos \theta$ and $\sin \theta$ are periodic functions with a period of 360°, $\cos \theta = \cos(\theta + k \cdot 360°)$ and $\sin \theta = \sin(\theta + k \cdot 360°)$, where k is an integer. Hence

$$[r \text{ cis } \theta]^{1/n} = \sqrt[n]{r} \text{ cis}\left(\frac{\theta + k \cdot 360°}{n}\right) \tag{9.8}$$

For a given positive integer n, the right-hand side of Equation (9.8) takes on n distinct values corresponding to $k = 0, 1, 2, \ldots, n - 1$. For $k > n - 1$, the result is merely a duplication of the first n values.

Example 2 Find the square roots of $4i$.

SOLUTION We first express $4i$ in polar form using

$$r = \sqrt{0^2 + 4^2} = 4 \qquad \text{and} \qquad \theta = 90°$$

Thus

$$4i = 4 \text{ cis } 90°$$

and the square roots of $4i$ are given by

$$2 \text{ cis}\left(\frac{90° + k \cdot 360°}{2}\right)$$

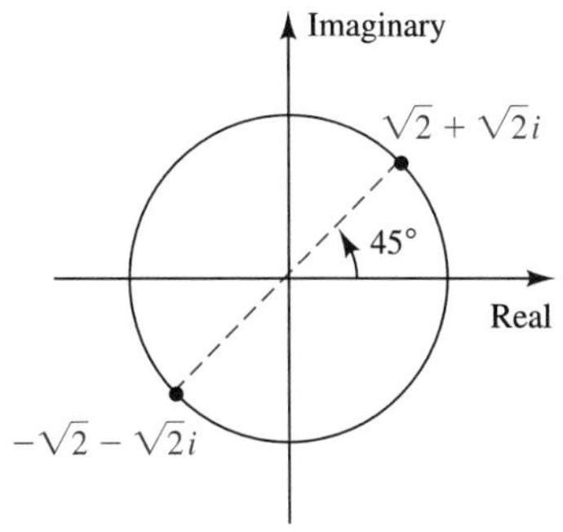

Figure 9.12

Therefore, we have, for $k = 0$,

$$2 \text{ cis } 45° = \sqrt{2} + \sqrt{2}i$$

and for $k = 1$,

$$2 \text{ cis } 225° = -\sqrt{2} - \sqrt{2}i$$

It is convenient and informative to plot these values in the complex plane, as shown in Figure 9.12. Notice that both roots are located on a circle of radius 2, but they are 180° apart. ▲

Example 3 Find the three cube roots of unity.

SOLUTION In polar form, the number 1 may be written 1 cis 0°. Thus

$$\sqrt[3]{1 \text{ cis } 0°} = 1 \text{ cis}\left(\frac{0° + k \cdot 360°}{3}\right)$$

For $k = 0$

$$1 \text{ cis } 0° = 1$$

For $k = 1$

$$1 \text{ cis } 120° = \frac{-1 + \sqrt{3}i}{2}$$

For $k = 2$

$$1 \text{ cis } 240° = \frac{-1 - \sqrt{3}i}{2}$$

Figure 9.13

These roots are graphed in Figure 9.13. Notice that they are located on a circle of radius 1 and are equally spaced at intervals of 120°. ▲

Example 4 Find the fourth roots of $-1 + \sqrt{3}i$.

SOLUTION First we write $-1 + \sqrt{3}i$ in polar form:

$$-1 + \sqrt{3}i = 2 \text{ cis } 120°$$

Then

$$[-1 + \sqrt{3}i]^{1/4} = \sqrt[4]{2} \text{ cis } \frac{120° + k \cdot 360°}{4}$$

The four roots corresponding to $k = 0, 1, 2, 3$ are as follows.

For $k = 0$

$$\sqrt[4]{2} \text{ cis } 30° = \sqrt[4]{2}\left(\frac{\sqrt{3}}{2} + \frac{1}{2}i\right)$$

For $k = 1$

$$\sqrt[4]{2} \text{ cis } 120° = \sqrt[4]{2}\left(-\frac{1}{2} + \frac{\sqrt{3}}{2}i\right)$$

For $k = 2$

$$\sqrt[4]{2} \text{ cis } 210° = \sqrt[4]{2}\left(-\frac{\sqrt{3}}{2} - \frac{1}{2}i\right)$$

For $k = 3$

$$\sqrt[4]{2} \text{ cis } 300° = \sqrt[4]{2}\left(\frac{1}{2} - \frac{\sqrt{3}}{2}i\right)$$

▲

▲▼ Exercises for Section *9.3*

In Exercises 1–10 use DeMoivre's theorem to evaluate each power. Leave the answer in polar form.

1. $(-1 + \sqrt{3}i)^3$
2. $(1 + i)^4$
3. $(\sqrt{3} \text{ cis } 60°)^4$
4. $(\sqrt{3} - i)^6$
5. $(-2 + 2i)^5$
6. $(-1 + 3i)^3$
7. $(-\sqrt{3} + i)^7$
8. $(2 \text{ cis } 20°)^5$
9. $(2 + 5i)^4$
10. $(3 + 2i)^{10}$

In Exercises 11–20 find each root indicated and sketch its location in the complex plane.

11. Fifth roots of 1
12. Cube roots of 64
13. Fourth roots of i
14. Fourth roots of -16
15. Square roots of $1 + i$
16. Fifth roots of $\sqrt{3} + i$
17. Sixth roots of $-\sqrt{3} + i$
18. Square roots of $-1 + i$
19. Fourth roots of $-1 + \sqrt{3}i$
20. Sixth roots of $-i$
21. Use DeMoivre's theorem to obtain expressions for $\cos 2\theta$ and $\sin 2\theta$ in terms of trigonometric functions of θ.
22. Find all roots of the equation $x^4 + 81 = 0$.
23. Find all roots of $x^3 + 64 = 0$.

24. A physicist studying the theory of optics needs to know the $\frac{3}{2}$ power of $2 + i$. This can be found by taking the cube of $2 + i$ and then finding the two square roots of that result. Of these two numbers, the physicist wants the one with the positive real part. Find it for her.

25. The function $1/(s^4 + 1)$ is used in the theory of Laplace transforms. Determine the values of s for which the denominator is equal to zero.

9.4 Polar Equations and Their Graphs

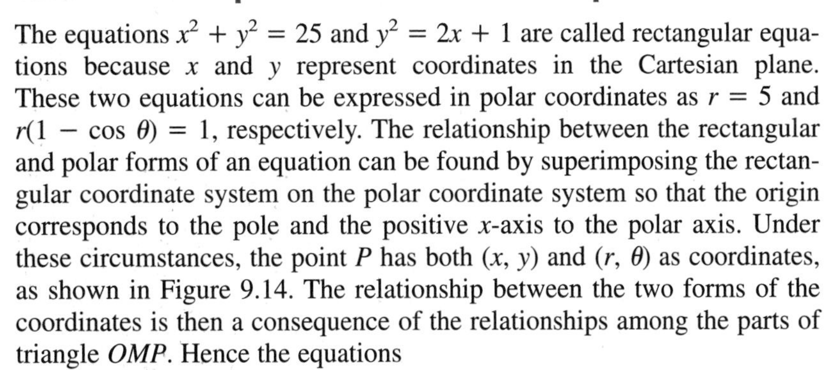

The equations $x^2 + y^2 = 25$ and $y^2 = 2x + 1$ are called rectangular equations because x and y represent coordinates in the Cartesian plane. These two equations can be expressed in polar coordinates as $r = 5$ and $r(1 - \cos\theta) = 1$, respectively. The relationship between the rectangular and polar forms of an equation can be found by superimposing the rectangular coordinate system on the polar coordinate system so that the origin corresponds to the pole and the positive x-axis to the polar axis. Under these circumstances, the point P has both (x, y) and (r, θ) as coordinates, as shown in Figure 9.14. The relationship between the two forms of the coordinates is then a consequence of the relationships among the parts of triangle OMP. Hence the equations

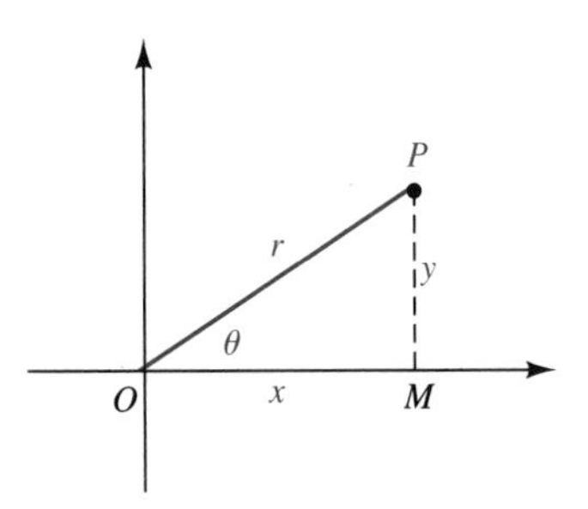

Figure 9.14

$$x = r\cos\theta \tag{9.9}$$

and

$$y = r\sin\theta \tag{9.10}$$

can be used to transform a rectangular equation into a polar equation.

Example 1 Find the polar equation of the circle whose rectangular equation is $x^2 + y^2 = a^2$.

SOLUTION Substituting Equations (9.9) and (9.10) into the given equation, we have

$$\begin{aligned} r^2\cos^2\theta + r^2\sin^2\theta &= a^2 \\ r^2(\cos^2\theta + \sin^2\theta) &= a^2 \\ r^2 &= a^2 \\ r &= a \end{aligned}$$

Hence $r = a$ is the polar equation of the given circle. ▲

To make the transformation from polar coordinates into rectangular coordinates, we use the following equations.

$$r = \sqrt{x^2 + y^2} \tag{9.11}$$

$$\sin \theta = \frac{y}{\sqrt{x^2 + y^2}} \tag{9.12}$$

$$\cos \theta = \frac{x}{\sqrt{x^2 + y^2}} \tag{9.13}$$

These equations are derived from the triangle illustrated in Figure 9.14.

Example 2 Transform the following polar equation into a rectangular equation:

$$r = 1 - \cos \theta$$

SOLUTION Substituting Equations (9.11) and (9.13) in the polar equation gives

$$\sqrt{x^2 + y^2} = 1 - \frac{x}{\sqrt{x^2 + y^2}}$$

$$x^2 + y^2 = \sqrt{x^2 + y^2} - x$$

Therefore

$$x^2 + y^2 + x = \sqrt{x^2 + y^2}$$

is the rectangular form of the original equation. ▲

Example 3 Show that

$$r = \frac{1}{1 - \cos \theta}$$

is the polar form of the parabola $y^2 = 2x + 1$.

SOLUTION Here our work is simplified if we multiply both sides of the given equation by $1 - \cos \theta$ before making the substitution. Thus,

$$r - r \cos \theta = 1$$

Substituting for r and $\cos \theta$ using Equations (9.11) and (9.13), we get

$$\sqrt{x^2 + y^2} - x = 1$$

Transposing x to the right and squaring both sides of the resulting equation gives

$$x^2 + y^2 = x^2 + 2x + 1$$

$$y^2 = 2x + 1$$

$$y^2 = 2(x + \tfrac{1}{2})$$

▲

A polar equation has a graph in the polar coordinate plane, just as a rectangular equation has a graph in the rectangular coordinate plane. To draw the graph of a polar equation, we start by assigning values to θ and finding the corresponding values of r. The graph is then generated by plotting the ordered pairs (r, θ) and connecting them with a smooth curve. Graphing polar equations will be easier if you use polar coordinate paper (available commercially). Polar coordinate paper is marked with equally spaced concentric circles with radial lines extending at equal angles through the pole. Several points are plotted in Figure 9.15 as examples.

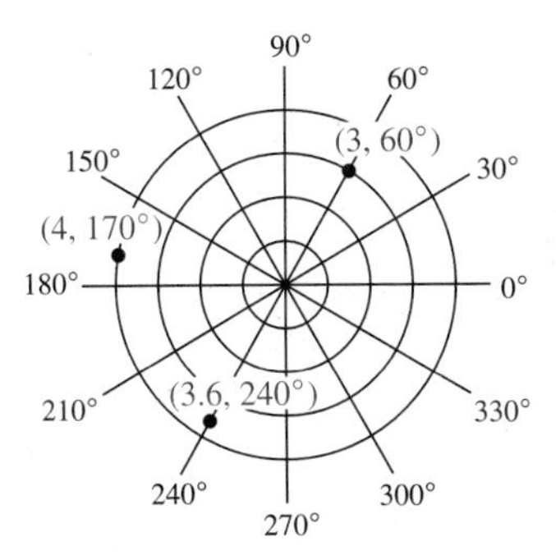

Figure 9.15

Example 4 Sketch the graph of the equation $r = 1 + \cos\theta$.

SOLUTION Using increments of 45° for θ, we obtain the following table.

θ	0	45°	90°	135°	180°	225°	270°	315°	360°
r	2.00	1.71	1.00	0.29	0.00	0.29	1.00	1.71	2.00

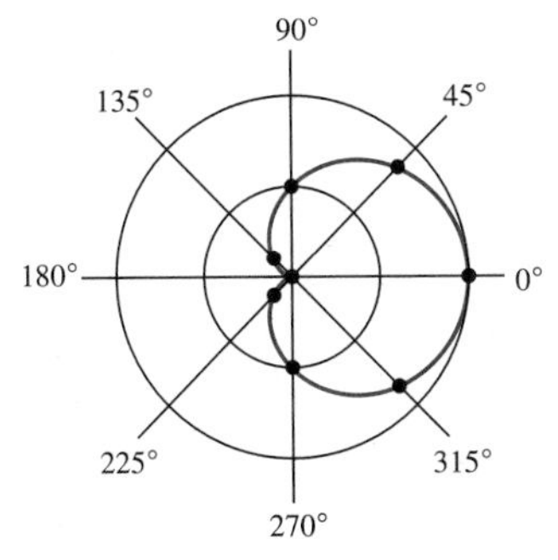

Figure 9.16

We then plot each point. The curve obtained by connecting these points with a smooth curve is called a *cardioid*. (See Figure 9.16.) ▲

Example 5 Sketch the graph of $r = \sin 3\theta$ for $0° \leq \theta \leq 180°$, using increments of 15° for θ.

SOLUTION The following table gives values of r corresponding to the indicated values of θ. Notice that the values of r corresponding to $\theta = 75°$, 90°, and 105° are negative and are plotted accordingly in Figure 9.17. This curve, which is called a *three-leaved rose*, is traced again for $180° \leq \theta \leq 360°$.

θ	0	15°	30°	45°	60°	75°	90°	105°	120°	135°	150°	165°	180°
r	0	0.7	1.0	0.7	0	−0.7	−1.0	−0.7	0	0.7	1.0	0.7	0

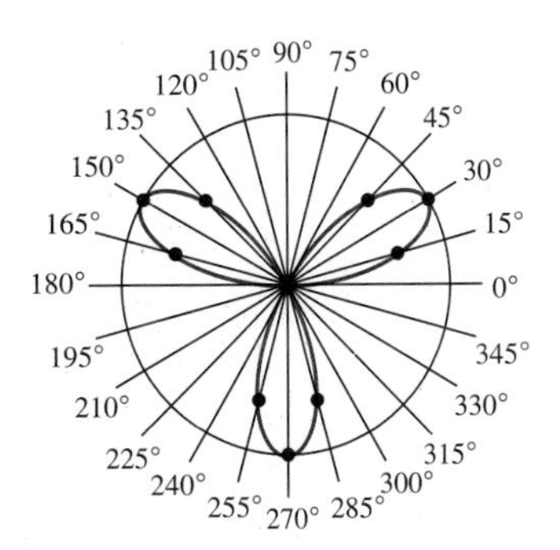

Figure 9.17

▲

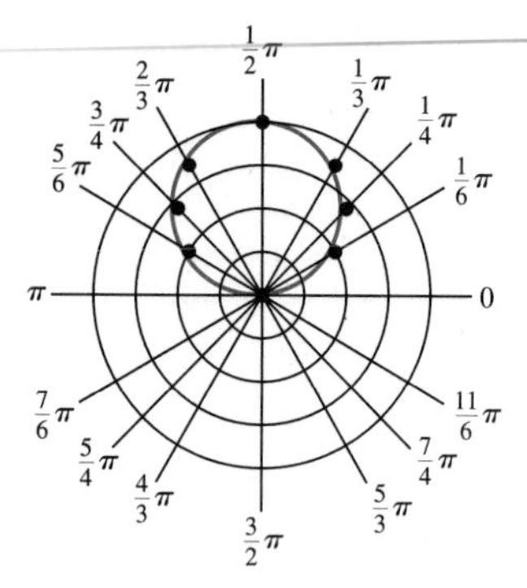

Figure 9.18

Example 6 Sketch the graph of the equation $r = 4 \sin \theta$ for $0 \le \theta \le \pi$.

SOLUTION The following table gives values of r corresponding to the indicated values of θ. Drawing a smooth curve through the plotted points, we obtain the *circle* shown in Figure 9.18. Notice that θ varies only from 0 to π radians. If we allow θ to vary from 0 to 2π radians, the graph will be traced out twice; once for $0 \le \theta \le \pi$ and again for $\pi < \theta \le 2\pi$. Demonstrate this by plotting points in the interval $\pi < \theta \le 2\pi$.

θ	0	$\frac{1}{6}\pi$	$\frac{1}{4}\pi$	$\frac{1}{3}\pi$	$\frac{1}{2}\pi$	$\frac{2}{3}\pi$	$\frac{3}{4}\pi$	$\frac{5}{6}\pi$	π
r	0	2	2.82	3.46	4	3.46	2.82	2	0

▲

To graph polar equations on graphing calculators, use the parametric mode with θ as the parameter. Some models have a built-in polar mode. Then covert polar equations of the form $r = f(\theta)$ into parametric equations with $\theta = t$ as the parameter to obtain

$$x = f(t)\cos t$$
$$y = f(t)\sin t$$

Note that these are formulas (9.9) and (9.10) with r replaced by $f(t)$ and t playing the role of θ.

Example 7 Use a graphing calculator to sketch the cardioid $r = 1 + \cos \theta$.

SOLUTION The parametric form of $r = 1 + \cos \theta$ with the parameter t is

$$x = (1 + \cos t)\cos t$$
$$y = (1 + \cos t)\sin t$$

The graph is shown below and is the same as that shown in Figure 9.16.

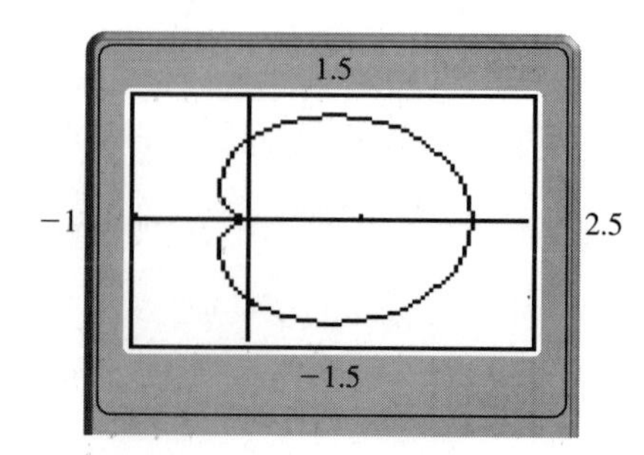

▲

Example 8 Use a graphing calculator to sketch the three-leaved rose $r = \sin 3\theta$.

SOLUTION The parametric form of $r = \sin 3\theta$ is

$$x = \sin 3t \cos t$$
$$y = \sin 3t \sin t$$

The calculator graph shown below is the same as the graph shown in Figure 9.17.

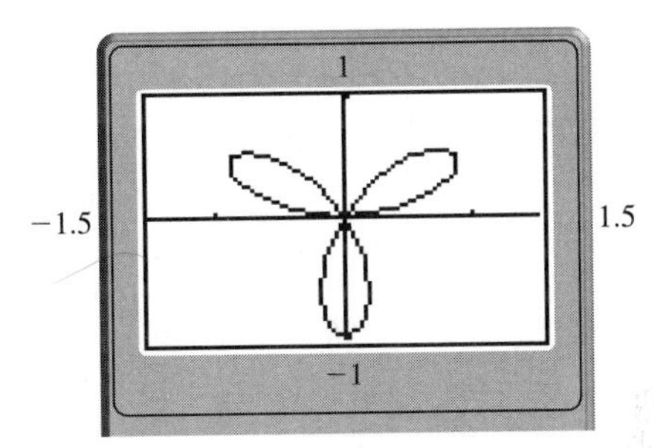

▲

▲▼ Exercises for Section 9.4

In Exercises 1–10 plot each point on polar coordinate paper.

1. $(5, 30°)$
2. $(3.6, -45°)$
3. $(12, \frac{2}{3}\pi)$
4. $(0.5, 220°)$
5. $(-7.1, 14°)$
6. $(-2, \frac{7}{3}\pi)$
7. $(1.75, -200°)$
8. $(\sqrt{2}, -311°)$
9. $(-5, -30°)$
10. $(5, 150°)$

In Exercises 11–18 convert each rectangular equation into an equation in polar coordinates.

11. $2x + 3y = 6$
12. $y = x$
13. $x^2 + y^2 - 4x = 0$
14. $x^2 - y^2 = 4$
15. $x^2 + 4y^2 = 4$
16. $xy = 1$
17. $x^2 = 4y$
18. $y^2 = 16x$

In Exercises 19–26 convert each polar equation into an equation in rectangular coordinates.

19. $r = 5$
20. $r = \cos \theta$
21. $r = 10 \sin \theta$
22. $r = 2(\sin \theta - \cos \theta)$
23. $r = 1 + 2 \sin \theta$
24. $r \sin \theta = 10$
25. $r = \dfrac{5}{1 + \cos \theta}$
26. $r(1 - 2 \cos \theta) = 1$

In Exercises 27–40 sketch the graph of the given equation.

27. $r = 5.6$
28. $r = \sqrt{2}$
29. $\theta = \frac{1}{3}\pi$
30. $\theta = 170°$
31. $r = 2 \sin \theta$
32. $r = 0.5 \cos \theta$

33. $r \sin \theta = 1$
34. $r \cos \theta = -10$
35. $r = 1 + \sin \theta$
36. $r = 1 - \cos \theta$
37. $r = \sec \theta$
38. $r = -\sin \theta$
39. $r = 4 \sin 3\theta$
40. $r = \sin 2\theta$

41. The radiation pattern of a particular two-element antenna is a cardioid of the form $r = 100(1 + \cos \theta)$. Sketch the graph of the radiation pattern of this antenna.
42. The radiation pattern of a certain antenna is given by

$$r = \frac{1}{2 - \cos \theta}$$

Plot this pattern.
43. Transform the polar equation in Exercise 42 into a rectangular equation.
44. The feedback diagram of a certain electronic tachometer can be approximated by the curve $r = \frac{1}{2}\theta$. Sketch the feedback diagram of this tachometer from $\theta = 0$ to $\theta = \frac{7}{6}\pi$.

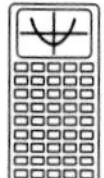

Graphing Calculator Exercises

In Exercises 1–8 use a graphing calculator to sketch the graph of the given polar equations. How does the calculator graph compare with the graph you sketched by hand in the indicated exercise?

1. $r = 5.6$ (Exercise 27)
2. $r = \sqrt{2}$ (Exercise 28)
3. $r = 2 \sin \theta$ (Exercise 31)
4. $r = 0.5 \cos \theta$ (Exercise 32)
5. $r = 1 + \sin \theta$ (Exercise 35)
6. $r = 1 - \cos \theta$ (Exercise 36)
7. $r = 4 \sin 3\theta$ (Exercise 39)
8. $r = \sin 2\theta$ (Exercise 40)

▲▼ Key Topics for Chapter 9

Define and/or discuss each of the following.

Complex Numbers
Graphical Representation of a Complex Number
Rectangular Form of a Complex Number
Complex Conjugate
Polar Coordinates
Polar Form of a Complex Number
DeMoivre's Theorem
Graphical Representation of a Polar Equation

▲▼ Review Exercises for Chapter 9

In Exercises 1–15 perform the indicated operations and express the results in the form $a + bi$. (Notice that $\overline{a + bi} = a - bi$.)

1. $(3 - 2i) + (6 - i)$
2. $(3 - 2i) - (i + 2)$
3. $(i + 7) - (i + 2)$
4. $(2 + i)(2 - i)$
5. $(3 - i)\overline{(3 + i)}$
6. $(i - 2)(2 + i)i$

7. $i(i^2 - 1)(i^2 + 1)$

8. $(6 - i)^2\overline{(6 + i)}$

9. $(3 + 2i)(2 - i) + \overline{(3 + 2i)(2 - i)}$

10. $\dfrac{i}{2 + i}$

11. $\dfrac{2i + 1}{i - 1}$

12. $\dfrac{\overline{6 - i}}{i^2}$

13. $\dfrac{i - 1}{\overline{i - 1}}$

14. $\dfrac{(9 - i)(9 - 2i)}{\overline{i - 2}}$

15. $\dfrac{(4 - i)(3 + 2i)}{i + 1}$

In Exercises 16–20 express each number in polar form.

16. $1 + i\sqrt{3}$

17. i

18. $i - 1$

19. $1 + i$

20. 4

In Exercises 21–25 express each number in rectangular form.

21. $2 \text{ cis } 45°$

22. $-3 \text{ cis } 75°$

23. $4 \text{ cis}(-20°)$

24. $(-2 \text{ cis } 30°)^2$

25. $(3 \text{ cis } 10°)^3$

In Exercises 26–30 convert each equation to polar coordinates and sketch.

26. $x^2 + y^2 + y = 0$

27. $y = 2x$

28. $y^2 + x^2 - 3x = 1$

29. $4x^2 + y^2 = 1$

30. $x = 4y^2$

In Exercises 31–40 convert each equation to rectangular coordinates and sketch.

31. $r = 2$

32. $r = 2 + 3 \cos \theta$

33. $r = \dfrac{2}{1 + \sin \theta}$

34. $r \cos \theta = 3$

35. $r = 3 \cos \theta$

36. $r = \theta$

37. $r = \sin 2\theta$

38. $r^2 = \sin \theta$

39. $r^2 - r = 0$

40. $r^2 - 3r + 2 = 0$

In Exercises 41–45 evaluate each expression. Leave the answer in polar form.

41. $(1 + i)^3$

42. $(-2 + 2i)^6$

43. $(-\sqrt{3} + i)^5$

44. $(\sqrt{3} \text{ cis } 60°)^4$

45. $(3 - i)^{10}$

In Exercises 46 and 47 find the indicated roots and sketch their locations in the complex plane. Leave the answers in polar form.

46. Square roots of i

47. Fifth roots of -1

Practice Test Questions for Chapter 9

In Exercises 1–10 answer *true* or *false*.

1. $3 + i$ is the complex conjugate of $3 - i$.
2. A number of the form $2 + 3i$ is called an imaginary number.
3. The sum of a complex number and its complex conjugate is a real number.

4. Two complex numbers are added by adding the real part to the imaginary part.
5. $a + bi$ is called the polar form of a complex number.
6. To multiply two complex numbers, multiply their magnitudes and add their arguments.
7. There is a one-to-one correspondence between the complex numbers and the points in the plane.
8. The complex number $z = 3 \text{ cis } 35°$ has a modulus of 3.
9. DeMoivre's theorem is valid for all real values of n.
10. DeMoivre's theorem is used to find the powers and roots of complex numbers.

In Exercises 11–20 fill in the blank to make the statement true.

11. The number i such that $i^2 = -1$ is called an __________ number.
12. The number $5 + 2i$ is the __________ of the number $5 - 2i$.
13. $z = r \text{ cis } \theta$ is called the __________ form of the complex number z.
14. The product of a complex number and its __________ is a nonnegative real number.
15. In the polar coordinate system, the distance from the pole to a point P is called the __________.
16. Given the complex number $z = r \text{ cis } \theta$, then θ is called the __________ of z.
17. The modulus of the product of two complex numbers is the __________ of the individual moduli.
18. DeMoivre's theorem states that if $z = r \text{ cis } \theta$, then $z^n =$ __________.
19. If $z = a + bi$, the modulus of z is given by $r =$ __________.
20. The nth roots of any number can be found using __________ theorem.

In the following exercises solve the stated problem. Show all your work.

21. Given $z_1 = 4 - i$ and $z_2 = 5 + 2i$, calculate $z_1 + z_2$ and $z_1 - z_2$.
22. Find the product $(-2 - i)(-7 + 2i)$.
23. Draw the graph of $z = -3 + 4i$. Draw the graph of the complex conjugate on the same axes.
24. Find the quotient $(4 - i)/(i - 3)$. Express your answer in the form $a + bi$.
25. Find the modulus of z, if $z = 1 - 2i$.
26. Plot $1 - \sqrt{3}i$ and then express it in polar form.
27. Plot $7 \text{ cis } 315°$ and then express it in rectangular form.
28. Evaluate $(\sqrt{5} \text{ cis } 120°)(7 \text{ cis } 80°)$. (Express in polar form.)
29. Find the cube roots of $-1 + i$.
30. Find all the roots of $x^5 + 32 = 0$.

Exponential and Logarithmic Functions

Logarithms were "invented" at the turn of the 17th century as an aid to the many arithmetic computations required by astronomers. Now modern calculators make logarithms obsolete as far as their computational task is concerned. However, a knowledge of the functional and graphical properties of logarithms is a requirement for many applications. We learn the fundamental properties of logarithms and their graphs by considering a logarithm as the inverse of the exponential. In this chapter there are some interesting applications of exponential and logarithmic functions.

10.1 Exponential Functions

In this section we describe *exponential functions*. The variable in an exponential function is in the exponent of some constant base, such as 2^x, 5^{-x}, 3^{-3x+2}, or a^x. In general we make the following definition.

DEFINITION 10.1 If a is any positive fixed constant, $a \neq 1$, then the exponential function with base a is defined by $f(x) = a^x$ where x is any real number.

But how do we give meaning to a variable exponent? When x is a rational number, there is no problem. When $x =$ the rational number p/q in lowest terms, then $a^{p/q}$ means the pth power of the qth root of a.

Example 1

(a) If $f(x) = 81^x$, then $f(\frac{1}{4}) = 3$ because $3^4 = 81$.

(b) If $f(x) = 8^x$, then $f(\frac{4}{3}) = 16$ because $(8^{1/3})^4 = 2^4 = 16$.

(c) If $f(x) = 32^x$, then $f(1.2) = 64$ because $32^{1.2} = 32^{6/5} = (32^{1/5})^6 = 2^6 = 64$

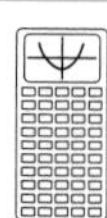

Exponentiation with rational numbers can be performed with a calculator. The keystroke process depends on the model of calculator. Two popular methods are given for evaluating $6.3^{1.43}$.

- **The use of a [y^x] button to obtain the value of $6.3^{1.43}$. We use the sequence 6.3 [y^x] 1.43 [=] to obtain the readout 13.90135.**
- **The use of a [^] key. In this case simply put in 6.3 [^] 1.43 and then [=] (or *enter*) to obtain the same result.**

The exponential function obeys the following laws of exponents.

RULE 1 $a^m a^n = a^{m+n}$
RULE 2 $a^m \div a^n = a^{m-n}$
RULE 3 $(a^m)^n = a^{mn}$

Example 2 Use the laws of exponents to solve the equation $4^{x+1} = 8$.

SOLUTION Since $4 = 2^2$ and $8 = 2^3$, the given equation can be written in the equivalent form

$$(2^2)^{x+1} = 2^3 \quad \text{or} \quad 2^{2(x+1)} = 2^3$$

For these two expressions to be equal, the exponents must be equal. Therefore

$$2(x+1) = 3$$
$$2x + 2 = 3$$
$$x = \frac{1}{2}$$

▲

To extend the concept of an exponent to include all real numbers, we can give some idea of the meaning of an irrational number exponent. First consider the meaning of 2^π. Since π is an irrational number, it has the unending and nonrepeating decimal expansion

$$3.141592653\ldots$$

Now consider the sequence of rational numbers 3, 3.1, 3.14, 3.141, 3.1415, and so forth. Each of these rational numbers is an approximation of the irrational number π; the more decimal places, the better the approximation. In the same way the corresponding rational powers of 2, namely,

$$2^3, 2^{3.1}, 2^{3.14}, 2^{3.141}, 2^{3.1415}$$

and so forth become better approximations of 2^π as more accurate estimates of π are used. In this way irrational exponents have a definite interpretation to which the exponential laws apply.

If $y = f(x)$ is an exponential function, then y is said to *vary exponentially* as x. The next example shows the characteristic shape of the graph of an exponential function.

Example 3 Sketch the graph of $y_1 = 2^x$ and $y_2 = (\frac{1}{2})^x$.

SOLUTION Figure 10.1 shows a table of values for each of the functions along with the graph.

x	$(\frac{1}{2})^x$
-2	4
-1	2
0	1
1	$\frac{1}{2}$
2	$\frac{1}{4}$

x	2^x
-2	$\frac{1}{4}$
-1	$\frac{1}{2}$
0	1
1	2
2	4

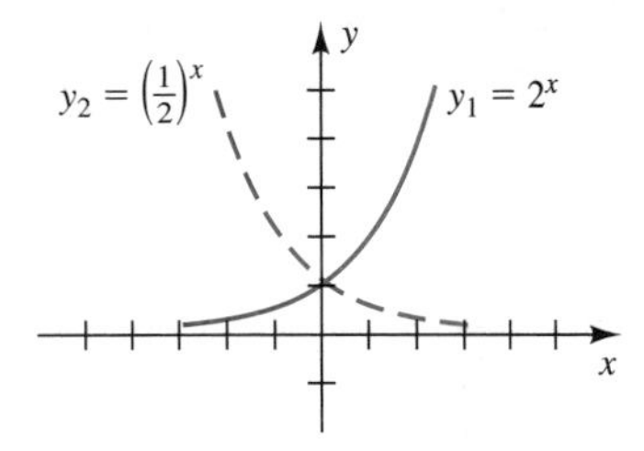

Figure 10.1

Notice that if we let $f(x) = 2^x$, then $f(-x) = 2^{-x} = (\frac{1}{2})^x$. Thus, the graphs of $y_1=2^x$ and $y_2=(\frac{1}{2})^x$ are reflections in the y-axis. ▲

We make the following observations about the function $f(x) = a^x$ and its graph.

PROPERTIES OF $f(x) = a^x$, $a > 1$

1. The domain of the function consists of all real numbers, and since $a^x > 0$ for all x, the range consists of all positive real numbers.
2. When $x = 0$, $a^0 = 1$. Hence the graph has a y-intercept at $(0, 1)$.
3. When $x = 1$, $a^1 = a$. Hence the graph passes through $(1, a)$.
4. As x increases without bound, a^x increases without bound.
5. As x decreases without bound, a^x approaches 0. Hence the graph is asymptotic to the negative x-axis.
6. If $x_1 < x_2$, then $a^{x_1} < a^{x_2}$. Hence a^x is an **increasing** function.

If $0 < a < 1$, then properties (4)–(6) become

4′. As x increases without bound, a^x approaches 0. In this case the positive x-axis is a horizontal asymptote.

5′. As x decreases without bound, a^x increased without bound.

6′. If $x_1 < x_2$, then $a^{x_1} > a^{x_2}$. Hence a^x is a decreasing function.

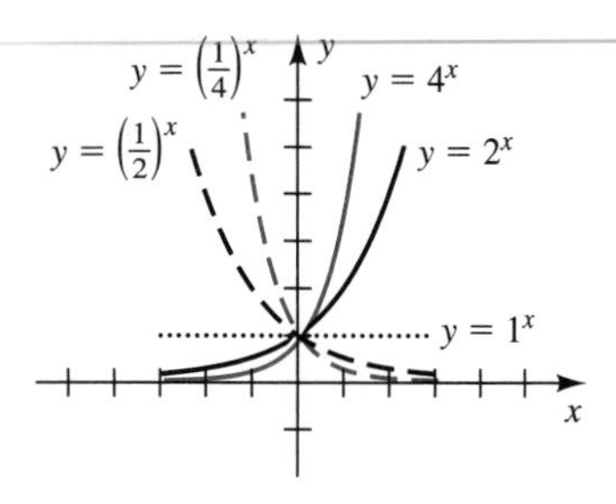

Figure 10.2

A different exponential function is obtained for each value of a, although the shape of the graph remains basically the same and the same kind of functional properties as (1)–(6) continue to hold.

Figure 10.2 shows the graph of a^x for $a = \frac{1}{4}, \frac{1}{2}$, 1, 2 and 4. We exclude the value of $a = 1$ because not only is the graph of 1^x trivial, but it is not of exponential shape. Furthermore, $f(x) = 1^x$ does not obey the functional properties (1) through (6), which are ordinarily associated with exponential functions. Hence the only acceptable bases are $a > 0$, $a \neq 1$.

WARNING Be aware of the fact that -3^x and $(-3)^x$ have different meanings [for example, $-3^2 = -9 \neq (-3)^2 = 9$]. The latter is an "unacceptable" exponential function since $-3 < 0$, while -3^x is merely the negative of 3^x. (See Figure 10.3.)

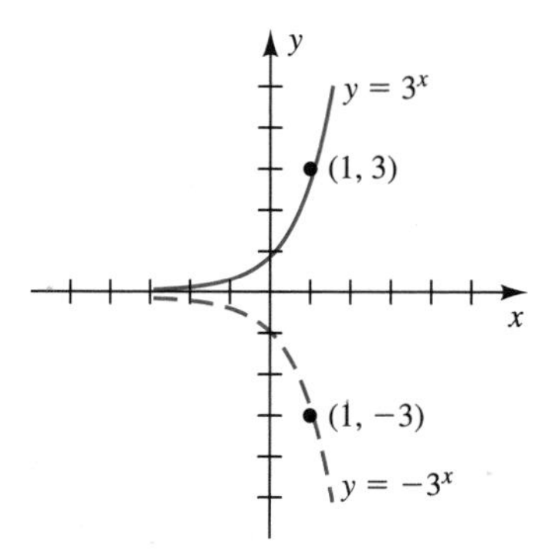

Figure 10.3

Exponential Functions with Base *e*

Exponential functions are essential to the description and understanding of growth and decay processes such as population size, radioactive decay, and compound interest. The most important base for exponential functions is the irrational number e, which is approximated to three decimal places by 2.718. Remember that irrational numbers do not have an exact decimal representation, so 2.718 is only an approximation of e. The purpose of this section is to give you some idea of why the number e is chosen as a base for exponential functions.

Most calculators have an [e^x] button. To find e^x for any number, enter x and push the [e^x] button. Thus you find that $e^{3.1}$ = [22.19795128] by entering 3.1 and pushing the [e^x] button. Note also that if you enter 1 and then push [e^x], the display will show [2.718281828], which is e approximated to nine decimal places.

Example 4 Sketch the graph of $y = 2 + e^x$.

SOLUTION The indicated table was constructed using a calculator. The graph is shown in Figure 10.4.

x	e^x	$2 + e^x$
−3	.05	2.05
−2	.14	2.14
−1	.37	2.37
0	1.00	3.00
1	2.72	4.72
2	7.39	9.39

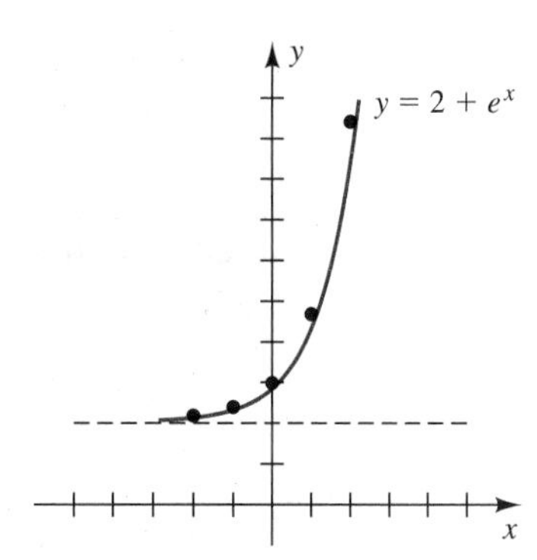

Figure 10.4

Note that as x decreases, $2 + e^x$ approaches 2. Thus $y = 2$ is an asymptote and is indicated in the figure by the dashed line. ▲

Applications of Exponential Functions

The exponential function is one of the most useful functions for describing physical and social phenomena, of which the following are just a few.

- Electric current
- Atmospheric pressure
- Decomposition of uranium
- Population growth
- Learning process
- Compound interest

If a physical quantity obeys a law that is described by an exponential function and is increasing, it is said to *increase exponentially*. If it is exponential in character but is decreasing, it is said to *decrease or decay exponentially*.

Example 5 A company finds that the net sales of its product double every year. Write this fact as an exponential function.

SOLUTION Let $s(t)$ represent sales at any time t and let s_0 be the initial sales. Then we can write

$$s(0) = s_0$$
$$s(1) = 2s_0$$
$$s(2) = 2s(1) = 2^2 s_0$$
$$s(3) = 2s(2) = 2^3 s_0$$

In general

$$s(t) = s_0 2^t$$

is the desired function. ▲

Example 6 An unmanned satellite has a radioisotope power supply whose power output in watts is given by the equation

$$P = 50e^{-t/260}$$

where t is the time in days that the battery has been in operation. How much power will be available at the end of one year? Give the answer to one decimal place.

SOLUTION Applying the given formula with $t = 365$,

$$P = 50e^{-365/260} \approx \boxed{12.28251678} \approx 12.3 \text{ W}$$ ▲

Example 7 Population growth is described by an exponential function of the form

$$P = P_0 e^{kt}$$

where P_0 is the initial population size, t is the elapsed time, and k is a constant. Estimate the population of the asteroid Malthus in 2025 and 2075 if the population in 1975 was 500. Assume $k = 0.032$.

SOLUTION The growth equation for this population is

$$P = 500e^{0.032t}$$

Using this formula with $t = 50$, we have

$$P = 500e^{1.6} \approx \boxed{2476.516212} \approx 2477$$

If $t = 100$, the population is

$$P = 500e^{3.2} \approx \boxed{12266.2651} \approx 12{,}267$$

Thus in 2025 the population of the asteroid is 2477, or almost five times what it was in 1975. By 2075 the population has increased to 12,267, which is about 25 times the population in 1975. (Some people think the population in 1975 was 750, but the census data shows 500.) ▲

Example 8 An approximate rule for atmospheric pressure at altitude less than 50 mi is the following: *Standard atmospheric pressure,* 14.7 *lb/in.*^{2}, *is halved for each* 3.5 *mi of vertical ascent.*

(a) Write an exponential function to express this rule.

(b) Compute the atmospheric pressure at an altitude of 19.5 mi.

SOLUTION

(a) If we let P denote the atmospheric pressure at less than 50 mi and h the altitude in miles, the general expression will be of the form

$$P = P_0 a^{kh}$$

where P_0 is the value of P when $h = 0$; that is, 14.7 lb/in.^{2}. Since the pressure is halved for each 3.25 mi of ascent, $a = \frac{1}{2}$. Since the halving is accomplished every 3.25 mi, $k = \frac{1}{3.25}$. Hence

$$P = 14.7\left(\frac{1}{2}\right)^{h/3.25}$$

This function is shown graphically in Figure 10.5.

(b) Using the expression for P and letting $h = 19.5$, we get

$$\begin{aligned} P &= 14.7\left(\frac{1}{2}\right)^{19.5/3.25} \\ &= 14.7\left(\frac{1}{2}\right)^{6} = \frac{14.7}{64} \approx 0.23 \text{ lb/in}^2 \end{aligned}$$

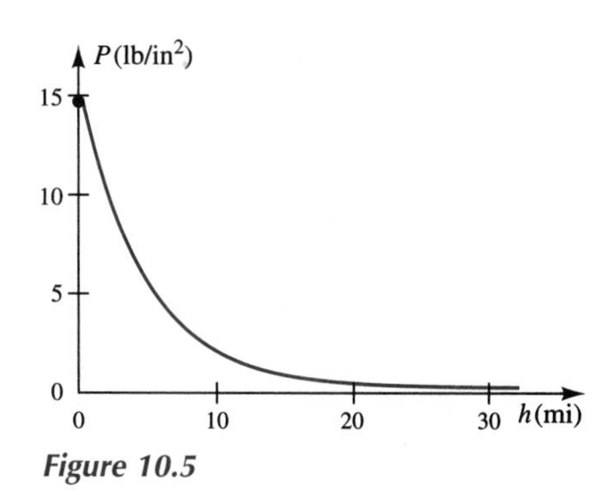

Figure 10.5 ▲

Example 9 A psychologist finds that the number of different mazes, R, a laboratory animal can learn to run is represented by the exponential function $R = 5(1 - e^{-t})$, where t is the time in days spent training the animal. Graph this function for $t \geq 0$, assuming that its domain is the set of all positive real numbers.

SOLUTION Letting $t = 0$ in the expression for R, we obtain

$$R(0) = 5(1 - e^{-0}) = 5(1 - 1) = 0$$

As t increases, e^{-t} decreases, and hence $1 - e^{-t}$ approaches the asymptote $y = 1$; consequently R approaches 5. The graph in Figure 10.6 shows that the ability of the animals to learn increases rapidly at first and then levels off at 5 as t becomes large. This curve is called a **learning curve**.

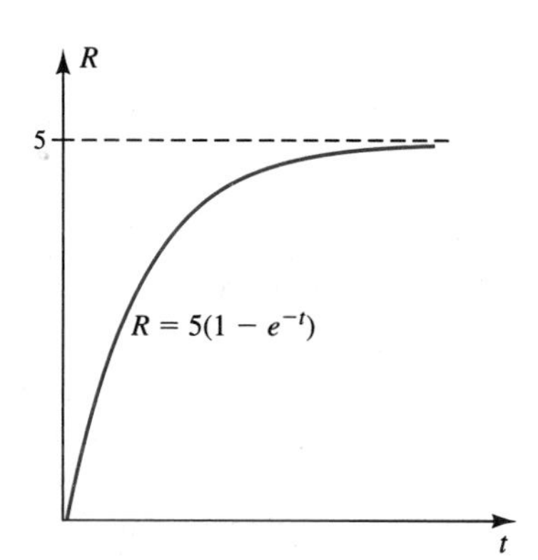

Figure 10.6 ▲

Exercises for Section 10.1

Solve for x in Exercises 1–10.

1. $x^{-3/2} = \frac{27}{8}$
2. $27^{-2/3} = x$
3. $27^{x+2} = 9^{2-x}$
4. $2^{x+4} = 8$
5. $(-2)^{x+2} = -8$
6. $(-x)^{-3} = \frac{27}{64}$
7. $[(3)^x]^x = 1$
8. $(2^x)^{x-1} = 64$
9. $3^{x^2+4x} = \frac{1}{81}$
10. $4^{x^2-3x} = 256$

Sketch the graph of each exponential function in Exercises 11–26.

11. $y = 5^x$
12. $h = 10^x$
13. $y = 3^{-x}$
14. $p = 10^{-t}$
15. $s = 3(2^x)$
16. $y = 2 + 2^x$
17. $y = 2 - 2^x$
18. $y = (\sqrt{2})^x$
19. $m(t) = 3^{3-t}$
20. $y = 1 + e^x$
21. $y = e^{x/2}$
22. $y = 10 + e^{3x/2}$
23. $f(t) = e^{1-t}$
24. $z = 2e^T$
25. $y = e^2e^x$
26. $g(x) = \dfrac{e}{e^x}$

27. Sketch the graph of $y = 2^x$. Using this graph, approximate the values of **(a)** $2^{\sqrt{2}}$; **(b)** 2^{π}; **(c)** $2^{3/2}$; **(d)** $2^{-3/2}$; **(e)** $2^{1/3}$.

28. Why is the base of exponential functions restricted to positive numbers?

29. Discuss the exponential function with base 1.

In Exercises 30–35 tell how the graph of the given exponential function compares with the graph of $y = 2^x$. In each case make a sketch to show the comparison.

30. 2^{-x}
31. -2^x
32. $2^{|x|}$
33. $|2^x|$
34. $3 + 2^x$
35. 2^{x+3}

36. The population of water spiders on a pond is estimated to be 1000. Using the population growth equation in Example 7, estimate the population for $t = 25$ if $k = 0.02$.

37. The mass m (in kilograms), of a certain radioactive substance decays with time according to $m = 100e^{-0.1t}$, where t is elapsed time in days. Draw an accurate sketch of mass versus time for $t > 0$. Using this graph, estimate the half-life of the substance. (*Hint:* The *half-life* is the time in which the mass of a substance decays to one-half its original value.)

38. A company has found that a person new to the assembly line produces items according to the function

$$N = 50 - 50e^{-0.5t}$$

where t is the number of days he or she has worked on the line. How many items will a new worker produce on the fifth day? Graph this function for $t > 0$, assuming it to be a function for real numbers, not just the positive integers.

39. If P dollars is deposited in a bank paying interest i compounded annually for t years, the amount of money A in the account at the end of that time is given by the formula $A = P(1 + i)^t$. Find the amount of money in an account after 5 yr for an initial investment of \$1000 at an interest rate of 5%.

40. With no promotional activity, the sales of a product decrease at a rate described by the exponential function $A(t) = A_0e^{-kt}$, where A_0 is the amount of the sales when $t = 0$ and k is the *sales decay constant*. What will the sales be after 1 yr if $A_0 = 10{,}000$ for a sales decay constant of 0.5?

Use a calculator to evaluate Exercises 41–48.

41. Find the velocity (in m/sec) of a rocket for $t = 2.75$ sec and $v = 10.772e^{2.032t}$.

42. If the mass (in grams, or g) of a radioactive material varies with time according to $m = 558.76e^{-0.01749t}$, what is the mass when (a) $t = 3.95$ and (b) $t = 5.50$?

43. What is the temperature of an object when $t = 7.5$ min if the temperature T varies according to $T = 198.6 + 95.3e^{-0.147t}$?

44. Given $F(x) = 16.79x^2e^{-3.11x}$, find $F(1.665)$.

45. Given $f(t) = \sqrt{t + 3.346}\,e^{0.0178t}$, find $f(30.542)$.

46. Given $g(t) = t^3e^{t^2} + 156.79$, find $g(2.069)$.

47. Given $H(x) = (e^{0.15x} + e^{-0.15x})e^{3.19-x}$, find $H(0.7683)$.

48. Given $r(n) = 0.00156(8013.6 - 13.607e^{1.116n})/n$, find $r(9.0075)$.

10.2 The Logarithm Function

A horizontal line will intersect the graph of the exponential function $y = a^x$ only once. Hence the function is one-to-one; that is, for each value of x, there is at most one value of y and for each value of y, there is a unique x. Using this property we can define another important function, the inverse of the exponential function.

DEFINITION 10.2

Let $a > 0$ and $a \neq 1$. Then for $x > 0$,

$$y = \log_a x \quad \text{means} \quad x = a^y$$

The function $\log_a x$ is called the **logarithm function** of x to the base a.

Thus the value of $\log_a x$ is the exponent to which the base a must be raised to give the value x. In other words, **every logarithm is an exponent of the base.** The next example shows how the logarithm of a number is found directly from the definition.

Example 1

(a) $\log_2 16 = 4$ because $2^4 = 16$.

(b) $\log_2 (\frac{1}{8}) = -3$ because $2^{-3} = \frac{1}{8}$.

(c) $\log_{10} (1/1000) = -3$ because $10^{-3} = 1/1000$.

In each of the cases, note that the logarithm is an exponent. ▲

Example 2 Find $\log_2 32$.

SOLUTION Let $y = \log_2 32$. Then $32 = 2^y$, or $y = 5$. ▲

Example 3 Find b, given that $\log_b 4 = -\frac{1}{2}$.

SOLUTION Writing this in exponential form, we have $4 = b^{-1/2}$. Solving $4 = 1/b^{1/2}$ for b, we find that $16 = 1/b$ and $b = \frac{1}{16}$. ▲

Example 4 Solve $3 = \log_4 (x^2 + 2)$ for x.

SOLUTION

$$3 = \log_4 (x^2 + 2)$$

$$4^3 = x^2 + 2 \qquad \text{Definition of the logarithm}$$

$$x^2 = 62 \qquad \text{Subtracting 2 from both sides}$$

$$x = \pm \sqrt{62} \qquad \text{Taking the square root}$$

▲

Example 5 Determine the values of x for which

$$1 < \log_3 (2x + 5) < 2$$

SOLUTION We use the fact that

$$1 < \log_3 u < 2 \qquad \text{when} \qquad 3^1 < u < 3^2$$

Thus we are to solve the inequality

$$3 < 2x + 5 < 9$$

Subtracting 5 from each member, we have

$$-2 < 2x < 4$$

which gives

$$-1 < x < 2$$

▲

Although the base of the logarithm may be any positive number except 1, in practice we seldom use bases other than 10 and e. Logarithms with base 10 are called **common logarithms** and are usually written just $\log x$, with the base understood. Logarithms with base e have a special notation.

DEFINITION 10.3 The function defined by $f(x) = \log_e x$ is called the **natural logarithmic function** and is denoted by $f(x) = \ln x$, $x > 0$.

Example 6 The intensity (energy) level β of a soundwave with intensity I is defined to be

$$\beta = 10 \log(I/I_0) \text{ db (decibels)}$$

where I_0 is the minimum intensity detectable by the human ear. (When two sounds differ in intensity by a factor of 10, they differ in loudness by 1 bel; a difference of 100 means a loudness difference of 2 bels. In practice the unit used is the decibel, one-tenth of a bel.) Solve for I.

SOLUTION Dividing both sides of $\beta = 10 \log (I/I_0)$ by 10, we have

$$\frac{\beta}{10} = \log(I/I_0)$$

Using the fact that $y = \log_b x$ means $x = b^y$, we have

$$\frac{I}{I_0} = 10^{\beta/10}$$

from which

$$I = I_0\, 10^{\beta/10}$$ ▲

Example 7 The relationship between the velocity, v, of a space vehicle, its exhaust velocity, c, and its mass ratio, R, is given by the exponential relationship.

$$R = e^{v/c}$$

Write this expression logarithmically by solving for v.

SOLUTION Using the fact that $y = e^x$ may also be written as $x = \ln y$, we have

$$\frac{v}{c} = \ln R$$

from which

$$v = c \ln R$$ ▲

Because of the definition, the domain and the range of the logarithm function are the same as the range and the domain of the exponential function, respectively. From Chapter 8, this means these two functions are inverses. Thus the domain of $y = \log_a x$ is the set of all positive reals (sometimes we say that the logarithm of a nonpositive number is undefined) and the range is the set of all real numbers. A diagram of these relationships is shown below.

Exponential function		Logarithm function
$y = a^x$		$y = \log_a x$
Domain	$\longrightarrow$ all real numbers $\longleftarrow$	Range
Range	$\longrightarrow$ all positive reals $\longleftarrow$	Domain

We may sketch the graph of the logarithm function by first constructing a table of values. The tables for $y = \ln x$ and $y = \log x$ are included on the next page, and the corresponding graphs are presented in Figure 10.7.

x	$\frac{1}{8}$	$\frac{1}{4}$	$\frac{1}{2}$	1	e	e^2	e^3
$\ln x$	−2.08	−1.39	−.69	0	1	2	3

x	$\frac{1}{1000}$	$\frac{1}{100}$	$\frac{1}{10}$	1	10	100
$\log x$	−3	−2	−1	0	1	2

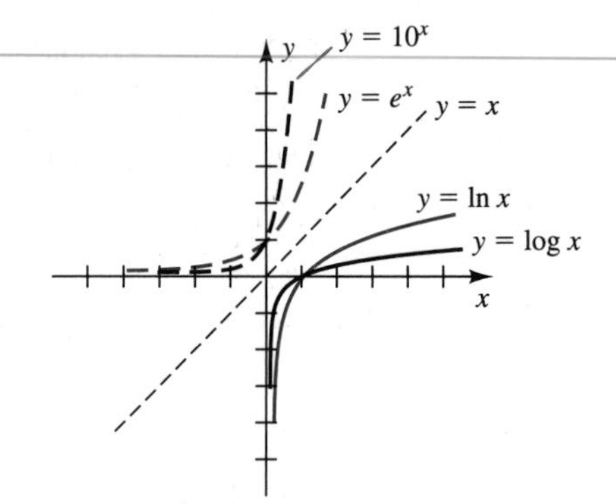

Figure 10.7

COMMENT The graphs of $y = e^x$ and $y = 10^x$ are also shown in Figure 10.7. Notice that the graph of $y = e^x$ is the mirror image of $y = \ln x$ in the line $y = x$ and that the same observation can be made about $y = 10^x$ and $y = \log x$. In fact for $a > 0$, $a \neq 1$, the graphs of $y = a^x$ and $y = \log_a x$ are mirror reflections of each other in $y = x$. This demonstrates graphically the inverse nature of the exponential and logarithmic functions.

Each of the solid curves of Figure 10.7 is characteristic of what is called the **logarithmic shape.** The figure clearly demonstrates the following properties of the logarithm function for $a > 1$. Any function that has these properties is said to **behave logarithmically.**

PROPERTIES OF $f(x) = \log_a x$, $a > 1$

1. $\text{Log}_a\, x$ is not defined for $x \leq 0$, so the domain is the set of positive real numbers.
2. $\text{Log}_a 1 = 0$.
3. $\text{Log}_a a = 1$.
4. The value of $\log_a x$ is negative for $0 < x < 1$ and positive for $x > 1$. Hence the range is the set of all real numbers.
5. As x approaches 0, $\log_a x$ decreases without bound. The graph is asymptotic to the negative y-axis.
6. As x increases without bound, $\log_a x$ increases without bound.
7. If $x_1 < x_2$, then $\log_a x_1 < \log_a x_2$. Hence $\log_a x$ is an increasing function.

More generally, we must work with logarithmic functions of the type $y = \log\,[g(x)]$, where $g(x)$ is some function of x.

Example 8 Sketch the graph of the function $y = \ln(-x)$.

SOLUTION At first glance you might conclude that this function is undefined since the logarithm of a negative number is undefined. However, $-x$ is positive when $x < 0$, so $\ln(-x)$ has a domain of $x < 0$. The graph passes through $(-e, 1)$ and $(-1, 0)$ since $\ln(-(-e)) = \ln e = 1$ and $\ln(-(-1)) = \ln 1 = 0$. The graph is shown in Figure 10.8. Notice that the characteristic logarithmic shape of $\ln(-x)$ is the mirror reflection of $\ln x$ in the y-axis.

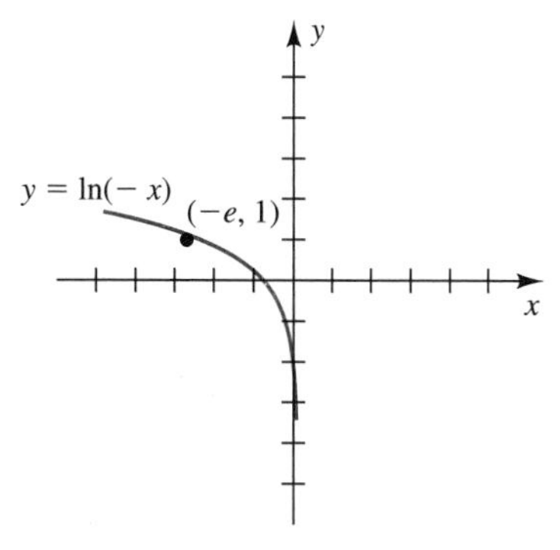

Figure 10.8 ▲

Example 9 Sketch the graph of $y = \ln(2x - 1)$.

SOLUTION Here we note that the function is *not* defined for values of $x \le \frac{1}{2}$. Furthermore, when $x = 1$, $y = \ln 1 = 0$, and when $x = \frac{1}{2}e + \frac{1}{2}$, $y = \ln e = 1$. These facts are summarized in the table and the graph drawn in Figure 10.9.

$y = \ln(2x - 1)$

x	y
$\frac{1}{2}$	undef.
1	0
$\frac{1}{2}e + \frac{1}{2} \approx 1.859$	1
increasing w/o bound	increasing w/o bound

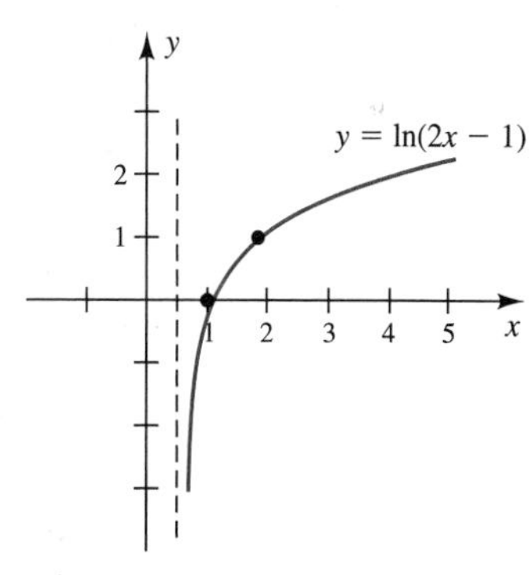

Figure 10.9 ▲

Exercises for Section 10.2

Write a logarithmic expression equivalent to the exponential expression given in Exercises 1–5.

1. $x = 2^3$ **2.** $y = 3^8$ **3.** $M = 5^{-3}$

4. $N = 10^{-2}$ **5.** $L = 7^2$

In Exercises 6–10 write an exponential expression equivalent to the given logarithmic expression. In each case find the base of the logarithm function.

6. $\log_b 8 = 3$ **7.** $\log_b 4 = 2$ **8.** $\log_b 0.25 = -2$
9. $\log_b 100 = 2$ **10.** $\log_b b = 1$

In Exercises 11–30 solve for the unknown.

11. $\log_{10} x = 4$
12. $\log_5 N = 2$
13. $\log_x 10 = 1$
14. $\log_x 25 = 2$
15. $\log_x 64 = 3$
16. $\log_{16} x = 2$
17. $\log_{27} x = \frac{2}{3}$
18. $\log_2 \frac{1}{8} = x$
19. $\log_3 9 = x$
20. $\log_{10} 10^7 = x$
21. $\log_b b^a = x$
22. $\log_b x = b$
23. $\log_x 2 = \frac{1}{3}$
24. $\log_x 0.0001 = -2$
25. $\log_x 6 = \frac{1}{2}$
26. $\log_b x = 0$
27. $6^{\log_6 x} = 6$
28. $x^{\log_x x} = 3$
29. $\log_2 (x + 3) = -1$
30. $\log_2 (x - 1) = 3$

In Exercises 31–34 solve the given logarithmic inequalities.

31. $\log_3 (x + 1) < 2$
32. $\log_2 (2 - x) > -1$
33. $2 \leq \log_2 x \leq 3$
34. $0 < \log x < 1$

Sketch a graph of the functions in Exercises 35–43. (Remember that $\log x$ means $\log_{10} x$.)

35. $f(x) = \ln x$
36. $f(x) = -\log x$
37. $f(x) = \log |x|$
38. $f(x) = \log (x - 1)$
39. $f(x) = \log 2x$
40. $g(x) = \log_3 (-x)$
41. $f(x) = \log_2 (1 - x)$
42. $f(x) = \log_2 (1 + x)$
43. $f(x) = \log_2 |1 + x|$

In Exercises 44–49 describe how the graph of the given logarithmic function is related to the graph of $y = \log x$. In each case make a sketch to show the comparison.

44. $\log (-x)$ **45.** $-\log x$ **46.** $\log |x|$
47. $|\log x|$ **48.** $3 + \log x$ **49.** $\log (x + 3)$

Find the base of the logarithm function $y = \log_a x$ whose graphs contain the points in Exercises 50–54.

50. $(100, 2)$ **51.** $(4, 2)$ **52.** $(64, 3)$
53. $(0.1, -1)$ **54.** $(0.5, -1)$

55. Let $f_b(x) = \log_b x$. For any acceptable base the graph of $f_b(x)$ passes through a point independent of b. What is that point?

56. Let $f(x) = \log_3 x$. Find $f(9)$, $f(\frac{1}{27})$, and $f(81)$.

57. Let $f(x) = \log_2 x$. By example show that
(a) $f(x + y) \neq f(x) + f(y)$
(b) $f(ax) \neq af(x)$

58. On the same set of axes, sketch $\log_2 x$ and $\log_3 x$ and
 (a) solve the equation $\log_2 x = \log_3 x$.
 (b) solve the inequality $\log_2 x > \log_3 x$.

59. How are the graphs of $y = \log_2 x$ and $y = 2^x$ related?

In Exercises 60–67 use a calculator to approximate the given logarithm.

60. $\log 2.67$
61. $\log 0.5$
62. $\log 5000$
63. $\log 5 + \log 2$
64. $\ln 10$
65. $\ln e^{-1}$
66. $\ln 1000$
67. $\ln 0.5$

68. A power supply has a power output in watts approximated by the equation

$$P = 64e^{-3t}$$

where t is in days. Solve this expression for t.

69. A certain radioactive material decays exponentially by the equation

$$A(t) = A_0e^{-t/5}$$

where A_0 is the initial amount of material. Find the time t_h required for the radioactive material to decay to one-half its initial value. The time t_h is called the "half-life" of the material.

10.3 Basic Properties of Logarithms

Since the logarithm function and the exponential function are inverses, the laws of exponents given in Section 10.1 have their analogous consequences in the rules for logarithms.

RULE 1

$$\log_a(MN) = \log_a M + \log_a N$$

(i.e., the logarithm of a product is the sum of the individual logarithms)

RULE 2

$$\log_a(M/N) = \log_a M - \log_a N$$

(i.e., the logarithm of a quotient is the difference of the logarithms of the numerator and denominator)

RULE 3

$$\log_a M^c = c \log_a M$$

(i.e., the logarithm of a number to a power is the power times the logarithm of the number)

RULE 4

$$\log_a x = \log_a y \leftrightarrow x = y$$

(i.e., if the logarithms of x and y are equal, then x and y are equal)

Examine these rules carefully and notice where they *do* apply as well as where they do *not* apply. For example, there is no rule for simplifying expressions of the formula $\log_a (x + y)$.

WARNING $\log_a (x + y)$ is NOT equal to $\log_a x + \log_a y$. The following example shows why.

$$\log_2 (8 + 8) = \log_2 16 = 4$$
$$\log_2 8 + \log_2 8 = 3 + 3 = 6 \neq 4$$

Example 1

(a) $\log_2 [(8)(64)] = \log_2 8 + \log_2 64 = \log_2 2^3 + \log_2 2^6 = 3 + 6 = 9$

(b) $\log_3 \sqrt{243} = \log_3 243^{1/2} = \frac{1}{2} \log_3 243 = \frac{1}{2}(5) = 2.5$

(c) $\log_2 (\frac{3}{5}) = \log_2 3 - \log_2 5$

(d) $\log (4 \cdot 29/5) = \log 4 + \log 29 - \log 5$ ▲

Example 2 Given that $\log 2 = 0.3010$ and $\log 3 = 0.4771$, find $\log 45$.

SOLUTION Since $45 = 9 \cdot 5$, we have

$$\begin{aligned} \log 45 &= \log 9 + \log 5 \\ &= \log 3^2 + \log(\tfrac{10}{2}) \\ &= 2 \log 3 + \log 10 - \log 2 \\ &= 2(0.4771) + 1 - 0.3010 \\ &= 1.6532 \end{aligned}$$ ▲

Example 3 Write the expression $\log x - 2 \log x + 3 \log (x + 1) - \log (x^2 - 1)$ as a single term.

SOLUTION Proceed as follows:

$$\log x - 2 \log x + 3 \log (x + 1) - \log (x^2 - 1)$$
$$= \log x - \log x^2 + \log (x + 1)^3 - \log (x^2 - 1) \qquad \text{Rule 3}$$
$$= \log \frac{x(x + 1)^3}{x^2(x^2 - 1)} = \log \frac{(x + 1)^2}{x(x - 1)} \qquad \text{Rules 1 and 2 and the cancellation law}$$ ▲

Example 4 Given $\ln x = 3$, find $\ln (1/x)$.

SOLUTION
$$\begin{aligned} \ln (1/x) &= \ln x^{-1} \\ &= -\ln x \\ &= -3 \end{aligned}$$ ▲

Exercises for Section 10.3

Evaluate the logarithms in Exercises 1–8.

1. $\log_2 (32 \cdot 16)$
2. $\log_2 16^5$
3. $\log_5 25^{1/4}$
4. $\log_3 27$
5. $\log_3 (27 \cdot 9 \cdot 3)$
6. $\log_2 (64 \cdot 32 \cdot 8)$
7. $\log_2 (8 \cdot 32)^3$
8. $\log_3 (9 \cdot 81)^8$

Given log 2 = 0.3010, log 3 = 0.4771, and log 7 = 0.8451, find the logarithms in Exercises 9–20.

9. $\log \frac{3}{2}$
10. $\log 4$
11. $\log 12$
12. $\log 30$
13. $\log 90$
14. $\log \sqrt{2}$
15. $\log \sqrt{5}$
16. $\log 21^{1/3}$
17. $\log 2400$
18. $\log 0.00018$
19. $\log 0.0014$
20. $\log 42000$

In Exercises 21–28 write the given expression as a single logarithmic term.

21. $\log_2 x^2 - \log_2 x$
22. $\log_2 (x^2 - 1) - \log_2(x - 1)$
23. $\log x + \log (1/x)$
24. $\log 3x + 3 \log(x + 2) - \log(x^2 - 4)$
25. $\log 5t + 2 \log(t^2 - 4) - \frac{1}{2} \log(t + 3)$
26. $\log z - 3 \log 3z - \log(2z - 9)$
27. $3 \log u - 2 \log(u + 1) - 5 \log(u - 1)$
28. $\log t + 7 \log(2t - 8)$
29. Let $\ln I = (-R/L)t + \ln I_0$. Show that $I = I_0 e^{-Rt/L}$.
30. If y is directly proportional to x^p, what relation exists between $\log y$ and $\log x$?
31. Compare the functions $f(x) = \log x^2$ and $g(x) = 2 \log x$. In what way are they the same? In what way different?
32. Let $f(x) = \log_a x$ and $g(x) = \log_{1/a} x$. Show that $g(x) = f(1/x)$.
33. If $\log_a x = 2$, find $\log_{1/a} x$ and $\log_a (1/x)$.
34. Compare the graphs of the functions $f(x) = \log_2 2x$, $g(x) = \log_2 x$, $h(x) = \log_2 \sqrt{x}$, and $m(x) = \log_2 x^2$.
35. Given the graph of $y = \log x$, explain a convenient way to obtain the following graphs. Assume $p > 0$.
    (a) $\log x^p$ **(b)** $\log px$
    (c) $\log(x + p)$ **(d)** $\log(x/p)$
36. If $f(x) = \log_b x$, is $f(x + y) = f(x) + f(y)$?

In Exercises 37–47 use a calculator to verify the given statement.

37. $\ln 6 = \ln 2 + \ln 3$
38. $\ln 100 = \ln 50 + \ln 2$
39. $\ln 9 = 2 \ln 3$

40. $\ln [16\sqrt{2}] = 2 \ln 4 + \frac{1}{2} \ln 2$

41. $\ln [(62.3)(28.6)] = \ln 62.3 + \ln 28.6$

42. $\ln e^{\pi} = \pi$

43. $\ln 12 = \ln 36 - \ln 3$

44. $\log [25 \cdot 34] = \log 25 + \log 34$

45. $\log [16 \cdot \pi] = \log 16 + \log \pi$

46. $\log \frac{125}{73} = \log 125 - \log 73$

47. $\log \frac{17}{35} = \log 17 - \log 35$

10.4 Exponential and Logarithmic Equations

Equations in which the variable occurs as an exponent are called **exponential equations.** To solve these equations, we use the fact that the logarithm is a one-to-one function. Thus $\log x = \log y$ if and only if $x = y$. Hence if both sides are positive, taking the logarithm of both sides yields an equivalent equation.

Example 1 Solve the exponential equation $3^x = 2^{2x+1}$

SOLUTION Taking the common logarithm of both sides, we write

$$\log 3^x = \log 2^{2x+1}$$

$$x \log 3 = (2x + 1) \log 2 \qquad \text{Using Rule 3}$$

$$x(\log 3 - 2 \log 2) = \log 2 \qquad \text{Expanding and collecting like terms}$$

$$x = \frac{\log 2}{\log 3 - 2 \log 2} \qquad \text{Solving for } x$$

$$\approx \boxed{-2.40942084} \qquad \text{Using a calculator}$$

$$x \approx -2.41 \qquad \text{Rounding off to two decimals}$$

▲

Example 2 The expression $(e^x - e^{-x})/2$ is called the hyperbolic sine of x and is denoted by $\sinh x$. Solve the equation $\sinh x = 3$. Round off the answer to two decimal places.

SOLUTION

$$\frac{e^x - e^{-x}}{2} = 3 \qquad \text{Replacing } \sinh x \text{ by its equal}$$

$$e^x - e^{-x} = 6 \qquad \text{Multiplying both sides by 2}$$

$$e^{2x} - 6e^x - 1 = 0 \qquad \text{Adding } -6 \text{ to both sides and multiplying by } e^x$$

$$u^2 - 6u - 1 = 0 \qquad \text{Letting } u = e^x$$

$$u = \frac{6 \pm \sqrt{40}}{2} = 3 \pm \sqrt{10} \qquad \text{Using the quadratic formula}$$

Since $u = e^x$ is always positive, discard the root $3 - \sqrt{10}$, which is negative. Thus

$$e^x = 3 + \sqrt{10}$$

Then

$$x = \ln(3 + \sqrt{10}) \approx \boxed{1.818446459} \approx 1.82$$

is the desired solution. ▲

Equations involving logarithms are called **logarithmic equations.** The use of one of the rules of logarithms frequently gives the needed simplification to allow you to solve such an equation.

Example 3 Solve the logarithmic equation $\log(x^2 - 1) - \log(x - 1) = 3$.

SOLUTION Simplifying the left-hand side by combining the two logarithm terms, we get

$$\log \frac{x^2 - 1}{x - 1} = 3$$

$$\log(x + 1) = 3 \qquad \text{Canceling } x - 1 \text{ in } x^2 - 1$$

$$x + 1 = 10^3 \qquad \log M = N \text{ means } M = 10^N$$

$$x = -1 + 10^3 = 999 \qquad \text{Adding } -1 \text{ to both sides}$$

▲

Example 4 Solve the equation $\log x + \log(x - 3) = 1$.

SOLUTION Writing the left-hand side as the logarithm of a product, we have

$$\log[x(x - 3)] = 1$$

$$x(x - 3) = 10^1 \qquad \text{Definition of a logarithm}$$

$$x^2 - 3x - 10 = 0 \qquad \text{Expanding on the left and subtracting 10 from both sides}$$

$$(x - 5)(x + 2) = 0 \qquad \text{Factoring}$$

From the factored form it appears that the solutions are $x = -2$ and 5. However, when we check these solutions in the given equation, we find that there is a problem with $x = -2$ because $\log(-2)$ is undefined. The value $x = -2$ is called an **extraneous solution** since it appears to be a solution but it is not. Therefore the equation has only $\boldsymbol{x} = 5$ as a solution. ▲

Example 5 Solve the equation

$$x^{\log x} = \frac{x^3}{100}$$

SOLUTION Taking the common logarithm of both sides, we have

$$\log x^{\log x} = \log \frac{x^3}{100}$$

$$(\log x)(\log x) = \log x^3 - \log 100 \qquad \text{Using Rules 2 and 3}$$

$$(\log x)^2 - \log x^3 + \log 100 = 0 \qquad \text{Collecting terms on the left}$$

$$(\log x)^2 - 3\log x + 2 = 0 \qquad \log x^3 = 3\log x \text{ and } \log 100 = 2$$

$$(\log x - 2)(\log x - 1) = 0 \qquad \text{Factoring}$$

$$\log x = 2, \text{ or } \log x = 1 \qquad \text{Solving for } \log x$$

Thus, $x = 100$ or 10. ▲

Example 6 A certain power supply has a power output in watts governed by the equation

$$P = 50e^{-t/250}$$

where t is the time in days. If the equipment aboard a satellite requires 10 W of power to operate properly, what is the operational life of the satellite?

SOLUTION Letting $P = 10$, we solve the equation $10 = 50e^{-t/250}$ for t. Dividing both sides by 50, we have

$$e^{-t/250} = 0.2$$

$$\frac{-t}{250} = \ln 0.2 \qquad e^y = x \text{ means } y = \ln x$$

$$t = -250 \ln 0.2 \qquad \text{Multiplying both sides by } -250$$

$$\approx \boxed{402.3594781} \qquad \text{Using a calculator}$$

$$\approx 402 \qquad \text{To three significant figures}$$

Hence the operational life of the satellite is 402 days. ▲

▲▼ Exercises for Section *10.4*

Solve for x in Exercises 1–26.

1. $7^{x+1} = 2^x$
2. $3^x 2^{2x+1} = 10$
3. $10^{x^2} = 2^x$
4. $8^x = 10^x$
5. $2^{1+x} = 3$
6. $3^{1-x} = 7^x$
7. $2^{x+5} = 3^{x-2}$
8. $(1/2)^x > 3$
9. $2^{x+1} > 5$
10. $5^{-(1+x)} < 8$
11. $2^{\log x} = 2$
12. $\log \log \log x = 1$
13. $(\log x)^{1/2} = \log \sqrt{x}$
14. $(\log x)^2 = \log x^2$
15. $(\log x)^3 = \log x^3$
16. $x^{\log x} = 10$
17. $\log (x + 15) + \log x = 2$
18. $\log 3x - \log 2x = \log 3 - \log x$
19. $\log (x - 2) - \log (2x + 1) = \log (1/x)$
20. $\log (x^2 + 1) - \log (x - 1) - \log (x + 1) = 1$
21. $\log (x + 2) - \log (x - 2) - \log x + \log (x - 3) = 0$
22. $\log x + \log (x - 99) = \log 2$
23. $\log x + \log (x^2 - 4) - \log 2x = 0$
24. $3 \log (x - 1) = \log 3(x - 1)$
25. $\log (x + 1) - \log x < 1$
26. $\log (x + 1) + \log (x - 1) < 2$
27. The expression $(e^x + e^{-x})/2$ is called the hyperbolic cosine of x and is denoted by $\cosh x$. Solve the equation $\cosh x = 2$. *Hint*: Let $u = e^x$.
28. Solve the two equations $y = 50e^{-2x}$ and $y = 2^x$ simultaneously by making a sketch of each of the two equations on the same coordinate system.

29. Explain why if $\log u(x) = v(x)$, then $u(x) = 10^{v(x)}$.
30. Show that if $y = e^{\ln f(x)}$, then $y = f(x)$.
31. The radioactive chemical element strontium 90 has a half-life of approximately 28 years. The element obeys the radioactive decay formula, $A(t) = A_0e^{-kt}$, where A_0 is the original amount and t is the time in years. Find the value of k.
32. Repeat Exercise 31 for the element iodine, whose decay formula is the same type. Express t in days. The half-life is 8 days.
33. What is the half-life of the power supply of Example 6?
34. The difference in intensity level of two sounds with intensities I and I_0 is defined by $10 \log(I/I_0)$ db. Find the intensity level in decibels of the sound produced by an electric motor that is 175.6 times greater than I_0.
35. As previously pointed out, the population growth curve is given by $P = P_0e^{kt}$, where P_0 is the initial size, t is time in hours, and k is a constant. If $k = 0.0132$ for a bacteria culture, how long does it take for the culture to double in size?

Key Topics for Chapter 10

Define and/or discuss each of the following.

Definition of Exponential and Logarithmic Functions
Graphs of Exponential and Logarithmic Functions
Properties of Exponential and Logarithmic Functions
Laws of Exponentials
Rules of Logarithms
Applications of Exponential and Logarithmic Functions
Exponential and Logarithmic Equations

Review Exercises for Chapter 10

In Exercises 1–14 solve for x.

1. $2^{x-5} = 3^x$
2. $x^{-1.2} = 18$
3. $2^{x^2 - 3x} = 16$
4. $3^{\log x} = 2$
5. $3^{1-x} = 5$
6. $\log 2^x = x^2 - 2$
7. $\log x = 3$
8. $\log_2 x = 5$
9. $\log_3 81 = x$
10. $\log_5 (\frac{1}{25}) = x$
11. $\log_x 64 = 6$
12. $\log_x 27 = 3$
13. $x^{4.3} = 2.1$
14. $\log_x (0.01) = -2$

In Exercises 15–24 make a sketch of the graph of the given function.

15. $y = e^{-(1-x)}$
16. $y = 2e^{-x} + 3$
17. $y = 2^2 2^t$
18. $y = (0.5)^{-x}$
19. $y = (2/3^x)$
20. $y = 2 - 5e^{-(x-1)}$
21. $y = -\log (x - 2)$
22. $y = 1 + \log x$
23. $y = 2 - \log x^3$
24. $y = \log (-x) - 3$

In Exercises 25–30 write the given expression as a single logarithmic term.

25. $\log x + \log x^2$

26. $\log 2x - \log x$

27. $\log_2 2x + 3 \log_2 x$

28. $3 \log_2 x - \log_2 x$

29. $3 \log_5 x - \log_5 (2x - 3)$

30. $5 \log (x + 2) - 2 \log x$

Solve each of the equations in Exercises 31–40.

31. $2^{x+2} = 10$

32. $5^{3-x} = 8$

33. $3^{x+1} = 4^x$

34. $2^x = 3^{x-1}$

35. $\log (\log x) = 1$

36. $(\log x)^2 = \log x$

37. $\log x + \log (x + 1) = \log 6$

38. $\log (x - 2) + \log x = \log 8$

39. $\log x + \log (x - 3) = 1$

40. $\log 5x + \log 2x = 2 + \log x$

41. Find the half-life of a radioactive material that decays exponentially according to the equation $A(t) = A_0e^{-t/4}$, where t is the time in years.

42. A colony of bacteria increases according to the law

$$N(t) = N(0)e^{kt}$$

If the colony doubles in five hours, find the time needed for the colony to triple.

43. A bacteria colony population is given by $N(t) = 2000\, e^{1.3t}$. Plot a graph of N as a function of t.

44. A principal of \$5000 is invested at the rate of 9.2% per year **compounded continuously.** The law governing continuous compounding is $P = P_0e^{rt}$, where t is the time in years, P_0 is the principal, and r is the rate. What will be the amount after 20 years?

45. In the dye-dilution procedure for measuring cardiac output, the amount in milligrams of dye in the heart at any time, t, in minutes is given by

$$D(t) = D(0)e^{-rt/V}$$

where $D(0)$ is the amount in milligrams of dye injected, r is a constant representing the outflow of blood and dye in liters per minute, and V is the volume of the heart in liters. Find the amount of dye in the heart after 5 sec, given that $V = 450$ mL, $r = 1.4$ L/min, and $D(0) = 2.3$ mg. (*Hint*: Use consistent units.)

▲▼ Practice Test Questions for Chapter *10*

In Exercises 1–10 answer *true* or *false*.

1. $f(x) = x^3$ is an exponential function.

2. The logarithm of x to the base a is an exponent of a.

3. For all $x > 0$, $y = \log x$ is an increasing function.

4. The domain of $y = e^x$ is $-\infty < x < \infty$.

5. The domain of $y = \ln x$ is $-\infty < x < \infty$.

6. If $f(x) = 2^x$, then $f(x + y) = f(x) \cdot f(y)$.

7. $\text{Log}(x^c) = c \log x$.
8. $\text{Log}(x + y) = \log x + \log y$.
9. The logarithm function is the reciprocal of the exponential function.
10. The exponential function is defined by $y = a^x$, where a is any real number.

In Exercises 11–20 fill in the blank to make the statement true.

11. If $\log_b 16 = 2$, then $b =$ __________.
12. The __________ of x to the base a is the power to which a must be raised to give x.
13. The exponential function is the __________ of the logarithm function.
14. The domain of $y = \log(-x)$ is __________.
15. The notation $\ln x$ is used to indicate the __________ logarithm of x.
16. $\log(m) - \log(n) = \log$ (__________).
17. A function of the form $y = 2^x$ is called $a(n)$ __________ function.
18. If $y = \ln x$, then $x =$ __________.
19. If $y = b^{2x}$, then $x =$ __________.
20. $c \ln (x) = \ln$ (__________).

In the following exercises solve the stated problem. Show all your work.

21. Make a sketch of $y = 2^{x-3}$ and $y = 2^x$ on the same coordinate plane. Compare the two graphs.
22. Make a sketch of $y = \log_2 x$ and $y = \log_2(x + 2)$ on the same coordinate plane. Compare the two graphs.
23. If $\log_x 8 = -3$, find x.
24. Express $3 \log x - \log(x^2 - 2) + 2 \log(x + 1)$ as a single logarithm term.
25. Solve $10^{x+1} = 13$ for x.
26. Solve $3^x = 2^{x+1}$ for x.
27. Solve $\log x + \log(x - 3) = 1$ for x.
28. Solve $\ln x + \ln(x + 2) = \ln 8$ for x.
29. Solve $x^{4.3} = 2.1$ for x.
30. The population of deer in a region is given by $P = P_0 e^{0.02t}$, where P_0 is the initial population and t is time in years. How long does it take for the population to double in size?

Analytic Geometry

Fundamental curves such as straight lines, circles, parabolas, ellipses, and hyperbolas have been studied by mathematicians for centuries. In this chapter we show how these curves are related to algebraic equations as well as some of their many modern applications. Section 11.1 covers the straight line, with an emphasis on the concept of slope. Parabolas, ellipses, circles, and hyperbolas, which are called the conic sections, are then covered in sequence with reference to the origin. Section 11.5 shows how to describe the conic sections when they are translated in the plane, and Section 11.6 covers their description under a rotation of axes. The approach in this chapter is to sketch lines and the conic sections without resorting to a point plotting technique.

11.1 Lines

By a **linear function**, we mean a function expressible in the form

$$f(x) = ax + b \tag{11.1a}$$

where a and b are constants. If we replace $f(x)$ with the letter y, the linear function can be written in the form

$$y = ax + b \tag{11.1b}$$

In this form, it is called a **linear equation in the two variables x and y**.

More generally, any equation in the variables x and y that can be arranged into the form

$$ax + by = c \tag{11.2}$$

where a, b, and c are constants is a linear equation since, if $b \neq 0$, it can be written $y = (-a/b)x + (c/b)$. If $b = 0$, we get $ax = c$, which is not a linear function but is still considered to be a linear equation.

The graph of every linear function is a straight line,* and since a line is determined by two distinct points, we usually choose points that correspond to the intersections of the line with the x- and y-axes, called the **x-intercept** and **y-intercept**, respectively.

- To find the x-intercept of the graph of a linear function, let $y = 0$ and solve for x.
- To find the y-intercept of the graph of a linear function, let $x = 0$ and solve for y.

Example 1 Graph $y = 2x - 4$.

SOLUTION Letting $x = 0$ gives $y = -4$ and letting $y = 0$ gives $x = 2$. Thus the graph passes through $(0, -4)$ on the y-axis and $(2, 0)$ on the x-axis. The graph of this equation is shown in Figure 11.1.

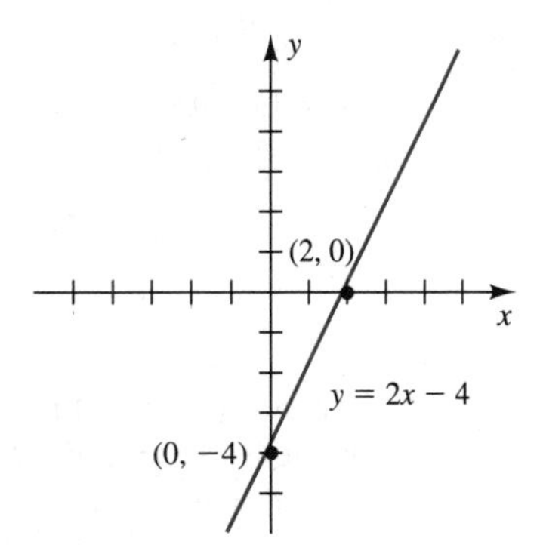

Figure 11.1 ▲

Slope of a Straight Line

If you were asked to describe the graph in Figure 11.1 you would say that the graph is a line that crosses the x-axis at $x = 2$ and the y-axis at $y = -4$. In addition, you might say that the line rises from left to right. Another bit of information that would be useful is a measure of the steepness of the

* In the future we will use *line* with the understanding that it is straight.

graph. The line in Figure 11.1 rises two units for every one unit moved to the right. Implicit in this description of steepness is a quantity called the **slope** of the line.

The slope of a line is defined as the ratio of the vertical rise of the line to the corresponding horizontal run; that is,

$$\text{slope} = \frac{\text{vertical rise}}{\text{horizontal run}}$$

Applying this definition to the line segment P_1P_2 in Figure 11.2, the slope

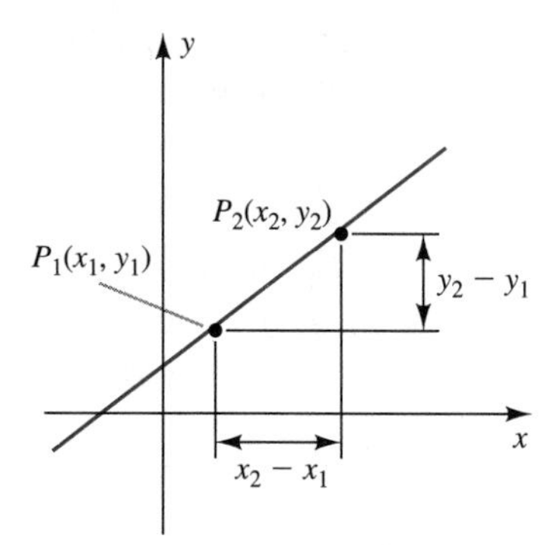

Figure 11.2

m is given by the equation

$$m = \frac{y_2 - y_1}{x_2 - x_1} \quad \textbf{(11.3)}$$

where $y_2 - y_1$ is the vertical distance between the given points and $x_2 - x_1$ is the horizontal distance between the points. Because of similar triangles, the choice of points in the slope formula is immaterial.

The Inclination of a Line

The inclination of a line is the angle θ, $0° \leq \theta < 180°$, measured clockwise from the positive x-axis to the line as shown in Figure 11.3. From the definition of the tangent of an angle, it is easy to see that the inclination angle, θ, and the slope m of nonvertical lines are related by the equation

$$m = \tan \theta$$

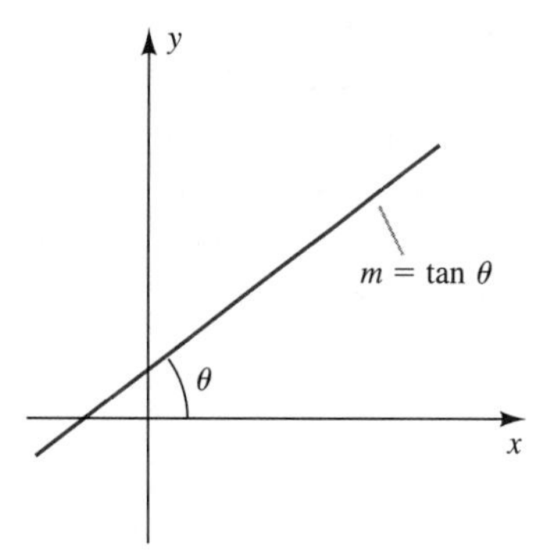

Figure 11.3

Example 2 Find the slope and inclination of the line passing through the points $(-5, 1)$, and $(2, -3)$ in Figure 11.4.

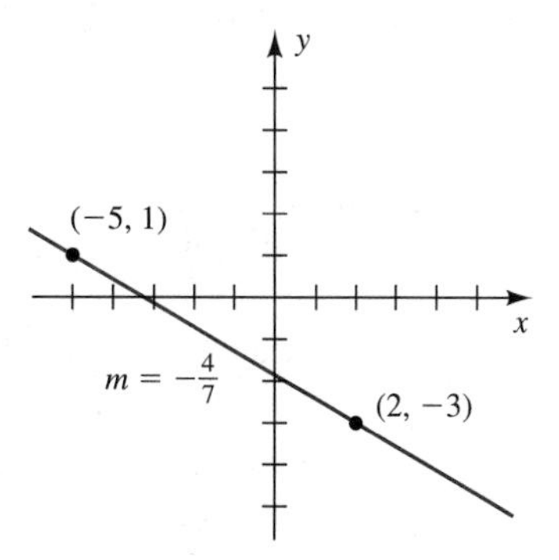

Figure 11.4

SOLUTION Letting $(x_1, y_1) = (-5, 1)$ and $(x_2, y_2) = (2, -3)$ and using Equation 11.3, we obtain the desired slope:

$$m = \frac{y_2 - y_1}{x_2 - x_1} = \frac{-3 - 1}{2 - (-5)} = \frac{-4}{7} = -\frac{4}{7}$$

If we interchange the labels of the given points and let $(x_1, y_1) = (2, -3)$ and $(x_2, y_2) = (-5, 1)$, the result is

$$m = \frac{y_2 - y_1}{x_2 - x_1} = \frac{1 - (-3)}{-5 - 2} = \frac{4}{-7} = -\frac{4}{7}$$

Hence the order in labeling the given points is immaterial. We interpret a slope of $\frac{-4}{7}$ to mean that for every 7 units moved to the right, the straight line moves down 4 units, or, for every 7 units moved to the left, the straight line moves up 4 units. The inclination of the line is determined from the equation: $m = \frac{-4}{7} = \tan\theta$. Solving for θ, $\theta = \arctan\left(\frac{-4}{7}\right) + 180° = 169.1°$. ▲

In working with lines, the following generalizations about slopes are usefull.

- The slope of a line is positive if, as you follow the curve from left to right, you move up. See Figure 11.5(a).
- The slope of a line parallel to the x-axis is zero because the rise is zero for any run. See Figure 11.5(b).

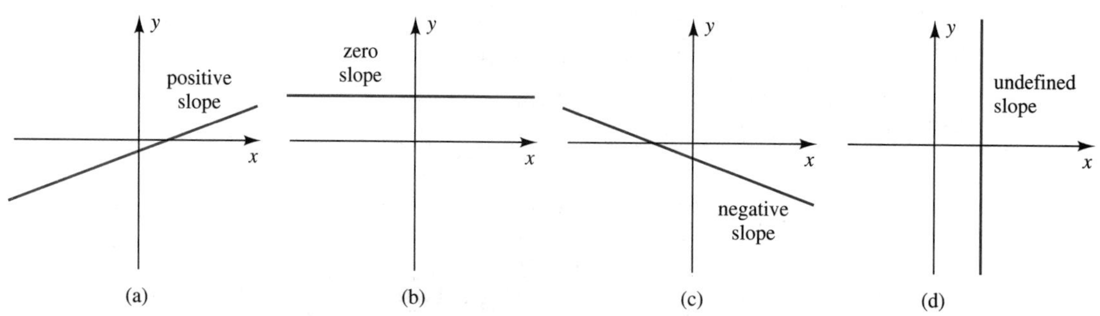

Figure 11.5

- The slope of a line is negative if, as you follow the curve from left to right, you move down. From Figure 11.5(c) we see that when the run is positive, the rise is negative, and when the run is negative, the rise is positive. The ratio of rise to run (that is, slope) in this case must be negative.
- The slope of a line parallel to the y-axis is undefined because the run is zero for any rise. Therefore to apply Equation 11.3, we would have to divide by zero. (Remember, division by zero is an undefined operation.) See Figure 11.5(d).
- Parallel lines have equal slopes.
- Perpendicular lines have slopes that are negative reciprocals. That is, $\boldsymbol{m_1 = -1/m_2}$. $(m_2 \neq 0)$

Example 3

(a) Find the slope of a line passing through (2, 1) and (−4, 6).

(b) Find the slope of a line drawn perpendicular to the given line at (2, 1) (see Figure 11.6).

SOLUTION

(a) The slope m of the given line is

$$m = \frac{y_2 - y_1}{x_2 - x_1} = \frac{6 - 1}{-4 - 2} = -\frac{5}{6}$$

(b) The slope m' of a line perpendicular to the given line is

$$m' = -\frac{1}{m} = -\frac{1}{-5/6} = \frac{6}{5}$$

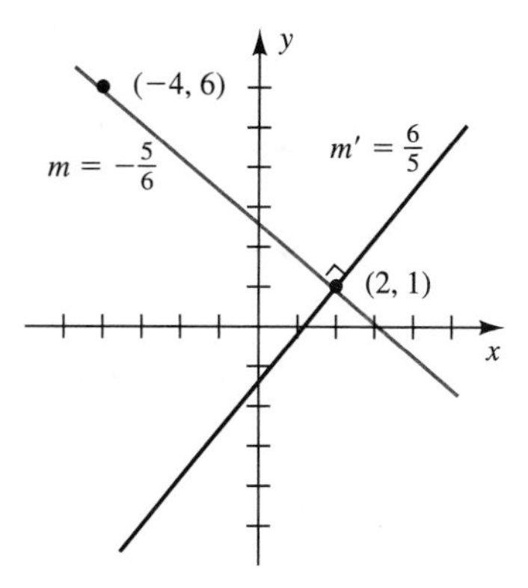

Figure 11.6

▲

Methods of Describing a Line

Whereas every line can be represented by a linear equation of the general form $ax + by = c$, sometimes certain other forms of representation are significant. Two important forms of a line are the point-slope form and the slope-intercept form. Each form shows two properties of a line that can be determined by inspection of the equation.

The **point-slope form** of a line is used when we know the slope and one point on the line. See Figure 11.7. To obtain the point-slope form of a

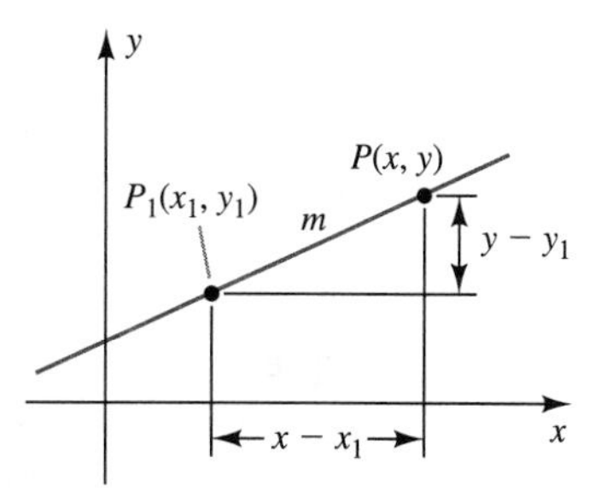

Figure 11.7

line, choose an arbitrary point $P(x, y)$ different from $P(x_1, y_1)$ on the line. Then, by the definition of the slope of a line, we have

$$\frac{y - y_1}{x - x_1} = m$$

Rearranging the terms yields

THE POINT-SLOPE FORM OF A LINE The equation of a line passing through (x_1, y_1) with slope m is

$$y - y_1 = m(x - x_1) \tag{11.4}$$

COMMENT Note that, since *any* fixed point (x_1, y_1) may be used in Equation 11.4, the equation will *appear* different for various choices of the point, but the resulting equations are equivalent.

Example 4 The point-slope form of the equation representing the line that passes through (2, 1) with slope $\frac{1}{3}$ is

$$y - 1 = \frac{1}{3}(x - 2)$$

However, the two points $(0, \frac{1}{3})$ and $(-1, 0)$ are also on the line, so that two other point-slope forms of the same line are

$$y - \frac{1}{3} = \frac{1}{3}(x - 0) \qquad \text{and} \qquad y = \frac{1}{3}(x + 1)$$

Infinitely many other point-slope forms of this equation are possible. Note that the general form of this line is

$$x - 3y = -1$$ ▲

If the point used in the point-slope form is $(0, b)$, we have the special case of the **slope-intercept form**. Substituting $x_1 = 0$ and $y_1 = b$ into the point-slope form yields

$$\frac{y - b}{x} = m$$

Solving for y, we have

THE SLOPE-INTERCEPT FORM OF A LINE The equation of a line passing through $(0, b)$ with slope m is

$$y = mx + b \tag{11.5}$$

COMMENT When an equation is in slope-intercept form, the constant on the right-hand side is the y-intercept and the coefficient of x is the slope. Given any linear equation, a few simple manipulations can represent the line in any desired form.

Example 5 Rearrange the linear equation $3x + 2y = -5$ into slope-intercept form and draw its graph.

SOLUTION To rearrange into slope-intercept form, solve for y. Thus

$$2y = -3x - 5$$

and

$$y = \underbrace{-\frac{3}{2}}_{m}x + \underbrace{\left(-\frac{5}{2}\right)}_{b}$$

from which we recognize the slope as $-\frac{3}{2}$ and the y-intercept as $-\frac{5}{2}$. (See Figure 11.8.)

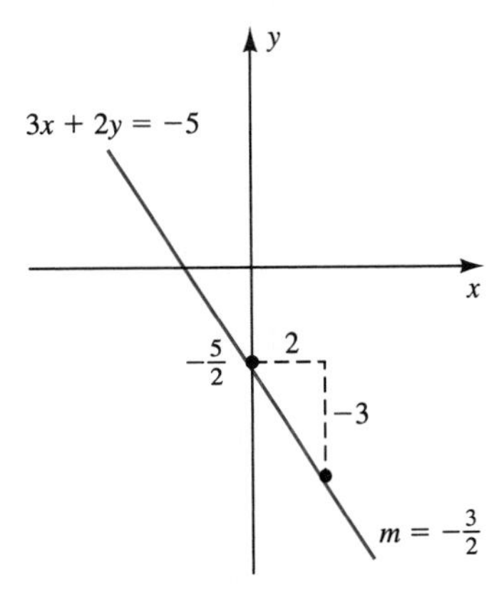

Figure 11.8 ▲

Example 6 Write the equation of the line perpendicular to $x - 3y = 7$ at the point $(4, -1)$.

SOLUTION First, rewrite the given equation in slope-intercept form; that is,

$$y = \frac{1}{3}x - \frac{7}{3}$$

From this form of the given equation, we see that the slope of the line is $m = \frac{1}{3}$. Denoting the slope of the perpendicular line by m', we have

$$m' = -\frac{1}{m} = -\frac{1}{\frac{1}{3}} = -3$$

Finally, using the point-slope form of a line, the equation of the perpendicular line is

$$\begin{aligned} y - (-1) &= -3(x - 4) \\ y + 1 &= -3x + 12 \\ y &= -3x + 11 \end{aligned}$$

The general form of the line is $3x + y = 11$. ▲

The general relationship between equations of lines that are perpendicular to each other is given in Exercise 53 of the following exercise set. Before looking at Exercise 53, see if you can find any relationship between the given equation and that of the perpendicular line.

Exercises for Section 11.1

In Exercises 1–12 sketch the graph of the given linear equation.

1. $x + y = 1$

2. $x - y = 1$

3. $x = 5$

4. $y = -2$

5. $2x - y = 5$

6. $3x + 2y = 5$

7. $4x - y + 1 = 0$

8. $x = 7 - y$

9. $3y - 4x = 4$

10. $y = \frac{1}{2}x - 3$

11. $y = \dfrac{x - 2}{3}$

12. $15(x + y) = 10$

In Exercises 13–20 sketch the line through the pairs of points and compute the slope and angle of inclination.

13. $(1, 2), (5, 4)$

14. $(-5, 2), (3, -7)$

15. $(-1, -1), (3, -6)$

16. $(7, 3), (0, 5)$

17. $(-2, -3), (-5, -7)$

18. $(3, -2), (7, 6)$

19. $(-2, 3), (5, 3)$

20. $(\frac{1}{2}, \frac{1}{2}), (\frac{1}{2}, -\frac{2}{5})$

In Exercises 21–24 sketch the line passing through the given point with the given slope and then determine its equation.

21. $(2, 5), m = \frac{1}{2}$

22. $(-1, -3), m = 3$

23. $(5, -2), m = -7$

24. $(3, 4), m = -\frac{2}{5}$

In Exercises 25–30 draw the line through the given points and then find its equation.

25. $(1, 3), (6, 2)$

26. $(2, 5), (-3, -7)$

27. $(-1, -1), (1, 2)$

28. $(0, 0), (3, -2)$

29. $(0, 2), (-5, 0)$

30. $(\frac{1}{2}, \frac{1}{3}), (-\frac{1}{2}, \frac{1}{3})$

Find the slope and the y-intercept of the graph of the equations in Exercises 31–42 and then draw the line.

31. $2x - 3y = 5$

32. $3x + 4y = 0$

33. $x + y = 2$

34. $5x + 2y = -3$

35. $4y - 2x + 8 = 0$

36. $-5x - y - 2 = 0$

37. $2y = x + 5$

38. $3x + 6 = 2y$

39. $y = 5$

40. $x + y = 1$

41. $x - 5y + 7 = 0$

42. $y = 1 - x$

In Exercises 43–52 write the equation of the line perpendicular to the given line at the indicated point.

43. $3x + 2y = 7$ at $(1, 2)$

44. $x + 3y = 11$ at $(2, 3)$

45. $x - y = 2$ at $(5, 3)$

46. $2x - 5y = 2$ at $(-4, -2)$

47. $2y - 3x = 1$ at $(1, 2)$

48. $y = x$ at $(0, 0)$

49. $5x - 7y = 0$ at $(0, 0)$

50. $5x - 7y = 3$ at $(2, 1)$

51. $-x + 3y = 5$ at $(-2, 1)$

52. $-2y - x = 3$ at $(-1, -1)$

53. Show that the equation of the line perpendicular to $Ax + By = C$ at (x_1, y_1) is

$$Bx - Ay = Bx_1 - Ay_1$$

(*Hint:* Use the fact that the product of the slopes equals -1.)

54. Show that the equation of a line with x-intercept a and y-intercept b may be written in the form

$$\frac{x}{a} + \frac{y}{b} = 1 \qquad (a \neq 0,\ b \neq 0)$$

This form of a line is known as the **intercept form** of a line.

55. Write the equation $x + 3y = 2$ in intercept form.

56. Write the equation $2x - y = 5$ in intercept form.

11.2 The Parabola

Curves that can be formed by cutting a right circular cone with a plane are called **conic sections**. Greek mathematicians in 400–300 B.C. discovered that they could obtain four distinct curves by cutting a right circular cone with a plane: the circle, the ellipse, the hyperbola, and the parabola. The properties of the conics discovered by the Greeks include those that we use as definitions in this chapter.

If two right circular cones are placed vertex to vertex with a common axis, the resulting figure is referred to as a cone with two *nappes*. To see how the four conics can be generated from a cone with two nappes, refer to Figure 11.9.

(a) A **circle** is obtained when the cutting plane is perpendicular to the axis, provided it does not pass through the vertex.

(b) An **ellipse** is obtained when the cutting plane is inclined so as to cut entirely through one nappe of the cone without cutting the other nappe.

(c) A **parabola** is obtained when the cutting plane is parallel to one element in the side of the cone.

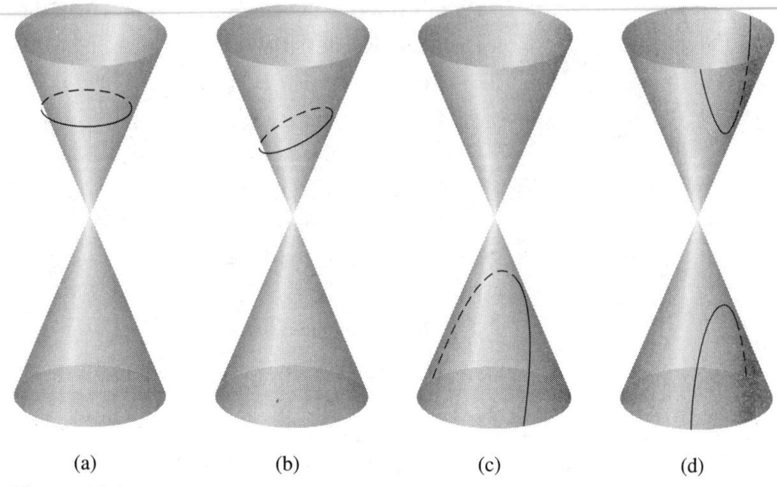

Figure 11.9

(d) A **hyperbola** is obtained when the cutting plane is parallel to the axis of the cone so that it cuts through both nappes.

Analytic geometry is the study of the relationship between algebra and geometry. In this section we define and discuss the **parabola**.

DEFINITION 11.1 A parabola is the set of all points in a plane that are equidistant from a fixed point (**focus**) and a fixed line (**directrix**).

The line through the focus and perpendicular to the directrix is called the **axis** of the parabola. The midpoint between the focus and the directrix is a point on the parabola and is known as the **vertex**.

Figure 11.10 shows a parabola with vertex at the origin, focus at $F(a, 0)$, and directrix perpendicular to the x-axis at $D(-a, 0)$. The distance from the vertex to the focus is sometimes referred to as the **focal distance**.

Consider a point $P(x, y)$ on the parabola. Then by definition

$$\overline{GP} = \overline{FP} \tag{11.6}$$

But from Figure 11.10 we see that

$$\overline{GP} = |x + a|$$

and

$$\overline{FP} = \sqrt{(x - a)^2 + y^2}$$

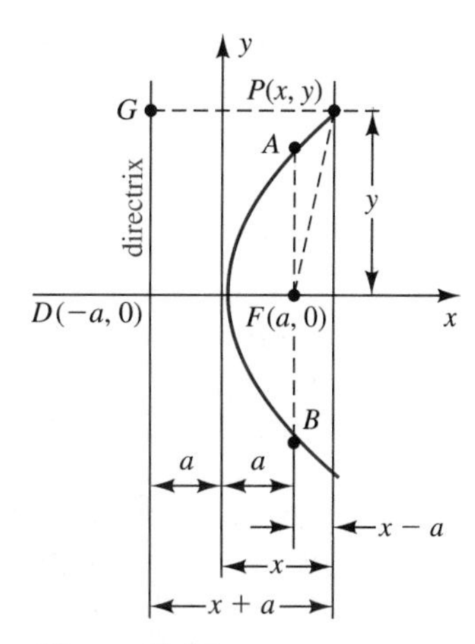

Figure 11.10

Substituting these expressions in Equation 11.6 yields

$$|x + a| = \sqrt{(x - a)^2 + y^2}$$

or, squaring both sides,

$$(x + a)^2 = (x - a)^2 + y^2$$

Expanding and collecting like terms, the equation of the parabola becomes

$$y^2 = 4ax \tag{11.7}$$

Equation 11.7 is referred to as the **standard form** of the equation of a parabola with vertex at the origin and focus on the x-axis. It is clearly symmetric about its axis. The focus is at $(a, 0)$ if the coefficient of x is positive. In this case the parabola opens to the right. If the coefficient of x is negative, the focus is at $(-a, 0)$ and the parabola opens to the left.

The chord AB through the focus and perpendicular to the axis is called the **right chord**, or the **latus rectum**. The length of the right chord is found by letting $x = a$ in Equation 11.7. Making this substitution, we find

$$y^2 = 4a^2$$

from which

$$y = \pm 2a$$

The length of the right chord is, therefore, equal numerically to $4|a|$. This fact is useful because it helps define the shape of the parabola by giving us an idea of the "width" of the parabola.

The standard form of a parabola with vertex at the origin and focus on the y-axis is

$$x^2 = 4ay \tag{11.8}$$

The derivation of this formula parallels that of Equation 11.7. The parabola represented by Equation 11.8 is symmetric about the y-axis. It opens upward if the coefficient of y is positive and downward if it is negative.

The equation of a parabola is characterized by one variable being linear and the other being squared. *The linear variable indicates the direction of the axis of the parabola.*

A rough sketch of the parabola can be drawn if the location of the vertex and the extremities of the right chord are known. This information can be obtained from the standard form of the equation.

Example 1 Discuss and sketch the graph of the equation $x^2 = 8y$.

SOLUTION This equation has the form of Equation 11.8 with $4a = 8$ or $a = 2$. The y-axis is the axis of the parabola, the focus is at (0, 2), and the directrix is the line $y = -2$. The endpoints of the right chord are then $(-4, 2)$ and $(4, 2)$. Figure 11.11 shows the parabola.

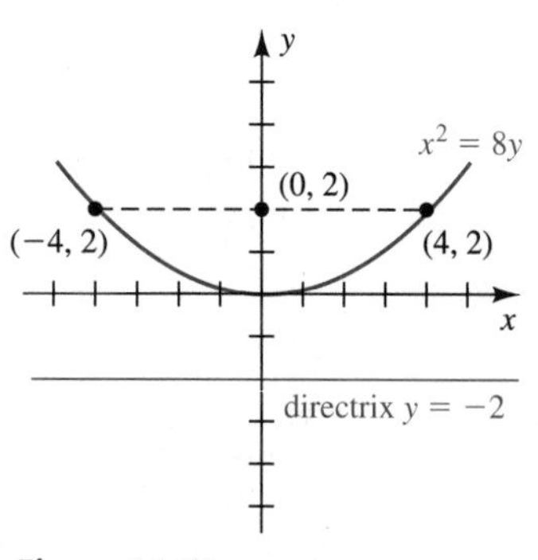

Figure 11.11 ▲

Example 2 Find the equation of the parabola with focus at $(-1, 0)$ and directrix $x = 1$ and sketch the curve.

SOLUTION The focus lies on the x-axis to the left of the directrix, so the parabola opens to the left. The desired equation is then the form in Equation 11.7 with $a = -1$; that is,

$$y^2 = -4x$$

To sketch the parabola, note that the length of the right chord is $\overline{AB} = 4$. Its extremities are therefore $(-1, 2)$ and $(-1, -2)$. The curve appears in Figure 11.12.

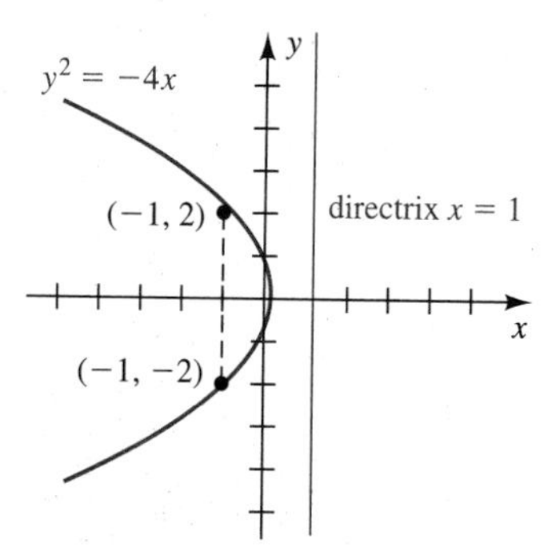

Figure 11.12

▲

COMMENT A unique physical property of the parabola is that it will reflect all rays emitted from the focus such that they travel parallel to the axis of the parabola. This feature makes the parabola a particularly desirable shape for reflectors in spotlights and reflecting telescopes and also for radar antennas. See Figure 11.13.

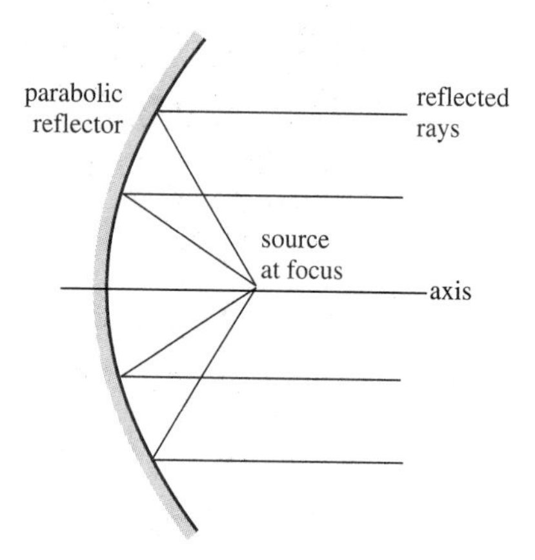

Figure 11.13

Example 3 A parabolic reflector is to be built with a focal distance of 2.25 ft. What is the diameter of the reflector if it is to be 1 ft deep at its axis?

SOLUTION We need the equation of the parabola used to generate the reflector. Referring to Figure 11.14, the vertex of the parabola is located at the origin and the focus on the x-axis. (We could have just as well placed the focus on the y-axis.) Then using Equation 11.2 with $a = 2.25$ yields the equation

$$y^2 = 4(2.25)x = 9x$$

The diameter of the reflector can be found by substituting $x = 1$ into this equation. Thus,

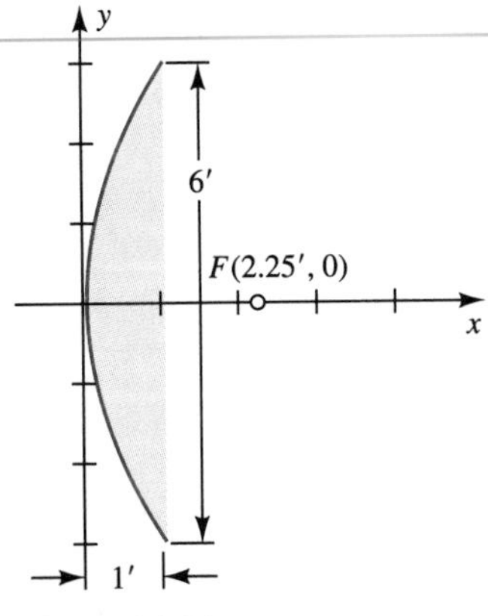

Figure 11.14

$$y^2 = 9 \qquad \text{and} \qquad y = \pm 3$$

which means that the diameter of the reflector is 6 ft. ▲

Exercises for Section 11.2

In Exercises 1–10 find the coordinates of the focus, the endpoints of the right chord, and the equation of the directrix of each of the parabolas. Sketch the graph of each parabola.

1. $y^2 = -8x$
2. $x^2 = 12y$
3. $2x^2 = 12y$
4. $x^2 = -24y$
5. $y^2 + 16x = 0$
6. $y + 2x^2 = 0$
7. $y^2 = 3x$
8. $3x^2 = 4y$
9. $y^2 = -2x$
10. $y^2 = 10x$

In Exercises 11–18 find the equation of the parabolas having the given properties. Sketch each curve.

11. Focus at $(0, 2)$, directrix $y = -2$.
12. Focus at $(0, -\frac{1}{2})$, directrix $y = \frac{1}{2}$.
13. Focus at $(\frac{3}{2}, 0)$, directrix $x = -\frac{3}{2}$.
14. Focus at $(-10, 0)$, directrix $x = 10$.
15. Endpoints of right chord $(2, -1)$ and $(-2, -1)$, vertex at $(0, 0)$.
16. Endpoints of right chord $(3, 6)$ and $(3, -6)$, vertex at $(0, 0)$.
17. Vertex at $(0, 0)$, vertical axis, one point of the curve $(2, 4)$.
18. Vertex at $(0, 0)$, horizontal axis, one point of the curve $(2, 4)$.

In Exercises 19–22 solve the given system of equations graphically.

19. $2x + 4y = 0$
    $x^2 - 4y = 0$
20. $y = e^x$
    $y^2 = -3x$

21. $y^2 = 12x$
$y = \log x$

22. $y = x^3$
$x^2 = 8y$

23. In the accompanying figure, the supporting cable of a suspension bridge hangs in the shape of a parabola.* Find the equation of a cable hanging from two 400-ft-high supports that are 1000 ft apart, if the lowest point of the cable is 250 ft below the top of the supports. Choose the origin in the most convenient location.

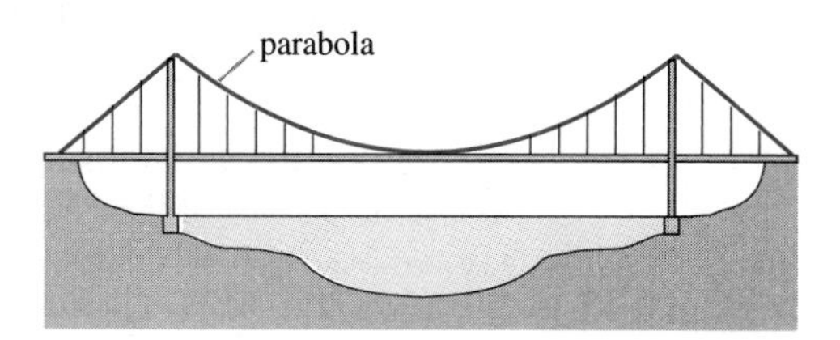

* *Free*-hanging cables hang in the shape of a curve called a catenary. Supporting cables such as the one described here are more closely approximated by a parabola.

24. A parabolic antenna is to be constructed by revolving the parabola $y^2 = 24x$. Sketch the cross-section of the antenna if the diameter of the circular front is to be 12 ft. Locate the focus.

11.3 The Ellipse and the Circle

The Ellipse

An ellipse can be constructed from a loop of string in the following way. Place two pins at F and F', as shown in Figure 11.15, and place the loop

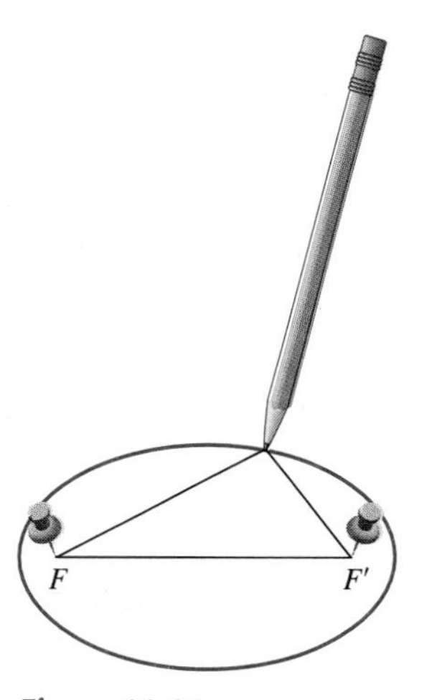

Figure 11.15

of string over them. Pull the string taut with the point of a pencil and then move the pencil, keeping the string taut. The figure generated is an ellipse. From this construction observe that the sum of the distances from the two fixed points to the point P is always the same, since the loop of string is kept taut. This property characterizes the ellipse.

DEFINITION 11.2 An **ellipse** is the set of all points in a plane the sum of whose distances from two fixed points (**foci**) in the plane is constant.

The midpoint of a line through the foci is called the **center** of the ellipse. We use the center of the ellipse to locate the ellipse in the plane in the same way we used the vertex to locate the parabola.

To obtain the equation of an ellipse, consider an ellipse with foci located on the x-axis such that the origin is midway between them, as in Figure 11.16. Let the foci be the points $F(c, 0)$ and $F'(-c, 0)$ and let the sum of the distances from a point $P(x, y)$ of the ellipse to the foci be $2a$, where $a > c$. Then

$$PF + PF' = 2a$$

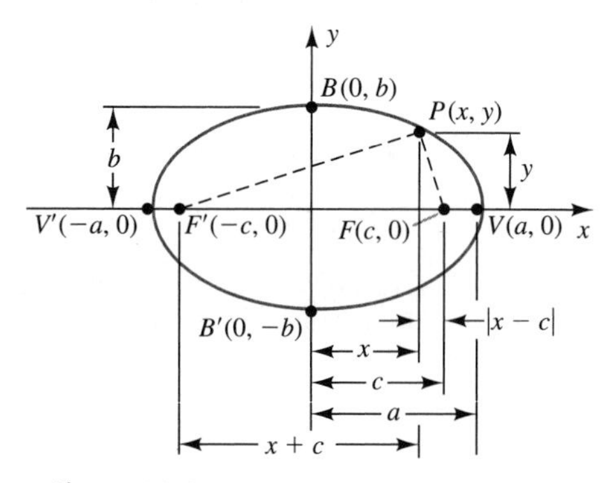

Figure 11.16

From Figure 11.16

$$PF = \sqrt{(x - c)^2 + y^2}$$

and

$$PF' = \sqrt{(x + c)^2 + y^2}$$

so that

$$\sqrt{(x - c)^2 + y^2} + \sqrt{(x + c)^2 + y^2} = 2a$$

Transposing the first radical and squaring,

$$(x + c)^2 + y^2 = 4a^2 - 4a\sqrt{(x - c)^2 + y^2} + (x - c)^2 + y^2$$

Expanding and collecting like terms,

$$a\sqrt{(x-c)^2 + y^2} = a^2 - cx$$

Squaring again and simplifying,

$$(a^2 - c^2)x^2 + a^2y^2 = a^2(a^2 - c^2)$$

If we substitute $b^2 = a^2 - c^2$, this equation becomes

$$b^2x^2 + a^2y^2 = a^2b^2$$

Finally, dividing through by the nonzero quantity a^2b^2, we can write the equation of an ellipse with its center at the origin and its foci on the x-axis as

$$\frac{x^2}{a^2} + \frac{y^2}{b^2} = 1 \tag{11.9}$$

The graph of Equation 11.9 is symmetric about both axes and the origin. Letting $y = 0$ in Equation 11.9, we see that the x-intercepts of the ellipse are $(a, 0)$ and $(-a, 0)$. The segment of the line through the foci from $(a, 0)$ to $(-a, 0)$ is called the **major axis** of the ellipse. The length of the major axis is $2a$, which is also the value chosen for the sum of the distances PF and PF'. The endpoints of the major axis are called the **vertices** of the ellipse.

The y-intercepts of the ellipse are found to be $(0, b)$ and $(0, -b)$ by letting $x = 0$ in Equation 11.9. The segment of the line perpendicular to the major axis from $(0, b)$ to $(0, -b)$ is called the **minor axis**. The graph of an ellipse can readily be sketched once the endpoints of the axes are known.

The foci of an ellipse are located by solving the equation $b^2 = a^2 - c^2$ for the focal distance c. Thus the expression for c is

$$c = \sqrt{a^2 - b^2}$$

A similar derivation will show that

$$\frac{x^2}{b^2} + \frac{y^2}{a^2} = 1 \tag{11.10}$$

is the equation of an ellipse with its center at the origin and its foci on the y-axis. [Note that the vertices are $(0, a)$ and $(0, -a)$.]

Example 1 Find the equation of an ellipse centered at the origin with foci on the x-axis if the major axis is 10 and the minor is 4.

SOLUTION In this case the major axis is on the x-axis with $a = 5$ and $b = 2$. Substituting these values into Equation 11.9 yields

$$\frac{x^2}{5^2} + \frac{y^2}{2^2} = 1$$

or

$$\frac{x^2}{25} + \frac{y^2}{4} = 1$$

which is the required equation. ▲

Example 2 Find the equation of the ellipse with vertices at (0, 5) and (0, −5) and foci at (0, 4) and (0, −4).

SOLUTION From the given information, the foci are on the y-axis, the center of the ellipse is at the origin, and $a = 5$. To find b, use the relation $b^2 = a^2 - c^2$. Thus $b = \sqrt{25 - 16} = \sqrt{9} = 3$. Substituting $a = 5$ and $b = 3$ into Equation 11.10 yields

$$\frac{x^2}{9} + \frac{y^2}{25} = 1$$

The ellipse is sketched in Figure 11.17.

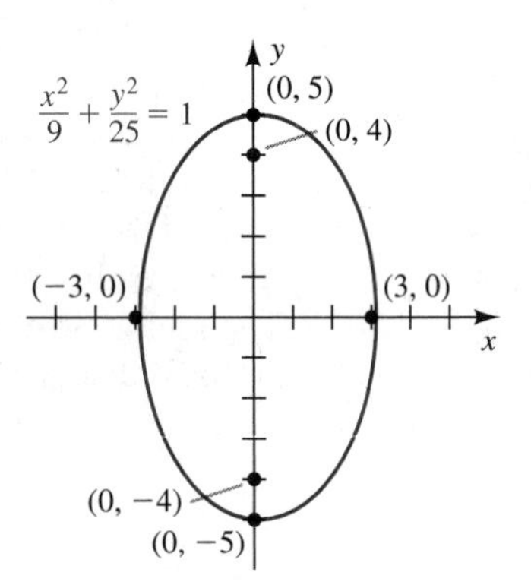

Figure 11.17 ▲

Example 3 Sketch the graph of the ellipse $4x^2 + 16y^2 = 64$.

SOLUTION The given equation divided by 64 can be written in the form

$$\frac{x^2}{16} + \frac{y^2}{4} = 1$$

The major axis lies along the x-axis since the denominator of the x term in the equation is larger than the denominator of the y term. Consequently $a = 4$ and $b = 2$. The vertices are then (4, 0) and (−4, 0) and the endpoints of the minor axis are (0, 2) and (0, −2). The ellipse is sketched in Figure 11.18.

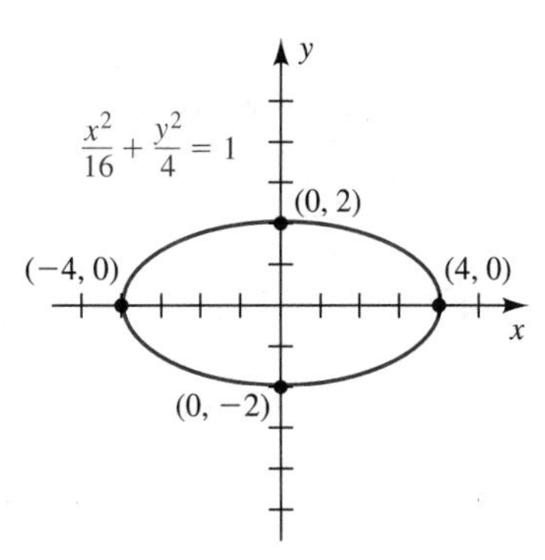

Figure 11.18

COMMENT One of the first scientific applications of the ellipse was in astronomy. The astronomer Johannes Kepler (c. 1600) discovered that the planets moved in elliptical orbits about the sun with the sun at one focus. Artificial satellites also move in elliptical orbits about the earth. Another application is in the design of machines, in which elliptical gears are used to obtain a slow, powerful movement with a quick return. A third application of the ellipse is found in electricity, in which the magnetic field of a single-phase induction motor is elliptical under normal operating conditions. The ellipse also has the property that all rays emitted from one focus are reflected through to the other focus, as shown in the figure.

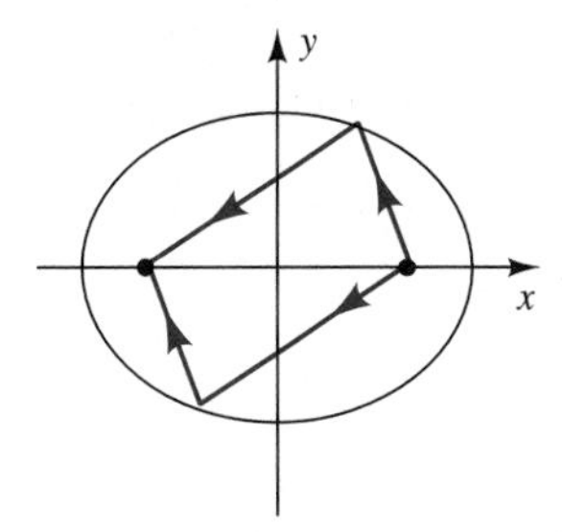

The Circle

When a plane intersects a cone perpendicular to the axis of the cone, other than at the vertex, a circle is formed. See Figure 11.9(a). If the plane intersects the vertex, the result is a single point.

DEFINITION 11.3 A circle is the set of points in a plane equidistant from a fixed point.

The equation of a circle centered at the origin is obtained from an ellipse which becomes a circle when both foci are located at the origin. Replacing a and b in Equation 11.9 with r, we find that the equation of a circle is

$$\frac{x^2}{r^2} + \frac{y^2}{r^2} = 1$$

Multiplying both sides of this equation by r^2, we obtain

$$x^2 + y^2 = r^2 \qquad \textbf{(11.11)}$$

as the standard form of the equation of a circle with center at the origin and radius equal to r.

Example 4 Write the equation of a circle centered at the origin with radius 4.

SOLUTION Substituting $r = 4$ into Equation 11.11, we get the desired equation:

$$x^2 + y^2 = 16$$

See Figure 11.19.

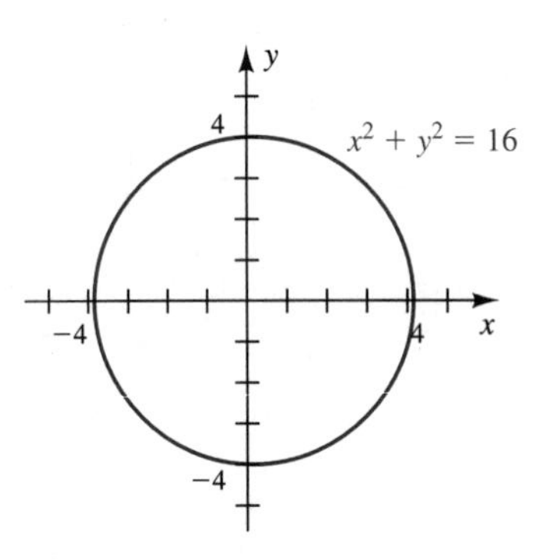

Figure 11.19 ▲

▲▼ Exercises for Section *11.3*

In Exercises 1–10 discuss each of the given equations and sketch their graphs.

1. $5x^2 + y^2 = 25$

2. $4x^2 + 9y^2 = 36$

3. $16x^2 + 4y^2 = 16$

4. $3x^2 + 9y^2 = 27$

5. $x^2 + y^2 = 9$

6. $25x^2 + 4y^2 = 100$

7. $2x^2 + 3y^2 - 24 = 0$

8. $5x^2 + 20y^2 = 20$

9. $9x^2 + 4y^2 = 4$

10. $4x^2 + y^2 = 25$

In Exercises 11–20 find the equation of the ellipses having the given properties. Sketch each curve.

11. Vertices at $(\pm 4, 0)$, minor axis 6.
12. Vertices at $(0, \pm 1)$, minor axis 1.
13. Vertices at $(0, \pm 5)$, minor axis 3.
14. Vertices at $(\pm 6, 0)$, minor axis 4.
15. Major axis 10, foci at $(\pm 4, 0)$.
16. Major axis 10, foci at $(0, \pm 3)$.
17. Foci at $(\pm 1, 0)$, major axis 8.
18. Vertices at $(0, \pm 7)$, foci at $(0, \pm\sqrt{28})$.
19. Vertices at $(\pm\frac{5}{2}, 0)$, one point of the curve at $(1, 1)$.
20. Vertices at $(\pm 3, 0)$, one point of the curve at $(\sqrt{3}, 2)$.

In Exercises 21–24 solve the given system of equations graphically.

21. $\frac{x^2}{4} + y^2 = 4$
$2y + 3x = 0$

22. $\frac{x^2}{9} + \frac{y^2}{9} = 1$
$y = e^{x+2}$

24. $y = x^2$
$x^2 + 4y^2 = 4$

24. $y^2 - 12x = 0$
$y^2 + 9x^2 = 9$

25. An elliptical cam with a horizontal major axis of 10 in. and a minor axis of 3 in. is to be machined by a numerically controlled vertical mill (see the figure).

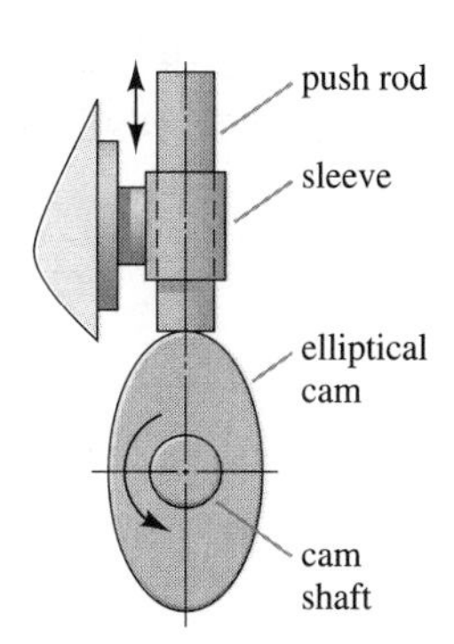

Find the equation of the ellipse to be used in programming the control device.

26. An elliptical cam having the equation $9x^2 + y^2 = 81$ revolves against a push rod. What is the maximum travel of the push rod? Assume the cam revolves about its center.

27. The magnetic field curves of a single-phase induction motor are given by the set of ellipses $x^2 + 4y^2 = c^2$. Sketch some of the curves.

28. A "ripple" tank is a tank of water in the form of an ellipse. When water is disturbed at one focus, ripples radiate outward and, eventually, a drop of water spurts up at the other focus. Suppose such a tank is of the shape given by the equation

$$3.1x^2 + 4.5y^2 = 15.6$$

Determine where to poke your finger into the water and where the water will spurt up.

29. The elliptical orbit of the earth is very nearly circular. In fact, it is much like the ellipse $x^2 + (y/1.1)^2 = 1$. Make a sketch of this ellipse.

11.4 The Hyperbola

The final conic that we consider is the hyperbola.

DEFINITION 11.4 A hyperbola is the set of all points in a plane the difference of whose distances from two fixed points (**foci**) in the plane is constant.

Figure 11.20 shows a hyperbola with foci at $F(c, 0)$ and $F'(-c, 0)$. The origin is at the midpoint between the foci, which corresponds to the **center** of the hyperbola. The points $V(a, 0)$ and $V'(-a, 0)$ are called the **vertices**, and the segment $\overline{VV'}$ is called the **transverse axis** of the hyperbola. The length of the transverse axis is $2a$. The segment BB', which is perpendicular to the transverse axis at the center of the hyperbola, is called the **conjugate axis** and has a length of $2b$. The conjugate axis has an important relation to the curve even though it does not intersect the curve.

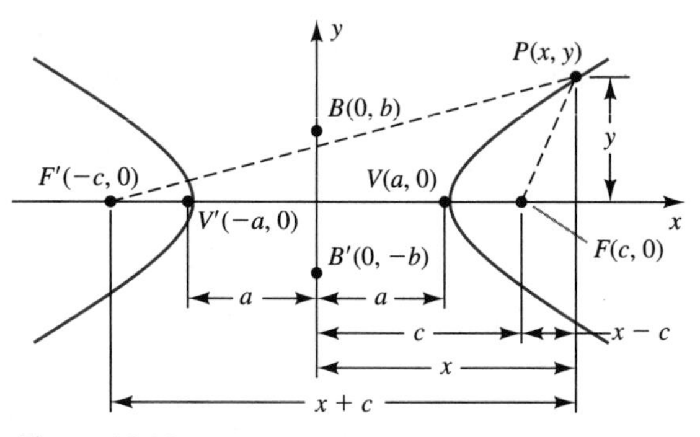

Figure 11.20

Consider a point $P(x, y)$ on the hyperbola. Then by definition

$$\overline{F'P} - \overline{FP} = 2a$$

which in turn can be written

$$\sqrt{(x + c)^2 + y^2} - \sqrt{(x - c)^2 + y^2} = 2a$$

Using the same procedure here that we used for the ellipse, we can reduce the radical equation to

$$\frac{x^2}{a^2} - \frac{y^2}{c^2 - a^2} = 1$$

Or, if we let $b^2 = c^2 - a^2$, the equation of the hyperbola becomes

$$\frac{x^2}{a^2} - \frac{y^2}{b^2} = 1 \tag{11.12}$$

which, like the ellipse, is symmetric about both axes and the origin. Letting $y = 0$, we find the x-intercepts of the graph of Equation 11.12 to be $x = \pm a$. Additional information on the shape of the hyperbola can be obtained by solving Equation 11.12 for y:

$$y = \pm\frac{b}{a}\sqrt{x^2 - a^2}$$

This equation shows that there are no points on the curve for which $x^2 < a^2$. Consequently the hyperbola consists of two separate curves, or *branches*—one to the right of $x = a$ and a similar one to the left of $x = -a$.

The shape of the hyperbola is constrained by two straight lines called the **asymptotes** of the hyperbola. The asymptotes of a hyperbola are the extended diagonals of the rectangle formed by drawing lines parallel to the coordinate axes through the endpoints of both the transverse axis and the conjugate axis. Referring to Figure 11.21, we see that the slope of the diagonals of this rectangle are

$$m = \pm\frac{b}{a}$$

Therefore the asymptotes are given by the lines

$$y = \pm\frac{b}{a}x$$

Now we show that the lines $y = \pm(b/a)x$ are asymptotes of the hyperbola; that is, that the hyperbola approaches arbitrarily close to the lines as

x increases without bound. Solving Equation 11.12 for y^2, we can write it in the form

$$y^2 = \frac{b^2x^2}{a^2}\left(1 - \frac{a^2}{x^2}\right)$$

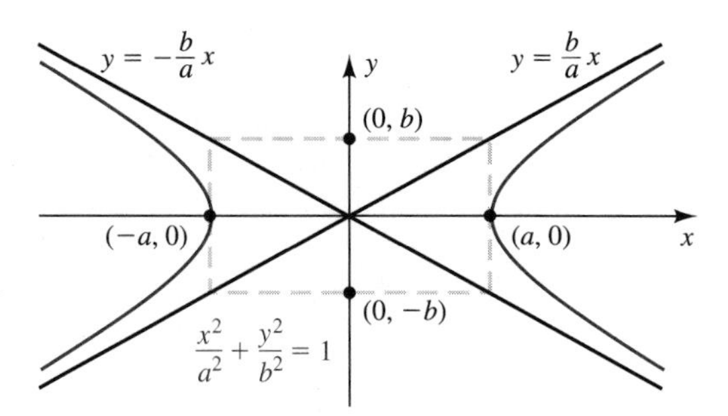

Figure 11.21

or, taking the square root of both sides,

$$y = \pm\frac{b}{a}x\sqrt{1 - \frac{a^2}{x^2}}$$

Now consider the value of the right member as x becomes large. The quantity a^2/x^2 becomes small and, therefore,

$$\pm\frac{b}{a}x\sqrt{1 - \frac{a^2}{x^2}}$$

approaches $\pm(b/a)x$, which means that the hyperbola $(x^2/a^2) - (y^2/b^2) = 1$ is asymptotic to the lines $y = \pm(b/a)x$.

If we begin with the foci of the hyperbola on the y-axis and the center at the origin, the standard form of the equation of a hyperbola becomes

$$\frac{y^2}{a^2} - \frac{x^2}{b^2} = 1 \qquad \textbf{(11.13)}$$

The vertices of this hyperbola are on the y-axis; **the positive term always indicates the direction of the transverse axis.** Notice that the standard form of the hyperbola, like that for the ellipse, demands that the coefficients of x^2 and y^2 be in the denominator and that the number of the right-hand side be 1. In the case of the hyperbola the *sign* of the term, *not the magnitude* of the denominator, determines the transverse axis.

To sketch the hyperbola, draw the rectangle through the extremities of the transverse and conjugate axes and extend the diagonals of the rectangle. Then draw the hyperbola so that it passes through the vertex and comes closer to the extended diagonals as x moves away from the origin.

Example 1 Discuss and sketch the graph of $4x^2 - y^2 = 16$.

SOLUTION Dividing by 16, we have

$$\frac{x^2}{4} - \frac{y^2}{16} = 1$$

which is the equation of a hyperbola with center at the origin and foci on the x-axis. It has vertices at $(\pm 2, 0)$, and its conjugate axis extends from $(0, 4)$ to $(0, -4)$. The foci are found from the equation $b^2 = c^2 - a^2$. Thus

$$c = \sqrt{a^2 + b^2} = \sqrt{4 + 16} = \sqrt{20} = 2\sqrt{5}$$

and the foci are located at $(\pm 2\sqrt{5}, 0)$. Plotting these points and drawing the rectangle and its extended diagonals, we obtain the hyperbola shown in Figure 11.22.

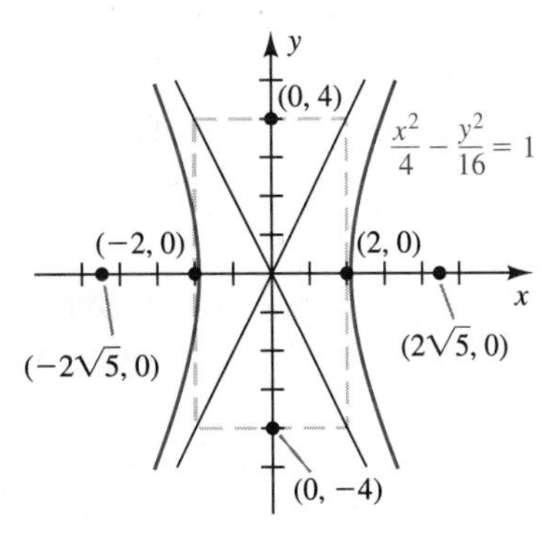

Figure 11.22 ▲

Example 2 Discuss and sketch the graph of $\frac{y^2}{13} - \frac{x^2}{9} = 1$.

SOLUTION This hyperbola has a vertical transverse axis with vertices at $(0, \sqrt{13})$ and $(0, -\sqrt{13})$. The extremes of the conjugate axis are then $(3, 0)$ and $(-3, 0)$. The foci are located at $(0, \sqrt{22})$ and $(0, -\sqrt{22})$. This information is used to sketch the hyperbola in Figure 11.23.

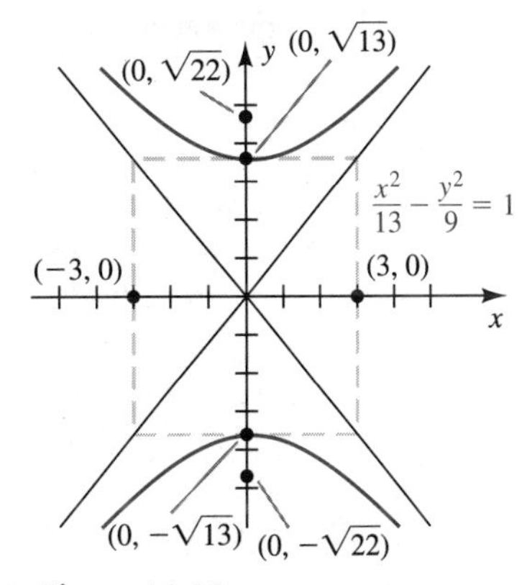

Figure 11.23 ▲

Example 3 Determine the equation of the hyperbola centered at the origin with foci at $(\pm 6, 0)$ and a transverse axis 8 units long.

SOLUTION Here $a = 4$ and $c = 6$. Since $c^2 = a^2 + b^2$, we have $b^2 = c^2 - a^2 = 36 - 16 = 20$. Substituting $a^2 = 16$ and $b^2 = 20$ into Equation 11.12, we get

$$\frac{x^2}{16} - \frac{y^2}{20} = 1$$ ▲

Exercises for Section *11.4*

In Exercises 1–10 discuss the properties of the graph of each of the given equations and then sketch the graph.

1. $x^2 - y^2 = 16$

2. $y^2 - x^2 = 9$

3. $4x^2 - 9y^2 = 36$

4. $9x^2 - y^2 = 9$

5. $4y^2 - 25x^2 = 100$

6. $3x^2 - 3y^2 = 9$

7. $4x^2 - 16y^2 = 25$

8. $4y^2 - x^2 = 9$

9. $y^2 + 1 = x^2$

10. $x^2 - 25 = 5y^2$

In Exercises 11–20 find the equation of the hyperbolas having the given properties. Sketch each curve.

11. Vertices at $(\pm 4, 0)$, foci at $(\pm 5, 0)$.

12. Vertices at $(0, \pm 3)$, foci at $(0, \pm 5)$.

13. Conjugate axis 4, vertices at $(0, \pm 1)$.

14. Conjugate axis 1, vertices at $(\pm 4, 0)$.

15. Transverse axis 6, foci at $(\pm \frac{7}{2}, 0)$.

16. Transverse axis 3, foci at $(\pm 2, 0)$.

17. Vertices at $(0, \pm 4)$, asymptotes $y = \pm(\frac{1}{2})x$.

18. Vertices at $(\pm 3, 0)$, asymptotes $y = \pm 2x$.

19. Vertices at $(0, \pm 3)$, one point of the curve $(2, 7)$.

20. Vertices at $(\pm 3, 0)$, one point of the curve $(7, 2)$.

21. In the study of electrostatic potential with particular boundary conditions, the equipotential curves are found to be

$$\frac{x^2}{\sin^2 c} - \frac{y^2}{\cos^2 c} = 1$$

Show that every member of this family has foci at $(-1, 0)$ and $(1, 0)$.

22. Curves of the form $xy = c$ are hyperbolas, but their foci lie along the lines $y = \pm x$ instead of along the coordinate axes. Make a sketch of the hyperbola $xy = 2$.

23. Any two variables x and y that are related by the equation $xy = c$ are said to vary inversely with each other. Sketch the inverse variation $xy = -1$ and note the hyperbolic shape.

24. Show that the curve defined parametrically by

$$x = \frac{e^t - e^{-t}}{2}, \; y = \frac{e^t + e^{-t}}{2}$$

is a hyperbola. (*Hint:* Show that x and y satisfy the equation $y^2 - x^2 = 1$. The two given functions are called the **hyperbolic functions** and are denoted respectively by $\sinh t$ and $\cosh t$.)

11.5 Translation of Axes

In the previous three sections of this chapter we developed the equations of the conic sections relative to the origin. Now we wish to write the equations of the conics referenced to some point (h, k) in the plane. To see how this is done, consider a point $P(x, y)$ in the xy-plane and an $x'y'$-coordinate system whose origin is at (h, k), as shown in Figure 11.24. Then the coordinates of P may be given with respect either to the xy-plane or to the $x'y'$-plane. From Figure 11.24, we see that the relationship between (x, y) and (x', y') is given by the equations

$$x = x' + h \qquad y = y' + k$$

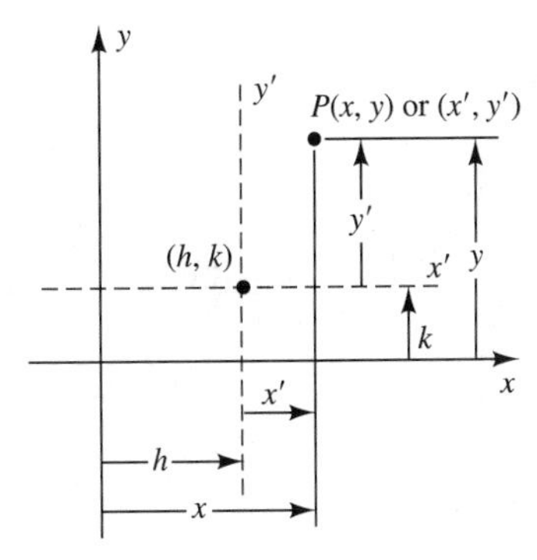

Figure 11.24

or

$$x' = x - h \qquad y' = y - k$$

These equations are called **translation equations**.

To discover the equation of a circle with its center at (h, k), we note that since (h, k) is the origin of the $x'y'$-coordinate system, the equation of the circle may be written as

$$(x')^2 + (y')^2 = r^2$$

The translation equations $x' = x - h$ and $y' = y - k$ are then used to give

$$(x - h)^2 + (y - k)^2 = r^2 \tag{11.14}$$

as the standard form of a circle with center at (h, k) and radius r. Similarly, the equation of a parabola with vertex at (h, k) is

$$(x - h)^2 = 4a(y - k) \tag{11.15}$$

if the axis of symmetry is parallel to the y-axis and

$$(y - k)^2 = 4a(x - h) \tag{11.16}$$

if the axis of symmetry is parallel to the x-axis. By a similar procedure, if $a > b$ the standard equations of an ellipse centered at a point (h, k) are seen to be

$$\frac{(x - h)^2}{a^2} + \frac{(y - k)^2}{b^2} = 1 \tag{11.17}$$

if the major axis is parallel to the x-axis and

$$\frac{(x - h)^2}{b^2} + \frac{(y - k)^2}{a^2} = 1 \tag{11.18}$$

if it is parallel to the y-axis. Finally, the standard equations of the hyperbola centered at a point (h, k) are seen to be

$$\frac{(x - h)^2}{a^2} - \frac{(y - k)^2}{b^2} = 1 \tag{11.19}$$

if the transverse axis is parallel to the x-axis and

$$\frac{(y - k)^2}{a^2} - \frac{(x - h)^2}{b^2} = 1 \tag{11.20}$$

if it is parallel to the y-axis.

Example 1 Write the equation of the ellipse centered at (2, −3) with horizontal axis 10 units and vertical axis 4 units. (See Figure 11.25.)

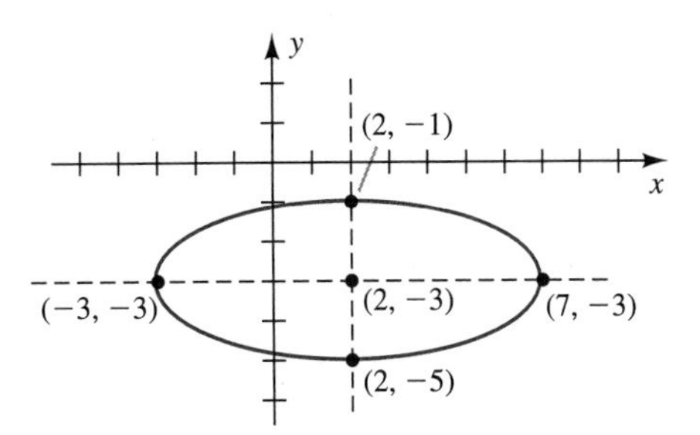

Figure 11.25

SOLUTION Since the longer of the two axes is the horizontal axis, use Equation 11.17 with $a = 5$ and $b = 2$. Also, $h = 2$ and $k = -3$. Making these substitutions, we have

$$\frac{(x-2)^2}{25} + \frac{(y+3)^2}{4} = 1$$

as the equation of the ellipse. ▲

Example 2 Write the equation of the parabola whose directrix is the line $y = -2$ and whose vertex is located at (3, 1). (See Figure 11.26).

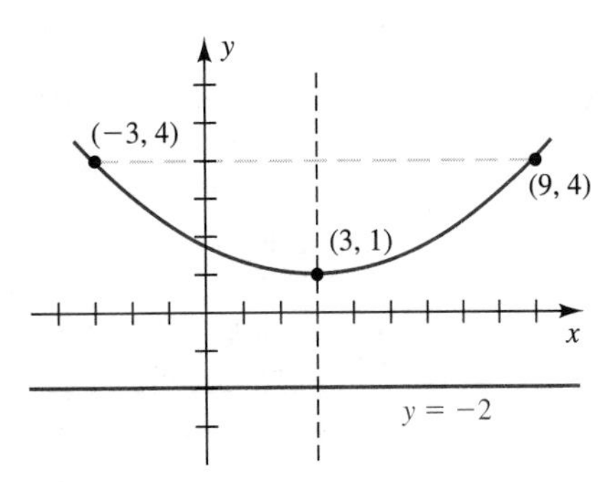

Figure 11.26

SOLUTION This parabola has a vertical axis and opens upward because the directrix is horizontal and lies below the vertex. The vertex lies midway between the focus and the directrix so that $a = 3$. If we let $h = 3$, $k = 1$, and $a = 3$ in Equation 11.15, the equation of the parabola is

$$(x-3)^2 = 12(y-1)$$ ▲

Example 3 Discuss and sketch the graph of the hyperbola

$$\frac{(x+5)^2}{16} - \frac{(y-2)^2}{9} = 1$$

SOLUTION The hyperbola is centered at $(-5, 2)$. By Equation 11.19, it has a horizontal transverse axis with vertices at $(-9, 2)$ and $(-1, 2)$. The endpoints of the conjugate axis are located at $(-5, 5)$ and $(-5, -1)$. Also, the foci are at $(-10, 2)$ and $(0, 2)$ since $c = \sqrt{9 + 16} = 5$. The graph appears in Figure 11.27.

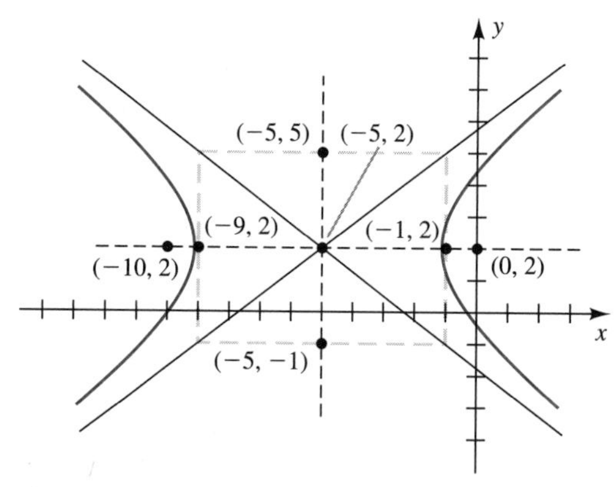

Figure 11.27

Exercises for Section 11.5

In Exercises 1–6 write the equations of the *parabolas* having the given properties. Sketch each graph.

1. Vertex at $(3, 1)$, focus at $(5, 1)$.
2. Vertex at $(-2, 3)$, focus at $(-2, 0)$.
3. Directrix $y = 2$, vertex at $(1, -1)$.
4. Directrix at $x = -1$, vertex at $(0, 4)$.
5. Endpoints of right chord $(2, 4)$ and $(2, 0)$, opening to the right.
6. Endpoints of right chord $(-1, -1)$ and $(5, -1)$, opening upward.

In Exercises 7–12 write the equations of the *ellipses* having the given properties. Sketch each graph.

7. Major axis 8, foci at $(5, 1)$ and $(-1, 1)$.
8. Minor axis 6, vertices at $(2, -1)$ and $(10, -1)$.
9. Minor axis 2, vertices at $(\frac{1}{2}, 0)$, and $(\frac{1}{2}, -8)$.
10. Major axis 3, foci at $(1, 1)$ and $(1, -1)$.
11. Vertices at $(-6, 3)$ and $(-2, 3)$, foci at $(-5, 3)$ and $(-3, 3)$.
12. Center at $(1, -3)$, major axis 10, minor axis 6, vertical axis.

In Exercises 13–18 write the equations of the *hyperbolas* having the given properties. Sketch each graph.

13. Center at $(-1, 2)$, transverse axis 7, conjugate axis 8, vertical axis.
14. Center at $(3, 0)$, transverse axis 6, conjugate axis 2, horizontal axis.
15. Vertices at $(5, 1)$ and $(-1, 1)$, foci at $(6, 1)$ and $(-2, 1)$.

16. Vertices at $(2, \pm 4)$, conjugate axis 2.

17. Vertices at $(-4, -2)$ and $(0, -2)$, slope of the asymptotes $m = \pm\frac{1}{2}$.

18. Vertices at $(3, 3)$ and $(5, 3)$, slope of asymptotes $m \pm 3$.

Write the equation of the family of curves indicated in Exercises 19–24.

19. Circles with center on the x-axis.

20. Parabolas with vertical axis and vertex on the x-axis.

21. Parabolas with vertex and focus on the x-axis.

22. Ellipses with center on the y-axis and horizontal major axis.

23. Circles passing through the origin with center on the x-axis.

24. Circles tangent to the x-axis.

25. The path of a projectile is given by $y = 20x - \frac{1}{10}x^2$, where y is the vertical distance and x the horizontal distance away from the initial point. Locate the vertex and sketch the path.

11.6 Rotation of Axes

In the previous section we saw how an analytic representation of a curve can be considerably simplified by moving the origin of the Cartesian coordinate system. This section shows how to *rotate* the axes to yield a new rectangular system of coordinates.

Figure 11.28 shows two rectangular coordinate systems with the x', y'-axes at an angle θ with the xy-axes. We say that the coordinates of a point (x, y) are **transformed** into the coordinates (x', y') by rotating the axes through an angle θ.

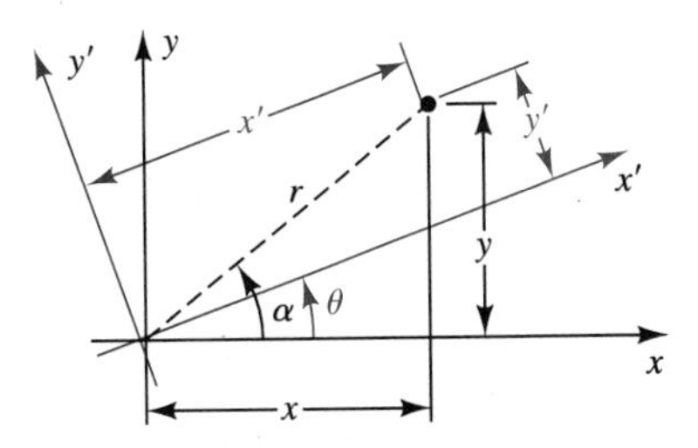

Figure 11.28

The equations of rotation are obtained by expressing (x', y') in terms of (x, y) and the rotation angle θ. From Figure 11.28,

$$x' = r\cos(\alpha - \theta)$$
$$y' = r\sin(\alpha - \theta) \qquad \textbf{(11.21)}$$

where α is the angle made by the x-axis and the line drawn from the origin to the point P. Using the trigonometric identities for the cosine and sine of the difference of two angles, we may write Equations 11.21 as

$$\begin{aligned} x' &= r \cos \alpha \cos \theta + r \sin \alpha \sin \theta \\ y' &= r \sin \alpha \cos \theta - r \cos \alpha \sin \theta \end{aligned} \qquad \textbf{(11.22)}$$

From Figure 11.28, $x = r \cos \alpha$ and $y = r \sin \alpha$; hence the required relationship between the two coordinate systems is given by

$$\begin{aligned} x' &= x \cos \theta + y \sin \theta \\ y' &= -x \sin \theta + y \cos \theta \end{aligned} \qquad \textbf{(11.23)}$$

Solving these equations for x and y in terms of x', y', and θ, we obtain the **equations of the inverse transformation:**

$$\begin{aligned} x &= x' \cos \theta - y' \sin \theta \\ y &= x' \sin \theta + y' \cos \theta \end{aligned} \qquad \textbf{(11.24)}$$

You can use Equations 11.23 to find the coordinates of any point in the rotated system if you know the coordinates in the original system.

Example 1 If the coordinates of a point in a coordinate system are $x = 4$, $y = 3$, find the coordinates of the same point in a rectangular system whose axes are rotated at 30° to the original.

SOLUTION Using the equations of rotation, we get

$$\begin{aligned} x' &= 4 \cos 30° + 3 \sin 30° \\ y' &= -4 \sin 30° + 3 \cos 30° \end{aligned}$$

from which $x' = 2\sqrt{3} + 3/2$ and $y' = -2 + (3\sqrt{3})/2$. ▲

To express an equation of a curve given in xy-coordinates in terms of $x'y'$-coordinates, merely use the rotation equations in Equations 11.24 to substitute for x and y in terms of x', y', and θ.

Example 2 Find the equation of the straight line $x + y = 1$ in $x'y'$-coordinates if the prime coordinate system is rotated 45° to the xy-system.

SOLUTION From Equations 11.24 the expressions for x and y in terms of x' and y' are

$$x = x'\frac{\sqrt{2}}{2} - y'\frac{\sqrt{2}}{2}$$

$$y = x'\frac{\sqrt{2}}{2} + y'\frac{\sqrt{2}}{2}$$

Hence the equation $x + y = 1$ becomes

$$\frac{\sqrt{2}}{2}(x' - y') + \frac{\sqrt{2}}{2}(x' + y') = 1$$

After simplification this equation becomes

$$\sqrt{2}x' = 1$$

▲

The foregoing example is typical of the actual use of rotation of coordinates; that is, the resulting equation should be in a simpler form. In the case of Example 2, the straight line is parallel to the y'-coordinate axis, which can be considered a simplification. Sometimes the curve itself becomes recognizable only after rotation. This simplification is done by judiciously choosing the rotation angle.

Example 3 Consider the equation $x^2 + xy + y^2 = 1$. Choose a rotated coordinate system in which the "product term" (that is, the product of the coordinates) is not present. Identify and sketch the curve that the equation represents.

SOLUTION Using the equations of rotation in Equations 11.24 and substituting into the given equation, we have

$$(x' \cos\theta - y' \sin\theta)^2 + (x' \cos\theta - y' \sin\theta)(x' \sin\theta + y' \cos\theta) + (x' \sin\theta + y' \cos\theta)^2 = 1$$

Simplifying yields

$$(x')^2(1 + \sin\theta\cos\theta) + (\cos^2\theta - \sin^2\theta)x'y' + (1 - \sin\theta\cos\theta)(y')^2 = 1$$

Note that the coefficient of $x'y'$ vanishes if $\cos^2\theta = \sin^2\theta$, so that we choose $\theta = 45°$, and the equation of the curve becomes

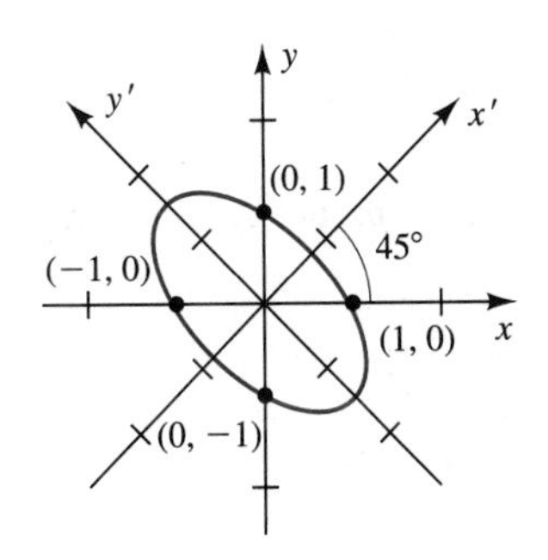

Figure 11.29

$$\left(1 + \frac{\sqrt{2}}{2}\frac{\sqrt{2}}{2}\right)(x')^2 + \left(1 - \frac{\sqrt{2}}{2}\frac{\sqrt{2}}{2}\right)(y')^2 = 1$$

or

$$\frac{(x')^2}{\frac{2}{3}} + \frac{(y')^2}{2} = 1$$

Hence the equation represents an ellipse whose axes are at 45° to the xy-axes. Figure 11.29 displays the graph.

▲

Exercises for Section 11.6

Find the coordinates of the points in Exercises 1–5 in a coordinate system rotated 30° from the x, y coordinate system.

1. (3, 1)
2. (−5, 3)
3. (−1, −1)
4. (2, 1)
5. (2, −1)

6. Find the equations of rotation for a system rotated first through 30° and then 45°. Compare with the equations obtained by one rotation of 75°.

Find the equations of the curves in Exercises 7–13 after the given rotation.

7. The line $y = 2x$ after a 45° rotation.
8. The line $y = x + 5$ after a 45° rotation.
9. The line $2x + 7y = 3$ after a 60° rotation.
10. The circle $x^2 + y^2 = 4$ after a rotation of any angle θ.
11. The ellipse $x^2 + 4y^2 = 4$ after a rotation of 90°.
12. The parabola $y = x^2$ after a rotation of 30°.
13. The circle $x^2 + 2x + y^2 = 0$ after a rotation of 30°.

14. What is the slope of the line $y = mx$ in a system of coordinates rotated at 45° from the x, y system?

15. Given the transformation equations

$$x' = 0.6x + 0.8y$$
$$y' = -0.8x + 0.6y$$

what is the angle of rotation?

16. Consider the curve $(x - y)^2 - (x + y) = 0$. Rotate the coordinate system 45° and identify the curve.

17. Consider the curve $3x^2 - 2xy + 3y^2 = 2$. Rotate the coordinate system 45° and identify the curve.

Eliminate the product term in Exercises 18–21 by using a proper rotation and then sketch.

18. $xy = 1$
19. $x^2 + xy = 1$
20. $y - xy = 1$
21. $x^2 - xy + y^2 = 1$

22. What is the form of the equation of the hyperbola whose equation is $x^2 - y^2 = a^2$ in a system rotated 45°?

11.7 The General Second-Degree Equation

The general second-degree equation is of the form.

$$Ax^2 + Bxy + Cy^2 + Dx + Ey + F = 0 \qquad \textbf{(11.25)}$$

where A, B, C, D, E, and F are constants. Each conic described in Sections 11.2 through 11.5 can be expressed in the form of Equation 11.25 with $B = 0$. This is seen by expanding the standard form of each conic. Assuming that $B = 0$, we make the following statements.

- If $A = C$, Equation 11.25 represents a circle.
- If $A \neq C$ and A and C have the same numerical sign, Equation 11.25 represents an ellipse.
- If A and C have different numerical signs, Equation 11.25 represents a hyperbola.
- If A or $C = 0$ (but not both), then Equation 11.25 represents a parabola.
- Special cases such as a single point or no graph may result.

If $B = 0$, the general form of a conic can be reduced to one of the standard forms by completing the square on x and y. Several examples of this technique follow.

Example 1 Discuss and sketch the graph of $x^2 - 4y^2 + 6x + 24y - 43 = 0$.

SOLUTION This is the equation of a hyperbola, since the coefficients of the x^2 and y^2 terms have unlike signs. To sketch the hyperbola, reduce the given equation to standard form by rearranging the terms and completing the square on the x-terms and the y-terms. Thus

$$x^2 - 4y^2 + 6x + 24y - 43 = 0$$

can be written

$$(x^2 + 6x) - 4(y^2 - 6y) = 43$$

Completing the square on each variable, we get

$$(x^2 + 6x + 9) - 4(y^2 - 6y + 9) = 43 + 9 - 36$$

$$(x + 3)^2 - 4(y - 3)^2 = 16$$

$$\frac{(x + 3)^2}{16} - \frac{(y - 3)^2}{4} = 1$$

The center of the hyperbola is the point $(-3, 3)$. The transverse axis is horizontal with vertices at $(1, 3)$ and $(-7, 3)$. The endpoints of the conjugate axis are located at $(-3, 5)$ and $(-3, 1)$. Finally, the foci of the hyperbola are at $(-3 + 2\sqrt{5}, 3)$ and $(-3 - 2\sqrt{5}, 3)$, since $c = \sqrt{16 + 4} = 2\sqrt{5}$. The graph appears in Figure 11.30. ▲

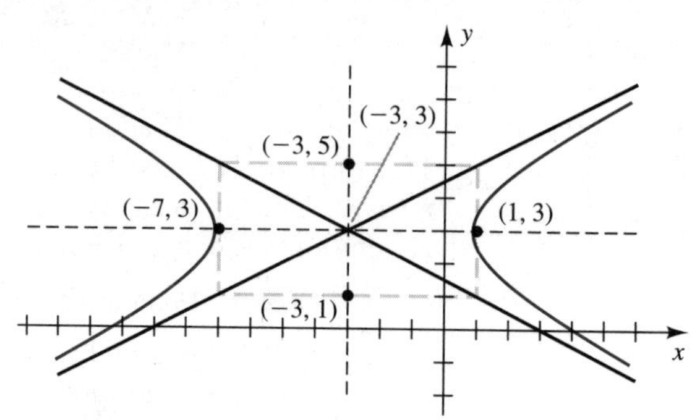

Figure 11.30

Example 2 Discuss and sketch the graph of $2y^2 + 3x - 8y + 9 = 0$.

SOLUTION By completing the square on the y-variable, this equation can be reduced to the form of Equation 11.16. Thus

$$\begin{aligned} 2y^2 + 3x - 8y + 9 &= 0 \\ 2(y^2 - 4y) &= -3x - 9 \\ 2(y^2 - 4y + 4) &= -3x - 9 + 8 \\ 2(y-2)^2 &= -3x - 1 \\ 2(y-2)^2 &= -3\left(x + \frac{1}{3}\right) \\ (y-2)^2 &= -\frac{3}{2}\left(x + \frac{1}{3}\right) \end{aligned}$$

This is the standard form of the equation of a parabola with horizontal axis and vertex at $(-\frac{1}{3}, 2)$. We see that $4a = -\frac{3}{2}$, so $a = -\frac{3}{8}$. Therefore the focus is at $(-\frac{17}{24}, 2)$ and the endpoints of the right chord are $(-\frac{17}{24}, \frac{11}{4})$ and $(-\frac{17}{24}, \frac{5}{4})$. The parabola, which opens to the left, is shown in Figure 11.31.

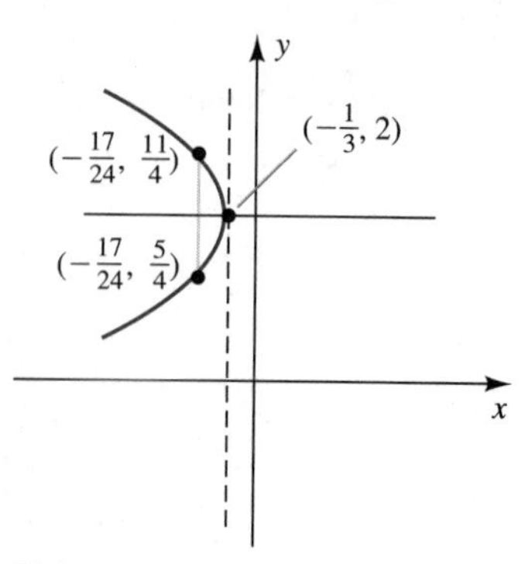

Figure 11.31 ▲

If the axis of the conic is not parallel to one of the coordinate axes, then a product term is present, $B \neq 0$, and the nature of the conic is not immediately obvious. By properly choosing θ, we can remove this product in a rotated $x'y'$-system and reduce the problem essentially to the case of $B = 0$.

By substituting the general rotation equations into Equation 11.25, we can show that the product term can be made to vanish if we choose a rotation angle θ so that

$$\cot 2\theta = \frac{A - C}{B}$$

Unfortunately, this formula does not give the angle of rotation directly but requires a little unraveling using familiar trigonometric identities. (See Example 3.)

Example 3 Eliminate the xy term in $3x^2 - 4xy = 20$ and sketch the graph.

SOLUTION Here $A = 3$, $B = -4$, and $C = 0$. Hence

$$\cot 2\theta = \frac{3 - 0}{-4} = -\frac{3}{4}$$

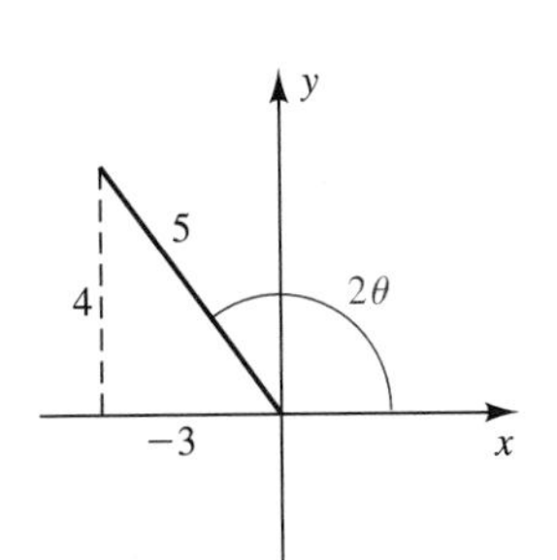

Figure 11.32

which implies the angle shown in Figure 11.32. Assuming that 2θ lies in the second quadrant and noting that $\sin\theta = \sqrt{(1 - \cos 2\theta)/2}$ and $\cos\theta = \sqrt{(1 + \cos 2\theta)/2}$, we find that

$$\sin\theta = \sqrt{\frac{1 + \frac{3}{5}}{2}} = \frac{2\sqrt{5}}{5}$$

$$\cos\theta = \sqrt{\frac{1 - \frac{3}{5}}{2}} = \frac{\sqrt{5}}{5}$$

Substituting these values into the rotation Equations 11.24, we get

$$x = \frac{\sqrt{5}}{5}x' - \frac{2\sqrt{5}}{5}y'$$

$$y = \frac{2\sqrt{5}}{5}x' + \frac{\sqrt{5}}{5}y'$$

The equation $3x^2 - 4xy = 20$ can then be written

$$3\left(\frac{\sqrt{5}}{5}x' - \frac{2\sqrt{5}}{5}y'\right)^2 -$$

$$4\left(\frac{\sqrt{5}}{5}x' - \frac{2\sqrt{5}}{5}y'\right)\left(\frac{2\sqrt{5}}{5}x' + \frac{\sqrt{5}}{5}y'\right) = 20$$

This equation reduces (with some effort) to

$$20y'^2 - 5x'^2 = 100$$

or

$$\frac{y'^2}{5} - \frac{x'^2}{20} = 1$$

which is a hyperbola in standard form with respect to the $x'y'$-coordinate system. (See Figure 11.33.) The rotation angle, θ, is given by

$$\theta = \sin^{-1}\left(\frac{2\sqrt{5}}{5}\right) = 63.4°$$

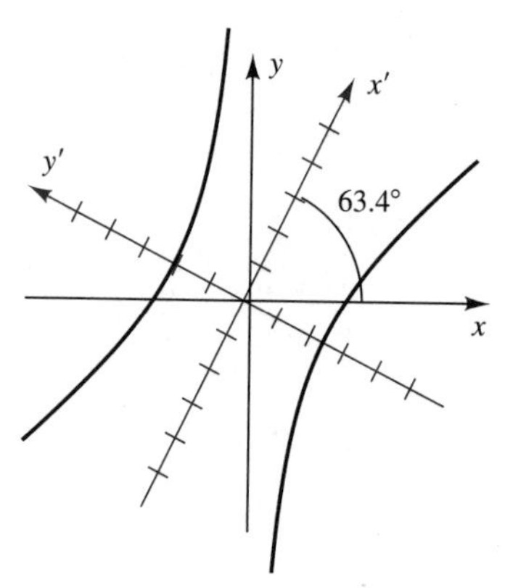

Figure 11.33

The asymptotes are

$$y' = \pm\frac{1}{2}x'$$

or, in terms of the original coordinates,

$$x = 0 \text{ (the } y \text{ axis)} \qquad \text{and} \qquad y = \frac{3}{4}x$$

▲

▲▼ Exercises for Section 11.7

Transform each of the equations in Exercises 1–20 into standard form and sketch their graphs.

1. $x^2 + y^2 + 4x + 6y + 4 = 0$

2. $x^2 - 2x - 8y + 25 = 0$

3. $9x^2 + 4y^2 + 18x + 8y - 23 = 0$

4. $2x^2 + 2y^2 - 4x - 16 = 0$

5. $x^2 - y^2 - 4x - 21 = 0$

6. $y^2 + 4y + 6x - 8 = 0$

7. $x^2 - 6x - 3y = 0$

8. $x^2 + 4y^2 + 8x = 0$

9. $2y^2 - 2y + x - 1 = 0$

10. $3x^2 + 4y^2 - 18x + 8y + 19 = 0$
11. $4x^2 - y^2 + 8x + 2y - 1 = 0$
12. $y^2 - 2x - 4y + 10 = 0$
13. $9x^2 + 4y^2 - 18x + 16y - 11 = 0$
14. $y^2 - 25x^2 + 50x - 50 = 0$
15. $x^2 - 2y^2 + 2y = 0$
16. $y^2 - y - \frac{1}{2}x + \frac{1}{4} = 0$
17. $x(x + 4) = y^2 + 3$
18. $x^2 + y(4 + 2y) = 0$
19. $y = x^2 + 5x + 7$
20. $y = 2x^2 + 10x$

Use rotation of axes to identify and sketch the conics in Exercises 21–24.

21. $5x^2 + 12xy = 9$
22. $5x^2 + 3xy + 5y^2 = 7$
23. $5x^2 + 24xy - 2y^2 = 44$
24. $x^2 + 2xy + y^2 = 1$
25. In the analysis of the heat distribution on a hotplate, the isotherms are found to be included in the family of curves

$$c^2x^2 - 8c^2x + 4y^2 + 12c^2 = 0, \text{ (c is a constant)}$$

What type of curve is each of the isotherms? Sketch a few of them for $y > 0$.

26. A space vehicle is scheduled to take off from the earth and follow the curved path given by the equation $y^2 - 4y - x^2 + 2x + 7 = 0$. Assuming that the earth is represented by the vertex in the part of the plane to the right of center of the conic, find the coordinates of the earth. Will the vehicle ever return to the neighborhood of the earth?
27. A garden is dug in the shape of a right triangle so that the hypotenuse is always 5 ft larger than one of the legs. Find the equation relating the legs. What kind of curve does it represent?
28. The outline of a lens of a camera has the equation $2x^2 - 3y - 4x + 2 = 0$. Where is the focus? What kind of curve describes the outline of the lens?

Review Exercises for Chapter 11

Sketch the graph of each of the equations given in Exercises 1–10. Identify all important parts.

1. $x^2 = -3y$
2. $x = -y^2 + 2y$
3. $x^2 = -y^2 - 4y + 5$
4. $x^2 - 3y^2 - 4 = 0$
5. $(x + 1)^2 - 2y = y^2$
6. $2x^2 - 3y^2 - x = 4$
7. $x^2 = y^2 - 100$
8. $y = 2x^2 + 3$
9. $y^2 + 2x^2 + 6x = 7$
10. $y^2 - y = x^2$
11. Write the equation of the parabola with vertex at (2, 0) and focus at the origin.
12. Write the equation of the parabola with vertex at (7, −2) and focus at (10, −2).

13. Write the equation of the circle with center at $(-1, 2)$ and radius equal to 8.
14. Write the equation of the circle with center at $(1, 3)$ that passes through $(6, 4)$.
15. Write the equation of the ellipse with center at the origin, focus at $(2, 0)$, and major axis equal to 8.
16. Write the equation of the hyperbola with center at $(2, -2)$, ends of transverse axis at $(2, 0)$ and $(2, -4)$, and one focus at $(2, 1)$.
17. Write the equation of the hyperbola with vertices at $(7, 4)$ and $(-1, 4)$ and whose asymptotes have slope ± 2.
18. Show that each ellipse in the family of ellipses

$$\frac{x^2}{\sin^2 c} + \frac{y^2}{\cos^2 c} = 1$$

has its foci at the same two points. What are the foci? Sketch a few ellipses from the family. These ellipses are said to be **confocal.**
19. "Whispering" galleries have cross sections that are ellipses. In such galleries a whisper at one focus will be heard distinctly at the other focus. Find the two foci in a gallery whose cross-section is a semiellipse with a height of 12 ft and length of 30 ft.
20. Certain navigational systems use the set of hyperbolas

$$\frac{x^2}{e^c - e^{-c}} - \frac{y^2}{e^c + e^{-c}} = 2$$

as references in a coordinate system. Show that each hyperbola pair has the same foci. Sketch a few hyperbolas from this family.
21. Solve the system of equations $2x^2 + x + y^2 + 2y = 1$ and $y + 2x = 1$ graphically.
22. Show graphically that $x^2 - 4y^2 = 4$ and $y = 3x + 2$ do not have any common real solutions.
23. Explain how each of the conic sections can be obtained from a right circular cone.
24. The trajectory of a body projected at an angle θ with the horizontal is shown to be $x = tv_0 \cos\theta$, $y = (v_0 \sin\theta)t - \frac{1}{2}gt^2$. Eliminate the variable t in these two equations and show that the path is parabolic. Make a sketch of the trajectory from the time it leaves the muzzle until it returns to the same level at which it was fired.

▲▼ Key Topics for Chapter 11

Define and/or discuss each of the following.

Lines and Linear Functions
Quadratic Functions and the Parabola
The Focus/Directrix Property of the Conics
Ellipses
Hyperbolas
Asymptotes to the Hyperbola
The General Second Degree Equation

Practice Test Questions for Chapter 11

In Exercises 1–10 answer *true* or *false*.

1. The slope of a straight line is equal to the tangent of the angle of inclination.

2. The slope of a line parallel to the y-axis is undefined.

3. A parabola is the set of points in the plane equidistant from a fixed point and a fixed line.

4. An arbitrary ray emitted from the focus of a parabola is reflected parallel to the directrix.

5. The transverse axis of a hyperbola passes through the vertices.

6. The inclination of a straight line is a positive acute angle.

7. The slope of a line parallel to the x-axis is zero.

8. The right chord of a parabola is parallel to the directrix.

9. Perpendicular lines have slopes that are negative reciprocals of one another.

10. The variable in the positive term of the equation of a hyperbola indicates the direction of the conjugate axis.

In Exercises 11–20 fill in the blank to make the statement true.

11. An equation of the form $ax + by = c$ is called a(n) ________ equation.

12. The ________ of a line is the ratio of the vertical rise to the corresponding horizontal run.

13. Parallel straight lines have ________ slopes.

14. The equation of a straight line passing through (0, 3) with a slope of 2 is ________.

15. The chord of a parabola through the focus and parallel to the directrix is called the ________.

16. The midpoint between the focus and the directrix of a parabola is called the ________.

17. A ray emitted from the focus of a parabola is reflected ________ to the axis of the parabola.

18. An ellipse is the set of points in the plane, the ________ of whose distances to two fixed points is constant.

19. $Ax^2 + By^2 + Cx + Dy + E = 0$ represents a(n) ________ if A and B have different signs.

20. A ray emitted from the focus on an ellipse is reflected to ________ .

In the following exercises solve the stated problem. Show all your work.

21. Write the equation of the line that is perpendicular to $x - 3y = 7$ at the point $(1, -2)$.

22. Write the equation of the line that passes through (1, 3) and $(-1, -2)$. Express the equation in slope-intercept form.

23. Find the equation of the line with intercepts (2, 0) and (0, 3).

24. Find the equation of the parabola with focus at $(-2, 1)$ and directrix at $x = 3$.

25. Find the equation of the ellipse with center at (1, 3), major axis of 6 parallel to the y-axis, and a minor axis of 3.
26. Find the equation of the hyperbola with center at (0, 0), vertices at (± 3, 0), and foci at (± 4, 0).
27. Discuss and sketch the graph of $x^2 + y^2 + 4y = 0$.
28. Discuss and sketch the graph of $2x^2 + 8x + y^2 + 2y = 7$.
29. Discuss and sketch the graph of $x^2 - 4y^2 = 4$.
30. Discuss and sketch the graph of $x^2 - 6x + 3y = 0$.

Calculators

The word *calculator* was originally used to describe a person who could perform arithmetic calculations quickly and accurately. Today the same word means a small, hand-held electronic device that will perform arithmetic calculations quickly and accurately. Electronic calculators have revolutionized the teaching of topics such as trigonometry, which involve relatively difficult arithmetic manipulations and the extensive use of tables. The tables that have been used in trigonometry for hundreds of years are now available to you with the push of a button. ▼

▲▼ A.1 Arithmetic Operations

In this section we shall cover some general information on arithmetic calculations. In descriptions of arithmetic computations performed using a calculator, we shall designate the arithmetic operation keys by [+], [−], [×], [÷], and [=]. You perform arithmetic operations by pressing the appropriate operation key. After all operations and data are entered, pressing the [=] key causes the result to be displayed in the register.

Example 1

(a) To find $5.2 + 6.32$ we enter 5.2, press [+], and then enter 6.32. The sum 11.52 is displayed when we press the [=] key. We will represent this sequence of keystrokes by

5.2 [+] 6.32 [=] [11.52]

(b) To find $3.42 - 6.14$ we use the following keystrokes.

3.42 [−] 6.14 [=] [−2.72]

The minus sign appears to the left of a negative number. ▲

COMMENT Calculators are designed to display an error message when an undefined operation is entered. For instance, $4 \div 0$ gives an error message. Try it on your calculator. Note that after an error message appears, you will have to reset your calculator by pushing the **clear** key.

Example 2 A certificate of deposit has a value of $10,450 on the first of January. Assume the CD pays simple interest and has an annual interest rate of 9.24%. How much interest does the CD earn for the month of January?

SOLUTION The interest rate for one day is $0.0924 \div 365$. Therefore the interest earned in 31 days is $31(0.0924 \div 365)(10{,}450)$. This is performed on a calculator as follows:

31 [×] 0.0924 [÷] 365 [×] 10450 [=] **82.00816438**

Thus the amount of interest is $82.01. ▲

COMMENT The number of digits displayed in the register of a calculator varies with the brand of calculator. The number of digits displayed is not always the number needed in the answer to a calculation. The problem of how to choose the correct number of digits in an answer was addressed in Section 1.1.

Special Function Keys

To facilitate more complicated computations, your calculator has a series of special function keys. For instance, most calculators have a key for squaring a number, [x^2]; a key for taking the square root of a number, [$\sqrt{\ }$]; and a key for taking the reciprocal of a number, [1/x] or [x^{-1}]. Determine how your calculator performs these operations; not all of them use the format we have described.

Example 3

If your calculator has a [$\sqrt{\ }$] key, you can obtain the square root of 42 by entering

[$\sqrt{\ }$] 42 [=] **6.480740698** ▲

Example 4

(a) 1/2.62 is found by entering 2.62 and pushing the [x^{-1}] key.

2.62 [x^{-1}] [=] **0.381679389**

(b) $1/(2.62)^2$ is found either by the sequence

2.62 [x^2] [x^{-1}] [=] **0.145679156**

or by the sequence

2.62 [x^{-1}] [x^2] [=] [0.145679156] ▲

Example 5

(a) $1/3.14 + 5.69^2$ can be found by the keystrokes

3.14 [x^{-1}] [+] 5.69 [x^2] [=] [32.69457134]

(b) $1/2.5 + 1/1.3$ can be found by the keystrokes

2.5 [x^{-1}] [+] 1.3 [x^{-1}] [=] [1.169230769] ▲

Exercises for Section *A.1*

In Exercises 1–20 perform the arithmetic computations using your calculator. Give the answer shown in the calculator display.

1. $2.1 + 3.2$

2. $5.66 + 2.8$

3. $6.2 - 8.4$

4. $45.3 - 82.7$

5. 7.2×8.5

6. $\pi \times 22$

7. 8.1×9.4

8. 6×4.52

9. $\pi \div 3$

10. $\pi + 2$

11. $\sqrt{3.1} - 1$

12. $5.3^2 - \frac{1}{3.4}$

13. $\sqrt{15} - \pi^2$

14. $\sqrt{2.5} + 9.2^2$

15. $\frac{1}{5.1} - \frac{1}{7.2}$

16. $\frac{1}{9.1} + \frac{1}{\sqrt{\pi}}$

17. $\frac{1}{\sqrt{2}} + \frac{1}{\sqrt{6}}$

18. $\frac{1}{\sqrt{5.6}} - \frac{1}{\sqrt{7.5}}$

19. $\sqrt{17} - (16)^{-1}$

20. $\sqrt{4.07} - \frac{1}{\sqrt{0.332}}$

21. Try calculating $\sqrt{-3}$ on your calculator. What happens? Why?

22. How much interest is earned on the certificate of deposit in Example 2 (page 400), if the rate is 10.25%?

23. Joe Jones, star basketball player for the High Jumpers, makes 78.3% of his free throws. If he shoots 185 free throws, how many free throws would you expect him to make?

A.2 Operating Systems and Memory

An expression such as $5 + 4 \div 2$ can be interpreted in two different ways that will give two different answers. One way of proceeding is to add 5 and 4 and then divide this sum by 2 to obtain 4.5. The other is to divide 4 by 2 and then add 5 to this quotient to obtain 7. Only the latter interpretation is acceptable and is used in calculators. Thus in sequences of additions, subtractions, multiplications, and divisions, the multiplications and divisions are performed first and the additions and subtractions are performed second. Calculators that use this priority of operations are said to have an **algebraic operating system** (AOS). See how your calculator handles $5 + 4 \div 2$.

A calculator that has an algebraic operating system automatically sorts the sequence of operations and numbers and applies the following priority of operations on the numbers from left to right.

Priority of Operations*

1. Special function keys such as [$\sqrt{\ }$], [x^2], [x^{-1}] operate immediately on the displayed number.
2. Multiplications and divisions are completed as they occur.
3. Additions and subtractions are completed in the order in which they occur after the operations described in steps 1 and 2 are completed.

COMMENT The priority of operations we have described assumes that parentheses are implied around numbers being multiplied or divided to separate them from the addition and subtraction operations. For instance, in the expression $5 + 4 \div 2$, parentheses around $4 \div 2$ are implied, and this expression is computed as though it were written $5 + (4 \div 2)$.

Example *1* Insert the implied parentheses in the expression

$$5 + 6^2 \div 2 + 4 - 2 \times 2$$

and evaluate the expression.

SOLUTION The square of 6 is calculated first, followed by the multiplications and divisions. The implied parentheses are around $6^2 \div 2$ and 2×2. Thus,

$$\begin{aligned} 5 + 6^2 \div 2 + 4 - 2 \times 2 &= 5 + (36 \div 2) + 4 - (2 \times 2) \\ &= 5 + 18 + 4 - 4 \end{aligned}$$

The additions and subtractions are calculated last to obtain 23. Check this result by entering the expression in your calculator. Remember to work from left to right in entering the numbers and operations. ▲

Parentheses

Suppose you want to use an order of operations other than the AOS of your calculator. You can override the AOS by using the parenthesis keys on the keyboard to identify special groupings of numbers and operations. Within each set of parentheses, the calculator operates according to the AOS priority.

* Remembering My Dear Aunt Sally (MDAS) may help you recall this priority.

Calculators have a left parenthesis key [( ] and a right parenthesis key [ )]. Keystrokes performed after pushing [( ] and before pushing [ )] are separated from the sequence of operations outside the grouping symbols. For example, $3 \cdot (6 + 4) + 7$ is evaluated by the following keystrokes:

3 [×] [( ] 6 [+] 4 [ )] [+] 7 [=] 37

If the parenthesis keys are not used, the display will show 29.

Example 2 Use a calculator to evaluate

$$12 \div \frac{2}{3}$$

SOLUTION Since the rule for dividing by a fraction is to invert the fraction and multiply, we know that the correct quotient is

$$12 \div \frac{2}{3} = 12 \times \frac{3}{2} = 18$$

To get this result using a calculator, we must enter the problem as

$$12 \div (2 \div 3)$$

If we enter $12 \div 2 \div 3$, we get 2, which is incorrect. ▲

Example 3 Use a calculator to evaluate

$$\frac{9 + 3 - 2}{3 + 4 \div 2}$$

SOLUTION Here it is important to evaluate the entire numerator and the entire denominator before doing the division. We accomplish this sequence of operations by using parentheses and writing the expression as

$$(9 + 3 - 2) \div (3 + 4 \div 2)$$

The keystrokes are

[( ] 9 [+] 3 [−] 2 [ )] [÷]
[( ] 3 [+] 4 [÷] 2 [ )] [=] [2] ▲

Memory

Your calculator has a "memory"—that is, the ability to store numbers and recall them later. The two basic memory keys are [STO] and [RCL]. The [STO] key stores the displayed number and at the same time leaves the number in the display register for immediate use. The [RCL] key retrieves a number from memory and displays it in the register for immediate use; memory maintains the displayed number until it is cleared or a new number is stored there.

COMMENT Many calculators have more than one memory. Such calculators must be given instructions as to where to store numbers and from which memory to recall them. For example, if a calculator has two memories labeled A and B, then [STO A] would put the number in memory A. Likewise, [RCL A] would recall and display a number from memory A.

Example 4 Use a calculator to solve for z if $w = \frac{2}{5}(28 - 13)$ and $z = 3w + 7\sqrt{w}$.

SOLUTION The value of w is given by

2 [÷] 5 [×] [(] 28 [−] 13 [)] [=] [6] [STO]

The value of w is 6. By pushing the [STO] key when this value is displayed, we store the number 6 in memory. Since the value of w is still in the display register after [STO] is pushed, we continue the calculation by multiplying by 3.

Displays $w = 6$ ↓

6 [×] 3 [+] 7 [×] [√] [RCL] [=] [35.1464282]

↑ Value of w

Thus the result is $z \approx 35.1464282$. ▲

Exercises for Section A.2

In Exercises 1–10 insert the implied parentheses to indicate how a calculator will perform the computation.

1. $2 + 3 \times 4$
2. $2 \times 4 + 3$
3. $2 + 3 \div 6 + 2$
4. $5 \div 2 + 3 \div 2$
5. $\sqrt{5} + 5 - 3 \times 4$
6. $3 \times 6 \div 2 + 4$
7. $6 \div 4 \div 2 - 1$
8. $8 + 1 \times 5 - 2 \times 3^2$
9. $6 - \frac{4}{3} \times 28$
10. $\sqrt{37} + 3/4 + 2$

In Exercises 11–17 use a calculator to evaluate each expression. Give the answer shown in the calculator display.

11. $\dfrac{2.4 + 5.1 - 2}{2.1}$
12. $\dfrac{6.3 \times 4 - 5 \times 2.8}{1.4}$
13. $\dfrac{\pi + \sqrt{2}}{\sqrt{3} + \sqrt{\pi}}$
14. $\dfrac{1/2 + 1/3}{1/4 + 1/5}$
15. $\dfrac{\pi^2 + \sqrt{80}}{\frac{1}{7} + \frac{1}{11}}$
16. $4 + \frac{1}{3} \div \frac{7}{4}$
17. $\sqrt{15} \div \frac{6}{11} - \sqrt{8}$

In Exercises 18–22 use a calculator to solve for z. Give the answer shown in the calculator display.

18. $w = \dfrac{(52 - 7)}{3}$

$z = 3w + 2/w$

19. $w = 0.261(0.20 + 0.527)$

$z = \sqrt{2w} + w$

20. $w = \dfrac{1}{5.1} + \dfrac{1}{6.3}$

$z = \dfrac{1}{w} + w$

21. $w = \dfrac{\sqrt{5} + \dfrac{1}{0.24}}{5.28 - 3.12}$

$z = w^2 + \dfrac{1}{w^2}$

22. $w = \dfrac{3 - \dfrac{1}{0.3}}{2 + \dfrac{1}{2}}$

$z = w^2 + w - 2$

A.3 Scientific Notation

Scientists often express positive real numbers in an abbreviated form called **scientific notation**. The number is written as the product of a number between 1 and 10 and a power of 10. If x is a positive real number, then

$$x = m \cdot 10^c$$

where m is a number between 1 and 10 and c is an integer equal to the number of places the decimal point of x must be moved to produce a number between 1 and 10. The number m is called the **mantissa**; the integer c is called the **power**. The exponent c is positive if the given number x is greater than 1 and negative if x is less than 1.

Example 1 Write the following numbers in scientific notation:

(a) 37,910,000 **(b)** 0.000172

SOLUTION

(a) The given number is greater than 1, so the decimal point must be moved seven digits to the *left* to produce a number between 1 and 10. We write

$$37{,}910{,}000 = 3.791 \times 10^7$$

(b) The given number is less than 1, so the decimal point must be moved four digits to the *right* to produce a number between 1 and 10. We write

$$0.000172 = 1.72 \times 10^{-4}$$

▲

Example 2 Convert the following numbers from scientific to standard notation:

(a) 4.76×10^4 **(b)** 9.93×10^{-2}

SOLUTION

(a) The exponent c is positive, so we move the decimal point four digits to the *right* to get a number greater than 1.

$$4.76 \times 10^4 = 47{,}600$$

(b) The exponent c is negative, so we move the decimal point two digits to the *left* to produce a number less than 1.

$$9.93 \times 10^{-2} = 0.0993$$ ▲

To multiply or divide numbers written in scientific notation, we add the exponents of 10, as demonstrated in the next example.

Example 3 Evaluate **(a)** $(2 \times 10^{-3})(6 \times 10^8)$ and **(b)** $\dfrac{8.4 \times 10^6}{3 \times 10^{14}}$

SOLUTION

(a) $(2 \times 10^{-3})(6 \times 10^8) = 2 \times 6 \times 10^{-3+8} = 12 \times 10^5 = 1.2 \times 10^6$

(b) $\dfrac{8.4 \times 10^6}{3 \times 10^{14}} = \dfrac{8.4}{3} \times 10^{6-14} = 2.8 \times 10^{-8}$ ▲

Scientific notation is especially valuable for expressing either very large or very small numbers. For example, 0.000000000000000000000053 represents the weight of an oxygen molecule. In scientific notation this is 5.3×10^{-23}; the decimal point is moved to the right 23 places to obtain the number 5.3, which is between 1 and 10.

Your calculator will display numbers in scientific notation with a key labeled **EE** (for "enter exponent"). To enter a number in scientific notation:

1. Enter the mantissa (as either a positive or a negative number between 1 and 10).
2. Press **EE** .
3. Enter the power of ten.

Example 4

(a) When entered in exponential notation, the number 2,654,000 will be displayed as

2.654 E 06

(b) The number 3.62×10^{-51} will be displayed as

3.62 E −51 ▲

COMMENT Numbers in scientific notation can be mixed with numbers in standard form for any calculation.

Example 5 The speed of light is 3×10^5 km/sec. How many kilometers does light travel in one hour?

SOLUTION To find the distance light travels in one hour, we use the formula

$$\text{distance} = \text{speed} \times \text{time}$$

Since the speed of light is given in kilometers per second, we express one hour as 3600 sec in the formula. The calculator keystrokes for this computation are

distance = 3 [EE] 5 [×] 3600 [=] [1.08 E 9]

Thus light travels 1.08×10^9 km in one hour. ▲

COMMENT Some calculators have a key that will change a number from standard format to scientific format.

Scientific notation is often used to get a quick estimate of the results of a computation. Each number in an expression is approximated by an integer between 1 and 10 in scientific format so that the calculation can be done mentally. Such an estimate is made to give some idea of the order of magnitude of the solution so that gross errors can easily be detected. The technique is illustrated in Example 6.

Example 6 Estimate the value of $16{,}021 \times 0.0286 \div 0.322$.

SOLUTION The estimates of the three numbers are

$$16{,}021 \approx 2 \times 10^4$$
$$0.0286 \approx 3 \times 10^{-2}$$
$$0.322 \approx 3 \times 10^{-1}$$

Thus we may estimate the order of our result to be

$$2 \times 10^4 \times 3 \times 10^{-2} \div 3 \times 10^{-1} = 2 \times 10^3 = 2000$$

The calculator value of this expression is [1422.98323]. Our mental estimation leads us to believe the answer is reasonable. Had we obtained [142.298323] in the display register, we would have known something was wrong because our estimate said the answer should be about 2000. Calculators generally don't make errors, but the people using them do. That is why it is advisable to spend a little time to make sure the answer is at least reasonable. ▲

Exercises for Section A.3

In Exercises 1–9 write each number in scientific notation.

1. 8,243,765,000,000
2. 8 followed by 100 zeros
3. 93,410,000
4. 0.000000000000000054
5. 0.0477
6. 0.0000003155

7. 0.975
8. 265.7
9. 31,110

In Exercises 10–17 write each number in standard notation.

10. 3.4852×10^9
11. 1.4239×10^{12}
12. 9.385×10^{-7}
13. 6.43×10^3
14. 8.993×10^{-11}
15. 2.285×10^{-1}
16. 5.36×10^{15}
17. 3.117×10^{-10}

18. Evaluate $\dfrac{(8{,}000{,}000)}{(0.0000000002)(400{,}000)}$ without using a calculator.

19. Evaluate $\dfrac{0.0004(0.0000015)}{5000(0.002)}$ without using a calculator.

20. The speed of light is 30,000,000,000 cm/sec. Write this number in scientific notation.

21. The modulus of elasticity for steel is 6,000,000. Write this number in scientific notation.

22. The expression

$$\frac{(2 \times 10^{-3})(2.5 \times 10^{-2})}{(100)^2}$$

is used to compute the force of attraction of two objects. Simplify this expression.

In Exercises 23–28 use a calculator with scientific notation to compute each expression. Write the answer shown in the calculator display.

23. (2,354,000)(19,380)(0.087689)

24. (887,350)(0.755318)(2,533)(35,679)

25. $\dfrac{(9005)(343{,}230)}{(0.00000003715)}$

26. $\dfrac{(0.00232445)(0.891153)}{(0.000000000039645)}$

27. $\dfrac{(39{,}555{,}000)(7{,}508.122)}{(90.758)(0.000065656)}$

28. $\dfrac{(7598.3248)(66{,}789)}{(0.3966)(0.0046872)}$

Review of Elementary Geometry

In this appendix we provide a review of some of the terminology and results of plane geometry that are important in the study of trigonometry.

Geometry is the study of the properties of points, lines, planes, angles, surfaces, and volumes. Certain geometric concepts, such as those of a point, a line, and a plane, are accepted without definition. Technically a dot on a paper is not really a point, since it has dimension; however, by general agreement we make visible marks (dots) to represent points and denote them with capital letters such as A, B, C, and so forth, as shown in Figure B.1.

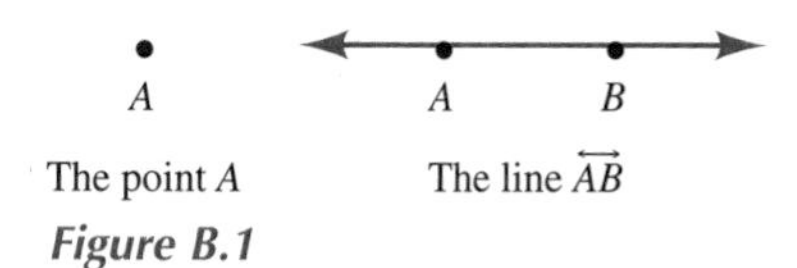

The point A The line $\overleftrightarrow{AB}$

Figure B.1

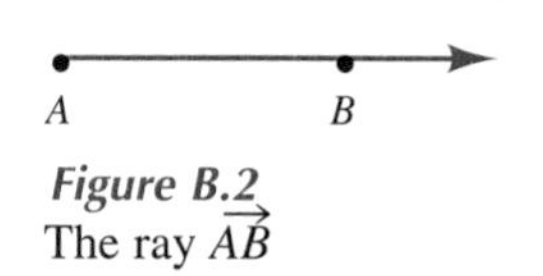

Figure B.2
The ray $\overrightarrow{AB}$

Similarly, a straight line drawn with a ruler is technically not a line, since a line has no width. In geometry we think of a **line*** as extending indefinitely in both directions. Two points determine a line; thus a line passing through the points A and B is denoted by $\overleftrightarrow{AB}$. (See Figure B.1.)

A **ray** consists of a point A on a line and that part of the line that is on one side of the point. A ray is denoted by $\overrightarrow{AB}$, as shown in Figure B.2.

A **line segment**, $\overline{AB}$, is the part of a line between the two points A and B. (See Figure B.3.) Depending on the application, the line segment may or may not include the points A and B.

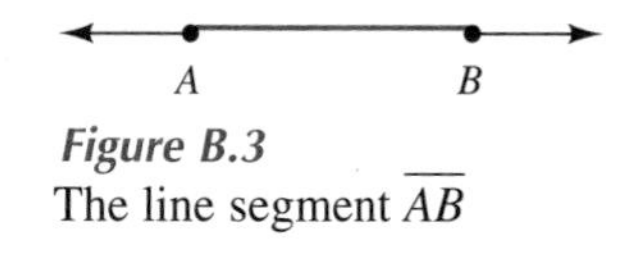

Figure B.3
The line segment $\overline{AB}$

 *B.1* Angles

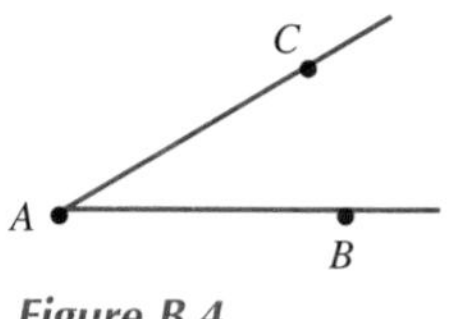

Figure B.4
$\angle BAC$ or $\angle A$

Two rays with a common endpoint form an **angle**. The common point is called the **vertex** of the angle, and the rays are called the **sides**. (See Figure B.4.) We generally refer to an angle by mentioning a point on each of its sides and the vertex, although sometimes only the vertex is mentioned. An

* We shall use the word *line* to mean a straight line.

angle whose two rays form a straight line is called a **straight angle**. (See Figure B.5.)

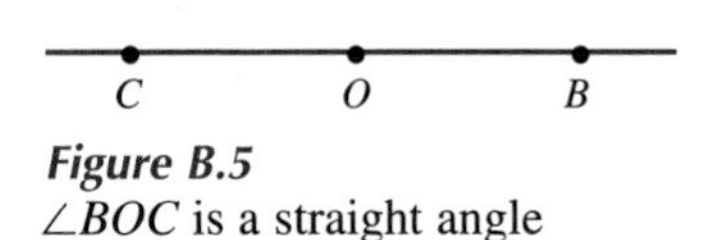

Figure B.5
$\angle BOC$ is a straight angle

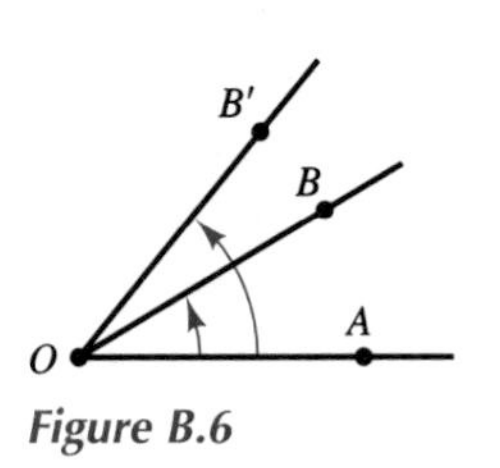

Figure B.6

Sometimes we conceive of an angle as being formed as one of the sides rotates about the angle's vertex while the other side remains fixed, as shown in Figure B.6. If *OA* is fixed and *OB* is rotated about the vertex, *OA* is called the **initial side** and *OB* the **terminal side** of the generated angle. The size of the angle depends on the amount of rotation of the terminal side. Thus $\angle AOB$ is considered smaller than $\angle AOB'$.

Angles with the same initial and terminal sides are said to be **coterminal**. However, two angles may not be equal even though the same rays form their sides. Figure B.7 shows two angles that have the same sides but are not equal because they are formed by different rotations.

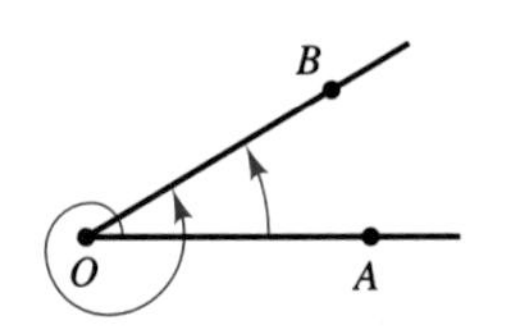

Figure B.7

The most common unit of angular measure is the **degree**. The measure of a straight angle is 180°. One-half of a straight angle is called a **right angle**; a right angle has a measure of 90°. (See Figure B.8.) An angle is **acute** if its measure is greater than 0° and less than 90° and **obtuse** if its measure is more than 90° but less than 180°. (See Figure B.9.)

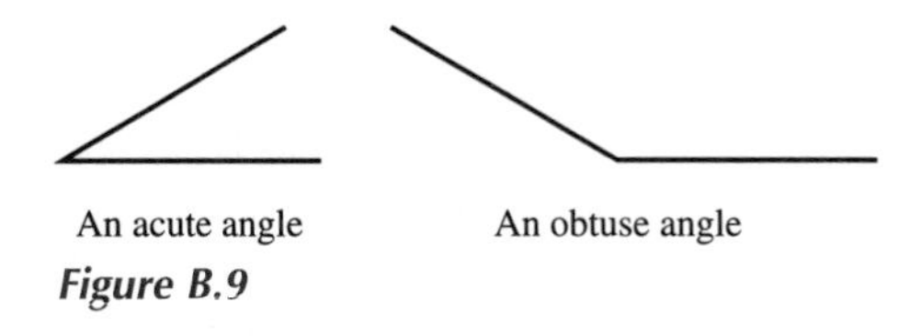

Figure B.9

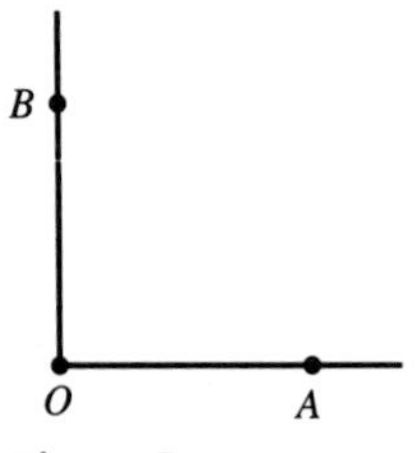

Figure B.8
A right angle

Two positive angles that together form a right angle are **complementary**. If two angles form a straight angle when combined, they are **supplementary** angles. (See Figure B.10.)

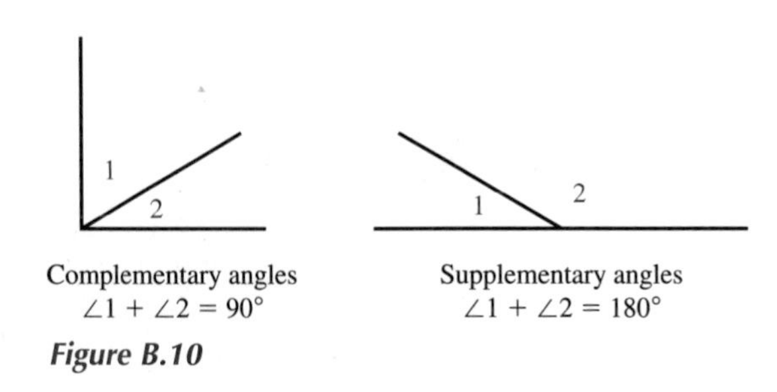

Figure B.10

B.2 Intersection of Lines

If two lines in the same plane have no points in common, they are said to be **parallel.** (See Figure B.11.) Two nonparallel lines intersect at a point to form four angles. In any such intersection, **opposite** angles are equal. Thus in Figure B.12, $\angle 1$ and $\angle 3$ are equal, as are $\angle 2$ and $\angle 4$.

Figure B.11
Parallel lines

If $\angle 1$, $\angle 2$, $\angle 3$, and $\angle 4$ are all right angles formed by intersecting lines, the two lines are perpendicular. (See Figure B.13.)

1 2 4 3

Figure B.12
Intersecting lines

A line drawn through two parallel lines forms eight angles. The following list summarizes the relationships among these angles. (See Figure B.14.)

- **Corresponding** angles such as $\angle 1$ and $\angle 5$ or $\angle 4$ and $\angle 8$ are equal.
- **Alternate interior** angles such as $\angle 3$ and $\angle 6$ or $\angle 5$ and $\angle 4$ are equal.
- **Alternate exterior** angles such as $\angle 1$ and $\angle 8$ or $\angle 2$ and $\angle 7$ are equal.

The relationship of parallelism and perpendicularity in the plane are such that

- Two lines perpendicular to the same line are parallel.
- Two lines parallel to the same line are parallel.

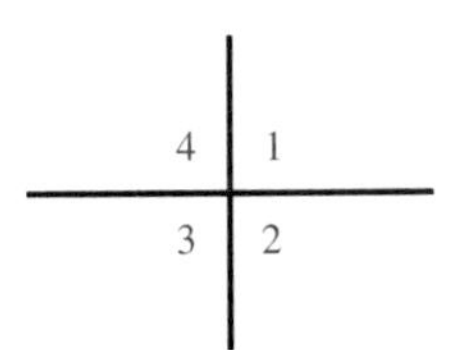

Figure B.13
Perpendicular lines

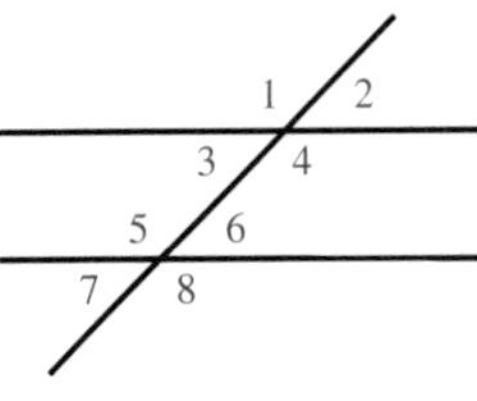

Figure B.14

B.3 Polygons

A **polygonal path** is one obtained by connecting line segments, as shown in Figure B.15. A polygonal path of more than two segments that begins and ends at the same point, and that has no other points of intersection, is called a **polygon**. (See Figure B.16.) A polygon has an interior and an exterior. The interior of a polygon is called its **area**. The **sides** of the polygon are the line segments, and the **angles** of the polygon are those angles on the interior of the polygon formed by the sides. Polygons are named for the number of sides.

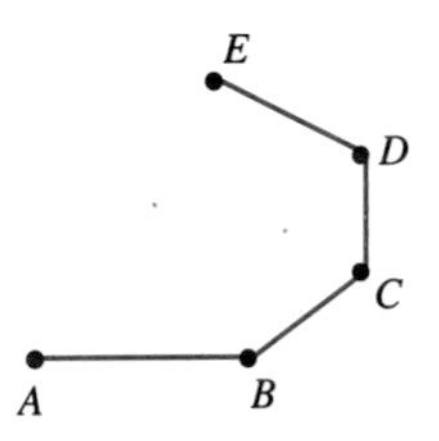

Figure B.15
Polygonal path

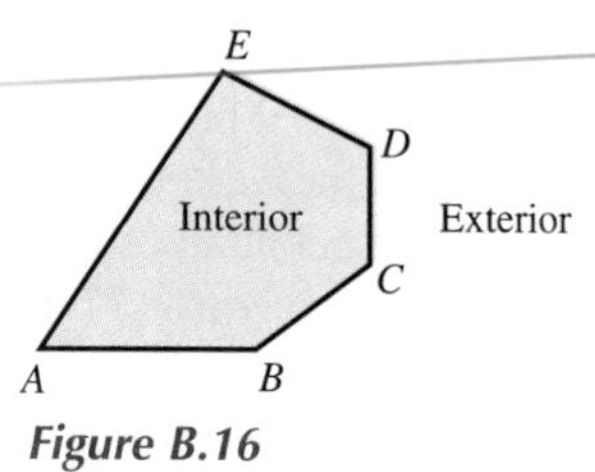

Figure B.16
A polygon

- A **triangle** has three sides.
- A **quadrilateral** has four sides.
- A **pentagon** has five sides.

A **regular** polygon is one in which all sides have the same length and all angles are equal.

- A regular triangle is said to be **equilateral**. See Figure B.17(a).
- A regular quadrilateral is a **square**. See Figure B.17(b).

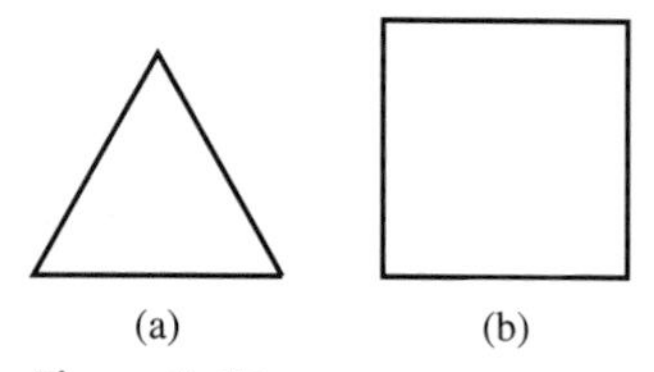

Figure B.17
Regular polygons

We compare polygons by comparing their line segments and vertices in a specific order. Thus the triangle ABC may be compared with the triangle $A'B'C'$ using any of the six orderings ABC, ACB, BAC, BCA, CAB, or CBA. When a comparison is made for a particular ordering, the sides and angles are said to **correspond**. For example, when $\triangle ABC$ and $\triangle A'C'B'$ are compared, the corresponding angles are $\angle A$ and $\angle A'$, $\angle B$ and $\angle C'$, and $\angle C$ and $\angle B'$; the corresponding sides are $\overline{AB}$ and $\overline{A'C'}$, $\overline{AC}$ and $\overline{A'B'}$, and $\overline{BC}$ and $\overline{C'B'}$. (See Figure B.18.) The notion of corresponding angles and sides allows us to compare the sizes and shapes of two or more polygons.

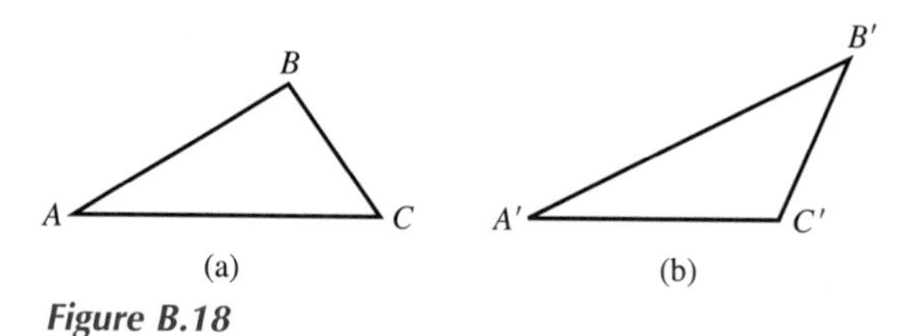

Figure B.18

Two polygons are **similar** if their corresponding angles are equal. In effect, similar polygons have the same shape but not necessarily the same size. The symbol for similarity is $\sim$. In Figure B.19, $\triangle ABC \sim \triangle A'B'C'$.

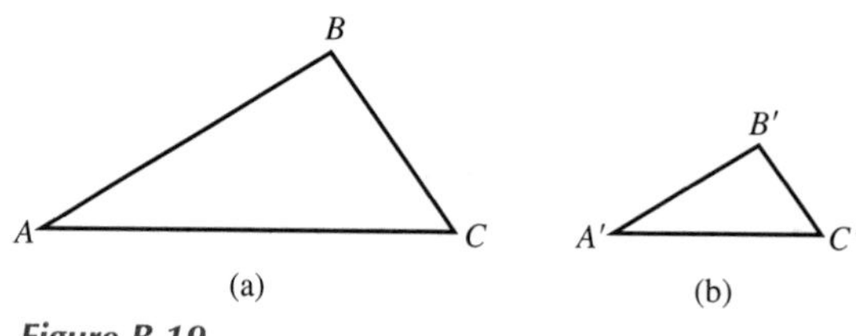

Figure B.19
Similar triangles

Two polygons are **congruent** if their corresponding angles and sides are equal. The parts of congruent polygons coincide exactly if the polygons are placed properly one upon the other. The symbol for congruence is $\cong$. In Figure B.20, quadrilateral $ABCD$ is congruent to quadrilateral $A'B'C'D'$.

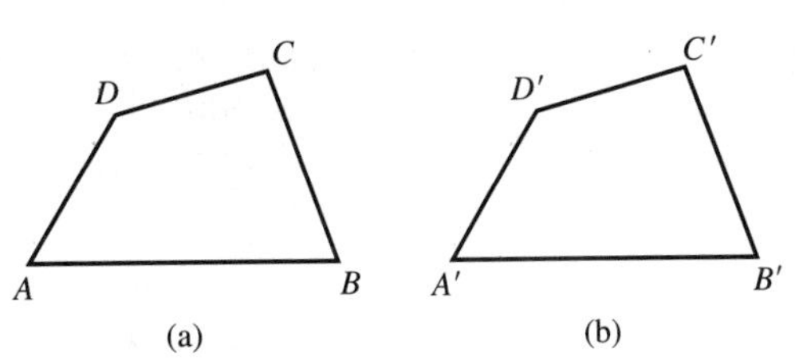

Figure B.20
Congruent quadrilaterals

B.4 Triangles

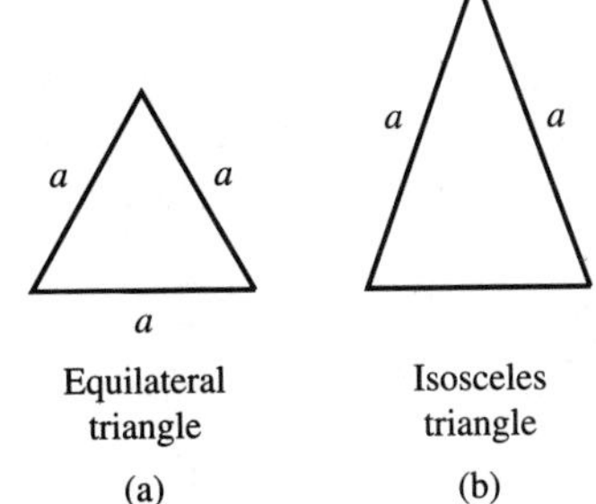

Figure B.21

A triangle is said to be **equiangular** if all of its three angles are exactly the same, and **equilateral** if all three sides have the same length.

- A triangle is equilateral if and only if it is equiangular. See Figure B.21(a).
- A triangle is **isosceles** if two of its angles are equal. See Figure B.21(b). In an isosceles triangle, the sides opposite the two equal angles are equal.

The perpendicular distance from a vertex to the opposite side of a triangle is called an **altitude**, and the side opposite the vertex is called the **base** for that altitude. Any two triangles with the same base and altitude have the same area. (See Figure B.22.)

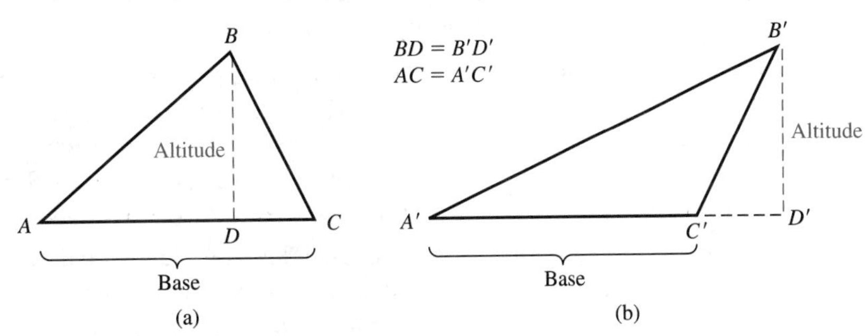

Figure B.22
Since $\overline{BD} = \overline{B'D'}$ and $\overline{AC} = \overline{A'C'}$, the areas of these two triangles are equal.

A **right** triangle is one in which one of the angles is a right angle. An **oblique** triangle is one without a right angle. In any triangle, the sum of the measures of the angles is 180°. Thus in an equilateral triangle each of the angles measures 60°. In a right triangle each of the non-right angles is acute and the sum of their measures is 90°.

There is a relatively standard method for referring to the sides and angles of any triangle. For example, consider a triangle with vertices A, B, and C. (See Figure B.23 on page 414.) The sides $\overline{AB}$ and $\overline{AC}$ are called the sides **adjacent** to the angle at vertex A. The side: $\overline{BC}$ is called the side

opposite angle A. The sides opposite and adjacent to angle B and those opposite and adjacent to angle C are referred to similarly. In the special case of a right triangle, the side opposite the right angle is called the **hypotenuse**.

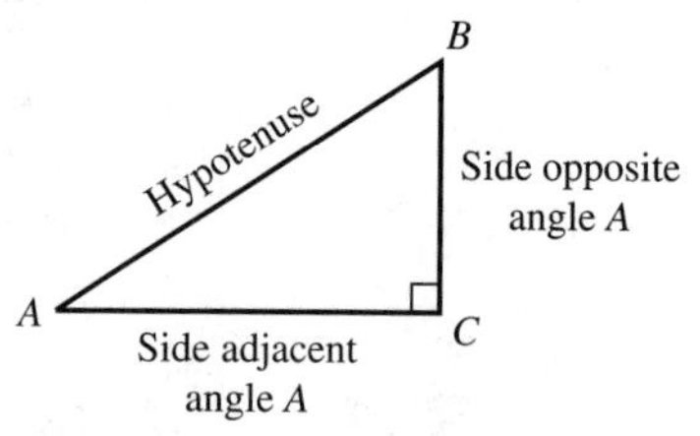

Figure B.23

Similar Triangles

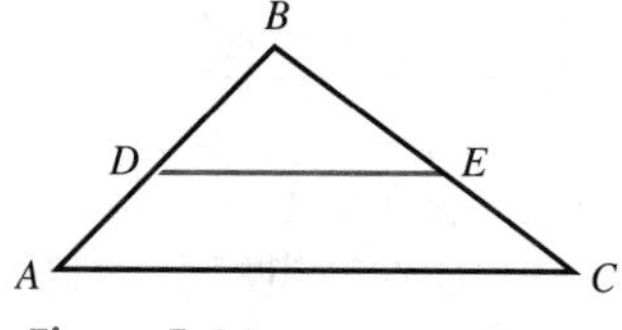

Figure B.24

In a triangle ABC, if $\overline{DE}$ is parallel to $\overline{AC}$, then $\triangle DBE$ is similar to $\triangle ABC$. (See Figure B.24.) The corresponding sides of similar triangles are proportional; that is, the ratios of corresponding sides are equal. Thus from Figure B.24,

$$\frac{\overline{AB}}{\overline{DB}} = \frac{\overline{BC}}{\overline{BE}} = \frac{\overline{AC}}{\overline{DE}}$$

Congruent Triangles

The following three theorems summarize the three ways in which two triangles can be congruent.

1. Two triangles are congruent if and only if two sides and the *included* angle of one are equal, respectively, to two sides and the *included* angle of the other. (See Figure B.25.) For instance, $\triangle ABC$ and $\triangle A'B'C'$ are congruent if $\overline{AB} = \overline{A'B'}$, $\overline{AC} = \overline{A'C'}$, and $\angle A = \angle A'$.
2. Two triangles are congruent if and only if three sides of one are equal, respectively, to three sides of the other. For instance, in Figure B.25, $\triangle ABC$ and $\triangle A'B'C'$ are congruent if $\overline{AB} = \overline{A'B'}$, $\overline{AC} = \overline{A'C'}$, and $\overline{BC} = \overline{B'C'}$.
3. Two triangles are congruent if and only if two angles and the *included* side of one are equal, respectively, to two angles and the

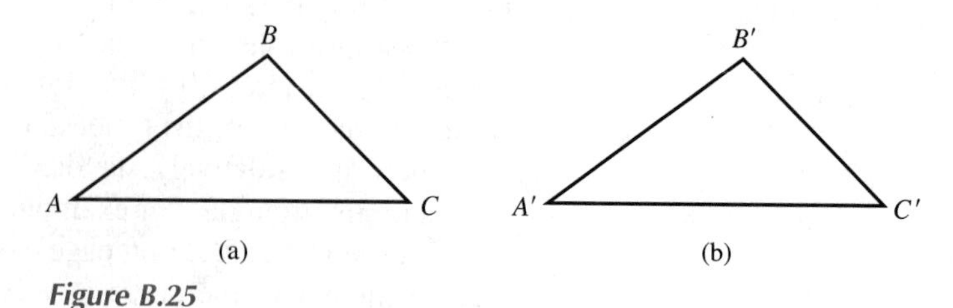

Figure B.25

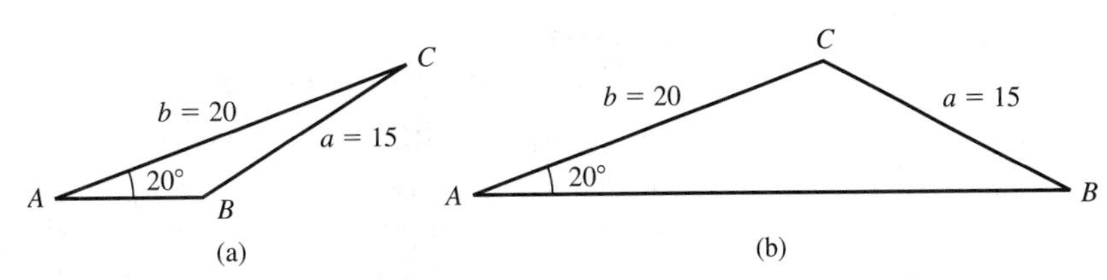

Figure B.26

included side of the other. For instance, $\triangle ABC$ and $\triangle A'B'C'$ are congruent if $\angle A = \angle A'$, $\angle B = \angle B'$, and $\overline{AB}$ and $\overline{A'B'}$.

Note that two triangles cannot be considered congruent merely because two sides of one triangle and the angle opposite one of them are equal, respectively, to two sides of the other triangle and the angle opposite one of them. Figure B.26 shows two triangles; two sides and the angle opposite $\overline{BC}$ are equal, but the two triangles are obviously not congruent.

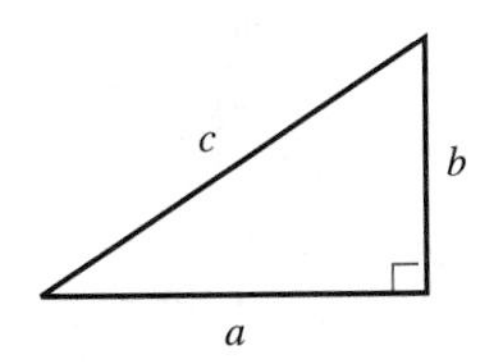

Figure B.27
Pythagorean theorem
$a^2 + b^2 = c^2$

Right Triangles

Right triangles are at the heart of the study of trigonometry. They have the following special properties:

1. The square of the hypotenuse is equal to the sum of the squares of the other two sides. (See Figure B.27.) This rule is called the *Pythagorean theorem.*
2. In a 30°–60° right triangle, the length of the side opposite the 30° angle is equal to one-half the length of the hypotenuse. If the hypotenuse is 2 units, the side opposite the 30° angle is 1 unit. Then, by the Pythagorean theorem, the side opposite the 60° angle is $\sqrt{3}$. (See Figure B.28.)
3. In a 45°–45° right triangle, the sides opposite the 45° angles are equal. If the sides opposite the 45° angles have a length of 1 unit, then by the Pythagorean theorem the length of the hypotenuse is $\sqrt{2}$. (See Figure B.29.)

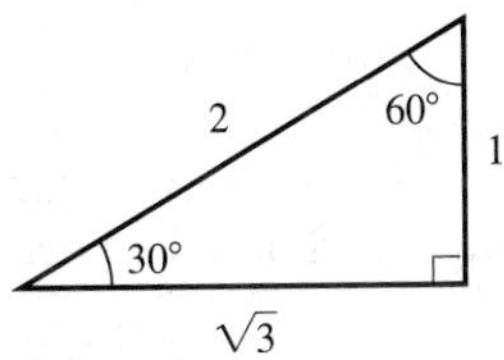

Figure B.28
30°-60° right triangle

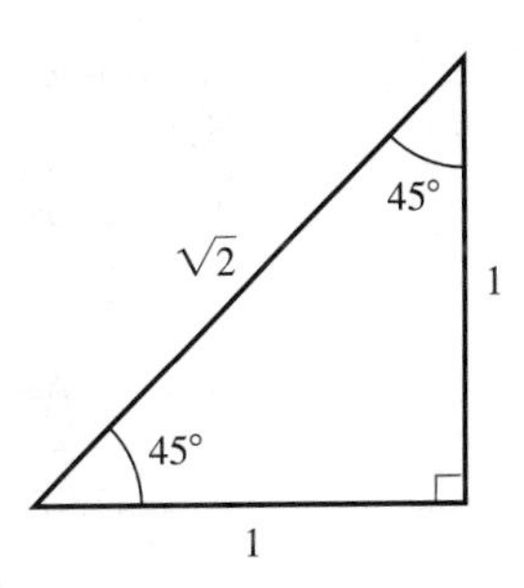

Figure B.29
45°-45° right triangle

B.5 Quadrilaterals

A **quadrilateral** is a polygon with four sides. (See Figure B.30.) A quadrilateral that has two and only two parallel sides is called a **trapezoid**. If both pairs of opposite sides are parallel line segments, the quadrilateral is called a **parallelogram**. A **rectangle** is a parallelogram with four right angles.

The perpendicular distance between the parallel sides of a trapezoid is called the **altitude** of the trapezoid. The parallel sides are called the **bases**.

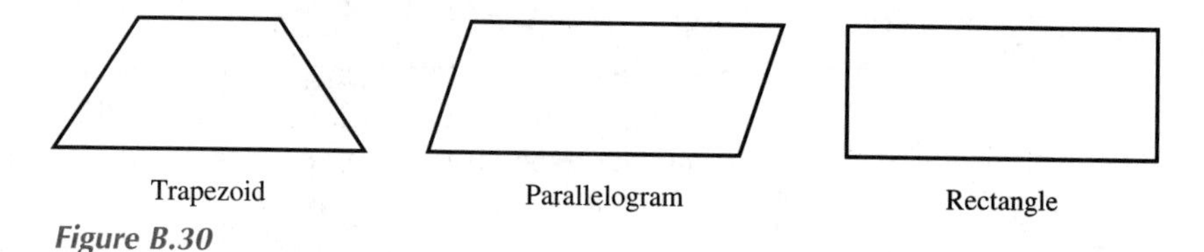

Figure B.30

B.6 Circles

A **circle** is a set of points in a plane that are the same distance from a fixed point. (See Figure B.31.) The fixed point is called the **center** of the circle, and the distance from the center to any point on the circle is called the **radius**. A line segment connecting any two points on a circle is called a **chord**. The **diameter** of the circle is the length of any chord that passes through the center. The total distance around the circle is its **circumference**.

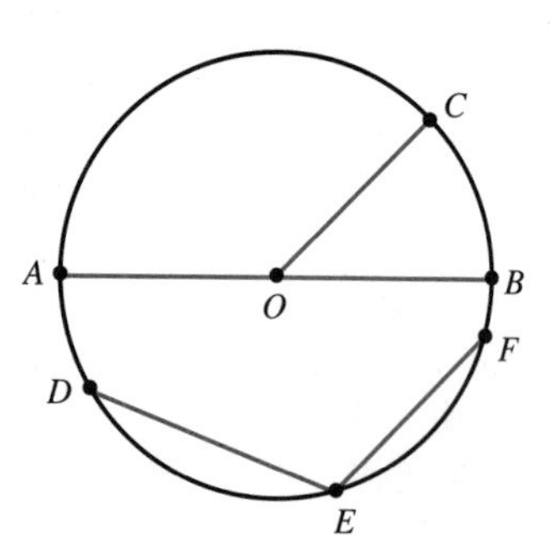

Figure B.31

An **inscribed angle** is an angle formed by two chords that meet at a common point on the circle. A **central angle** is formed at the center of a circle by two radii. A central angle **subtends** an arc on the circle. A **sector** of a circle is the closed figure formed by a central angle and its subtended arc.

In Figure B.31, $\overline{OA}$, $\overline{OB}$, and $\overline{OC}$ are radii, $\overline{AB}$ is a diameter; $\overline{AB}$, $\overline{DE}$, and $\overline{EF}$ are chords; BOC is a central angle; BOC subtends arc BC; and DEF is an inscribed angle.

Circles have the following properties:

- An angle inscribed in a semicircle is a right angle. (See Figure B.32.)
- The ratio of the circumference to the diameter is the same for all circles. This ratio is equal to π.

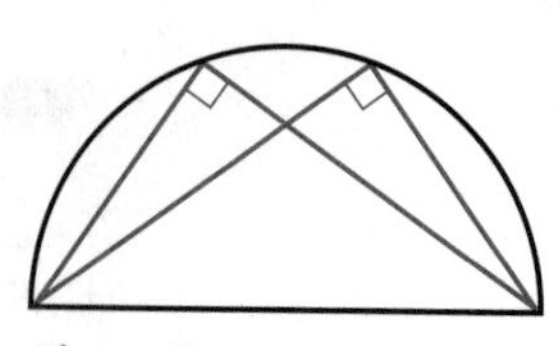

Figure B.32

B.7 Formulas from Geometry

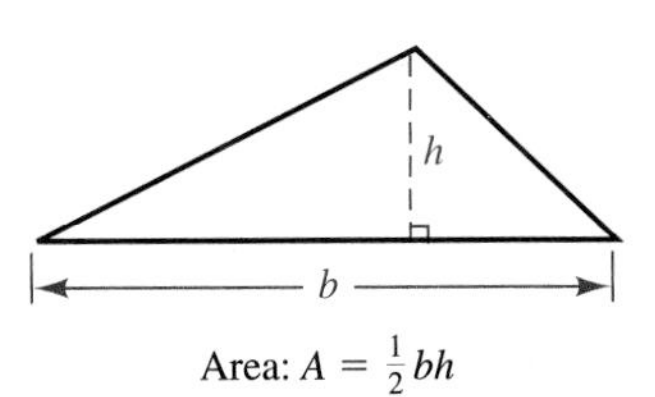

Area: $A = \frac{1}{2}bh$

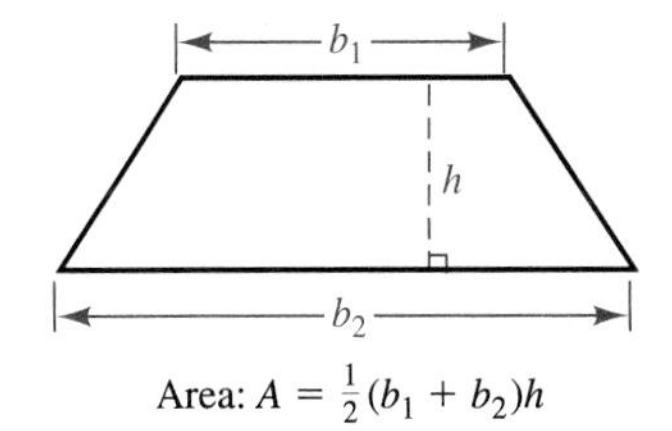

Area: $A = \frac{1}{2}(b_1 + b_2)h$

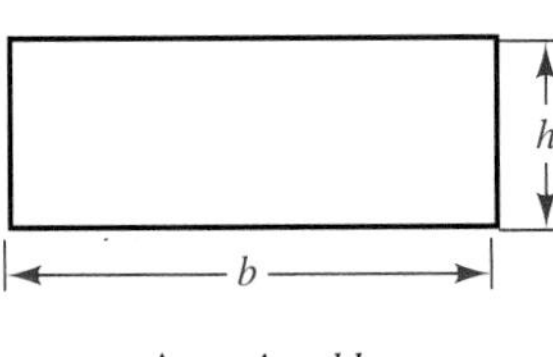

Area: $A = bh$

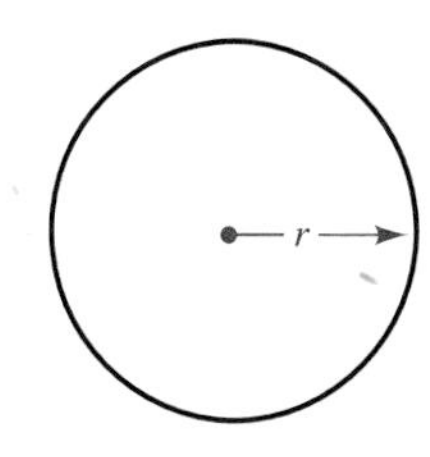

Area: $A = \pi r^2 = \frac{1}{4}\pi D^2 \quad (D = 2r)$
Circumference: $C = 2\pi r = \pi D$

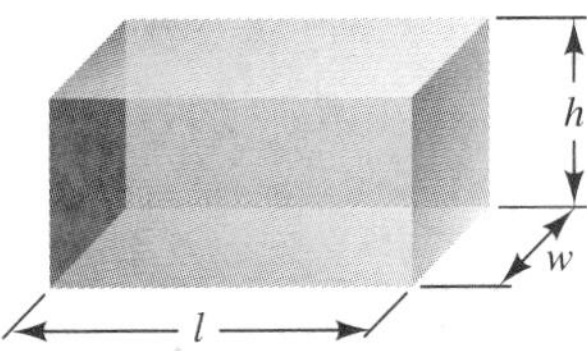

Surface area: $A = 2(lw + lh + wh)$
Volume: $V = hwl$

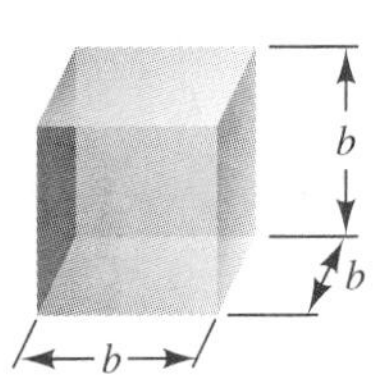

Surface area: $A = 6b^2$
Volume: $V = b^3$

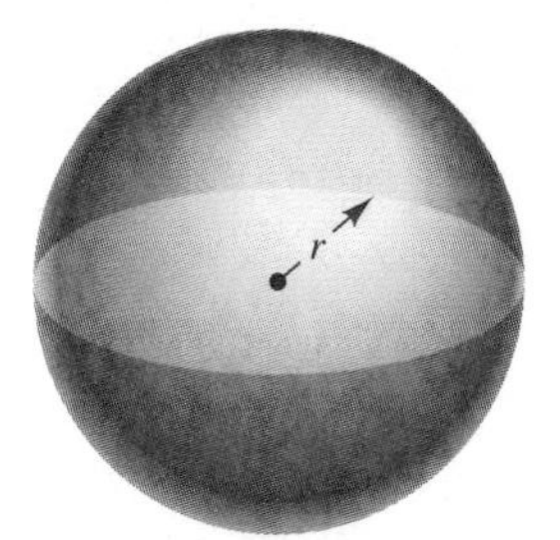

Surface area: $A = 4\pi r^2$
Volume: $V = \frac{4}{3}\pi r^3$

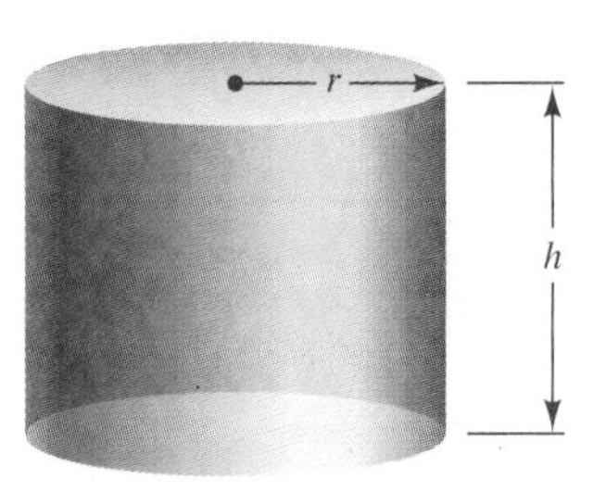

Surface area: $A = 2\pi rh + 2\pi r^2$
Volume: $V = \pi r^2 h$

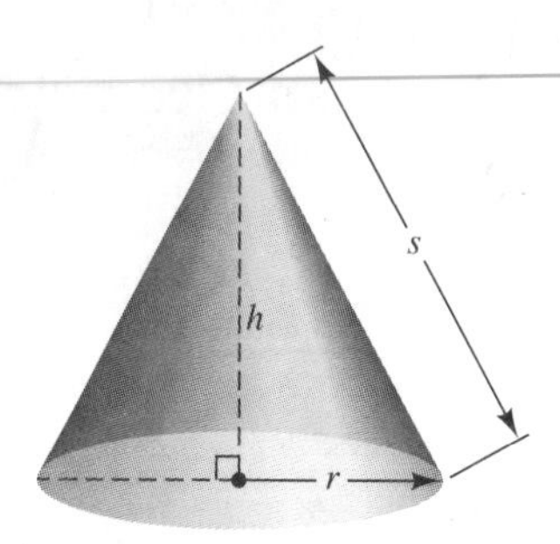

Lateral area: $A = \pi rs$
Volume: $V = \frac{1}{3}\pi r^2 h$

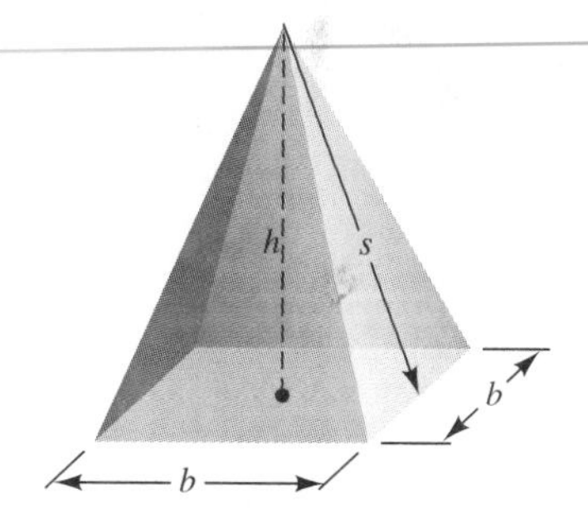

Lateral area: $A = 2bs$
Volume: $V = \frac{1}{3}b^2 h$

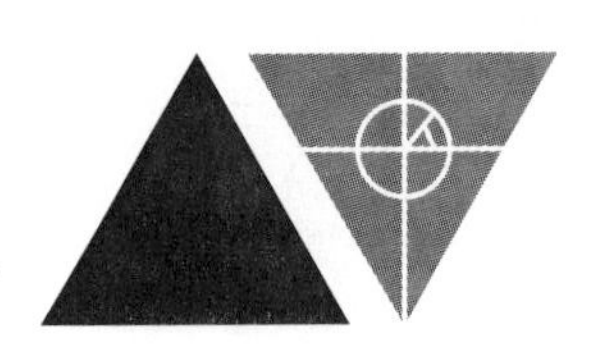

Answers to Odd-Numbered Exercises

Section 1.1

1. $(7 + 2) - 3$ **3.** $(6 \times 5) \div 3$ **5.** $(6 \times 7) - 8$
7. $\{[(5 + 4) - 3] + 2\}$ **9.** $(5 \div 4) + (3 \times 2)$ **11.** -17.4
13. 28 **15.** -63.732 **17.** 3; 2 **19.** 4; 2 **21.** 2 to 4; 0
23. 4; 4 **25.** 2; 6 **27.** 9820 **29.** 54.7 **31.** 25.0
33. 0.490 **35.** 900,000 **37.** 0.0 **39.** -7.9 **41.** 824
43. -4.32 **45.** 3.769 **47.** 4.4 **49.** 139.6 **51.** 0.0029

Section 1.2

1, 3, 5

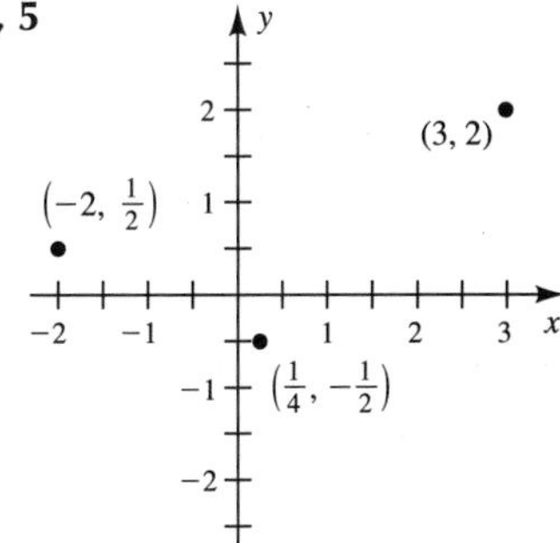

7. I, IV **9.** III **11.** 0

13. $\sqrt{2} \approx 1.414$ **15.** $\frac{5}{4}$ **17.** 8.2 **19.** 4.79
21. The y-axis
23.

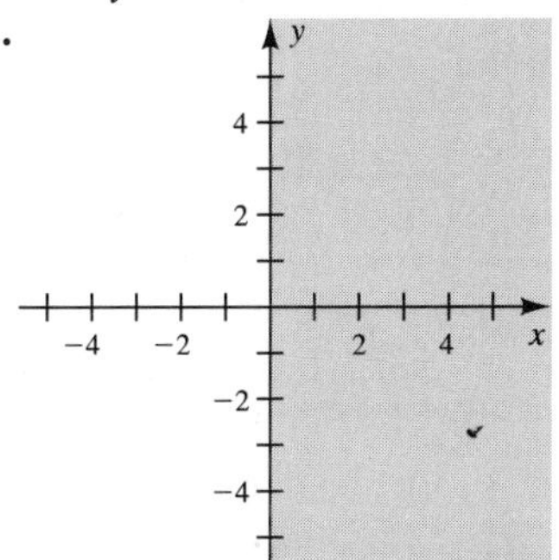

25.

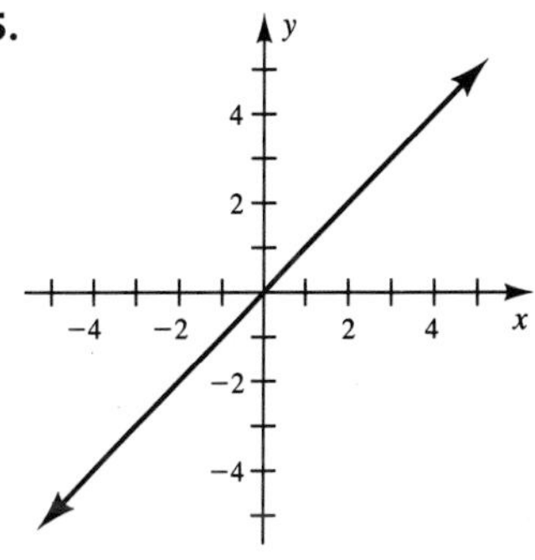

27.

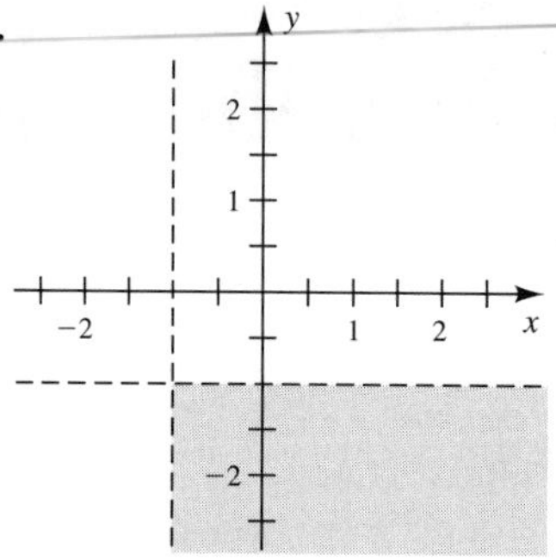

29. $5 \pm \sqrt{12}$ **31.** $2|x|$

33. $\sqrt{5}$ **35.** $\sqrt{13}$

Graphing Calculator Exercises

1.

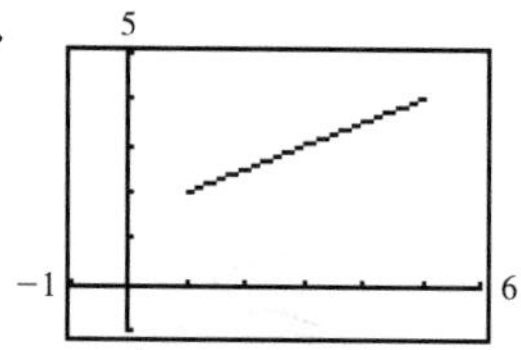

3.

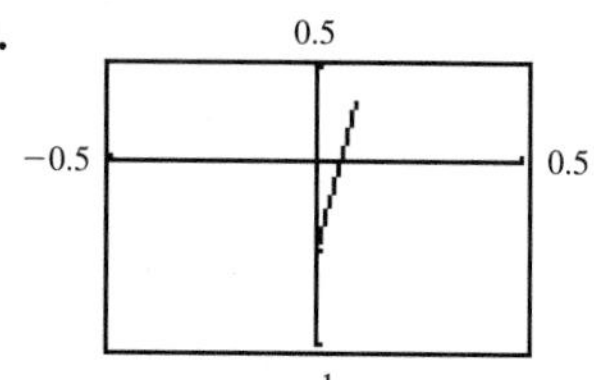

5.

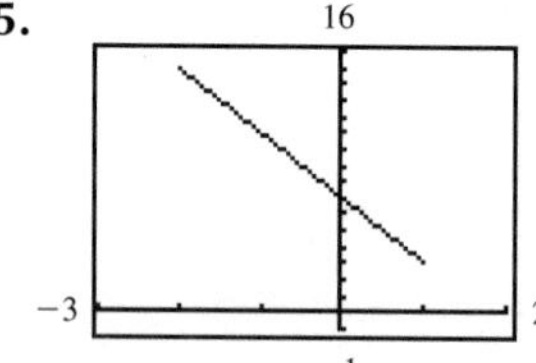

7.

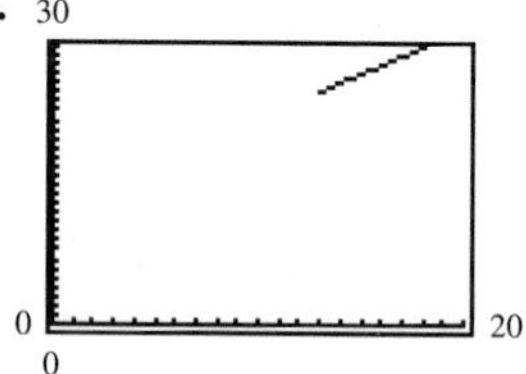

Section 1.3

1. Function **3.** Function **5.** Function **7.** Not a function
9. Function **11.** Not a function **13.** (a) and (b)
15. D: all reals; R: all reals **17.** D: $x \geq 0$; R: $y \geq 0$
19. D: all reals; R: all reals **21.** D: $x \leq 0$; R: $f(x) \geq 0$
23. D; $x \geq 1$; R: $f(x) \geq 0$
25. **(a)** 10 **(b)** $3\pi + 1$ **(c)** $3z + 1$ **(d)** $3(x - h) + 1$
(e) 3 **(f)** All reals
27. Odd **29.** Odd **31.** Neither **33.** Even
35. $4x^2 - 48x + 135$ **37.** $9x^4 - 54x^3 + 81x^2$ **39.** $x = 1, -3$
41. $x = 0, \frac{2}{3}$
43. **(a)** x_2/x_1 **(b)** $f(1/x) = kx$; $1/f(x) = x/k$
(c) $f(x^2) = k/x^2$; $[f(x)]^2 = k^2/x^2$
(d) $f(x) + 1 = (x + k)/x$; $f(x + 1) = k/(x + 1)$

(e) $f(x_1 + x_2) = k/(x_1 + x_2); f(x_1) + f(x_2) = k(x_1 + x_2)/x_1x_2$
(f) $af(x) = ak/x; f(ax) = k/ax$

45. -11.6918 **47.** 2.366 A **49.** $P = 0.705w - 150$

Section 1.4

1. (a) Function (b) Not a function
3. (a) Not a function (b) Function
5. (a) Increasing for all x (b) Decreasing for x
(c) Increasing for $x < -1$, decreasing for $x > -1$

7.

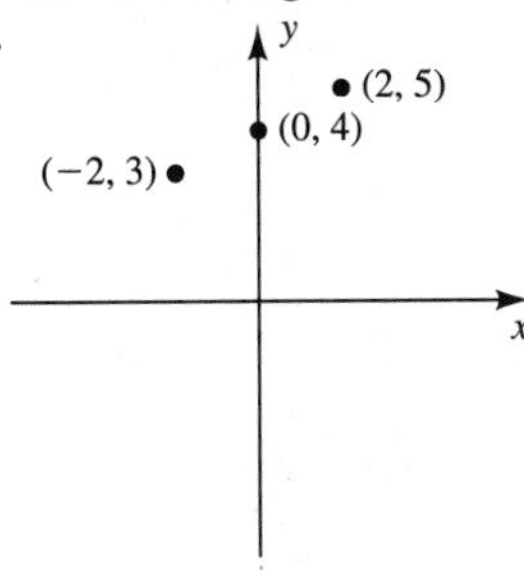

D: $-2, 0, 2$
R: $3, 4, 5$

9.

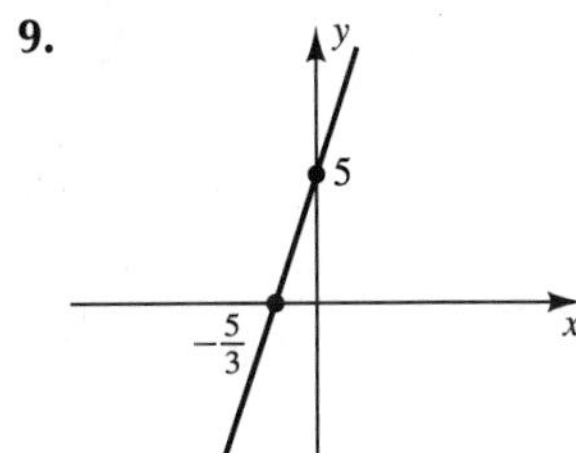

D: all reals
R: all reals

11.

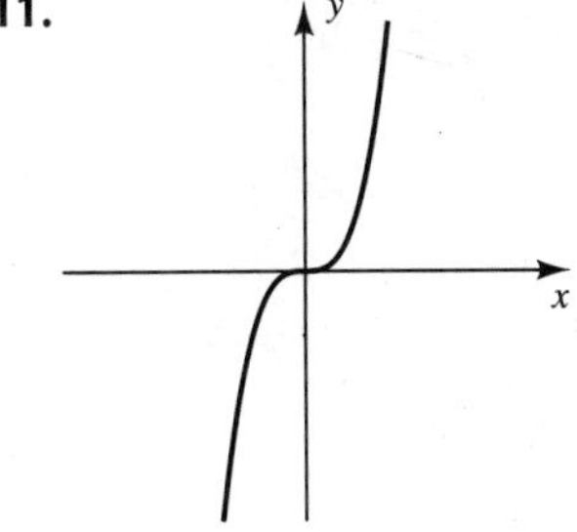

D: all reals
R: all reals

13.

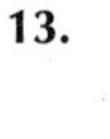

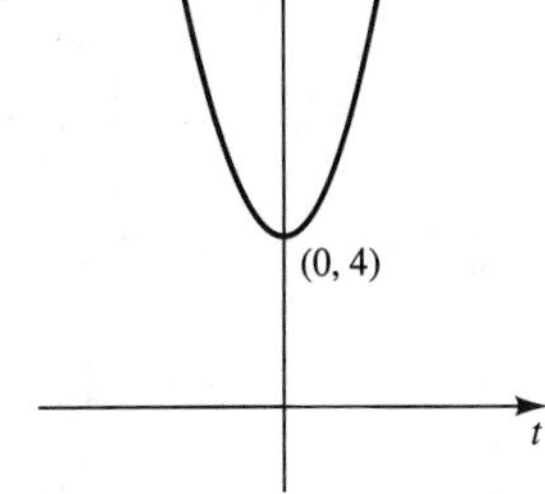

D: all reals
R: all reals $\geqslant 4$

15.

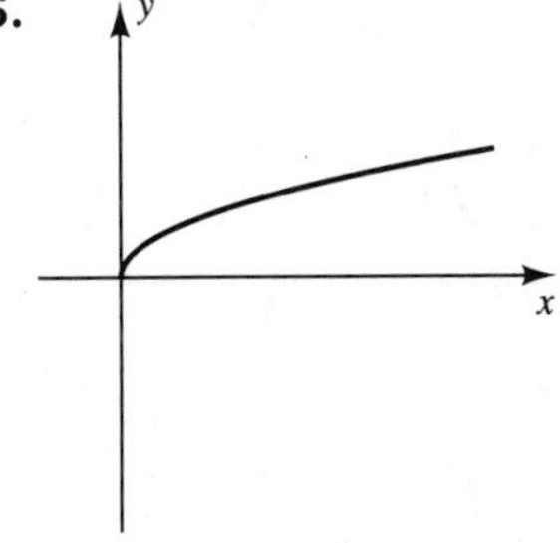

D: $x \geqslant 0$
R: $y \geqslant 0$

17.

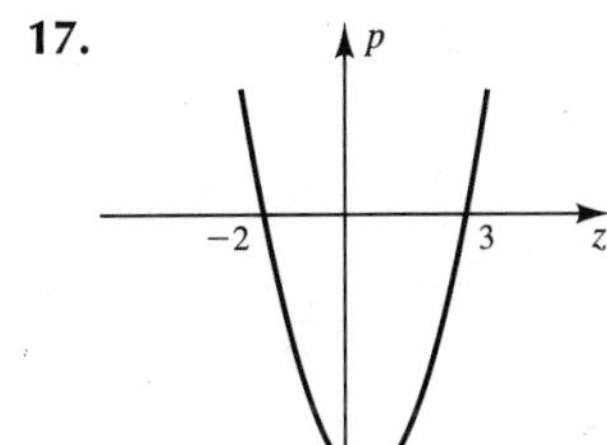

D: all z
R: $p \geqslant -6.25$

19.

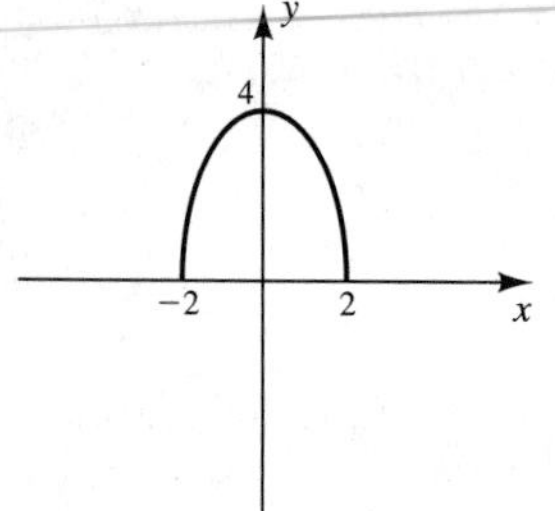

D: $-2 \leqslant x \leqslant 2$
R: $0 \leqslant y \leqslant 4$

21.

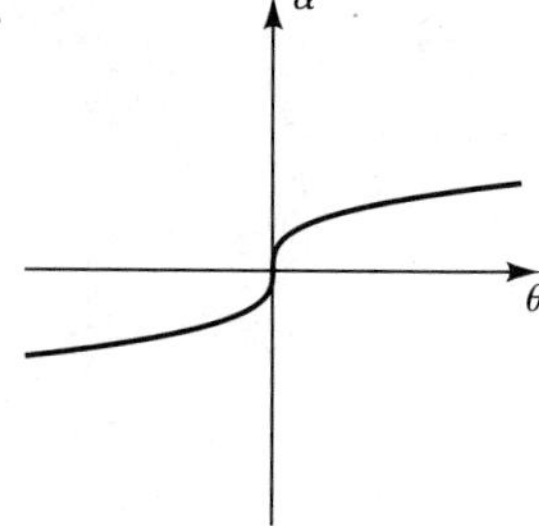

D: all θ
R: all α

23.

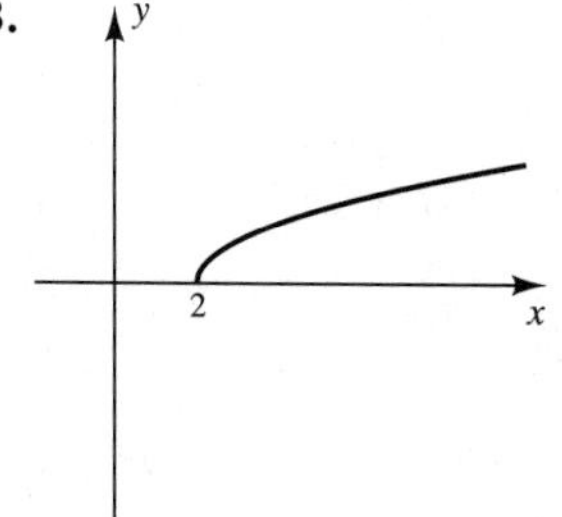

D: $x \geqslant 2$
R: $y \geqslant 0$

25.

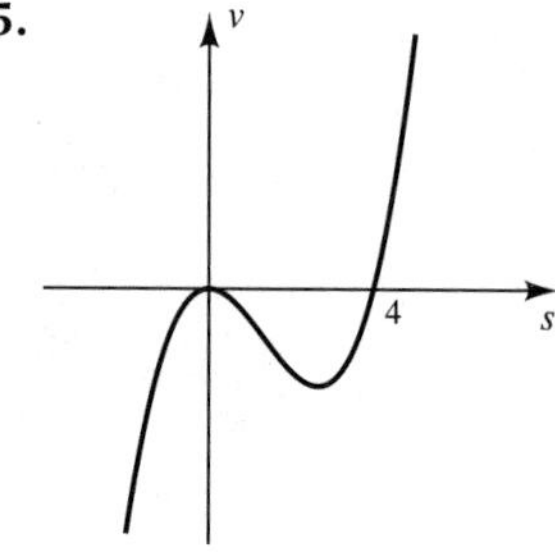

D: all s
R: all v

27.

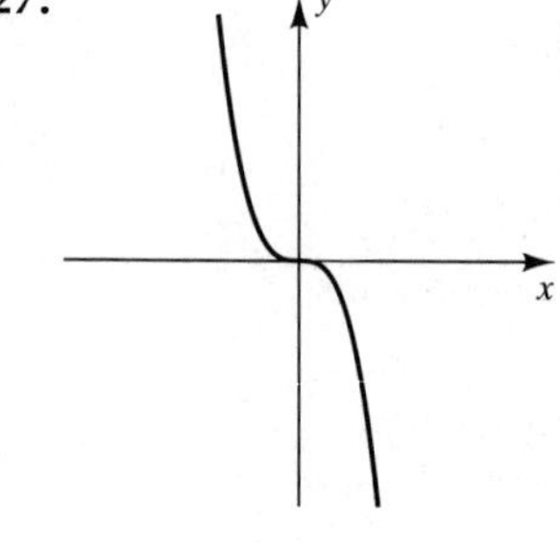

D: all x
R: all y

29.

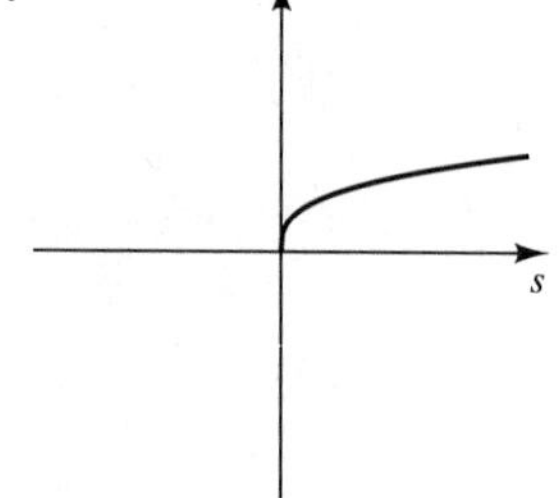

31.

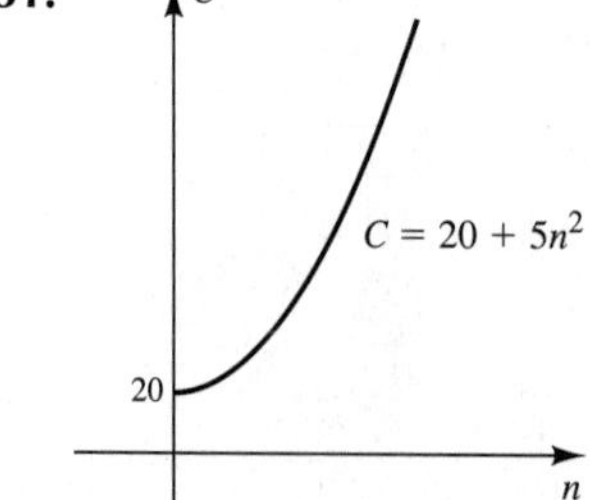

33.

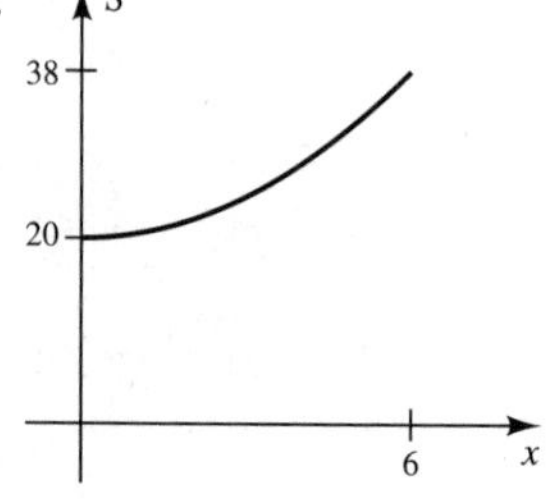

35.

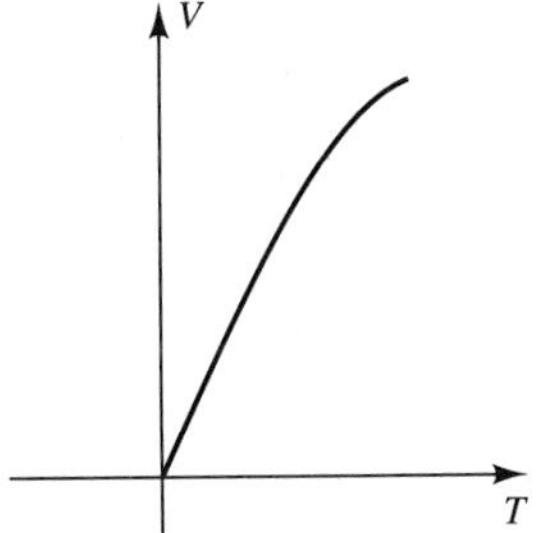

Section 1.5

1.

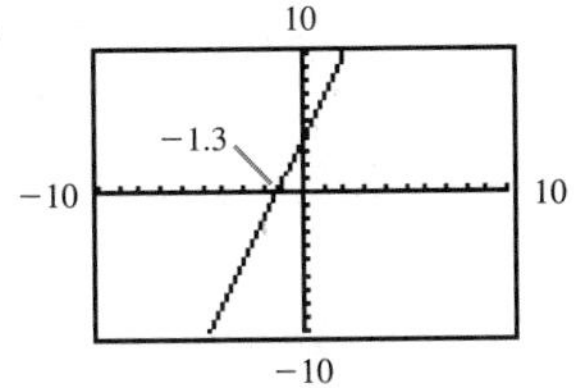

3.

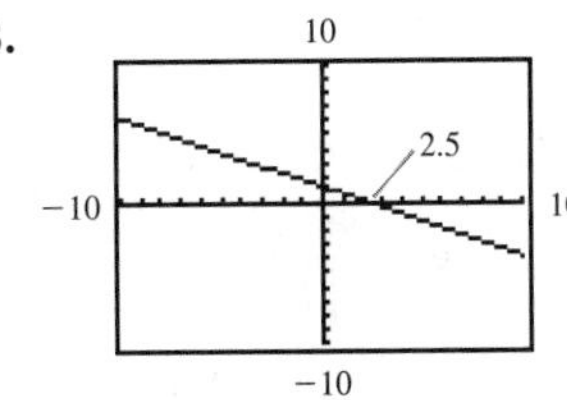

5.

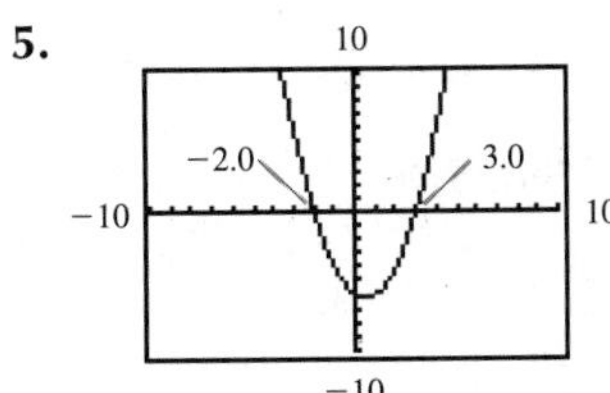

7.

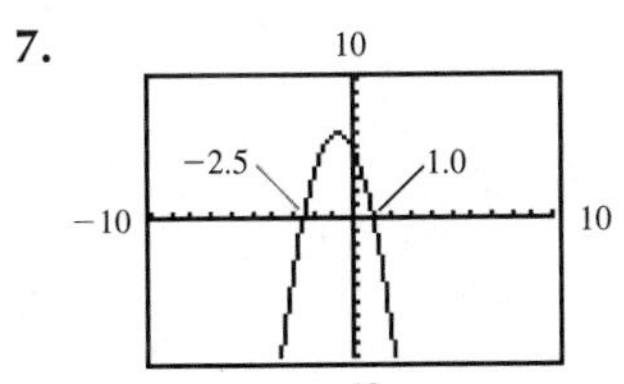

9.

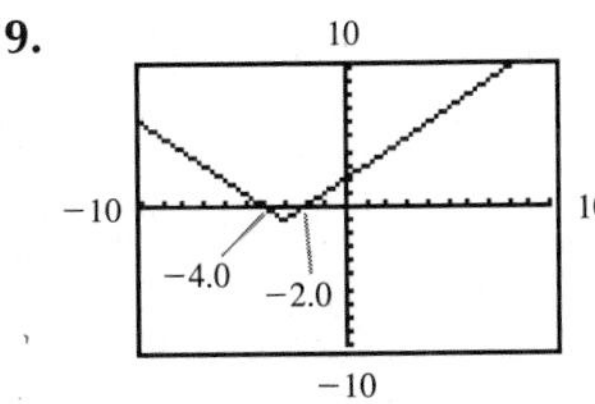

11.

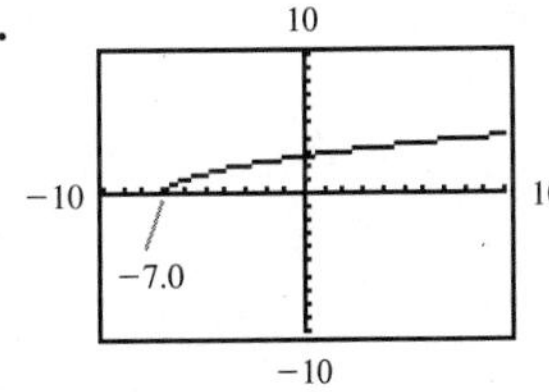

13.

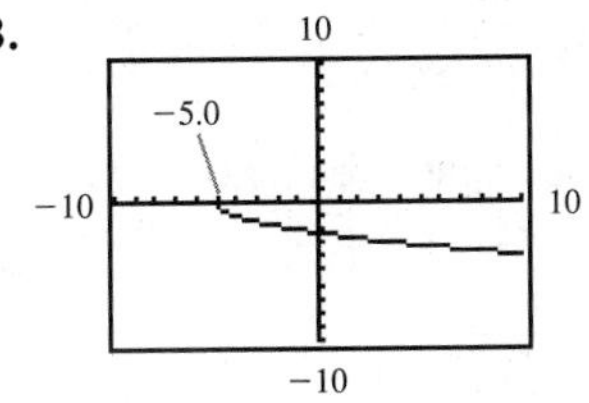

15.

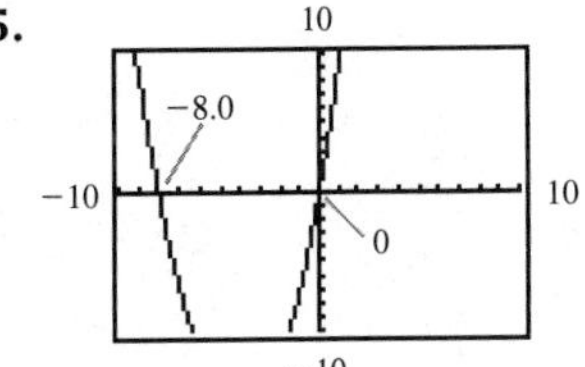

17.

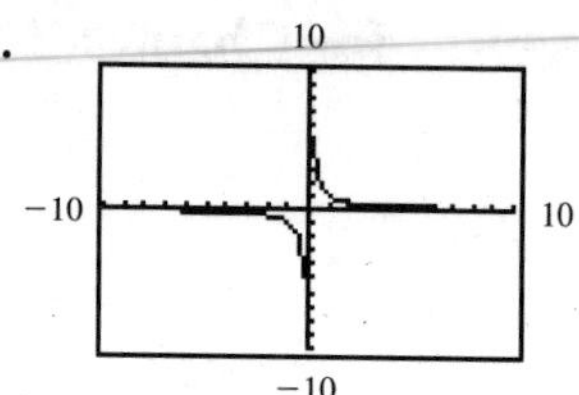

19.

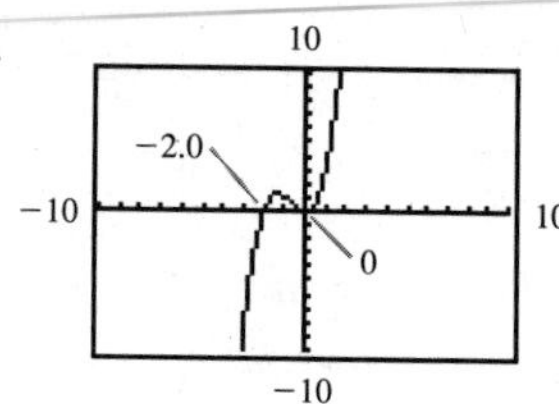

21. $0 \le x \le 20$, $-1 \le y \le 5$

23. $-5 \le x \le 15$, $-300 \le y \le 10$

25. $-10 \le x \le 15$, $-50 \le y \le 80$

27. $x = -0.48$, $y = -6.2$

29. $x = 1.1$, $y = 3.9$

31. $x = 2.24$

33. $x = -1.59$

35. $x = -0.59$ and -3.41

37.

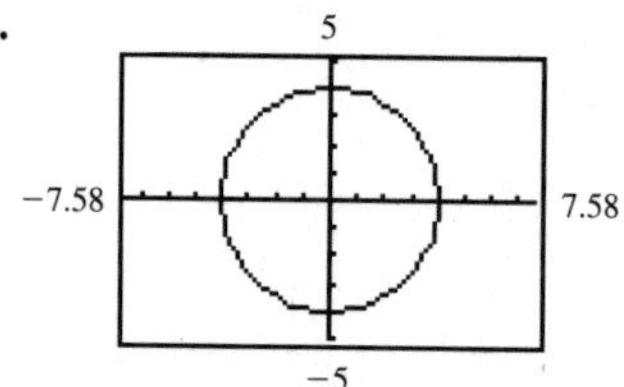

39.

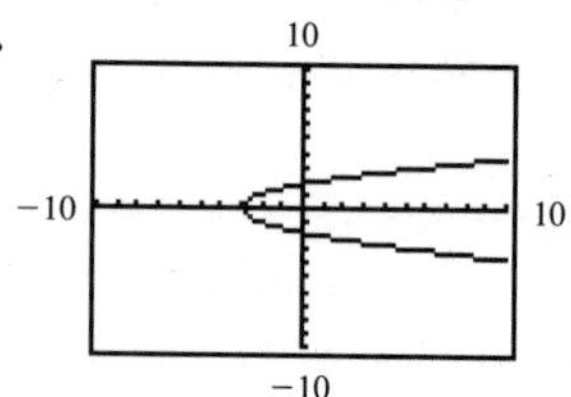

Section 1.6

1. 90° 3. 75°15′, 15°5′ 5. 109°10′, 8°1′20″
7. 130°55′15″, −10°34′45″ 9. 574°10′54″, −92°39′34″
11. −121°1′10″, 39°35′16″ 13. 18.4267° 15. 94.2856°
17. 283.6083° 19. 183.2444° 21. 48°15′26″
23. −235°27′0″ 25. 45°45′27″ 27. 15°15′27″ 29. 60°
31. 42.5° 33. 135° 35. 230° 37. 0° 39. 120°
41. 45° 43. 150° 45. 135° 47. −90° 49. 180°
51. 79°28′23″ 53. 0.08° 55. 4500°

Section 1.7

1. 74° 3. 127 ft 5. 3.7 m 7. 8.7 cm
15. 30°–60° right triangle; $A = 30°$ 17. Not a 30°–60° right triangle
19. 30°–60° right triangle; $B = 30°$ 21. Right triangle
23. Not a right triangle 25. Not a right triangle
27. No answer required. 29. 11, 60, 61 31. $a = 8.1$, $c = 10.3$
33. $a = 9.5$, $b = 3.5$ 35. 33 cm
37. Yes, because angles are all 60°. No, angle size may vary. 39. 80 m
41. $\sqrt{2}m$ 43. $\alpha = 7°2'$, $\phi = 58°46'$

Review Exercises for Chapter 1

1. 2, 1, −2

3.

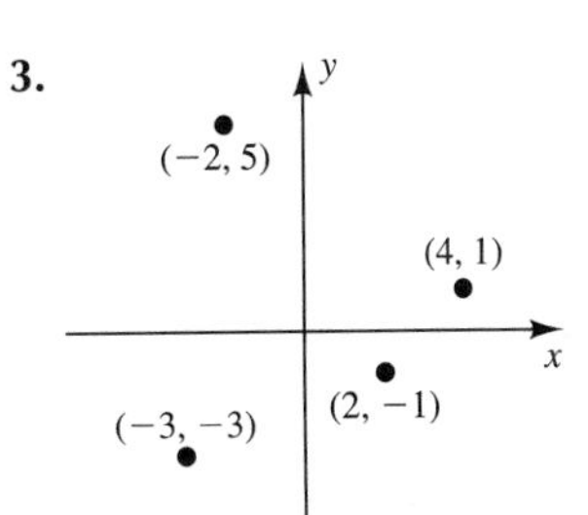

5.

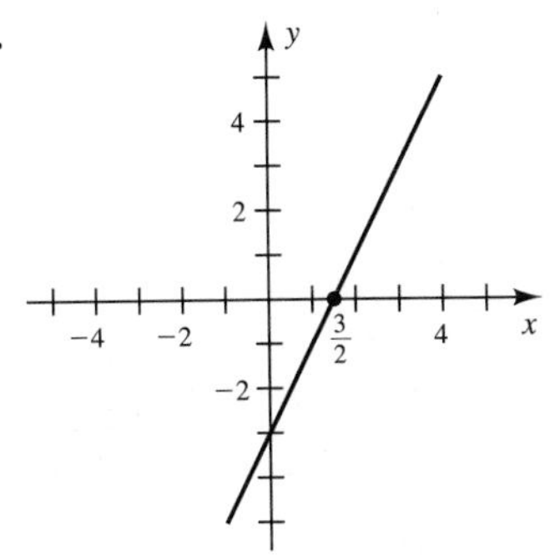

7. Function

9. Not a function **11.** Function **13.** Not a function
15. Function **17.** Not a right triangle **19.** 160°
21. 38.7231° **23.** 135° **25.** 9°47′20″ **27.** 280°
29. 51.8 ft **31.** 3.0 km
33.

Section 2.1

(The trigonometric functions are listed in order: sine, cosine, tangent, cotangent, secant, cosecant.)

1. $\frac{2}{\sqrt{5}}, \frac{1}{\sqrt{5}}, 2, \frac{1}{2}, \sqrt{5}, \frac{\sqrt{5}}{2}$

3. $\frac{16}{\sqrt{337}}, -\frac{9}{\sqrt{337}}, -\frac{16}{9}, -\frac{9}{16}, -\frac{\sqrt{337}}{9}, \frac{\sqrt{337}}{16}$

5. $-\frac{7}{\sqrt{53}}, \frac{2}{\sqrt{53}}, -\frac{7}{2}, -\frac{2}{7}, \frac{\sqrt{53}}{2}, -\frac{\sqrt{53}}{7}$

7. $-\frac{1}{\sqrt{10}}, \frac{3}{\sqrt{10}}, -\frac{1}{3}, -3, \frac{\sqrt{10}}{3}, -\sqrt{10}$

9. $-\frac{1}{2}, -\frac{\sqrt{3}}{2}, \frac{\sqrt{3}}{3}, \sqrt{3}, -\frac{2\sqrt{3}}{3}, -2$

11. **(a)** I, II **(b)** I, IV **(c)** I, III **13.** II **15.** III

17. IV **19.** $\frac{3}{5}, \frac{4}{5}, \frac{3}{4}, \frac{4}{3}, \frac{5}{4}, \frac{5}{3}$ **21.** $-\frac{3}{5}, -\frac{4}{5}, \frac{3}{4}, \frac{4}{3}, -\frac{5}{4}, -\frac{5}{3}$

23. Q I: $\frac{1}{2}, \frac{\sqrt{3}}{2}, \frac{1}{\sqrt{3}}, \sqrt{3}, \frac{2}{\sqrt{3}}, 2$; Q IV: $-\frac{1}{2}, \frac{\sqrt{3}}{2}, -\frac{1}{\sqrt{3}}, -\sqrt{3}, \frac{2}{\sqrt{3}}, -2$

25. $\frac{2}{3}, -\frac{\sqrt{5}}{3}, -\frac{2}{\sqrt{5}}, -\frac{\sqrt{5}}{2}, -\frac{3}{\sqrt{5}}, \frac{3}{2}$

27. Q III: $-\frac{1}{2}, -\frac{\sqrt{3}}{2}, \frac{\sqrt{3}}{3}, \sqrt{3}, -\frac{2}{\sqrt{3}}, -2$;

Q IV: $-\frac{1}{2}, \frac{\sqrt{3}}{2}, -\frac{\sqrt{3}}{3}, -\sqrt{3}, \frac{2}{\sqrt{3}}, -2$

29. $10/\sqrt{101}, 1/\sqrt{101}, 10, 1/10, \sqrt{101}, \sqrt{101}/10$

31. Q I: $\frac{5}{13}, \frac{12}{13}, \frac{5}{12}, \frac{12}{5}, \frac{13}{12}, \frac{13}{5}$; Q IV: $-\frac{5}{13}, \frac{12}{13}, -\frac{5}{12}, -\frac{12}{5}, \frac{13}{12}, -\frac{13}{5}$

33. Q II: $\frac{\sqrt{3}}{2}, -\frac{1}{2}, -\sqrt{3}, -\frac{1}{\sqrt{3}}, -2, \frac{2}{\sqrt{3}}$;

Q IV: $-\frac{\sqrt{3}}{2}, \frac{1}{2}, -\sqrt{3}, -\frac{1}{\sqrt{3}}, 2, -\frac{2}{\sqrt{3}}$

35. $\frac{u}{v}, \frac{\sqrt{v^2-u^2}}{v}, \frac{u}{\sqrt{v^2-u^2}}, \frac{\sqrt{v^2-u^2}}{u}, \frac{v}{\sqrt{v^2-u^2}}, \frac{v}{u}$

37. $\sqrt{1-u^2}, u, \frac{\sqrt{1-u^2}}{u}, \frac{u}{\sqrt{1-u^2}}, \frac{1}{u}, \frac{1}{\sqrt{1-u^2}}$ **39.** $\frac{3}{2} = 1.5$

41. $\frac{128}{\sqrt{191{,}552}}$ **43.** $\frac{255}{\sqrt{70650}}$

Section 2.2

1. 2 **3.** $\frac{1}{3}$ **5.** $-\frac{1}{2}$ **7.** $\frac{1}{\sqrt{2}}$ **9.** $\frac{\sqrt{3}}{2}$ **11.** $-\frac{5}{12}$

13. $-\dfrac{1}{\sqrt{5}}$ **15.** $\dfrac{1}{3}$ **17.** 2 **19.** $-\dfrac{\sqrt{2}}{\sqrt{7}}$

21. $\dfrac{1}{3}, \dfrac{\sqrt{8}}{3}, \dfrac{1}{\sqrt{8}}, \sqrt{8}, \dfrac{3}{\sqrt{8}}, 3$ **23.** $\dfrac{1}{\sqrt{5}}, -\dfrac{2}{\sqrt{5}}, -\dfrac{1}{2}, -2, -\dfrac{\sqrt{5}}{2}, \sqrt{5}$

25. $-\dfrac{\sqrt{2}}{\sqrt{3}}, -\dfrac{1}{\sqrt{3}}, \sqrt{2}, \dfrac{1}{\sqrt{2}}, -\sqrt{3}, -\dfrac{\sqrt{3}}{\sqrt{2}}$ **27.** 8.658 **29.** 0.5258

31. $\dfrac{1}{\sqrt{1.6}}$

Section 2.3

5. cos 70° **7.** cot 69°45′ **9.** csc 9° **11.** sin 72.5°
13. cot 29.6° **15.** $(5\sqrt{2} - 2)/2$ **17.** −1 **19.** 4/3
21. Undefined **23.** 0 **25.** 30° **27.** 60° **29.** 30°
31. 90° **33.** 45° **35.** 45° **37.** $5\sqrt{3}$ ft

Section 2.4

1. 80° **3.** 60° **5.** 30° **7.** 60° **9.** 45° **11.** 45°

13. 30° **15.** $\dfrac{\sqrt{2}}{2}$ **17.** $\dfrac{1}{2}$ **19.** $-\sqrt{3}$ **21.** −2 **23.** $\sqrt{2}$

25. $-\dfrac{1}{2}$ **27.** $\sqrt{3}$ **29.** $\dfrac{\sqrt{3}}{2}$ **31.** $\dfrac{-2\sqrt{3}}{3}$ **33.** $\dfrac{2\sqrt{3} - \sqrt{2}}{2}$

35. $90° + n \cdot 360°$ **37.** $90° + n \cdot 360°, 270° + n \cdot 360°$

39. $30° + n \cdot 360°, 150° + n \cdot 360°$ **41.** $0, \dfrac{3}{2}, \dfrac{3\sqrt{3}}{2}, \dfrac{3\sqrt{3}}{2}, \dfrac{3}{2}, 0$

Section 2.5

1. 0.2250 **3.** 0.3115 **5.** −0.1771 **7.** 0.7771
9. −5.6713 **11.** −1.2690 **13.** 1.0778 **15.** −1.0612
17. −0.8337 **19.** 1.2290 **21.** 1.9781 **23.** −0.1772
25. 33.8° **27.** 203.8° **29.** 334.8° **31.** 18.2° **33.** 318.0°
35. 161.3°, 341.3° **37.** 41.6°, 138.4° **39.** 117.2°, 297.2°
41. 335.6 ft **43.** 76.6 m **45.** 37.1° **47.** 802 ft/sec
49. 68.1°

Review Exercises for Chapter 2

1. $\dfrac{5}{\sqrt{29}}, -\dfrac{2}{\sqrt{29}}, -\dfrac{5}{2}, -\dfrac{2}{5}, -\dfrac{\sqrt{29}}{2}, \dfrac{\sqrt{29}}{5}$ **3.** $-\dfrac{3}{5}, -\dfrac{4}{5}, \dfrac{3}{4}, \dfrac{4}{3}, -\dfrac{5}{4}, -\dfrac{5}{3}$

5. 44°42′ **7.** 61.9° **9.** 87.8° **11.** $-\frac{1}{3}\sqrt{3}$ **13.** $-\frac{1}{2}\sqrt{2}$
15. $-\sqrt{2}$ **17.** 0.4245 **19.** −0.3843 **21.** −6.9273
23. −0.9659 **25.** 60°, 120° **27.** 120°, 300° **29.** 270°
31. 234°7′ and 305°53′ **33.** 292.6° **35.** 322.2° **37.** 231 mph

Section 3.1

1. $A = 31.1°$, $B = 58.9°$, $c = 82.4$ **3.** $A = 36.9°$, $B = 53.1°$, $b = 16.0$
5. $B = 80°40'$, $b = 30.4$, $c = 30.8$ **7.** 41 ft **9.** 45 ft
11. 143 yd **13.** 5.1 m **15.** 57.5°, 32.5° **17.** 13.9 mm
19. 0.93 in. **21.** 110.3 ft **23.** 2 min 25 sec **25.** 7.97 cm
27. 48.5 ft **29.** 920 m **31.** $r = 1200$ mi **33.** 19,000 mi
35. 84.3° **37.** 617 ft **39.** 1.4 km **41.** 270 ft/sec

Section 3.2

1. $c = 39.7$ **3.** $c = 46.9$ **5.** $a = 74.0$ **7.** $C = 80.6°$
9. $A = 95.7°$ **11.** $B = 118.1°$ **13.** $A = 129.7°$, $B = 19.6°$, $c = 2.79$
15. $A = 37°48'$, $B = 47°47'$, $c = 195.2$
17. $A = 29.0°$, $B = 46.5°$, $C = 104.5°$ **19.** 3.81 km
21. **(a)** 66.8 ft **(b)** 63.7 ft **23.** 30.8°
25. 13.6° **27.** $A = 66.6°$, $C = 53.4°$, $\overline{AC} = 3775$ m **29.** 346 m
31. 6.7 mi

Section 3.3

1. $C = 100.0°$, $b = 14.0$, $c = 18.6$ **3.** $B = 24.5°$, $C = 110.3°$, $c = 11.7$
5. $B = 23.9°$, $C = 120.5°$, $c = 25.9$ **7.** $B = 41°20'$, $C = 17°50'$, $c = 2.35$
9. $B = 28.9°$, $A = 98.1°$, $a = 37.4$ **11.** $A = 45.0°$, $C = 13.0°$, $c = 7.98$
13. $A = 20.0°$, $a = 17.9$, $c = 49.1$ **15.** $C = 88.6°$, $a = 42.1$, $b = 45.6$
17. 5.72 ft **19.** 1 hr 31 min
21. **(a)** 0.43×10^8 km **(b)** 46.53° **23.** 4.72 ft
25. Antenna, 545 ft; hill, 847 ft **27.** 31.8°

Section 3.4

1. No solution **3.** Two solutions **5.** One solution
7. No solution **9.** $B = 14.5°$, $C = 15.5°$, $b = 7.53$ **11.** No solution
13. $A = 38.8°$, $B = 113.2°$, $b = 29.4$
$A = 141.2°$, $B = 10.8°$, $b = 5.96$
15. $B = 38.08°$, $C = 111.88°$, $c = 602.6$
$B = 141.92°$, $C = 8.04°$, $c = 90.82$
17. $A = 40.0°$, $B = 70.0°$, $a = 68.4$
19. $B = 47°50'$, $C = 60°00'$, $c = 0.819$ **21.** $15 \sin 25° < b < 15$

Section 3.5

1. $C = 45.0°$, $b = 669$, $c = 490$ **3.** $A = 60.0°$, $b = 1.20$, $c = 4.46$
5. No solution **7.** $A = 32.0°$, $B = 118.0°$, $c = 283$ **9.** No solution
11. $A = 77.8°$, $C = 43.9°$, $c = 7.95$
$A = 102.2°$, $C = 19.5°$, $c = 3.83$
13. $C = 20.0°$, $a = 1.76$, $c = 0.695$ **15.** $C = 50.5°$, $B = 29.5°$, $c = 1.57$
17. $A = 53.9°$, $B = 29.8°$, $C = 96.3°$ **19.** 1.93 mi

Section 3.6

1. 2.24, 63.4° **3.** 3.00, 61.9° **5.** 5.66, 225° **7.** 5.00, 323.1°
9. 6.24, 106.1° **11.** (6.43, 7.66) **13.** (11.3, 7.69)
15. (−90.6, 129) **17.** (−33.3, −28.0) **19.** (9.43, −4.40)
21. 25.0, 36.9° **23.** 71.5, 75.8° **25.** 0.309, 65.1°
27. 10.7, 318.8° **29.** 8.03, 254.8° **31.** 44.3, 34.5°
33. 360, 37.2° **35.** 4.81, 68.5° **37.** 179.4, 174.1°
39. 1.22, 86°3′

51.

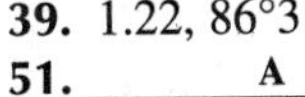

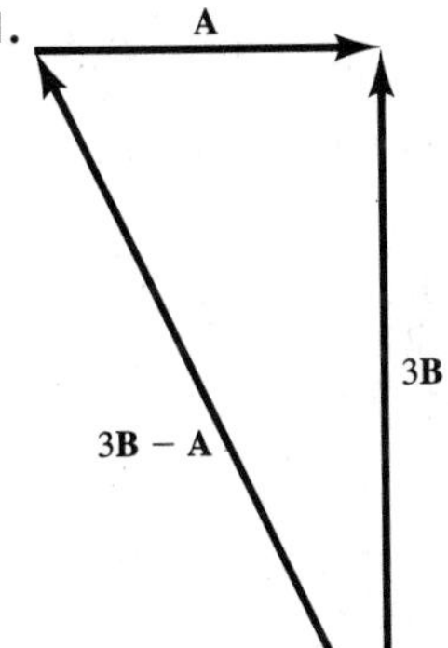

53.

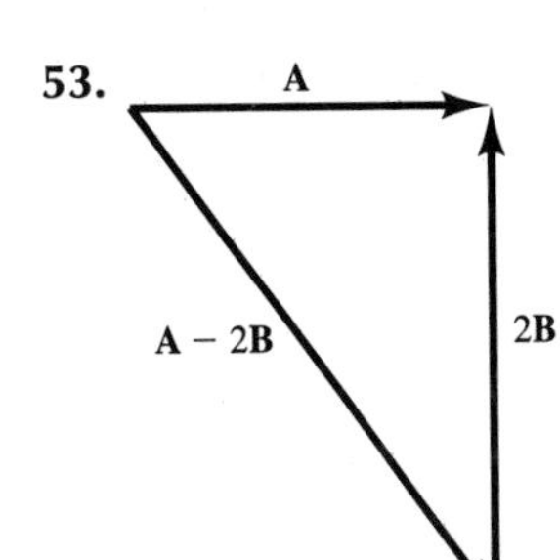

55.

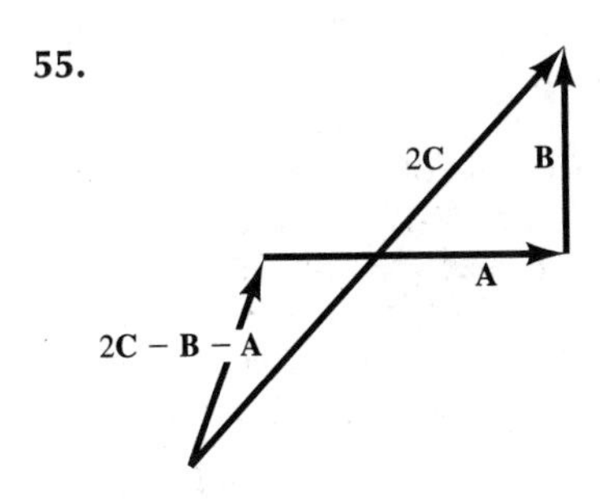

57.

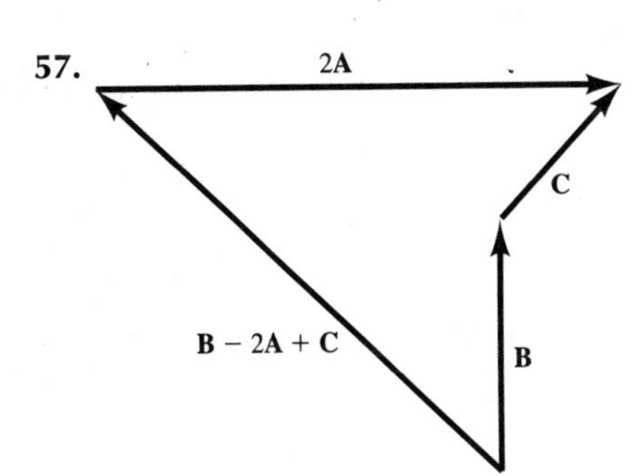

Section 3.7

1. 76.6 ft/sec, 64.3 ft/sec
3. 37° upstream from a line straight across the stream
5. 41.8 ft/sec, 61.4° from vertical
7. 1680 ft/sec, 14.2° below the horizontal
9. 304 mph, 350.5° clockwise from north **11.** 172 mph
13. 353.6°, 1 hr 46 min **15.** Horizontal, 73.3 lb; vertical, 15.8 lb
17. 13.5 lb **19.** 26.7° **21.** 82.8° **23.** 36.9° from vertical
25. 14° **27.** 157.3 lb, 36.8° with the horizontal
29. $R = 152$ lb, $\theta = 25.3°$
31. 4 hr, 7 min, 30 sec. Southbound: 188.1°; Northbound: 351.9°
33. 94.8°, 203 mph **35.** 62.7 mi
37. $X = 52.0$, $\theta = 60.0°$ **39.** $Z = 224$, $\theta = 26.6°$
41. $R = 43.3$, $Z = 50.0$

Section 3.8

1. 18.8 **3.** 39.4 **5.** 6.0 **7.** 0.448 **9.** 2.24
11. Doing so does not define a unique triangle; similar triangles may have unequal sides.
13. 85.3 in.^{2} **15.** 573 ft^2 **17.** 301 m or 500 m
19. Other side is 5.55 m and angles are 16.8°, 36.1°, 127.1°; or other side is 26.3 m and angles are 163.2°, 9.7°, 7.1°
21. 120 m^2

Review Exercises for Chapter 3

1. $B = 58°$, $b = 4.8$, $c = 5.7$ **3.** $A = 45.0°$, $B = 45.0°$, $b = 29.0$
5. Horizontal, 40.1 lb; vertical, 29.9 lb **7.** Magnitude 83.9 lb, angle 59.2°
9. 67.9° **11.** 472.0 ft **13.** (13.8, 5.86)
15. $C = 118.9°$, $B = 32.1°$, $a = 15.5$
17. $A = 44.3°$, $B = 29.5°$, $C = 106.2°$
19. $A = 148.0°$, $B = 7.8°$, $C = 24.2°$
21. $B = 40.2°$, $b = 2180$, $c = 1560$
23. $B = 19.8°$, $C = 135.2°$, $c = 25.0$ **25.** No triangle exists.
27. 10.1 **29.** $A = 24.3°$, $C = 17.7°$, $b = 14.7$, area $= 20.2$
31. 90.3 m **33.** 123 ft **35.** 3.6° south of east **37.** 16.7 ft

Section 4.1

1. An integral multiple of 2π **3.** $\frac{5}{12}\pi$ **5.** $\frac{8}{3}\pi$ **7.** $-\frac{4}{3}\pi$
9. $\frac{19}{36}\pi$ **11.** $\frac{25}{6}\pi$ **13.** 1.61 **15.** 0.00161 **17.** 4.426
19. 0.0079 **21.** $\frac{1}{4}\pi$ **23.** π **25.** $-\frac{3}{4}\pi$ **27.** $-\frac{1}{2}\pi$
29. 229.18° **31.** 180° **33.** π, $-540°$ **35.** 3.717, 572.96°
37. 0, 18,000° **39.** 2.8198 **41.** 7.0153 **43.** 1.3476
45. -0.4161 **47.** 2.69 **49.** 3.66 **51.** 0.45 **53.** 5.52
55. $\dfrac{\sqrt{2}}{2}$ **57.** $-\sqrt{3}$ **59.** -1 **61.** $\dfrac{\sqrt{3}}{2}$ **63.** $\sqrt{2}$
65. -1 **67.** $a = 19.3$, $b = 24.1$, $C = 0.97$ rad
69. $b = 159$, $c = 273$, $A = 1.28$ radians **73.** 4.60

Section 4.2

1. 34.7 cm **3.** 0.44 rad **5.** 20.5 ft **7.** 2.62 ft **9.** 11.4 in.
11. 4.2 in **13.** 1.47 in. **15.** 57.1° **17.** 293 in.
19. 12.4 in. **21.** 108 ft^2 **23.** 5.53 in. **25.** 9.5°, 34.8 ft
27. 292° **29.** 6.4 ft^3

Section 4.3

1. 12.5 ft/sec **3.** 25.7 ft/sec **5.** 15.3 ft/sec **7.** 0.011 rad/sec
9. 29.9 rad/sec **11.** 628 cm **13.** 8.73 ft/sec, 47.6 mi
15. 0.63 in. **17.** 16.2 rad/sec **19.** $\frac{20}{3}\pi$ ft/sec, 200 rpm
21. 860 mph

Review Exercises for Chapter 4

1. $\frac{\pi}{5}$ **3.** $\frac{11\pi}{18}$ **5.** $-\frac{17\pi}{36}$ **7.** 10° **9.** 24,923.7°
11. $-\frac{1}{2}$ **13.** $\frac{\sqrt{3}}{3}$ **15.** −1 **17.** −0.9165 **19.** −0.0715
21. $\frac{5\pi}{3}$ **23.** 1.25 **25.** $\frac{2\pi}{3}$ **27.** 4.02 **29.** 3.605
31. 7.39 or 10.53 **33.** 6.7 cm **35.** 2.6 ft **37.** 68.4 cm^2
39. 18.3 rad/sec; 174.3 rpm **41.** 44.2 rad/sec

Section 5.1

1. **(a)** $\cos 1 = 0.54, \sin 1 = 0.84$ **(b)** $\cos(-2) = -0.42, \sin(-2) = -0.91$
(c) $\cos 3 = -0.99, \sin 3 = 0.14$ **(d)** $\cos 10 = -0.84, \sin 10 = -0.54$
(e) $\cos 3\pi = -1, \sin 3\pi = 0$ **(f)** $\cos(-4) = -0.65, \sin(-4) = 0.76$
(g) $\cos(-4\pi) = 1, \sin(-4\pi) = 0$ **(h)** $\cos \frac{1}{3}\pi = 0.5, \sin \frac{1}{3}\pi = 0.87$
(i) $\cos \frac{1}{3} = 0.95, \sin \frac{1}{3} = 0.33$ **(j)** $\cos \frac{1}{2} = 0.88, \sin \frac{1}{2} = 0.48$
(k) $\cos \sqrt{7} = -0.88, \sin \sqrt{7} = 0.48$
(l) $\cos 5.15 = 0.42, \sin 5.15 = -0.91$

5. 0 **7.** $\frac{\sqrt{3}}{2}$ **9.** $\frac{1}{2}$ **11.** 0.8415 **13.** 0.7055
15. 0.5440 **17.** $\frac{(4n+1)\pi}{2}$ **19.** $\frac{\pi}{6} + 2n\pi, \frac{5\pi}{6} + 2n\pi$
21. No solution **23.** $0.10 + 2n\pi, 3.04 + 2n\pi$
25. $3.44 + 2n\pi, 5.98 + 2n\pi$ **27.** 1 **29.** $\frac{1}{2}$ **31.** $-\frac{\sqrt{3}}{2}$
33. 0.5403 **35.** 0.8776 **37.** 0.9450 **39.** $2n\pi$
41. $\frac{5\pi}{6} + 2n\pi, \frac{7\pi}{6} + 2n\pi$ **43.** No solution
45. $1.37 + 2n\pi, 4.91 + 2n\pi$ **47.** $0.54 + 2n\pi, 5.74 + 2n\pi$
49. $1.23 + 2n\pi, 5.05 + 2n\pi$
51. **(a)** $n\pi$, n an integer **(b)** $\frac{1}{2}\pi + n\pi$, n an integer **53.** 0.56 ft/sec
55. 0.79 sec

Section 5.2

1. Domain: all real numbers; range $[-1, 1]$
3. Unbounded for large values of x **5.** Bounded by +3 and −3
7. Unbounded near 4 **9.** Unbounded near integral multiples of π
11. Unbounded near odd multiples of $\pi/2$
13. Unbounded for large values of x **15.** Even **17.** Odd
19. Neither **21.** Even **23.** Even
25. Sine wave: period $= 2\pi/3$ **27.** Cosine wave: period $= 4\pi$

29.

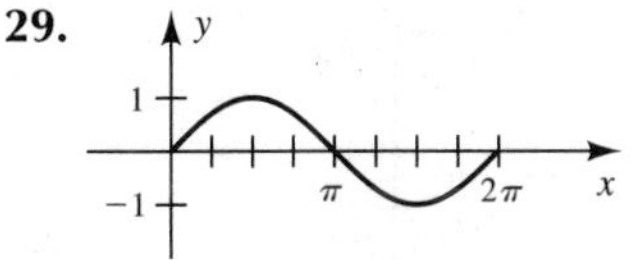

31.

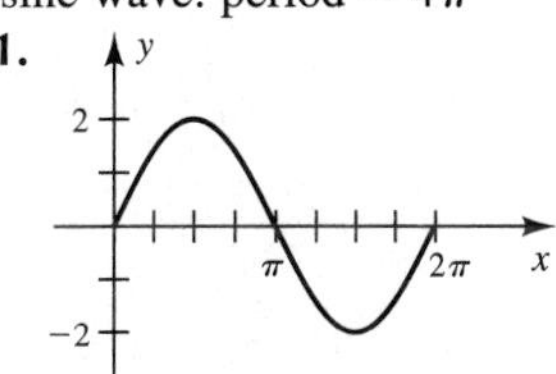

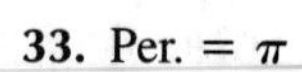

33. Per. $= \pi$

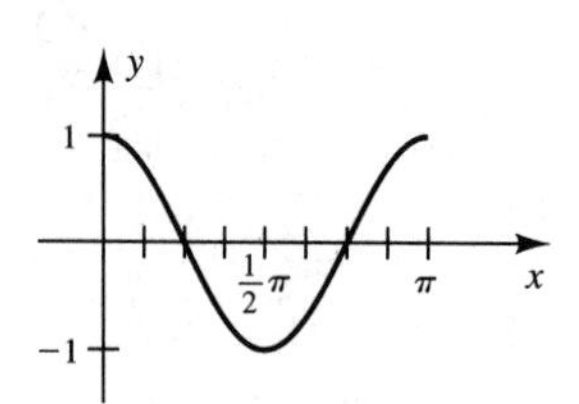

35.

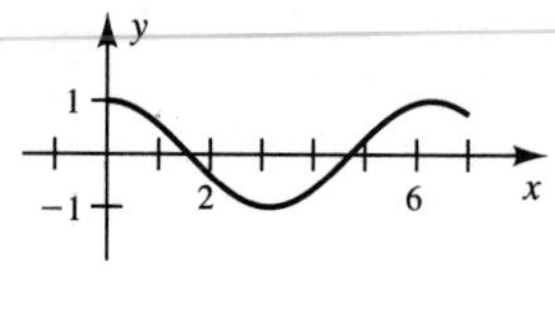

37.

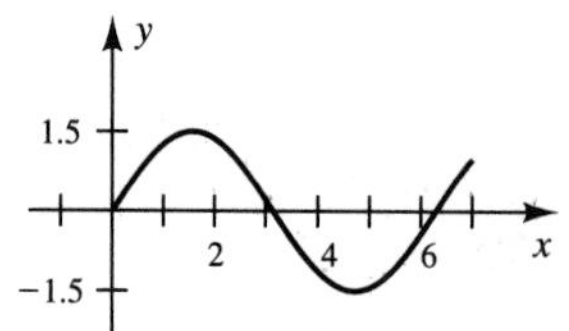

39.

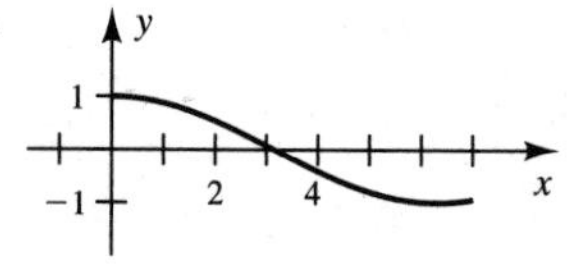

41. Per. $= \pi$

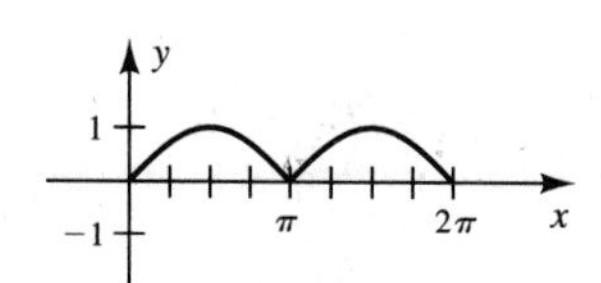

43.

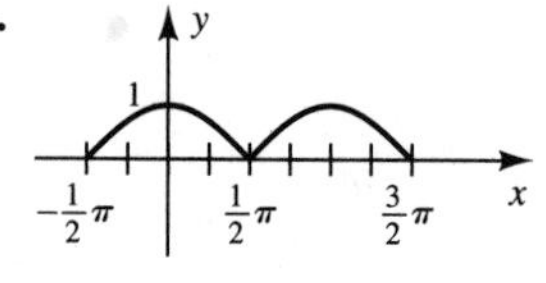

45. Initial velocity $= 1$, $v = 0$ for $t = \frac{1}{2}\pi + n\pi$
Points where the graph crosses the t-axis are where $v = 0$.

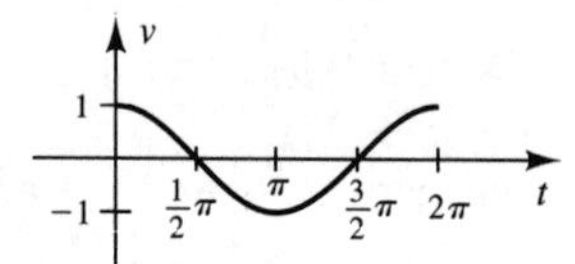

47.

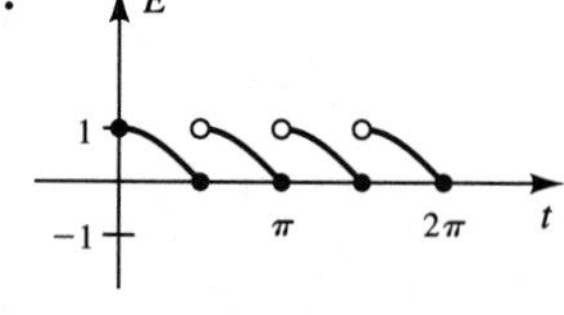

Graphing Calculator Exercises

1.

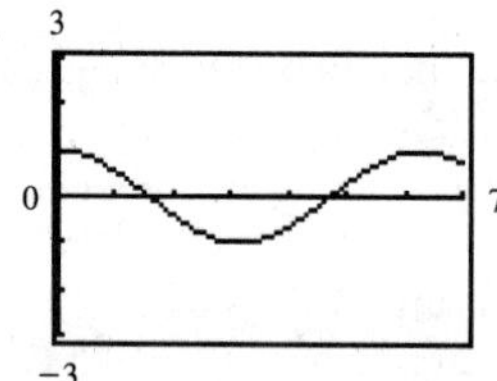

3.

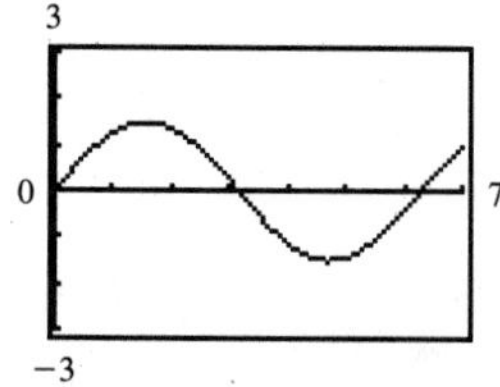

5.

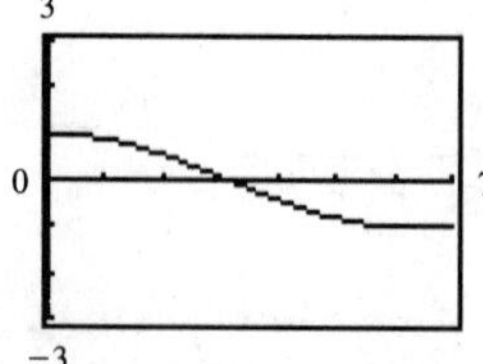

11. The graph of $y = |\sin x|$ is the same as that of $y = \sin x$ when $\sin x \geq 0$, and it is the mirror image in the x-axis of $y = \sin x$ when $\sin x < 0$. The graph of $y = \sin |x|$ is the same as that of $y = \sin x$ for $x \geq 0$, and for $x < 0$ it is the mirror image in the y-axis of the graph of $y = \sin x$ for $x \geq 0$.

Section 5.3

1. $A = 3$, per. $= 2\pi$

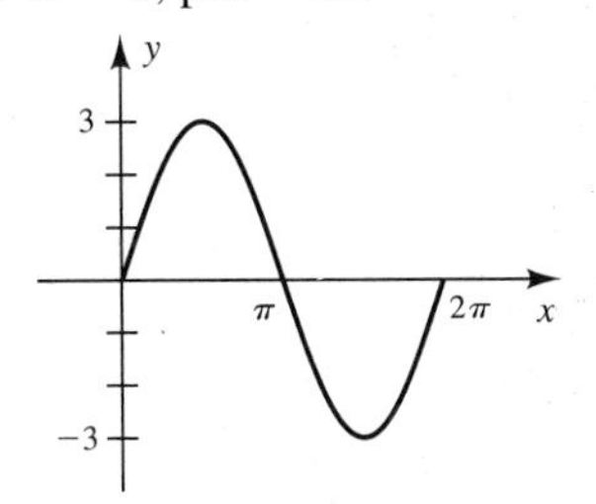

3. $A = 6$, per. $= 2\pi$

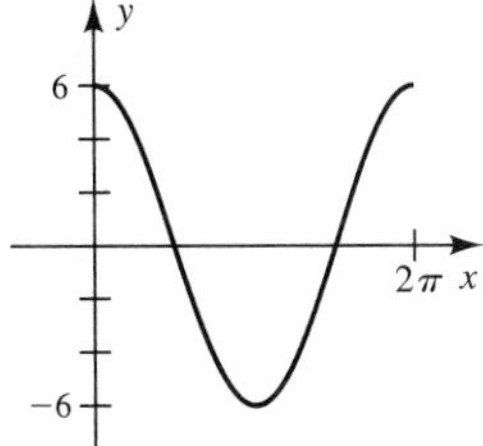

5. $A = 1$, per. $= 3\pi$

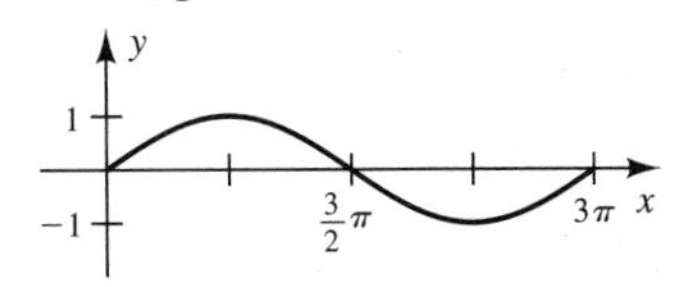

7. $A = 1$, per. $= 2$

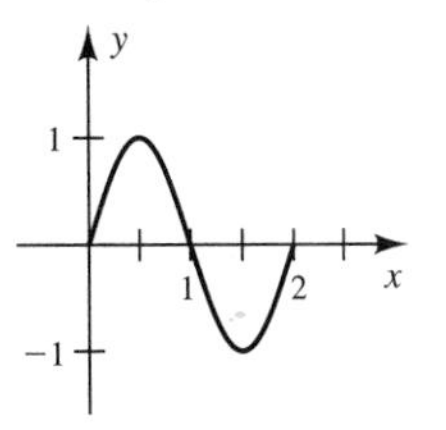

9. $A = \frac{1}{2}$, per. $= \pi$

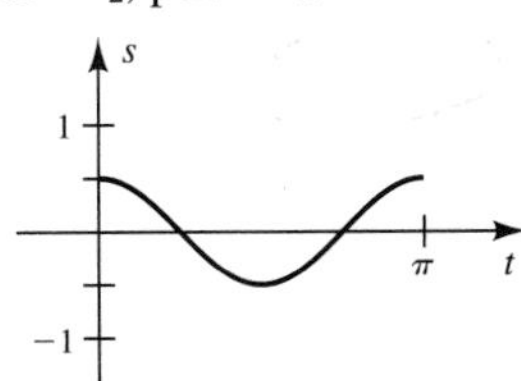

11. $A = 8.2$, per. $= 5\pi$

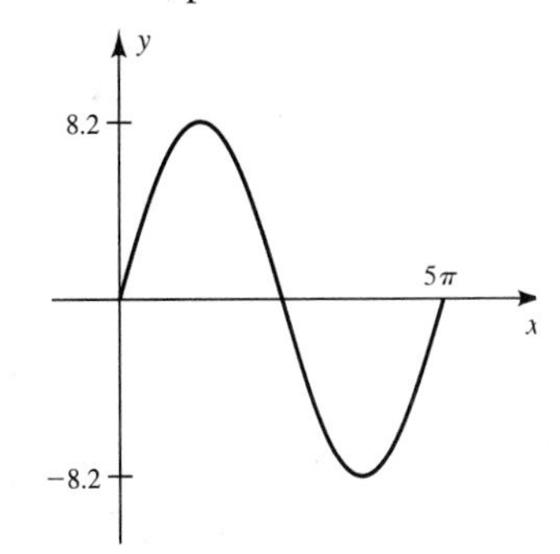

13. $A = 1$, per. $= 5\pi$

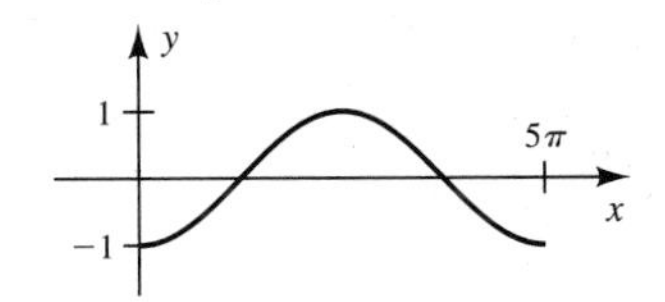

15. $A = \pi$, per. $= \frac{1}{50}\pi$

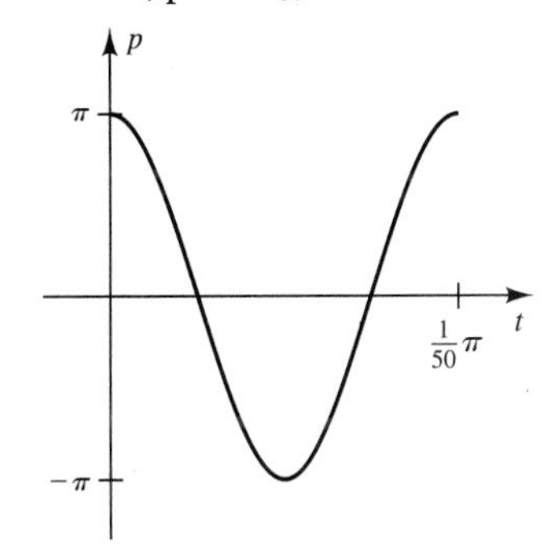

17. $A = 12$, per. $= 10\pi$

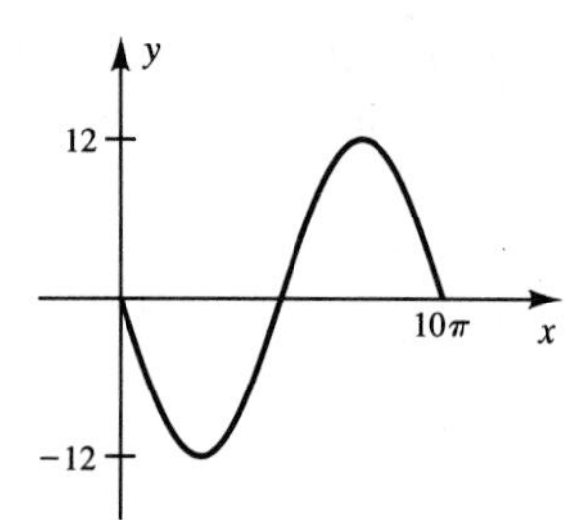

19. $A = \frac{1}{50}$, per. $= \frac{1}{500}$

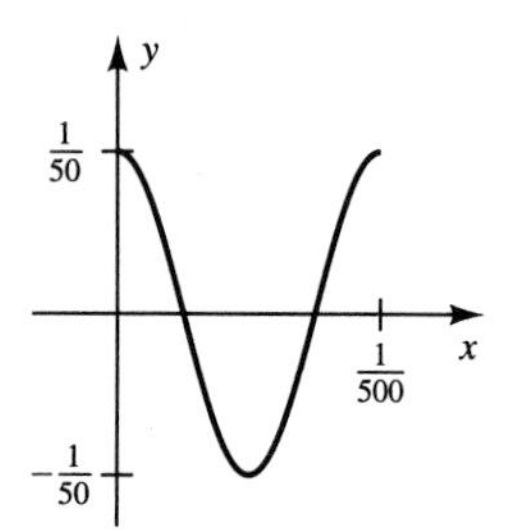

21. $y = \frac{1}{3} \sin \frac{1}{6}\pi x$ **23.** $y = 20 \sin \frac{16}{3}\pi x$ **25.** $y = 2.4 \sin 6x$

27. $y = \pi \sin \frac{2}{3}x$ **29.** $x = 3.2 \sin \frac{4}{5}\pi i$

31.

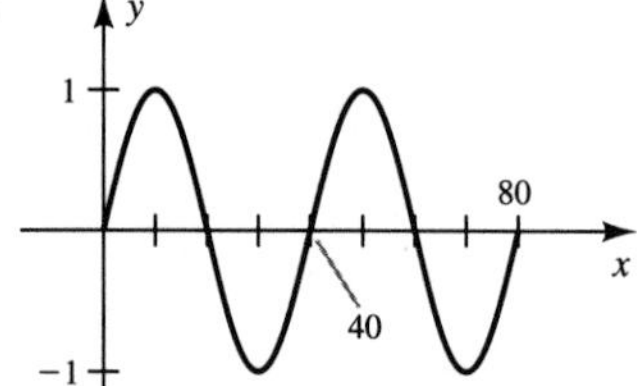

Graphing Calculator Exercise

3.

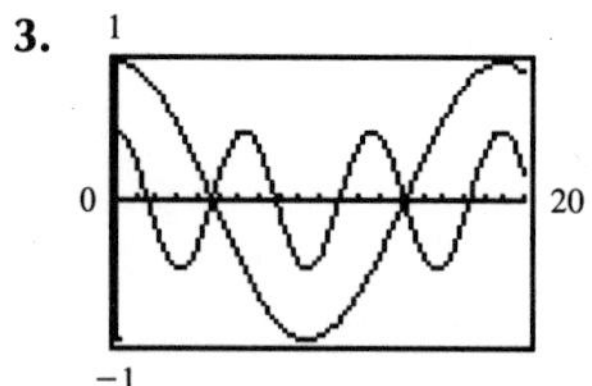

5.

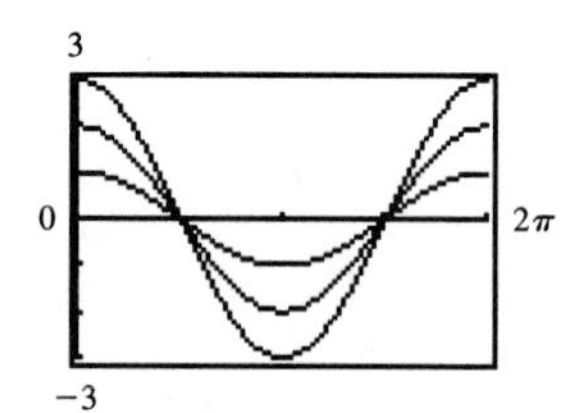

Section 5.4

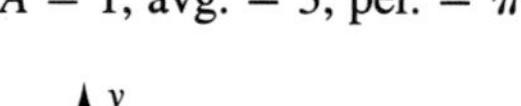

1. $A = 1$, avg. $= 3$, per. $= \pi$

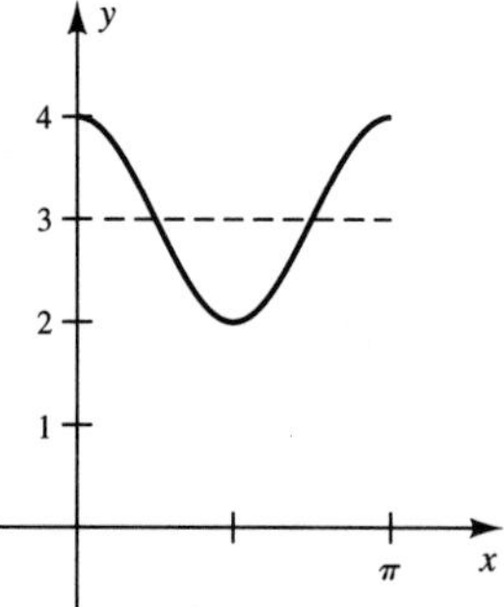

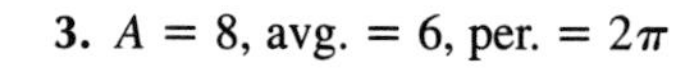

3. $A = 8$, avg. $= 6$, per. $= 2\pi$

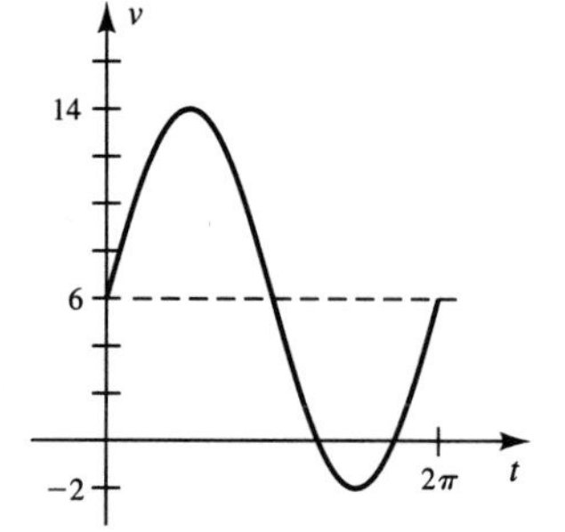

5. $A = 1$, avg. $= -2$, per. $= 4\pi$

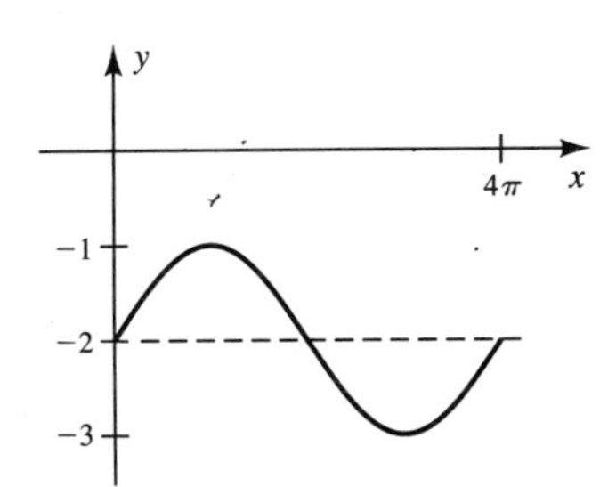

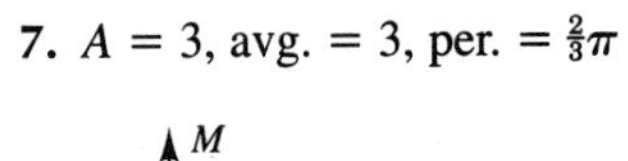

7. $A = 3$, avg. $= 3$, per. $= \frac{2}{3}\pi$

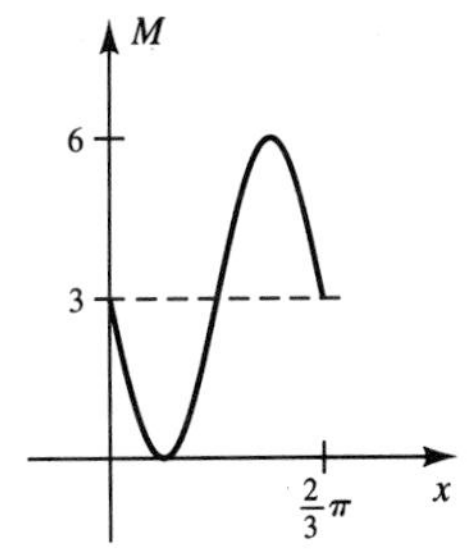

9. $A = 1$, ps $= -\frac{1}{3}\pi$, per. $= 2\pi$

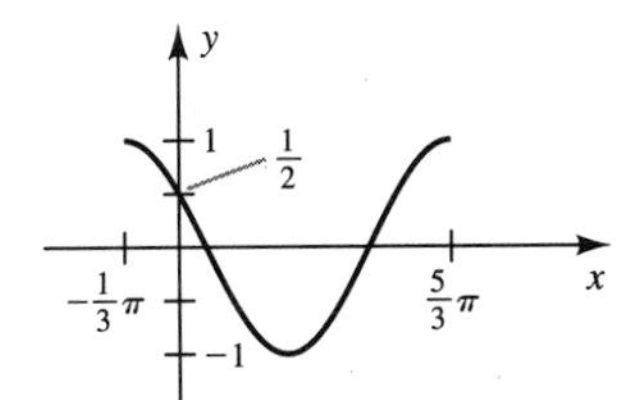

11. $A = 2$, ps $= \pi$, per. $= 4\pi$

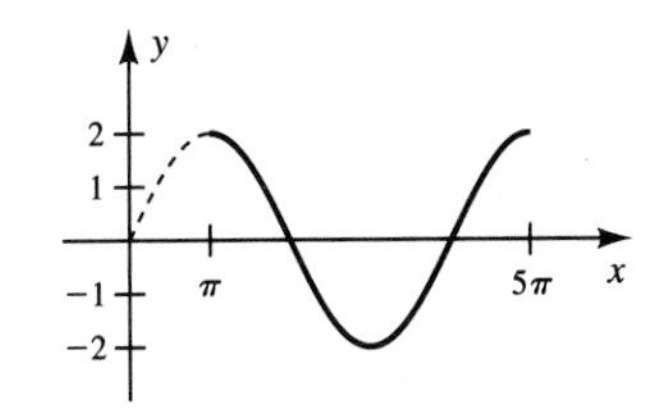

13. $A = 1$, ps $= -\frac{1}{2}\pi$, per. $= \pi$

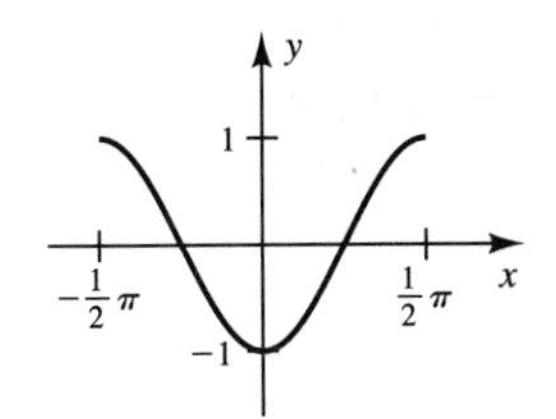

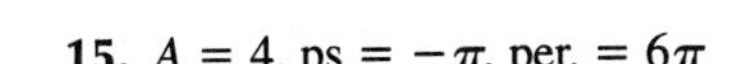

15. $A = 4$, ps $= -\pi$, per. $= 6\pi$

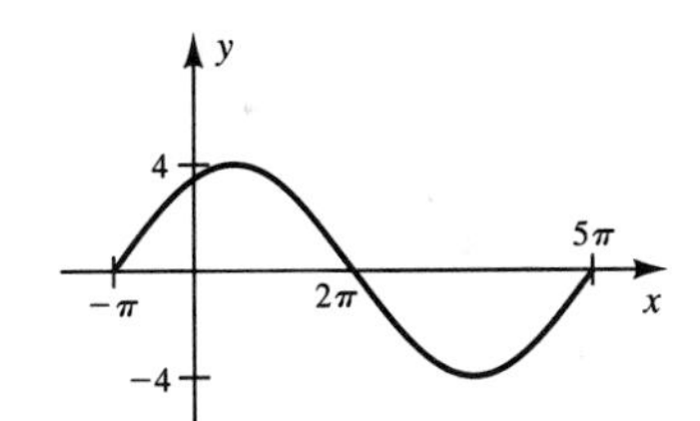

17. $A = 1$, ps $= \frac{1}{4}$, per. $= 2$

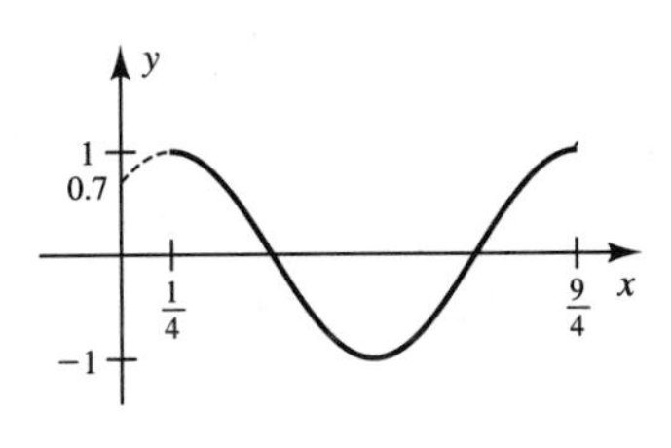

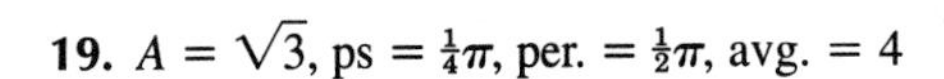

19. $A = \sqrt{3}$, ps $= \frac{1}{4}\pi$, per. $= \frac{1}{2}\pi$, avg. $= 4$

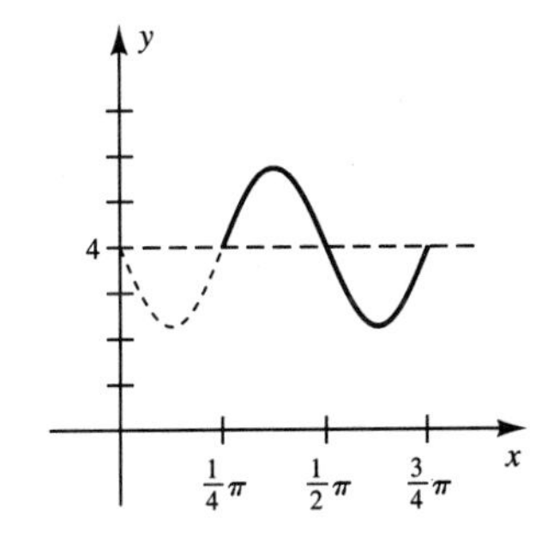

21. $\sin[x + (1 + \pi)]$

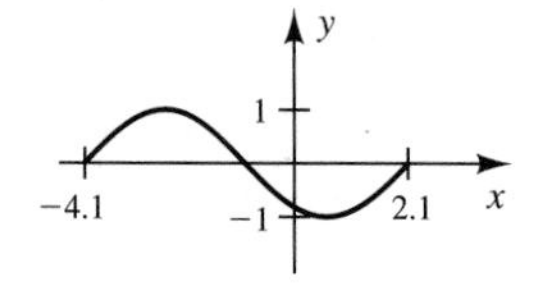

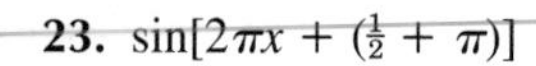
23. $\sin[2\pi x + (\frac{1}{2} + \pi)]$

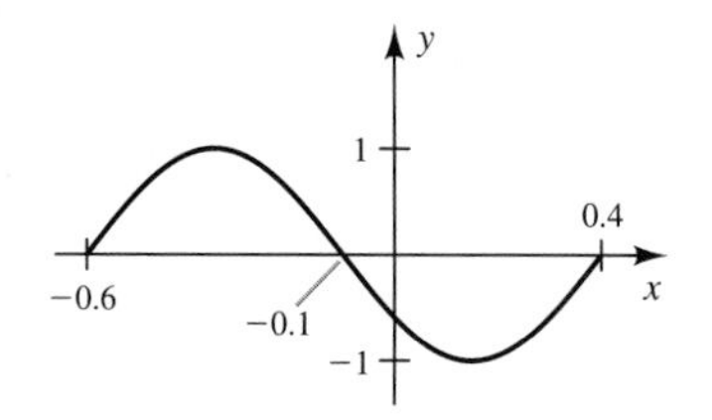

25. $\sin[\pi x + (1 - \frac{1}{2}\pi)]$

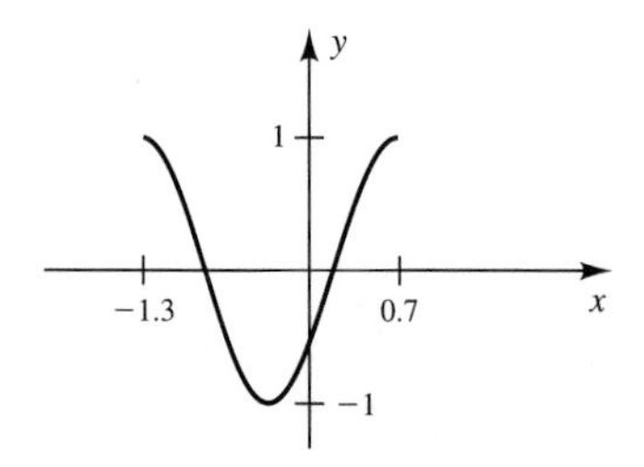

27. $y = \cos 2(x + \frac{1}{6}\pi)$

29. $y = -2 \sin \frac{1}{2}x$

31. They are mirror images in the t-axis.

33. They are the same.

35.

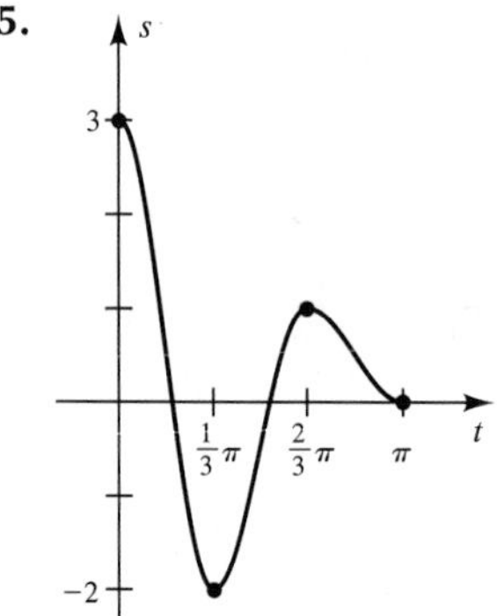

37.

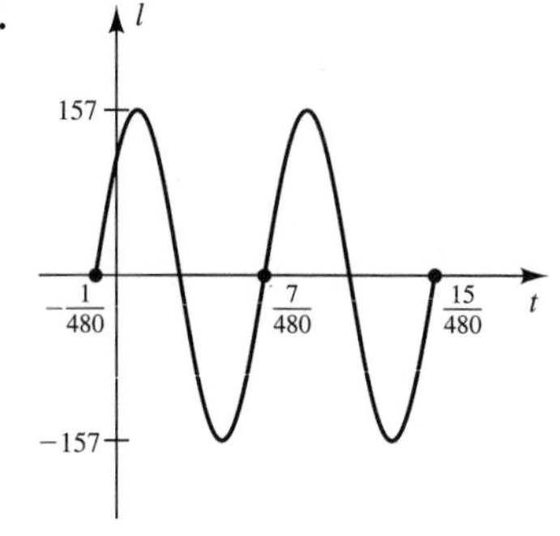

39. **(a)** 0.6 sec

(b)

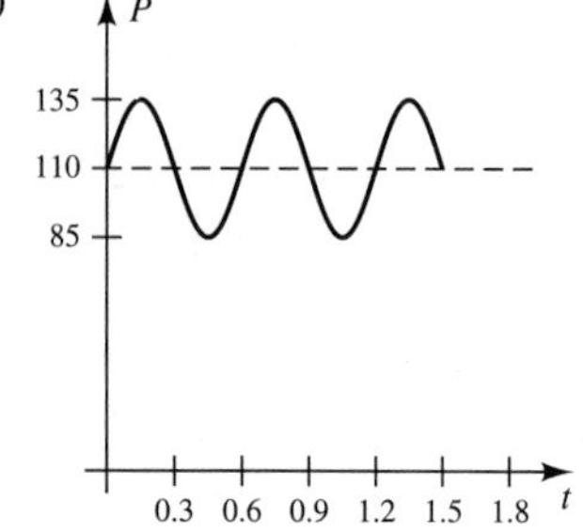

Graphing Calculator Exercises

1.

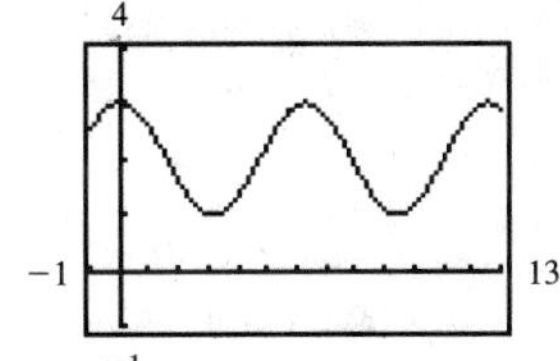

3.

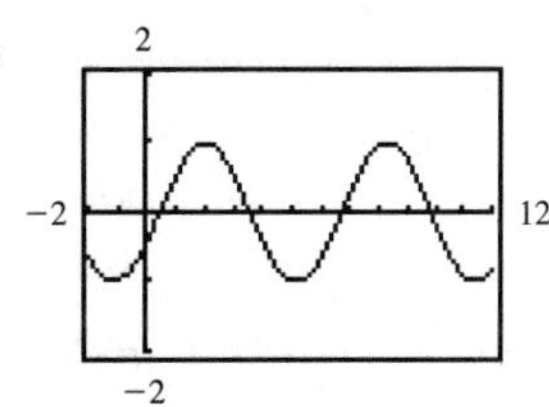

5.

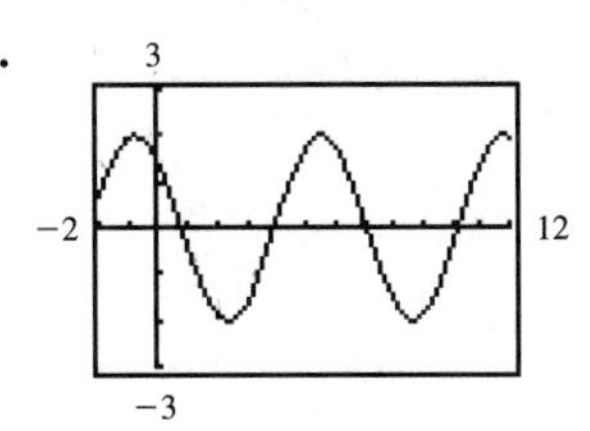

7.

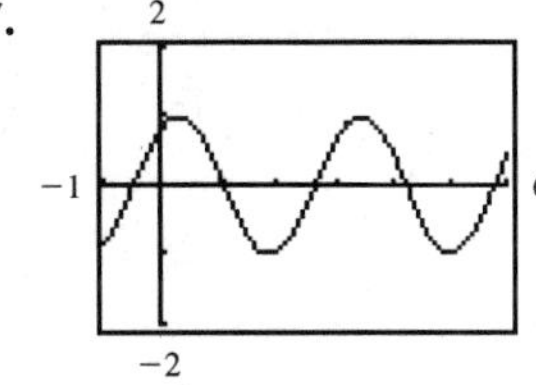

9. $\sin(x + 2) \neq \sin x + 2$ since they have different graphs.
11. $-\sin x = \sin(x - \pi)$ since they have the same graphs.

Section 5.5

1. **(a)** 3.14 sec **(b)** −6.24 cm
3. **(a)** $h = 6 + 7 \cos \frac{1}{5}\pi(t - 2)$ **(b)**

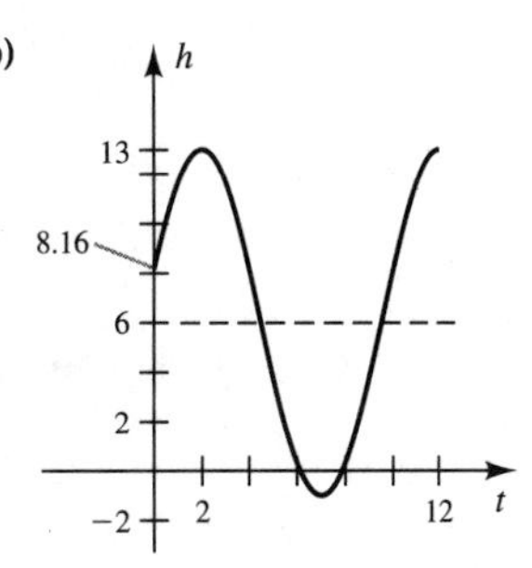

(c) $t = 7.86$ sec
5. **(a)** $y = -3 - 20 \cos \frac{1}{3}\pi(t - 2)$

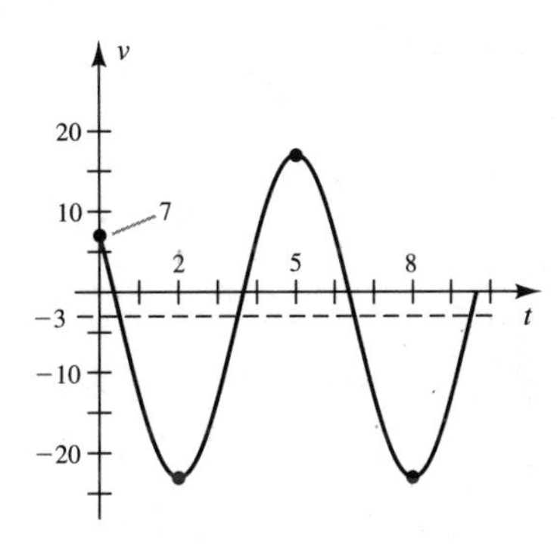

(b) −16.4, 1.2, −13
(c) At $y = 7$
(d) $t = 3.64$ sec

7.

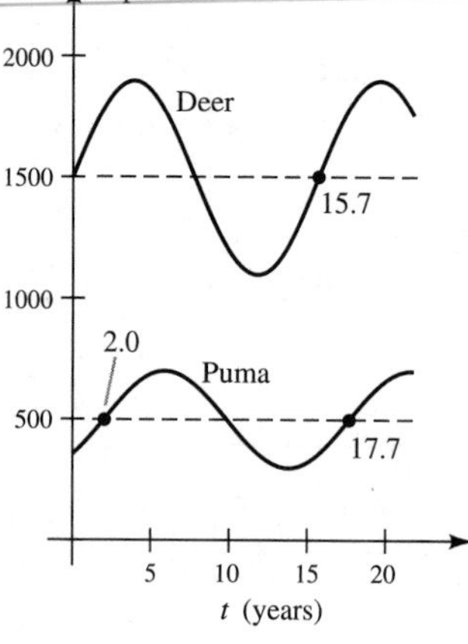

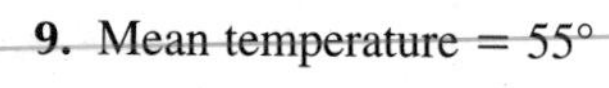

9. Mean temperature = 55°

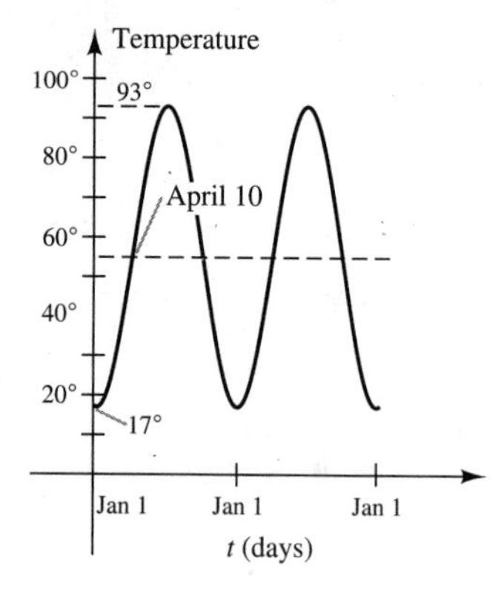

11. **(a)** 11 years **(b)**

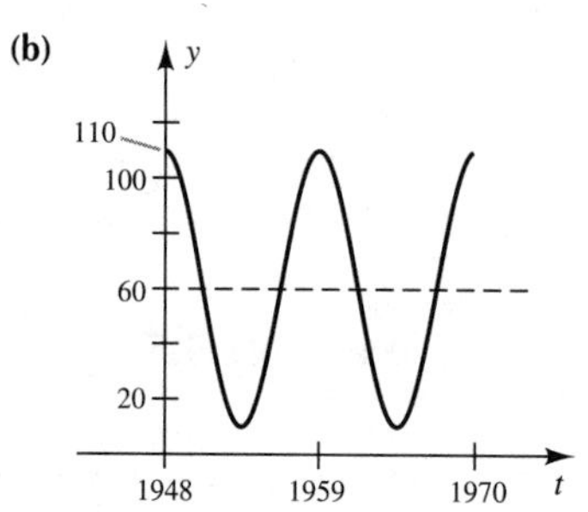

(c) $y = 60 + 50 \cos \frac{2}{11}\pi(t - 1948)$

(d) Answers will depend on the year; 52.9 **(e)** 2003

Section 5.6

1. Per. $= 2\pi$

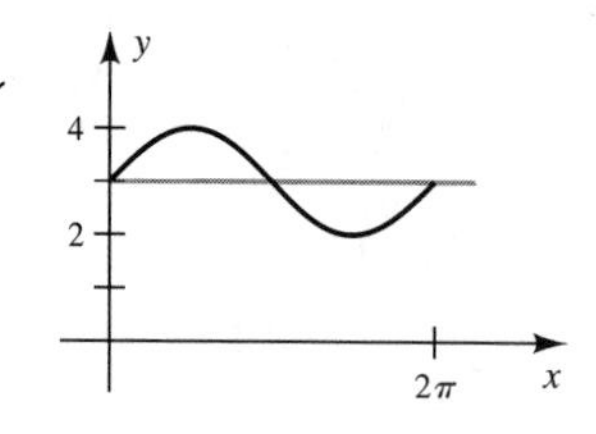

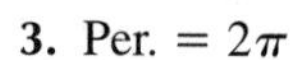

3. Per. $= 2\pi$

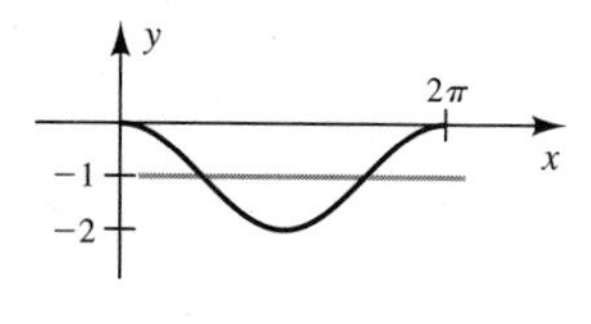

5. Per. $= 2\pi$

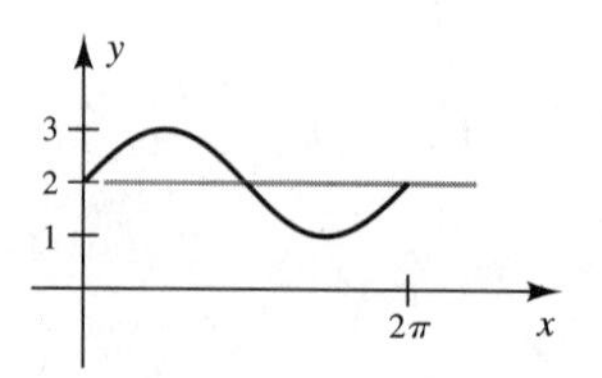

7. Not periodic

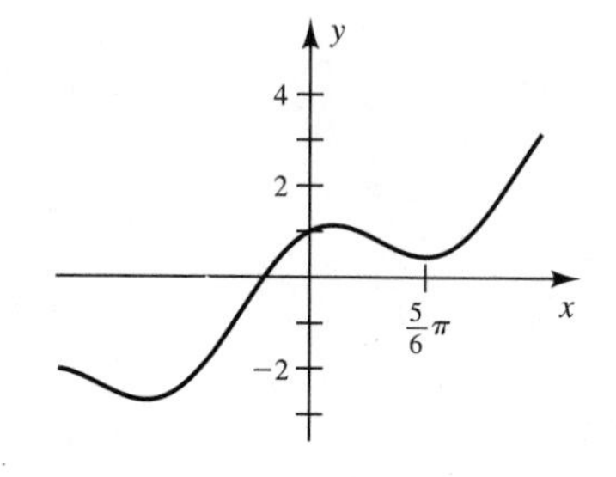

9. Not periodic

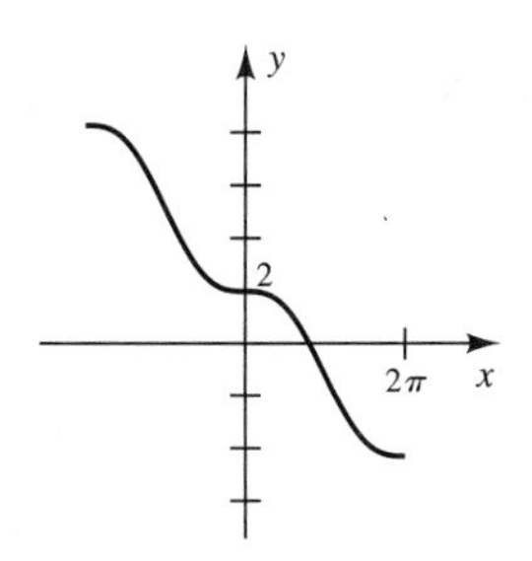

11. Not periodic

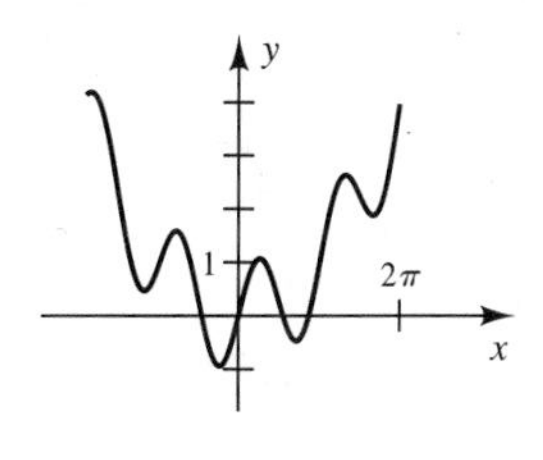

13. Per. = 2π

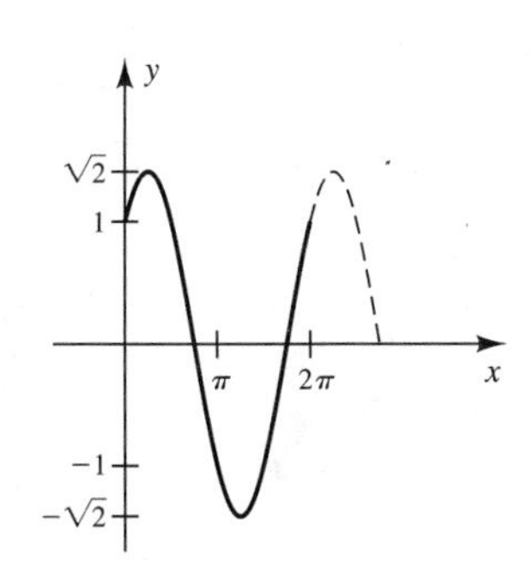

15. Per. = 4π

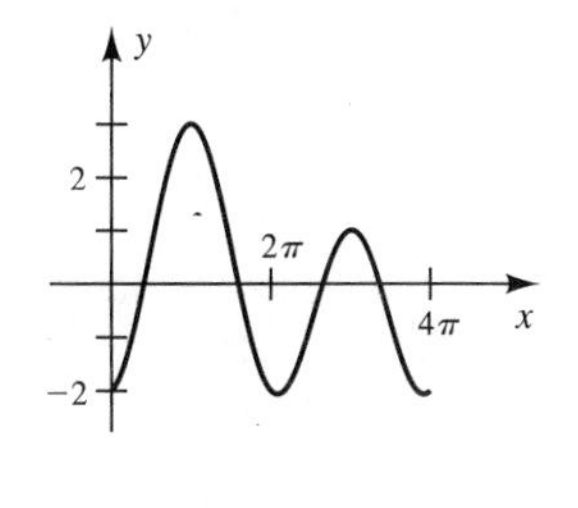

17. Per. = 2π

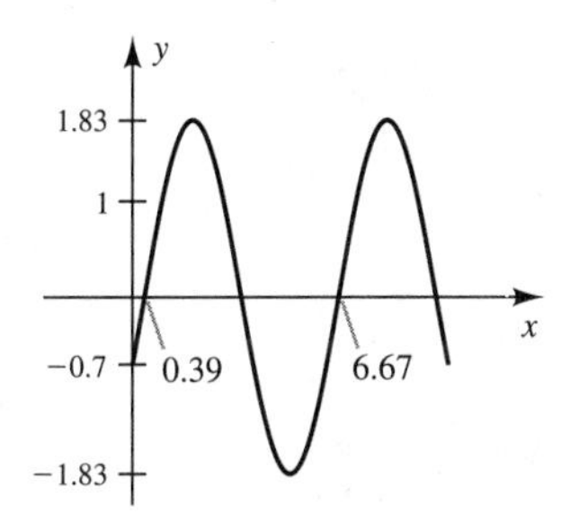

19. Not periodic

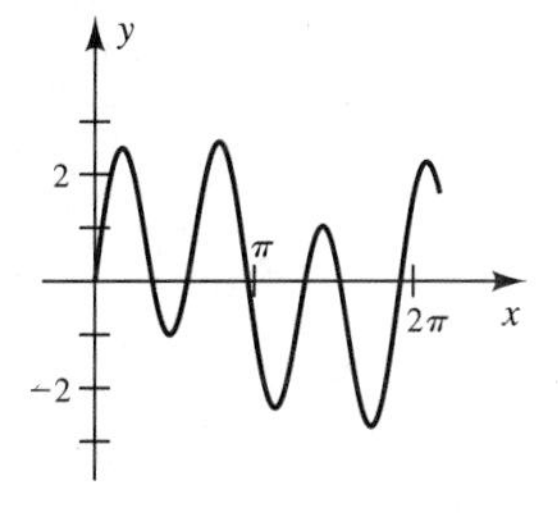

21. They are the same.

23.

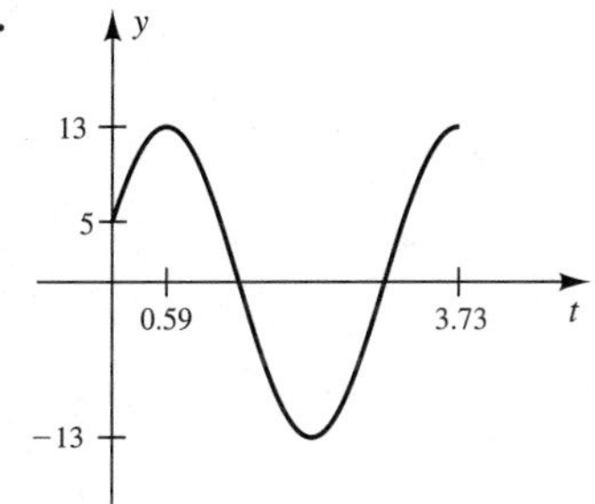

25.

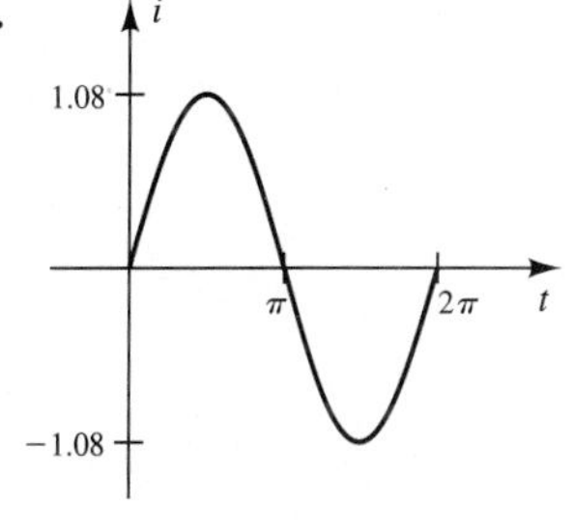

Graphing Calculator Exercises

1.

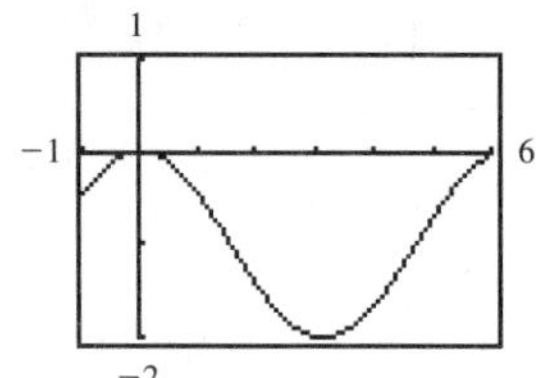

3.

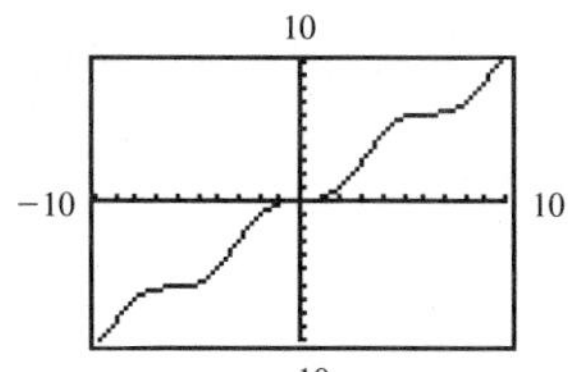

5.

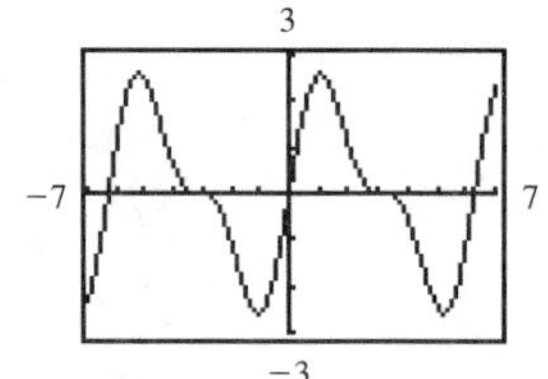

7.

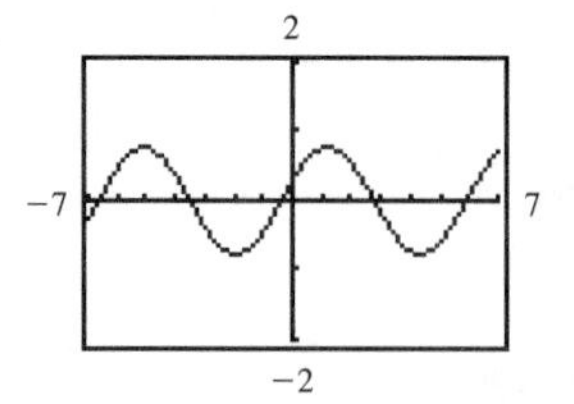

Sections 5.7 and 5.8

1. 2π **3.** 2 **5.** 2 **7.** $\frac{6}{5}\pi$

9. Per. $= \frac{1}{2}\pi$, ps $= 0$,
asym. at $x = \frac{1}{4}\pi + \frac{1}{2}n\pi$

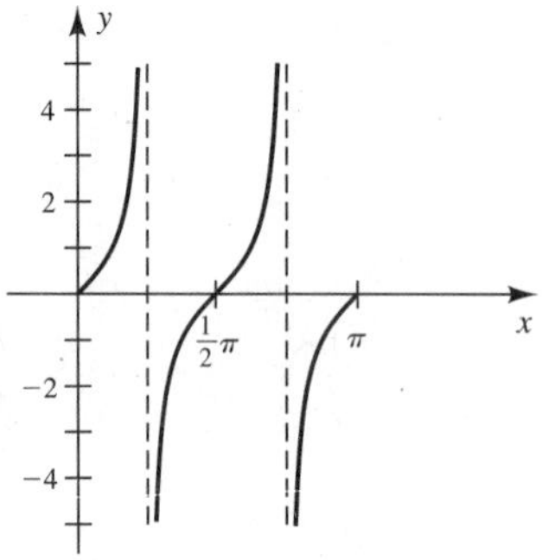

11. Per. $= \pi$, ps $= \frac{1}{4}\pi$,
asym. at $x = \frac{1}{4}\pi + n\pi$

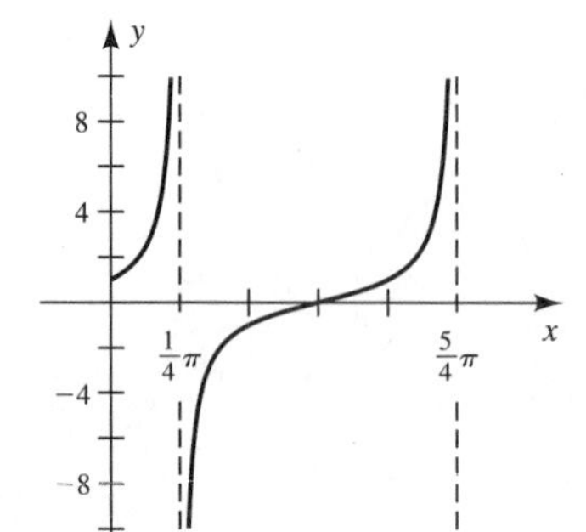

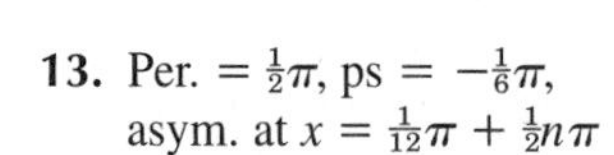

13. Per. $= \frac{1}{2}\pi$, ps $= -\frac{1}{6}\pi$,
asym. at $x = \frac{1}{12}\pi + \frac{1}{2}n\pi$

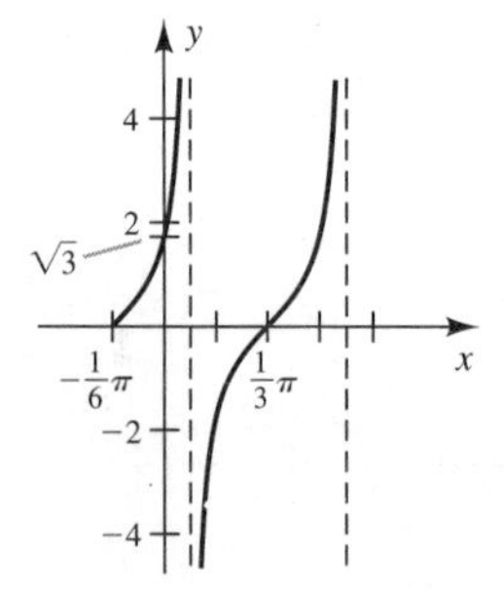

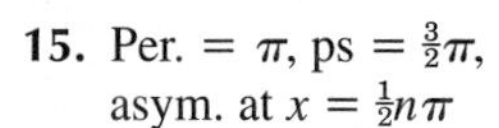

15. Per. $= \pi$, ps $= \frac{3}{2}\pi$,
asym. at $x = \frac{1}{2}n\pi$

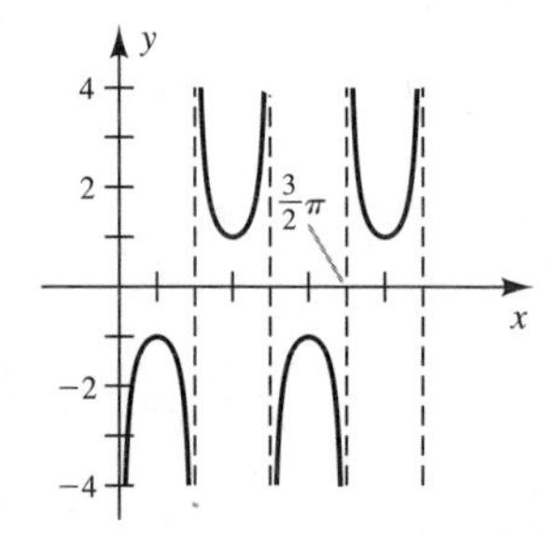

17. Per. $= \pi$, ps $= \frac{1}{4}\pi$, asym. at $x = \frac{3}{4}\pi + n\pi$

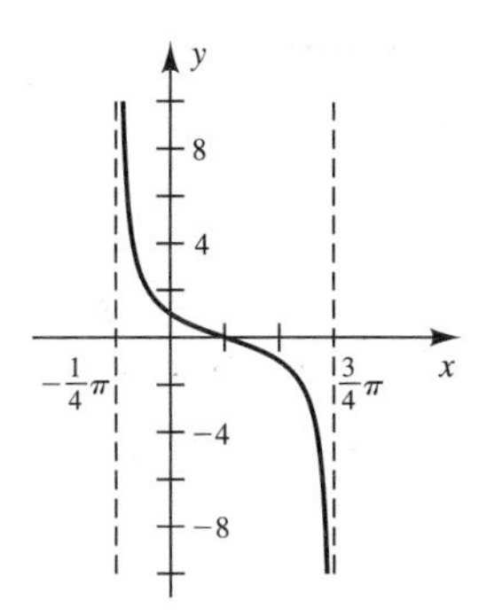

19. $\tan x = -\cot(x + \frac{1}{2}\pi)$

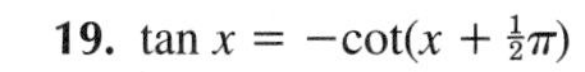

21. They are the same.

23. They are mirror reflections in the y-axis.

25.

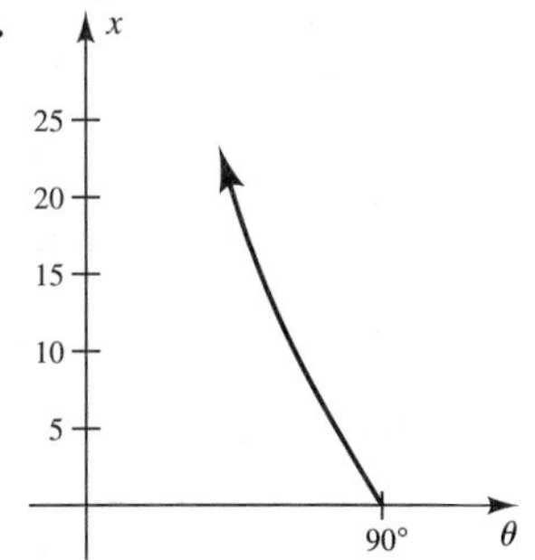

27.

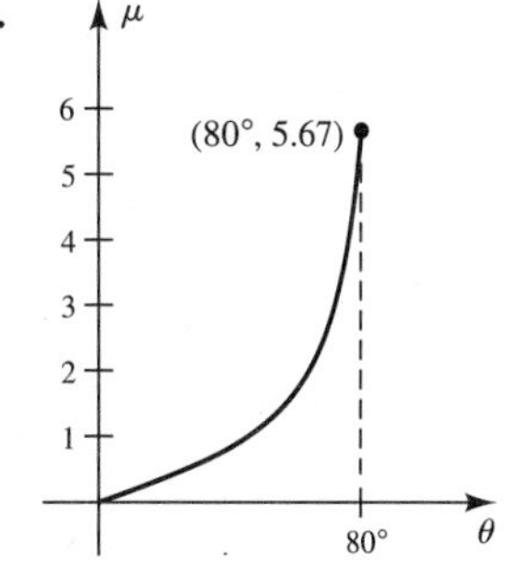

Graphing Calculator Exercises

1.

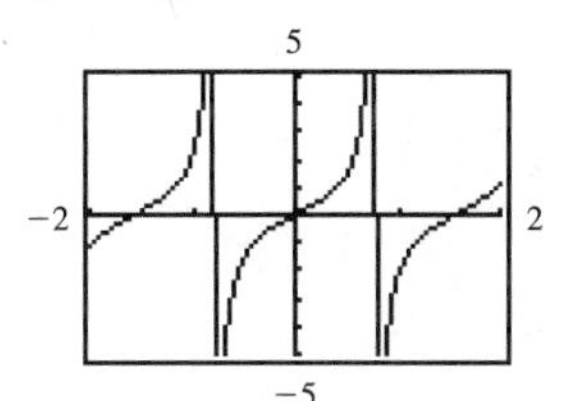

3.

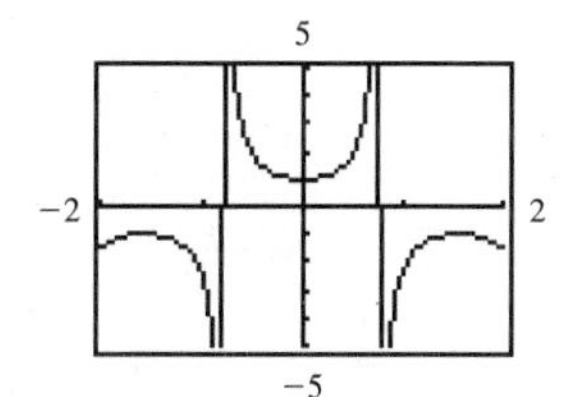

5.

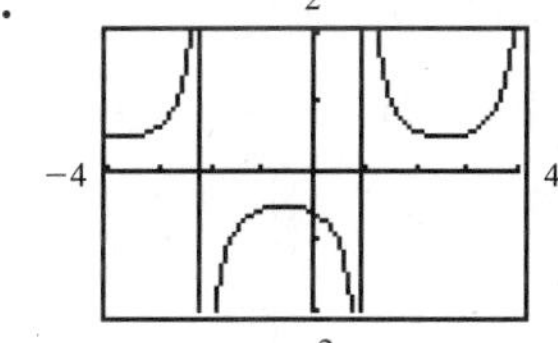

7.

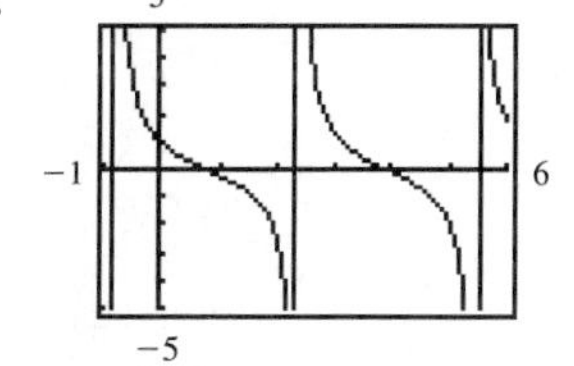

9.

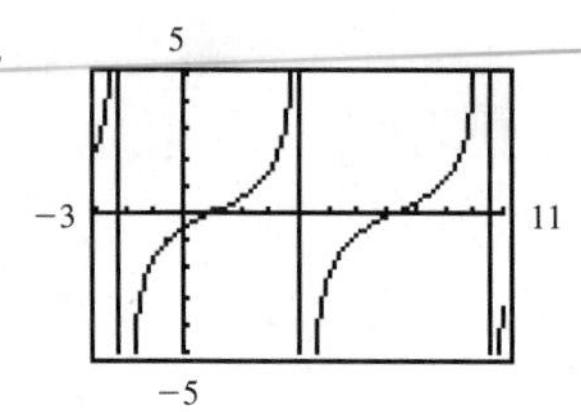

Section 5.9

3. Approaches 2

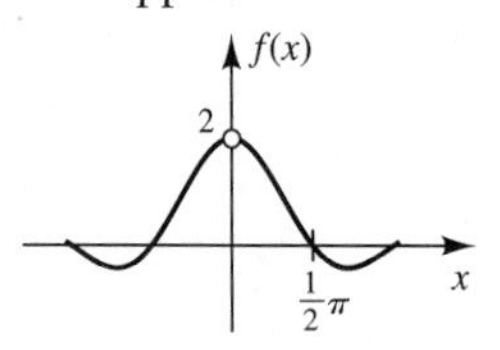

$f(0)$ is undefined.

Review Exercises for Chapter 5

1. **(a)** -1.683 **(b)** 0 **(c)** -1.683

3. $\frac{1}{18}\pi + \frac{1}{3} + \frac{2}{3}n\pi$; $\frac{5}{18}\pi + \frac{1}{3} + \frac{2}{3}n\pi$

5. $A = 1$, ps $= \frac{1}{3}\pi$, per. $= 2\pi$, avg. $= 2$

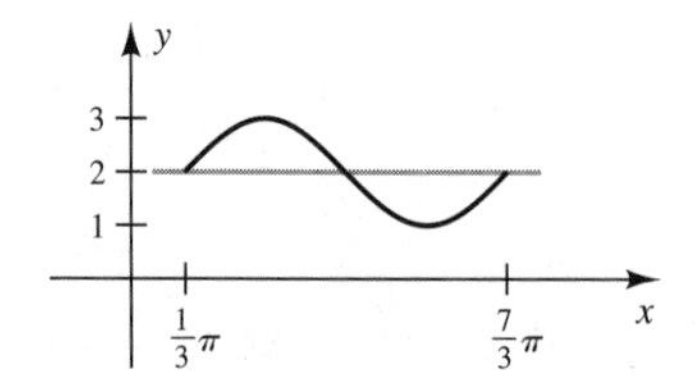

7. ps $= -\frac{1}{4}\pi$, per. $= 2\pi$

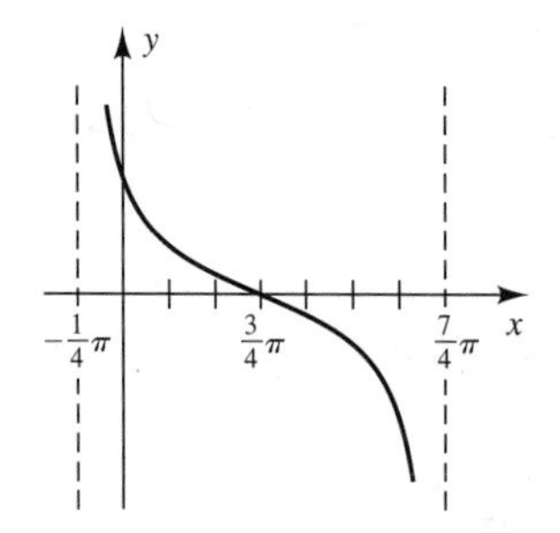

9. ps = π, per. = 6π, avg. = 2

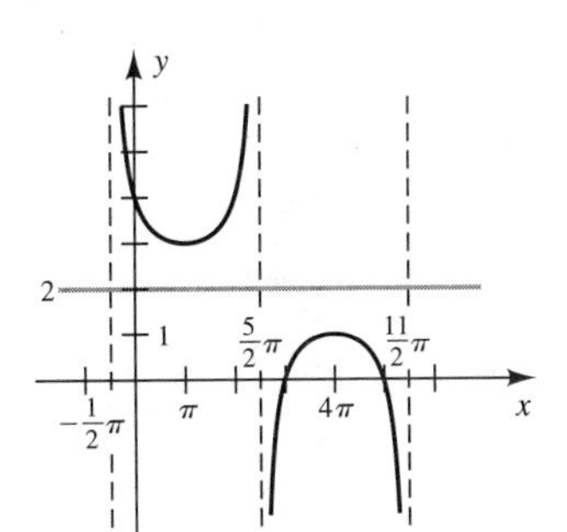

11. $A = 9$, per. = $\frac{2}{3}\pi$

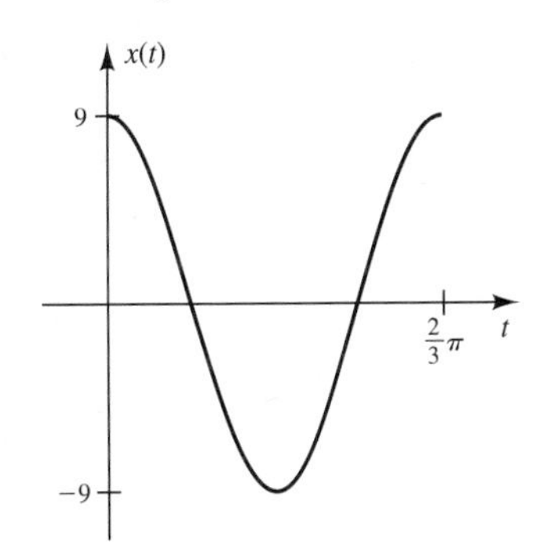

13. $x = -3 + (2n + 1)(\frac{1}{2}\pi)$

15.

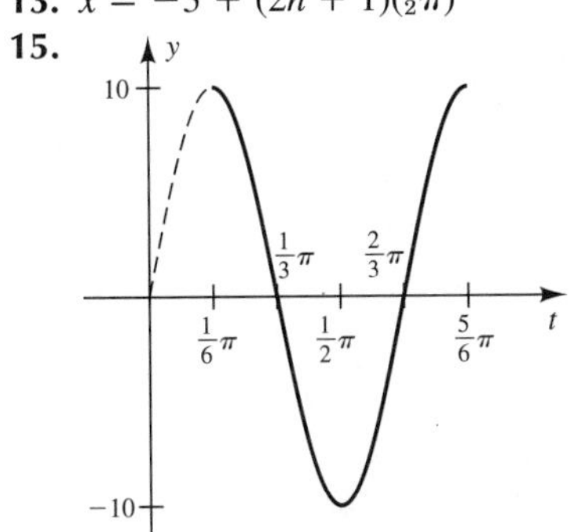

17.

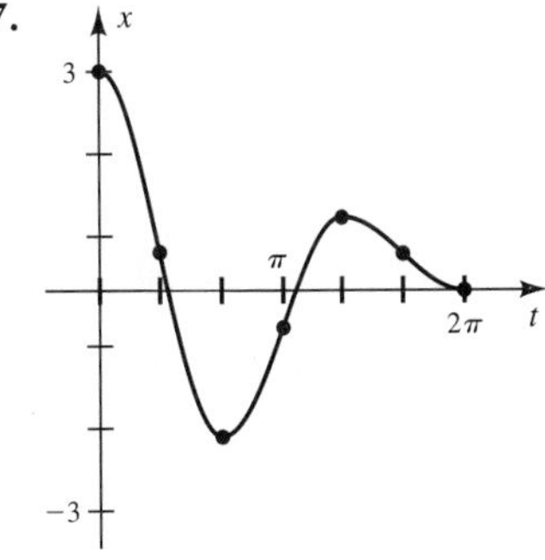

19. -1.13 cm **21.** $n\pi$, where n is an integer

23. $x = 2 \cos 2\pi t$, $y = 2 \sin 2\pi t$, $(-2, 0)$, $(0, -2)$, $(2, 0)$

Section 6.1

1. $\sec\theta$ **3.** $\sec x$ **5.** $\cot x$ **7.** $\cot x$ **9.** $-\tan^2 x$
11. $\sin x$ **13.** 1 **15.** 1 **17.** $\sec x$ **19.** $\cos x$
21. $a \sec\theta, \dfrac{x}{\sqrt{a^2 + x^2}}$ **23.** $\sin\theta$ **25.** $\sqrt{3}\cos\theta$ **29.** No

Section 6.2

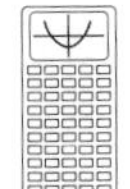

Graphing Calculator Exercises

13. Identity **15.** Not an identity **17.** Identity **19.** Identity
21. Identity

Section 6.3

1. $\frac{1}{6}\pi, \frac{5}{6}\pi$

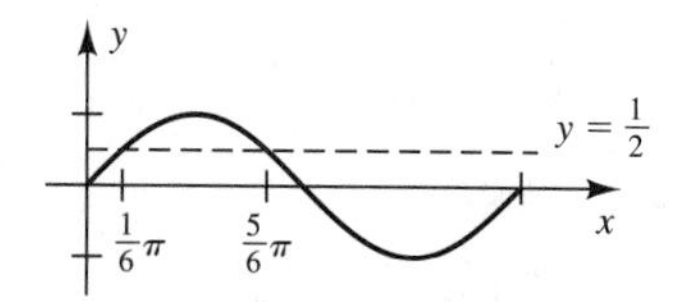

3. $\frac{1}{3}\pi$

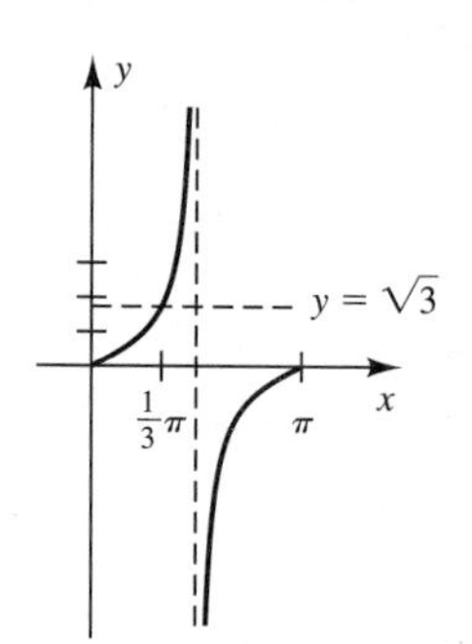

5. $\frac{1}{3}\pi, \frac{2}{3}\pi$

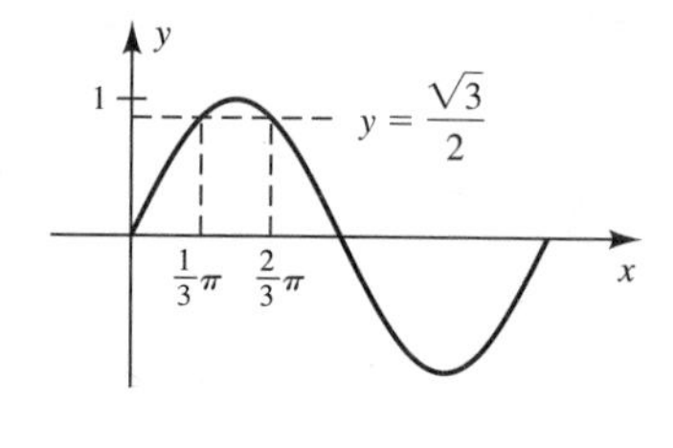

7. $\frac{7}{12}\pi, \frac{11}{12}\pi$

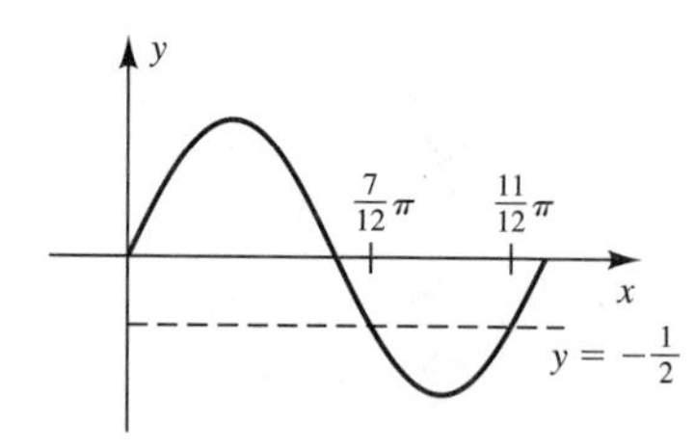

9. $\frac{1}{4}\pi, \frac{3}{4}\pi$

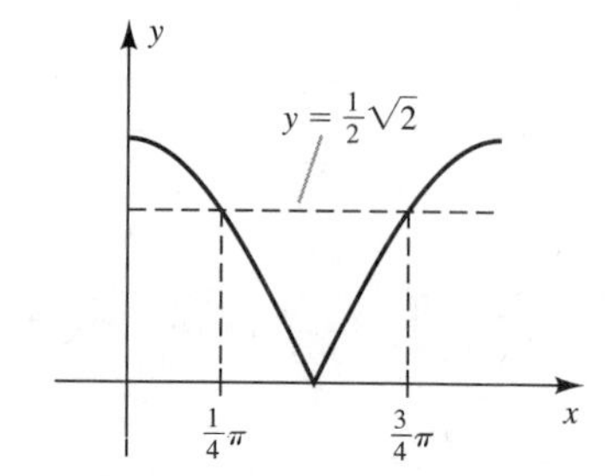

11. $\frac{7}{6}\pi, \frac{11}{6}\pi$ **13.** 0 **15.** π **17.** $0, \pi, \frac{1}{6}\pi, \frac{5}{6}\pi$
19. No solution **21.** $\frac{1}{4}\pi, \frac{5}{4}\pi$ **23.** No solution **25.** $\frac{1}{6}\pi, \frac{5}{6}\pi, \frac{3}{2}\pi$
27. $\frac{1}{2}\pi$ **29.** $0, \frac{1}{4}\pi, \pi, \frac{5}{4}\pi$ **31.** $0, \pi, \frac{1}{4}\pi, \frac{3}{4}\pi, \frac{5}{4}\pi, \frac{7}{4}\pi$ **33.** 0
35. $\frac{1}{2}\pi, \pi$ **37.** $\frac{1}{4}\pi, \frac{3}{4}\pi, \frac{5}{4}\pi, \frac{7}{4}\pi$ **39.** 45°, 135°, 225°, 315°
41. 90°, 270° **43.** 90°, 150°, 210° **45.** 0°, 180°, 240°, 300°
47. 0°, 45°, 180°, 225° **49.** 30°, 150° **51.** 120°, 300°

53. $\dfrac{5\pi}{12} + 2n\pi, \dfrac{25}{12}\pi + 2n\pi$; n an integer

55. $\dfrac{-\pi}{6} + \dfrac{2n\pi}{3}$; n an integer

57. $\dfrac{\pi}{8} - \dfrac{1}{2} + n\pi, \dfrac{3\pi}{8} - \dfrac{1}{2} + n\pi$; n an integer

59. $\dfrac{\pi}{18} - \dfrac{2}{3} + \dfrac{2n\pi}{3}, \dfrac{5\pi}{18} - \dfrac{2}{3} + \dfrac{2n\pi}{3}, \dfrac{7\pi}{18} - \dfrac{2}{3} = \dfrac{2n\pi}{3}, \dfrac{11\pi}{18} - \dfrac{2}{3} + \dfrac{2n\pi}{3};$ n an integer

61. $-1.114 + 2n\pi, 0.256 + 2n\pi; 2.028 + 2n\pi, 3.397 + 2n\pi$, n an integer

63. 0.58, 2.56, 3.72, 5.70 **65.** 0, 3.14, 3.48, 5.94

67. 0.875, 2.267, 3.591, 5.834 **69.** 0.67, 2.48

71. $\frac{2}{3} + 2n, \frac{5}{3} + 2n$, where n is an integer

73. $-\dfrac{\phi}{\omega_c} + \dfrac{n\pi}{\omega_c}$, where n is an integer

Section 6.4

1.

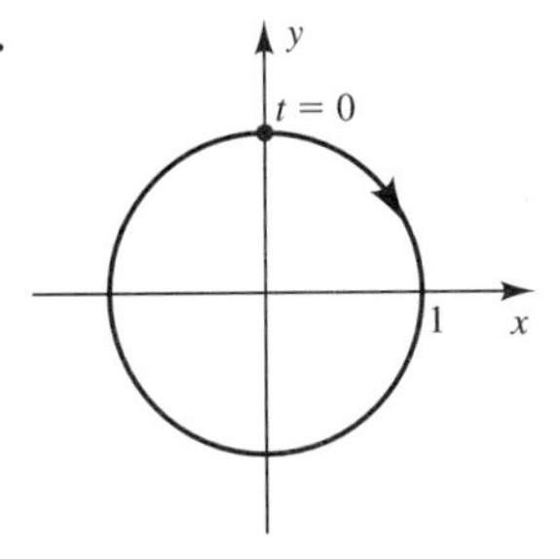

3.

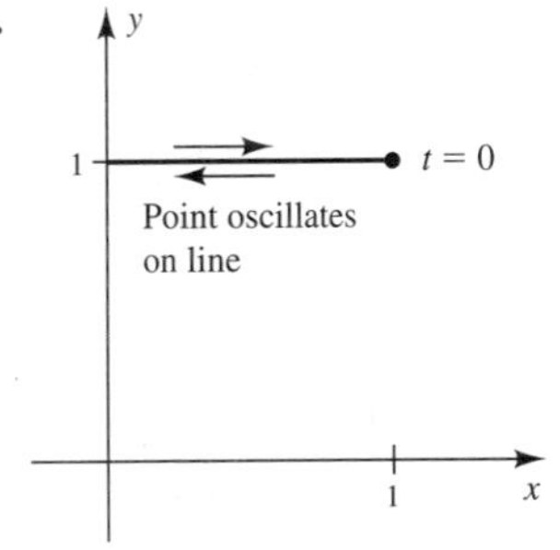

5.

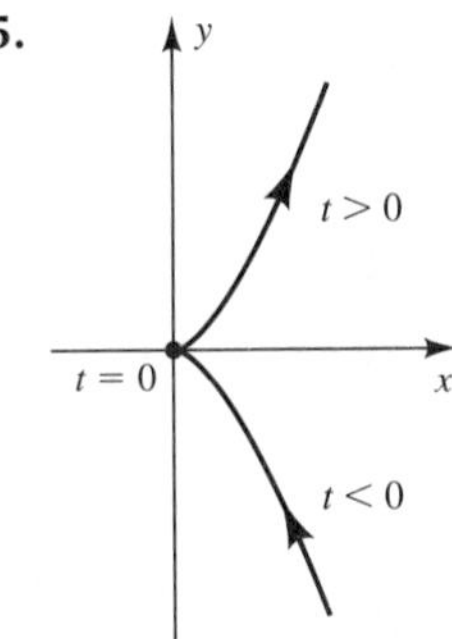

7.

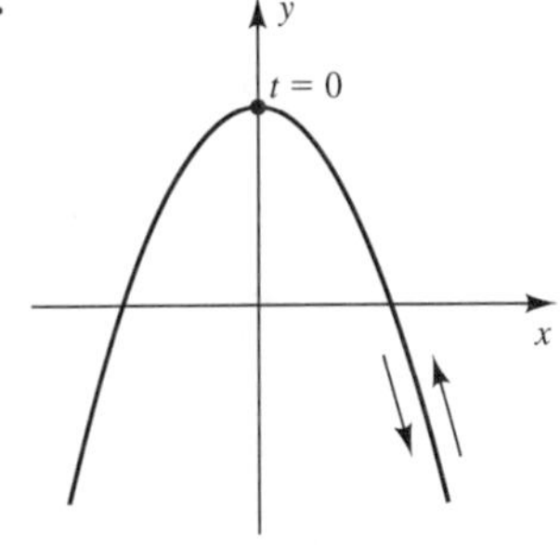

9.

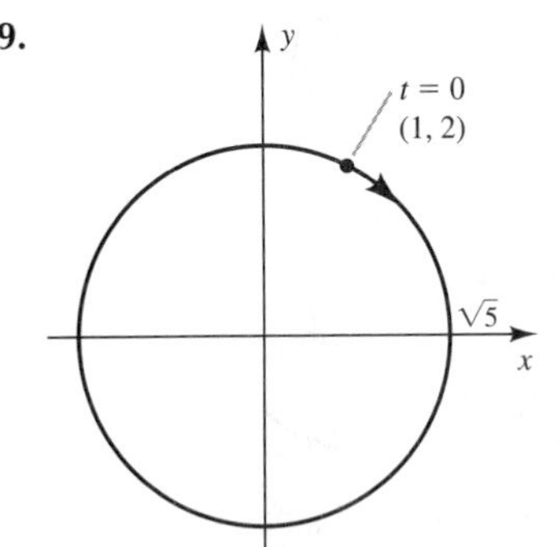

11. $x^2 + y^2 = 1$

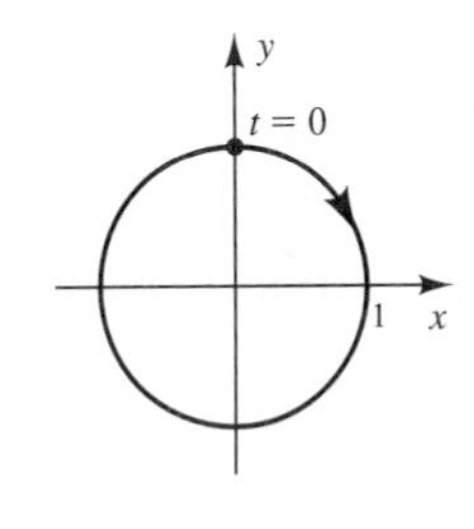

13. $y = x + 1$

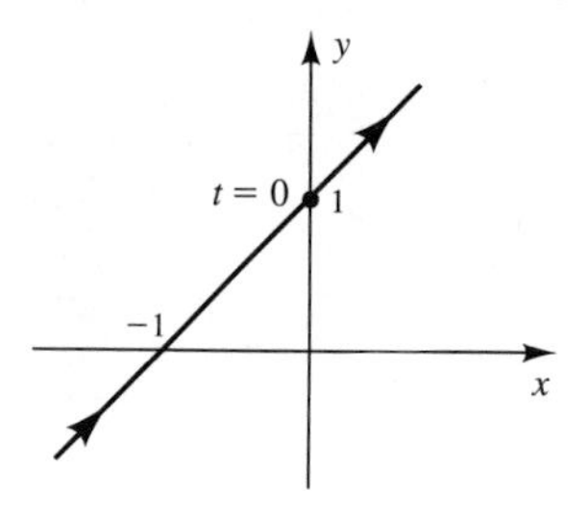

15. $y = x, -1 \leq x \leq 1$

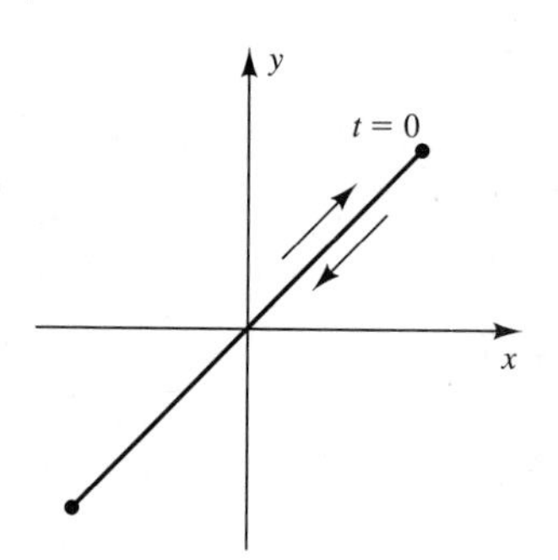

17. $x = \sin^2 y$

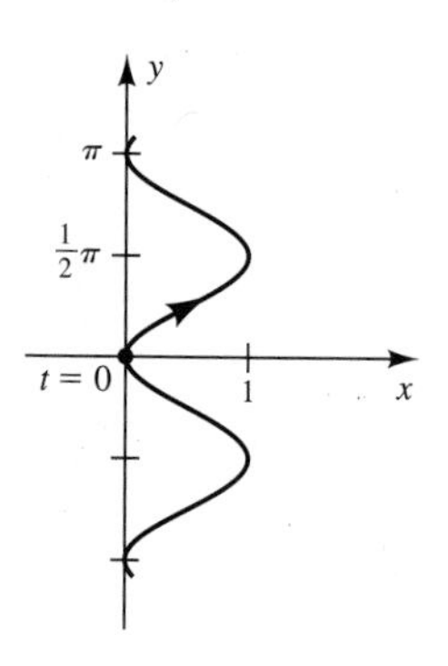

19. $(y - 1) = (x + 3)^2$

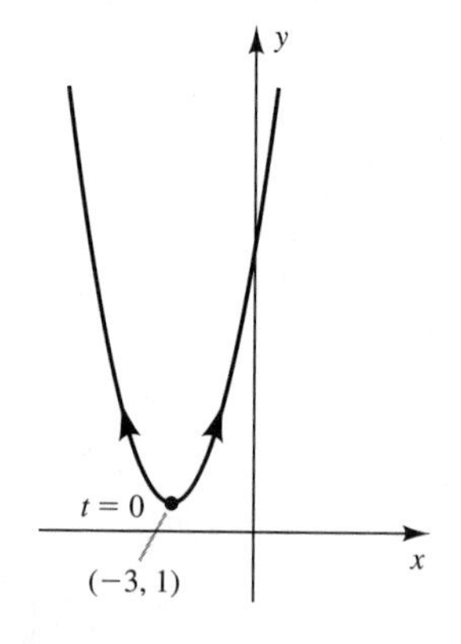

21.

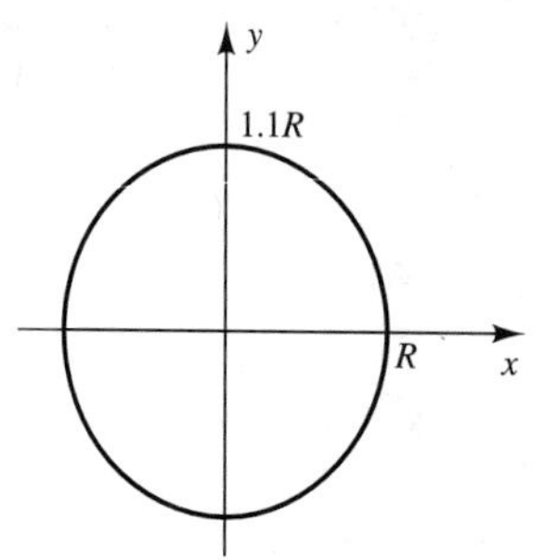

23.

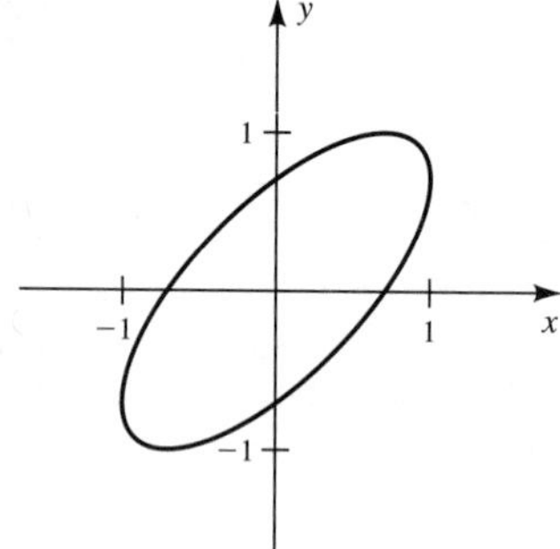

25.

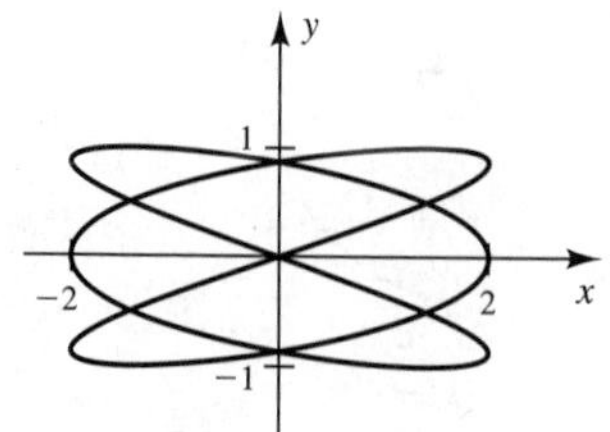

Graphing Calculator Exercises

1. See Exercise 1. **3.** See Exercise 3. **5.** See Exercise 7.
7. See Exercise 23. **9.** He should increase the angle by about 4°.

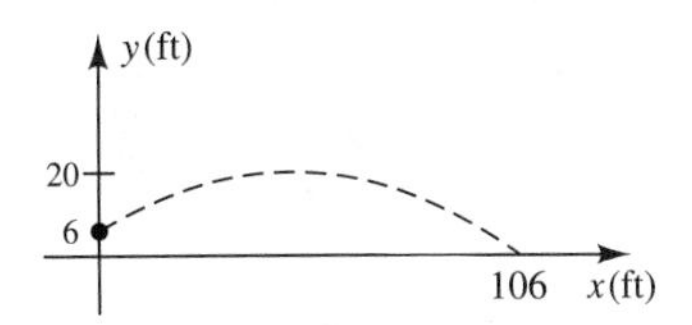

Section 6.5

1. 0

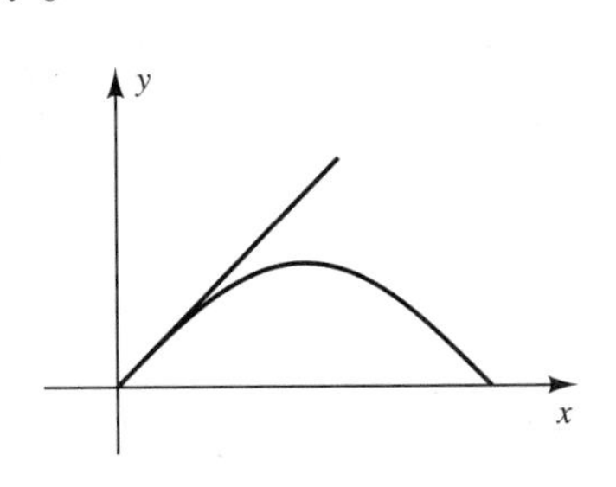

3. 1.1, 2.8

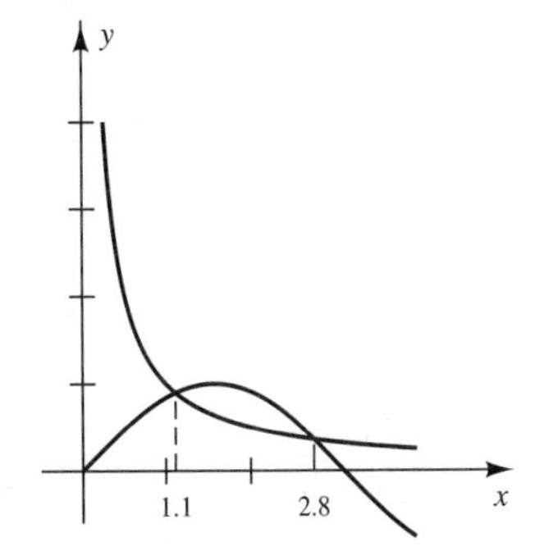

5. 0.8

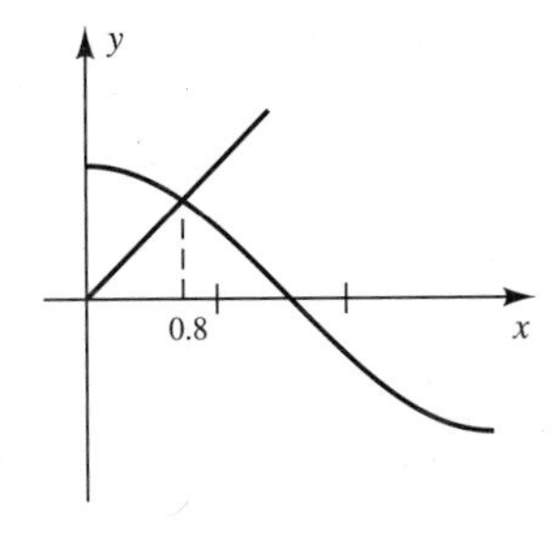

7. 0, 4.67

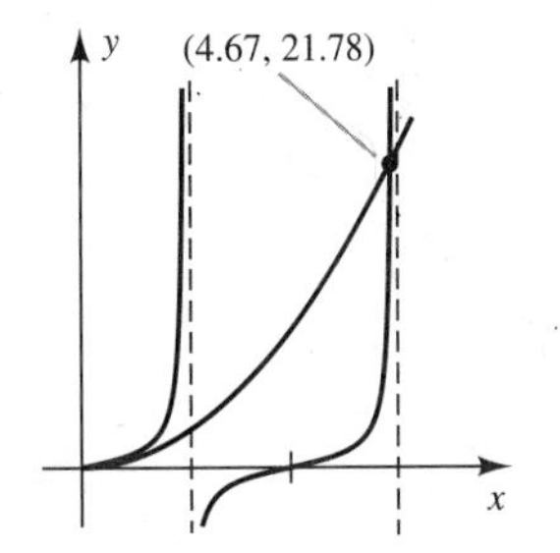

9. 0.7, 2.5

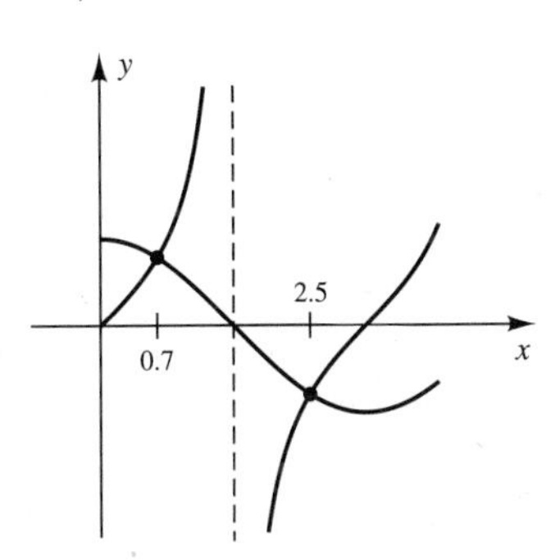

11. $\frac{1}{6}\pi \leq x \leq \frac{5}{6}\pi$

13. $\frac{1}{4}\pi < x \le \frac{3}{4}\pi, \frac{5}{4}\pi \le x \le \frac{7}{4}\pi$
15. $\frac{1}{4}\pi < x < \frac{1}{2}\pi,\ \pi < x < \frac{5}{4}\pi, \frac{3}{2}\pi < x < 2\pi$
17. $-0.4 < x < 0, x > 0.4$ **19.** $0 \le x \le \frac{1}{4}\pi, \frac{3}{4}\pi \le x \le \frac{5}{4}\pi, \frac{7}{4}\pi \le x \le 2\pi$
21. $0 < x < 1.16$

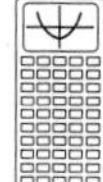

Graphing Calculator Exercises

1. 2, 4.28 **3.** 1.25, 5.01 **5.** 1.26, 5.04 **7.** 1.26, 5.04
9. 0.74 **11.** 0.87 **13.** $0 < x < 0.78, 3.9 < x < 6.3$
15. $1.11 < x < 2.77$

Review Exercises for Chapter 6

11. $\frac{4}{3}\pi, \frac{5}{3}\pi$ **13.** $\frac{7}{6}\pi, \frac{11}{6}\pi, \frac{1}{2}\pi$ **15.** $\frac{1}{4}\pi, \frac{3}{4}\pi, \frac{5}{4}\pi, \frac{7}{4}\pi$
17. $\frac{1}{4}y^2 - x^2 = 1$

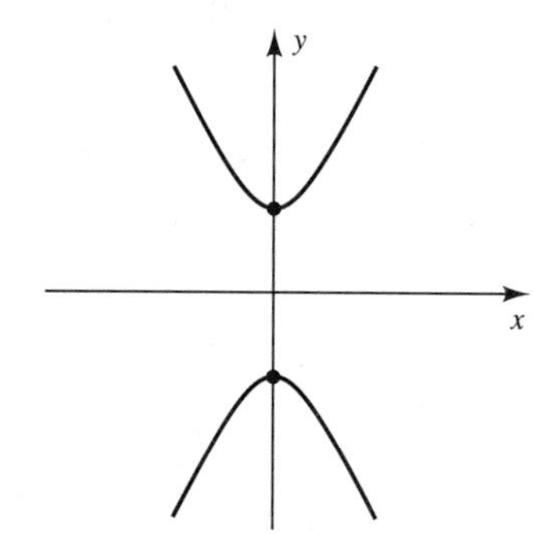

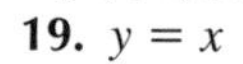
19. $y = x$

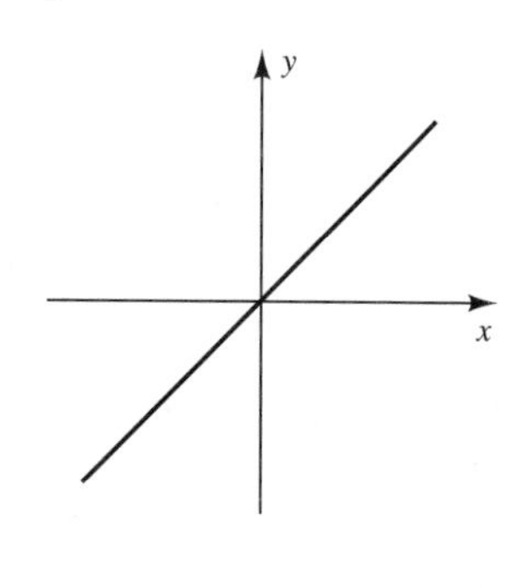

21. 0, 0.88 **23.** $0 \le x < \frac{1}{4}\pi, \frac{5}{4}\pi < x \le 2\pi$
25.

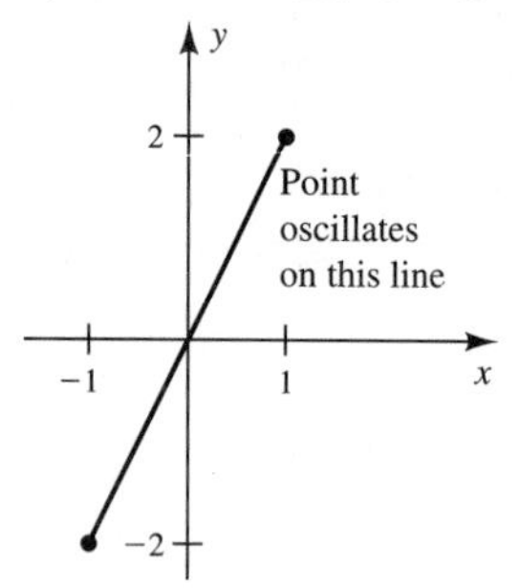

29. $4 \tan \theta \sec \theta$ **31.** 0.81

33.

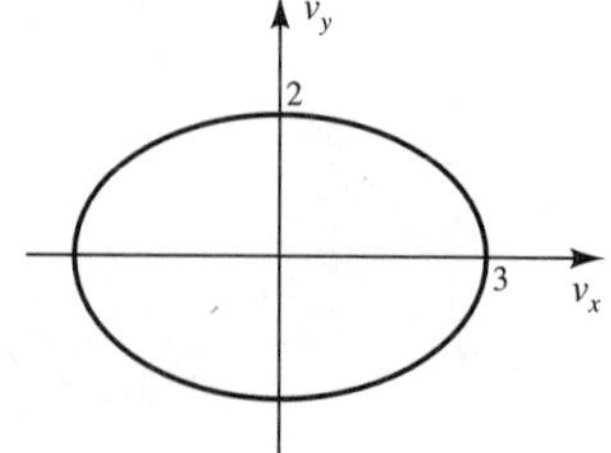

Section 7.1

7. $\frac{\sqrt{2}}{4}(1 + \sqrt{3})$ **9.** $\frac{-\sqrt{2}}{4}(1 + \sqrt{3})$ **15.** $\cos 8x$ **17.** $\cos 2\theta$

29. $\frac{\sqrt{3} + \sqrt{8}}{6}$ **31.** $\frac{3}{5}$ **33.** $\frac{5\sqrt{8} + \sqrt{11}}{18}$

35. $y = \sqrt{2} \cos(x + \frac{1}{4}\pi)$, $A = \sqrt{2}$, $\text{ps} = -\frac{1}{4}\pi$

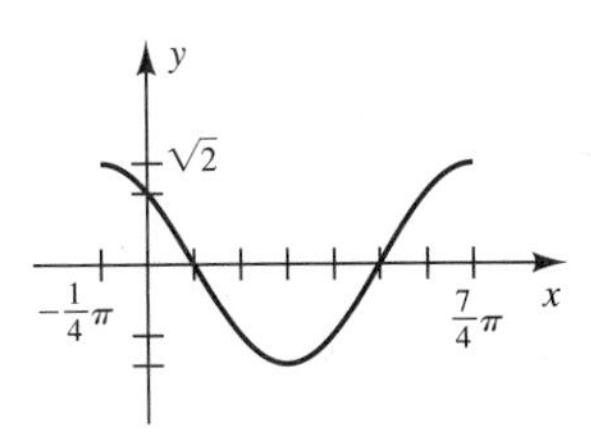

37. $y = 2 \cos 2(x - \frac{1}{6}\pi)$, $A = 2$, $\text{ps} = \frac{1}{6}\pi$

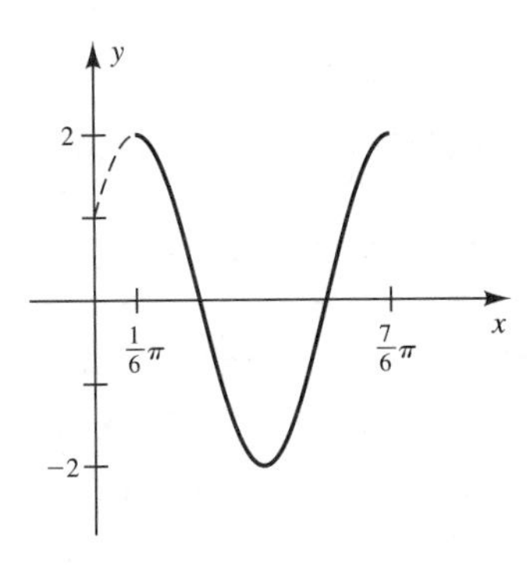

45. $78 \cos 2\pi ft + 78\sqrt{3} \sin 2\pi ft$

47. $c_1 = A \cos \frac{2\pi v}{\lambda}$

$c_2 = A \sin \frac{2\pi v}{\lambda}$

49. $y = 3 \cos(2\pi x - ct)$

Graphing Calculator Exercise

1. **(b)**, **(c)**, and **(f)** are not identities. The others are identities.

Section 7.2

1. $\frac{\sqrt{2}}{4}(1 + \sqrt{3})$ **3.** $\frac{\sqrt{2}}{4}(\sqrt{3} + 1)$ **5.** $\frac{\sqrt{3} - 1}{\sqrt{3} + 1}$

25. $\tan x$ **27.** $\tan(x + y + z)$ **29.** 36/325, 36/323

31. $\sqrt{2}\sin(2x + \frac{1}{4}\pi)$, ps $= -\frac{1}{8}\pi$

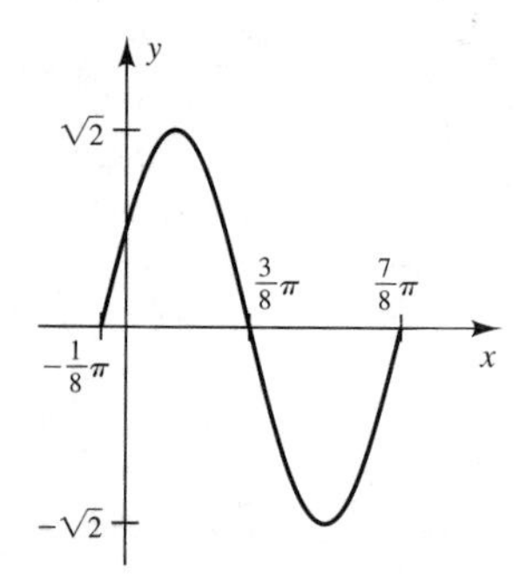

33. $2\sin(\pi x + \frac{1}{6}\pi)$, ps $= -\frac{1}{6}$

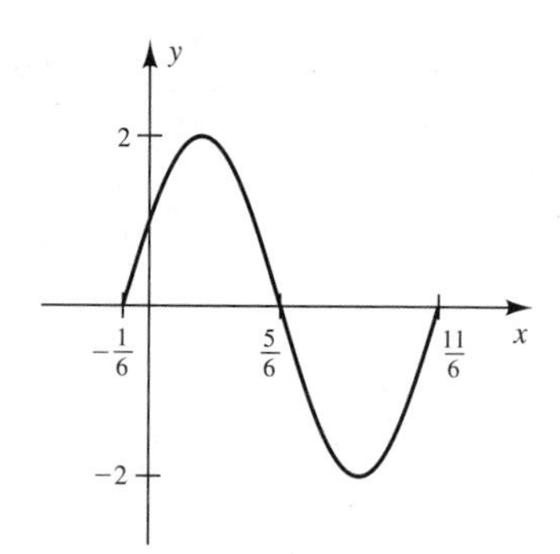

41. $25\sqrt{2}(\sqrt{3} + 1)$ **43.** $v(t) = 30\sqrt{3}\cos 2\pi ft + 30\sin 2\pi ft$

45. $\tan\beta = \dfrac{m_2 - m_1}{1 + m_1 m_2}$

Graphing Calculator Exercise

1.

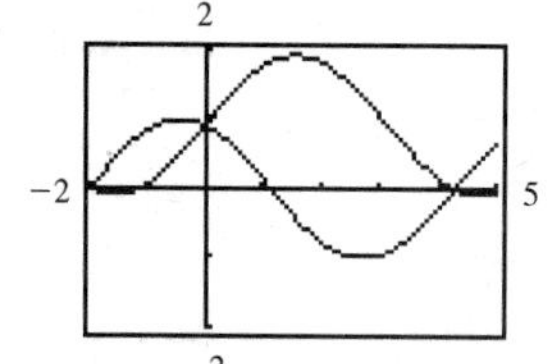

Section 7.3

1. $\sin 6x$ **3.** $-\cos 8x$ **5.** $\tan \frac{1}{3}x$

7. max $= \frac{1}{2}$ at $\frac{1}{8}\pi$

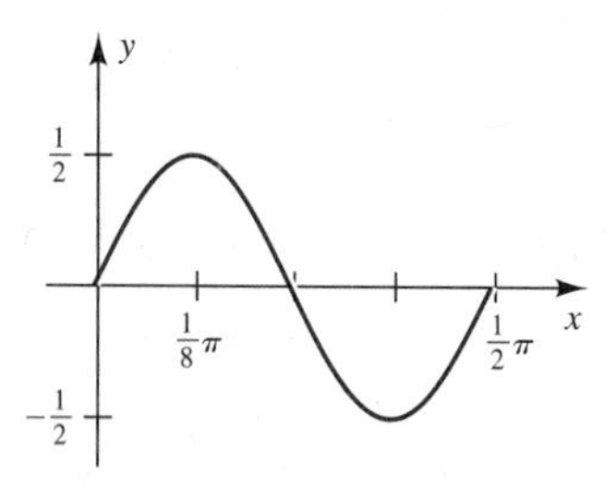

9. Undefined at $0, \frac{1}{2}\pi, \pi$

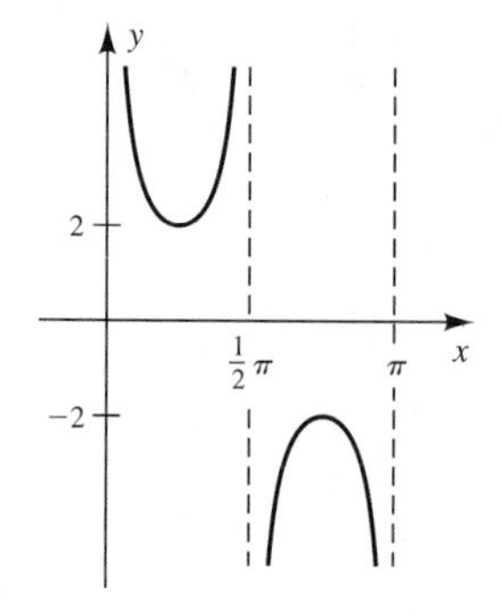

11. Undefined at $x = 0, \frac{1}{2}\pi, \pi$; per. $= \pi$ (The graph is the same as the one for Exercise 9.)

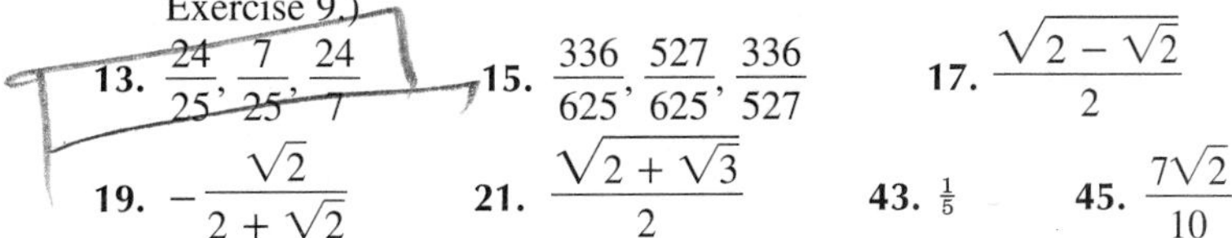

13. $\frac{24}{25}, \frac{7}{25}, \frac{24}{7}$ **15.** $\frac{336}{625}, \frac{527}{625}, \frac{336}{527}$ **17.** $\frac{\sqrt{2-\sqrt{2}}}{2}$

19. $-\frac{\sqrt{2}}{2+\sqrt{2}}$ **21.** $\frac{\sqrt{2+\sqrt{3}}}{2}$ **43.** $\frac{1}{5}$ **45.** $\frac{7\sqrt{2}}{10}$

47. $\frac{1}{4}\pi, \frac{5}{4}\pi$ **49.** $0, \pi, 2\pi$ **51.** $\frac{3}{8}\pi, \frac{7}{8}\pi, \frac{11}{8}\pi, \frac{15}{8}\pi$

53. $\frac{1}{6}\pi, \frac{1}{3}\pi, \frac{2}{3}\pi, \frac{5}{6}\pi, \frac{4}{3}\pi, \frac{7}{6}\pi, \frac{5}{3}\pi, \frac{11}{6}\pi$

65. $\cos\theta = \sqrt{\frac{\sqrt{5}+1}{2\sqrt{5}}}$, $\sin\theta = \sqrt{\frac{\sqrt{5}-1}{2\sqrt{5}}}$ **67.** $R = 16v_0^2 \sin 2\theta$

Section 7.4

1. $2\sin 2\theta\cos\theta$ **3.** $2\sin 5x\cos 3x$ **5.** $-2\sin 40°\sin 10°$

7. $2\cos\frac{1}{2}\sin\frac{1}{4}$ **9.** $\frac{1}{2}[\sin(\frac{3}{2}x) - \sin(\frac{1}{2}x)]$ **11.** $\frac{1}{2}(\cos 8x + \cos 4x)$

19. $\frac{-2\sin\frac{1}{2}(2x+h)\sin\frac{1}{2}h}{h}$ **21.** $0, \frac{1}{6}\pi, \frac{1}{2}\pi, \frac{5}{6}\pi, \pi$ **23.** $0, \frac{2}{5}\pi, \frac{4}{5}\pi$

25. $A = 2\cos 1$, per. $= \frac{2}{3}\pi$ **31.** $2\cos x\cos\frac{1}{2}x$

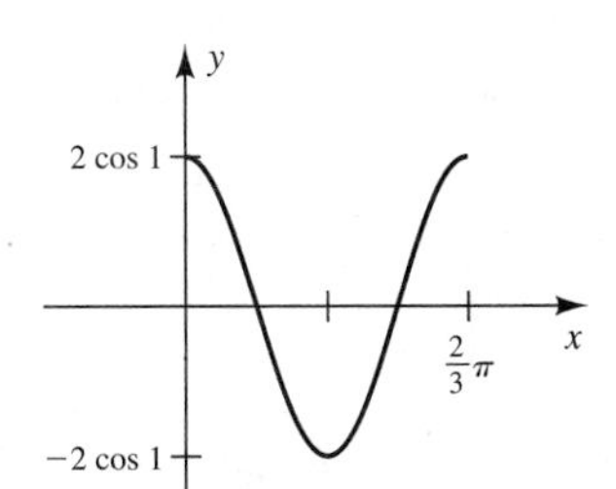

Review Exercises for Chapter 7

1. $\frac{6-4\sqrt{5}}{15}$ **3.** $-\frac{24}{7}$

21. $f(x) = 13\cos(2x + 1.176)$; $A = 13$, per. $= \pi$, ps $= -0.588$

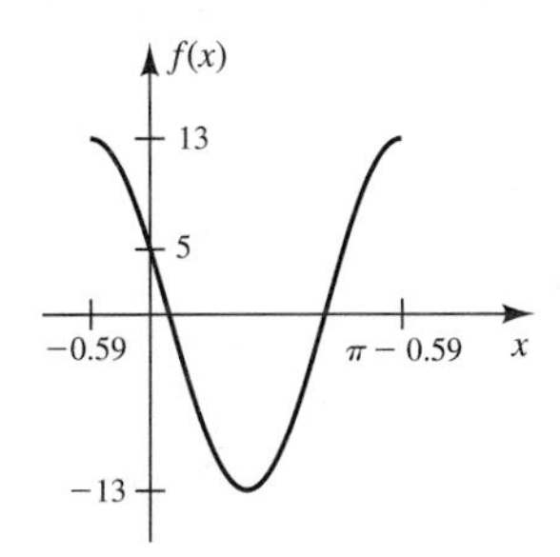

23. $f(x) = \sqrt{2}\cos(3x + \frac{1}{4}\pi)$; $A = \sqrt{2}$, per. $= \frac{2}{3}\pi$, ps $= -\frac{1}{12}\pi$

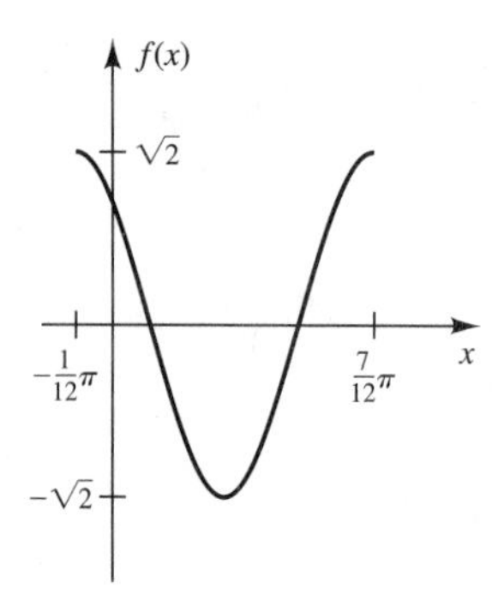

25. $f(x) = 2\cos(x - \frac{1}{3}\pi)$; $A = 2$, per. $= 2\pi$, ps $= \frac{1}{3}\pi$

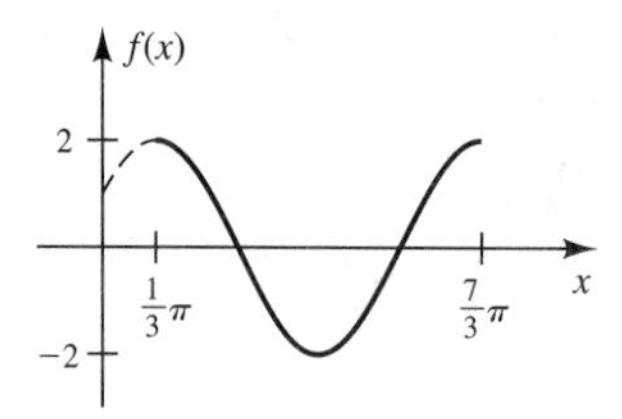

27. $-\sqrt{\dfrac{3 - \sqrt{8}}{6}}$

29. $\frac{24}{25}$ **31.** $\sin\theta = \dfrac{2\sqrt{5}}{5}$, $\cos\theta = \dfrac{\sqrt{5}}{5}$ **33.** $2\sin\frac{3}{2}x\cos\frac{1}{2}x$

Section 8.1

1. $\frac{3}{2}$ **3.** $\frac{5}{2}$ **5.** 3; none **7.** $\{(7, 3), (9, 5), (3, 7), (5, 9)\}$
9. No inverse function **11.** $\{(3, -2), (4, -1), (0, 0)\}$
13. No inverse function **15.** $y = x + 3$ **17.** $y = \dfrac{1 - x}{x}$
19. No inverse function **21.** $y = \dfrac{1 + x}{1 - x}$
23. The inverse is a function. **25.** The inverse is a function.
27. Inverse functions **29.** Not inverse functions
31. Inverse functions **33.** Inverse functions

Section 8.2

1. $\frac{1}{6}\pi$ **3.** $\frac{1}{4}\pi$ **5.** $\frac{5}{6}\pi$ **7.** $\frac{2}{3}\pi$ **9.** 0 **11.** $-\frac{1}{3}\pi$
13. $\frac{12}{13}$ **15.** $\dfrac{2}{\sqrt{5}}$ **17.** $\dfrac{\sqrt{15}}{4}$ **19.** $\frac{7}{8}$ **21.** $\dfrac{1}{\sqrt{82}}$
23. $\dfrac{2\sqrt{3} - 1}{2 + \sqrt{3}}$ **25.** $-\frac{4}{3}$ **27.** $\dfrac{x}{(1 - x^2)^{1/2}}$
29. $y\sqrt{1 - x^2} + x\sqrt{1 - y^2}$ **31.** 0.9711
33. No solution because -0.5 is not in $[0, \pi]$ **35.** 2.5722
37. 0.2350 **39.** No solution because 2.8947 is not in $\left[-\dfrac{\pi}{2}, \dfrac{\pi}{2}\right]$

49.

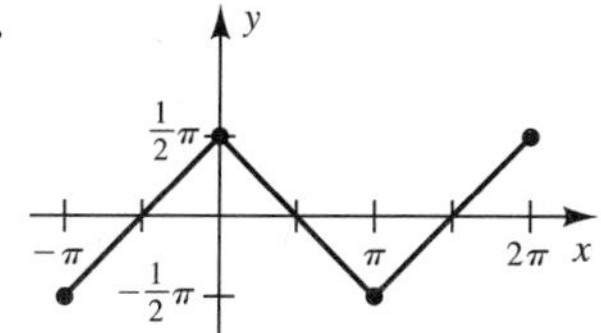

51. If x is the cosecant of a number, then $1/x$ is the sine of that number. Hence to find $\csc^{-1}x$, enter x, press the 1/x key and then the inv sin keys. A similar procedure is used to find $\sec^{-1}x$ and $\cot^{-1}x$.

53. $\theta = \frac{1}{2} \arcsin \dfrac{R}{16v_0^2}$ **55.** 1.211 **57.** 1.397 **59.** 1.407

61. 0.840 **63.** 0.8143 **65.** 0.4636

Section 8.3

1.

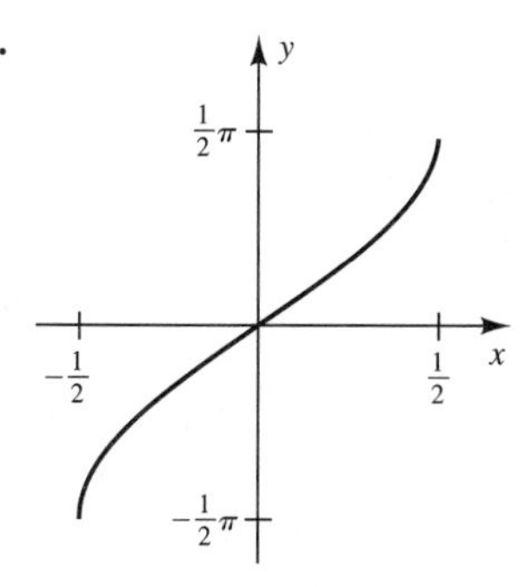

3.

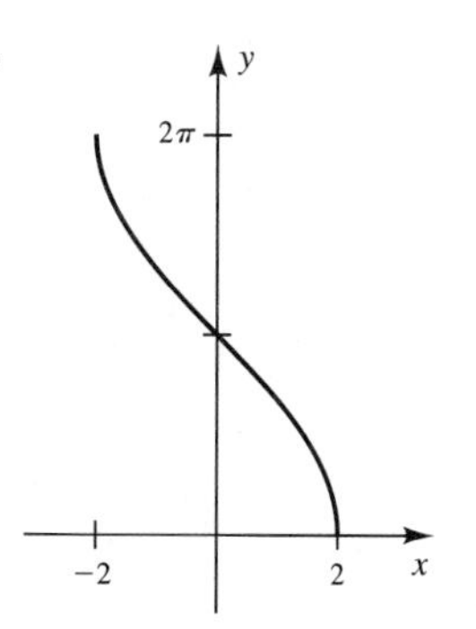

5.

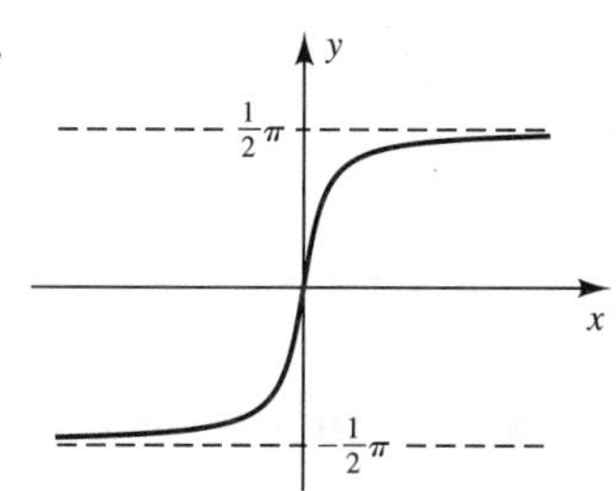

7.

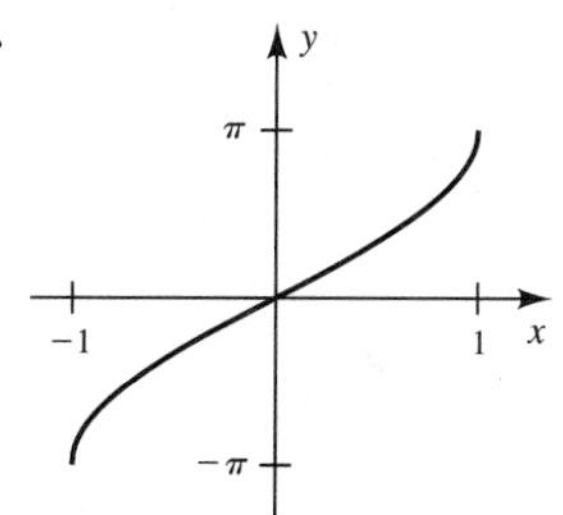

9.

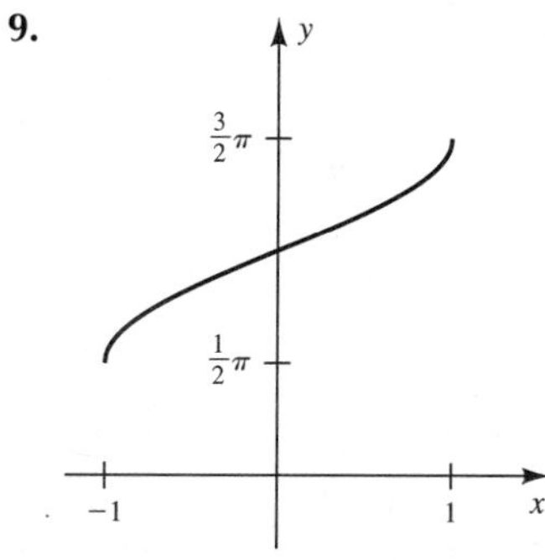

11.

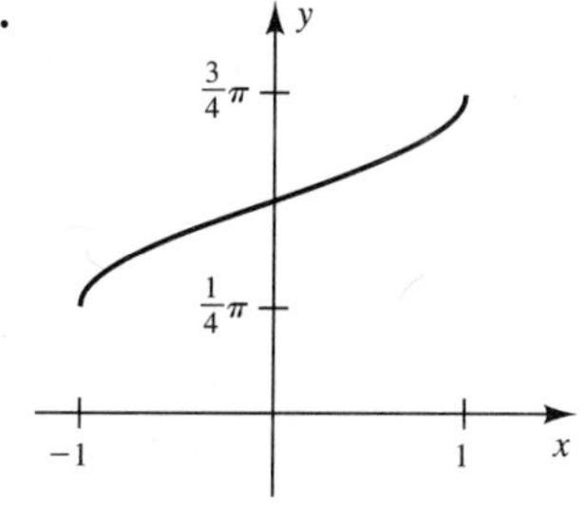

13.

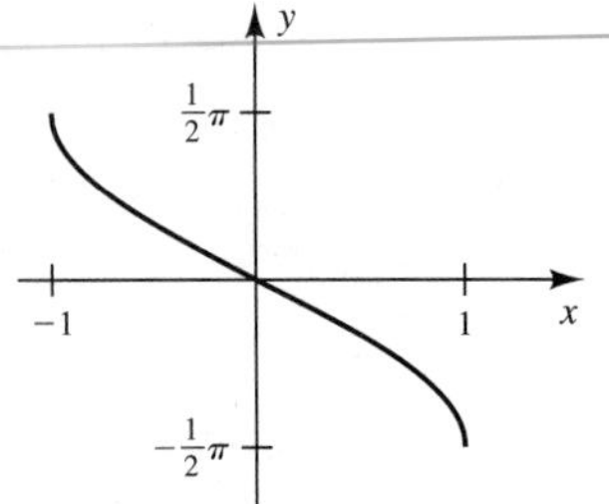

15.

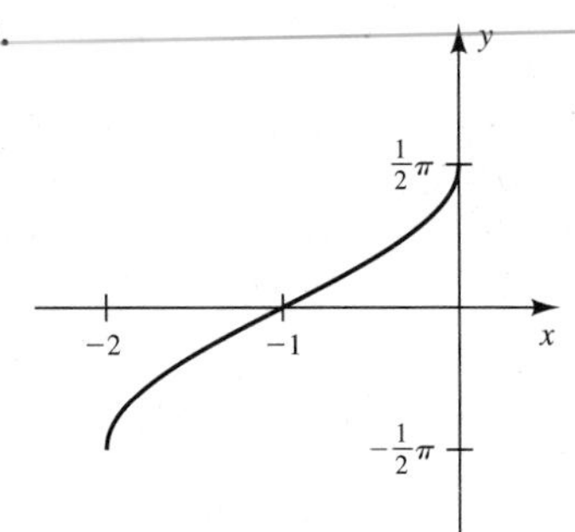

17.

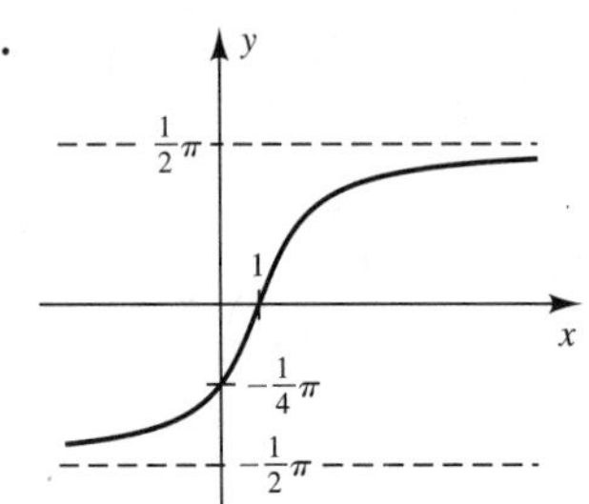

19.

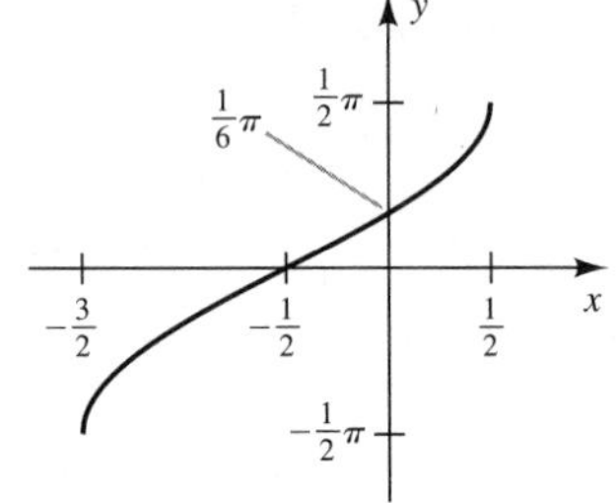

21.

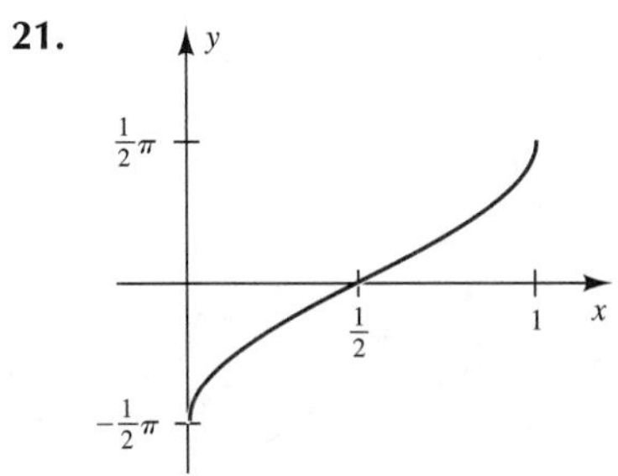

23.

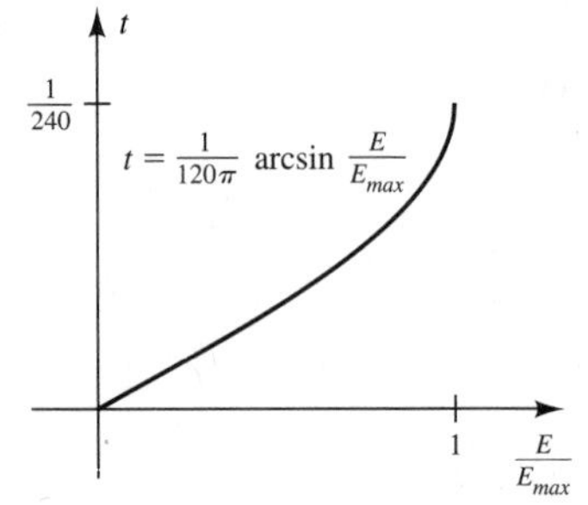

Graphing Calculator Exercises

1.

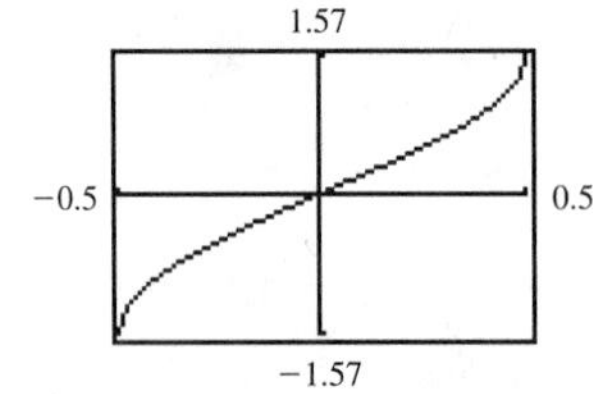

3.

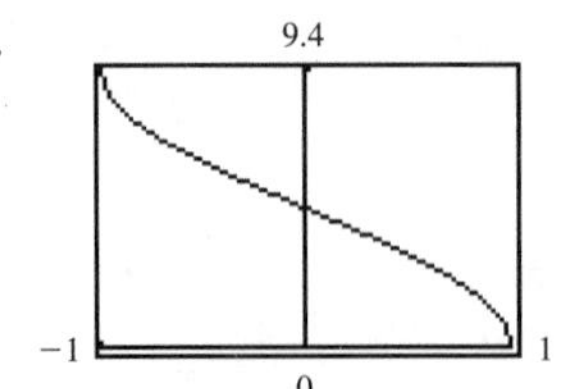

5.

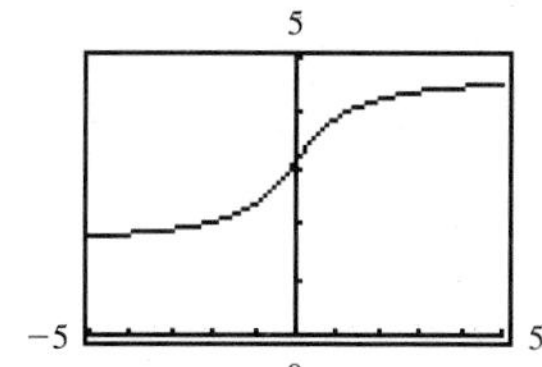

7.

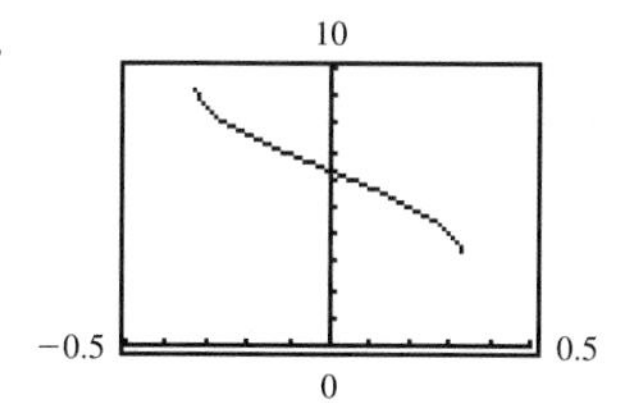

Review Exercises for Chapter 8

1. 4 **3.** $y = \dfrac{x - 5}{2}$ **5.** $y = \frac{1}{2}[5 + \arcsin(x - 3)]$ **7.** $\frac{1}{3}\pi$

9. $\dfrac{-2}{\sqrt{21}}$ **11.** $-\frac{24}{25}$ **13.** $\dfrac{-(1 + \sqrt{120})}{12}$ **15.** 1

17.

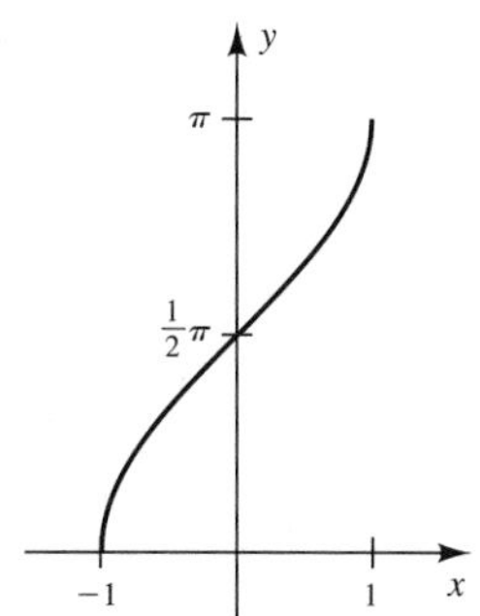

19.

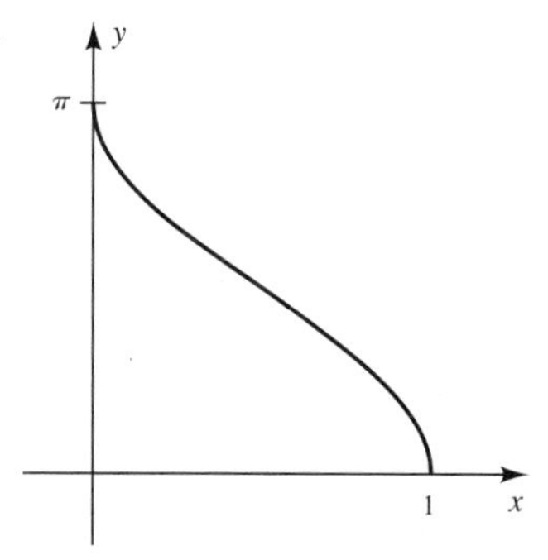

21.

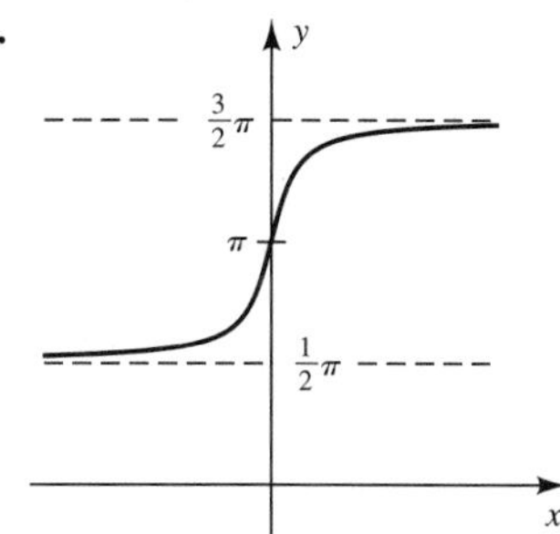

23.

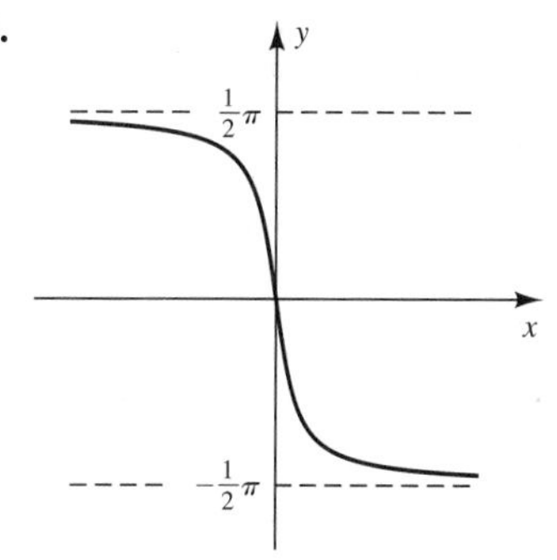

25.

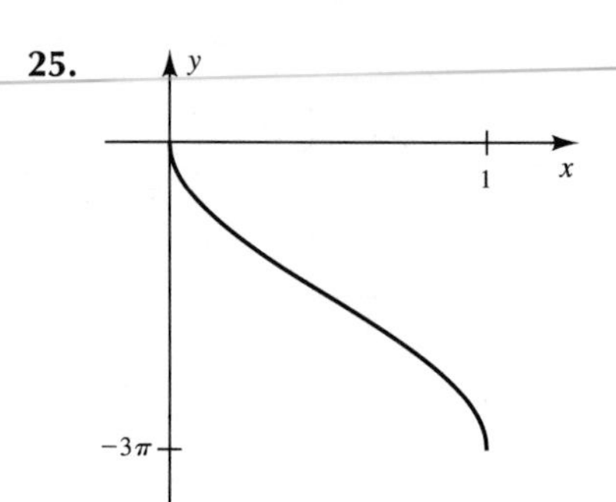

Section 9.1

1.

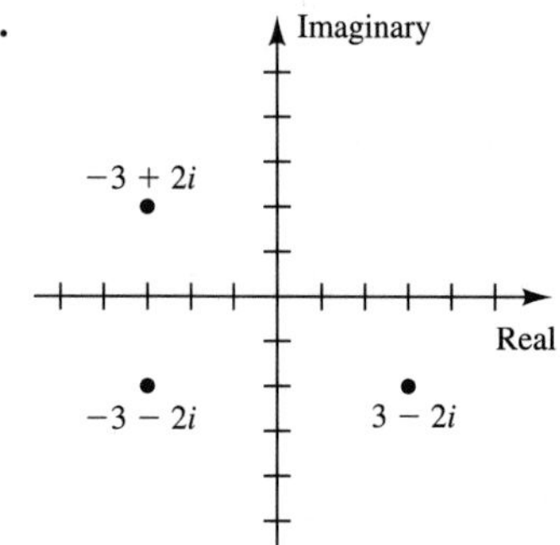

3.

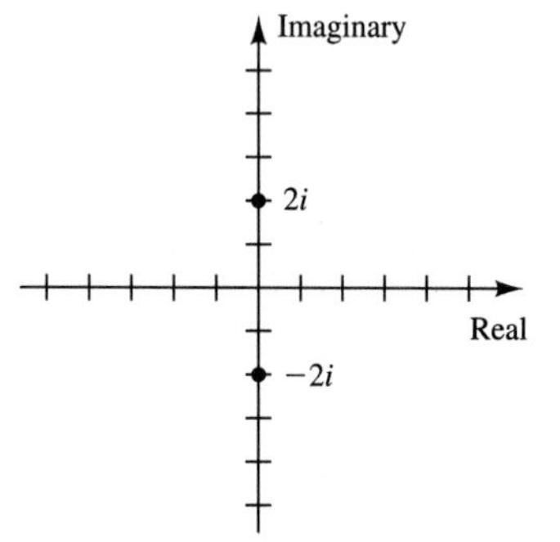

5.

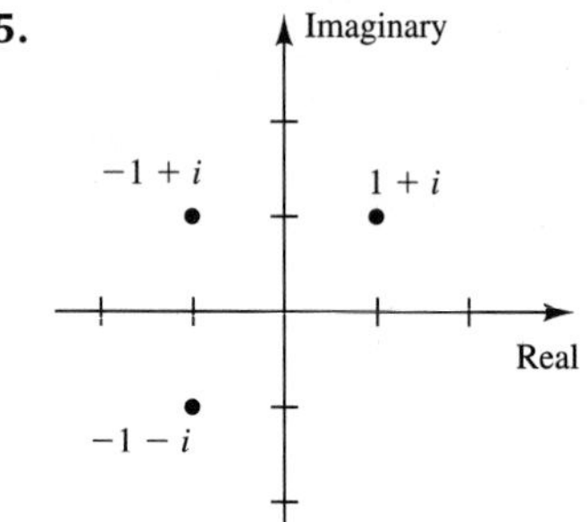

7.

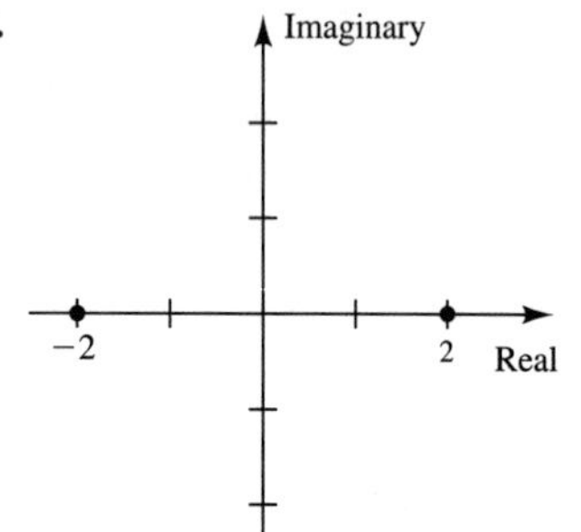

9.

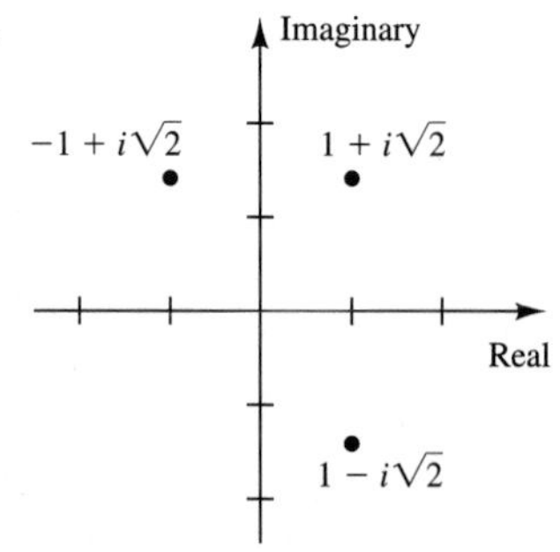

11. $7 + 5i$

13. $-2 + 3i$ **15.** 4 **17.** $3 + i$ **19.** $-7 + 22i$ **21.** 26

23. $13 + 8\sqrt{3}i$ **25.** $18 + 24i$ **27.** $\dfrac{5 - i}{2}$ **29.** $\dfrac{6 + 9i}{13}$

31. $\dfrac{-i}{5}$ **33.** $\dfrac{(-4 + 3\sqrt{2}) - (12 + \sqrt{2})i}{18}$ **37.** $5 - 2i$

39.

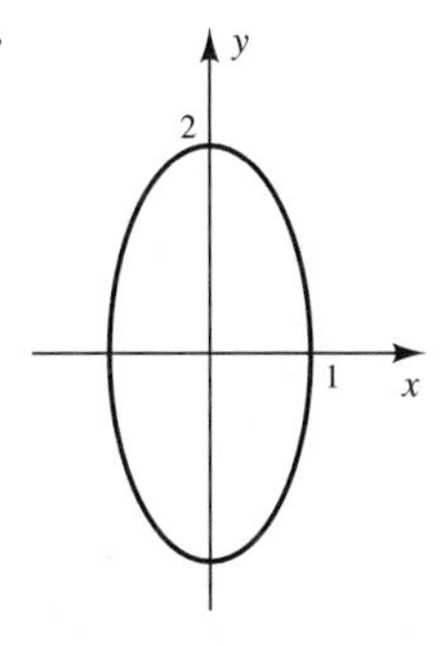

Section 9.2

1. $2 \text{ cis}(-60°)$ **3.** $3 \text{ cis } 41°49'$ **5.** $9 \text{ cis } 0°$ **7.** $5 \text{ cis}(-53.1°)$

9. $\sqrt{61} \text{ cis}(-50.2°)$ **11.** $\sqrt{3} + i$ **13.** $\dfrac{5\sqrt{2}}{2}(-1 + i)$

15. $\dfrac{-3 - \sqrt{3}i}{2}$ **17.** $\dfrac{3 - 3\sqrt{3}i}{2}$ **19.** $9.397 + 3.40i$

21. $12 \text{ cis } 90°$ **23.** $2 \text{ cis } 330°$ **25.** $20 \text{ cis } 135°$ **27.** $10 \text{ cis } 23.1°$

29. $5 \text{ cis}(-60°)$ **31.** $2 \text{ cis } 7.5°$ **33.** $\dfrac{\sqrt{2}}{2} \text{ cis}(-75°)$

35. $\dfrac{4}{\sqrt{2}} \text{ cis}(-45°)$

39. $p = 1000 \text{ cis}(-1.2t)$; $p = 1000 \text{ cis}(-1.2)$; $1000 \cos(-1.2)$; $1000 \sin(-1.2)$

Section 9.3

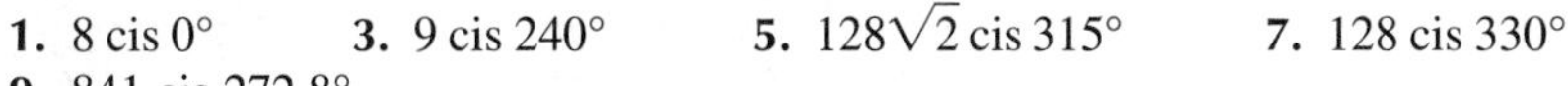

1. $8 \text{ cis } 0°$ **3.** $9 \text{ cis } 240°$ **5.** $128\sqrt{2} \text{ cis } 315°$ **7.** $128 \text{ cis } 330°$

9. $841 \text{ cis } 272.8°$

11. $1 \text{ cis } 0° = 1$
$1 \text{ cis } 72° = .3090 + .9511i$
$1 \text{ cis } 144° = -.8090 + .5878i$
$1 \text{ cis } 216° = -.8090 - .5878i$
$1 \text{ cis } 288° = .3090 - .9511i$

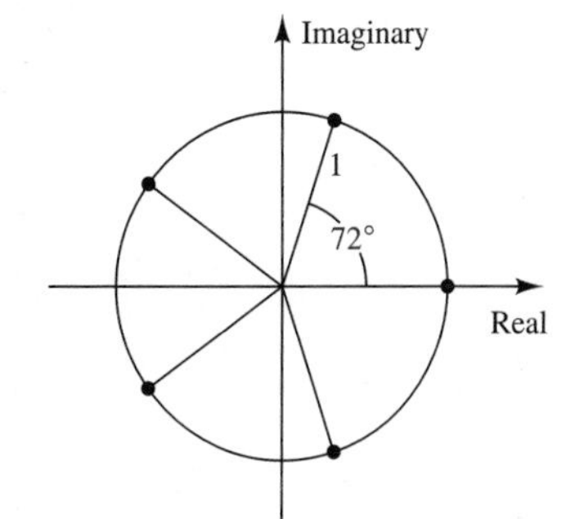

13. $1 \text{ cis } 22.5° = .9239 + .3827i$
$1 \text{ cis } 112.5° = -.3827 + .9239i$
$1 \text{ cis } 202.5° = -.9239 - .3827i$
$1 \text{ cis } 292.5° = .3827 - .9239i$

15. $\sqrt{2} \text{ cis } 22.5°$
$\sqrt{2} \text{ cis } 202.5°$

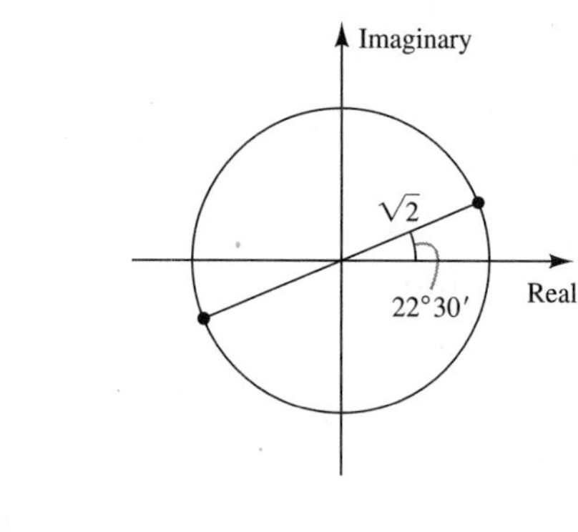

17. $\sqrt[6]{2} \text{ cis}(25° + M \cdot 60°), M = 0, 1, 2, 3, 4, 5$

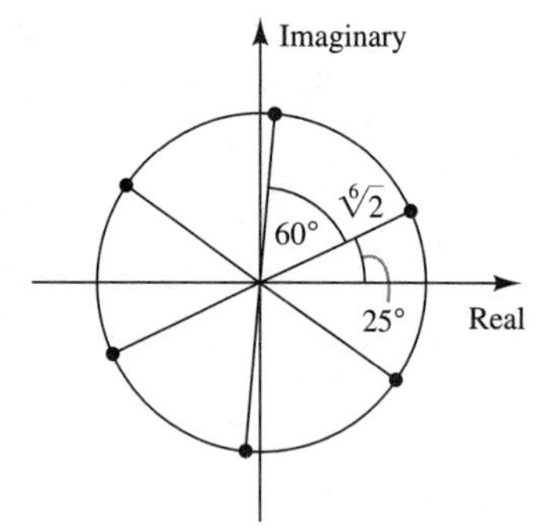

19. $\sqrt[4]{2}\left(\frac{\sqrt{3}}{2} + \frac{i}{2}\right)$
$\sqrt[4]{2}\left(\frac{-1}{2} + \frac{i\sqrt{3}}{2}\right)$
$\sqrt[4]{2}\left(-\frac{\sqrt{3}}{2} - \frac{i}{2}\right)$
$\sqrt[4]{2}\left(\frac{1}{2} - \frac{i\sqrt{3}}{2}\right)$

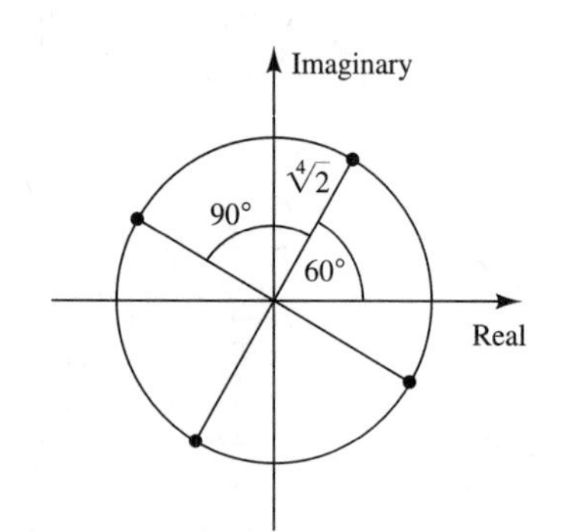

23. $4 \text{ cis } 60° = 2 + 2\sqrt{3}i$
$4 \text{ cis } 180° = -4$
$4 \text{ cis } 300° = 2 - 2\sqrt{3}i$

25. $\frac{1 \pm i}{\sqrt{2}}, \frac{-1 \pm i}{\sqrt{2}}$

Section 9.4

1.

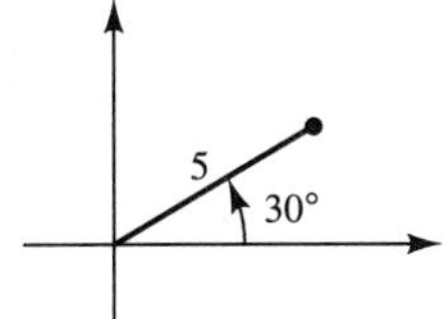

3.

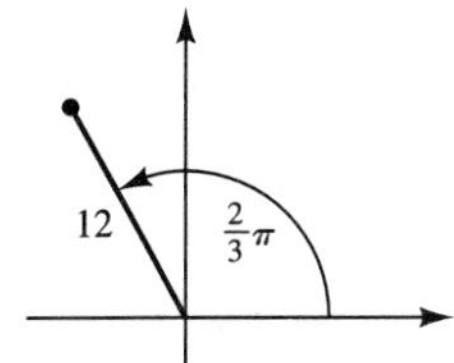

5.

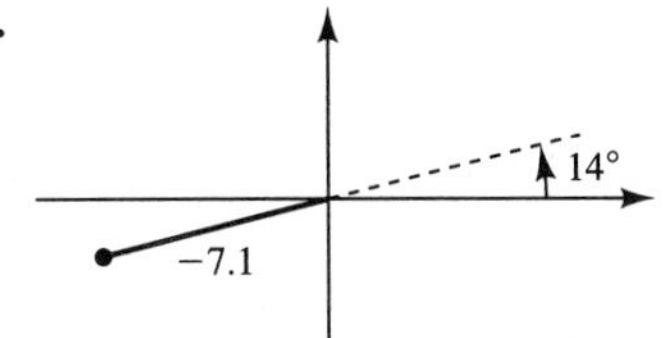

7.

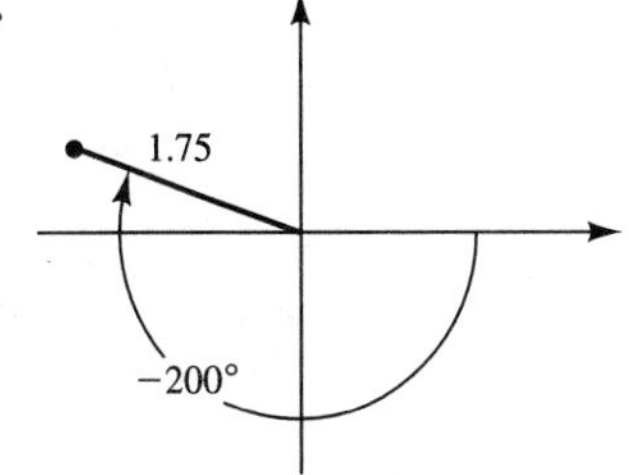

9.

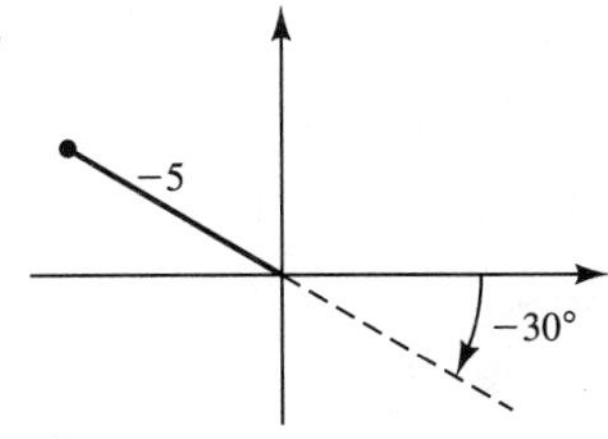

11. $r(2\cos\theta + 3\sin\theta) = 6$ **13.** $r = 4\cos\theta$

15. $r^2(\cos^2\theta + 4\sin^2\theta) = 4$ **17.** $r\cos^2\theta = 4\sin\theta$

19. $x^2 + y^2 = 25$

21. $x^2 + y^2 = 10y$ **23.** $x^2 + y^2 = (x^2 + y^2)^{1/2} + 2y$

25. $y^2 = 25 - 10x$

27.

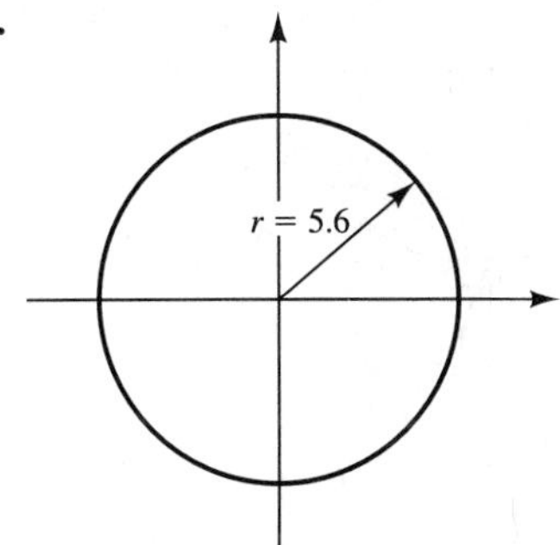

29.

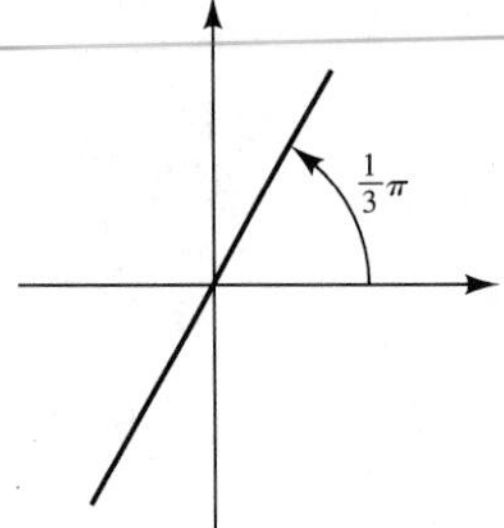

31.

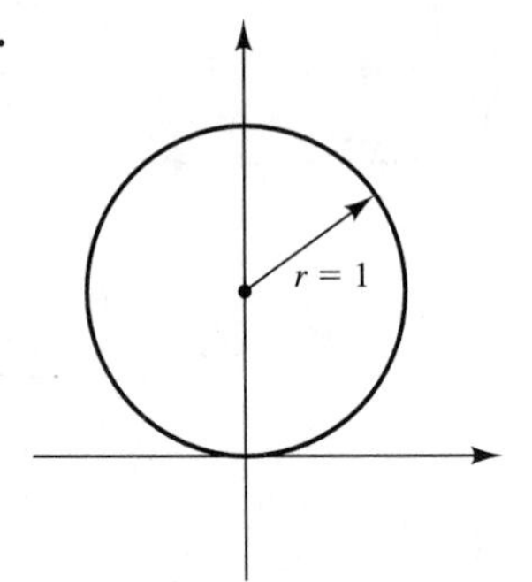

33.

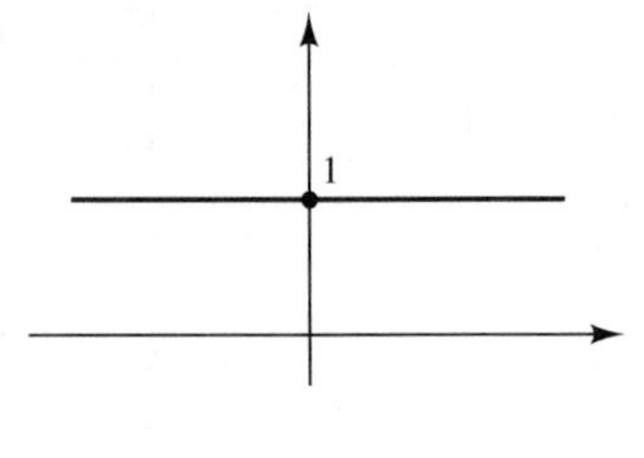

35.

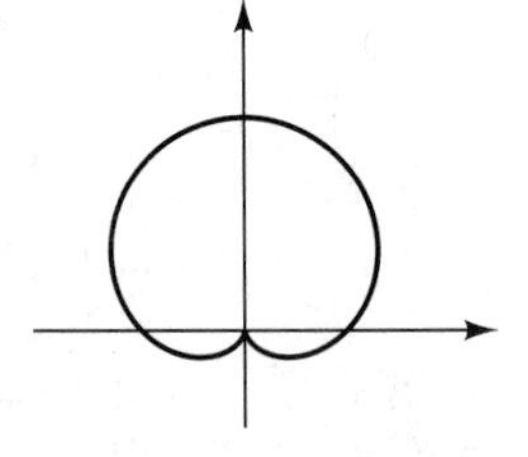

37.

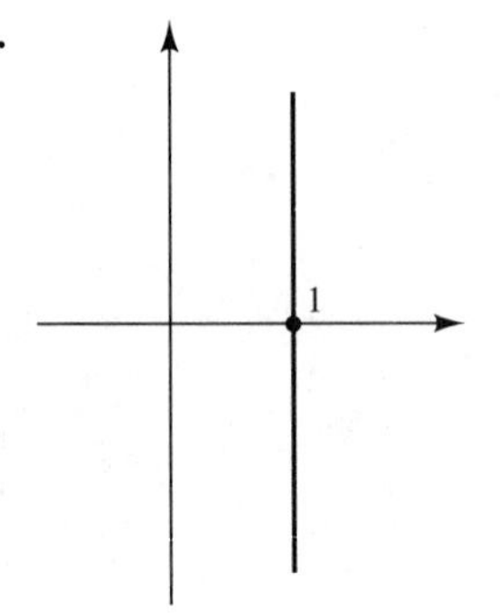

39.

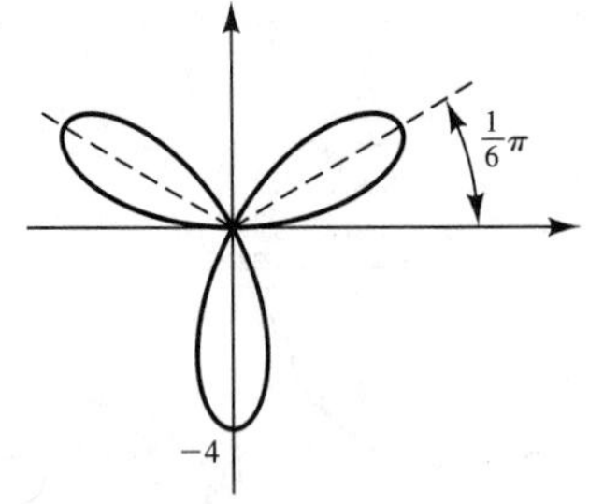

41.

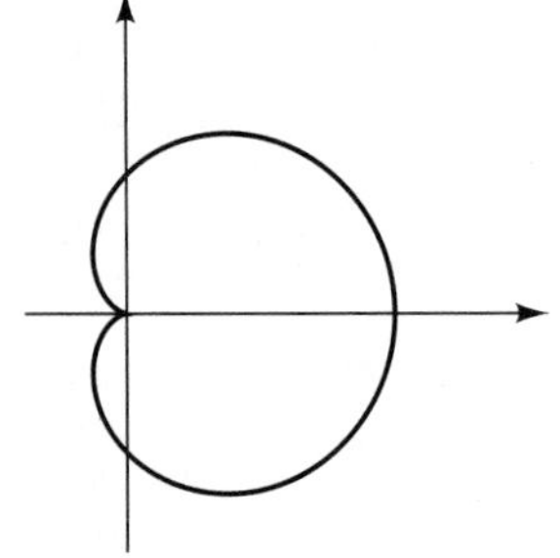

43. $3x^2 + 4y^2 - 2x - 1 = 0$

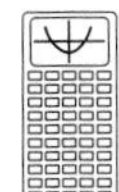

Graphing Calculator Exercises

1. See graph in Exercise 27 on page 459.
3. See graph in Exercise 31 on page 460.
5. See graph in Exercise 35 on page 460.
7. See graph in Exercise 39 on page 460.

Review Exercises for Chapter 9

1. $9 - 3i$ **3.** 5 **5.** $8 - 6i$ **7.** 0 **9.** 16

11. $\frac{1 - 3i}{-2}$ **13.** $-i$ **15.** $\frac{19}{2} - \frac{9}{2}i$ **17.** 1 cis 90°

19. $\sqrt{2}(\text{cis } 45°)$ **21.** $\sqrt{2}(1 + i)$

23. $4(\cos 20° - i \sin 20°) = 3.8 - 1.4i$ **25.** $\frac{27}{2}(\sqrt{3} + i)$

27. $\tan \theta = 2$

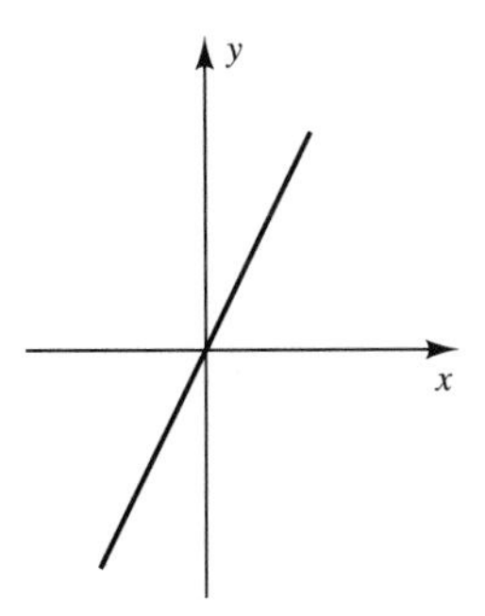

29. $r^2 = \dfrac{1}{4 \cos^2\theta + \sin^2\theta}$

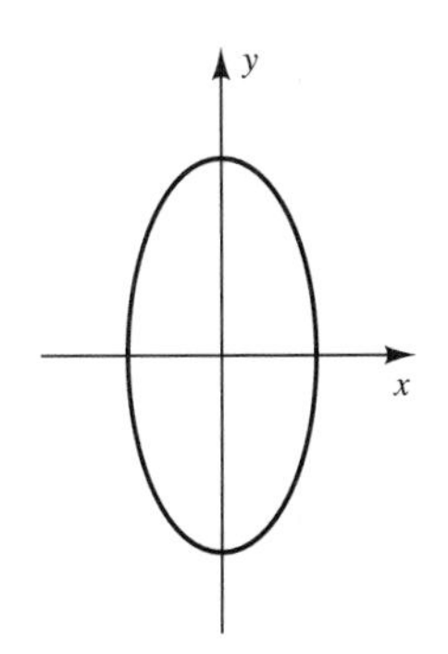

31. $x^2 + y^2 = 4$

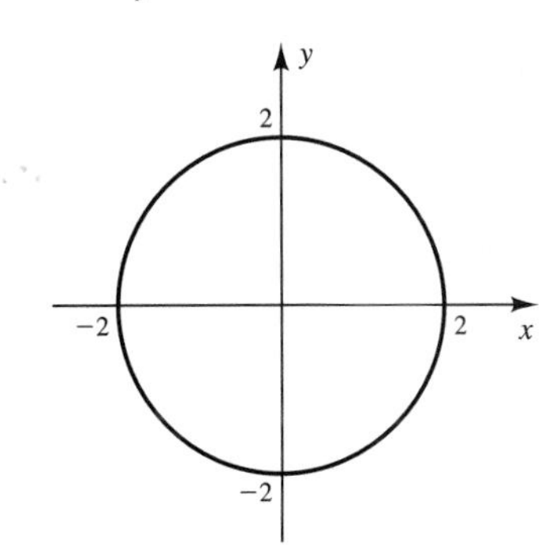

33. $x^2 + 4y = 4$

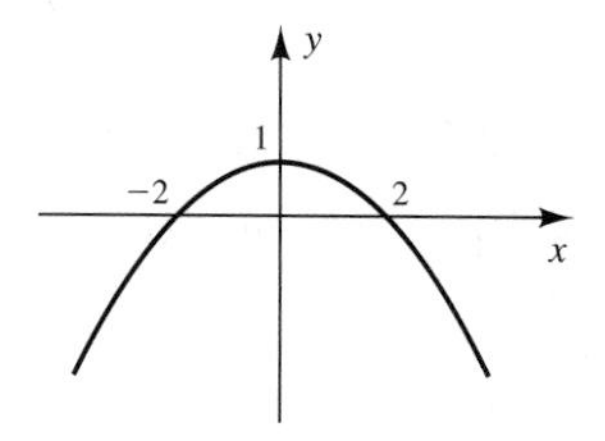

35. $x^2 + y^2 - 3x = 0$

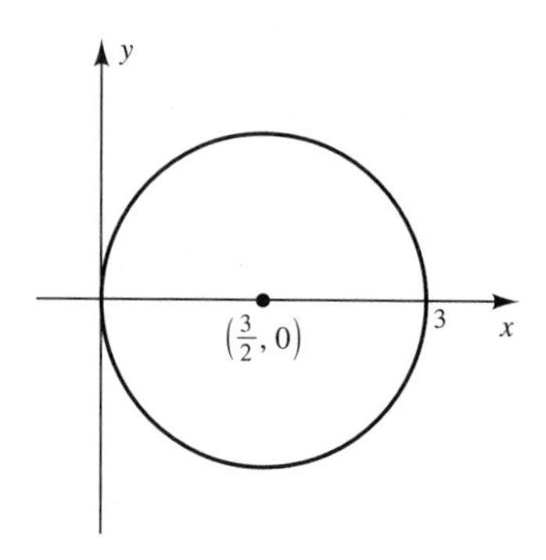

37. $(x^2 + y^2)^{3/2} = 2xy$

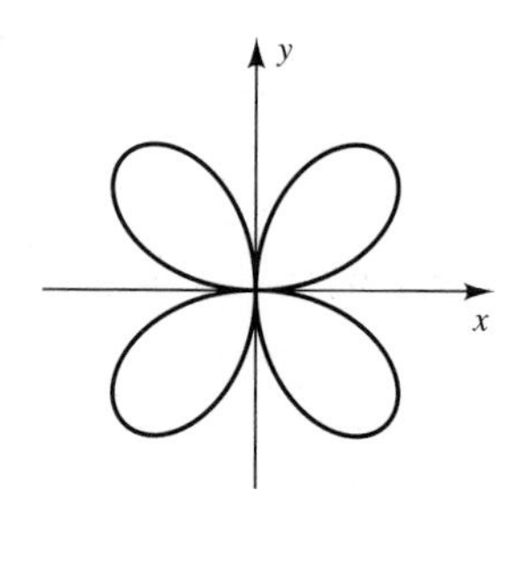

39. $x^2 + y^2 = 1$

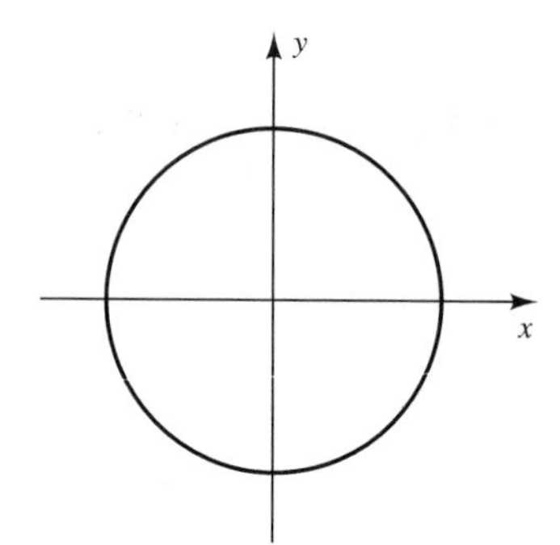

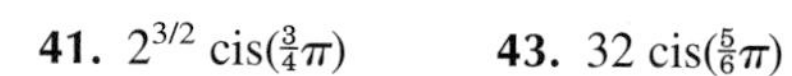

41. $2^{3/2}\ \text{cis}(\frac{3}{4}\pi)$ **43.** $32\ \text{cis}(\frac{5}{6}\pi)$

45. $10^5\ \text{cis}\ 3.06$

47. $1\ \text{cis}(\frac{1}{5}\pi)$, $1\ \text{cis}(\frac{3}{5}\pi)$, $1\ \text{cis}(\pi)$, $1\ \text{cis}(\frac{7}{5}\pi)$, $1\ \text{cis}(\frac{9}{5}\pi)$

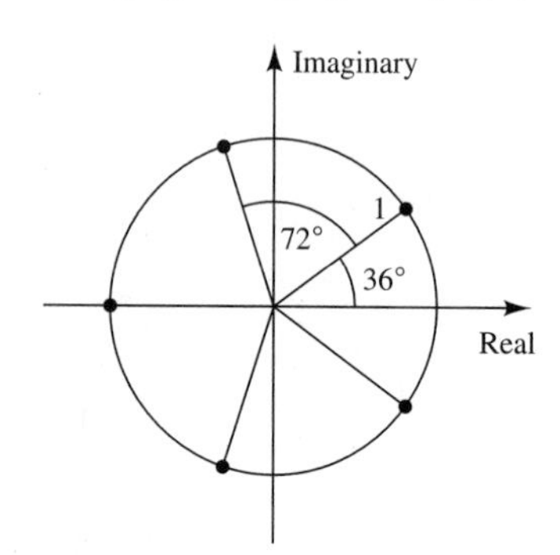

Section 10.1

1. $\frac{4}{9}$ **3.** $-\frac{2}{5}$ **5.** 1 **7.** 0 **9.** -2

11.

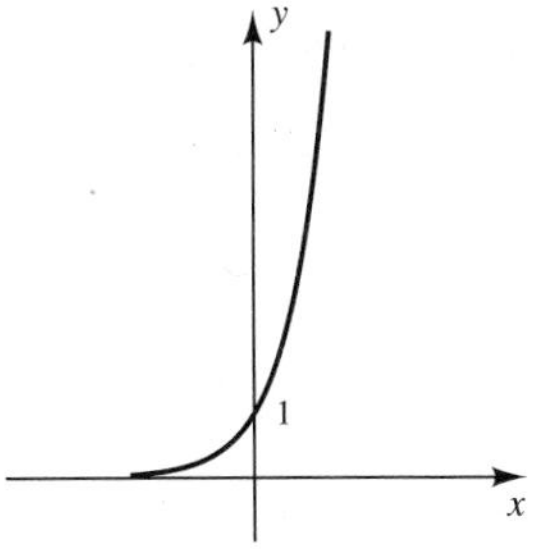

13.

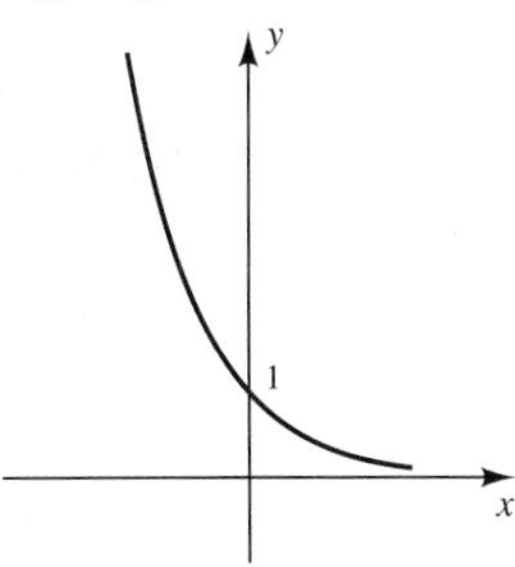

15.

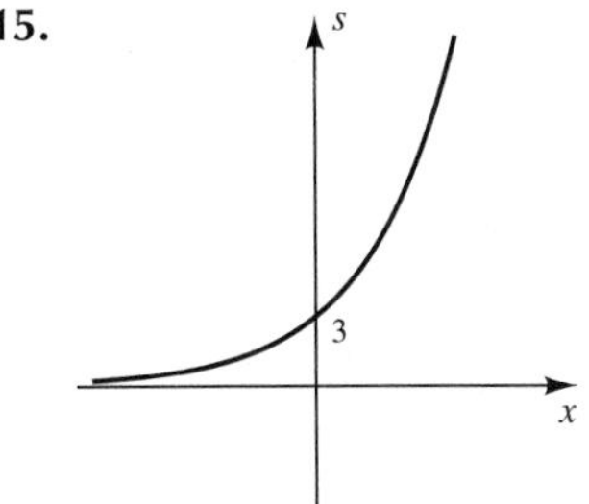

17.

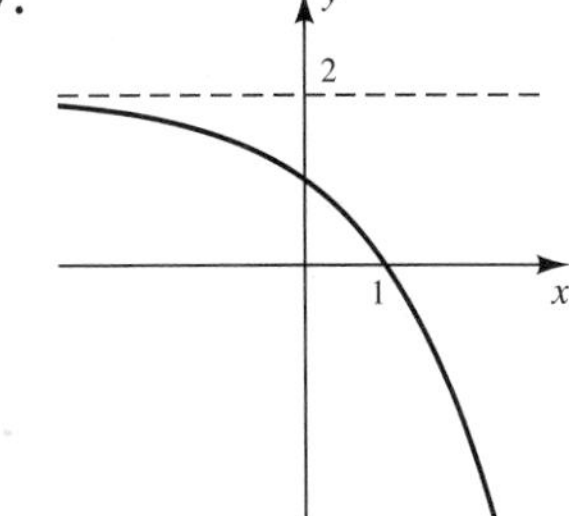

19.

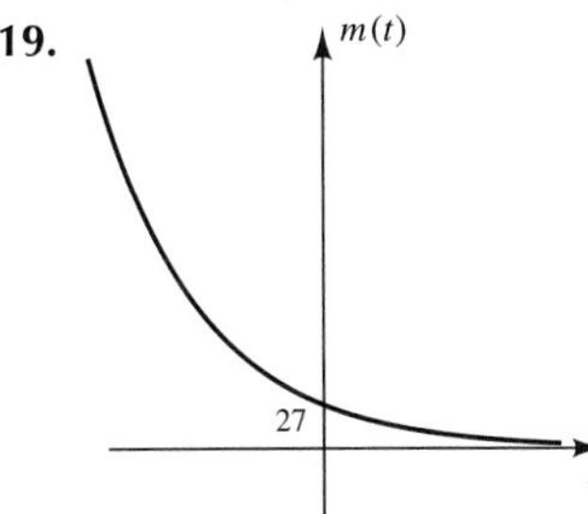

21.

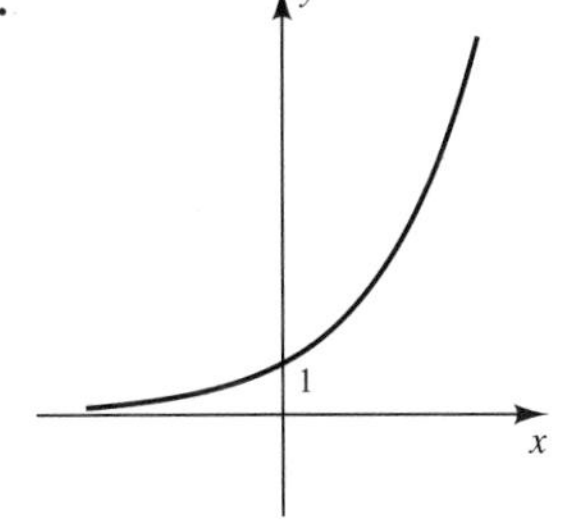

23.

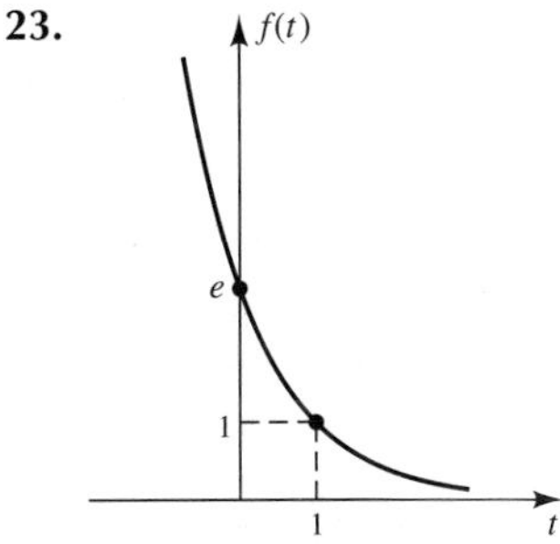

25.

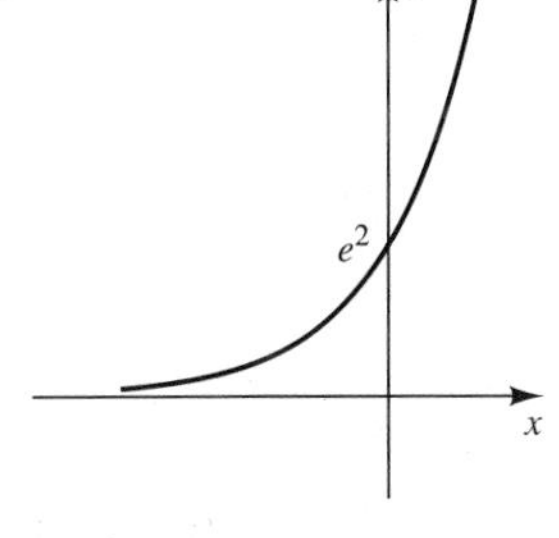

27. (a) 2.7 (b) 8.8 (c) 2.8 (d) 0.35 (e) 1.3
29. $y = 1$ for all x **31.** Reflection across the x-axis
33. The same curve
35. The same curve shifted 3 units in the negative x direction **37.** 7 days
39. \$1276.28 **41.** 2878.29 m/sec **43.** 230.24° **45.** 10.026
47. 22.68

Section 10.2

1. $\log_2 x = 3$ **3.** $\log_5 M = -3$ **5.** $\log_7 L = 2$ **7.** $b = 2$
9. $b = 10$ **11.** 10,000 **13.** 10 **15.** 4 **17.** 9
19. 2 **21.** a **23.** 8 **25.** 36 **27.** 6 **29.** $-\frac{5}{2}$
31. $-1 < x < 8$ **33.** $4 \leq x \leq 8$

35.

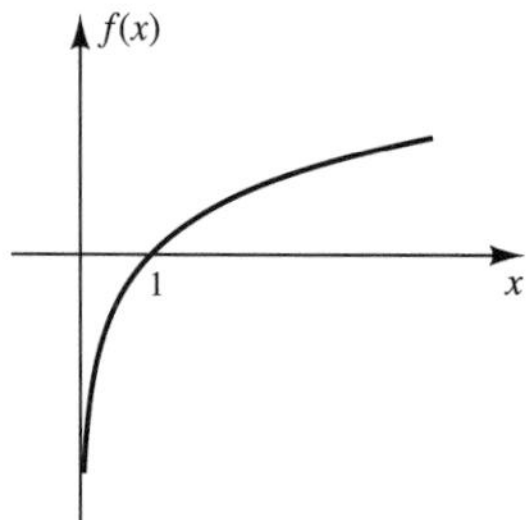

37.

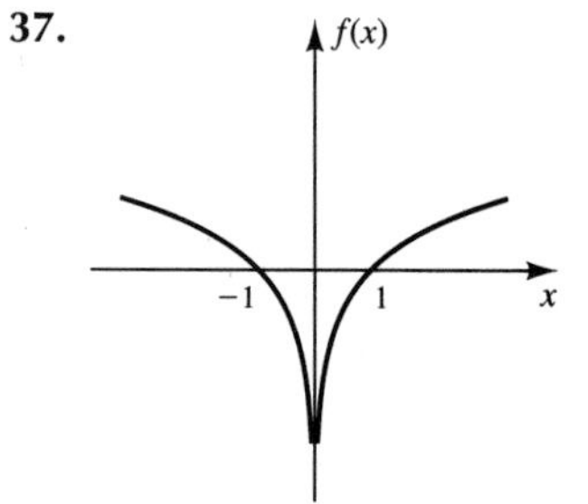

39.

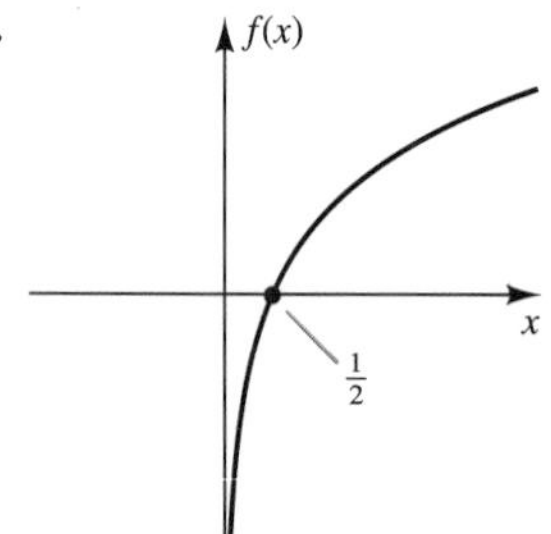

41.

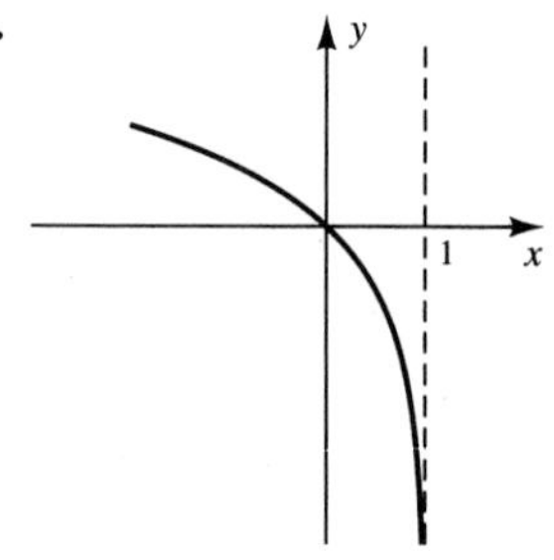

43.

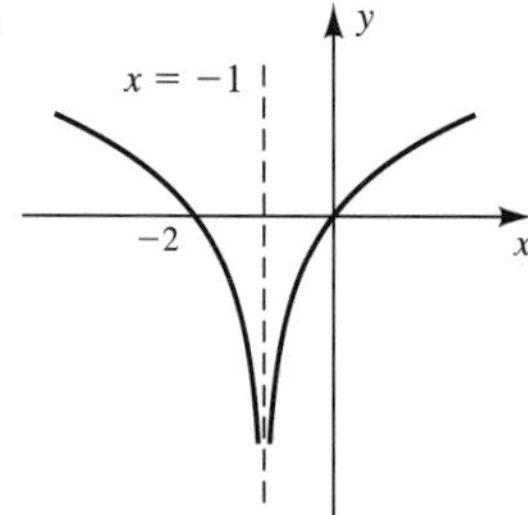

45. Reflection in the x-axis
47. Two branches: the negative branch is the reflection of $y = \log x$ in the x-axis.

49. Curve moved 3 units in the negative x direction
51. 2 **53.** 10 **55.** $(1, 0)$
57. No answer required **59.** Reflections of each other in $y = x$
61. -0.3010 **63.** 1 **65.** -1 **67.** -0.69315
69. 3.4657 time units

Section 10.3

1. 9 **3.** $\frac{1}{2}$ **5.** 6 **7.** 24 **9.** 0.1761 **11.** 1.0791
13. 1.9542 **15.** 0.3495 **17.** 3.3801 **19.** -2.8539
21. $\log_2 x$ **23.** 0 **25.** $\log \dfrac{5t(t^2-4)^2}{\sqrt{t+3}}$
27. $\log \dfrac{u^3}{(u+1)^2(u-1)^5}$
31. The same for $x > 0$. For $x < 0$, $\log x^2$ is defined, $2 \log x$ is not.
33. Both are -2.
35. **(a)** Multiply each ordinate value by p. **(b)** Move curve up $\log p$ units.
(c) Move curve p units to the left **(d)** Move curve down $\log p$ units.

Section 10.4

1. -1.553 **3.** $x = 0, x = \log 2 \approx 0.301$ **5.** 0.5850 **7.** 13.97
9. $x > \ln 2.5/\ln 2 \approx 1.32$ **11.** $x = 10$ **13.** $x = 1, 10^4$
15. $x = 1$
$x = 10^{\sqrt{3}} \approx 53.96$
$x = 10^{-\sqrt{3}} \approx 0.0185$
17. $x = 5$ **19.** $x = 2 + \sqrt{5} \approx 4.236$
21. $x = 6$ **23.** $x = \sqrt{6}$ **25.** $x > \frac{1}{9}$
27. $x = \ln(2 + \sqrt{3}) \approx 1.317$
$x = \ln(2 - \sqrt{3}) \approx -1.317$
29. The logarithm is a 1-to-1 function.
31. $k = 0.02476$ **33.** 173.3 days **35.** 52.2 hr

Review Exercises for Chapter 10

1. -8.55 **3.** $4, -1$ **5.** -0.465 **7.** 1000 **9.** 4
11. 2 **13.** 1.188
15.

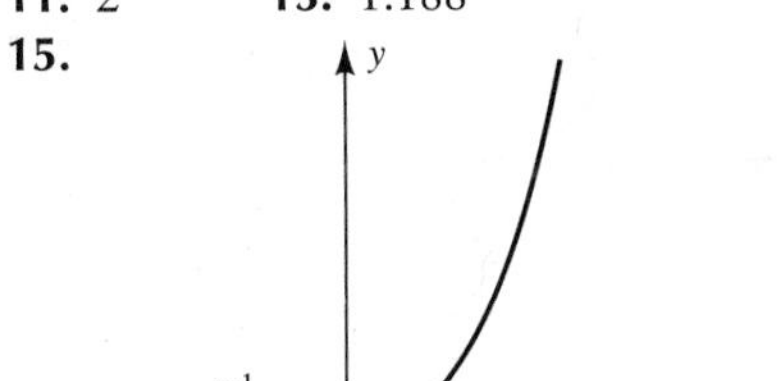

17.

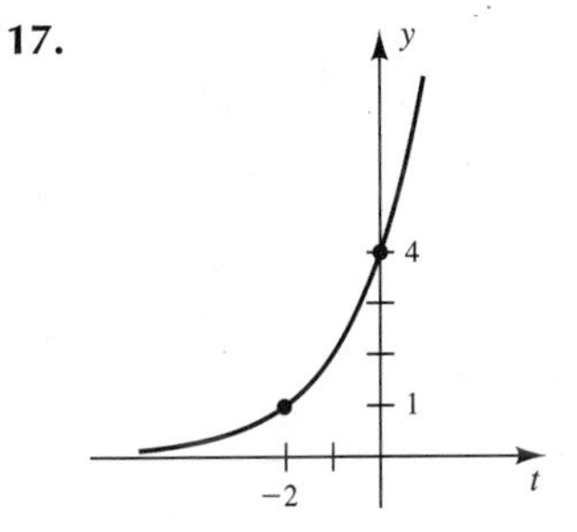

19.

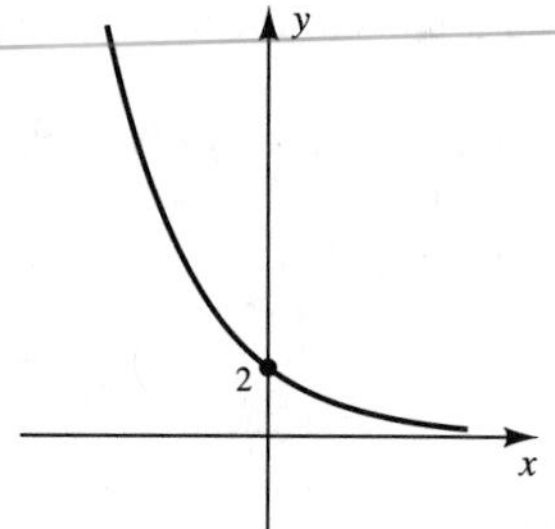

21.

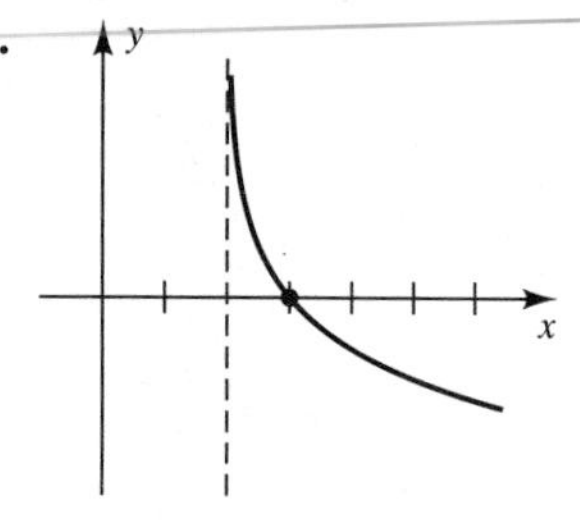

23.

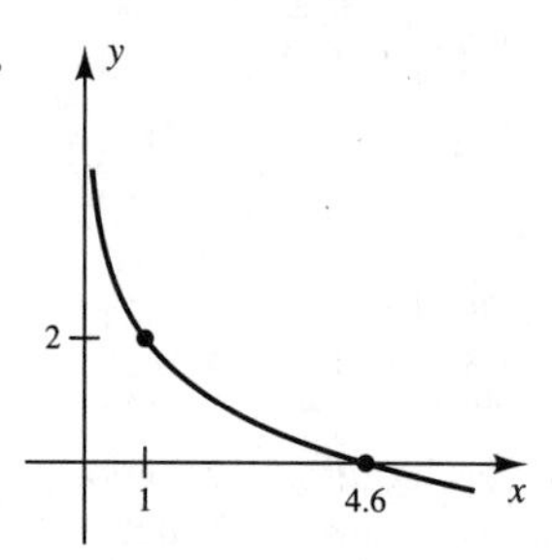

25. $3 \log x$ **27.** $\log_2(2x^4)$ **29.** $\log_5[x^3/(2x - 3)]$
31. $x = 1.322$ **33.** $x = 3.819$ **35.** $x = 10^{10}$ **37.** $x = 2$
39. $x = 5$ **41.** 2.77 years
43.

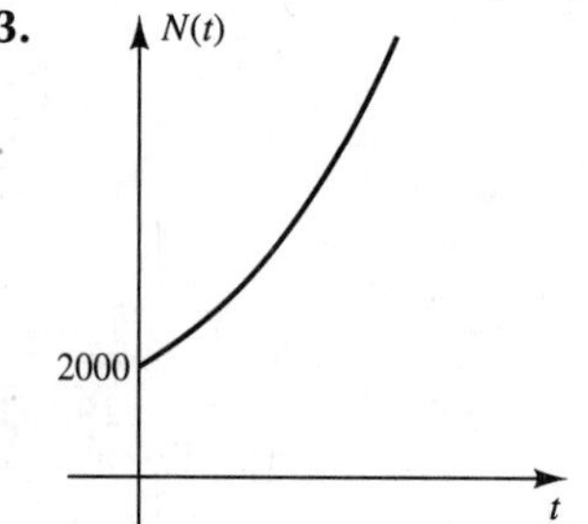

45. 1.77 mg

Section 11.1

1.

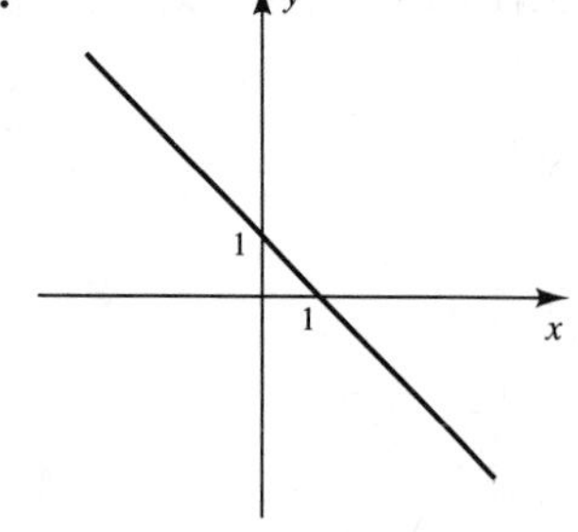

3.

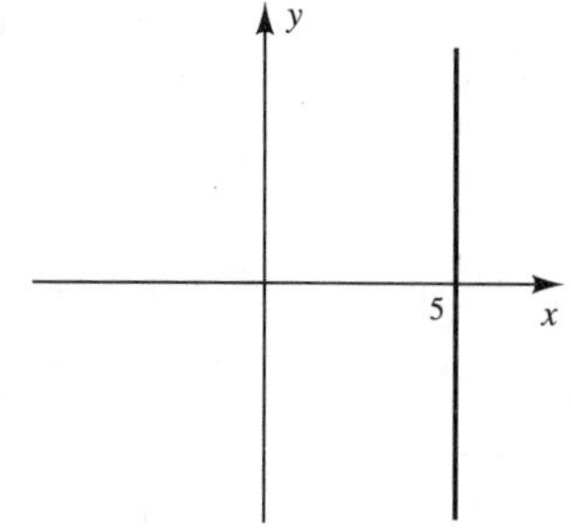

5.

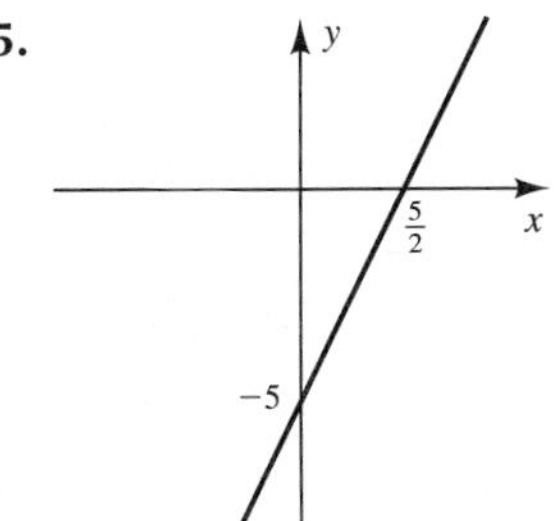

7.

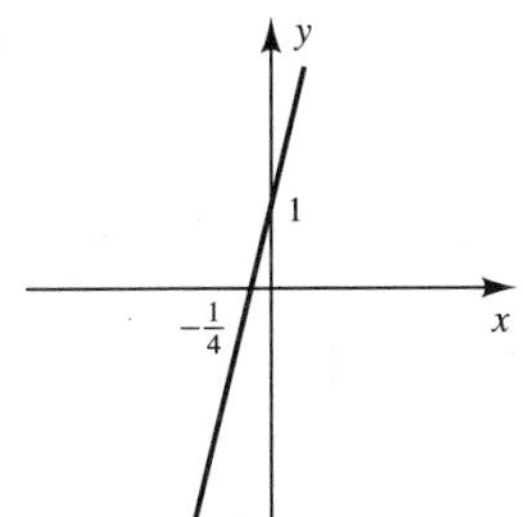

9.

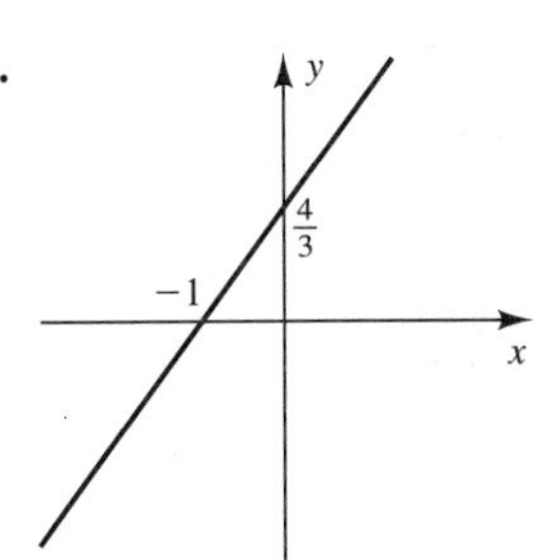

11.

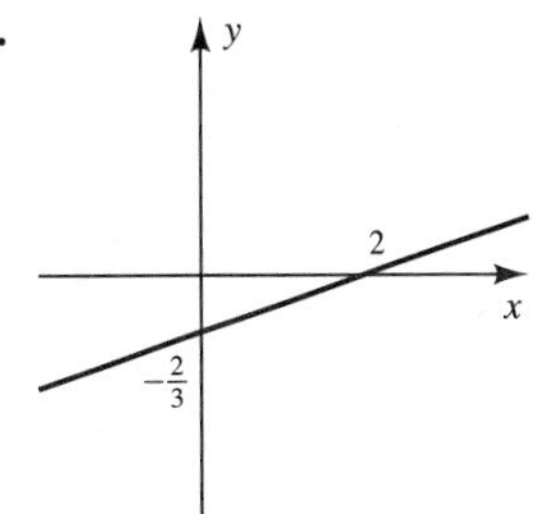

13. $m = \frac{1}{2}, \theta = 26.6°$ **15.** $m = -\frac{5}{4}, \theta = 128.7°$
17. $m = \frac{4}{3}, \theta = 53.1°$ **19.** $m = 0, \theta = 0$
21. $x - 2y = -8$ **23.** $7x + y = 33$ **25.** $x + 5y = 16$
27. $2y - 3x = 1$ **29.** $5y - 2x = 10$ **31.** $m = \frac{2}{3}, b = -\frac{5}{3}$
33. $m = -1, b = 2$ **35.** $m = \frac{1}{2}, b = -2$ **37.** $m = \frac{1}{2}, b = \frac{5}{2}$
39. $m = 0, b = 5$ **41.** $m = \frac{1}{5}, b = \frac{7}{5}$ **43.** $2x - 3y = -4$
45. $x + y = 8$ **47.** $2x + 3y = 8$ **49.** $7x + 5y = 0$
51. $3x + y = -5$ **53.** No answer required **55.** $\frac{x}{2} + \frac{y}{\frac{2}{3}} = 1$

Section 11.2

1. Focus $(-2, 0)$
Directrix $x = 2$
Right chord $(-2, 4), (-2, -4)$

3. Focus $(0, \frac{3}{2})$
Directrix $y = -\frac{3}{2}$
Right chord $(-3, \frac{3}{2}), (3, \frac{3}{2})$

5. Focus $(-4, 0)$
Directrix $x = 4$
Right chord $(-4, 8), (4, 8)$

7. Focus $(\frac{3}{4}, 0)$
Directrix $x = -\frac{3}{4}$
Right chord $(\frac{3}{4}, \frac{3}{2}), (\frac{3}{4}, -\frac{3}{2})$

9. Focus $(-\frac{1}{2}, 0)$
Directrix $x = \frac{1}{2}$
Right chord $(-\frac{1}{2}, 1), (-\frac{1}{2}, -1)$

11. $x^2 = 8y$ **13.** $y^2 = 6x$ **15.** $x^2 = -4y$

17. $x^2 = y$ **19.** $(0, 0), (-2, 1)$ **21.** $(.09, -1.05)$
23. $x^2 = 1000y$

Section 11.3

1. Ellipse
$a = 5$
$b = \sqrt{5}$
Foci $(0, \pm 2\sqrt{5})$

3. Ellipse
$a = 2$
$b = 1$
Foci $(0, \pm\sqrt{3})$

5. Circle
$r = 3$

7. Ellipse
$a = \sqrt{12}$
$b = \sqrt{8}$
Foci $(\pm 2, 0)$

9. Ellipse
$a = 1$
$b = \frac{2}{3}$
Foci $(0, \pm\frac{1}{3}\sqrt{5})$

11. $9x^2 + 16y^2 = 144$

13. $100x^2 + 9y^2 = 225$

15. $9x^2 + 25y^2 = 225$

17. $15x^2 + 16y^2 = 240$

19. $4x^2 + 21y^2 = 25$

21. $(1.3, -1.9), (-1.3, 1.9)$

23. $(.94, .88), (-.94, .88)$

25. $9x^2 + 100y^2 = 225$

27.

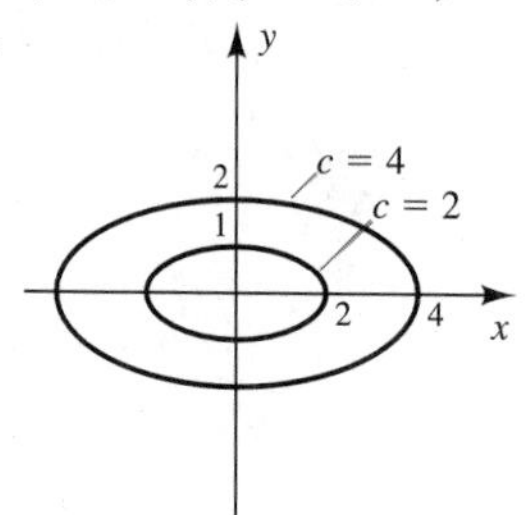

29.

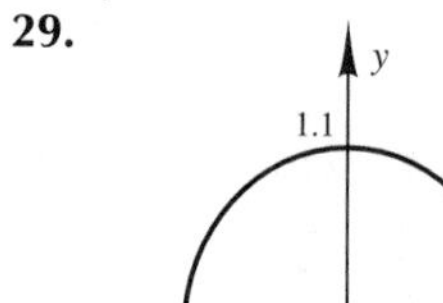

Section 11.4

1. $a = 4$
$b = 4$
Foci $(\pm\sqrt{32}, 0)$

3. $a = 3$
$b = 2$
Foci $(\pm\sqrt{13}, 0)$

5. $a = 5$
$b = 2$
Foci $(0, \pm\sqrt{29})$

7. $a = \frac{5}{2}$
$b = \frac{5}{4}$
Foci $(\pm\frac{5}{4}\sqrt{5}, 0)$

9. $a = 1$
$b = 1$
Foci $(\pm\sqrt{2}, 0)$

11. $9x^2 - 16y^2 = 144$

13. $4y^2 - x^2 = 4$

15. $13x^2 - 36y^2 = 117$

17. $4y^2 - x^2 = 64$

19. $y^2 - 10x^2 = 9$

23.

y

x

Section 11.5

1. $8x = y^2 - 2y + 25$

3. $12y = -13 + 2x - x^2$

5. $4x = y^2 - 4y + 8$

7. $7x^2 + 16y^2 - 28x - 32y - 68 = 0$

9. $16x^2 + y^2 - 16x + 8y + 4 = 0$ **11.** $3x^2 + 4y^2 + 24x - 24y + 72 = 0$
13. $64y^2 - 49x^2 - 98x - 256y = 577$ **15.** $7x^2 - 9y^2 - 28x + 18y = 44$
17. $x^2 - 4y^2 + 4x - 16y - 16 = 0$ **19.** $(x - h)^2 + y^2 = r^2$
21. $y^2 = 4a(x - h)$ **23.** $(x - r)^2 + y^2 = r^2$
25.

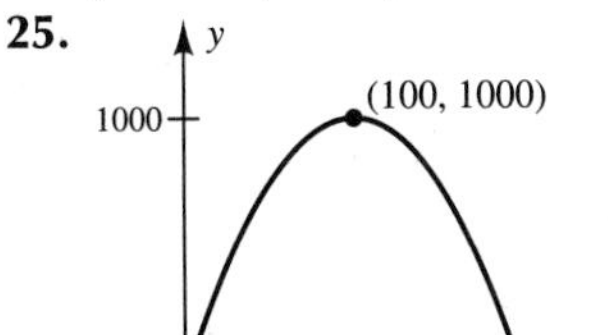

Section 11.6

1. $[(3\sqrt{3} + 1)/2, (-3 + \sqrt{3})/2]$ **3.** $[(-\sqrt{3} - 1)/2, (1 - \sqrt{3})/2]$
5. $[(2\sqrt{3} - 1)/2, (-2 - \sqrt{3})/2]$ **7.** $y' = \frac{1}{3}x'$
9. $(2 + 7\sqrt{3})x' + (7 - 2\sqrt{3})y' = 6$ **11.** $4x'^2 + y'^2 = 4$
13. $x'^2 + y'^2 + \sqrt{3}x' - y' = 0$ **15.** $53°8'$
17. $x'^2 + 2y'^2 = 1$
19. $\dfrac{x'^2}{.828} - \dfrac{y'^2}{4.828} = 1$ **21.** $\dfrac{x'^2}{2} + \dfrac{y'^2}{\frac{2}{3}} = 1$

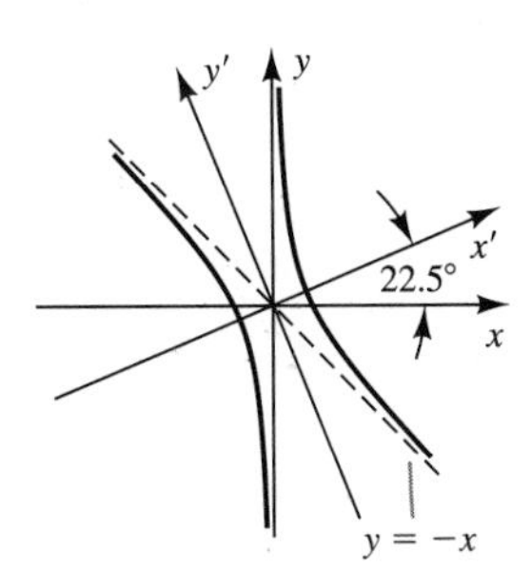

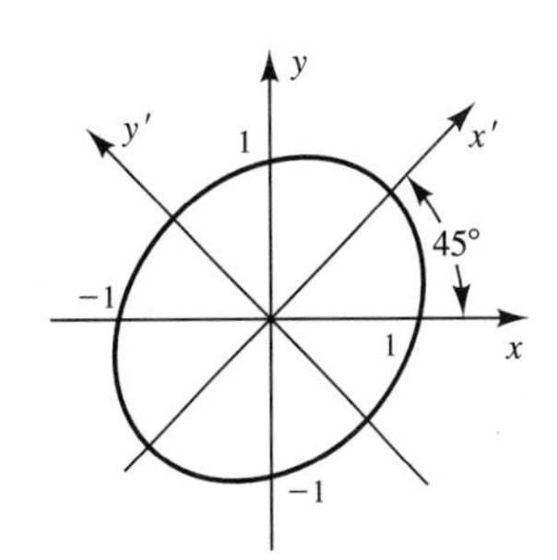

Section 11.7

1. $(x + 2)^2 + (y + 3)^2 = 3^2$ **3.** $\dfrac{(x + 1)^2}{4} + \dfrac{(y + 1)^2}{9} = 1$

5. $\dfrac{(x - 2)^2}{25} - \dfrac{y^2}{25} = 1$ **7.** $(x - 3)^2 = 3(y + 3)$

9. $(y - \frac{1}{2})^2 = -\frac{1}{2}(x - \frac{3}{2})$ **11.** $\dfrac{(x + 1)^2}{1} - \dfrac{(y - 1)^2}{4} = 1$

13. $\dfrac{(x - 1)^2}{4} + \dfrac{(y + 2)^2}{9} = 1$ **15.** $\dfrac{(y - \frac{1}{2})^2}{\frac{1}{4}} - \dfrac{x^2}{\frac{1}{2}} = 1$

17. $\dfrac{(x + 2)^2}{7} - \dfrac{y^2}{7} = 1$ **19.** $(x + \frac{5}{2})^2 = y - \frac{3}{4}$

21. $\dfrac{x'^2}{1} - \dfrac{y'^2}{\frac{9}{4}} = 1$

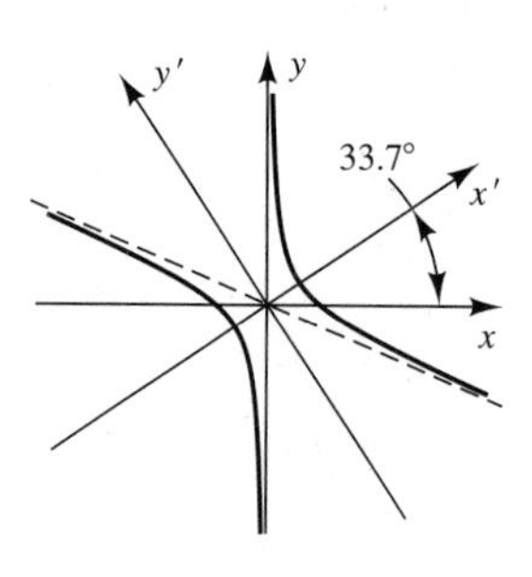

23. $\dfrac{x'^2}{\frac{22}{7}} - \dfrac{y'^2}{4} = 1$

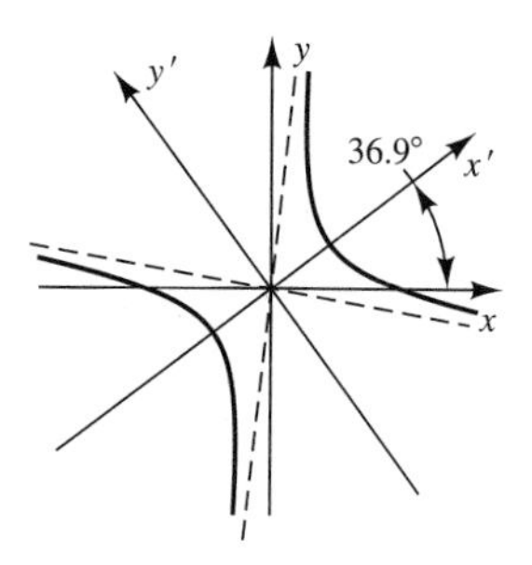

25. Ellipse

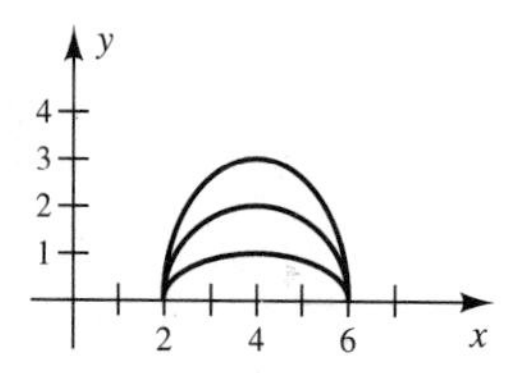

27. Parabola: $y^2 = 10x + 25$

Review Exercises for Chapter 11

1. Parabola
Vertex (0, 0)
Focus $(0, -\frac{3}{4})$

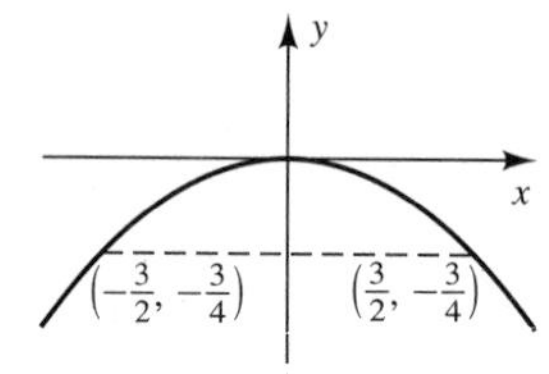

3. Circle
Center (0, −2)
Radius = 3

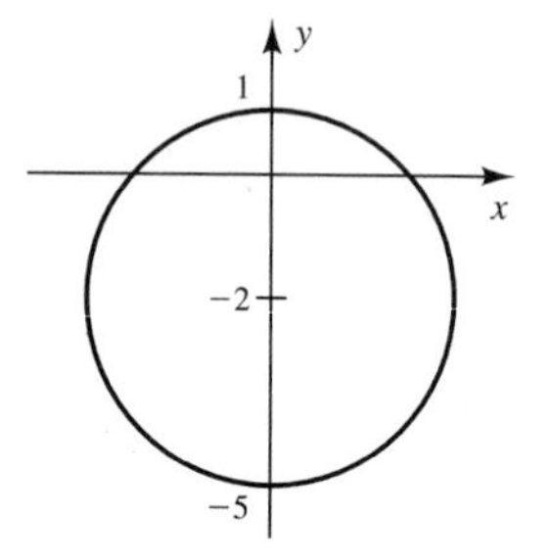

5. Hyperbola
Center (−1, −1)
$a = 1$
$b = 1$
$c = \sqrt{2}$

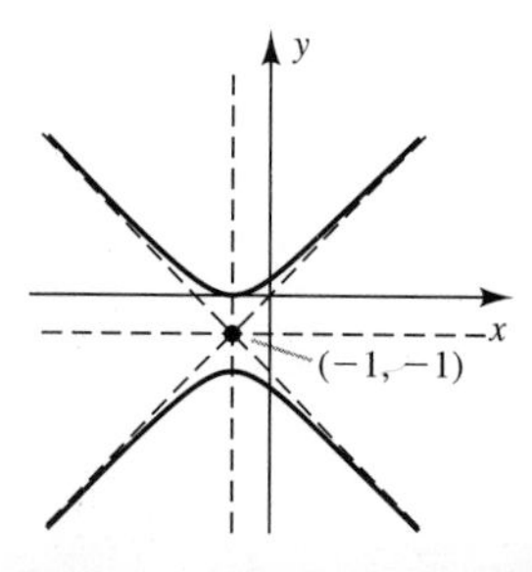

7. Hyperbola
Center (0, 0)
$a = 10$
$b = 10$
$c = 10\sqrt{2}$

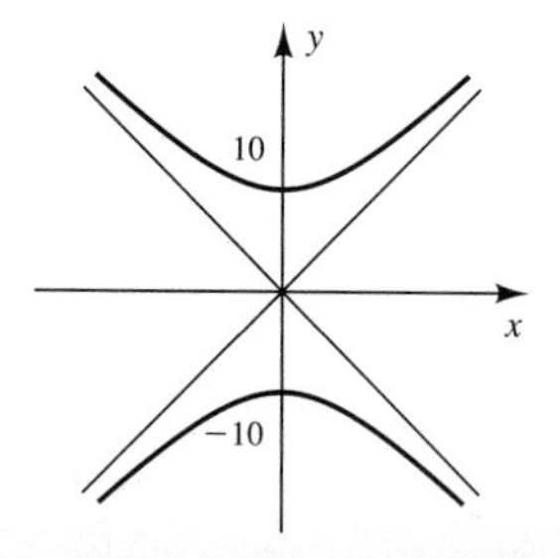

9. Ellipse
Center $(-\frac{3}{2}, 0)$
$a = \sqrt{\frac{23}{2}} \approx 3.4$
$b = \sqrt{\frac{23}{4}} \approx 2.4$
$c = \sqrt{\frac{23}{4}} \approx 2.4$

11. $y^2 = -8(x - 2)$

13. $(x + 1)^2 + (y - 2)^2 = 64$ **15.** $x^2/16 + y^2/12 = 1$
17. $(x - 3)^2/16 - (y - 4)^2/64 = 1$ **19.** 9 ft from center
21. $(0.67, -0.33)$ and $(0.5, 0)$

Appendix A.1 Exercises

1. 5.3 **3.** −2.2 **5.** 61.2 **7.** 76.14 **9.** 1.047197551
11. 0.760681686 **13.** −5.996621055 **15.** 0.0571895425
17. 1.115355072 **19.** 4.060605626
21. An error message occurs because you cannot take the square root of a negative number.
23. About 145

Appendix A.2 Exercises

1. $2 + (3 \times 4)$ **3.** $2 + (3 \div 6) + 2$ **5.** $\sqrt{5} + 5 - (3 \times 4)$
7. $[(6 \div 4) \div 2] - 1$ **9.** $6 - [(4 \div 3) \times 28]$ **11.** 2.619047619
13. 1.299985778 **15.** 80.481582 **17.** 4.272042343
19. 0.805777843 **21.** 8.900462547

Appendix A.3 Exercises

1. 8.243765×10^{12} **3.** 9.341×10^7 **5.** 4.77×10^{-2}
7. 9.75×10^{-1} **9.** 3.111×10^4 **11.** 1,423,900,000,000
13. 6430 **15.** 0.2285 **17.** 0.0000000003117 **19.** 6×10^{-11}
21. 6.0×10^6 **23.** 4.0004×10^9 **25.** 8.3197×10^{16}
27. 4.9839×10^{13}

Index